Advances in Neurobiology

Volume 34

Series Editor

Arne Schousboe, Department of Drug Design & Pharmacology
University of Copenhagen
Copenhagen, Denmark

Advances in Neurobiology covers basic research in neurobiology and neurochemistry. It provides in-depth, book-length treatment of some of the most important topics in neuroscience including molecular and pharmacological aspects. The main audiences of the series are basic science researchers and graduate students as well as clinicians including neuroscientists (neurobiologists and neurochemists) and neurologists. *Advances in Neurobiology* is indexed in PubMed, Google Scholar, and the Thompson Reuters Book Citation Index.

Alberto A. Rasia-Filho
Maria Elisa Calcagnotto
Oliver von Bohlen und Halbach
Editors

Dendritic Spines

Structure, Function, and Plasticity

Springer

Editors
Alberto A. Rasia-Filho
Department of Basic Sciences/Physiology and Graduate Program in Biosciences
Universidade Federal de Ciências da Saúde de Porto Alegre
Porto Alegre, RS, Brazil

Graduate Program in Neuroscience
Universidade Federal do Rio Grande do Sul
Porto Alegre, RS, Brazil

Oliver von Bohlen und Halbach
Institut für Anatomie und Zellbiologie
Universitätsmedizin Greifswald
Greifswald, Germany

Maria Elisa Calcagnotto
Graduate Program in Neuroscience
Universidade Federal do Rio Grande do Sul
Porto Alegre, RS, Brazil

Department of Biochemistry
Universidade Federal do Rio Grande do Sul
Porto Alegre, RS, Brazil

Graduate Program in Biochemistry
Universidade Federal do Rio Grande do Sul
Porto Alegre, RS, Brazil

Graduate Program in Psychiatry and Behavioral Science
Universidade Federal do Rio Grande do Sul
Porto Alegre, RS, Brazil

ISSN 2190-5215 ISSN 2190-5223 (electronic)
Advances in Neurobiology
ISBN 978-3-031-36161-6 ISBN 978-3-031-36159-3 (eBook)
https://doi.org/10.1007/978-3-031-36159-3

This Springer imprint is published by the registered company Springer Nature Switzerland AG
The registered company address is: Gewerbestrasse 11, 6330 Cham, Switzerland

Paper in this product is recyclable.

Foreword

The history of the discovery and elucidation of dendritic spine structure and function is one of the most exciting stories in neuroscience. Our knowledge progressed because of the ingenuity of neuroscientists taking advantage of new techniques. This is true since over a century ago when Santiago Ramón y Cajal (1888) used the Golgi technique, chronicled in detail in this volume by the editors, Dr. Alberto A. Rasia-Filho, Dr. Maria E. Calcagnotto and Dr. Oliver von Bohlen und Halbach. This technique, using chrome silver impregnation, is employed today accompanied by advanced stereological tools to decipher changes in the structure, number, and distribution of the spines along the dendritic shaft with various physiological and behavioral paradigms. Cajal's drawings are revered for their detailed and novel information, as well as for their artistic aesthetic. A classic example of the Golgi technique in brain disorders is a study by Dominick P. Purpura from 1974, whereby the Golgi technique revealed dendritic spine dysgenesis in the brains of children with intellectual developmental disorder. That is, abnormally long, thin spines together with the absence of short thick spines, as well as tangled spines, were visualized in dendrites of cortical neurons of affected children.

The elation felt by Sanford L. Palay (1958) in the mid-1950s when he visualized the structure of the synapse for the first time by electron microscopy is palpable even today. In addition to synaptic vesicles, Dr. Palay described pre- and postsynaptic thickenings in close apposition to the pre- and postsynaptic membranes, respectively. In 1959, Edward G. Gray (1959) described synaptic contacts on dendritic spines of the cerebral cortex using electron microscopy. Subsequently, many investigators have revealed comprehensive ultrastructural analyses of dendritic spines. Kristen M. Harris' (2020) elegant reconstructions detail structure-function relationships, in particular, structural synaptic plasticity, following long-term potentiation (LTP) with the hippocampal slice paradigm. The spine, itself, is a complex body containing many organelles, the precise function of which in nerve conduction and synaptic plasticity remains an enigma. The evolution of our knowledge of dendritic spine structure and function is detailed in the first chapter of this volume.

The postsynaptic density (PSD) – the dense, submembranous, filamentous array beneath the postsynaptic membrane – figures prominently at the synaptic contact and is evident and complex in dendritic spines (Fig. 1). Investigators, including Alan Peters and Ita R. Kaiserman-Abramof (1969), demonstrated the variability of PSD size, shape, and perforations within it. The significance of the perforations remains unknown. Our knowledge of the complexity of the PSD has advanced considerably from initial biochemical characterizations of PSDs in the 1970s and 1980s by the Carl W. Cotman (1974), Philip Siekevitz (1977), Andrew Matus (1980), and James W. Gurd (1982) groups. Despite its apparent dense and impenetrable structure as seen with electron microscopy, the PSD contains a wealth of molecules that mediate communication of signaling events in space and time. These molecules permit temporal and spatial communication of signaling events with signal transduction pathways in the cytoplasm and modulation of these events by regulatory proteins, such as kinases and phosphatases. The PSD is a dynamic structure, whereby its component proteins are arranged in a precise and hierarchical manner as detailed by Mary B. Kennedy (2008), one of the pioneers in revealing the molecular architecture of the PSD. The constituents, which can be grouped into three major tiers of proteins, include: (1) receptors, ion channels, and adhesion proteins shared with the postsynaptic membrane; (2) scaffold proteins, which connect receptors to each other, to other membrane components, and to the actin cytoskeleton; and (3) the actin-based cytoskeleton. In this way, the PSD has special importance in dendritic spine function, especially because of its structural complexity in dendritic spines.

Dendritic spines contain an actin cytoskeleton, which occupies most of the interior of the spine. Whereas actin filaments appear randomly arranged within the spine, there is a complex compartmentalization of actin, which may impact or be a result of dendritic spine function. Thomas A. Blanpied's group (2010) proposed that the dendritic spine cytoskeleton is a "network of networks" regulating synapse function in a coordinated manner by way of several mechanisms that are located within different areas of the spine. These include receptor positioning, remodeling of the PSD, regulation of endocytosis, trafficking of organelles, neck structure and function regulation, and control of morphology.

Organelles present within the dendritic spine include membranous sacs in various configurations known as the spine apparatus (Fig. 1). Some of the sacs appear as organized sacs, separated by dense bands, although some membranous sacs can appear individually. As detailed in the introduction to this volume, the spine apparatus and other membranous organelles may be involved in the sequestration and release of Ca^{2+}. Menahem Segal and colleagues (2010) suggest that synaptopodin, an actin-binding protein that co-localizes with the spine apparatus, plays an essential role in the relationship between calcium stores and long-term plasticity.

Adding to the complexity of activity-dependent synaptic plasticity may be its dependence on local regulation of protein translation. Oswald Steward and William B. Levy (1982) first described the presence of polyribosomes at the base of dendritic spines. Several issues emerged from this discovery: that is, the processes by which RNA translation is repressed or silenced during its transport to its target synaptic site; the nature of activating stimuli controlling translation at the destination site; and the effects of synaptic deficits in disease states on local translation (cf. Wang et al. (2010) for review).

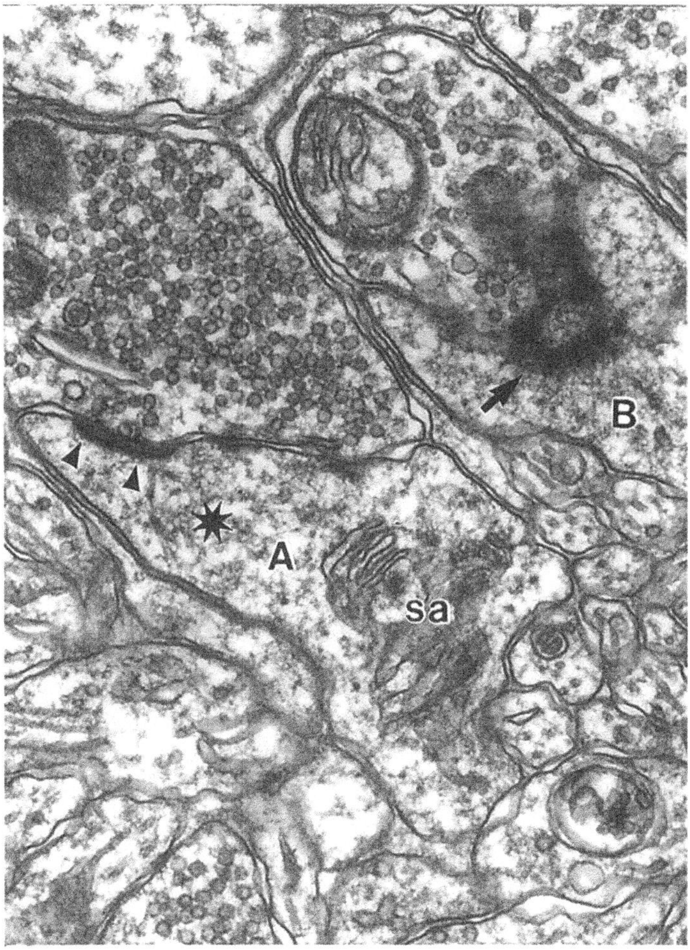

Fig. 1 An electron micrograph of a *mushroom-shaped* spine displaying a spine apparatus. The *mushroom-shaped* postsynaptic spine labeled (A) has a large bulbous head and displays a spine apparatus (*sa*). The postsynaptic density is indicated by *arrowheads* on its *left side*, although some dense material can also be seen beneath the postsynaptic membrane of the rest of the synaptic contact. The actin cytoskeleton is indicated by an *asterisk*. The neighboring synapse (B) shows a PSD (*arrow*) cut parallel to the surface of the postsynaptic membrane. (Cohen RS (2021) Cell biology of the synapse. In: *Neuroscience in the 21st century: from basic to clinical*, 3rd edn. Springer Nature, p 22)

The present volume is a unique and comprehensive map amplifying our current understanding of structure-function relationships within the dendritic spine. This structure is complex in terms of factors that impact it, and these are addressed in this volume. Chapters include those that consider morphology, preservation techniques, and innovative visualization approaches; dendritic spine development at various stages; trophic factors and dendritic spines; electrophysiology; steroid hormones, dendritic spines, and neuropsychiatric disorders; and interactions of dendritic spines with astrocytes and the extracellular milieu. As noted above, the introduction by the editors gives an expansive view of the history and evolution of our knowledge of dendritic spine structure and function.

Many studies on dendritic spines demonstrate correlative or causal phenomena. One of the main challenges to some studies involving morphology is a faithful representation of the living tissue in the fixed state. This issue is addressed in two of the following chapters by Dr. Floris G. Wouterlood and Dr. Alberto A. Rasia-Filho (below). Dr. Wouterlood presents a compendium of various approaches used to visualize dendritic spines, ranging from Golgi metallic silver staining to genetic engineering and transfection by a virus. Specific details of various procedures are included, making this chapter invaluable for approaching structure-function relationships of dendritic spines. These structures project in all directions from a dendrite, making them especially difficult to visualize in their entirety, and more advanced stereological techniques are essential. Advances in microscopy are detailed, for example, volume scanning electron microscopy and 3D reconstruction by field ion beam scanning electron microscopy (FIB-SEM), simulated emission depletion microscopy (STED), and other cutting-edge microscopy techniques, whereby dendritic spines can be visualized in greater detail than with conventional confocal laser scanning microscopy (CLSM) and without some of its limitations.

Dr. Joseane Righes Marafiga and Dr. Maria E. Calcagnotto present the electrophysiology of dendritic spines in regard to information processing, dynamic compartmentalization, and synaptic plasticity. Of particular interest is their discussion of backpropagating action potentials and synaptic potentials in dendritic spines as shown by novel technologies. The authors describe the temporal and spatial dynamics of receptors on spines related to synaptic plasticity and examine studies of synaptic potentials mediated in dendritic spines in different biochemical and electrophysiological compartments and in different neuronal subpopulations. How spine structure is related to function and structural and biochemical relationships within the spine, itself, is a special analysis within this chapter. Also, noteworthy is a discussion of the influence of the extracellular matrix on spine synaptic plasticity and the formation and stabilization of new dendritic spines and even receptor trafficking. Spine neck resistance, associated with the electrical compartmentalization of spines, is another intriguing consideration in this chapter.

The elucidation of mechanisms involved in dendritic spine development requires a two-fold approach: one, its role in brain development, itself, and, second, its development in the adult as a consequence of internal and external inputs, such as trophic factors, drugs, electrophysiological changes, or experiential conditions. Dr. Zdravko Petanjek, Dr. Ivan Banovac, Dr. Dora Sedmak, and Dr. Ana Hladnik focus on human brain development with emphasis on spine and synapse overproduction and synaptic pruning for the development of brain circuits. The significance of overproduction and elimination of dendritic spines allows for dynamic reorganization of synaptic circuitry at different stages of development before stabilizing in the adult. These studies impact our understanding of cognitive aptitudes during various life phases, as well as vulnerability to the formation of abnormal circuitry seen in neuropsychiatric disorders.

Neurotrophic factors, such as brain-derived neurotrophic factor (BDNF), appear to have many effects on various brain areas and specific brain structures and, consequently, on neural plasticity. Dr. Oliver von Bohlen und Halbach presents beautifully illustrated, all-inclusive, and detailed studies on the role of the neurotrophin family and ephrin family in mediating neural plasticity via dendritic spines in the adult brain. Dendritic spines exhibit variability in morphology and numbers, and this diversity may reflect their ability to adapt to changing conditions, thereby resulting in long-lasting effects on higher processes, such as learning and memory and affective behaviors. In addition to the roles of neurotrophins (particularly BDNF) and ephrins in detail in dendritic spine plasticity, the roles of other neurotrophic factors that impact dendritic spines are given, including epidermal growth factors (EGFs), fibroblast growth factors (FGFs), ghrelin and insulin, glial cell derived neurotrophic factor, insulin-like growth factor, leptin, pituitary adenylate-cyclase-activating polypeptide (*PACAP*), and the transforming growth factor beta (TGFß) family. Studies described range from the cell biology of dendritic spines to their impact on behaviors and behavioral disorders.

Surrounding dendritic spines are glia and the extracellular space. The historical reputation of glia as a "glue" in the nervous system has been superseded and revolutionized by numerous investigators. Glia are now thought to play an active role in various key aspects of brain function, including synaptic modulation. The comprehensive and informative chapter by Dr. Alberto A. Rasia-Filho, Dr. Maria E. Calcagnotto, and Dr. Oliver von Bohlen und Halbach addresses the morphology of astrocytes and microglia in relationship to dendritic spines, their functional associations, and how their interactions affect synaptic plasticity. Specifically, the chapter includes a discussion of the structural interface between glial cells and dendritic spines, electrophysiological implications of associations and factors that may influence neurotransmission, such as hormones, neurotransmitters, cell adhesion molecules, and the extracellular matrix and neurotrophic factors. These studies are highly significant in that they show that glia play a role in neuron excitability, synaptic transmission and as a possible substrate for long-term plasticity.

Dr. Nicolas Heck and Dr. Marc Dos Santos present a comprehensive perspective of dendritic spines and their role in learning and memory. They take us on a fascinating and thought-provoking journey starting with the early visualization of dendritic spines using light and electron microscopy through state-of-the-art and novel imaging methods that permit the integration of structure-function and cause-and-effect relationships. Included are discussions of dendritic spine formation relative to developmental stages versus adult spinogenesis and the role of presynaptic input and other spines that cluster within the vicinity. The importance of spine elimination and spine formation is discussed, as well as a consideration of their role in new memory formation versus stabilization of pre-existing memories. Molecular signaling mechanisms are also considered. Disorders of the brain involving dendritic spines are reviewed. The exciting combination of molecular techniques and behavioral models are profound innovations in our understanding of dendritic spine structure and function.

Dr. Maya Frankfurt, Dr. Zeinab Nassrallah, and Dr. Victoria Luine present their seminal and impactful work on steroid hormone effects on dendritic spines. Hormonal fluctuations may lead to behavioral alterations at various timepoints during human life stages and these changes may involve hormonal effects on dendritic spines. Estrogen has been shown to have powerful growth effects in neurons via both genomic and non-genomic pathways. Activation of these pathways may lead to the transcription of growth-related genes, which in turn may result in persistent alterations in cellular and synaptic substrates of growth, including dendritic spines in brain areas, such as the hippocampus and prefrontal cortex, areas involved in cognitive processing. Evidence for estrogen effects on the cell biological mechanisms of dendritic spine formation is presented. Androgens, also, appear to affect some of these processes. Sex differences in neuropsychiatric disorders and the possible role of steroid effects on dendritic spines in these disorders are discussed. Of special interest, the authors consider the potential of steroid hormone actions in the brain in the amelioration of some of these disorders.

Human brain tissue presents a recurring and problematic issue for light and electron microscopic studies of dendritic spines. The chapter by Dr. Alberto A. Rasia-Filho and Josué Renner includes a comprehensive discussion of dendritic structure for synaptic organization, phylogenetic and ontogenetic differences of dendritic spines between humans and other species, and examples of spiny neurons from the spinal cord, and various brain areas. Reconstructions using Golgi technique and bright-field microscopy detail the *continuum* of spine shapes and sizes, using confocal microscopy with extracellular (DiI) and intracellular (Lucifer Yellow) fluorescent dye, and transmission electron microscopy. A review of pathological changes from the literature is presented. Other perspectives, including high-resolution and multi-patch clamp recordings and biocytin for morphology, are discussed. This chapter will allow an in depth understanding of normal and pathological specimens in human disease states. It, also, serves as a compendium for referencing dendritic spines in various parts of the nervous system.

The following chapters present cellular, molecular, and electrophysiological aspects of dendritic spines that have emerged since the innovative discoveries of the nineteenth and early to mid-twentieth centuries. The swift development of key technologies together with the combined behavioral, imaging, electrophysiological, and molecular strategies as presented here give a front-row view of how these approaches bridge dendritic spine and brain function. A special and unique feature of this collection is its inclusion of data regarding human dendritic spines, as well as technical approaches to study human brain tissue. Deficits in brain function may result from dendritic spine abnormalities, which may impact disorders, such as cognitive decline, addiction, anxiety, and depression. Whereas experimental designs and emphases may differ, they all seek to understand the basis of our motivations, higher cognitive processes, and behaviors.

Rochelle S. Cohen
Professor Emerita, Department of Anatomy
and Cell Biology
University of Illinois at Chicago,
Chicago, IL, USA

References

Cohen RS, Blomberg F, Berzins K, Siekevitz P (1977) The structure of postsynaptic densities isolated from dog cerebral cortex. I. Overall morphology and protein composition. J Cell Biol 74(1):181–203. https://doi.org/10.1083/jcb.74.1.181

Cotman CW, Banker G, Churchill L, Taylor D (1974) Isolation of postsynaptic densities from rat brain. J Cell Biol 63(2 Pt 1):441–455. https://doi.org/10.1083/jcb.63.2.441

Frost NA, Kerr JM, Lu, HE, Blanpied TA (2010) A network of networks: cytoskeletal control of compartmentalized function within dendritic spines. Curr Opin Neurobiol 20(5):578–587. https://doi.org/10.1016/j.conb.2010.06.009

Gray EG (1959) Electron microscopy of synaptic contacts on dendritic spines of the cerebral cortex. Nature 183:1592–1593

Gurd JW, Gordon-Weeks P, Evans WH (1982) Biochemical and morphological comparison of postsynaptic densities prepared from rat, hamster, and monkey brains by phase partitioning. J Neurochem 39(4):1117–1124. https://doi.org/10.1111/j.1471-4159.1982.tb11504.x. PMID: 7119784

Harris KM (2020) Synaptic odyssey. J Neurosci 40(1):61–80

Kennedy MB, Marcora E, Carlisle HJ (2008) Scaffold proteins in the postsynaptic density. In: Hell JW, Ehlers MD (eds) Structural and functional organization of the synapse. Springer, New York, pp 407–440

Matus A, Pehling G, Ackermann M, Maeder J (1980) Brain postsynaptic densities: the relationship to glial and neuronal filaments. J Cell Biol 87(2 Pt 1):346–359. https://doi.org/10.1083/jcb.87.2.346

Palay SL (1958) The morphology of synapses in the central nervous system. Exp Cell Res 5(Suppl 5):275–293

Peters A, Kaiserman-Abramof IR (1969) The small pyramidal neuron of the rat cerebral cortex. The synapses upon dendritic spines. Z Zellforsch Mikrosk Anat 100(4):487–506. https://doi.org/10.1007/BF00344370

Purpura DP (1974) Dendritic spine "dysgenesis" and mental retardation. Science 186:1126–1128

Ramón y Cajal S (1888) Estructura de los centros nerviosos de las aves. Rev Trim Histol Norm Patol 1:1–10

Segal M, Vlachos A, Korkotian E (2010) The spine apparatus, synaptopodin, and dendritic spine plasticity. Neuroscientist 16(2):125–131. https://doi.org/10.1177/1073858409355829

Steward O, Levy WB (1982) Preferential localization of polyribosomes under the base of dendritic spines in granule cells of the dentate gyrus. J Neurosci 2(3):284–191. https://doi.org/10.1523/JNEUROSCI.02-03-00284.1982

Wang DO, Martin KC, Zukin S (2010) Spatially restricting gene expression by local translation at synapses. Trends Neurosci 33(4):173–182

Preface

Dendritic spines are fascinating neuronal elements that modulate synaptic transmission, strength, and plasticity of neural networks. Spines are found in very different species and multiple areas of the nervous system in normal and altered, pathological conditions. Recently, research in the field of dendritic spines has undergone dramatic advances in terms of techniques and experimental findings from *in vitro* and *in vivo* data, from animal models to human neurons, also including computational models. To address these cutting-edge findings, we aim to provide a state-of-the-art comprehensive book with chapters that can serve as a reference for current and future research with implications in broadly related areas in neuroscience. This work is suitable for undergraduate as well as graduate students, researchers, and leaders in fields including morphology, neurophysiology, development, neurochemistry, neuroendocrinology, neuropathology, electrophysiology, microscopy, psychology, neurology, psychiatry, and artificial intelligence/computational neural modeling.

We would like to thank the excellent editorial support for this book from William Lamsback and Vishnu Prakash (Springer Nature, New York) and the outstanding participation, guidance, and inspiration of Prof. Rochelle Cohen (Professor Emerita, University of Illinois at Chicago). We are indebted to the distinguished colleagues that dedicated their time, knowledge, and care to preparing the relevant chapters that compose this book. We also acknowledge Prof. Juan A. De Carlos (Cajal Institute/Cajal Legacy, Madrid) for the kind permission to reproduce the original drawings of Santiago Ramón y Cajal in this book, and Prof. U. Valentin Nägerl (Université de Bordeaux) for his efforts to be with us and for permitting us to include his work in the first chapter.

This book is dedicated to my beloved "encantos," cheerful daughter and son, and my sunny nieces (Alberto); to my beloved husband (Maria Elisa); to my beloved wife and my wonderful children (Oliver); to the patients and their relatives; and to scientists worldwide.

Porto Alegre, RS, Brazil — Alberto A. Rasia-Filho
Greifswald, Germany — Maria Elisa Calcagnotto
Oliver von Bohlen und Halbach

Contents

Chapter 1
Introduction: What Are Dendritic Spines?

Alberto A. Rasia-Filho, Maria Elisa Calcagnotto, and Oliver von Bohlen und Halbach

Abstract Dendritic spines are cellular specializations that greatly increase the connectivity of neurons and modulate the "weight" of most postsynaptic excitatory potentials. Spines are found in very diverse animal species providing neural networks with a high integrative and computational possibility and plasticity, enabling the perception of sensorial stimuli and the elaboration of a myriad of behavioral displays, including emotional processing, memory, and learning. Humans have trillions of spines in the cerebral cortex, and these spines in a *continuum* of shapes and sizes can integrate the features that differ our brain from other species. In this chapter, we describe (1) the discovery of these small neuronal protrusions and the search for the biological meaning of dendritic spines; (2) the heterogeneity of shapes and sizes of spines, whose structure and composition are associated with the fine-tuning of synaptic processing in each nervous area, as well as the findings that support the role of dendritic spines in increasing the wiring of neural circuits and their functions; and (3) within the intraspine microenvironment, the integration and activation

A. A. Rasia-Filho (✉)
Department of Basic Sciences/Physiology and Graduate Program in Biosciences, Universidade Federal de Ciências da Saúde de Porto Alegre, Porto Alegre, RS, Brazil

Graduate Program in Neuroscience, Universidade Federal do Rio Grande do Sul, Porto Alegre, RS, Brazil

M. E. Calcagnotto
Graduate Program in Neuroscience, Universidade Federal do Rio Grande do Sul, Porto Alegre, RS, Brazil

Department of Biochemistry, Universidade Federal do Rio Grande do Sul, Porto Alegre, RS, Brazil

Graduate Program in Biochemistry, Universidade Federal do Rio Grande do Sul, Porto Alegre, RS, Brazil

Graduate Program in Psychiatry and Behavioral Science, Universidade Federal do Rio Grande do Sul, Porto Alegre, RS, Brazil

O. von Bohlen und Halbach
Institut für Anatomie und Zellbiologie, Universitätsmedizin Greifswald, Greifswald, Germany

A. A. Rasia-Filho et al. (eds.), *Dendritic Spines*, Advances in Neurobiology 34,
https://doi.org/10.1007/978-3-031-36159-3_1

of signaling biochemical pathways, the compartmentalization of molecules or their spreading outside the spine, and the biophysical properties that can affect parent dendrites. We also provide (4) examples of plasticity involving dendritic spines and neural circuits relevant to species survival and comment on (5) current research advancements and challenges in this exciting research field.

Keywords Synapse · Postsynaptic processing · Synaptic plasticity · Biochemical compartmentalization · Morphology and function · Neural circuits · Behavior

1.1 The Discovery of Small Neuronal Protrusions and the Search for the Biological Meaning of Dendritic Spines

In 1906, the Nobel Prize in Physiology or Medicine was awarded to Camillo Golgi and Santiago Ramón y Cajal "in recognition of their work on the structure of the nervous system." Cajal[1] conceived that "The nerve cells are morphological entities, neurons, to use the word brought into use by the authority of Professor Waldeyer. My celebrated colleague Professor Golgi has already demonstrated this property with respect to the dendritic or protoplasmic processes of the nerve cells; but at the beginning of our research there were only vague conjectures as regards the behaviour of the axon branches and collaterals. We applied Golgi's method, firstly in the cerebellum and then in the spinal cord, the cerebrum, the olfactory bulb, the optic lobe, the retina and so on of embryos and young animals, and our observations revealed, in my opinion, the terminal arrangement of the nerve fibres. These fibres, ramifying several times, always proceed towards the neuronal body, or towards the protoplasmic expansions around which arise plexuses or very tightly bound and rich nerve nests" (Cajal 1906;[2] Fig. 1.1).[3]

Using the Golgi technique, Cajal discovered thin protrusions emerging from the surface of dendrites of Purkinje cells in birds and described them as "espinas dendríticas" (dendritic spines, from the monograph "Estructura de los centros nerviosos de las

---

[1] We use "Cajal" to refer to Santiago Ramón y Cajal. It is also usual to find "Ramón y Cajal" in the literature. We are following the same way Cajal cited his own works in the book "¿Neuronismo o reticularismo?" (1933, reprinted 1952), the way Rafael Lorente de Nó refered to him, and based on the preface of the translated edition of Cajal's book of 1909–1911 (by Swanson and Swanson, Oxford University Press, 1995).

[2] Award ceremony speech. NobelPrize.org. Nobel Prize Outreach AB 2022 at https://www.nobelprize.org/prizes/medicine/1906/ceremony-speech/

[3] In this chapter, we use quotation marks and footnotes to include parts of selected works aiming to present original descriptions for data on dendritic spines. However, dendritic spines research has been subject to constant advancement. The references presented along the text are some examples for each topic or experimental finding, but more references can be found in each article or book cited here.

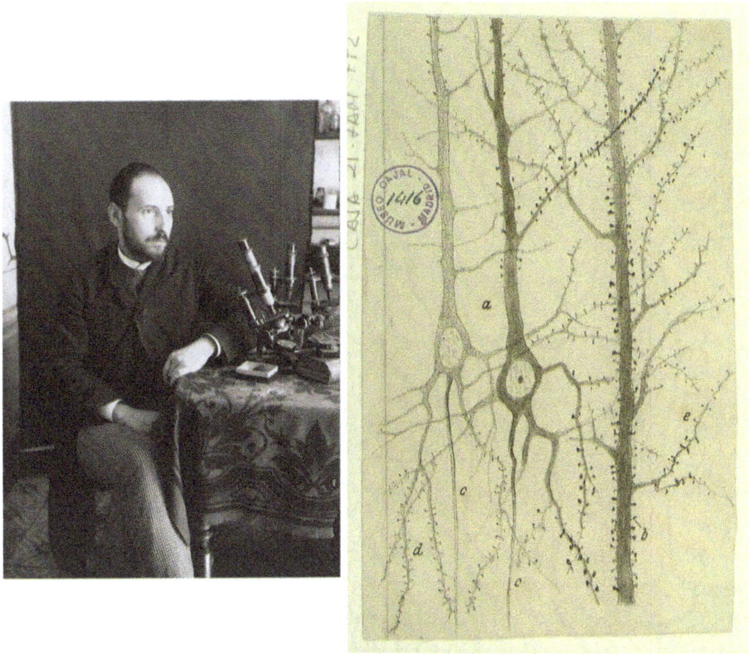

Fig. 1.1 (left) Santiago Ramón y Cajal (1852–1934). (right) Cajal's drawing of dendritic spines stained using the Ehrlich methylene blue method under light microscopy. Pyramidal cells in the cerebral cortex of the guinea pig. Note the presence of spines protruding from dendrites. *a*, two medium-sized pyramidal cells; *b*, spines on a dendritic trunk belonging to a large pyramidal cell; *c*, axons; *d*, basal dendrites with their spines; *e*, collateral branches of a dendritic trunk with their spines (Cajal 1909–1911). (Courtesy of the Cajal Legacy, Cajal Institute (CSIC), Madrid, Spain)

aves," dated 1888; Yuste 2010, see Pannese 2015 for a reference to Owsjannikov in 1864). Cajal's description of dendritic spines as actual elements of neurons was: "… the surface of the Purkinje cells´ dendrites appears ruffled with thorns or short spines, which on the terminal dendrites look like light protrusions. Early on we thought that these eminences were the result of a tumultuous precipitation of the silver; but the constancy of their existence and its presence even in preparations where the staining appears with great delicacy in the remaining elements, incline us to consider them as a normal disposition" (translated from the Spanish by Yuste 2010; Fig. 1.2). Indeed, dendritic spines were further depicted in cortical pyramidal cells (Fig. 1.3), as when describing the appearance of "psychic cells" in different vertebrates and humans (Cajal 1894) or the cerebral cortex of guinea pigs and cats using another staining, the Ehrlich methylene blue one (Cajal 1909–1911, 1933; Fig. 1.1).

These major achievements unraveled fundamental principles for the functioning of the nervous system, which demanded skilled interpretation of the results from staining procedures in different species and developmental stages over the lifetime of each individual (Pannese 1996; Cimino 1999; Rodríguez de Romo 2005; García-López et al. 2010; Yuste 2010; DeFelipe 2011; de Castro 2019). From that time to

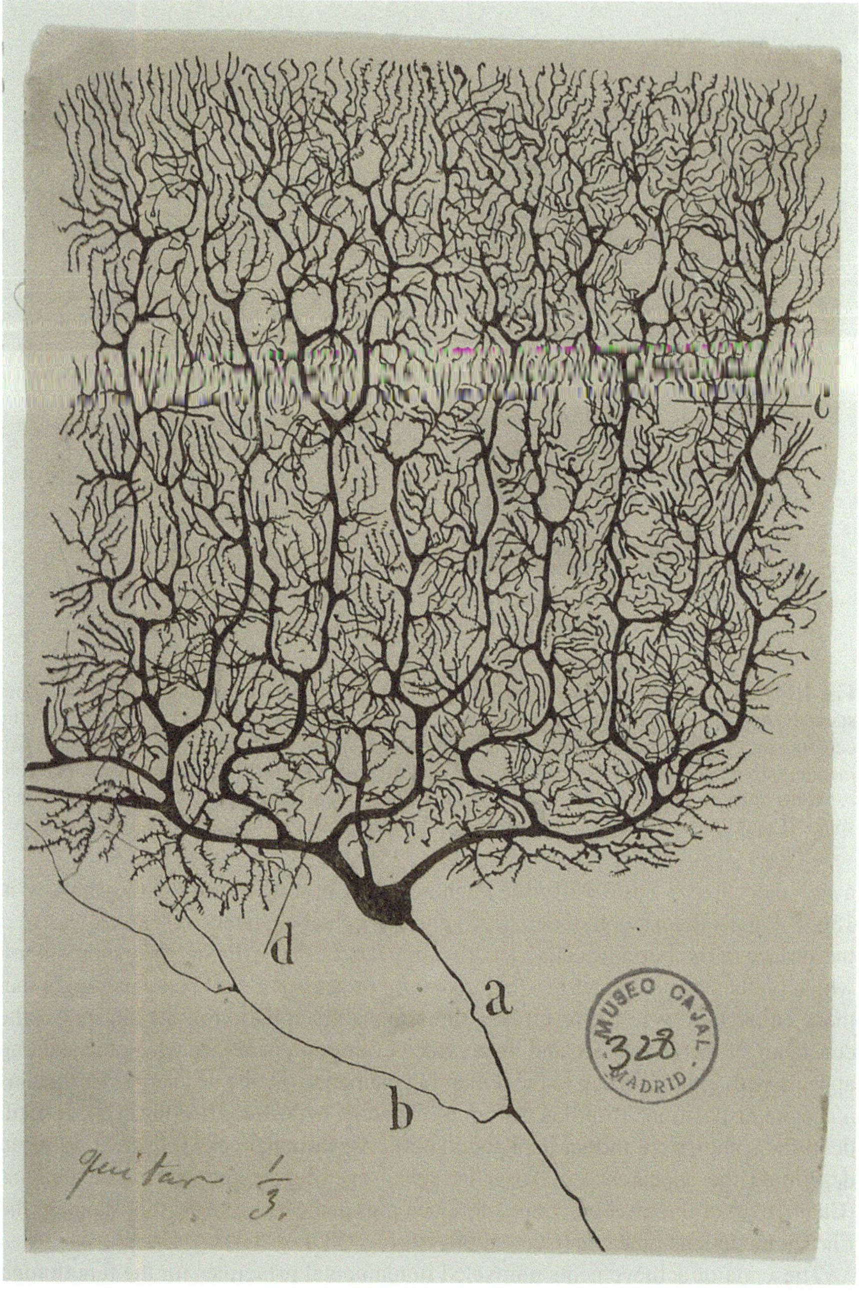

Fig. 1.2 Cajal's drawing of a Golgi-impregnated Purkinje cell from the human cerebellar cortex under light microscopy. Note the presence of protruding spines distributed in dendrites of multiple orders and along proximal to distal branches. a, axon; b, recurrent collateral; c and d, spaces in the dendritic foliage for capillaries, stellate, or basket cells (Cajal 1909–1911). (Courtesy of the Cajal Legacy, Cajal Institute (CSIC), Madrid, Spain)

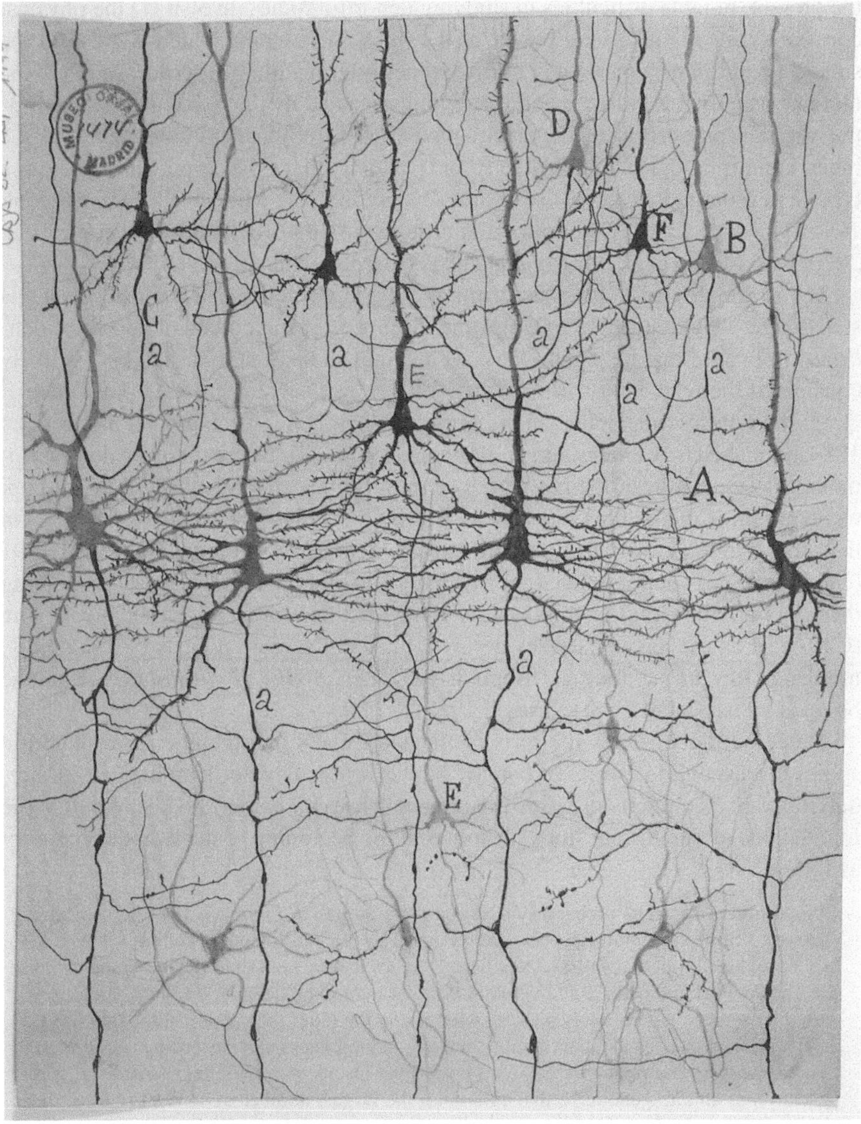

Fig. 1.3 Cajal's drawing of Golgi-impregnated pyramidal neurons from inner layers of the cerebral cortex under light microscopy. *A*, giant pyramidal cell layer; *B*, deep part of the stellate cell layer; *C, D*, small pyramidal cell with an axon (*a*) that bifurcates into two arcuate ascending branches; *E*, medium-sized pyramidal cell (at the middle) with a long, descending axon, and a small pyramidal cell (at the bottom) with a descending axon in a deeper layer; *F*, small pyramidal cell with an axon that bifurcates into recurrent and descending collaterals (description by comparison with Figs. 388, 389, and 399 in Cajal 1909–1911). Note the presence of various protruding spines in basal and apical dendrites. Short, gnarled twigs projecting from axonal branches are terminal swellings to innervate nearby structures (from middle to bottom). (Courtesy of the Cajal Legacy, Cajal Institute (CSIC), Madrid, Spain)

the present, notable difficulties in studying dendritic spines involve (1) the physical constraints for the resolving power of the light microscopy, which is close to the smallest and thinnest parts of dendritic spines after histological processing, (2) the fact that usually few cells are completely impregnated by the Golgi method, and (3) the variable proportion of dendritic spines due to phylogeny, ontogeny, and the inherent morphological heterogeneity and plasticity that exists even within the same type of neuron in the same layer or area (Valverde 1967; Miller 1988; Kasper et al. 1994; Tjia et al. 2017; Cembrowski and Spruston 2019; Benavides-Piccione et al. 2020; Rasia-Filho et al. 2021; Fuentealba-Villarroel et al. 2022).

The originality of Cajal's descriptions of the existence of dendritic spines is stated here: "Held attribute to Golgi the discovery of the collateral spines of dendrites. I am sure that he should have seen spines in his first material, but he likely considered them, as later surmised by Kolliker, as a silver chromate precipitate… Golgi did not describe spines in his prior works… Note, for example, his work of 1883, published in *Archives ital. de Biologie*, volume III, where there are no descriptions and no drawings of spines not even in Purkinje cells. Likewise, Forel did not name nor draw them in his report from Arch. J. Psychiatrie, etc. Bd 18, 1887… On the other hand, we described and illustrated them in 1888 (cerebellum) and 1891 (cerebrum), not counting that we provided a specific monograph in 1896, staining them by both the Golgi method and by some modification of the Ehrlich technique (CAJAL: Las espinas colaterales de las células del cerebro teñidas por el azul de metileno. Rev. trim. microgr. Tomo I, pág. 151, 1896)…" (translated from the Spanish by us; Cajal 1933, reprinted 1952).

Time and much research efforts confirmed Cajal's interpretation that dendritic spines correspond to points of contact between neurons, not histological artifacts due to fixative solutions or silver precipitate (Figs. 1.2 and 1.3). Cajal detailed the foundations of our current study on spines at the beginning of the twentieth century as follows:

> "Spines are very short processes arising at right angles from any part of the dendritic surface... Spines generally appear as very tightly stretched filaments ending freely in a spherical or elliptical swelling. Nowhere are they better seen than on the branches of cerebellar Purkinje cells… and cerebral cortical pyramidal cells…; on the former, they are thick, stubby, and densely packed, whereas on the latter they appear to be thin, long, and widely spaced. Thus, their number, length, and thickness vary according to cell type. They even vary between species; if we compare neurons in homologous structures, it is usually possible to generalize that cells with the largest number of spines are from the brains of more highly evolved animals. To cite just one example from among the vertebrates, Purkinje cells dendrites in birds have fewer spines than those in mammals (we should point out that the neuronal processes of invertebrates generally are bare and thus appear to lack spines)

1. Spines are... demonstrated with the original Golgi method as with Cox's variant and with the Ehrlich technique.
2. They always arise in the same way, and from the same regions of the dendritic tree.
3. They are never found on certain other parts of the neuron, such as the axon, cell body, larger dendritic trunks, or, more correctly, from initial parts of dendrites.
4. When viewed with a high-power apochromatic objective, spines do not in any way resemble crystals or irregular deposits. Instead, they appear to be either simple, or often complex, thread-

like structures that are in direct continuity with the main part of the dendritic process from which they arise, and there is no indication that they are separated from these processes.

5. When the Ehrlich technique is used correctly, it too impregnates spines clearly, with their pedicles appearing pale blue and their terminal spherules appearing deep blue...
6. Finally, it should be noted that spines are never stained with neurofibrillary methods...

> The fact that spines end completely freely is both supplementary evidence and, in a sense, decisive testimony for the equally free termination of dendrites themselves. And the general theory of neuronal independence and individuality gains further support from these invaluable new arguments.
>
> ... we showed that the receptive surface of dendrites is increased tremendously by the presence of spines, and that, because of them, contacts between terminal arborizations of axons and dendrites could be more intimate" (Cajal 1909–1911, translated by Swanson and Swanson 1995).

In a personal letter dated 1934, Lorente de Nó ascertained to Cajal that "collateral spines are doubtless points of contact" (Fairén 2007). Twenty-five years after Cajal's death, Gray (1959) used transmission electron microscopy (EM) to determine that dendritic spines are postsynaptic structures forming contacts with axon terminals (see examples in Fig. 1.4). The first paragraph of Gray's seminal two-page article is: "When stained by the Golgi or methylene blue method for light microscopy, certain dendrites of the cerebral cortex and elsewhere appear to have numerous spinous projections. The nature of these spines has long been disputed. For example, it has been suggested that they are simply "nutritive" expansions, or presynaptic end-feet, or postsynaptic processes of the dendrite - the presynaptic component remaining unstained. Electron microscopy shows that the spines are in fact sites of synaptic contact." By the data originally obtained in the rat visual cortex (Gray 1959), spines have been reported as postsynaptic elements under various methodological approaches. An exception occurs for the dendritic spines ("gemmules") of granule cells in the mammalian olfactory bulb that form reciprocal synapses with mitral cell dendrites and can be presynaptic elements as well (Rall et al. 1966; Pannese 2015; Shepherd et al. 2020 - e.g., see Fig. 2A in this latter article). Furthermore, spines are key to the evolution of the nervous system functioning and species-specific skills (Yuste 2010; DeFelipe 2011). Remarkably, spines change their shape when the first orientation flight occurs in honeybees (Brandon and Coss 1982; Fig. 1.5). Spines can be found in polyclad flatworms, and evident morphological spine plasticity was also reported in fishes and birds submitted to ecologic relevant behavioral activity or experimental manipulations (Coss and Perkel 1985). That is, the spine's "exterior geometry and variation in size are remarkably similar in animals with evolutionary histories that diverged as long ago as Precambrian and late Cambrian times. In light of this long evolutionary history of conservation or convergence, spines must play a fundamentally important functional role" (Coss and Perkel 1985).

The study of dendritic spines developed with the Golgi method, its variants, and additional techniques over decades. The characterization of spiny neurons was expanded with EM, immunocytochemical, axonal retrograde tracing markers, electrophysiological, genetic, and optical tools providing new experimental

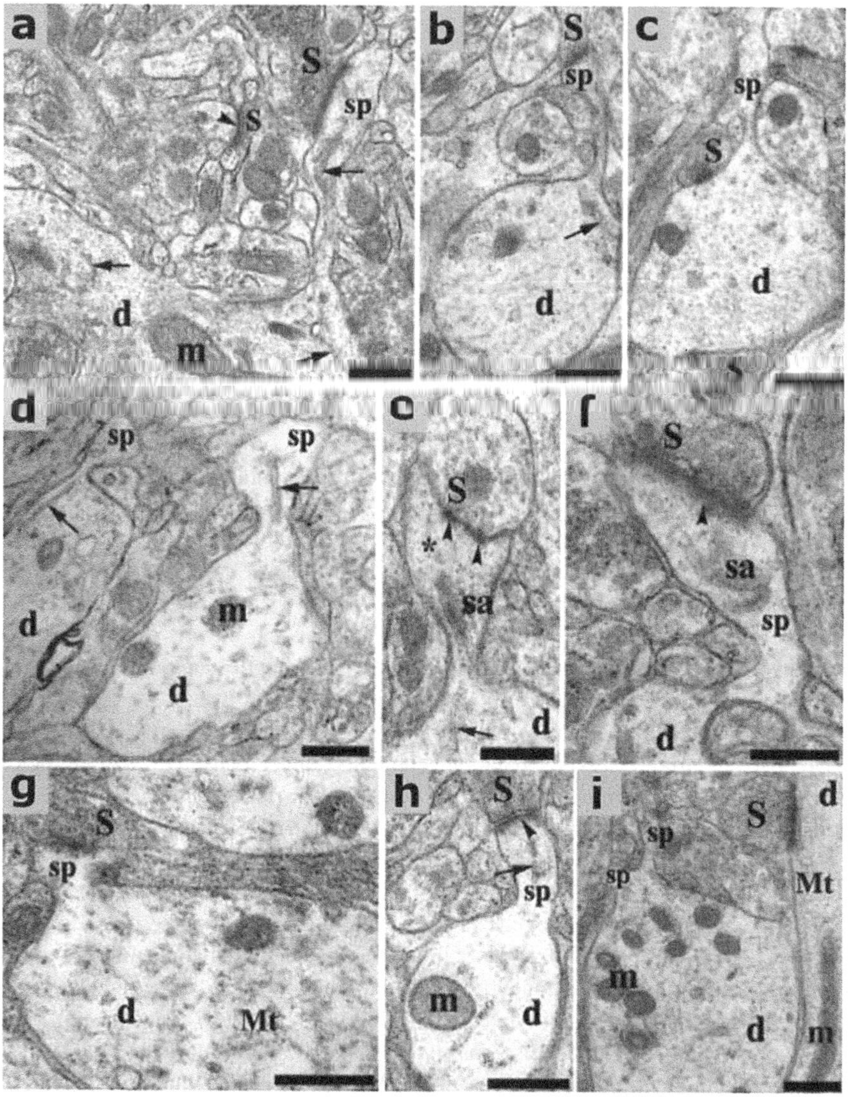

Fig. 1.4 Transmission electron micrographs showing the ultrastructure of dendritic spines of thin (**a**–**c**), mushroom-like (**d**–**f**), and stubby/wide (**g**–**i**) shapes. Spines are specialized postsynaptic units. Note the presence of synapses (S) and the postsynaptic density (arrowheads), smooth endoplasmic reticulum (arrows), spine apparatus (sa), and actin filaments (asterisk) in different spines. Microtubules (Mt) and mitochondria (m) are observed in dendritic processes (**d**). Data from the posterodorsal medial amygdala neurons of adult rats. Scale bar = 500 nm (**a**–**d**, and **g**–**i**), 400 nm (**e**), and 250 nm (**f**). (Legend adapted and figure reprinted from Hermel et al. (2006), under CCC RightsLink® license #5279390654924, originally published by John Wiley & Sons, Inc)

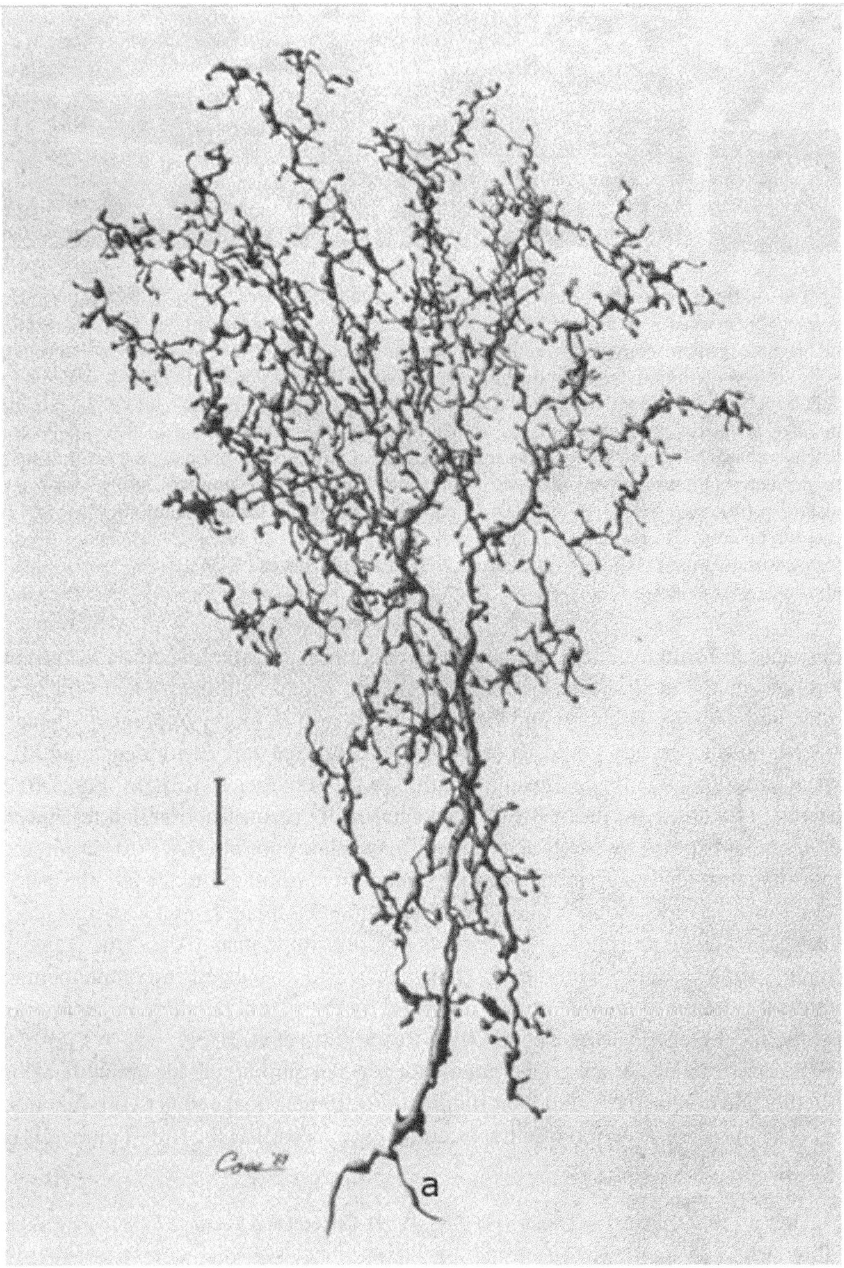

Fig. 1.5 Dendritic spines are found in different species along phylogeny and show structural and functional plasticity. Camera lucida drawing of the Golgi-impregnated dendritic tree of a spiny Kenyon cell in the calyx of the honeybee *corpora pedunculata*. Illustration was made from a large scale (1 μm = 6 mm) tracing of a 6-day-old flyer interneuron. Axon is labeled by the letter *a* (adapted from the original, in the bottom). Rapid Golgi method. Calibration bar = 10 μm. (Legend adapted and figure reprinted from Brandon and Coss (1982), under CCC RightsLink® license #5282600876347, originally published by Elsevier)

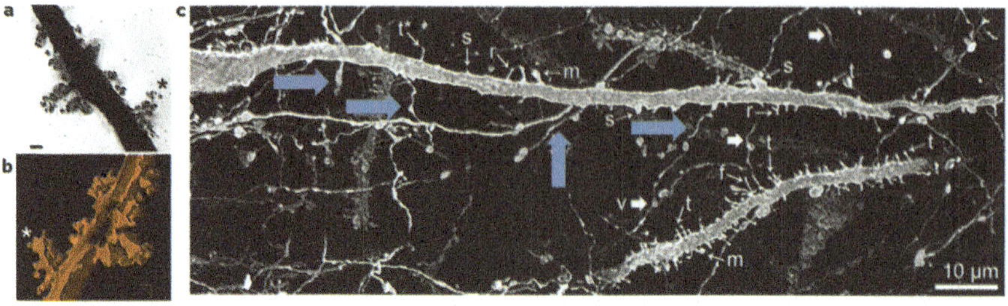

Fig. 1.6 (**a, b**) Higher magnification of a Golgi-impregnated dendritic segment showing a thorny excrescence (marked with an asterisk) from the human CA3 hippocampal field. In **a**, the original raw brightfield microscopy image and, in **b**, the same image after 3D reconstruction and rotated 180°. Scale = 2 μm. (Legend and figure reprinted from Reberger et al. (2018), under CCC RightsLink® license #5346040208162, data originally published by Elsevier). (**c**) Use of extracellular dye to obtain fluorescent images of neurons from the posterodorsal medial amygdaloid nucleus of rats. Cells were 3D reconstructed by confocal microscopy following the application of fine-powdered DiI in a coronal brain slice. Spines were classified as thin (t), stubby/wide (s), or mushroom (m) ones. Arrows point to axons with varicosities (v), which usually form *en passant* synapses or are close to dendritic shafts with a perpendicular to oblique orientation. (Figure reprinted from Brusco et al. (2010), under CCC RightsLink® license #5346050766588, originally published by Elsevier)

approaches.[4] From excellent drawings and diagrammatic, stereoscopic illustrations to represent the spatial distribution of dendritic spines within the neuropil (e.g., Szentágothai 1978; Brandon and Coss 1982), images of Golgi-impregnated spines under light microscopy could be reconstructed for their three-tridimensional (3D) aspect using specifically designed algorithm (e.g., Reberger et al. 2018; Fig. 1.6) or specific software (e.g., the software package for 3D reconstruction and to analyze dendrites and spines by MBF at www.mbfbioscience.com/blog/2017/03/complete-guide-imaging-analyzing-spines-neurons-neurolucida-360/; Dokter et al. 2015), including free, open-source one (e.g., Neuromantic, ImageJ, and ilastik; Guerra et al. 2023). Then, morphological features of Golgi-impregnated dendritic spines in human samples could be studied in more detail and visualized in complementary angles of rotation to include spines obscured by the parent dendrite thickness in a specific focal plane (Vásquez et al. 2018; Rasia-Filho et al. 2021).

The last decades witnessed a dramatic increase in our knowledge about the spine structure and connectivity and their implications for neuronal and network functioning.[5] The 3D reconstruction of ultrastructural data, as well as the use of intracellular

[4] Fairén et al. (1977), Freund and Somogyi (1983, 1989), Gabbott and Somogyi (1984), Peters et al. (1991), Fairén (2005), Harvey et al. (2008), Knott et al. (2008), Chen et al. (2011), Lanciego and Wouterlood (2011), Mancuso et al. (2013), Araya (2014), Sadakane et al. (2015), Lanciego and Wouterlood (2020), and Frankfurt and Bowman (2021).

[5] Scheibel and Scheibel (1955), Ramón-Moliner (1962), Valverde (1967), Fairén et al. (1977), Fifková and Van Harreveld (1977), Woolley et al. (1990), Freund and Somogyi (1983), Harvey et al. (2008), Bourne and Harris (2009), Kasai et al. (2010), Segal (2010), von Bohlen und Halbach (2010), Yuste (2010), Chen et al. (2011), Rochefort and Konnerth (2012), Spruston et al. (2013),

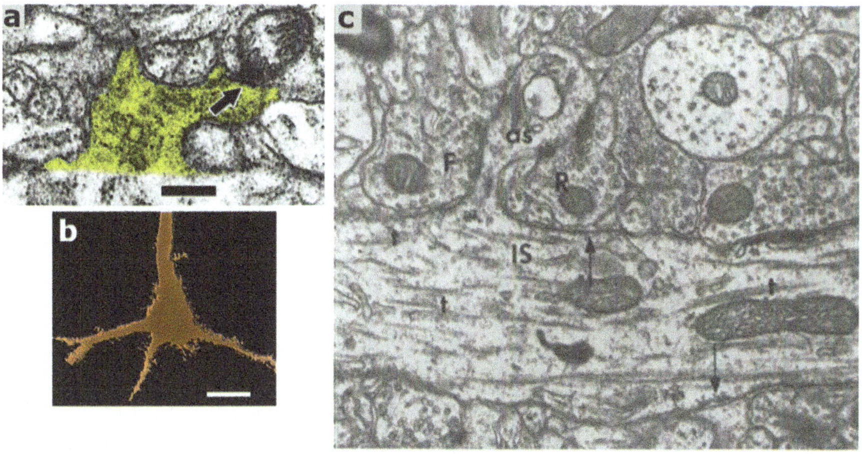

Fig. 1.7 (**a**) Digitized electron micrograph of the ultrastructure of a ramified spine (highlighted in yellow) arising from the neuronal cell body as observed in the posterodorsal medial amygdaloid nucleus of a female rat in late proestrus. Scale bar = 0.5 μm. (Legend and image reprinted from Zancan et al. (2015), under CCC RightsLink® license #5346080098217, originally published by John Wiley & Sons, Inc). (**b**) Golgi-impregnated pyramidal neuron from the adult human cerebral cortex. The 3D reconstruction was made using serial, z-sequence light microscopy images. Arrows point to spines protruding from the cell body and from proximal primary dendrites. Scale = 20 μm. (**c**) "An axonic spine (as) having a definite tip and stalk or stem in continuity with an initial segment (IS) which contains fascicles of microtubules (t) and the dense undercoating (arrows). Two presynaptic processes, one containing round vesicles (R) and the other flattened vesicles (F) form contacts with the axonic spine tip and stalk respectively. Image ×27,000." (Legend and image reprinted from Westrum (1970), under CCC RightsLink® license #5352501008129, originally published by John Wiley & Sons, Inc)

or extracellular fluorescent dyes in combination with laser scanning confocal microscopy (Fig. 1.6), two-photon laser scanning microscopy, and optogenetics or super-resolution STED microscopy added valuable findings to complement the scenario of spine morphology and function in various neuronal types and species. For example, to complement the original descriptions, EM images demonstrated that dendritic spines can also project from the neuronal perikarya in neurons from the medial amygdala of female rats (Zancan et al. 2015) and in a human cortical pyramidal neuron (Fig. 1.7a, b), from the initial segment of an axon in the prepyriform cortex of the rat (Westrum 1970; Peters et al. 1991; Fig. 1.7c), or in proximal dendrites of human cortical pyramidal neurons (i.e., along the initial 0–50 μm, Petanjek et al. 2008; Luengo-Sanchez et al. 2018; Rasia-Filho et al. 2021; Fig. 1.7c).

Sala and Segal (2014), Dall'Oglio et al. (2015), Hayashi-Takagi et al. (2015), Tønnesen and Nägerl (2016), Lanciego and Wouterlood (2020), Helm et al. (2021), and Cornejo et al. (2022).

1.1.1 Spiny Neurons and Synaptic Features

Dendritic spines enhance the connectivity and expand the possibilities of synaptic modulation in neurons[6] (Fig. 1.8). Henceforth, considering the existence, spatial distribution, and density of spines, neurons were classified as spiny, sparsely spiny, or non-spiny (also termed "aspiny") cells. Nevertheless, it is important to consider that the same neuron can have a heterogeneous density of spines along its different dendrites (Feldman 1984; Peters and Jones 1984), and those apparently aspiny dendrites can still display a few simple spines over their entire length (Fiala and Harris 1999).

Heterogeneity in spine density is also found in the same subset of neurons. For example, interneurons are much less spiny than the most abundant pyramidal cells in the human neocortex (Shapson-Coe et al. 2021). In the CA1 and CA3 hippocampal areas of mice, most of the interneurons are non-spiny and few of them are sparsely spiny; otherwise, a subset of parvalbumin-expressing GABAergic interneurons in the dentate gyrus show a high density of dendritic spines (Foggetti et al. 2019).

Cortical pyramidal neurons are examples of spiny cells. The number of neurons and synaptic profiles obtained from a 50 μm/side cube within layer V of the human anterolateral temporal cortex estimated the existence of 958,890 synapses for 21 local neurons, which gives about 30,000 synapses per neuron, 90% being asymmetric and 10% symmetric ones (DeFelipe 2011). These impressive numbers allow the estimation of approximately 100 trillion spines in the human cerebral cortex (Kasai et al. 2021). Near 99.5% of these spines lie in pyramidal neurons (Kubota et al. 2016) for the organization of the timely synaptic transmission using multiple neurochemical circuits (Palomero-Gallagher and Zilles 2019).

Ultrastructural data demonstrated that most spines establish a synaptic contact (Arellano et al. 2007b; Parajuli et al. 2020), and only 4% of the spines along the dendrites of neocortical neurons in rats did not contact an axon (Arellano et al. 2007b). It was usually considered that "most dendritic spines in the cerebral cortex receive only one axon terminal forming an asymmetric synapse" and, then, "the number of dendritic spines possessed by cortical neurons, and the distribution of those spines can be assumed to reflect the excitatory input impinging on a neuron"

---

[6] The dendritic spine capacity to form new synapses is exemplified here: "With the evolution of the brain, the increase in synaptic density had to be accompanied by a significant reorganization of the neuropil. Dendritic spines permit dendrites to synapse with neurons 1–2 mm away, which allows for increased synaptic density in densely packed neuropil (Fig. 1.8a). Even some invertebrate neurons exhibit spine-like structures, indicating that dendritic spines appeared well before the evolution of the complex mammalian brain. Consider the simple case of an orthogonal relationship between dendrites and axons (Fig. 1.8b). There can only be two synapses on either side of the dendrite in any given plane without the presence of spines. Dendrites with spines can reach beyond their immediate perimeter to connect with axons in nearby rows, thereby at least doubling the density of possible connections. In addition, the shape of dendritic spines allows efficient interdigitation between neighboring processes, thus achieving the high synapse packing density in the neuropil" (Bourne and Harris 2009).

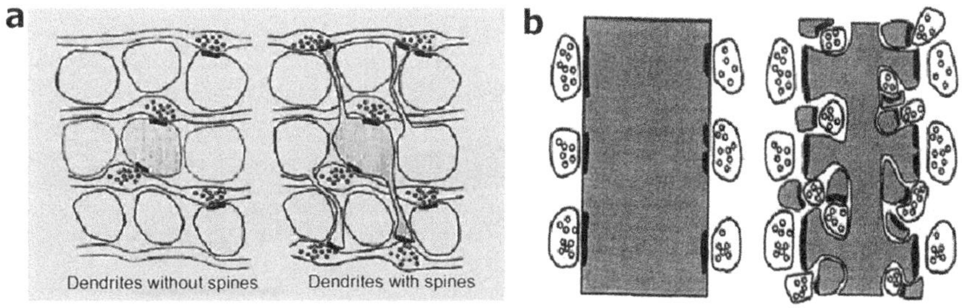

Fig. 1.8 (**a**) Spines increase the packing density of synapses. Convolution and interdigitation of dendrite, axon, and spine membranes support more synapses. Schematic figures illustrate a cross-section through two dendrites (shaded), one without and one with dendritic spines (Sorra and Harris, 2000). (**b**) The presence of spines also allows for an increase in synaptic density without increasing the overall volume of the brain. (Legend slightly adapted and figure reprinted from the chapter published by Bourne and Harris (2009), https://doi.org/10.1016/b978-008045046-9.01771-x, Copyright Elsevier (2009), reprinted under permission #220405-025798)

(Peters et al. 1991). Spines receive most (>90%) excitatory glutamatergic inputs (Rochefort and Konnerth 2012; Yuste 2013). This monosynaptic (one-to-one matching of spine and axon) aspect of spines can occur in subcortical structures as well (Fig. 1.9). Surprisingly, synapses on spines accounted for only one-quarter of dendritic synapses in a small sample of neurons in the rat medial amygdala (Brusco et al. 2014). In the human temporal cortex, an axon usually establishes one synapse with a postsynaptic cell, and few axonal inputs to a neuron make more than one synapse with a target cell (Shapson-Coe et al. 2021). These findings mean that the number of spines per cell adds a huge possibility for each neuron to compute and integrate inputs (with different time windows for synaptic processing) coming from many perpendicular axons crossing long dendritic segments (Yuste 2010). Those axons making contact with more than one spine can be running close along the longitudinal axis of the dendritic segment (compare the interesting axonal features in Figs. 1.6c and 1.10c). The cerebellum is an example where an incoming axon forms more than one synaptic contact with a Purkinje cell (Parajuli and Koike 2021). EM studies showed that approximately one-fifth to one-fourth of the spines had a presynaptic partner contacting as many as four different spines on a dendrite of a Purkinje cell (Parajuli and Koike 2021 and references therein).

A lower percentage of spines, but essential for neuronal circuit functioning, is contacted by inhibitory γ-aminobutyric acid (GABA)-containing axon terminals (Kubota et al. 2007; Brusco et al. 2014; Müllner et al. 2015; Kubota et al. 2016) usually at proximal dendritic segments (Peters et al. 1991; Kubota et al. 2016). Nevertheless, GABAergic inputs also contacted spines in distal dendritic segments and form symmetrical synaptic junctions (Kubota et al. 2007; for other data, see also Kwon et al. 2019). In these cases, most of targeted spines were also innervated by an asymmetrical terminal, with spine size and postsynaptic density (PSD) aspect differences depending on the distinct origin of the afferent innervation coming to

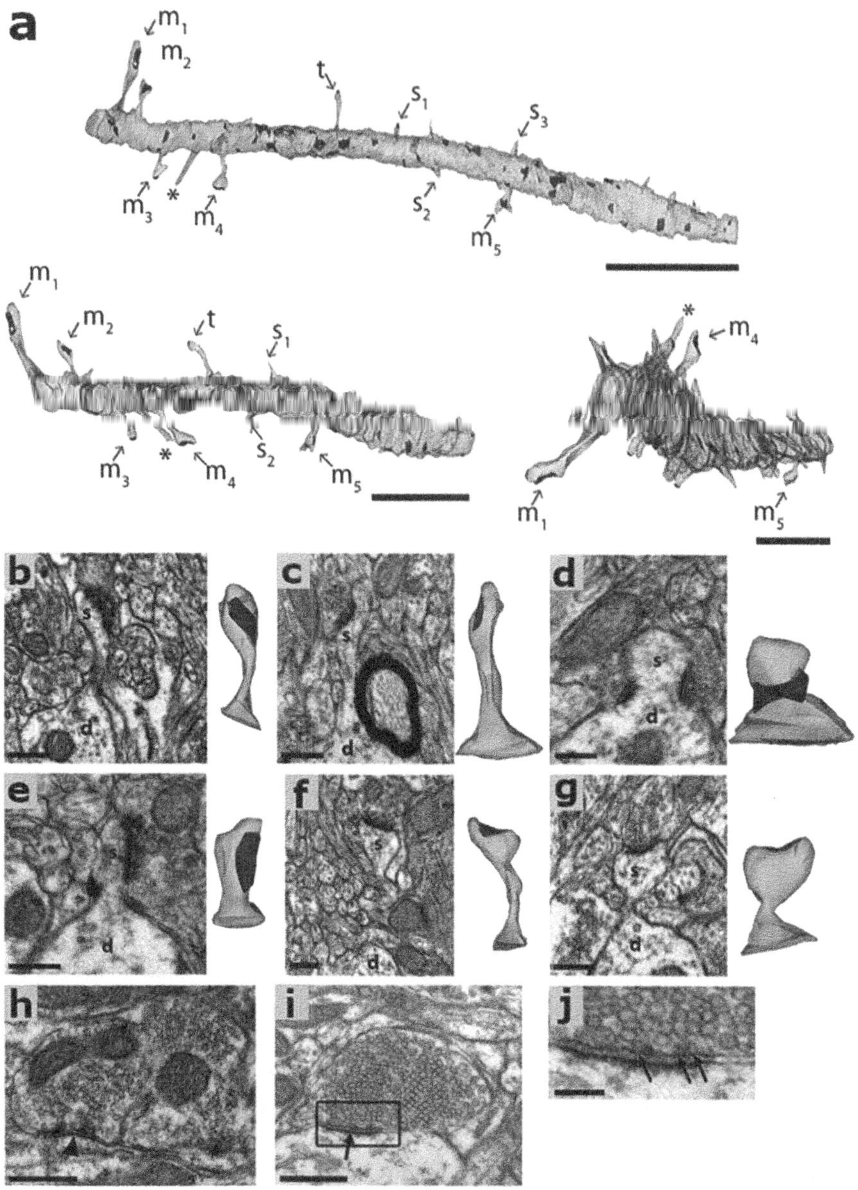

Fig. 1.9 Dendritic spines from the posterodorsal medial amygdaloid nucleus of rats. (**a**) Three-dimensional views of the same dendritic shaft and dendritic spines reconstructed by electron micrographs from 420 serial sections of 60–70 nm. Spines were classified as mushroom (m), stubby (s), and thin (t) ones, and numbers indicate the same spine in different views. The asterisk points to a filopodium with no synaptic contact. Scale bars = 5 μm. (**b**, **c**) Thin spines, (**d**) stubby and (**e**) wide spines, and (**f**, **g**) mushroom spines. Synaptic contacts are in dark gray on the reconstructed structures. Scale bars = 250 nm. Examples of synapses classified as (**h**) symmetric or inhibitory (arrowhead) and (**i**) asymmetric or excitatory (arrow). Scale bars = 500 nm. (**j**) Highlighted area in I to point examples of docked vesicles (arrows). Scale bar = 250 nm. (Legend adapted and figure reprinted from Brusco et al. (2014), under CCC RightsLink® license #5282610237568, originally published by John Wiley & Sons, Inc)

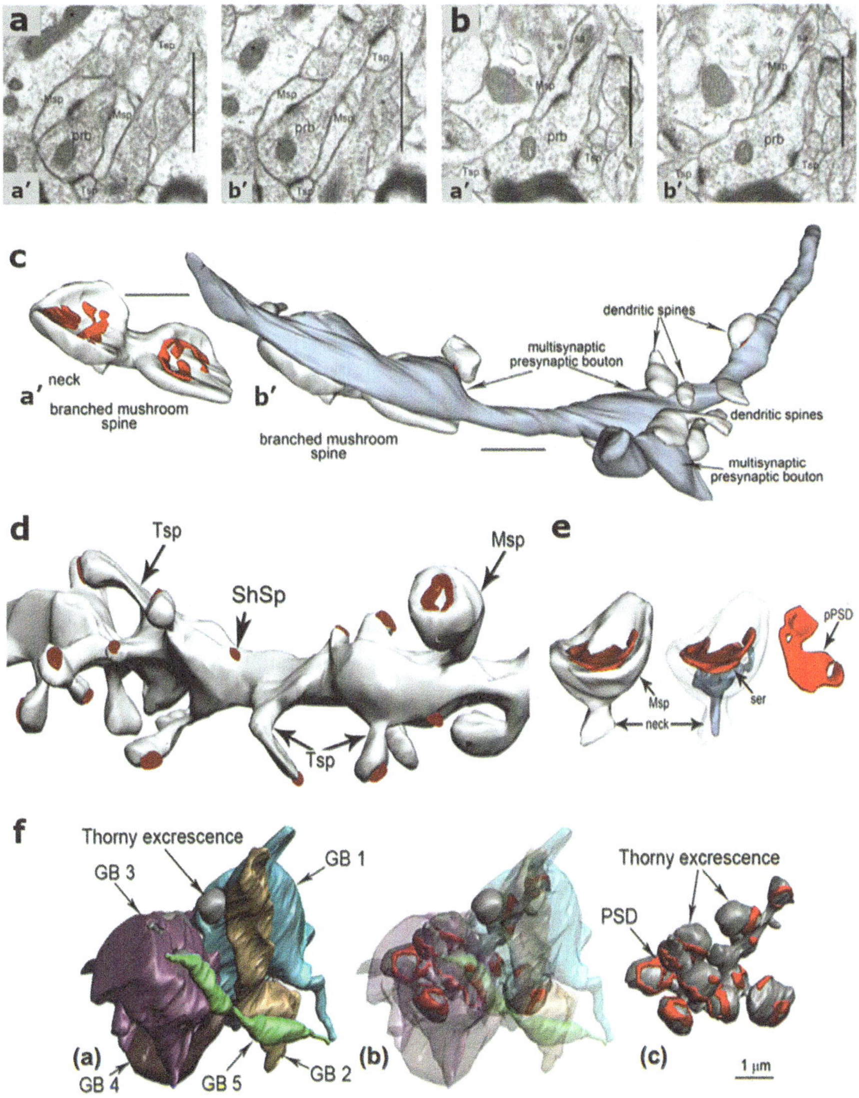

Fig. 1.10 Electron micrographs from serial sections of the adult rat dentate gyrus. The mushroom spine (Msp, with spine apparatus, sa) shown in both (**a** a′, b′) and (**b** a′, b′) is a branched mushroom spine (shown in **c** a′) where heads and postsynaptic density (PSD, in **c** a′) contact the same presynaptic bouton (prb, in **c** b′). The PSD is marked and painted in red on each dendritic spine. Note the presence of a multisynaptic presynaptic bouton (prb in **a** and **b**, painted in blue in **c**) contacting spines along the longitudinal axis of the dendritic segment. (**d**) Synapses on thin spines (Tsp) and Msp. Three-dimensional reconstruction of a segment of dendrite approximately 10 μm in length. Note the PSD on different spines, the dendritic shaft, and the base of a spine (ShSp). (**e**) Three-dimensional reconstruction of a large mushroom spine. Spine and synapse forms are similar throughout the hippocampus except in CA3 in *stratum lucidum* where the dendrites are covered with thorny excrescences, which are contacted by the giant boutons (GB) of the mossy fibers reconstructed in (**f**). ser, smooth endoplasmic reticulum; pPSD, perforated postsynaptic density. (**f** a′) a complete reconstruction of a thorny excrescence with contacting GB (colored and numbered 1–5); (**f** b′) with GBs translucent; and (**f** c′) without GBs, showing only the spiny thorny excrescence. All scale bars = 1 μm. (Legend adapted and figure reprinted from Stewart et al. (2014), https://doi.org/10.1016/b978-0-12-418675-0.00001-8, under CCC RightsLink® license #5279371390765, originally published by Elsevier)

the frontal cortex in rats (Kubota et al. 2007). Other neurotransmitters can modulate spine function. For example, thalamic and cortical glutamatergic inputs target separate dendritic spines in basolateral amygdala of mice (McDonald et al. 2019). Half of the spines targeted by vesicular glutamate transporter 1 (VGLUT1)- and VGLUT2-containing terminals express type 1 muscarinic receptors (M1R) that directly contact or are near the PSD (McDonald et al. 2019). Dopamine is also a modulator of spine structure and network function in a tone-reward, discrimination learning task, and classical conditioning in mice (Iino et al. 2020; Kasai et al. 2021). That is, from data obtained in nucleus accumbens slices, a transient reduction in dopamine concentration (the dopamine dip as short as 0.4 s) was detected by type 2 dopaminergic receptors (D2R) in spiny projection neurons to disinhibit adenosine A_{2A} receptor ($A_{2A}R$) mediated enlargement of dendritic spines (Iino et al. 2020).

There can be a variable number of spines in different neurons and in dendritic segments of the same neuron. Dendritic shafts and branches have preferred orientations within the neuropil related to the cytoarchitectonic organization and spatial distribution of local and extrinsic connections upon each neuron (Dall'Oglio et al. 2008; Clemo et al. 2016; Weiler et al. 2022). Dendrites can have a multitude of forms and were classified according to (1) branching pattern as radiate, tufted, wavy, atypical, or intermediate and (2) considering orientation within the neuropil, with no preponderant pattern (i.e., radiating in all directions), linear (uni-dimensional orientation with parallel dendrites in an imaginary elongated cylinder within the neuropil), planar (bi-dimensional orientation with dendrites within an imaginary flattened cylinder; Ramón-Moliner 1962). Alternatively, dendrites can be classified as spindle-shaped, spherical, laminar, cylindrical, conical, biconical, or with a fan aspect (Fiala and Harris 1999). These findings indicate that spine number, distribution, shape, synaptic, and integrated functional properties can vary according to each cell and nervous area. Indeed, spines have been studied in dendrites of different branching orders and spatial orientation within the neuropil (Woolley et al. 1990; Woolley and McEwen 1993; Barreto-Cordero et al. 2020). The giant pyramidal neuron in layer V of the monkey visual cerebral cortex (Meynert cell) is a notable example of that.[7]

Morphological data also helped to underscore the integrative role that each spiny neuron would carry out in neural networks and the relevance of intrinsic membrane

---

[7] The fundamental description of a spiny Meynert cell is the following: "The perikaryon and primary segments of all dendrites are spine-free; however, more distally a total of 36,000 spines are present, differentially disposed upon the dendritic surfaces. The basal dendrites bear over 77% of the spines on the Meynert cell, although they account for only 66% of the total length of the dendritic arborization. The first part of the apical dendrite is the most densely decorated with appendages, accounting for almost 10% of the spines on the whole dendritic tree. The apical dendrite becomes progressively less spiny as it passes through the superficial part of layers IV and III; less than 2.5% of the total number of spines of the Meynert cell project from this part of the apical dendrite. When the dendrite reaches layer II it bursts into an umbel of rapidly tapering branches. These are highly spinose, accounting for 8–13% of the cell's total, dispersed over only 23% of the linear dendritic length. It is suggested that this differential distribution of thorns can be correlated with the axonal inputs in the various cortical layers and that the Meynert cell is designed to receive maximal information from layers I and II and from layers V and VI, which are sources mainly of intracortical inputs" (Chan-Palay et al. 1974).

properties of dendrites and spines (Spruston et al. 2013; Ramaswamy and Markram 2015; Tønnesen and Nägerl 2016; Cornejo et al. 2022). In layer V of the rat primary visual cortex, morphological, intrinsic electrophysiological, and axonal target features characterize local spiny pyramidal neurons with their function. Those cells that project to the superior colliculus have thick apical dendrites with a notable terminal arborization in layer I, whereas those projecting to the contralateral visual cortex have a slender apical dendrite that terminates below layer I lacking a terminal tuft, a rounder and less conical cell body, and fewer basal dendrites than corticotectal neurons (Kasper et al. 1994).

Considering hippocampal and neocortical pyramidal cells as examples, the distribution of spines in different dendritic segments represents "domains" for information processing in the same neuron and across layers, indicating distinct synaptic inputs and weights (Konur et al. 2003; Andersen et al. 2007; Fairén 2007; Spruston et al. 2013; Larriva-Sahd 2014). Therefore, morphological data can form the basis to the study of synaptic processing and dendritic compartmentalization (depending on each cell, Fiala and Harris 1999) as well as the possibility of integration of synchronized incoming information along the dendritic tree (Fairén 2007; Larriva-Sahd 2014). Although the ability of cortical neurons to perform temporally accurate computations would be crucial for encoding information in the cortex, these neurons could not do that because of the dendritic filtering properties, the stochastic nature of synaptic transmission and ion channel gating, and/or background synaptic "noise" (Ariav et al. 2003). A step forward finding was the identification of synaptically evoked spikes in basal dendrites of CA1 pyramidal neurons. That is, these dendrites respond to coincidently activated glutamate-mediated excitatory postsynaptic potentials (EPSPs) innervating the same segment by the generation of local fast spikes. Whereas coactivated, spatially distributed synaptic inputs in dendrites would produce temporally imprecise output action potentials (APs) over several milliseconds, the integration of closely spaced basal inputs does initiate local dendritic spikes that amplify and sharpen the summed somatic potential (Ariav et al. 2003). This electrophysiological organization would allow precise timing of output APs (with submillisecond temporal jitter over a wide range of activation intensities and background synaptic noise) and serve as precise "timers" of output APs and temporal coding for cortical network activity states (Ariav et al. 2003).

1.1.2 Dendritic Spine Morphology

In spiny dendrites, there can be filopodia, simple spines (with sessile and pedunculated shapes), and branched spines (or ramified ones, with two or more branches; Fiala and Harris 1999). Reconstructed EM images and high-resolution microscopy made it possible to quantify with more accuracy the spine "head" with an ovoid, spherical sharp edge or a flat end-bulb and the narrow stalk (the "neck") attached to the parent dendrite (see relevant data in Peters et al. 1991; Fiala and Harris 1999; Arellano et al. 2007a, b; Pannese 2015; Basu et al. 2018). In contrast, filopodium is

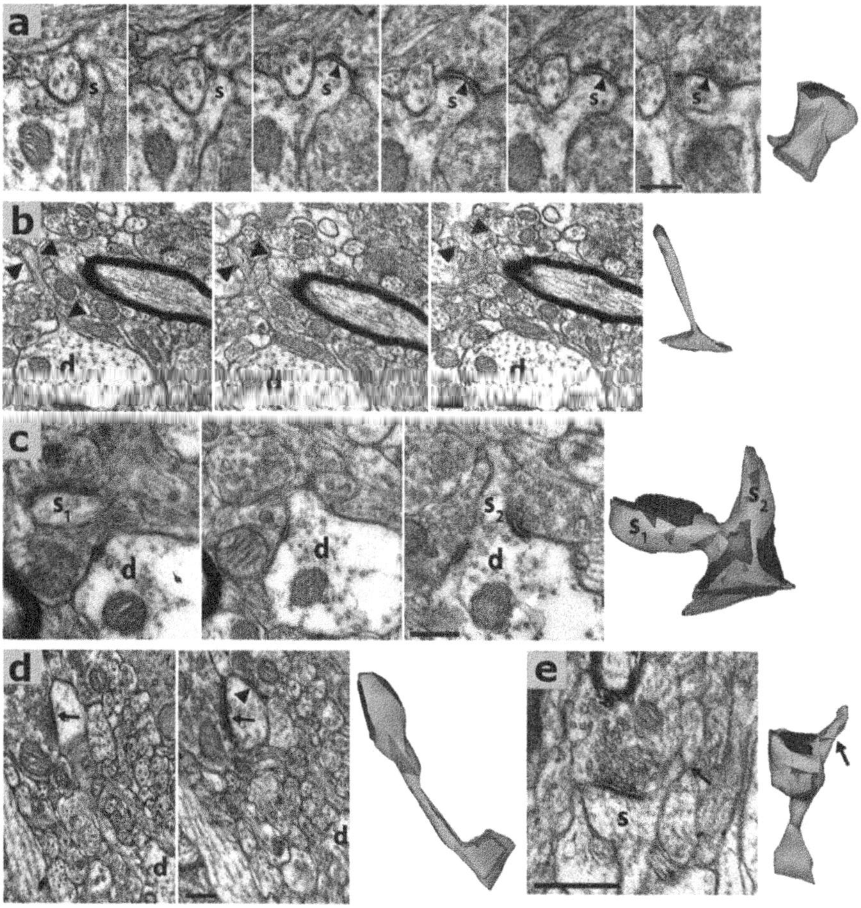

Fig. 1.11 Synapses on dendritic spines in the posterodorsal medial amygdaloid nucleus of rats. (**a**) Serial electron micrographs were used for the reconstruction of a dendritic spine receiving one symmetric contact (arrowheads). Scale bar = 200 nm. (**b**, **c**) Parent dendrites (d) and spines (s) were identified to reconstruct a filopodium (**b**, arrowheads) and (**c**) a ramified dendritic spine with two independent protrusions (s1 and s2) from the same neck. (**d**) Multisynaptic dendritic spine receiving one asymmetric (arrows) and one symmetric (arrowhead) synaptic contact. The site of asymmetric and symmetric contacts is represented in dark gray in the reconstructed spines. (**e**) Spine with a protruding spinule (arrow). Scale bar = 250 nm. (Legend adapted and figure reprinted from Brusco et al. (2014), under CCC RightsLink® license #5282610237568, originally published by John Wiley & Sons, Inc)

a thin and long protrusion (e.g., 2 µm) with no evident head (Brusco et al. 2014; Fig. 1.10b). Filopodia usually display no PSD, but have a role in synaptogenesis and are regarded as a sign of dynamic spine formation or elimination (García-López et al. 2010).

Dendritic spines are morphologically heterogeneous, ranging in a *continuum* of shapes and sizes (Figs. 1.6, 1.9, 1.10, and 1.11). Spines have been classified according to their head and neck characteristics as stubby/wide, thin, mushroom, ramified or "atypical" (including "intermediate" and "transitional" forms between classes, or shapes that cannot be classified in the preceding types, such as "double" spines, for example; based on Peters and Kaiserman-Abramof 1970; Harris et al. 1992; Arellano et al. 2007a; Bourne and Harris 2009; Yuste 2010; González-Burgos et al. 2012; Brusco et al. 2014; González-Ramírez et al. 2014; Stewart et al. 2014; Dall'Oglio et al. 2015; Correa-Júnior et al. 2020; e.g., Figs. 1.8, 1.9, and 1.10). Each classification requires the complete observation of the spine shape along the "z" axis and at various angles of microscopic image rotation (Arellano et al. 2007a; Brusco et al. 2014), considering the physics laws and limits of resolution for each type of microscope (e.g., 0.2 μm for light microscopy - see a critical discussion in Chapter 2 in this book; Tønnesen and Nägerl 2016; Parajuli and Koike 2021). There are different proportions of spines in each class among neurons. Moreover, the characterization of the 3D structure of dendritic spines advanced by volume EM techniques evidenced many "atypical" shapes, which provided pieces of evidence that the spine classification cannot be restrictive (Parajuli and Koike 2021, further examples in Maiti et al. 2015 and Rasia-Filho et al. 2021). The diversity in the morphological architecture of dendritic spines implies functional purposes linked to diverse input processing and computations (Parajuli and Koike 2021).

Some convoluted forms, which likely add more possibilities to the synaptic processing, include thorny excrescences (Dall'Oglio et al. 2013; Stewart et al. 2014), coralline excrescences, and racemose appendages (Fiala and Harris 1999). They would represent varying stages of development, retraction, or functional specialization for multiform spines. For example, thorny excrescences are found in CA3 hippocampal pyramidal neurons and in non-pyramidal neurons in the medial amygdaloid nucleus of humans (Lauer and Senitz 2006; Dall'Oglio et al. 2013, 2015; Lu et al. 2013). This type of specialized protrusion establishes multiple (as many as 37) synaptic contacts in various bulbous endings (Chicurel and Harris 1992; Stewart et al. 2014; Reberger et al. 2018; Fig. 1.10f), which contrasts with the common finding of monosynaptic spines (Peters et al. 1991; Fig. 1.9). Multisynaptic spines can show PSD at varied locations along the spine surface with both symmetric and asymmetric synapses (Brusco et al. 2014; Dall'Oglio et al. 2015, see further data in Kleinjan et al. 2022; Figs. 1.11 and 1.12). Multiple synaptic inputs converging onto a single spine is also exemplified by "toric" spines on the soma and proximal dendrites of "space-specific," sensory processing neurons in the barn owl inferior colliculus (Sanculi et al. 2020). These spines with unusually complex and variable morphology can potentially integrate multiple input sources and have up to 49 active zones derived from multiple axons (up to 11) distributed widely throughout the neuropil volume (Sanculi et al. 2020).

As occurs for the heterogeneity of shape, the dimensions of dendritic spines are not uniform and can vary even for the same type of spine along the same dendritic segment and neuron. The size of the spines displays differences depending on the studied area and species. For example, spines in the frontal cortex are bigger than in the visual cortex, and human spines are larger and occur at higher densities than those in corresponding brain areas of mice, for example (Yuste 2010; DeFelipe 2011 and references therein). Dimensions of individual spines (length, width, and

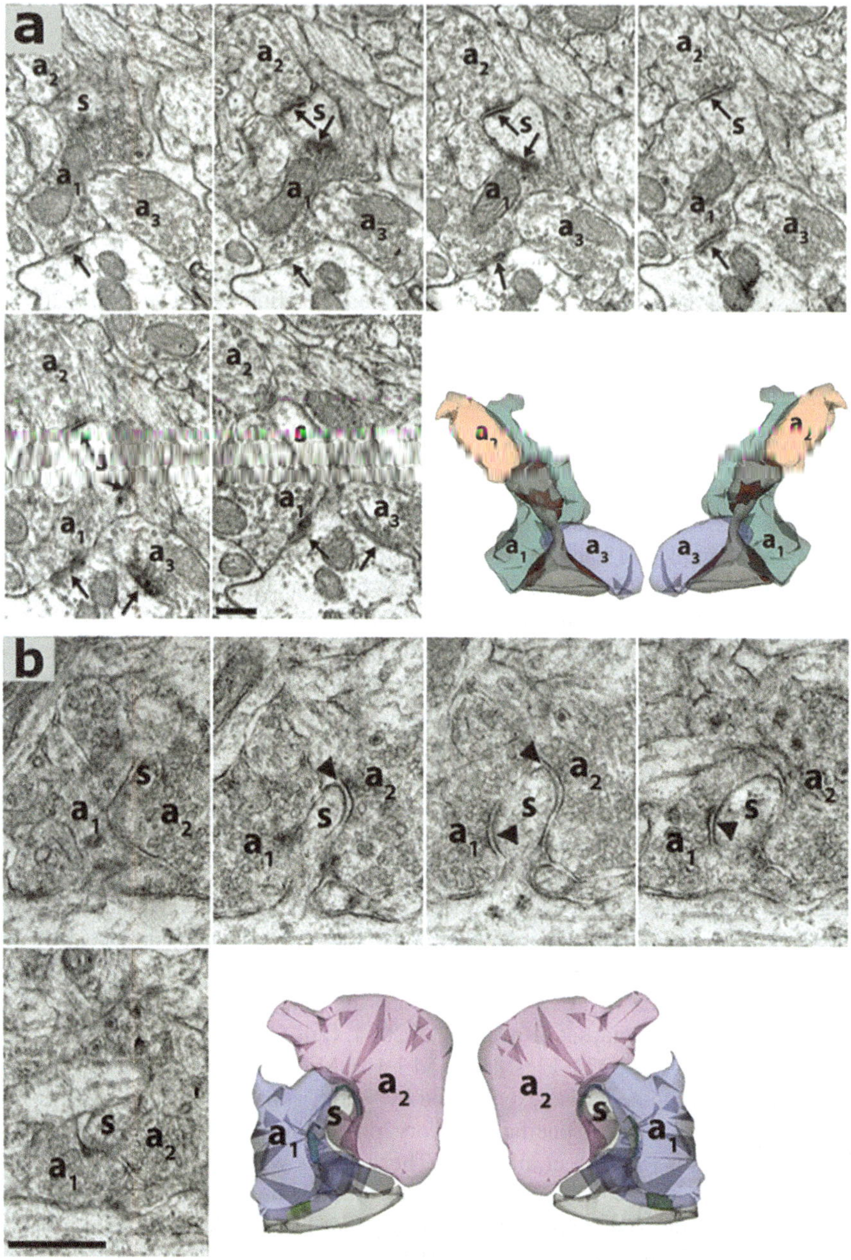

Fig. 1.12 Multisynaptic dendritic spines in the posterodorsal medial amygdaloid nucleus of rats. (**a**) Electron micrographs were used for the reconstruction of a dendritic spine receiving two asymmetric contacts (arrows). The first axon terminal (a_1), in light green, makes synapses with the spine (s) and the parent dendrite; the second axon (a_2), in orange, makes a synapse in this same spine (on the spine head); the third axon (a_3), in purple, contacts the parent dendrite. Asymmetric synaptic contacts are in red on the reconstructed spine. (**b**) Electron micrographs were used for the reconstruction of a dendritic spine (s) receiving two symmetric contacts (arrowheads) from different axon terminals (a_1 and a_2). The reconstructions show the terminals in purple (a_1) and pink (a_2) at different synaptic sites. Symmetric synaptic contacts are in green. Scale bar = 250 nm. (Legend slightly adapted and figure reprinted from Brusco et al. (2014), under CCC RightsLink® license #5282610237568, originally published by John Wiley & Sons, Inc)

volume) were measured using EM (fixed tissue) and STED (live tissue, Araya 2014). Values for the total spine length in cerebellar Purkinje cell and CA1 pyramidal neuron range from 0.2 to 3 μm, the lower value being at the limit of resolution using light microscopy. The neck width (diameter) and the neck length can range between 0.038–0.5 μm and 0.1–2.2 μm, respectively (data from serial EM, Fiala and Harris (1999), Sorra and Harris and references therein). Spine head volumes can vary from 0.01 to 1 μm^3, spine necks from 0.05 to 0.5 μm in diameter and approximately 3 μm in length, with little or no correlation between parameters, a finding that reinforces the variety of shapes and sizes of spines (Tønnesen and Nägerl 2016 and references therein). The spine head volume can be regulated by the interspine distance in the dendritic segment and according to the distance of this segment from the soma (Yuste 2010). This volume has up to a 204-fold difference in hippocampal, cerebral cortex, and striatum of mice (Parajuli et al. 2020). The average head volume of mushroom spines is much higher than of thin spines, which is one of the criteria for the morphological classification of these types (Medvedev et al. 2014).

Moreover, spinules can occur in all types of dendritic spines (Spacek and Harris 2004), across very different species (Brusco et al. 2014; Dall'Oglio et al. 2015; Gore et al. 2022) and across lifespan (Petralia et al. 2018). They are tiny vesicular or vermiform protrusions extending from a spine toward an adjacent cell (Brusco et al. 2014, Zaccard et al. 2020; Petralia et al. 2018, 2021; Gore et al. 2022; Fig. 1.11e). Invaginating spinules and related structures can be derived from the postsynaptic, presynaptic, or glial components of synapses (Petralia et al. 2018). For example, a spinule can protrude from a postsynaptic spine, invaginate, and be enveloped into a presynaptic axon terminal (Gore et al. 2022), as well as into adjacent axonal or glial processes (Spacek and Harris 2004; Petralia et al. 2018).

It is common to find a spine with one protruding spinule (Brusco et al. 2014), but six spinules per spine was already reported (Spacek and Harris 2004). In the stratum radiatum of an adult male rat, spinules emerged from spine heads or necks (62% and 24%, respectively), axons (13%), or dendritic shafts (1%, Spacek and Harris 2004). Some spinules do not have synaptic active zones within the invagination, but can be related to synaptic activity, strength, and plasticity (Petralia et al. 2018) or other forms of intercellular signaling (Spacek and Harris 2004). Interestingly, a higher number of spinules was found in hippocampal slice cultures following K^+-induced depolarization (Tao-Cheng et al. 2009). Activation of *N-methyl-D-aspartate* (NMDA) glutamate receptors increased spinule number, length, and contact with distal presynaptic elements (Zaccard et al. 2020). Being regulated by intracellular Ca^{2+} levels (Zaccard et al. 2020), transient increase in spinule size was identified 30 min after long-term potentiation (LTP; Toni et al. 1999). In CA1 hippocampal neurons of an adult male rat, spinules were found in approximately 74% of the sampled excitatory presynaptic terminals, being present in synaptic sites with larger boutons and larger PSD than in synapses devoid of these protrusions. Two types of spinules were observed in this area: small clathrin-coated spinules, and much larger spinules without clathrin (Gore et al. 2022). Spinules from mushroom spines usually invaginate their respective presynaptic axon (Spacek and Harris 2004). On the other hand, most of the spinules from thin spines were engulfed by neighboring axons, which would represent an ongoing axonal competition for synaptic sites and signaling via spinules on immature spines (Spacek and Harris 2004). The remaining spinules on thin spines were surrounded by presynaptic axons or astrocytic processes (Spacek and Harris 2004). The necks of thin or branched spines also showed both short and long spinules directed toward adjacent nonsynaptic axons (Spacek and Harris 2004). Recently, it was shown that spinules can be "short-lived," dynamic, exploratory, and originated near simple PSDs, whereas another type was "long-lived," elongated, and associated with complex PSD fragments (Zaccard et al. 2020). This latter subset can contact distal presynaptic terminals and form secondary synapses (Zaccard et al. 2020).

Classification or clusterization of dendritic spines is a matter of debate when considering spines with a wide range of shapes and sizes (Ruszczycki et al. 2012; Pchitskaya and Bezprozvanny 2020; Ammassari-Teule et al. 2021). There are statistically significant differences that allow the separation of spines in classes (Ruszczycki et al. 2012), but there are no unique values that determine the exact criteria for including spines in one specific predefined subtype or another (Arellano et al. 2007a; Yuste 2010; Parajuli and Koike 2021). Clusterization of spine shapes might represent a more realistic evaluation of the existence of spines within a *continuum* of shapes, avoiding subjective classifications or arbitrary subdivisions. However, an important issue for this latter approach is that different clustering algorithms can bring about different results, with clusters varying in shape and content on the same dataset, and influencing the interpretation of obtained data (Pchitskaya and Bezprozvanny 2020).

1.1.3 Spine Structure and Function

The number and shape of spines have been tested for their likely relationship with the ongoing synaptic transmission, input weight, and plasticity (Sorra and Harris 2000; Bourne and Harris 2007; Kasai et al. 2010, 2021; Segal 2010; Chen et al. 2011). Axospinous synapses functioning and strength depend on the composition of the PSD (Yuste 2013). The PSD is an electron-dense area behind the postsynaptic membrane made up of many proteins, including neurotransmitter receptors, ion channels, adhesion proteins, cytoskeletal, and scaffolding proteins (Cohen and Siekevitz 1978; Cohen 2013; Sala and Segal 2014). Spines of different shapes and sizes can differ in the PSD area available for receptor composition and trafficking (Yuste 2013 and references therein). Notably, the size of the spine head relates to the area of the PSD, the number and proportion of receptors and types [e.g., for glutamate NMDA and α-amino-3-hydroxy-5-methyl-4-isoxazolepropionic acid (AMPA) receptors], the number of presynaptic docked vesicles and the releasable pool of neurotransmitters, and the synaptic current and postsynaptic response (Segal 2010; Yuste 2010, 2013; Pannese 2015; Borczyk et al. 2019). These findings apply to mushroom spines, which can display PSD with macular or perforated aspects (Stewart et al. 2014; Fig. 1.10). In some networks, like the hippocampal one, the responsiveness of thin spines to synaptic activity would characterize them as "learning spines," whereas the stability of mushroom spines would let them serve as "memory spines" (Bourne and Harris 2007). In other words, in the hippocampus, thin spines would be more labile (plastic) elements than (more stable) mushroom spines (Bourne and Harris 2007). Caution has to be taken with generalizations for spine structure and function in other brain areas (Ammassari-Teule et al. 2021) because spine dimension, EPSP amplitudes, and putative functions are not always correlated (Sobczyk et al. 2005; Segal 2010). For example, apart from thin or mushroom spines, small spines can have a high density of NMDA receptors (Sobczyk et al. 2005), and stubby/wide spines can generate larger excitatory postsynaptic current (Segal 2010).

Spines are made as small as possible (e.g., the number of glutamate receptors in a synapse can approach single molecules and show a packing aspect that could help the neuron to maximize the sampled neuropil, enhancing its connectivity and the input integration (Yuste 2010)). In this same regard, it is relevant to consider that "the diversity in spine size tells us also something more: synapses have very different strengths, presumably as a consequence of their plasticity. Thus, in this distributed matrix of connectivity, each axon and each synapse could have a different weight on a postsynaptic neuron. While maximizing the sampling of axons, a neuron also keeps their functional influence different from one another.... combining maximum connectivity with the functional individuality and plasticity for each connection could provide spiny circuits with increased flexibility and computational power" (Yuste 2010).

Spine structure and function may have mutual influences (Chidambaram et al. 2019). The first experiments showing activity-driven changes in head and neck geometry were linked to the excitatory input received by the spine (e.g., Fifková and Van Harreveld 1977; Coss and Perkel 1985; Yuste 2010). However, the function is determined at a single spine level and may involve a structural *continuum* along the time. Spines with large heads and smaller necks would promote local electrical (Tønnesen and Nägerl 2016) and biochemical compartmentalization for calcium levels (Noguchi et al. 2011). Mushroom spines contain smooth endoplasmic reticulum for calcium storage or release (Hering and Sheng 2001). Shorter and wider spines (e.g., stubby or wide ones devoid of a spine neck) may transfer charge into the parent dendrite with lower resistance (Coss and Perkel 1985) at the same time that they can have a higher density of postsynaptic glutamate receptors than longer spines (Segal 2010). Thin spines have a neck that imposes higher resistance to synaptically dependent voltage changes than stubby spines (Tønnesen and Nägerl 2016). Ramified spines would confer different temporal and spatial signaling microdomains for synapses on each "branch" (Chen and Sabatini 2012). For a branched spine, separating the postsynaptic receptors on the surface of the spine may add diversity to the circuitry and, considering the surface area as a dynamic variable, the efficacy of the synapses may vary continuously over time (Verzi and Noris 2009).

The dendritic spine dynamics in different neural circuits are the result of phylogenetic, ontogenetic, epigenetic (with various microRNAs, miRNAs), ongoing, and learned functional events (García-López et al. 2010; DeFelipe 2011; Tjia et al.

2017; Reza-Zaldivar et al. 2020; see Table 1 in Sala and Segal 2014). Activity-dependent actions (afferent synaptic-dependent over both short and long timescales), but also activity-independent ones (Nagaoka et al. 2016), promote turnover, stabilization, differentiation, and remodeling with enlargement or shrinkage and pruning of spines (Hering and Sheng 2001; Oray et al. 2006; Bourne and Harris 2007; Segal 2010; Zancan et al. 2018; Runge et al. 2020; Kasai et al. 2021). Alterations in spine structure and stability may change or be affected by various molecules' dynamics and diffusivity within and outside the spine (Bourne and Harris 2009; Yasuda 2017; Obashi et al. 2021). In conjunction, the occurrence, the number and distribution, the varied shape and size, and the function of spines adjoin to the dendritic features increasing considerably the modulation of neuronal excitability (Spruston et al. 2013; Sala and Segal 2014; Obashi et al. 2021). The "mosaic of possibilities" for dendritic spines has to be studied and tested for each neuron type and area. For example, one has to consider the morphological heterogeneity of spiny pyramidal neurons along the subcortical-allocortical-neocortical *continuum* (Rasia-Filho et al. 2021) or the different types of spiny neurons in the retina and basal ganglia (Fiala and Harris 1999).

The dendritic geometry and complexity can reflect the possibility for a neuron to contact axons from a large number of afferent cells or establish various connections with the axons of just a few cells, for which it is pivotal to know the axonal organization pattern in the neuropil and the direction of axons relative to dendrites of each neuronal subpopulation (Fiala and Harris 1999; Weiler et al. 2022). In this regard, by allowing a dendrite to connect with a larger number of axons, spines could minimize the wiring and sample a wider choice of axons maximizing connectivity (Yuste 2010). This possibility would justify why dendritic spines are ubiquitous elements for synaptic processing and neural function in so many species, at the same that they have both general morphological features and particular functions for each neuron and circuit.

1.2 Dendritic Spines Increase the Neuronal Connectivity for Site-Specific Functions in Neural Circuits

An intriguing question has accompanied the study of spines: Why would spines increase the synaptic capacity of the neuron if the available aspiny dendritic segments can provide space enough and direct contact sites to alter the dendritic shaft voltage? The simplest answer is that both axospinous and axodendritic synapses in spine-free shafts are two complementary possibilities used for synaptic modulation. Nevertheless, there are various additional findings to be considered. For example, spines can be so numerous that they extend the available membrane for synapses to represent approximately half of the dendritic surface area of a hippocampal pyramidal neuron (Bannister and Larkman 1995). The nervous system is so densely packed with axonal bundles and boutons that only a short dendritic outgrowth would let an encounter with a new presynaptic partner (Fiala and Harris 1999). However, spines would not be only simple devices to increase the dendritic surface area available for synapses since spiny dendritic shafts also have areas between spines that are free of synapses (Pannese 2015). Wiring is a non-random, controlled process that organizes the network architecture (Kasthuri et al. 2015; Udvary et al. 2022), and spines would be specialized synaptic sites selective for afferent axons coming from particular sources, whereas distinct axons would form direct shaft contacts (Pannese 2015). Several lines of evidence reinforce the role of some specific spines in the connectional fine-tuning of neurons (e.g., see representative figures in Chen et al. 2011 and Hayashi-Takagi et al. 2015) as well as in metaplasticity, i.e., the "plasticity of synaptic plasticity" (Kasai et al. 2021).

Henceforth, spines have been considered multifunctional integrative units (Shepherd 1996), which evolved to increase the packing density of selected synapses by the convolution and interdigitation of cellular membrane and to support more synapses without increasing the overall volume of the brain within a restrictive skull (Bourne and Harris 2009). If not, a huge brain volume would need to change the head and neck anatomy of the offspring, adding more difficulties for labor and risks at birth, imposing the need of changes in the female anatomy and so on (more than occurred in the Homo evolution, Mayr 2001). Spines might be fundamental to improve the functional capacity of each neuron without the need to add much more cells to increase information processing and behaviors at the expense of the size of the brain, metabolic support, and energy expenditure. Spines provide additional modulatory possibilities to maximize the connectivity repertoire already governing the shape of dendritic arbors (Wen et al. 2009). It is like improving the functionality of the components of a computer giving them more computational power for information processing rather than putting more and more limited computers within a restricted space. Indeed, processing manifold stimuli from external and internal *milieux* engender specialization and functional integration of neural cells, areas, and networks (e.g., Rasia-Filho 2006; Freiwald 2020; Rasia-Filho et al. 2021; Fuentealba-Villarroel et al. 2022). Pyramidal cells are examples for the emergence and development of spines in neurons evolved to provide more elaborated functional sensory, motor, emotional, social behavior, and higher-order functions (Rasia-Filho et al. 2021 and Chapter 9 in this book). Cajal noted that cells from more highly evolved animals have more spines and argued that spines would be linked to "intelligence" (Yuste 2010). Interestingly, some human spines are similar in shape when compared to other animals, but dendritic spines would have reached higher levels of complexity in our species (Yuste 2010, 2013; DeFelipe 2011; Dall'Oglio et al. 2015; Vásquez et al. 2018; Benavides-Piccione et al. 2020, 2021; Rasia-Filho et al. 2021). Human spines can be more numerous, larger, and longer and show a diverse shape spectrum compared to other species or, in our species, even across the same cortical lobe (Benavides-Piccione et al. 2002, 2020, 2021; DeFelipe 2011; Yuste 2013; Dall'Oglio et al. 2015; Ofer et al. 2022). It is likely that spiny neurons with evolved complexity increased our capacity for synaptic processing and behavioral repertoire (DeFelipe 2011; Rasia-Filho et al. 2021). Otherwise, an aberrant number and immature morphology of spines are found in genetic syndromes associated with intellectual disability. The apical dendrites of layer V pyramidal neurons in the motor cortex of infants show sparse and longer spines in Patau syndrome whereas thinner, shorter, or longer spines are found in Down syndrome compared to controls (Kaufmann and Moser 2000; Dierssen and Ramakers 2006; Bączyńska et al. 2021 and references therein; see also Chapter 9 in this book).

Dendritic spines can undergo plastic changes in shape and size, appear and fade away over time (Holtmaat et al. 2005). Depending on the brain area, most spines can prune in youth and be stable in adulthood (Runge et al. 2020). The dynamics of spines can occur with a fast time course, as milliseconds for synaptically induced biophysical and biochemical events in the spine head, less than 1 min, or along few minutes (Engert and Bonhoeffer 1999; Fiala and Harris 1999; Toni et al. 1999; Yuste 2010). Spines can also change along hours to days (e.g., across the phases of the estrous cycle in adult rats; Woolley et al. 1990; Rasia-Filho et al. 2004; Brusco et al. 2008), days to a few weeks (during puberty; Zancan et al. 2018 and references therein), and days to months (during hibernation, Popov et al. 1992). Some spines can be very stable, but others may not last a lifetime and their turnover time varies in different structures (Ammassari-Teule et al. 2021). Although calculated as a few proportion, these remaining plastic spines would represent a quantity per cell significant enough to impact on neural network functions, learning and behavior (Runge et al. 2020, see further data and functional implications in Ma and Zuo 2022). In vivo studies showed that some spines change over time, while other spines are stable, indicating a balance between plasticity and stability in neural circuits (Spruston et al. 2013). In the case of stroke, 2-photon imaging has revealed dendritic swelling and dendritic spine loss within minutes of ischemic onset as well as increased synaptogenesis within peri-infarct tissues recovering from insult over longer time scales (Sigler and Murphy 2010). These latter findings prompt further caution when interpreting the function of dynamic structures assessed from static images (Pannese 2015) or when establishing the function of a type of spine in different brain areas submitted to distinct synaptic demands or experimental conditions (Nimchinsky et al. 2002; Segal 2010).

Dendritic spines' number, shape, and quantitative parameters (length, width, and volume) can also show changes in vivo or are modified by in vitro experimental and technical conditions. Experimental caveats may involve the following: sample size; age and developmental period; the composition of the experimental medium, buffer conditions and temperature; *postmortem* time interval until tissue fixation; ischemic/excitotoxic, osmotic, and pH changes in the studied tissue; uncontrolled morphological changes in the glia and extracellular matrix surrounding neurons; use of specific (aldehyde) fixatives or solutions that remove water and lipids from the tissue during histological processing (which causes shrinkage of varied degrees); use of overexpressed sensors that may change cellular properties when studied, for example (Coss and Perkel 1985; Sobczyk et al. 2005; Baryshnikova et al. 2006; Bourne and Harris 2009; Segal 2010; Pannese 2015; Yasuda 2017; Vásquez et al. 2018; Parajuli and Koike 2021 and references therein). Moreover, it is also essential to consider that spines can have species-specific, region-specific, subpopulation-specific, dendritic segment-specific, task-specific, age- and sex-specific features.

These characteristics can limit the possibility of some direct comparisons between data.

It is noteworthy that dendritic spines are associated with the processing of the ongoing synaptic demand for each neuron (Arellano et al. 2007b; Tjia et al. 2017). The development of particularly specialized neurons involves the structural remodeling of their dendritic branches, for which an optimal degree of connectivity could be associated with the dendritic length and branching pattern, spine features, and the intrinsic membrane properties of these elements (Papoutsi et al. 2014; Litwin-Kumar et al. 2017). That also involves sequential steps for spinogenesis and synaptogenesis, synaptic establishment and strength, and PSD remodeling. Reflecting the cellular function and the properties of the neural circuits where they lie within, the development of dendritic spines can occur at different rates after birth (Tjia et al. 2017; Runge et al. 2020). For example, dendritic spine formation in cortical pyramidal neurons of rats occurs after birth, but not in guinea pigs, which are born with a mature pattern of spines and synapses (Yuste 2010 and references therein). In the murine CA1 hippocampus there is an increase in spine densities between postnatal weeks 8 and 15 (von Bohlen und Halbach et al. 2006a). In monkeys, the number of dendritic spines of pyramidal neurons in the primary visual cortex reduces following the onset of visual experience, whereas in areas for sensory association in the inferotemporal cortex and for executive function in the granular prefrontal cortex grow more spines than they lose during the same period (Oga et al. 2017). In the rat visual cortex, the first surge in the production of spines occurs near the end of the first postnatal week and follows the afferent innervation by callosal and thalamic fibers (Miller 1988). The second surge occurs during postnatal week 3 when animals open their eyes; afterward, the spine density decreases significantly along the second and third months (Miller 1988 and references therein). Accordingly, apical dendritic spines of the visual cortex are reduced after light deprivation in the newborn mouse (Valverde 1967). In the apical tufts of mice neocortical pyramidal neurons (from layers II/III and V of the somatosensory and visual cortices), spine densities were higher at younger ages, showing a net loss of dendritic spines with developmental age at impressive daily rates (Holtmaat et al. 2005). At postnatal day 16, more than 30% of the spines disappeared, while 25% appeared from 1 day to the next in vivo (Holtmaat et al. 2005). In the mature brain, a subpopulation of likely thin spines appeared and disappeared from day to day compared to mushroom spines on the same dendrites, which tended to persist (Holtmaat et al. 2005). The fraction of persistent spines (lifetime ≥8 days) grew gradually during development and into adulthood (from 35% at postnatal days 16–25 to 73% at postnatal 175–225), indicating a balance between stability and plasticity in neural circuits that undergoes continuous changes even after early critical periods (Holtmaat et al. 2005; Chidambaram et al. 2019; Runge et al. 2020).

These structural and connectional reorganizations rendered neurons with spiny dendrites to receive multiple sources of inputs and to adapt, integrate, and fine-tune them for functional purposes in specific networks. This rationale applies to CA1 pyramidal neurons, which display numerous dendritic spines and form part of circuits for memory elaboration (Andersen et al. 2007). Along a total dendritic length of 12,000 μm, segments receive near 30,000 excitatory and 1700 inhibitory inputs (Megías et al. 2001). The integrative property of these neurons is exemplified by the morphology of their spiny dendrites in which (1) convergent excitation arriving onto distal dendrites of pyramidal cells is controlled by proximally located inhibition, and (2) there is a distinct organization for the excitatory and inhibitory inputs in layers receiving Schaffer collateral inputs (in *stratum radiatum* and *stratum oriens*) compared to perforant path inputs (reaching the *stratum lacunosum-moleculare*; Megías et al. 2001). In mice, dendrites of CA1 pyramidal neurons within the *stratum oriens* have numerous spines. In rats, proximal basal dendrites within *strata radiatum* and *oriens* as well as proximal apical dendrites are spine-free or sparsely spiny whereas intermediate to distal *strata radiatum* and *oriens* dendrites (most parts of the dendritic tree) are densely spiny (Megías et al. 2001). Excitatory inputs target dendritic spines, while inhibitory inputs terminate directly on dendritic shafts. Inhibitory inputs occur on proximal dendritic segments, cell body, and axon initial segment. Dendrites in the *stratum lacunosum-moleculare* show small-to-moderate density of spines, but their excitatory synapses are larger (frequently with a perforated aspect) than in other layers and can be located on dendritic shafts whereas the local relatively high inhibitory inputs can contact dendritic spines (Megías et al. 2001). Interestingly, human CA1 pyramidal neurons have large

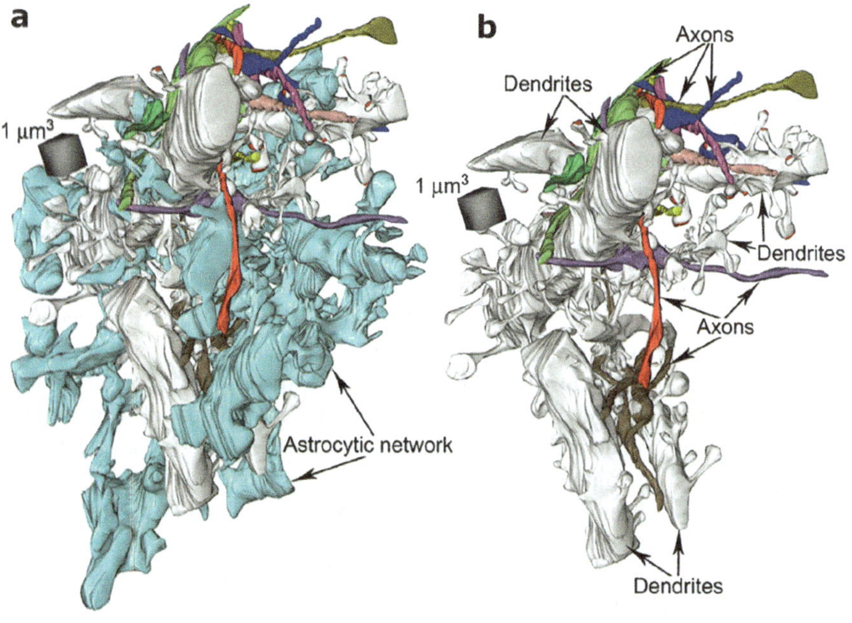

Fig. 1.13 Spatial geometry of a 3D reconstructed astrocytic network associated with dendritic shafts, spines, and axonal segments in the CA1 *stratum radiatum* of ground squirrel. (**a**) Six dendritic segments and 10 axonal segments are surrounded by the astrocytic network in (**b**), which comprised 9–12% of the tissue space. Scale = 1 μm³. (Legend adapted and image reprinted from Popov et al. (2007), under CCC RightsLink® license #5346430419699, originally published by Elsevier)

dendritic trees with more complex patterns of branching in the apical and basal dendrites than in mice, which implies the functional specializations of the neuronal networks in each species (Benavides-Piccione et al. 2020). Furthermore, the presence of different spines in human pyramidal neurons aligns well with the proposition that synapses with a gradation of states, each bridged by distinct metaplastic transitions, would provide neural networks with enhanced information processing and storage capacity (Lee et al. 2012; Rasia-Filho et al. 2021).

Axospinous synapses also differ in the extent to which they are surrounded by glial processes (Fiala and Harris 1999; Fig. 1.13). Data show an almost complete glial covering of spines in Purkinje cells (Spacek 1985) and approximately 40% of the rat CA1 *stratum radiatum* synapses without astroglial ensheathment (Ventura and Harris 1999, see further data in Bernardinelli et al. 2014). Perisynaptic astrocytic processes adjust the extracellular ionic concentration, control the glutamate uptake and spillover from the synaptic cleft, release gliotransmitters and signaling molecules, and participate in energy metabolism that, together, modulate synaptic transmission and strength, circuits, memory, learning, and behavior (Araque et al. 1999; Ostroff et al. 2014; Gavrilov et al. 2018; Lyon and Allen 2022; see Chapter 6 in this book).

Glia can modulate spine development, shape, and synaptic function (Seil 2001), and activity can modulate spine and glial structure. For example, astrocytic processes are motile (Haber et al. 2006) and change ensheathing of spines in the rat lateral amygdala after fear conditioning or conditioned inhibition. The former procedure induces a transient increase in the density of synapses and a long astrocyte-free perimeter on enlarged spines (Ostroff et al. 2014). In dentate granule cells of rats, astrocytic processes may selectively approach synapses on thin dendritic spines compared with those on mushroom spines (Medvedev et al. 2014). Another study showed that synaptic potentiation enlarges spine head volume, PSD area, presynaptic terminals, and glial coverage of both pre- and postsynaptic structures, indicating an activity-induced structural synapse remodeling in hippocampal organotypic slice cultures (Lushnikova et al. 2009). Similar astrocytic coverage of both small and larger spines was observed forming glutamatergic synapses in hippocampal CA1 *stratum radiatum* (Gavrilov et al. 2018, see illustrative EM images in Peters et al. 1991 and Bernardinelli et al. 2014).

There is, also, an associated role for extracellular matrix proteins in spine formation, maturation, and plasticity (Sala and Segal 2014). The structural and functional heterogeneity of glial cells, with biochemical and functional modulatory routes for the synaptic processing, as well as the extracellular matrix in tetrapartite synapses bestow additional levels of complexity to the synaptic transmission, plasticity, and metaplasticity (Chelini et al. 2018; Mederos et al. 2018; Tønnesen et al. 2018; Nguyen et al. 2020; Klimczak et al. 2021). In conjunction, there could be an "active milieu" formed by neuronal and glial (astrocytes, oligodendrocytes, and microglia) compartments, extracellular space, extracellular matrix, and vasculature working dynamically in the perisynaptic microenvironment for information processing at each moment (Semyanov and Verkhratsky 2021; see Chapter 6 in this book).

1.3 Dendritic Spines as Specialized Postsynaptic Units: Integrating Molecules, Biochemical Compartmentalization, and Local Biophysical Properties

Even being tiny neuronal elements (few femtoliters in volume, Sobczyk et al. 2005), spines have a remarkable molecular composition that can differentiate the intraspine microenvironment from the parent dendrite (Hering and Sheng 2001; Yuste 2010; Kasai et al. 2010; Sala and Segal 2014; Helm et al. 2021). Spines display ionic channels and receptors (Sala and Segal 2014) and transform the synaptic input into a complex biochemical coupling of downstream signaling cascades (Calabrese et al. 2006; Harvey et al. 2008; Sala and Segal 2014). Spines can compartmentalize this information (Tønnesen and Nägerl 2016; Cornejo et al. 2022) or propagate the impact of synaptically induced changes onward (Yasuda 2017). Some spines can also show plastic changes to adapt circuitry function to new demands, such as those coming from stimuli perceived as new experiences (Hayashi-Takagi et al. 2015), or be modulated by systemic variables (e.g., the variations in the circulating levels of hormones, Woolley and McEwen 1993; Rasia-Filho et al. 2004; Barreto-Cordero et al. 2020 and references therein). Still, dendritic spines show region, neuronal subpopulation, and dendritic segment-specific properties that impose difficulties to determine a specific role for each spine type valid for all studied areas and all orchestrated series of activity-dependent biochemical cascades (Nimchinsky et al. 2002; Spruston et al. 2013). It is noteworthy that individual spines can regulate the subunit composition of their NMDA receptors and the effective fractional Ca^{2+} current through these receptors (Sobczyk et al. 2005). There is no obvious morphological feature that predicts which spine responds to glutamate with high or low NMDA-mediated Ca^{2+} transient amplitude (Sobczyk et al. 2005). On the other hand, the number of AMPA receptors scale with the PSD size and the dendritic spine geometry (Matsuzaki et al. 2001). On a spine-by-spine basis (Oray et al. 2006; Woolfrey and Dell'Acqua 2015) within the design of each circuit, dendritic spines would enable input-specific plasticity, provide an individual tuning for each synaptic strength, and compose distributed elements for connectivity along dendritic segments (Yuste 2010). Spines can then integrate chemically mediated synaptic inputs with their intrinsic properties and those of the parent dendrite to promote both linear and nonlinear summation of synaptic inputs. As key elements for the functioning of neural circuits, spines have biochemical and biophysical features that enable them to perform a multitude of functions (Yasuda 2017; Chidambaram et al. 2019; Helm et al. 2021; Kasai et al. 2021).

Spine function involves a rich and dynamic local actin organization, signal transduction pathways (Fig. 1.14), and constitutive organelles. Smooth endoplasmic reticulum (the "spine apparatus" in spines possessing a perforated synapse, Spacek – https://synapseweb.clm.utexas.edu/spine-apparatus), endosomes, polyribosomes related to locally regulated protein synthesis (Spacek and Harris 2004; Yuste 2010; Sala and Segal 2014; Miermans et al. 2017; Okabe 2020), and mitochondria (during

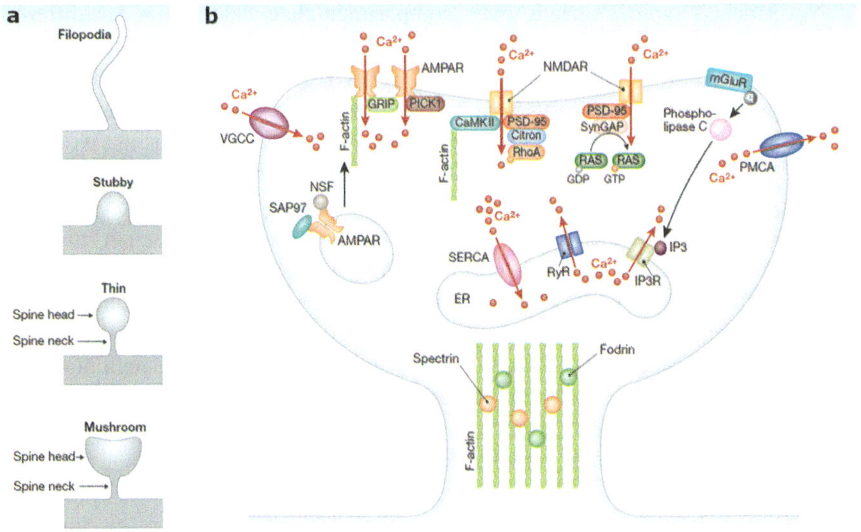

Fig. 1.14 Structural and molecular organization of spines. (**a**) Schematic drawings of spine morphologies based on the most common four-category classification. Note that on the same dendrite a *continuum* of shapes can be observed, and that the morphology of a spine can change rapidly. (**b**) Receptors and molecules related to calcium (Ca^{2+}) signaling in spines. Red arrows indicate flux of calcium ions. AMPAR α-amino-3-hydroxy-5-methyl-4-isoxazolepropionic acid receptor, CaMKII Ca^{2+}/calmodulin-dependent kinase II, ER endoplasmic reticulum, GAP GTPase-activating protein, GRIP glutamate-receptor-interacting protein, IP3(R) inositol trisphosphate (receptor), mGluR metabotropic glutamate receptor, NMDA *N*-methyl-D-aspartate, NSF N-ethylmaleimide sensitive factor, PICK1 protein interacting with C kinase, PMCA plasma membrane Ca^{2+}-ATPase, PSD postsynaptic density, RyR ryanodine receptor, SAP97 synapse-associated protein 97, SERCA sarco/endoplasmic reticulum Ca^{2+}-ATPase, VGCC voltage-gated calcium channel. (Legend adapted and image reprinted from Rochefort and Konnerth (2012), under CCC RightsLink® license #5346470498540, originally published by John Wiley & Sons, Inc)

development, Li et al. 2004), but not microtubules (Peters et al. 1991; Pannese 2015), are observed in spines. Large synaptic inputs impinging directly on the dendritic shafts might lead to the eventual death of highly activated neurons, whereas spines would buffer calcium and help to maintain synaptic activity-induced calcium rise and duration within physiological ranges (Segal 2010). In this regard, the number of synapses on dendritic shafts corresponds to only 5% of all synapses on pyramidal cells in the CA1 *stratum radiatum* (Fiala and Harris 1999). However, there are many examples of aspiny neurons or cells with a very few number of spines that normally participate in active circuits and execute their physiological roles likely dealing with synaptically induced increases in intracellular calcium levels (see examples in Chapter 9 in this book). Calcium compartmentalization can also occur following the activation of single synapses on aspiny dendrites of neocortical interneurons (Yuste 2010 and references therein).

Nevertheless, spines' heads may compartmentalize calcium under different regimes by using multiple mechanisms for calcium influx, concentration, and removal (Yuste 2010; Spruston et al. 2013; Araya 2014).[8] Compartmentalization was documented after auditory-evoked calcium transients in single dendritic spines using high-resolution two-photon imaging of mouse cortical neurons in vivo (Chen et al. 2011). It is worth noting that a notable heterogeneity was found between neighbor spines, that is, active spines were found widely distributed on basal and apical dendrites and spines tuned for different frequencies were interspersed on the same dendrites (Chen et al. 2011; Fig. 1.15). Moreover, the spine biochemical compartmentalization can go beyond calcium and involve other ions, second messengers, signaling enzymes, mRNA, small proteins, or voltage (Yuste 2010). This "control" of diffusive signals can be subject to activity-dependent regulation and depend on organelles, cytoskeletal structures, and geometry of the spine head and neck (Tønnesen and Nagerl 2016).

Uncaging-evoked transient amplitudes of Ca^{2+} vary from spine to spine (Sobczyk et al. 2005). When it occurs, Ca^{2+} increases manifold in the dendritic spine after synaptic stimulation and can trigger the activation of intraspine Ca^{2+}-binding proteins and a sequence of biochemical events (Yasuda 2017). Once started, "the activities of these proteins have a variety of spatiotemporal patterns, which orchestrate signaling activity in different subcellular compartments at different timescales. The diffusion and the decay kinetics of signaling molecules play important roles in determining the degree of their spatial spreading, and thereby the degree of the spine specificity of the signaling pathway" (Yasuda 2017). For sequentially integrated functions, various structural proteins and molecules involved with biochemical signaling pathways have been identified in dendritic spines (Calabrese et al. 2006; Kasai et al. 2010; Yuste 2010; Danielson et al. 2012; Rochefort and Konnerth 2012; Sala and Segal 2014; Helm et al. 2021; Suratkal et al. 2021; Fig. 1.14). They comprise actin binding and cytoskeletal proteins (including microtubule-binding proteins), small GTPases (e.g., the Rho family ones) and associated proteins, cell surface receptors and adhesion molecules, postsynaptic scaffold proteins and adaptor proteins, miRNAs, mRNA binding protein, and transcription factors relevant for the spinogenesis and for the function of mature spines (Sala and Segal

[8] A "useful feature of the spine is the relatively small volume of the spine head, which allows large changes in intra-spine calcium levels in response to the opening of a small number of receptors or channels. For example, it is estimated that individual spines contain only 1–20 voltage-sensitive calcium channels, depending on their size... as several different types of calcium-permeable receptor/channels are colocalized in spines, the spine head can act as an efficient integrator of different postsynaptic signals" (Hering and Sheng 2001). Furthermore, "The exquisite sensitivity of spine $[Ca^{2+}]_i$ to the timing of input or output of the cell... strongly suggests that spine $[Ca^{2+}]_i$ dynamics are involved in computational tasks. In fact, the pairing of input and output that generates the supralinear $[Ca^{2+}]_i$ increases in spines drives synapse-specific LTP (or spike-timing dependent plasticity) in a variety of systems in vitro and in vivo... In the case of spines, then, calcium compartmentalization would restrict plasticity to individual inputs and thus implement local learning rules" (Yuste 2010 and references therein).

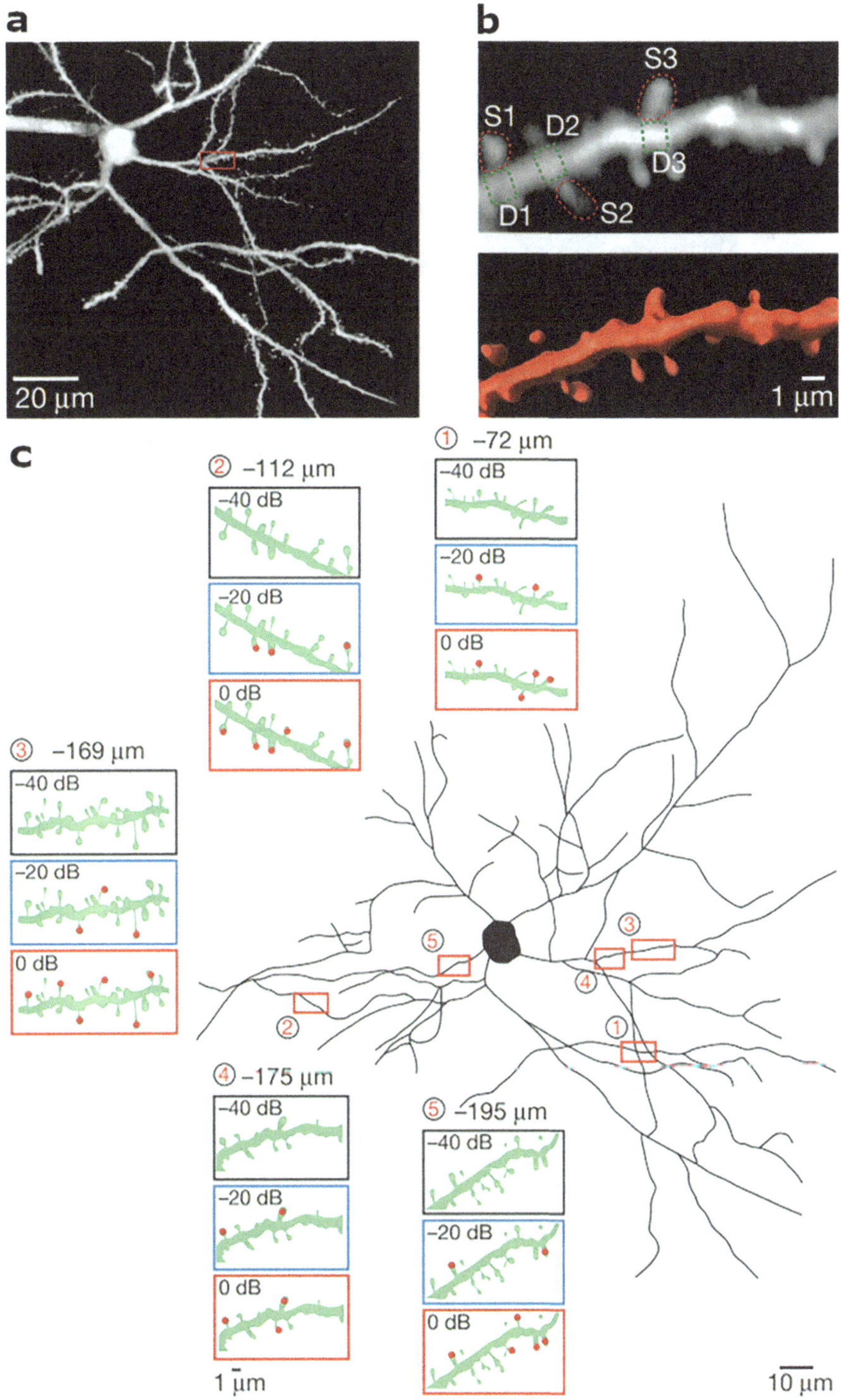

Fig. 1.15 Calcium signals in dendritic spines of mouse cortical neurons in vivo using two-photon microscopy. (**a**) z-projection of a layer II/III neuron of the primary auditory cortex. The red rectangle indicates the area magnified in **b**. (**b**) Upper panel: image at high magnification of the den

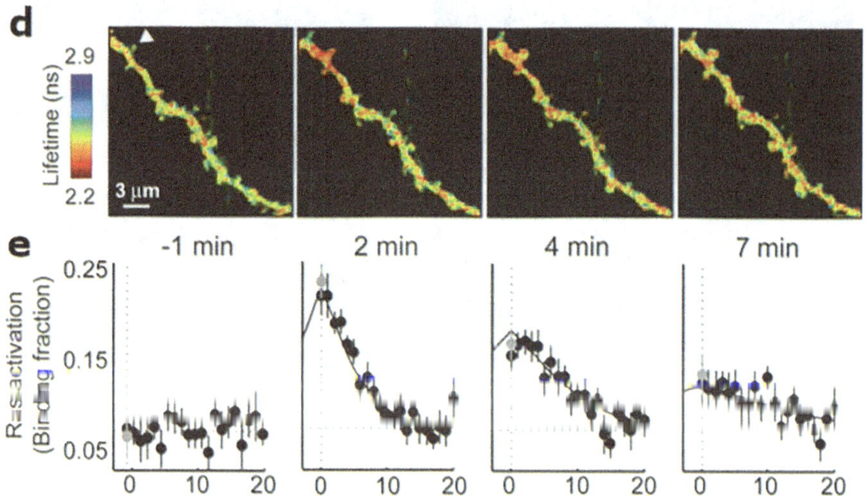

Fig. 1.15 (continued) dritic segment indicated in **a** (average of 6250 frames). Three spines of interest (S1–S3) and the adjacent dendritic shaft regions (D1–D3) are indicated by dashed lines. Lower panel: 3D image reconstruction of the same dendritic segment. (**c**) Image of a reconstructed neuron with dendritic recording sites marked and numbered, surrounded by corresponding insets indicating spine activity. Note the widespread dendritic distribution and sound-intensity-dependent recruitment of active spines. Insets illustrate the corresponding dendritic segments at high magnification (green), indicating with red dots the spines that were activated at the three sound intensities tested: 240 dB (black frame), 220 dB (blue), and 0 dB (red). The depth (under the cortical surface) of the imaged dendritic segments is indicated above each inset. The soma of the neuron was located 2192 mm under the cortical surface. This is the same neuron as that shown in (**a**, **b**). (Legend slightly adapted and figures reprinted from Chen et al. (2011), under CCC RightsLink® license #5352031487410, originally published by Springer-Nature). (**d**) Dendritic spread of active Ras triggered by activation of a single dendritic spine. Fluorescence lifetime images of Ras activity are shown at different time points and color intensity along the dendrite. At time = 0, LTP was induced at the spine marked by an arrowhead and Ras spreads inside the parent dendrite and to neighbor spines. (**e**) The spatial spread (in micrometers from the site of origin) of Ras activation is shown at different time points. Black circles indicate distances along the dendrite relative to the stimulated spine (gray circle). The solid line shows the fitted profile of Ras activation derived from a 1-D diffusion-reaction model. (Legend slightly adapted and figure reprinted from Harvey et al. (2008), under CCC RightsLink® license #5347091263742 and permission from The American Association for the Advancement of Science)

2014). Furthermore, synaptic strength and plasticity can be controlled through intraspine balanced protein phosphorylation and dephosphorylation by postsynaptic serine/threonine kinase and phosphatase signalling by scaffold proteins (Woolfrey and Dell'Acqua 2015).

1.3.1 Molecules and Biochemical Signaling Pathways

A multitude of spine molecules has been described in the last years.[9] Some of them are as follows:

1. Proteins associated with glutamate NMDA, AMPA, kainate, and metabotropic receptors and their subunits compositions (evidenced by the spine response to a glutamate uncaging pulse on specific spines).
2. NMDA-PSD-95 complexes, which cluster glutamate receptors to link with intracellular signaling pathways, and compose the membrane-associated guanylate kinases family, including PSD-95/SAP90, PSD-93/chapsyn-110, SAP102, SAP97, SAPAP/GKAP, Shank family, Homer scaffold proteins, SynGAP (a Ras GTPase-activating protein) and SPAR (which also regulate actin polymerization directly or via cofilin function), and respective PDZ domains, some associated with inositol trisphosphate (IP_3) function.
3. AMPA-GRIP complexes/AMPAR-binding proteins and S-SCAM/MAGI-2, which relate to the regulation of trafficking and stabilization of AMPA receptors and their modulation by neuronal activity.
4. Ionic pump, endogenous buffers, and release mechanisms for calcium homeostasis and actions (involving IP_3 receptors, the smooth endoplasmic reticulum, and synaptopodin in some spines).
5. Actin, whose polymerization relates to spine shape, maturation, stabilization, function, and plasticity, and actin-associated proteins, including spinophilin/neurabin II, myosins II and VI, Abi-1, Abi-2, Abi-3/NESH, Arp2/3, acidic calponin, drebrin, α-actinin-2, adducin, synaptopodin, ankyrin-G, Arp 2/3, profilin, ADF/cofilin, gelsolin, Ras, SPAR, SPIN90, VASP, WAVE, and cortactin, also involving local Ca^{2+} levels, the activity-dependent regulation of actin turnover, and spreading of molecules from activated spines to neighboring dendrites and spines (e.g., Ras).
6. Adhesion molecules, including N-cadherin, cadherin-associated proteins (e.g., αN-catenin), integrins, the polysialylated form of the neural cell adhesion molecule (PSA-NCAM), neuroligin 1, and syndecan-2, for example, to regulate the synaptic formation, modulate actin, spine development, and plasticity.
7. Kinases, such as the subunit α of the Ca^{2+}-calmodulin-dependent kinase II (CAMKIIα), phosphatidylinositol 3-kinase (PI3K), and receptor tyrosine kinase EphB2 interacting with presynaptic ephrin and postsynaptic NMDA receptors and syndecan-2, among others.
8. Phosphatases, including protein phosphatase 1 and calcineurin/protein phosphatase 2B.

[9] Segal (2010), Yuste (2010), Murakoshi et al. (2011), Danielson et al. (2012), Andres et al. (2013), Araya (2014), Guirado et al. (2014), Sala and Segal (2014), Lee et al. (2015), Woolfrey and Dell' Acqua (2015). Harward et al. (2016), Liu et al. (2016), Spence et al. (2016), Smith and Penzes (2018), Chang et al. (2019), Yang et al. (2019), Ben Zablah et al. (2020), Costa et al. (2020), Kesaf et al. (2020), Obashi et al. (2021), Suratkal et al. (2021), and references therein.

9. Neurotrophin receptors, from which either high specific trk receptors or the pan-neurotrophin receptor p75NTR have modulatory roles on spine number and synaptic plasticity.
10. Small GTPase and associated proteins (e.g., RhoA, which can also spread to adjacent spines following stimulation, Cdc42, and Rac1) and various other related proteins that regulate spine density, motility, and shape, including the modulation of Rac1-Trio8 signaling pathway by SESTD1(SEC14 and spectrin domains 1).
11. Proteases, such as the Ca^{2+}-dependent enzyme calpain.
12. Ionic transporters (i.e., K-Cl cotransporter KCC2 and channels, including voltage-dependent Na^+ channels, voltage-gated Ca^{2+} channels, and different subtypes of K^+ channels, which have effects on active membrane conductance and the synaptic input impact on the parent dendrite.

The interaction of these abovementioned molecules relates to several functional possibilities of modulation of synaptically mediated responses by every dendritic spine[10] (Fig. 1.14). Different routes of signaling cascades are activated by the level of intra-spine Ca^{2+} concentration, which can induce the enlargement, shrinkage, or pruning of dendritic spines caused by the presynaptic activity (Kasai et al. 2021).[11] Furthermore, following LTP and advancing over the course of hours, the turnover, function, and plasticity of spines involve structural modifications of the PSD and multiple possibilities for the reorganization of intraspine actin cytoskeleton, and molecular composition (Fifková and Van Harreveld 1977; Muller et al. 2000; Kasai et al. 2010; Suratkal et al. 2021). For example, in apical dendrites of CA1 pyramidal neurons of mice in vitro, binding of CaM to CaMKIIα occurs rapidly (within 128

[10] The basic functional possibility for this arrangement is exemplified by the following condition: a dendritic spine begins the postsynaptic response to glutamate released from a presynaptic axon terminal. Glutamate binds to AMPA and NMDA receptors clustered at the PSD, where "subregions of the spine membrane contain G protein-coupled glutamate receptors (mGluR) and endocytic zones for recycling of membrane proteins. Receptors, in turn, connect to scaffolding molecules, such as PSD-95, which recruit signaling complexes (e.g., regulators of RhoGTPases, or protein kinases). Actin filaments provide the main structural basis for spine shape. Via a network of protein interactions, actin filaments indirectly link up with the neurotransmitter receptors and other transmembrane proteins that regulate spine shape and development, including Eph receptors, cadherins, and neuroligins. Actin-regulatory molecules such as profilin, drebrin, cofilin, and gelsolin control the extent and rate of actin polymerization. These, in turn, are regulated by signaling cascades through the engagement of the transmembrane receptors" (Calabrese et al. 2006).

[11] That is, "large increases in intracellular Ca^{2+} concentration activate CaMKII (within about 1 s), resulting in the activation of a serine/threonine-protein kinase, PAK, and the phosphorylation of many proteins, including LIM-domain kinase (LIMK), slingshot (SSH) and cofilin, which are responsible for the stable enlargement of filamentous actin (F-actin) networks, and are constituents of stress fibers. By contrast, moderate increases in intracellular Ca^{2+} concentration selectively activate calcineurin (in about 1 s), but not CaMKII, leading to the dephosphorylation of SSH and cofilin. A lateral spread of cofilin induces shrinkage of neighboring spines. In addition to the global increases in intracellular Ca^{2+} concentration, Ca^{2+} nanodomains beneath the cytosolic mass of synaptic NMDA receptors are necessary for synaptic plasticity. The involvement of a small, representative subset of (other) molecules in spine remodeling" also occurs during this modulatory action (Kasai et al. 2021 and references therein).

ms) and plateaus with the first glutamate uncaging pulse on a spine, decaying faster (0.4 ± 0.5 s at near physiological temperature) or maintaining a constant level under repetitive stimulation (Chang et al. 2019). In contrast, a high basal level of intra-spine CaMKIIα activity can occur before glutamate uncaging, CaMKIIα activity increases rapidly after single glutamate stimulation of a single spine, shows a slower decay in response to a single uncaging pulse (approximately 10 s), or accumulates to higher levels during trains of uncaging pulses (decaying over 6 s afterward; Chang et al. 2019).

Other neurotransmitters than glutamate can affect spine activity as well. Dopamine promotes spine enlargement when it is released within a time window of 0.3–2 s after the onset of glutamatergic input, tracking the temporal profiles of intracellular cAMP levels (Kasai et al. 2021). This intracellular cAMP increase leads to CaMKII disinhibition via protein kinase A that, when summating effects, can induce spine enlargements within 1 min (Kasai et al. 2021). Noradrenaline can also affect spine size in similar time windows likely involving changes of adenylate cyclase within the spines (Kasai et al. 2021 and references therein).

1.3.2 Biophysical Properties of Spines

Spines can also modulate the magnitude of the EPSPs associating them with the "dendroarchitecture" and the passive and active properties for voltage propagation in dendritic segments (Coss and Perkel 1985; Segev et al. 1995; Rall 1999; Vetter et al. 2001; Spruston et al. 2013). These integrative properties relate to the transformation of synaptic inputs into a specific AP output (as a cellular code within its network), involving the spatiotemporal processing of inputs received to be summed and transformed into a variety of voltage signals by dendrites and spines (Losonczy and Magee 2006). Depending upon the specific form of dendritic integration (i.e., from summated compound EPSPs to brief or prolonged dendritic spikes), dendritic voltage signals can evoke different patterns of AP output such as a single AP, trains, or bursts of multiple APs (Losonczy and Magee 2006). Based on a modeling procedure, even a partial fusion of active spines or changes in spine branch positions would be able to increase synaptic signal transfer (Rusakov et al. 1996).

Recently, it was shown the occurrence of voltage compartmentalization in dendritic spines in vivo. Data were obtained using two-photon microscopy, a genetically encoded voltage indicator (fluorescent postASAP) plus PSD95.FingR nanobody domain for enriched expression in spines, and somatic whole-cell recording (Cornejo et al. 2022). Membrane potentials were studied in basal dendrites of layers II/III pyramidal neurons from the somatosensory cortex of mice (4–8 weeks old). During spontaneous activity (in darkness) and sensory stimulation (air puffs to activate the whiskers in anesthetized mice), spines and adjacent dendrites were (1) synchronously depolarized during backpropagation of trains of axonal APs (i.e., the invasion of APs into spines occurred without failures or decrement) or, when in the absence of APs (during transient subthreshold potentials or in the absence of pronounced somatic depolarization), (2) dendritic segments and their spines depolarized together, or (3) spines activated independently, likely representing isolated synaptic potentials (Cornejo et al. 2022). When spine heads were photostimulated to mimic synaptic potentials, depolarizations attenuated while spreading from the spine into the adjacent dendrite, which suggests that individual spines can compartmentalize voltage in awake animals (Cornejo et al. 2022).

The geometry of spiny dendritic branches is relevant for cellular excitability. In pyramidal neurons, most arbors (~95%) are composed of fairly short (~100 μm) and thin (~0.5 μm) terminal dendrites that branch off the trunk or soma (radial oblique, apical tuft, and basal branches; Bannister and Larkman 1995; Megías et al. 2001; Losonczy and Magee 2006). The use of multisite two-photon glutamate uncaging on spines, simultaneous two-photon Ca^{2+} imaging, and somatic whole-cell voltage recording revealed the integrative properties of radial oblique (thin collaterals) dendrites in hippocampal CA1 pyramidal neurons in slices from adult rats. Accordingly, synchronous and sized input patterns (~20 inputs within ~6 ms) produced a supralinear summation and dendritic spikes (Losonczy and Magee 2006). It was also tested what responses could be obtained if all inputs were tightly clustered on a 20 μm long dendritic segment (e.g., 20 synapses would represent ~30% of all synapses within 20 μm) or if the inputs were spread across the entire dendritic branch and onto spines randomly covering approximately 67% of a single oblique dendrite (~60 μm long). Results showed that input patterns spatially distributed in both conditions were capable of producing a dendritic spike, which indicates that each oblique branch can serve as an integrative compartment (Losonczy and Magee 2006). These branches summate input currents, and the synchronous activity required to shift from linear to supralinear integration represents ~5% of the total excitatory input that can reach the branch (Losonczy and Magee 2006). Again, the dendritic geometry act in concert with cable and active membrane properties and the spatiotemporal summation features for the AP propagation in dendrites (Vetter et al. 2001). These data are relevant for identifying the extent (number of inputs, frequency, spatial distribution of the synaptic inputs, and EPSPs summation) of afferent activity on spines needed to modify the dendritic voltage and neuronal output in these cells. They also suggest that there are more spines available to make synapses than the minimum needed to generate a dendritic spike, which provides a "safety factor" and evidence of the high potential possibility for spatiotemporal input integration and processing by these neurons. From a study modeling human temporal layer II–III pyramidal neurons, it was predicted that a somatic Na^+ spike can be generated (with 50% probability) by 134 ± 28 simultaneously activated synapses with large AMPA- and NMDA-conductances per synaptic contact, EPSPs at the spine head of 12.7 ± 4.6 mV and at spine base of 9.7 ± 5.0 mV, among other parameters (Eyal et al. 2016). These data suggest that the number of synchronized activated dendritic spines needed for a somatic AP is low when compared to the estimated total number of spines per pyramidal cell (see additional relevant data in Mueller and Egger, 2020).

The scenario involving synapses and spines is notably complex. Approximately 10% of the proteins encoded in the human genome can be found composing synapses and their diversity, providing different compositions and functional possibilities for individual synapses (Grant and Fransén 2020) with an evolutionary perspective (Berg et al. 2021; Viscardi et al. 2021). Over 500 different types of molecules modulate the structure and function of dendritic spines (Ammassari-Teule et al. 2021). For example, in mice, the apical dendrites of pyramidal neurons in the CA1 *stratum radiatum* show a high synaptic diversity, with postsynaptic proteins organized into gradients and differentially distributed along the length of the dendritic tree (Grant and Fransén 2020 and references therein). This structural organization can have different functional implications considering that various axons reaching dendritic segments show pathway-specific synaptic responses (Chabrol et al. 2015). In other words, "unlike the homogeneous population where all the synapses either strengthen or weaken, the heterogeneous populations show physiological diversity at the single-synapse level: some synapses strengthen and others weaken" (Grant and Fransén 2020). Moreover, there might be a dynamic, multisubcellular plasticity involving AMPA receptor trafficking (from a relatively small perisynaptic compartment) in and around the spine PSD that could alter the activity of G protein Gs and Gq and promote LTP or long-term depression (LTD), respectively (Mihalas et al. 2021; see also Woolfrey and Dell'Acqua 2015). All these possibilities and their implications have to be further tested, including the mechanisms for synaptic strength maintenance despite protein turnover (Danielson et al. 2012; Helm et al. 2021).

Therefore, more than "passive" elements receiving synapses, spines provided a major increase in neuronal computational capabilities (Yuste 2010 and references therein). "Purely hypothetical at first, studies pointed to synaptic amplification, non-linear sensitivity to several parameters, and the possibility of chain-reactions, in which several firing spine heads depolarize the dendritic membrane; the dendritic membrane could be passive, but under favorable conditions local dendritic depolarization might suffice to bring some adjacent excitable spines to their firing threshold" (Rall 1999). Furthermore, the *raison d'être* for dendritic spines would be the ability to show activity-dependent structural and molecular changes, modifying the synaptic strength and altering the gain of the linearly integrated subthreshold depolarizations before the generation of a dendritic spike (Araya 2014).

Some relevant questions were then raised: (1) what is the impact of differently shaped spines on processing synaptic inputs and corresponding synaptic strength ("weight"), and (2) because the electrical properties of dendrites vary with distance from the soma (Parajuli and Koike 2021), how are the spine-related postsynaptic currents along the length of the dendritic tree. The former issue involves head and neck morphological features, the PSD structure, and several intraspine components associated with intrinsic membrane properties of spines. The parallel resistance of the spine head is very large and the neck resistance is, by comparison, a path out and into the spine with multiple modulatory possibilities[12] (Spruston et al. 2013; Tønnesen and Nägerl 2016; Fig. 1.16).

[12] That is, "if the synapse operates in the current source regime, a change in neck resistance will only affect the voltage in the spine head, whereas if it acts as a voltage source, it will only influence the dendritic EPSP. In reality, most synapses are likely to occupy a middle ground between these

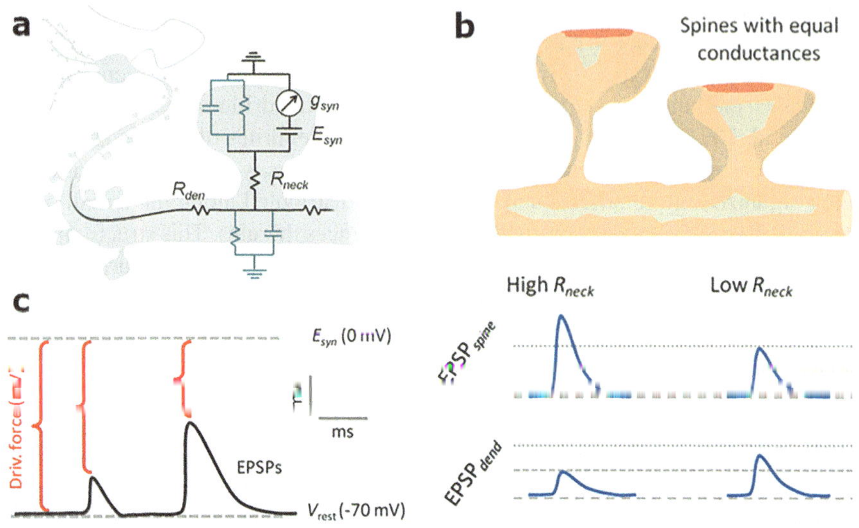

Fig. 1.16 Electrical compartmentalization can occur in dendritic spines. "(**a**) In the spine electrical circuit diagram, a variable current enters through the synaptic receptors, scaling with their conductance (*gsyn*) and with the electrical driving force, which is the difference between resting membrane potential and the reversal potential of the conductance (*Esyn*). The membrane resistance is so high that current will not escape, and it will instead pass first the neck resistance (*Rneck*) and then the dendritic input resistance (*Rdendrite*) on the way to the soma. The EPSP that the synaptic current generates along the way is defined by Ohm's law and follows voltage divider law. (**b**) As the synaptic current scales with driving force, the depolarizing EPSPs produced by the current will have a self-dampening effect as they approach the glutamate receptor reversal potential (*Esyn*). (**c**) A thin and long spine neck will have a high *Rneck*, which will locally boost the EPSP in the spine head. This in turn causes a loss of driving force, so that less current will flow over the synaptic conductance. While the EPSP in the spine head sees both the boosting and the loss of driving force, the corresponding EPSP in the dendrite only experiences the loss of driving force. Conversely, a spine with a low *Rneck* will see less boosting of the spine head EPSP and less current attenuation, so the spine and dendritic EPSPs are more similar. Beyond the illustrated passive effects of morphology, the boosted spine head EPSP may locally recruit voltage-gated conductances on the spine, which may in turn increase or decrease the synaptic current." (Legend and figure reprinted from Tønnesen and Nägerl (2016), https://doi.org/10.3389/fpsyt.2016.00101, under CC BY 4.0 license and copyright 2016)

two extreme regimes, so that spine neck plasticity might simultaneously influence synaptic signals in the spine and dendrite... Hence, a reduction in neck resistance is likely to have at the same time differential effects on the EPSP on either side of the spine neck, lowering it in the spine head, while elevating it in the dendrite. Conversely, an increase in neck resistance will boost the voltage in the spine head and lower it in the dendrite. The actual magnitude of the effects will depend on the relative sizes of the parameters synaptic conductance, neck resistance, and dendrite resistance... By influencing the spine head EPSP, changes in neck resistance might strongly affect the activation of voltage-gated ion channels in the spine head, such as voltage-sensitive calcium and sodium channels, which in turn shape the EPSP... Likewise, the voltage-dependent block of the NMDA receptor by extracellular magnesium will be directly affected by changes in the spine head EPSP... Beyond the immediate effects on EPSPs, large changes in neck resistance might effectively shift the operat-

Spines would also form helixes indicating that their locations along the dendrites might not be set at random (Yuste 2010). There is no clear dependence of the spine morphologies on their distance to the soma in layer II/III pyramidal cells of the mice visual cortex (Arellano et al. 2007a). On the other hand, the number of dendritic spines is organized in a structured pattern along the dendrite that affects the PSD area density (expressed as the PSD area per unit length of the dendrite) in the CA1 *stratum radiatum* of mice (Parajuli et al. 2020; Parajuli and Koike 2021). A positive correlation was observed between dendritic diameter and the density of dendritic spines, and between dendritic diameter and total PSD area per unit length of dendrite in this brain area (Parajuli et al. 2020). Pyramidal neurons in the mice visual cortex have another heterogeneous distribution of spines along the length of the apical dendrite, that is, a few spines in the thicker proximal dendrite and more numerous spines in the intermediate segment (Valverde 1967). In both cases, the morphological diversity of spines' shapes suggests a likely variability for the synaptic strength along the dendritic arbor (Arellano et al. 2007a). Passive and active biophysical properties of spines have to be considered for functional purposes involving spines in dendrites of different calibers (Spruston et al. 2013; Sala and Segal 2014; Gidon et al. 2020), which, again, can vary from neuron to neuron and brain area. Spines located at different distances from the soma would standardize local postsynaptic potentials throughout the dendritic tree, reducing the location-dependent variability of local excitatory properties (Gulledge et al. 2012; Harnett et al. 2012; van der Zee 2015). Hypothetically, using simulations of dendritic electrical properties, if the heads of distal dendritic spines have excitable/active membrane properties, spines could reach the threshold for AP and bring about a possibility of saltatory propagation into more proximal branches, increasing the efficacy of distal synaptic inputs (Shepherd et al. 1985). Synaptic potentials from perforant path inputs at distal dendrites would still induce LTP at the CA1 proximal Schaffer collateral synapses when the two inputs are paired at a precise interval on local pyramidal neurons (Dudman et al. 2007). This subthreshold form of heterosynaptic plasticity can occur in the absence of somatic spiking but needs synaptically induced activation of NMDA receptors and the release of Ca^{2+} from internal stores mediated by IP_3 (Dudman et al. 2007). Then, the synaptic processing made by spines would influence the neuronal activity depending upon its distribution along proximal to distal dendritic branches and considering the pattern of spatiotemporal integration of different synaptic inputs (Spruston et al. 2013).

In addition, spines can have their functioning changed by the influence of the parent dendrite voltage either for onward propagation of synaptic inputs or for back-propagation of APs or dendritic spikes (Yuste 2010; Spruston et al. 2013).[13] In this

ing regime of the synapse, acting more like a voltage or current source. Such a major 'parametric' change would modify the voltage transformation of the synapse and may, thus, affect dendritic integration and the computational performance of the neuron" (Tønnesen and Nägerl 2016 and references therein).

[13] Let "the spine be considered electrically as a very high-resistance spine head connected to a relatively low-resistance dendrite via an intermediate-resistance spine neck. This electrical circuit consists of multiple impedance mismatches, which means that voltage drops are highly asymmetrical: voltage drops considerably from the spine head to the dendritic shaft, but it should hardly change

regard, it is crucial to take into account the densities, subtypes, and kinetics of sodium and potassium channels in spines. For example, backpropagation of APs in spines is possible when the voltage-gated sodium channels on spines differ in density or inactivation features compared to those in the same dendrite, and potassium channels on the spine head and neck would shunt synaptic potentials (Yuste 2010). Trains of EPSPs would saturate the spine sodium gradient and serve to filter high-input frequencies (Yuste 2010). These data clearly indicate further complex and dynamic coupling of spines and dendrites in the spatiotemporal synaptic integration, providing multiple "functional codes" for the information processing.

When in clusters, spines can also show spike-timing-dependent cooperativity and plasticity (Tazerart et al. 2020). That is, synaptic amplification between close dendritic spines would enhance input cooperativity among coactive inputs (Harnett et al. 2012) and a timely postsynaptic depolarization to relieve the Mg^{2+} block of NMDA receptors and influx of Ca^{2+} in adjacent synapses. These interactions relate to the detection of temporally coincident presynaptic and postsynaptic activity by spines (Yuste 2010). Likewise, small molecules (e.g., Ras and Rho) can diffuse out of the activated spine and carry the "synaptic activation message" to adjacent spines (Sala and Segal 2014, see functional implications in Ma and Zuo 2022). Various diffusional properties affect the distribution of molecules between the PSD and dendritic shaft (Obashi et al. 2021). Nevertheless, in hippocampal pyramidal neurons, Ca^{2+} influx through synaptic NMDA receptors activates Ras that, from a single dendritic spine, diffuses over approximately 10 μm along the parent dendrite, invades nearby spines, and regulates the LTP induction threshold at neighboring synapses (Harvey et al. 2008; Fig. 1.15). In response to stimulation of a single dendritic spine, the spreading of cytosolic protein kinase ERK was calculated to be even greater (~100 μm), perhaps playing a role in synapse-to-nuclear signaling (Yasuda 2017 and references therein).

Interestingly, in apical dendrites of layer V pyramidal neurons in the motor cortex of mice, a third of new dendritic spines formed during the acquisition phase of learning novel forelimb skills emerge in clusters, usually as neighboring pairs (Fu et al. 2012). These new spines persist by repetitive activation of the cortical circuitry during learning and even longer after training stops (compared to non-clustered counterparts), overcoming a spatial constraint and being formed in close vicinity to previously existing stable spines (Fu et al. 2012). In this same regard, imaging and

at all when spreading in the other direction. Functionally, this means that when an excitatory synapse on a spine head is activated, the voltage in the spine head should be considerably larger than the voltage in the dendrite to which it is attached. On the other hand, voltages in the dendrite are expected to transmit reliably (i.e., without significant attenuation) into the spine head. This is biophysically similar to the situation described earlier for the dendrites. Dendritic spikes propagate poorly in the forward direction (from smaller dendrites into larger dendrites), but backpropagating action potentials propagate more reliably in the reverse direction (from larger dendrites into smaller dendrites)... This makes it possible for a group of spines to influence one another, as well as for backpropagating action potentials or dendritic spikes to change the membrane potential in the spine" (Spruston et al. 2013, see also Yuste 2010).

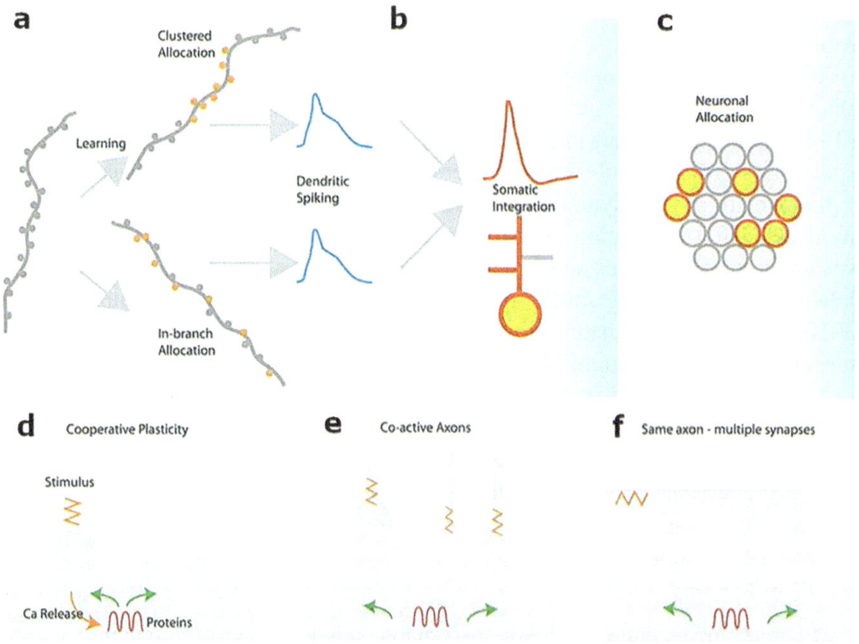

Fig. 1.17 Schematic representation of synaptic and cellular engram formation. "(**a**) Dendritic spikes can be induced either from localized activation of a coactive group of synapses (clustered synapse allocation) or from more dispersed activation of synapses within the same branch segment (In-branch allocation). (**b**) In both cases, the elicitation of dendritic spikes is integrated with the somatic compartment, controlling the action potential generation or bursting behavior of the neuron and enabling local or global plasticity to occur. (**c**) Neuronal populations activated for each memory are selected by mechanisms such as excitability and CREB activation leading to the selection of a population memory engram. Mechanisms for the generation of synaptic clusters: (**d**) cooperative sharing of plasticity-related resources such as proteins facilitates cooperative LTP in nearby synapses after LTP induction. The spreading of activation of plasticity-related proteins, enzymes, and mRNAs may prime nearby synapses for subsequent plasticity. (**e**) Coactive axons in nearby synapses can drive cooperative plasticity by initiating similar resource-sharing mechanisms as a result of coincident activation. (**f**) A single axon may make multiple contacts in a short segment of a dendrite, thereby driving cooperative plasticity via synchronized synapse activation." (Legend and figure reprinted from Kastellakis and Poirazi (2019), https://doi.org/10.3389/fnmol.2019.00300, under CC BY 4.0 license and copyright 2019)

molecular findings indicate that the recruitment of synapses that participate in the encoding and expression of memory is neither random nor uniform; rather they are in clusters[14] (Kastellakis and Poirazi 2019; Fig. 1.17; see further data in Ma and Zuo

[14] In this regard, "a hallmark observation is the emergence of groups of synapses that share similar response properties and/or similar input properties and are located within a stretch of a dendritic branch. This grouping of synapses has been termed 'synapse clustering' and has been shown to emerge in many different memory-related paradigms, as well as in in vitro studies. The clustering of synapses may emerge from synapses receiving similar input, or via many processes, which allow for cross-talk between nearby synapses within a dendritic branch, leading to cooperative

2022 and the possibility of higher-order interactions in dendrites in Hodassman et al. 2022).

In conjunction, the complex molecular composition of spines associated with the shape of the spine neck and head can lead to (1) biochemical compartmentalization, (2) an input-specific and independent regulation of synaptic strength in different spines, and (3) the possibility of cooperativity and summation of the electrical signaling weight of each synapse in the dendritic segment. Plasticity lies in the activity-dependent modulation of membrane channels and receptors and the spatiotemporal dynamics and kinetics of the actin cytoskeleton and other intracellular components of the spines, including the PSD and associated proteins (Yuste 2010). Neuronal and synaptic plasticity can be associated with changes in the structure of spines (e.g., number, shape, and connectivity) whose activity will widen neural network properties for physiological, innate, and adaptive purposes (see Heck and Benavides-Piccione 2015; Rasia-Filho and Cohen 2016; Calagnotto et al. 2019; Ammassari-Teule et al. 2021).

Many questions remain unsettled on the structure and function of dendritic spines (Ammassari-Teule et al. 2021). At least at this moment, adapting the general ideas for a spiny neuron and its network (Yuste 2010), one would consider that spines were likely designed

1. To maximize connectivity as postsynaptic elements accommodating a great number of distributed synapses along the dendritic surface, but using the minimal volume and molecules per spine as possible
2. To modulate EPSPs using passive and active biophysical properties to provide a different voltage regime toward parent dendrites (or be affected by them) and can also promote biophysical compartmentalization in single spines
3. To integrate AMPA and NMDA receptors with compositions and functions (considering only the most frequent excitatory input at principle) to amplify intraspine responses by orchestrating different biochemical routes as signaling pathways at different timescales
4. To define biochemical compartments that modulate local calcium levels, actin phosphorylation, and second messengers availability, among other possible routes
5. To sum various inputs along dendritic segments (with different properties along proximal to distal branches, main shafts, and collateral ones) to integrate them within a spatiotemporal window for linear and nonlinear impacts on the somatic voltage and firing output

plasticity... The mechanisms that lead to the nonrandom, clustered positioning of synapses (would) require at least one of the following: (a) convergence of axonal projections carrying similar information onto the same dendrite; (b) the presence of active ionic conductances in dendrites; (c) activity-dependent axonal rewiring; and (d) local protein synthesis. Clustered synapses can act in concert to exploit maximally the nonlinear integration potential of the dendritic branches in which they reside. Their main contribution is to facilitate the induction of dendritic spikes and dendritic plateau potentials, which provide advanced computational and memory-related capabilities to dendrites and single neurons" (Kastellakis and Poirazi 2019).

6. To maintain a certain degree of isolation for each input or form small clusters for each input to determine an activity-dependent synaptic weight and plasticity for the network in which these spines and neurons lie within.

1.4 Some Examples of Plasticity Involving Dendritic Spines and Neural Circuits

Dendritic spines considerably increase the computational repertoire of cells and circuits. It is not surprising that spines can be plastic elements, as abovementioned. Experience-dependent structural synaptic plasticity of cortical neurons includes changes in dendritic spine number by the generation of new spines or the elimination of existing ones. The balance between spine number, structure, and function might relate to the mechanisms for long-term changes in synaptic strength, connectivity, memory, and cognition in selective synaptic ensembles (Yuste and Bonhoeffer 2001; Knafo et al. 2005; Bourne and Harris 2007, 2009; Holtmaat and Svoboda 2009; Fu et al. 2012; Hayashi-Takagi et al. 2015). For example, optogenetic manipulation allowed the identification and erasure of specific synaptic memory traces in potentiated spines of the mouse motor cortex (Hayashi-Takagi et al. 2015; Fig. 1.18).

Within the hippocampus, the phenomenon of LTP, an electrophysiological model of neuronal plasticity (Bliss and Lomo 1973; Bliss and Collingridge 1993), is associated with enlargement of the spine head and shortening of the spine neck or the induction of short, stubby spines and shaft contacts, and thus changes in synaptic strength and EPSPs (reviewed in Yuste 2010; Parajuli and Koike 2021). LTP could also be associated with the transient formation of new, mature, and probably functional synapses involving an increase in spine density and in the proportion of axon terminals contacting two or more dendritic spines (Toni et al. 1999; Muller et al. 2000). However, whether these both processes require the same signaling pathways is still a matter of debate. The hippocampus is involved in spatial learning tasks, and aged humans and rodents exhibit spatial memory deficits (Barnes 1987). LTP induction and maintenance are reduced by aging (see details in Rosenzweig and Barnes 2003). At least in mice, there is an age-related impairment in spatial learning as well as a decrease in the densities of basal dendrites in CA1 neurons (von Bohlen und Halbach et al. 2006a, b). Interestingly, it has been shown that an increase in spine densities appeared exclusively in basal dendrites of CA1 area after spatial learning in rats (Moser et al. 1994). This finding indicates that altering neuronal plasticity affects electrophysiological properties, as well as leads to morphological changes in dendritic spines (at least in the hippocampus) that are accompanied by altered learning and memory capacities. On the other hand, LTD, another form of synaptic plasticity, reduces various morphological parameters of spines (Tønnesen and Nägerl 2016).

It is important to determine when dendritic spines are more susceptible to plastic changes and how changes can occur in spine number and shape besides genetic determination. Plasticity is more evident during early development, but modulatory

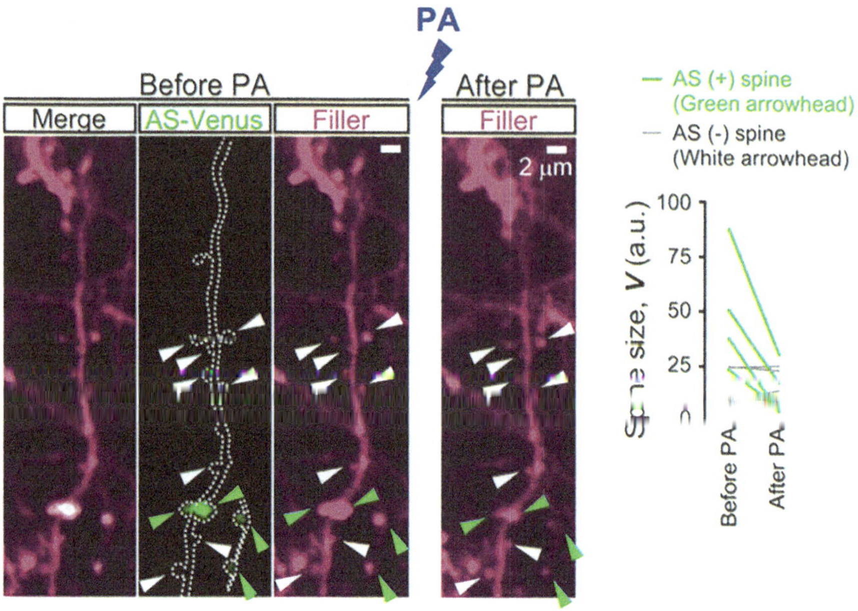

Fig. 1.18 Spine shrinkage in the motor cortex (M1) of mice promotes an erasure of synaptic memory traces. (left) Representative images of spine shrinkage in the M1 cortex upon photoactivation (PA) in vivo. Potentiated spines were labeled with the synaptic optoprobe AS-PaRac1 (activated synapse targeting photoactivatable Rac1). In vivo two-photon imaging of AS-PaRac1 revealed that motor learning can induce substantial synaptic remodeling. This acquired motor learning was disrupted by the optical shrinkage of the spines labeled with AS-PaRac1 following PA (right). (Legend adapted and Figure reprinted from Hayashi-Takagi et al. (2015), under CCC RightsLink® license #5352040706317, originally published by Springer-Nature)

actions upon spine structure and function can occur along the lifespan (Runge et al. 2020). For example, a plastic NMDA receptor-dependent mechanism modulates the estrogen regulation of hippocampal spine and synapse density in female rats (Woolley and McEwen 1994), although some spines are not indefinitely plastic and a certain number and function of stable spines should remain over time. A functional example of this condition occurs with region-specific features in the central control of reproductive behavior (McCarthy 2010; Rasia-Filho et al. 2012a, b). Dendritic spines can be sexually dimorphic or modulated by gonadal steroids (Woolley and McEwen 1993; Rasia-Filho et al. 2004; McCarthy and Konkle 2005; Frankfurt and Luine 2015; Hansberg-Pastor et al. 2015; Luine and Frankfurt 2020) and neurosteroids (Fester and Rune 2021). Spines can also be affected by sexual experience (Zancan et al. 2018) and motherhood (Rasia-Filho et al. 2004) in the posterodorsal medial amygdaloid nucleus (MePD) of rats, for example.

The MePD is an important area for puberty and sexual behavior display in rodents (Rasia-Filho et al. 1991, 2012b; Newman 1999; Cooke and Woolley 2009; Li et al. 2015; Johnson et al. 2021). This subnucleus has one of the brain highest concentrations of androgen receptors, both estrogen receptor-α and -β, and progesterone receptors, some coexpressed in the same neurons (Simerly et al. 1990; Gréco et al. 2001, 2003; Shughrue et al. 1997; De Vries and Simerly 2002; Portillo et al. 2006). Male-to-female differences in the dendritic spines in the MePD of prepubertal rats suggest a sex-specific synaptic processing that was initially genetically determined and modulated by gonadal hormones during the intrauterine period and early postnatal days to be further modified by puberty and at adulthood (Nishizuka and Arai 1981; Rasia-Filho et al. 2004, 2012a; Cooke and Woolley 2005, 2009; Cooke et al. 2007; Zancan et al. 2018). In this regard, the MePD of prepubertal male rats have nearly 80% more excitatory synapses per neuron than females (Cooke and Woolley 2005). Male prepubertal gonadectomy impaired the expression of sexually dimorphic behavior at the same time that reduced both the excitatory synaptic transmission and the density of dendritic spines, with no reductions in the overall dendritic length and branching (Cooke and Woolley 2009). Castration of prepubertal male rats reduced dendritic spine number by 30–45% along the first 70 μm of primary dendrites or along the last 70 μm of terminal dendrites (Cooke and Woolley 2009). In the adult rat MePD, the number of proximal dendritic spines reduced around 20% following castration of males (de Castilhos et al. 2008) whereas, in ovariectomized females, the dendritic spine density was increased after replacement therapy with estradiol and progesterone (de Castilhos et al. 2008).

Sexual activity requires the integration of conspecific sensory information (e.g., olfactory and vomeronasal, auditory, visual, and tactile), the ongoing actions of sex steroids on their receptors, and the processing of dynamic synaptic information in interconnected areas of the social behavior brain network (Newman 1999; Rasia-Filho et al. 2004, 2012a, b; Petrulis 2020; Gutierrez-Castellanos et al. 2022). Therefore, it is conceivable that part of the synaptic contacts and spines are developed to establish a "hardwired" connection with sex-specific features relevant "to secure" species reproduction and survival (Zancan et al. 2018, see a relevant discussion in Gutierrez-Castellanos et al. 2022). This proportion of dendritic spines would serve as a "reference and baseline point" for their density. From this value, the effects of sex steroids can occur in another proportion of spines aiming for the excitation of sexual behavior in males. Plastic spines modulate the disinhibition of sexual behavior and the timing regulation of neuroendocrine secretion and ovulation in females (Rasia-Filho et al. 2004, 2012a; Zancan et al. 2018; Dalpian et al. 2019).

These ideas need additional discussion. Based on morphological and electrophysiological features, there are two coexisting subpopulations of spiny neurons that are sexually dimorphic in the MePD of adult rats (Rasia-Filho et al. 2012a; Dalpian et al. 2019). In these cells, the density of Golgi-impregnated proximal dendritic spines is (1) approximately 30% higher in males than in females in proestrus, and (2) affected by ovarian hormones fluctuations in normally cycling females, decreasing around 30% from diestrus to proestrus (Rasia-Filho et al. 2004, 2012a). Ultrastructural data showed that, in the right MePD of females in proestrus, the proportion of inhibitory synapses on dendritic shafts is higher than in the left MePD, and higher than in the right MePD of males or females in diestrus or estrus (Brusco et al. 2014). Electrophysiological recordings provided congruent results: a higher excitatory input onto MePD neurons of males compared to cycling females and between estrous phases (higher in diestrus compared to proestrus or estrus; Dalpian et al. 2019). In proestrus, there is an increase in the inhibitory input onto MePD neurons compared to males and females in diestrus or estrus (Dalpian et al. 2019). In conjunction, these data indicate what would be the percentage of "labile," dynamic spines modulated by sex steroids compared with the percentage of more "stable" spines in the MePD of both sexes. A balance between stability and plasticity of spines is a major elaboration in areas related to social behavior and reproduction. That is, stable spines would provide steady properties for the neural circuit, whereas plastic spines would adapt the circuitry to new synaptic demands (Zancan et al. 2018). The plasticity of these spines of adult male rats was further tested following sexual intercourse. Sexual experience plays an essential role in inducing neurons that express androgen receptors to be activated by female cues (olfactory and vomeronasal stimuli) and by ejaculation (Greco et al. 1998; Li et al. 2017). Naïve males have comparatively less pronounced sexual behavior than sexually experienced animals (Swaney et al. 2012). It is highly likely that the afferent synaptic inputs to the MePD spiny neurons code the processing of learned social cues and the contextual memories of sexual encounters of the animal (Stark et al. 1998; Stark 2005; Becker et al. 2017; Zancan et al. 2018). This possibility implies that the synaptic plasticity in the MePD would improve the execution of an innate behavior and undergo adaptive modulatory changes (Zancan et al. 2018). At the same time, an experience-dependent lasting structural remodeling of local dendritic spines would improve the execution of sexual behavior that can promote learning and have a positive effect on reinforcement of ongoing and future social contacts (Ågmo 1999; Ferguson et al. 2001; Zancan et al. 2018).

The study of dendritic spines also opens research avenues for further hypotheses about synaptic and brain circuitries changes in pathological conditions. That is, considering multiple functional implications, spines are significant in that their morphology can change in various neurological and psychiatric disorders (Fiala et al. 2002; Penzes et al. 2011; Maiti et al. 2015; Herms and Dorostkar 2016; Forrest et al. 2018; Runge et al. 2020; Parajuli and Koike 2021). These disorders include epilepsy, Alzheimer's disease, Parkinson's disease, and major depression, among others (Wong and Guo 2013; Maiti et al. 2015; Smith and Penzes 2018; Chidambaram et al. 2019; Bączyńska et al. 2021). It is currently challenging to ascertain whether altered dendritic spines can be the cause, consequence, or compensatory response in each pathological condition (Fiala et al. 2002; Penzes et al. 2011). In epilepsy, abnormalities in dendrites and spines have been reported in both patients with refractory epilepsy and animal models, including altered density and morphology of dendritic spines in isolation or associated with varicose swelling of dendritic branches (González-Burgos et al. 2004; Ampuero et al. 2007). Direct examination of the epileptogenic zone of neocortical or hippocampal tissue from patients with epilepsy and animal models has identified spine loss (Zeng et al. 2007; Freiman et al. 2011; Kitaura et al. 2011; Santos et al. 2011; Guo et al. 2012; Rossini et al. 2021; Puhahn-Schmeiser et al. 2022), or increased density (Freiman et al. 2011; Puhahn-Schmeiser et al. 2022), decreased or enlarged spine head size, elongated spine necks, or increased spine length (Meikle et al. 2008; Chapleau et al. 2009), which are also the common findings observed in genetic disorders with a high prevalence of epilepsy. Extensive evidence from preclinical models supports the idea that seizures may directly contribute to the dendritic and spine structural pathology in epilepsy, but other studies indicate that the abnormal dendritic arbors and spines are likely to be etiology-driven instead of seizure-dependent (Wong and Guo 2013). Cutting-edge high-resolution imaging and molecular technologies will determine whether seizures are a cause or a consequence of the dendritic spine pathology in refractory epilepsies in humans. Likewise, the mechanisms of major depression, as well as the neurobiological basis of antidepressant therapy, are still enigmatic, but it is likely that the pathogenesis and the treatment of major depression involve changes in neuronal plasticity, which is accompanied by changes in dendritic spines in the hippocampus (Serafini 2012; Jiang and Salton 2013). Alterations in the morphology or density of dendritic spines can be seen in various models of neurodevelopmental disorders or in cases of intellectual disability (von Bohlen und Halbach 2010). The analysis of dendritic spines in mouse models of genes that, in their mutated form, leads to these conditions in humans helps to discover new proteins that would be involved in the structure and plasticity of synapses. Unraveling brain microcircuits, dendritic spines, and synaptic processing could ultimately help to understand the mechanisms of neuronal plasticity and offer the opportunity to develop specific therapeutics. However, they would require multiple innovative approaches and technological advancements, such as super-resolution microscopy (Urban et al. 2011; Berning et al. 2012) or optogenetics and intra-vital microscopy (Roth et al. 2020). A note of caution is also needed when considering that neuronal subpopulations of mice (as animal model) and humans can differ in their transcriptomic profile and phenotypic expressions (Hodge et al. 2019).

1.5 Dendritic Spines and Current Research Advancements

Several lines of research have been describing the dendritic spines in terms of structure, development and motility, biochemical composition and calcium compartmentalization, passive and active electrical modulation of synaptic inputs, synaptic strength and electrophysiology recorded at different segments of the dendritic tree, function assumed from theoretical approaches to in vitro and in vivo recordings along the time, plasticity involving activity-dependent changes in spines and circuits, age and sex effects, or abnormalities of spines in pathological conditions. Various experimental approaches and technological advancements have provided an unprecedented level of detail for the nervous system. They include the study of dendritic spines using high-resolution microscopy and computational resources in living animals, as mentioned above, and recent ion beam scanning and EM tomography in human tissue (e.g., Rollenhagen et al. 2020; Cano-Astorga et al. 2021). Neurons can be characterized using highly multiplexed, high-resolution brain-wide cell-type mapping, and high-throughput spatially resolved transcriptomics approaches to link cell types with connectomic and functional data (Hodge et al. 2019, 2020; Close et al. 2021; Ortiz et al. 2021). Cellular features were revealed on a nanoscale level with likely functional implications to circuits when identifying genes and constitutive proteins, subpopulations of cells and their networks with high-speed actions, higher-ordered mental states, and a multitude of disparate behaviors (Fuzik et al. 2016; Xu et al. 2018; Turcotte et al. 2019; Hodge et al. 2020; Close et al. 2021; Demas et al. 2021; Helm et al. 2021). Henceforth, each neuronal type across species will require single-cell transcriptomic data associated with morphology, electrophysiology, and connectivity for its probabilistic definition (Xu et al. 2018; Hodge et al. 2019; Yuste et al. 2020; Berg et al. 2021; Planert et al. 2021; Fig. 1.19). The dendritic spines have to be integrated into this current context with a coherent perspective associating relevant techniques (e.g., Vints et al. 2019; Berg et al. 2021; Planert et al. 2021; Rasia-Filho 2022).

The study of spines continues to represent a huge challenge considering neural networks and developmental periods, cell type-specific features and heterogeneity, and species-specific functional characteristics (DeFelipe 2011; Soltesz and Losonczy 2018; Cembrowski and Spruston 2019; Winnubst et al. 2019; Yang et al. 2019; Correa-Júnior et al. 2020; BRAIN Initiative Cell Census Network-BICCN 2021; Garin et al. 2022). Discrete and continuous variations may coexist and underlie cell-type diversity, forming a "combination of specification through evolutionarily driven and developmentally regulated genetic mechanisms, and refinement of cellular identities through intercellular interactions within the network in which the cells are embedded" (BICCN 2021, see also Peng et al. 2021). The magnitude of this endeavor is exemplified by the connectomic study of a fragment of the human temporal cortex (1 mm^3, >5000 slices cut at ~30 nm), imaged using a high-speed multi-beam scanning EM and 3D reconstructions. Samples exhibited 57,216 cells and approximately 133 million synapses in a 1.4-petabyte volume (Shapson-Coe et al. 2021). Furthermore, human cortical pyramidal neurons have the following: (1) larger dendritic length and branch complexity than macaque and mice (Mohan et al. 2015; Benavides-Piccione et al. 2020; additional data on the heterogeneity of human neurons is provided in Benavides-Piccione et al. 2021); (2) a class of calcium-mediated graded dendritic action potentials that would classify linearly nonseparable inputs (Gidon et al. 2020); and (3) membrane properties that significantly enhance synaptic charge-transfer from dendrites to soma and spike propagation along the axon (Eyal et al. 2016, see much more information in Chapter 9 in this book). These data and other transcriptomic features indicate that extrapolations on some neuronal functions from other species to the human brain have to be done with caution (Xu et al. 2018; Hodge et al. 2019) whereas, in the human brain, both transcriptional differences and cell-specific molecular responses can vary between individuals (Xu et al. 2018).

Finally, a large-scale nanoscopy and biochemistry analysis of postsynaptic dendritic spines showed that significant differences could exist for different types of neighbor dendritic spines (Helm et al. 2021). Stubby and mushroom spines show similar average protein copy number and topology for PSD composition identified after summing EM, stimulated emission depletion microscopy, mass spectrometry,

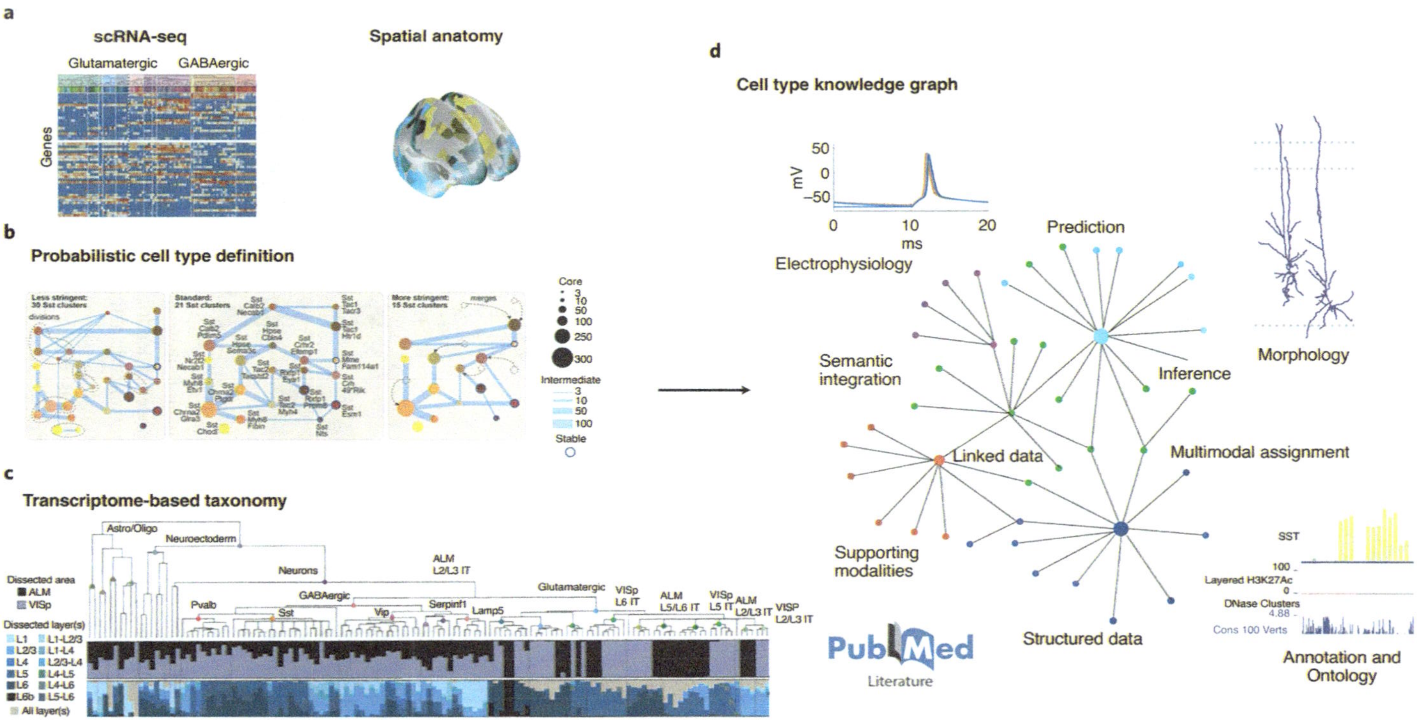

Fig. 1.19 Neuronal identification and classification may benefit from transcriptome-based taxonomy, identifying probabilistic cell types, and elaborating cell-type knowledge graphs. "(**a**) A transcriptome-based cell-type taxonomy is constructed from scRNA-seq data, related epigenomic datasets, and neuroanatomy. (**b**) Cell types are initially defined based on transcriptomic signatures in a probabilistic manner with multiresolution clustering and statistical analysis to identify robustness and variability. (**c**) Reproducible gene expression patterns identify hierarchies of putative cell types that are subject to further analyses and validation. (**d**) Transcriptomic cell-type taxonomies form a basis for constructing cell-type knowledge graphs that summarize the present state of definable cell types. Multimodal assignment of data, such as morphology, electrophysiology, and connectivity, is associated and reported with statistical variability over assigned types. A knowledge graph contains relevant and essential supporting information, such as supporting data for further analysis and mapping, descriptive annotation and ontology, and literature citations". (Legend and figure reprinted from Yuste et al. (2020), https://doi.org/10.1038/s41593-020-0685-8, under CC BY 4.0 license and copyright 2020)

fluorescence microscopy, and 3D reconstruction procedures in cultured hippocampal neurons of rats (Helm et al. 2021). However, proteins related to synaptic strength, spine dynamics, ion channels, endocytosis cofactors, cytoskeletal structure, signaling and trafficking, secretory proteins, and ribosomes are more evident in (and in the vicinity of) mushroom spines than in stubby ones (even when considering the time for the turnover rate of constitutive and inducible proteins in spines, Helm et al. 2021; Fig. 1.20).

There can be a high potential for synaptic processing involving a spatiotemporal window for dendritic integration and functional heterogeneity among individual synapses on spines of the same dendrite, between different neurons, and across and

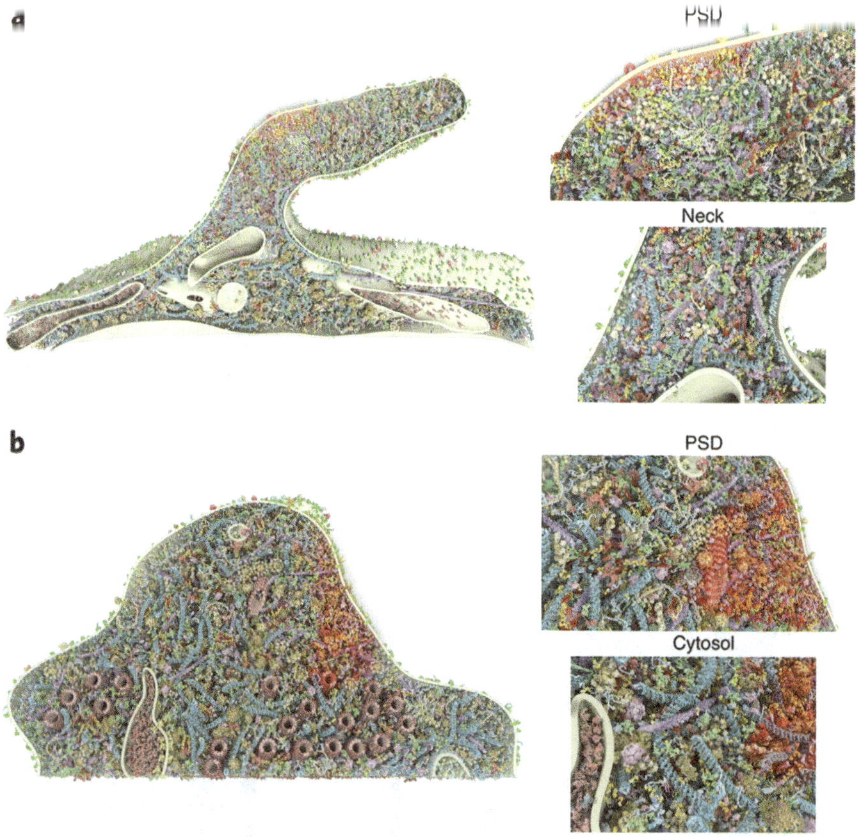

Fig. 1.20 Nanoscopy and biochemistry analysis can identify the composition of postsynaptic dendritic spines of different shapes. "A quantitative 3D model of dendritic spines is presented with all proteins shown to scale (except the highly abundant monomeric actin). (**a**) View into a mushroom spine. Magnifications into the PSD (highlighted with red glow) and neck are depicted. (**b**) View into a stubby spine. Again, a magnification of the PSD is shown and a zoom into the cytosolic region of the spine." (Legend and figure reprinted from Helm et al. (2021), under CCC RightsLink® license #5352050166807, data originally published by Springer Nature)

between brain regions (Konur et al. 2003; Grant and Fransén 2020; see also the temporal evolution of cortical ensembles underlying memory retrieval in the prelimbic subregion of the prefrontal cortex of mice in DeNardo et al. 2019). An instigating example is the huge spine density and shape variation in the human CA1 pyramidal neurons, which are related to personal memories and self-identity (Rasia-Filho et al. 2021). Synaptic diversity and strength are finely adjusted to code information, integrating synaptic processing and complex plasticity into evolved neuronal and glial circuits. These elaborations are related to the emergence of multiple sensorimotor, cognitive, emotional, abstract, creative, and conscious elaborations; visceral reactions; and behavioral displays using various brain networks (a parallel discussion can be found in Timo-Iaria and Valle 1995; DeFelipe 2011; Hodge et al. 2019; Freiwald 2020; Rasia-Filho et al. 2021; Garin et al. 2022, for example).

We look forward to the positive implications of the current and future knowledge about dendritic spines, connectivity, and brain networks to relieve harmful health conditions and restore functioning using neurotechnologies. For example, to be associated with current scientific efforts (safe genetic technology, improved imaging techniques, new pharmacological approaches, or protective strategies) aiming to alleviate the course of neurodevelopmental or neurodegeneration disorders that affect dendritic spines and synapses. These cases are represented by the progressive neurological and psychiatric disabilities that occur in Huntington's, Parkinson's, and Alzheimer's disease (Graveland et al. 1985; Dierssen and Ramakers 2006; Mancuso et al. 2013; Ammassari-Teule et al. 2021; Montero-Crespo et al. 2020, 2021; Johnson et al. 2023).

1.6 Conclusion

Dendritic spines are key elements for neural function at both network scale and synapse scale. From individual proteins to alterations in synaptic morphology and from individual synapses to a systemic level, dendritic spines participate in all these processes and with a notable integrative complexity (Mancuso et al. 2014; Rasia-Filho 2022).

The research on dendritic spines structure, function, and plasticity has undergone advances in terms of techniques and experimental findings from experimental animal models to human neurons. These approaches range from theoretical and computational models to in vivo imaging and recording techniques after stimulation of a single spine. The abovementioned data demonstrate how fascinating the dendritic spines are as neuronal components that modulate synaptic transmission, development, strength, and plasticity for the function of multiple areas of the nervous system along phylogeny and ontogeny.

This chapter also expresses our admiration for nature and life found in the following sentences:

> "… the cerebral cortex is similar to a garden filled with trees… which, thanks to intelligent culture, can multiply their branches, sending their roots deeper and producing more and more varied and exquisite flowers and fruits" (Cajal 1894).

"Perhaps histologists are following the correct path, in view of the fact that the static and dynamic are always inextricably intermixed in the minutiae that they study. In any event, microscopic anatomy has not evolved to the point that a division of labor between morphological neurocytology and physiological cytology is necessary. Even if such a division eventually comes about, we may speculate that it will be rather conventional because structure, which is constantly changing and moving, is often the only tangible manifestation of intimate cellular activities that we cannot observe directly. Thus, the appearance and relationships established by a mature neuron actually reflect a long series of movements and stimuli that have acted on it from within as well as from without during embryogenesis and after birth. Reasons for a particular shape are to be found entirely in past and present function. We might also point out that, in the distant future, when science has developed vast resources and chemistry and physics are no longer regarded as separate ways of approaching the same mechanisms at the level of atoms, anatomy will be an even more rigorous discipline. Practitioners will only be able to claim that a valid explanation of a histological observation has been provided if three questions can be answered satisfactorily: what is the functional role of the arrangement in the animal; what mechanisms underlie this function; and what sequence of chemical and mechanical events during evolution and development gave rise to these mechanisms? ... Thus, every hypothesis—no matter how incomplete—is plausible and at least temporarily acceptable as long as it takes a step in the direction of the truth and leads to new research... Never forget, though, that a theory is just a theory; it is simply a temporary, artificial edifice constructed by our minds to synthesize a particular group of observations and allow us to grasp the whole along with its dynamics" (Cajal 1909–1911, translated by Swanson and Swanson 1995).

"Spines would endow neural circuits with the ability to perform Boolean logic, to implement associative memory, to have multistable dynamics, to become self-organized, and to become veritable learning machines. Spines would be the way the brain builds neural networks and generates emergent functional properties. When considering the complexity of these apparently different aspects of the spine function and their elegant interactions toward this joint aim, it is truly remarkable that nature has found this clever solution. Indeed, spines are used widely in many types of nervous systems (Brandon and Coss 1982), so the 'invention' of spines could be one of the key advances in the evolution of the nervous system" (Yuste 2010).

References

Ågmo A (1999) Sexual motivation: an inquiry into events determining the occurrence of sexual behavior. Behav Brain Res 105:129–150

Ammassari-Teule M, Sala C, Segal M (2021) Editorial – dendritic spines: from biophysics to neuropathology. Front Synaptic Neurosci 13:652117. https://doi.org/10.3389/fnsyn.2021.652117

Ampuero E, Dagnino-Subiabre A, Sandoval R et al (2007) Status epilepticus induces region-specific changes in dendritic spines, dendritic length and TrkB protein content of rat brain cortex. Brain Res 1150:225–228

Andersen P, Morris R, Amaral D et al (2007) The hippocampus book. Oxford University Press, New York

Andres AL, Regev L, Phi L (2013) NMDA receptor activation and calpain contribute to disruption dendritic spines by the stress neuropeptide CRH. J Neurosci 33:16945–16960

Araque A, Parpura V, Sanzgiri RP et al (1999) Tripartite synapses: glia, the unacknowledged partner. Trends Neurosci 22:208–215

Araya R (2014) Input transformation by dendritic spines of pyramidal neurons. Front Neuroanat 8:141. https://doi.org/10.3389/fnana.2014.00141

Arellano JI, Benavides-Piccione R, DeFelipe J et al (2007a) Ultrastructure of dendritic spines: correlation between synaptic and spine morphologies. Front Neurosci 1:131–143

Arellano JI, Espinosa A, Fairén A et al (2007b) Non-synaptic dendritic spines in neocortex. Neuroscience 145:464–469

Ariav G, Polsky A, Schiller J (2003) Submillisecond precision of the input-output transformation function mediated by fast sodium dendritic spikes in basal dendrites of CA1 pyramidal neurons. J Neurosci 23:7750–7758

Bączyńska E, Pels KK, Basu S et al (2021) Quantification of dendritic spines remodeling under physiological stimuli and in pathological conditions. Int J Mol Sci 22:4053. https://doi.org/10.3390/ijms22084053

Bannister NJ, Larkman AU (1995) Dendritic morphology of CA1 pyramidal neurones from the rat hippocampus: II. Spine distributions. J Comp Neurol 360:161–171

Barnes CA (1987) Neurological and behavioral investigations of memory failure in aging animals. Int J Neurol 21:130–136

Barreto-Cordero LM, Ríos-Carrillo J, Roldán-Roldán G et al (2020) Cyclic changes and actions of progesterone and allopregnanolone on cognition and hippocampal basal (stratum oriens) dendritic spines of female rats. Behav Brain Res 379:112355. https://doi.org/10.1016/j.bbr.2019.112355

Baryshnikova LM, von Bohlen und Halbach O, Kaplan S et al (2006) Two distinct events, section compression and loss of particles ("lost caps"), contribute to z-axis distortion and bias in optical disector counting. Microsc Res Tech 69:738–756

Basu S, Saha PK, Roszkowska M et al (2018) Quantitative 3-D morphometric analysis of individual dendritic spines. Sci Rep 8:3545. https://doi.org/10.1038/s41598-018-21753-8

Becker RO, Rasia-Filho AA, Giovenardi M (2017) Selective deletion of the oxytocin gene remodels the number and shape of dendritic spines in the medial amygdala of males with and without sexual experience. Neurosci Lett 660:155–159

Ben Zablah Y, Merovitch N, Jia Z (2020) The role of ADF/cofilin in synaptic physiology and Alzheimer's disease. Front Cell Dev Biol 8:594998. https://doi.org/10.3389/fcell.2020.594998

Benavides-Piccione R, Ballesteros-Yáñez I, DeFelipe J et al (2002) Cortical area and species differences in dendritic spine morphology. J Neurocytol 31:337–346

Benavides-Piccione R, Regalado-Reyes M, Fernaud-Espinosa I et al (2020) Differential structure of hippocampal CA1 pyramidal neurons in the human and mouse. Cereb Cortex 30:730–752

Benavides-Piccione R, Rojo C, Kastanauskaite A et al (2021) Variation in pyramidal cell morphology across the human anterior temporal lobe. Cereb Cortex 31:3592–3609

Berg J, Sorensen SA, Ting JT et al (2021) Human neocortical expansion involves glutamatergic neuron diversification. Nature 598:151–158. Erratum in: Nature (2022) 601(7893):E12

Bernardinelli Y, Muller D, Nikonenko I (2014) Astrocyte-synapse structural plasticity. Neural Plast 2014:232105. https://doi.org/10.1155/2014/232105

Berning S, Willig KI, Steffens H et al (2012) Nanoscopy in a living mouse brain. Science 335:551. https://doi.org/10.1126/science.1215369

Bliss TV, Collingridge GL (1993) A synaptic model of memory: long-term potentiation in the hippocampus. Nature 361:31–39

Bliss TV, Lomo T (1973) Long-lasting potentiation of synaptic transmission in the dentate area of the anaesthetized rabbit following stimulation of the perforant path. J Physiol 232:331–356

Borczyk M, Śliwińska MA, Caly A et al (2019) Neuronal plasticity affects correlation between the size of dendritic spine and its postsynaptic density. Sci Rep 9:1693. https://doi.org/10.1038/s41598-018-38412-7

Bourne JN, Harris KM (2007) Do thin spines learn to be mushroom spines that remember? Curr Opin Neurobiol 17:381–386

Bourne JN, Harris KM (2009) Ultrastructural analysis of spine plasticity. In: Squire LR (ed) Encyclopedia of neuroscience. Elsevier, New York, pp 11–17

BRAIN Initiative Cell Census Network (BICCN) (2021) A multimodal cell census and atlas of the mammalian primary motor cortex. Nature 598:86–102

Brandon JG, Coss RG (1982) Rapid dendritic spine stem shortening during one-trial learning: the honeybee's first orientation flight. Brain Res 252:51–61

Brusco J, Wittmann R, de Azevedo MS et al (2008) Plasma hormonal profiles and dendritic spine density and morphology in the hippocampal CA1 stratum radiatum, evidenced by light microscopy, of virgin and postpartum female rats. Neurosci Lett 438:346–350

Brusco J, Dall'Oglio A, Rocha LB et al (2010) Descriptive findings on the morphology of dendritic spines in the rat medial amygdala. Neurosci Lett 483:152–156

Brusco J, Merlo S, Ikeda ÉT et al (2014) Inhibitory and multisynaptic spines, and hemispherical synaptic specialization in the posterodorsal medial amygdala of male and female rats. J Comp Neurol 522:2075–2088

Cajal SR (1894) The cronian lecture: la fine structure des centres nerveux. Proc Royal Soc Lond 55:444–468

Cajal SR (1906) The structure and connexions of neurons. https://www.nobelprize.org/uploads/2018/06/cajal-lecture.pdf

Cajal SR (1909–1911) Histologie du système nerveux de l'homme et des vertébrés. Maloine, Paris. English edition: Cajal SR (1995) Histology of the nervous system of man and vertebrates (trans. Swanson N, Swanson LW). Oxford University Press, New York

Cajal SR (1933) ¿Neuronismo o reticularismo? Arch Neurol 13. Reprinted: Ramón y Cajal S (1952) ¿Neuronismo o reticularismo? Las pruebas objetivas de la unidad anatómica de las células nerviosas. Consejo Superior de Investigaciones Científicas, Instituto Ramón y Cajal, Madrid

Calabrese B, Wilson MS, Halpain S (2006) Development and regulation of dendritic spine synapses. Physiology 21:38–47

Calcagnotto ME, Rasia-Filho AA, Segev I (2019) Editorial: dynamics and modulation of synaptic transmission in the mammalian CNS. Front Synaptic Neurosci 11:11. https://doi.org/10.3389/fnsyn.2019.00011

Cano-Astorga N, DeFelipe J, Alonso-Nanclares L (2021) Three-dimensional synaptic organization of layer III of the human temporal neocortex. Cereb Cortex 31:4742–4764. https://doi.org/10.1093/cercor/bhab120

Cembrowski MS, Spruston N (2019) Heterogeneity within classical cell types is the rule: lessons from hippocampal pyramidal neurons. Nat Rev Neurosci 20:193–204

Chabrol FP, Arenz A, Wiechert MT et al (2015) Synaptic diversity enables temporal coding of coincident multisensory inputs in single neurons. Nat Neurosci 18:718–727

Chang JY, Nakahata Y, Hayano Y et al (2019) Mechanisms of Ca2+/calmodulin-dependent kinase II activation in single dendritic spines. Nat Commun 10:2784. https://doi.org/10.1038/s41467-019-10694-z

Chan-Palay V, Palay SL, Billings-Gagliardi SM (1974) Meynert cells in the primate visual cortex. J Neurocytol 3:631–658

Chapleau CA, Calfa GD, Lane MC et al (2009) Dendritic spine pathologies in hippocampal pyramidal neurons from Rett syndrome brain and after expression of Rett-associated MECP2 mutations. Neurobiol Dis 35:219–233

Chelini G, Pantazopoulos H, Durning P et al (2018) The tetrapartite synapse: a key concept in the pathophysiology of schizophrenia. Eur Psychiatry 50:60–69

Chen Y, Sabatini BL (2012) Signaling in dendritic spines and spine microdomains. Curr Opin Neurobiol 22:389–396

Chen X, Leischner U, Rochefort NL et al (2011) Functional mapping of single spines in cortical neurons in vivo. Nature 475:501–505

Chicurel ME, Harris KM (1992) Three-dimensional analysis of the structure and composition of CA3 branched dendritic spines and their synaptic relationships with mossy fiber boutons in the rat hippocampus. J Comp Neurol 325:169–182

Chidambaram SB, Rathipriya AG, Bolla SR et al (2019) Dendritic spines: revisiting the physiological role. Prog Neuropsychopharmacol Biol Psychiatry 92:161–193

Cimino G (1999) Reticular theory versus neuron theory in the work of Camillo Golgi. Physis 36:431–472
Clemo HR, Lomber SG, Meredith MA (2016) Synaptic basis for cross-modal plasticity: enhanced supragranular dendritic spine density in anterior ectosylvian auditory cortex of the early deaf cat. Cereb Cortex 26:1365–1376
Close JL, Long BR, Zeng H (2021) Spatially resolved transcriptomics in neuroscience. Nat Methods 18:23–25
Cohen RS (2013) The postsynaptic density. In: Pfaff DW (ed) Neuroscience in the 21st century. Springer, New York, pp 403–437
Cohen RS, Siekevitz P (1978) Form of the postsynaptic density. A serial section study. J Cell Biol 78:36–46
Cooke BM, Woolley CS (2005) Sexually dimorphic synaptic organization of the medial amygdala. J Neurosci 25:10759–10767
Cooke BM, Woolley CS (2009) Effects of prepubertal gonadectomy on a male-typical behavior and excitatory synaptic transmission in the amygdala. Dev Neurobiol 69:141–152
Cooke BM, Stokas MR, Woolley CS (2007) Morphological sex differences and laterality in the prepubertal medial amygdala. J Comp Neurol 501:904–915
Cornejo VH, Ofer N, Yuste R (2022) Voltage compartmentalization in dendritic spines in vivo. Science 375:82–86
Correa-Júnior ND, Renner J, Fuentealba-Villarroel F et al (2020) Dendritic and spine heterogeneity of von Economo neurons in the human cingulate cortex. Front Synaptic Neurosci 12:25. https://doi.org/10.3389/fnsyn.2020.00025
Coss RG, Perkel DH (1985) The function of dendritic spines: a review of theoretical issues. Behav Neural Biol 44:151–185
Costa JF, Dines M, Lamprecht R (2020) The role of Rac GTPase in dendritic spine morphogenesis and memory. Front Synaptic Neurosci 12:12. https://doi.org/10.3389/fnsyn.2020.00012
de Castilhos J, Forti CD, Achaval M et al (2008) Dendritic spine density of posterodorsal medial amygdala neurons can be affected by gonadectomy and sex steroid manipulations in adult rats: a Golgi study. Brain Res 1240:73–81
de Castro F (2019) Cajal and the Spanish neurological school: neuroscience would have been a different story without them. Front Cell Neurosci 13:187. https://doi.org/10.3389/fncel.2019.00187
Dall'Oglio A, Xavier LL, Hilbig A et al (2013) Cellular components of the human medial amygdaloid nucleus. J Comp Neurol 521:589–611
Dall'Oglio A, Dutra AC, Moreira JE et al (2015) The human medial amygdala: structure, diversity, and complexity of dendritic spines. J Anat 227:440–459
Dall'Oglio A, Gehlen G, Achaval M et al (2008) Dendritic branching features of posterodorsal medial amygdala neurons of adult male and female rats: further data based on the Golgi method. Neurosci Lett 430:151–156
Dalpian F, Rasia-Filho AA, Calcagnotto ME (2019) Sexual dimorphism, estrous cycle and laterality determine the intrinsic and synaptic properties of medial amygdala neurons in rat. J Cell Sci 132:jcs227793. https://doi.org/10.1242/jcs.227793
Danielson E, Zhang N, Metallo J et al (2012) S-SCAM/MAGI-2 is an essential synaptic scaffolding molecule for the GluA2-containing maintenance pool of AMPA receptors. J Neurosci 32:6967–6980
DeNardo LA, Liu CD, Allen WE et al (2019) Temporal evolution of cortical ensembles prompting remote memory retrieval. Nat Neurosci 22:450–469
De Vries GJ, Simerly RB (2002) Anatomy, development, and function of sexually dimorphic neural circuits in the mammalian brain. In: Pfaff DW, Arnold AP, Etgen AM et al (eds) Hormones, brain and behavior. Academic Press, San Diego, pp 137–191
DeFelipe J (2011) The evolution of the brain, the human nature of cortical circuits, and intellectual creativity. Front Neuroanat 5:29. https://doi.org/10.3389/fnana.2011.00029

Demas J, Manley J, Tejera F et al (2021) High-speed, cortex-wide volumetric recording of neuroactivity at cellular resolution using light beads microscopy. Nat Methods 18:1103–1111

Dierssen M, Ramakers GJ (2006) Dendritic pathology in mental retardation: from molecular genetics to neurobiology. Genes Brain Behav 5(Suppl 2):48–60

Dokter M, Busch R, Poser R et al (2015) Implications of p75NTR for dentate gyrus morphology and hippocampus-related behavior revisited. Brain Struct Funct 220:1449–1462

Dudman JT, Tsay D, Siegelbaum SA (2007) A role for synaptic inputs at distal dendrites: instructive signals for hippocampal long-term plasticity. Neuron 56:866–879

Engert F, Bonhoeffer T (1999) Dendritic spine changes associated with hippocampal long-term synaptic plasticity. Nature 399:66–70

Eyal G, Verhoog MB, Testa-Silva G et al (2016) Unique membrane properties and enhanced signal processing in human neocortical neurons. elife 5:e16553. https://doi.org/10.7554/eLife.16553

Fairén A (2005) Pioneering a golden age of cerebral microcircuits: the births of the combined Golgi-electron microscope methods. Neuroscience 136:607–614

Fairén A (2007) Cajal and Lorente de Nó on cortical interneurons: coincidences and progress. Brain Res Rev 55:430–444

Fairén A, Peters A, Saldanha J (1977) A new procedure for examining Golgi impregnated neurons by light and electron microscopy. J Neurocytol 6:311–337

Feldman ML (1984) Morphology of the neocortical pyramidal neuron. In: Jones EG, Peters A (eds) Cerebral cortex. Plenum Press, New York, pp 107–121

Ferguson JN, Aldag JM, Insel TR et al (2001) Oxytocin in the medial amygdala is essential for social recognition in the mouse. J Neurosci 21:8278–8285

Fester L, Rune GM (2021) Sex neurosteroids: hormones made by the brain for the brain. Neurosci Lett 753:135849. https://doi.org/10.1016/j.neulet.2021.135849

Fiala JC, Harris KM (1999) Dendrite structure. In: Stuart G, Spruston N, Häusser M (eds) Dendrites. Oxford University Press, New York, pp 1–34

Fiala JC, Spacek J, Harris KM (2002) Dendritic spine pathology: cause or consequence of neurological disorders? Brain Res Rev 39:29–54

Fifková E, Van Harreveld A (1977) Long-lasting morphological changes in dendritic spines of dentate granular cells following stimulation of the entorhinal area. J Neurocytol 6:211–230

Foggetti A, Baccini G, Arnold P et al (2019) Spiny and non-spiny parvalbumin-positive hippocampal interneurons show different plastic properties. Cell Rep 27:3725–3732.e5

Forrest MP, Parnell E, Penzes P (2018) Dendritic structural plasticity and neuropsychiatric disease. Nat Rev Neurosci 19:215–234

Frankfurt M, Bowman R (2021) Rapid Golgi stain for dendritic spine visualization in hippocampus and prefrontal cortex. J Vis Exp 178. https://doi.org/10.3791/63404

Frankfurt M, Luine V (2015) The evolving role of dendritic spines and memory: interaction(s) with estradiol. Horm Behav 74:28–36

Freiman TM, Eismann-Schweimier J, Frotscher M (2011) Granule cell dispersion in temporal lobe epilepsy is associated with changes in dendritic orientation and spine distribution. Exp Neurol 229:332–338

Freiwald WA (2020) Social interaction networks in the primate brain. Curr Opin Neurobiol 65:49–58

Freund TF, Somogyi P (1983) The section-Golgi impregnation procedure. 1. Description of the method and its combination with histochemistry after intracellular iontophoresis or retrograde transport of horseradish peroxidase. Neuroscience 9:463–474

Freund TF, Somogyi P (1989) Synaptic relationships of Golgi-impregnated neurons as identified by electrophysiological or immunocytochemical techniques. In: Heimer L, Záborsky L (eds) Neuroanatomical tract-tracing methods. 2. Recent progress. Plenum Press, New York, pp 201–238

Fu M, Yu X, Lu J et al (2012) Repetitive motor learning induces coordinated formation of clustered dendritic spines in vivo. Nature 483:92–95

Fuentealba-Villarroel FJ, Renner J, Hilbig A et al (2022) Spindle-shaped neurons in the human posteromedial (precuneus) cortex. Front Synaptic Neurosci 13:769228. https://doi.org/10.3389/fnsyn.2021.769228

Fuzik J, Zeisel A, Máté Z et al (2016) Integration of electrophysiological recordings with single-cell RNA-seq data identifies neuronal subtypes. Nat Biotechnol 34:175–183

Gabbott PL, Somogyi J (1984) The "single" section Golgi-impregnation procedure: methodological description. J Neurosci Methods 11:221–230

García-López P, García-Marín V, Freire M (2010) Dendritic spines and development: towards a unifying model of spinogenesis – a present day review of Cajal's histological slides and drawings. Neural Plast 2010:769207. https://doi.org/10.1155/2010/769207

Garin CM, Hori Y, Everling S et al (2022) An evolutionary gap in primate default mode network organization. Cell Rep 39:110669. https://doi.org/10.1016/j.celrep.2022.110669

Gavrilov N, Golyagina I, Brazhe A et al (2018) Astrocytic coverage of dendritic spines, dendritic shafts, and axonal boutons in hippocampal neuropil. Front Cell Neurosci 12:248. https://doi.org/10.3389/fncel.2018.00248

Gidon A, Zolnik TA, Fidzinski P et al (2020) Dendritic action potentials and computation in human layer 2/3 cortical neurons. Science 367:83–87

González-Burgos I, López-Vázquez MA, Beas-Zárate C (2004) Density, but not shape, of hippocampal dendritic spines varies after a seizure-inducing acute dose of monosodium glutamate in rats. Neurosci Lett 363:22–24

González-Burgos I, Rivera-Cervantes MC, Velázquez-Zamora DA et al (2012) Selective estrogen receptor modulators regulate dendritic spine plasticity in the hippocampus of male rats. Neural Plast 2012:309494. https://doi.org/10.1155/2012/309494

González-Ramírez MM, Velázquez-Zamora DA, Olvera-Cortés ME et al (2014) Changes in the plastic properties of hippocampal dendritic spines underlie the attenuation of place learning in healthy aged rats. Neurobiol Learn Mem 109:94–103

Gore A, Yurina A, Yukevich-Mussomeli A, Nahmani M (2022) Synaptic spinules are reliable indicators of excitatory presynaptic bouton size and strength and are ubiquitous components of excitatory synapses in CA1 hippocampus. Front Synaptic Neurosci 14:968404. https://doi.org/10.3389/fnsyn.2022.968404

Grant SG, Fransén E (2020) The synapse diversity dilemma: molecular heterogeneity confounds studies of synapse function. Front Synaptic Neurosci 12:590403. https://doi.org/10.3389/fnsyn.2020.590403

Graveland GA, Williams RS, DiFiglia M (1985) Evidence for degenerative and regenerative changes in neostriatal spiny neurons in Huntington's disease. Science 227:770–773

Gray EG (1959) Electron microscopy of synaptic contacts on dendrite spines of the cerebral cortex. Nature 183:1592–1593

Gréco B, Edwards DA, Zumpe D et al (1998) Androgen receptor and mating-induced fos immunoreactivity are co-localized in limbic and midbrain neurons that project to the male rat medial preoptic area. Brain Res 781:15–24

Gréco B, Allegretto EA, Tetel MJ et al (2001) Coexpression of ERβ with ERα and progestin receptor proteins in the female rat forebrain: effects of estradiol treatment. Endocrinology 142:5172–5181

Gréco B, Blasberg ME, Kosinski EC et al (2003) Response of ERα -IR and ERβ-IR cells in the forebrain of female rats to mating stimuli. Horm Behav 43:444–453

Guerra KTK, Renner J, Vásquez CE et al (2023) Human cortical amygdala dendrites and spines morphology under open-source 3D reconstruction procedures. J Comp Neurol 531:344–365

Guirado R, Perez-Rando M, Sanchez-Matarredona D et al (2014) The dendritic spines of interneurons are dynamic structures influenced by PSA-NCAM expression. Cereb Cortex 24:3014–3024

Gulledge AT, Carnevale NT, Stuart GJ (2012) Electrical advantages of dendritic spines. PLoS One 7:e36007. https://doi.org/10.1371/journal.pone.0036007

Guo DJ, Arnspiger S, Rensing NR et al (2012) Brief seizures cause dendritic injury. Neurobiol Dis 45:348–355

Gutierrez-Castellanos N, Husain BFA, Dias IC et al (2022) Neural and behavioral plasticity across the female reproductive cycle. Trends Endocrinol Metab:S1043-2760(22)00176-X. https://doi.org/10.1016/j.tem.2022.09.001
Haber M, Zhou L, Murai KK (2006) Cooperative astrocyte and dendritic spine dynamics at hippocampal excitatory synapses. J Neurosci 26:8881–8889
Hansberg-Pastor V, González-Arenas A, Piña-Medina AG et al (2015) Sex hormones regulate cytoskeletal proteins involved in brain plasticity. Front Psych 6:165. https://doi.org/10.3389/fpsyt.2015.00165
Harnett MT, Makara JK, Spruston N et al (2012) Synaptic amplification by dendritic spines enhances input cooperativity. Nature 491:599–605
Harris KM, Jensen FE, Tsao B (1992) Three-dimensional structure of dendritic spines and synapses in rat hippocampus (CA1) at postnatal day 15 and adult ages: implications for the maturation of synaptic physiology and long-term potentiation. J Neurosci 12:2685–2705. Erratum in: J Neurosci (1992) 12(8)
Harvey CD, Yasuda R, Zhong H et al (2008) The spread of Ras activity triggered by activation of a single dendritic spine. Science 321:136–140
Harward SC, Hedrick NG, Hall CE et al (2016) Autocrine BDNF-TrkB signalling within a single dendritic spine. Nature 538:99–103
Hayashi-Takagi A, Yagishita S, Nakamura M et al (2015) Labeling and optical erasure of synaptic memory traces in the motor cortex. Nature 525:333–338
Heck N, Benavides-Piccione R (2015) Editorial: dendritic spines: from shape to function. Front Neuroanat 9:101. https://doi.org/10.3389/fnana.2015.00101
Helm MS, Dankovich TM, Mandad S et al (2021) A large-scale nanoscopy and biochemistry analysis of postsynaptic dendritic spines. Nat Neurosci 24:1151–1162
Hering H, Sheng M (2001) Dendritic spines: structure, dynamics and regulation. Nat Rev Neurosci 2:880–888
Hermel EE, Faccioni-Heuser MC, Marcuzzo S et al (2006) Ultrastructural features of neurons and synaptic contacts in the posterodorsal medial amygdala of adult male rats. J Anat 208:565–575
Herms J, Dorostkar MM (2016) Dendritic spine pathology in neurodegenerative diseases. Annu Rev Pathol 11:221–250
Hodassman S, Vardi R, Tugendhaft Y et al (2022) Efficient dendritic learning as an alternative to synaptic plasticity hypothesis. Sci Rep 12:6571. https://doi.org/10.1038/s41598-022-10466-8
Hodge RD, Bakken TE, Miller JA et al (2019) Conserved cell types with divergent features in human versus mouse cortex. Nature 573:61–68
Hodge RD, Miller JA, Novotny M et al (2020) Transcriptomic evidence that von Economo neurons are regionally specialized extratelencephalic-projecting excitatory neurons. Nat Commun 11:1172. https://doi.org/10.1038/s41467-020-14952-3
Holtmaat A, Svoboda K (2009) Experience-dependent structural synaptic plasticity in the mammalian brain. Nat Rev Neurosci 10:647–658
Holtmaat AJ, Trachtenberg JT, Wilbrecht L et al (2005) Transient and persistent dendritic spines in the neocortex in vivo. Neuron 45:279–291
Iino Y, Sawada T, Yamaguchi K et al (2020) Dopamine D2 receptors in discrimination learning and spine enlargement. Nature 579:555–560
Jiang C, Salton SR (2013) The role of neurotrophins in major depressive disorder. Transl Neurosci 4:46–58
Johnson CS, Hong W, Micevych PE (2021) Posterodorsal medial amygdala regulation of female social behavior: GABA versus glutamate projections. J Neurosci 41:8790–8800
Johnson GA, Tian Y, Ashbrook DG et al (2023) Merged magnetic resonance and light sheet microscopy of the whole mouse brain. Proc Natl Acad Sci USA 120(17):e2218617120
Kasai H, Fukuda M, Watanabe S et al (2010) Structural dynamics of dendritic spines in memory and cognition. Trends Neurosci 33:121–129
Kasai H, Ziv NE, Okazaki H et al (2021) Spine dynamics in the brain, mental disorders and artificial neural networks. Nat Rev Neurosci 22:407–422

Kasper EM, Larkman AU, Lübke J et al (1994) Pyramidal neurons in layer 5 of the rat visual cortex. II Development of electrophysiological properties. J Comp Neurol 339:475–494
Kastellakis G, Poirazi P (2019) Synaptic clustering and memory formation. Front Mol Neurosci 12:300. https://doi.org/10.3389/fnmol.2019.00300
Kasthuri N, Hayworth KJ, Berger DR et al (2015) Saturated reconstruction of a volume of neocortex. Cell 162:648–661
Kaufmann WE, Moser HW (2000) Dendritic anomalies in disorders associated with mental retardation. Cereb Cortex 10:981–991
Kesaf S, Khirug S, Dinh E et al (2020) The kainate receptor subunit GluK2 interacts with KCC2 to promote maturation of dendritic spines. Front Cell Neurosci 14:252. https://doi.org/10.3389/fncel.2020.00252
Kitaura H, Hiraishi T, Murakami H et al (2011) Spatiotemporal dynamics of epileptiform propagations: imaging of human brain slices. NeuroImage 58:50–59
Kleinjan MS, Buchta WC, Ogelman R et al (2022) Dually innervated dendritic spines develop in the absence of excitatory activity and resist plasticity through tonic inhibitory crosstalk. Neuron:S0896-6273(22)01001-7. https://doi.org/10.1016/j.neuron.2022.11.002
Klimczak P, Rizzo A, Castillo-Gómez E et al (2021) Parvalbumin interneurons and perineuronal nets in the hippocampus and retrosplenial cortex of adult male mice after early social isolation stress and perinatal NMDA receptor antagonist treatment. Front Synaptic Neurosci 13:733989. https://doi.org/10.3389/fnsyn.2021.733989
Knafo S, Libersat F, Barkai E (2005) Olfactory learning-induced morphological modifications in single dendritic spines of young rats. Eur J Neurosci 21:2217–2226
Knott G, Marchman H, Wall D et al (2008) Serial section scanning electron microscopy of adult brain tissue using focused ion beam milling. J Neurosci 28:2959–2964
Konur S, Rabinowitz D, Fenstermaker VL et al (2003) Systematic regulation of spine sizes and densities in pyramidal neurons. J Neurobiol 56:95–112
Kubota Y, Hatada S, Kondo S et al (2007) Neocortical inhibitory terminals innervate dendritic spines targeted by thalamocortical afferents. J Neurosci 27:1139–1150
Kubota Y, Karube F, Nomura M et al (2016) The diversity of cortical inhibitory synapses. Front Neural Circuits 10:27. https://doi.org/10.3389/fncir.2016.00027
Kwon T, Merchán-Pérez A, Rial Verde EM et al (2019) Ultrastructural, molecular and functional mapping of GABAergic synapses on dendritic spines and shafts of neocortical pyramidal neurons. Cereb Cortex 29:2771–2781
Lanciego JL, Wouterlood FG (2011) A half century of experimental neuroanatomical tracing. J Chem Neuroanat 42:157–183
Lanciego JL, Wouterlood FG (2020) Neuroanatomical tract-tracing techniques that did go viral. Brain Struct Funct 225:1193–1224
Larriva-Sahd JA (2014) Some predictions of Rafael Lorente de Nó 80 years later. Front Neuroanat 8:147. https://doi.org/10.3389/fnana.2014.00147
Lauer M, Senitz D (2006) Dendritic excrescences seem to characterize hippocampal CA3 pyramidal neurons in humans. J Neural Transm 113:1469–1475
Lee KFH, Soares C, Béïque JC (2012) Examining form and function of dendritic spines. Neural Plast 2012:704103. https://doi.org/10.1155/2012/704103
Lee CC, Huang CC, Hsu KS (2015) The phospholipid-binding protein SESTD1 negatively regulates dendritic spine density by interfering with Rac1-Trio8 signaling pathway. Sci Rep 5:13250. https://doi.org/10.1038/srep13250
Li Z, Okamoto K, Hayashi Y et al (2004) The importance of dendritic mitochondria in the morphogenesis and plasticity of spines and synapses. Cell 119:873–887
Li XF, Hu MH, Hanley BP et al (2015) The posterodorsal medial amygdala regulates the timing of puberty onset in female rats. Endocrinology 156:3725–3736
Li Y, Mathis A, Grewe BF et al (2017) Neuronal representation of social information in the medial amygdala of awake behaving mice. Cell 171:1176–1190.e17
Litwin-Kumar A, Harris KD, Axel R (2017) Optimal degrees of synaptic connectivity. Neuron 93:1153–1164

Liu A, Zhou Z, Dang R (2016) Neuroligin 1 regulates spines and synaptic plasticity via LIMK1/cofilin-mediated actin reorganization. J Cell Biol 212:449–463

Losonczy A, Magee JC (2006) Integrative properties of radial oblique dendrites in hippocampal CA1 pyramidal neurons. Neuron 50:291–307

Lu D, He L, Xiang W et al (2013) Somal and dendritic development of human CA3 pyramidal neurons from midgestation to middle childhood: a quantitative Golgi study. Anat Rec 296:123–132

Luengo-Sanchez S, Fernaud-Espinosa I, Bielza C et al (2018) 3D morphology-based clustering and simulation of human pyramidal cell dendritic spines. PLoS Comput Biol 14:e1006221. https://doi.org/10.1371/journal.pcbi.1006221

Luine V, Frankfurt M (2020) Estrogenic regulation of memory: the first 50 years. Horm Behav 121:104711. https://doi.org/10.1016/j.yhbeh.2020.104711

Lushnikova I, Skibo G, Muller D et al (2009) Synaptic potentiation induces increased glial coverage of excitatory synapses in CA1 hippocampus. Hippocampus 19:753–762

Lyon KA, Allen NJ (2022) From synapses to circuits, astrocytes regulate behavior. Front Neural Circuits 15:786293. https://doi.org/10.3389/fncir.2021.786293

Ma S, Zuo Y (2022) Synaptic modifications in learning and memory – a dendritic spine story. Semin Cell Dev Biol 125:84–90

Maiti P, Manna J, Ilavazhagan G, Rossignol J, Dunbar GL (2015) Molecular regulation of dendritic spine dynamics and their potential impact on synaptic plasticity and neurological diseases. Neurosci Biobehav Rev 59:208–237

Mancuso JJ, Chen Y, Li X et al (2013) Methods of dendritic spine detection: from Golgi to high-resolution optical imaging. Neuroscience 251:129–140

Mancuso JJ, Cheng J, Yin Z et al (2014) Integration of multi scale dendritic spine structure and function data into systems biology models. Front Neuroanat 8:130. https://doi.org/10.3389/fnana.2014.00130

Matsuzaki M, Ellis-Davies GC, Nemoto T et al (2001) Dendritic spine geometry is critical for AMPA receptor expression in hippocampal CA1 pyramidal neurons. Nat Neurosci 4:1086–1092

Mayr E (2001) What evolution is. Basic Books/Perseus Books Group, New York

McCarthy MM (2010) Sex differences in brain function. eLS. Wiley. https://doi.org/10.1002/9780470015902.a0022344

McCarthy MM, Konkle AT (2005) When is a sex difference not a sex difference? Front Neuroendocrinol 26:85–102

McDonald AJ, Jones GC, Mott DD (2019) Diverse glutamatergic inputs target spines expressing M1 muscarinic receptors in the basolateral amygdala: an ultrastructural analysis. Brain Res 1722:146349. https://doi.org/10.1016/j.brainres.2019.146349

Mederos S, González-Arias C, Perea G (2018) Astrocyte–neuron networks: a multilane highway of signaling for homeostatic brain function. Front Synaptic Neurosci 10:45. https://doi.org/10.3389/fnsyn.2018.00045

Medvedev N, Popov V, Henneberger C et al (2014) Glia selectively approach synapses on thin dendritic spines. Philos Trans R Soc Lond Ser B Biol Sci 369:20140047. https://doi.org/10.1098/rstb.2014.0047

Megías M, Emri Z, Freund T et al (2001) Total number and distribution of inhibitory and excitatory synapses on hippocampal CA1 pyramidal cells. Neuroscience 102:527–540

Meikle L, Pollizzi K, Egnor A et al (2008) Response of a neuronal model of tuberous sclerosis to mammalian target of rapamycin (mTOR) inhibitors: effects on mTORC1 and Akt signaling lead to improved survival and function. J Neurosci 28:5422–5432

Miermans CA, Kusters RP, Hoogenraad CC et al (2017) Biophysical model of the role of actin remodeling on dendritic spine morphology. PLoS One 12(2):e0170113. https://doi.org/10.1371/journal.pone.0170113

Mihalas S, Ardiles A, He K et al (2021) Multisubcellular compartment model of AMPA receptor trafficking for neuromodulation of Hebbian synaptic plasticity. Front Synaptic Neurosci 13:703621. https://doi.org/10.3389/fnsyn.2021.703621

Miller M (1988) Development of projection and local circuit neurons in neocortex. In: Peters A, Jones EG (eds) Cerebral cortex, vol VII. Plenum Press, New York, pp 133–175

Mohan H, Verhoog MB, Doreswamy KK et al (2015) Dendritic and axonal architecture of individual pyramidal neurons across layers of adult human neocortex. Cereb Cortex 25:4839–4853

Montero-Crespo M, Dominguez-Alvaro M, Rondon-Carrillo P et al (2020) Three-dimensional synaptic organization of the human hippocampal CA1 field. elife 9:e57013. https://doi.org/10.7554/eLife.57013

Montero-Crespo M, Domínguez-Álvaro M, Alonso-Nanclares L et al (2021) Three-dimensional analysis of synaptic organization in the hippocampal CA1 field in Alzheimer's disease. Brain 144:553–573

Moser MB, Trommald M, Andersen P (1994) An increase in dendritic spine density on hippocampal CA1 pyramidal cells following spatial learning in adult rats suggests the formation of new synapses. Proc Natl Acad Sci U S A 91:12673–12675

Mueller M, Egger V (2020) Dendritic integration in olfactory bulb granule cells upon simultaneous multispine activation: Low thresholds for nonlocal spiking activity. PLoS Biol 18(9):e3000873

Muller D, Toni N, Buchs PA (2000) Spine changes associated with long-term potentiation. Hippocampus 10:596–604

Müllner FE, Wierenga CJ, Bonhoeffer T (2015) Precision of inhibition: dendritic inhibition by individual GABAergic synapses on hippocampal pyramidal cells is confined in space and time. Neuron 87:576–589

Murakoshi H, Wang H, Yasuda R (2011) Local, persistent activation of Rho GTPases during plasticity of single dendritic spines. Nature 472:100–104

Nagaoka A, Takehara H, Hayashi-Takagi A et al (2016) Abnormal intrinsic dynamics of dendritic spines in a fragile X syndrome mouse model in vivo. Sci Rep 6:26651. https://doi.org/10.1038/srep26651

Newman SW (1999) The medial extended amygdala in male reproductive behavior: a node in the mammalian social behavior network. Ann N Y Acad Sci 877:242–257

Nguyen PT, Dorman LC, Pan S et al (2020) Microglial remodeling of the extracellular matrix promotes synapse plasticity. Cell 182:388–403.e15

Nimchinsky EA, Sabatini BL, Svoboda K (2002) Structure and function of dendritic spines. Annu Rev Physiol 64:313–353

Nishizuka M, Arai Y (1981) Sexual dimorphism in synaptic organization in the amygdala and its dependence on neonatal hormone environment. Brain Res 212:31–38

Noguchi J, Nagaoka A, Watanabe S et al (2011) In vivo two-photon uncaging of glutamate revealing the structure-function relationships of dendritic spines in the neocortex of adult mice. J Physiol 589:2447–2457

Obashi K, Taraska JW, Okabe S (2021) The role of molecular diffusion within dendritic spines in synaptic function. J Gen Physiol 153(4):e202012814. https://doi.org/10.1085/jgp.202012814

Ofer N, Benavides-Piccione R, DeFelipe J et al (2022) Structural analysis of human and mouse dendritic spines reveals a morphological continuum and differences across ages and species. eNeuro 9(3):ENEURO.0039-22.2022. https://doi.org/10.1523/ENEURO.0039-22.2022

Oga T, Elston GN, Fujita I (2017) Postnatal dendritic growth and spinogenesis of layer-V pyramidal cells differ between visual, inferotemporal, and prefrontal cortex of the macaque monkey. Front Neurosci 11:118. https://doi.org/10.3389/fnins.2017.00118

Okabe S (2020) Regulation of actin dynamics in dendritic spines: nanostructure, molecular mobility, and signaling mechanisms. Mol Cell Neurosci 109:103564. https://doi.org/10.1016/j.mcn.2020.103564

Oray S, Majewska A, Sur M (2006) Effects of synaptic activity on dendritic spine motility of developing cortical layer V pyramidal neurons. Cereb Cortex 16:730–741

Ortiz C, Carlén M, Meletis K (2021) Spatial transcriptomics: molecular maps of the mammalian brain. Ann Rev Neurosci 44:547–562

Ostroff LE, Manzur MK, Cain CK et al (2014) Synapses lacking astrocyte appear in the amygdala during consolidation of Pavlovian threat conditioning. J Comp Neurol 522:2152–2163

Palomero-Gallagher N, Zilles K (2019) Cortical layers: Cyto-, myelo-, receptor- and synaptic architecture in human cortical areas. NeuroImage S1053-8119:30682–30691

Pannese E (1996) The black reaction. Brain Res Bull 41:343–349
Pannese E (2015) Neurocytology. Springer International, Basel
Papoutsi A, Kastellakis G, Psarrou M et al (2014) Coding and decoding with dendrites. J Physiol 108:18–27
Parajuli LK, Koike M (2021) Three-dimensional structure of dendritic spines revealed by volume electron microscopy techniques. Front Neuroanat 15:627368. https://doi.org/10.3389/fnana.2021.627368
Parajuli LK, Urakubo H, Takahashi-Nakazato A et al (2020) Geometry and the organizational principle of spine synapses along a dendrite. eNeuro 7(6):ENEURO.0248-20.2020. https://doi.org/10.1523/ENEURO.0248-20.2020
Pchitskaya E, Bezprozvanny I (2020) Dendritic spines shape analysis -classification or clusterization? Perspective. Front Synaptic Neurosci 12:31. https://doi.org/10.3389/fnsyn.2020.00031
Peng H, Xie P, Liu L et al (2021) Morphological diversity of single neurons in molecularly defined cell types. Nature 598:174–181
Penzes P, Cahill ME, Jones KA et al (2011) Dendritic spine pathology in neuropsychiatric disorders. Nat Neurosci 14:285–293
Petanjek Z, Judas M, Kostovic I et al (2008) Lifespan alterations of basal dendritic trees of pyramidal neurons in the human prefrontal cortex: a layer-specific pattern. Cereb Cortex 18:915–929
Peters A, Jones EG (1984) Classification of cortical neurons. In: Peters A, Jones EG (eds) Cerebral cortex. Functional properties of cortical cells. Plenum Press, New York, pp 107–121
Peters A, Kaiserman-Abramof IR (1970) The small pyramidal neuron of the rat cerebral cortex. The perikaryon, dendrites and spines. Am J Anat 127:321–355
Peters A, Palay SL, Webster H (1991) The fine structure of the nervous system. Oxford University Press, New York
Petralia RS, Wang YX, Mattson MP et al (2018) Invaginating structures in mammalian synapses. Front Synaptic Neurosci 10:4. https://doi.org/10.3389/fnsyn.2018.00004
Petralia RS, Yao PJ, Kapogiannis D et al (2021) Invaginating structures in synapses – perspective. Front Synaptic Neurosci 13:685052. https://doi.org/10.3389/fnsyn.2021.685052
Petrulis A (2020) Structure and function of the medial amygdala. In: Urban JH, Rosenkranz JA (eds) Handbook of behavioral neuroscience. Elsevier, Amsterdam, pp 39–61
Planert H, Mittermaier FX, Grosser S et al (2021) Intra-individual physiomic landscape of pyramidal neurons in the human neocortex. bioRxiv. https://doi.org/10.1101/2021.11.08.467668
Popov VI, Bocharova LS, Bragin AG (1992) Repeated changes of dendritic morphology in the hippocampus of ground squirrels in the course of hibernation. Neuroscience 48:45–51
Popov VI, Medvedev NI, Patrushev IV et al (2007) Reversible reduction in dendritic spines in CA1 of rat and ground squirrel subjected to hypothermia-normothermia in vivo: a three-dimensional electron microscope study. Neuroscience 149:549–560. Erratum in: Neuroscience (2008) 152:1145
Portillo W, Díaz NF, Cabrera EA et al (2006) Comparative analysis of immunoreactive cells for androgen receptors and oestrogen receptor α in copulating and non-copulating male rats. J Neuroendocrinol 18:168–176
Puhahn-Schmeiser B, Kleemann T, Jabbarli R et al (2022) Granule cell dispersion in two mouse models of temporal lobe epilepsy and reeler mice is associated with changes in dendritic orientation and spine distribution. Hippocampus 32:517–528
Rall W (1999) An historical perspective on modeling dendrites. In: Stuart G, Spruston N, Häusser M (eds) Dendrites. Oxford University Press, New York, pp 193–204
Rall W, Shepherd GM, Reese TS et al (1966) Dendrodendritic synaptic pathway for inhibition in the olfactory bulb. Exp Neurol 14:44–56
Ramaswamy S, Markram H (2015) Anatomy and physiology of the thick-tufted layer 5 pyramidal neuron. Front Cell Neurosci 9:233. https://doi.org/10.3389/fncel.2015.00233
Ramón-Moliner E (1962) An attempt at classifying nerve cells on the basis of their dendritic patterns. J Comp Neurol 119:211–227
Rasia-Filho AA (2006) Is there anything "autonomous" in the nervous system? Adv Physiol Educ 30:9–12

Rasia-Filho AA (2022) Unraveling brain microcircuits, dendritic spines, and synaptic processing using multiple complementary approaches. Front Physiol 13:831568. https://doi.org/10.3389/fphys.2022.831568

Rasia-Filho AA, Peres TM, Cubilla-Gutierrez FH et al (1991) Effect of estradiol implanted in the corticomedial amygdala on the sexual behavior of castrated male rats. Braz J Med Biol Res 24:1041–1049

Rasia-Filho AA, Fabian C, Rigoti KM et al (2004) Influence of sex, estrous cycle and motherhood on dendritic spine density in the rat medial amygdala revealed by the Golgi method. Neuroscience 126:839–847

Rasia-Filho AA, Dalpian F, Menezes IC et al (2012a) Dendritic spines of the medial amygdala: plasticity, density, shape, and subcellular modulation by sex steroids. Histol Histopathol 8:985–1011

Rasia-Filho AA, Haas D, de Oliveira AP et al (2012b) Morphological and functional features of the sex steroid-responsive posterodorsal medial amygdala of adult rats. Mini Rev Med Chem 12:1090–1106

Rasia-Filho AA, Cohen RS, von Bohlen und Halbach O (2016) Editorial- frontiers in synaptic plasticity: dendritic spines, circuitries and behavior. Front Psych 7:112. https://doi.org/10.3389/fpsyt.2016.00112

Rasia-Filho AA, Guerra KTK, Vásquez CE et al (2021) The subcortical-allocortical-neocortical *continuum* for the emergence and morphological heterogeneity of pyramidal neurons in the human brain. Front Synaptic Neurosci 13:616607. https://doi.org/10.3389/fnsyn.2021.616607

Reberger R, Dall'Oglio A, Jung CR et al (2018) Structure and diversity of human dendritic spines evidenced by a new three-dimensional reconstruction procedure for Golgi staining and light microscopy. J Neurosci Methods 293:27–36

Reza-Zaldivar EE, Hernández-Sápiens MA, Minjarez B et al (2020) Dendritic spine and synaptic plasticity in Alzheimer's disease: a focus on microRNA. Front Cell Dev Biol 8:255. https://doi.org/10.3389/fcell.2020.00255

Rochefort NL, Konnerth A (2012) Dendritic spines: from structure to in vivo function. EMBO Rep 13:699–708

Rodríguez de Romo AC (2005) ¿Reticularismo o neuronismo?: diferente percepción de la misma circunstancia. Arch Neurocien (México, D.F.) 10(1):26–32. www.scielo.org.mx/scielo.php?script=sci_arttext&pid=S0187-47052005000100006&lng=es&tlng=es

Rollenhagen A, Walkenfort B, Yakoubi R et al (2020) Synaptic organization of the human temporal lobe neocortex as revealed by high-resolution transmission, focused ion beam scanning, and electron microscopic tomography. Int J Mol Sci 21:5558. https://doi.org/10.3390/ijms21155558

Rosenzweig ES, Barnes CA (2003) Impact of aging on hippocampal function: plasticity, network dynamics, and cognition. Prog Neurobiol 69:143–179

Rossini L, De Santis D, Mauceri RR et al (2021) Dendritic pathology, spine loss and synaptic reorganization in human cortex from epilepsy patients. Brain 144:251–265

Roth RH, Cudmore RH, Tan HL et al (2020) Cortical synaptic AMPA receptor plasticity during motor learning. Neuron 105:895–908

Runge K, Cardoso C, de Chevigny A (2020) Dendritic spine plasticity: function and mechanisms. Front Synaptic Neurosci 12:36. https://doi.org/10.3389/fnsyn.2020.00036

Rusakov DA, Stewart MG, Korogod SM (1996) Branching of active dendritic spines as a mechanism for controlling synaptic efficacy. Neuroscience 75:315–323

Ruszczycki B, Szepesi Z, Wilczynski GM et al (2012) Sampling issues in quantitative analysis of dendritic spines morphology. BMC Bioinf 13:213. https://doi.org/10.1186/1471-2105-13-213

Sadakane O, Watakabe A, Ohtsuka M et al (2015) In vivo two-photon imaging of dendritic spines in marmoset neocortex. eNeuro 2(4):ENEURO.0019-15.2015. https://doi.org/10.1523/ENEURO.0019-15.2015

Sala C, Segal M (2014) Dendritic spines: the locus of structural and functional plasticity. Physiol Rev 94:141–188

Sanculi D, Pannoni KE, Bushong EA et al (2020) Toric spines at a site of learning. eNeuro 7:ENEURO.0197-19.2019. https://doi.org/10.1523/ENEURO.0197-19.2019

Santos VR, de Castro OW, Pun RY et al (2011) Contributions of mature granule cells to structural plasticity in temporal lobe epilepsy. Neuroscience 197:348–357

Scheibel ME, Scheibel AB (1955) The inferior olive. A Golgi study. J Comp Neurol 102:77–131

Segal M (2010) Dendritic spines, synaptic plasticity and neuronal survival: activity shapes dendritic spines to enhance neuronal viability. Eur J Neurosci 31:2178–2184

Segev I, Rinzel J, Shepherd GM (1995) The theoretical foundations of dendritic function. The MIT Press, Cambridge

Seil FJ (2001) Interactions between cerebellar Purkinje cells and their associated astrocytes. Histol Histopathol 16:955–968

Semyanov A, Verkhratsky A (2021) Astrocytic processes: from tripartite synapses to the active milieu. Trends Neurosci 44:781–792

Serafini G (2012) Neuroplasticity and major depression, the role of modern antidepressant drugs. World J Psychiatry 2:49–57

Shapson-Coe A, Januszewski M, Berger DR et al (2021) A connectomic study of a petascale fragment of human cerebral cortex. bioRxiv 2021.05.29.446289. https://doi.org/10.1101/2021.05.29.446289

Shepherd GM (1996) The dendritic spine: a multifunctional integrative unit. J Neurophysiol 75:2197–2210

Shepherd GM, Brayton RK, Miller JP et al (1985) Signal enhancement in distal cortical dendrites by means of interactions between active dendritic spines. Proc Natl Acad Sci U S A 82:2192–2195

Shepherd GM, Hines ML, Migliore M et al (2020) Predicting brain organization with a computational model: 50-year perspective on lateral inhibition and oscillatory gating by dendrodendritic synapses. J Neurophysiol 124:375–387

Shughrue PJ, Lane MV, Merchenthaler IJ (1997) Comparative distribution of estrogen receptor-α and-β mRNA in the rat central nervous system. J Comp Neurol 388:507–525

Sigler A, Murphy TH (2010) In vivo 2-photon imaging of fine structure in the rodent brain: before, during, and after stroke. Stroke 41:S117–S123

Simerly RB, Chang C, Muramatsu M et al (1990) Distribution of androgen and estrogen receptor mRNA-containing cells in the rat brain: an *in situ* hybridization study. J Comp Neurol 294:76–95

Smith KR, Penzes P (2018) Ankyrins: roles in synaptic biology and pathology. Mol Cell Neurosci 91:131–139

Sobczyk A, Scheuss V, Svoboda K (2005) NMDA receptor subunit-dependent [Ca2+] signaling in individual hippocampal dendritic spines. J Neurosci 25:6037–6046

Soltesz I, Losonczy A (2018) CA1 pyramidal cell diversity enabling parallel information processing in the hippocampus. Nat Neurosci 21:484–493

Sorra KE, Harris KM (2000) Overview on the structure, composition, function, development, and plasticity of hippocampal dendritic spines. Hippocampus 10:501–511

Spacek J (1985) Three-dimensional analysis of dendritic spines. III. Glial sheath. Anat Embryol 171:245–252

Spacek J, Harris KM (2004) Trans-endocytosis via spinules in adult rat hippocampus. J Neurosci 24:4233–4241

Spence EF, Kanak DJ, Carlson BR et al (2016) The Arp2/3 complex is essential for distinct stages of spine synapse maturation, including synapse unsilencing. J Neurosci 36:9696–9709

Spruston N, Häusser M, Stuart G (2013) Information processing in dendrites and spines. In: Squire LR, Berg D, Bloom FE et al (eds) Fundamental neuroscience. Academic Press, Waltham, pp 231–260

Stark CP (2005) Behavioral effects of stimulation of the medial amygdala in the male rat are modified by prior sexual experience. J Gen Psychol 132:207–224

Stark CP, Alpern HP, Fuhrer J et al (1998) The medial amygdaloid nucleus modifies social behavior in male rats. Physiol Behav 63:253–259

Stewart MG, Popov VI, Kraev IV et al (2014) Structure and complexity of the synapse and dendritic spine. In: Pickel V, Segal M (eds) The synapse. Academic Press, New York, pp 1–20
Suratkal SS, Yen YH, Nishiyama J (2021) Imaging dendritic spines: molecular organization and signaling for plasticity. Curr Opin Neurobiol 67:66–74
Swaney WT, Dubose BN, Curley JP et al (2012) Sexual experience affects reproductive behavior and preoptic androgen receptors in male mice. Horm Behav 61:472–478
Szentágothai J (1978) The neuron network of the cerebral cortex: a functional interpretation. Proc R Soc Lond B 201:219–248
Tao-Cheng J-H, Dosemeci A, Gallant PE et al (2009) Rapid turnover of spinules at synaptic terminals. Neuroscience 160:42–50
Tazerart S, Mitchell DE, Miranda-Rottmann S et al (2020) A spike-timing-dependent plasticity rule for dendritic spines. Nat Commun 11:4276. https://doi.org/10.1038/s41467-020-17861-7
Timo-Iaria C, Valle AC (1995) The functional role of the conscious process. Ciência e Cultura J Braz Assoc Adv Sci 47:221–234
Tjia M, Yu X, Jammu LS et al (2017) Pyramidal neurons in different cortical layers exhibit distinct dynamics and plasticity of apical dendritic spines. Front Neural Circuits 11:43. https://doi.org/10.3389/fncir.2017.00043
Toni N, Buchs PA, Nikonenko I et al (1999) LTP promotes formation of multiple spine synapses between a single axon terminal and a dendrite. Nature 402:421–425
Tønnesen J, Nägerl V (2016) Dendritic spines as tunable regulators of synaptic signals. Front Psych 7:101. https://doi.org/10.3389/fpsyt.2016.00101
Tønnesen J, Inavalli VVGK, Nägerl UV (2018) Super-resolution imaging of the extracellular space in living brain tissue. Cell 172:1108–1121.e15
Turcotte R, Liang Y, Tanimoto M et al (2019) Dynamic super-resolution structured illumination imaging in the living brain. Proc Natl Acad Sci U S A 116:9586–9591
Udvary D, Harth P, Macke JH et al (2022) The impact of neuron morphology on cortical network architecture. Cell Rep 39:110677. https://doi.org/10.1016/j.celrep.2022.110677
Urban NT, Willig KI, Hell SW et al (2011) STED nanoscopy of actin dynamics in synapses deep inside living brain slices. Biophys J 101:1277–1284
van der Zee EA (2015) Synapses, spines and kinases in mammalian learning and memory, and the impact of aging. Neurosci Biobehav Rev 50:77–85
von Bohlen und Halbach O (2010) Dendritic spine abnormalities in mental retardation. Cell Tissue Res 342:317–323
von Bohlen und Halbach O, Krause S, Medina D et al (2006a) Regional- and age-dependent reduction in trkB receptor expression in the hippocampus is associated with altered spine morphologies. Biol Psychiatry 59:793–800
von Bohlen und Halbach O, Zacher C, Gass P et al (2006b) Age-related alterations in hippocampal spines and deficiencies in spatial memory in mice. J Neurosci Res 83:525–531
Valverde F (1967) Apical dendritic spines of the visual cortex and light deprivation in the mouse. Exp Brain Res 3:337–352
Vásquez CE, Reberger R, Dall'Oglio A et al (2018) Neuronal types of the human cortical amygdaloid nucleus. J Comp Neurol 526:2776–2801
Ventura R, Harris KM (1999) Three-dimensional relationships between hippocampal synapses and astrocytes. J Neurosci 19:6897–6906
Verzi DW, Noris OY (2009) A compartmental model for activity-dependent dendritic spine branching. Bull Math Biol 71:1048–1072
Vetter P, Roth A, Häusser M (2001) Propagation of action potentials in dendrites depends on dendritic morphology. J Neurophysiol 85:926–937
Vints K, Vandael D, Baatsen P et al (2019) Modernization of Golgi staining techniques for high-resolution, 3-dimensional imaging of individual neurons. Sci Rep 9:130. https://doi.org/10.1038/s41598-018-37377-x
Viscardi LH, Imparato DO, Bortolini MC et al (2021) Ionotropic receptors as a driving force behind human synapse establishment. Mol Biol Evol 38:735–744

Weiler S, Guggiana Nilo D, Bonhoeffer T et al (2022) Orientation and direction tuning align with dendritic morphology and spatial connectivity in mouse visual cortex. Curr Biol 32:1743–1753.e7

Wen Q, Stepanyants A, Elston GN et al (2009) Maximization of the connectivity repertoire as a statistical principle governing the shapes of dendritic arbors. Proc Natl Acad Sci U S A 106:12536–12541

Westrum LE (1970) Observations on initial segments of axons in the prepyriform cortex of the rat. J Comp Neurol 139:337–356

Winnubst J, Bas E, Ferreira TA et al (2019) Reconstruction of 1,000 projection neurons reveals new cell types and organization of long-range connectivity in the mouse brain. Cell 179:268–281.e13

Wong M, Guo D (2013) Dendritic spine pathology in epilepsy: cause or consequence? Neuroscience 251:141–150

Woolley CS, McEwen BS (1993) Roles of estradiol and progesterone in regulation of hippocampal dendritic spine density during the estrous cycle in the rat. J Comp Neurol 336:293–306

Woolley CS, McEwen BS (1994) Estradiol regulates hippocampal dendritic spine density via an N-methyl-D-aspartate receptor-dependent mechanism. J Neurosci 14:7680–7687

Woolley CS, Gould E, Frankfurt M et al (1990) Naturally occurring fluctuation in dendritic spine density on adult hippocampal pyramidal neurons. J Neurosci 10:4035–4039

Woolfrey KM, Dell'Acqua ML (2015) Coordination of protein phosphorylation and dephosphorylation in synaptic plasticity. J Biol Chem 290:28604–28612

Xu X, Stoyanova EI, LemieszJie AE et al (2018) Species and cell-type properties of classically defined human and rodent neurons and glia. elife 7:e37551. https://doi.org/10.7554/eLife.37551

Yang XY, Stanley RE, Ross AP et al (2019) Sestd1 encodes a developmentally dynamic synapse protein that complexes with BCR Rac1-GAP to regulate forebrain dendrite, spine and synapse formation. Cereb Cortex 29:505–516. Corrected in Cereb Cortex (2019) 29: 1217

Yasuda R (2017) Biophysics of biochemical signaling in dendritic spines: implications in synaptic plasticity. Biophys J 113:2152–2159

Yuste R (2010) Dendritic spines. MIT Press, Cambridge

Yuste R (2013) Electrical compartmentalization in dendritic spines. Ann Rev Neurosci 36:429–449

Yuste R, Bonhoeffer T (2001) Morphological changes in dendritic spines associated with long-term synaptic plasticity. Annu Rev Neurosci 24:1071–1089

Yuste R, Hawrylycz M, Aalling N et al (2020) A community-based transcriptomics classification and nomenclature of neocortical cell types. Nat Neurosci 23:1456–1468. Erratum in Nat Neurosci 2020 Dec 4; Erratum in Nat Neurosci 2021 24:612

Zaccard CR, Shapiro L, Martin-de-Saavedra MD et al (2020) Rapid 3D enhanced resolution microscopy reveals diversity in dendritic spinule dynamics, regulation, and function. Neuron 107:522–537

Zancan M, Dall'Oglio A, Sarzenski TM et al (2015) Glial and axonal perikaryal coverage and somatic spines in the posterodorsal medial amygdala of male and cycling female rats. J Comp Neurol 523:2127–2137

Zancan M, da Cunha RSR, Schroeder F et al (2018) Remodeling of the number and structure of dendritic spines in the medial amygdala: from prepubertal sexual dimorphism to puberty and effect of sexual experience in male rats. Eur J Neurosci 48:1851–1865

Zeng LH, Xu L, Rensing NR et al (2007) Kainate seizures cause acute dendritic spine loss and actin depolymerization in vivo. J Neurosci 27:11604–11613

Chapter 2
Techniques to Render Dendritic Spines Visible in the Microscope

Floris G. Wouterlood

Abstract A tiny detail visible on certain neurons at the limit of resolution in light microscopy went in 130 years of neuroscience research through a dazzling career from suspicious staining artifact to what we recognize today as a complex postsynaptic molecular machine: the dendritic spine.

This chapter deals with techniques to make spines visible. The original technique, Golgi silver staining, is still being used today. Electron microscopy and automated field ion beam scanning electron microscopy are ultrahigh resolution techniques, albeit specialized. Other methods are intracellular injection, uptake of dyes, and recently the exploitation of genetically modified animals in which certain neurons express fluorescent protein in all their processes, including the nooks and crannies of their dendritic spines.

Keywords Golgi silver impregnation · FIB-SEM · Intracellular injection · Carbocyanine dyes · Genetic engineering · Microscopy · 3D reconstruction

2.1 Introduction: Their Discovery

First of all, dendritic spines are what is being called "diffraction sensitive," i.e., their size hovers around the resolution limit of a microscope operating with visible light. These small, irregularly shaped protrusions studding the surfaces of numerous neurites and their branches that we easily dismiss today as "dendritic spines" were first reported in 1888, with a considerable amount of reservation, by Santiago Ramon y Cajal in a monograph on brains of birds (Ramón y Cajal 1888; Yuste 2015; illustration in DeFelipe 2015). The celebrated Spanish neurohistologist subsequently tagged such structures as "espinas," i.e., spines. Cajal was convinced that the appendages that he observed in the avian cerebellum on branches of dendrites of Purkyně cells were real, delicate excrescences of the neurons instead of artifactual

F. G. Wouterlood (✉)
Department of Anatomy & Neurosciences, Amsterdam UMC, Amsterdam, The Netherlands

A. A. Rasia-Filho et al. (eds.), *Dendritic Spines*, Advances in Neurobiology 34,
https://doi.org/10.1007/978-3-031-36159-3_2

silver metallic staining-introduced depositions such as suspected by the majority of Cajal's contemporaries. Although over time more and more colleagues became convinced, the dispute was finally settled by the first high-resolution images delivered in the 1960s by a revolutionary new instrument: the transmission electron microscope. Ramon y Cajal's "espinas" are a real feature of neurons.

2.2 Methods and Microscopes

If one wants to visualize dendritic spines, there are two barriers to take: methods and microscopes. These go hand in hand. Basically, there exist six ways to render dendritic spines visible:

- Electron microscopy
- Golgi metallic silver staining
- Intracellular or pericellular filling of neurons with some "stain"
- Staining with lipophilic dye
- Genetic engineering: create do-it-yourself fluorescent neurons
- Transfection by a virus, introducing a gene that codes for a fluorescent protein

Note that although electron microscopy arrived on the scene much later than Golgi metallic silver staining, we will discuss this combination of method and microscope first. Before we entered one decade ago the age of super-resolution microscopy, this combination of method and microscope provided unique insight into the dimensions and structural details of dendritic spines. Electron microscopy also revealed the enormous variation in spine morphology. While conventional light microscopes and super-resolution microscopes use light in the visible range with wavelengths measured in nanometers, an electron microscope operates with beam wavelengths in Ångströms. Diffraction sensitivity of spines is therefore not an issue in electron microscopy.

2.3 Electron Microscopy

Until the advent of super-resolution light microscopy, the best way to study the morphology of dendritic spines was to investigate them by electron microscopy (EM). We are dealing here with a tightly intertwined combination of technique and microscope. Electron microscopy requires rigid tissue fixation, specialized histological lab equipment, trained personnel, and access to an electron microscope. Briefly, brain tissue is initially fixed with a "strong" fixative that contains a certain amount of glutaraldehyde. Good preservation of ultrastructural detail depends strongly on how well the fixation procedure has been run. Best fixation is usually obtained through perfusion fixation using the brain's vascular system as delivery route. First, one must flush the brain with body temperature saline to remove erythrocytes and blood cells. Keeping the osmolarity of the fixative as close as possible to that inside the brain inhibits neurons from osmotically losing or accumulating water and thus from changing morphology and losing ultrastructural detail. Small tissue samples are post-fixed in 4% osmium tetroxide, block stained in a 2% uranyl acetate solution, and subsequently embedded in epoxy resin. After curing in an oven, ultrathin (100 nm) sections are cut on an ultramicrotome, mounted on copper or nickel grids, and post-stained with lead citrate. In an electron microscope, at comfortable magnifications of 5000× and higher the dimensions of dendritic spines as well as their content can be studied and recorded. Fixatives, embedding techniques, and staining procedures are extensively discussed by Glauert and Lewis (2016). Note that in electron microscopy, we always inspect radically fixed material (Wilson et al. 1983); in the living brain spine shape may vary continuously (Bonhoeffer and Yuste 2002; Nimchinsky et al. 2002; Segal 2010).

2.3.1 Transmission Electron Microscopy

The first images recorded with a transmission electron microscope that include dendritic spines were published more than 60 years ago by Gray (1959a, b). Two decades of incredible expansion of knowledge of the fine details of neurons followed after Gray's trailblazing publication. Almost all progress of that era has been summarized by in a book by Peters, Palay, and Webster, printed in its first edition in 1970 that surveys and lavishly illustrates ultrastructural details of central and peripheral nervous tissue found in the entire vertebrate nervous system. The third and most recent edition of this book was published in 1991 (Peters et al. 1991).

Focusing on dendritic spines, these are present at high densities on apical dendrites of cortical pyramidal neurons, cerebellar Purkyně cells, and hippocampal CA1 pyramidal cells. In addition, they have been reported elsewhere in the brain, e.g., on striatal medium-size spiny neurons and neurons in the amygdaloid complex. Spines have also been described to arise, albeit spuriously, from cell bodies (Peters et al. 1991) and from initial axon segments (Westrum 1970).

The shape of a dendritic spine is usually classified as "stubby," "mushroom," "thin," or "branched multiform" (Peters and Kaiserman-Abramof 1970; review in Knott and Holtmaat 2008). It should be remembered that in living brain, especially during development, spines are by no means rigid; they change shape and may be transient structures as well (Grutzendler et al. 2002). The electron micrographs shown in Fig. 2.1 were taken from rat brain: parahippocampal cortex (Fig. 2.1a, b) and dorsal cochlear nucleus (DCN, Fig. 2.1c). The dendritic spine in Fig. 2.1a is of the "stubby" type while Fig. 2.1b depicts a typical "thin" dendritic spine with a long neck. Especially this kind of spine is encountered rarely in transmission electron microscopy because necks of long spines are usually 100 nm or less in diameter (Gray 1959b), that is in the same dimension as the thickness of the ultrathin section

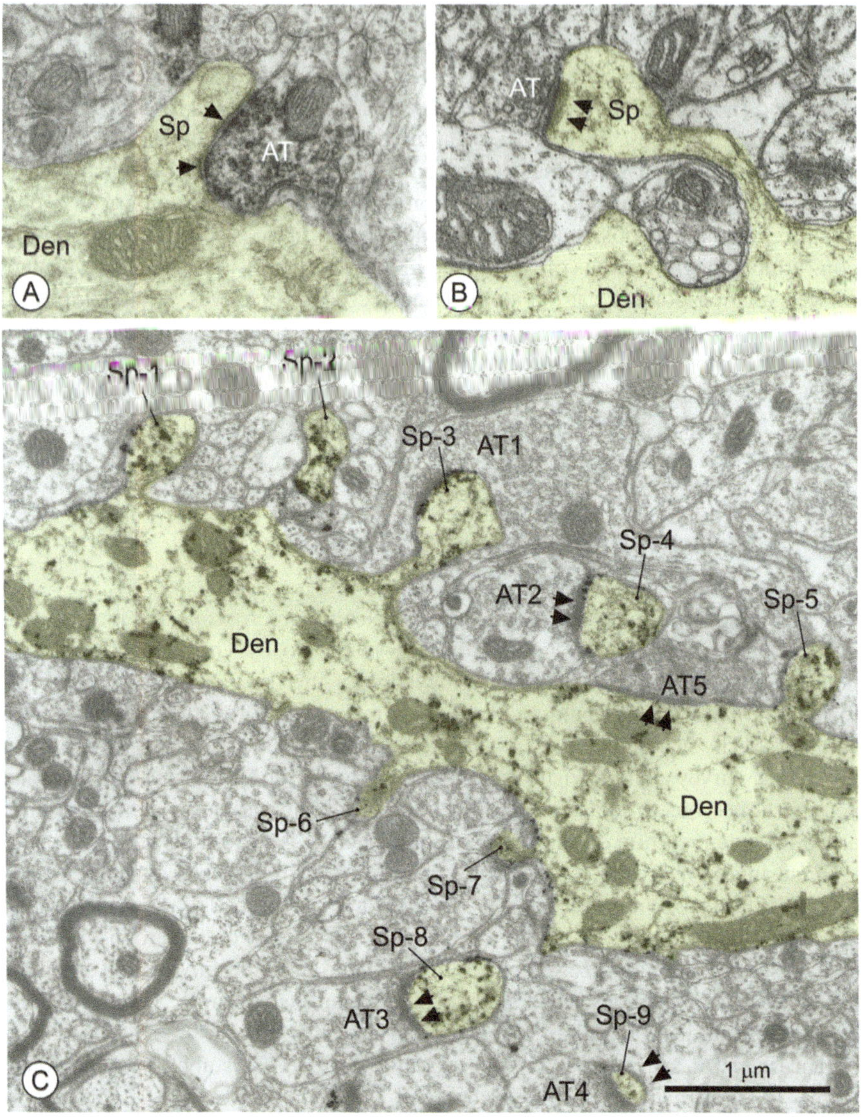

Fig. 2.1 Dendrites and their spines in various shapes revealed in transmission electron microscopy. (**a**, **b**) Standard transmission electron microscopy. (**c**) Golgi-electron microscopy of cartwheel neurons in the dorsal cochlear nucleus of rat. Scale marker in **c** also holds for **a** and **b**. In transmission electron microscopy, thin spine necks are often out of the sectioning plane; in-plane spines necks, such as in **b**, are rarely encountered. One needs an electron-dense marker, such as gold-toned Golgi silver impregnation (frame **c**) to become surprised to see the real amount of spine heads surrounding a main dendritic trunk. Arrowheads indicate synapses

itself (80–100 nm). A spine neck can thus easily be out of the cutting plane or, conversely, it is a matter of chance that a spine with a long, twisted neck is completely aligned with the cutting plane. For instance, the head of spine "9" in Fig. 2.1c is approximately one micron away from the axis of the parent dendrite. Spines with long, slender necks can therefore be best evaluated in 3D reconstructions. An alternative could be high-voltage electron microscopy on thick (1 µm) sections (Kang et al. 2017, review).

2.3.2 Volume Transmission Electron Microscopy and 3D Reconstruction

Dendritic spines project in all directions from a dendrite. A spine with its neck oriented in the cutting plane such as in Fig. 2.1b can be considered a "lucky hit." An electron-dense marker that homogeneously and completely fills the cytoplasmic compartment of single neurons would greatly assist the positive identification in the electron microscope of profiles, such as dendritic spines, belonging to these neurons. Fortunately, several electron-dense markers are available: diaminobenzidine precipitate (following horseradish peroxidase or biotin filling), Golgi metallic silver impregnation and carbocyanine or horseradish peroxidase-labeled immunocytochemical markers converted in the presence of diaminobenzidine into an electron-dense precipitate (Sandell and Masland 1989). A survey of fiduciary markers applicable in correlative light-electron microscopy can be found in Kageyama and Meyer (1987) and, more recently, in Van Hest et al. (2019).

Figure 2.1c shows a spiny dendrite whose cytoplasm contains gold substituted Golgi silver precipitate. The latter was produced via a technique that was pioneered by Blackstad (1965) and vastly improved by Fairén et al. (1977) (review in Fairén 2005). The image demonstrates nicely that in transmission electron microscopy an electron-dense cytoplasmic identification marker is required to make crystal clear which excrescences emerge from a particular dendrite and which do not, notably those spines that possess heads at some distance from the stem and necks that are not included in the ultrasectioning plane.

If a spine with a long neck comparable to spines Sp3 and Sp8 indicated in Fig. 2.1c projects a micron or so away from its stem dendrite and at the same time is oriented perpendicular to the cutting plane, a relatively long and completely intact ribbon of at least 10 ultrathin sections may be required to "catch" the complete spine. In the early days of electron microscopy, this type of work was rocket science while today we identify it as a normal application in the greater realm of what is called "volume electron microscopy" (Peddie and Collinson 2014). Until the advent of electronic image acquisition equipment in the 1980s, three-dimensional (3D) electron microscopy reconstructions were made in a tradition founded by the embryologist Karl Peter (1906). One would cut profiles out of contour drawings or from printed electron micrographs, stack them as good as possible depending on fiduciary markers, and coat the assembly with wax, clay, paraffin, or some other plastic material to manufacture a fancy three-dimensional shape. All this artisanal work was done manually where the main challenge was to stack "sections" properly on top of each other while rendering orientation and registration as realistic as possible (e.g., Bang and Bang 1957). An exquisite example of such a "heroic" early 3D EM reconstruction of dendritic spines of (hippocampal) neurons is that by Blackstad and Kjaerheim (1961). The introduction in the 1980s of electronic equipment like video grabbers and image memory eased the workload and set the stage for digital 3D EM reconstruction of for instance Purkyně dendrites and their spines (e.g., Harris and Stevens 1988). Today, the rendering of 3D EM reconstructions is within reach of everyone on a simple personal computer and with open source software (e.g., Fiji software by Schindelin et al. 2012; or the IMOD package at https://bio3d.colorado.edu/imod/) (Kremer et al. 1996). The preparation of long ribbons of ultra-thin sections remains a delicate performance, say a form of highly skilled craftsmanship (Harris et al. 2006).

2.3.3 Volume Scanning Electron Microscopy and 3D Reconstruction

In the early years of the twenty-first century, the first automated scanning electron microscopy-abrasion instruments were introduced in biomedicine, as an offspring of the semiconductor industry that developed these instruments to inspect the quality of lithographic layer deposition. Biomedicine scanning-abrasion instruments are either equipped with a focused ion beam (Denk and Horstmann 2004; FIB-SEM; Knott et al. 2008) or with a mechanical ultramicrotome inside the scanning chamber (serial block-face scanning electron microscopy, SBEM; Denk and Horstmann 2004), or even more ingenious, a tape-collecting ultramicrotomy system followed by scanning electron microscopy (ATLUM; Hayworth et al. 2006). This type of machinery and its accompanying software reduce the workload from the old heroic handwork to modern manageable workstation activity (Parajuli and Koike 2021; see also Peddie and Collinson 2014).

Field ion beam scanning electron microscopy (FIB-SEM) was originally developed for materials science to visualize in extremely high detail structures as varied as lithographically applied layers in semiconductor wafers and metal-coating interfaces. Work is performed inside the vacuum chamber of a scanning electron microscope and consists of controlled ablation of material ("milling") with a focused Gallium ion beam. Next, the surface of the preparation is scanned and backscattered electrons are detected to acquire an image. Then, the process is repeated: milling, scanning, milling, scanning, and so forth. The process and the image acquisition are fully automated: ablation and scanning are repeated until a specimen is completely "Z-sectioned." Each ion beam-ablated layer can be as thin as 5 nm. Post-acquisition image processing produces images (Figs. 2.2 and 2.3) with a quality that approaches that of transmission electron microscopy.

Because the ablation process proceeds along a fixed Z-axis, a stack of images is obtained with calibration in the *Z* direction. This makes fiduciary matching of

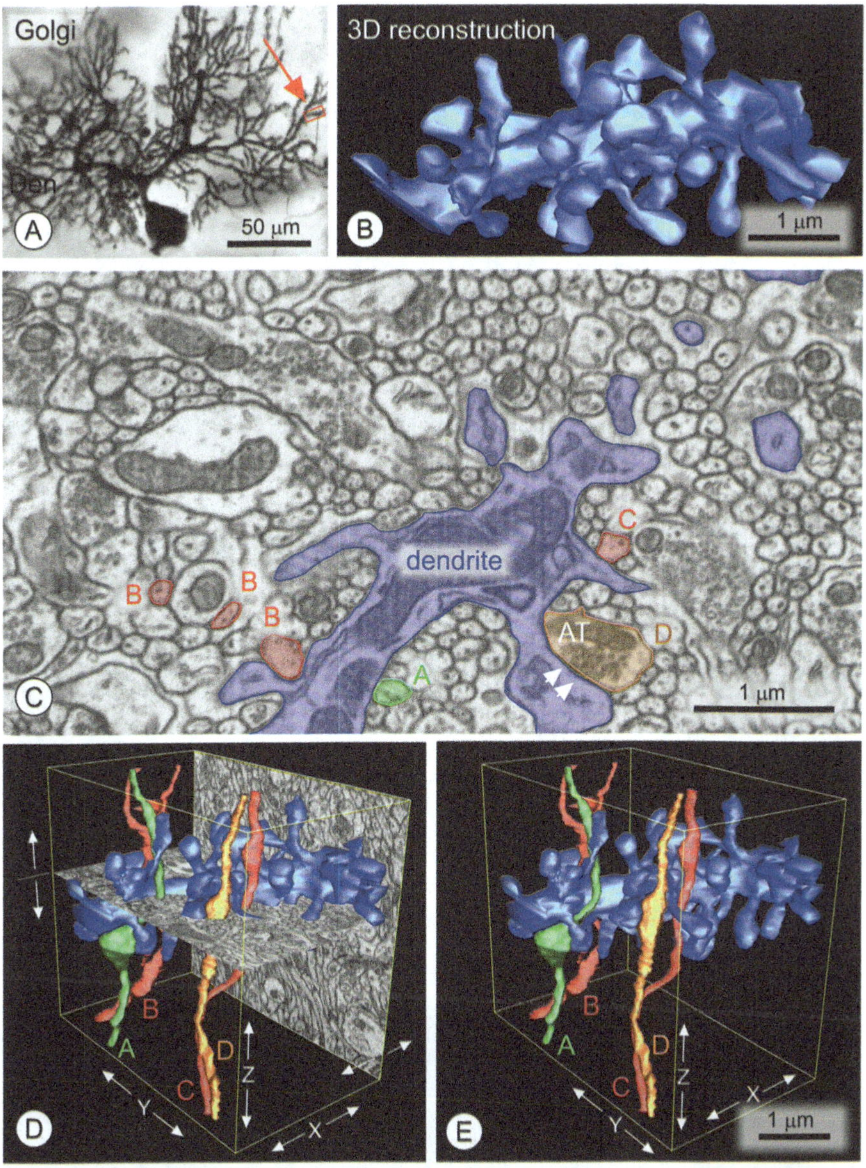

Fig. 2.2 FIB-SEM; 3D computer reconstructions made with the IMOD software package. (**a**) Golgi silver-stained rat cerebellar Purkyně cell—for illustrative purposes only—indicating which portion of the dendritic tree was subjected to FIB-SEM: a distal spiny branchlet in the molecular layer (red arrow). (**b**) 3D reconstructed Purkyně cell dendritic branchlet. The process is studded with an abundance of spines. (**c**) The dataset used here (kindly supplied by FEI company) is a stack of 360 image frames from mouse cerebellum, 25 nm abrasion depth. In this dataset, we outlined several Purkyně cell branchlets of which one is shown (blue), together with a number of parallel fibers. Several of these (fibers A, B, C, and D) form synaptic contacts with spines of this Purkyně dendrite. One synapse is visible in frame B: fiber D axon terminal (AT) (white arrowheads). (**d**, **e**) selected screen captures. IMOD allows 3D reconstructions to be blended with the original *XY* images but also with generated *XZ* and *YZ* image planes

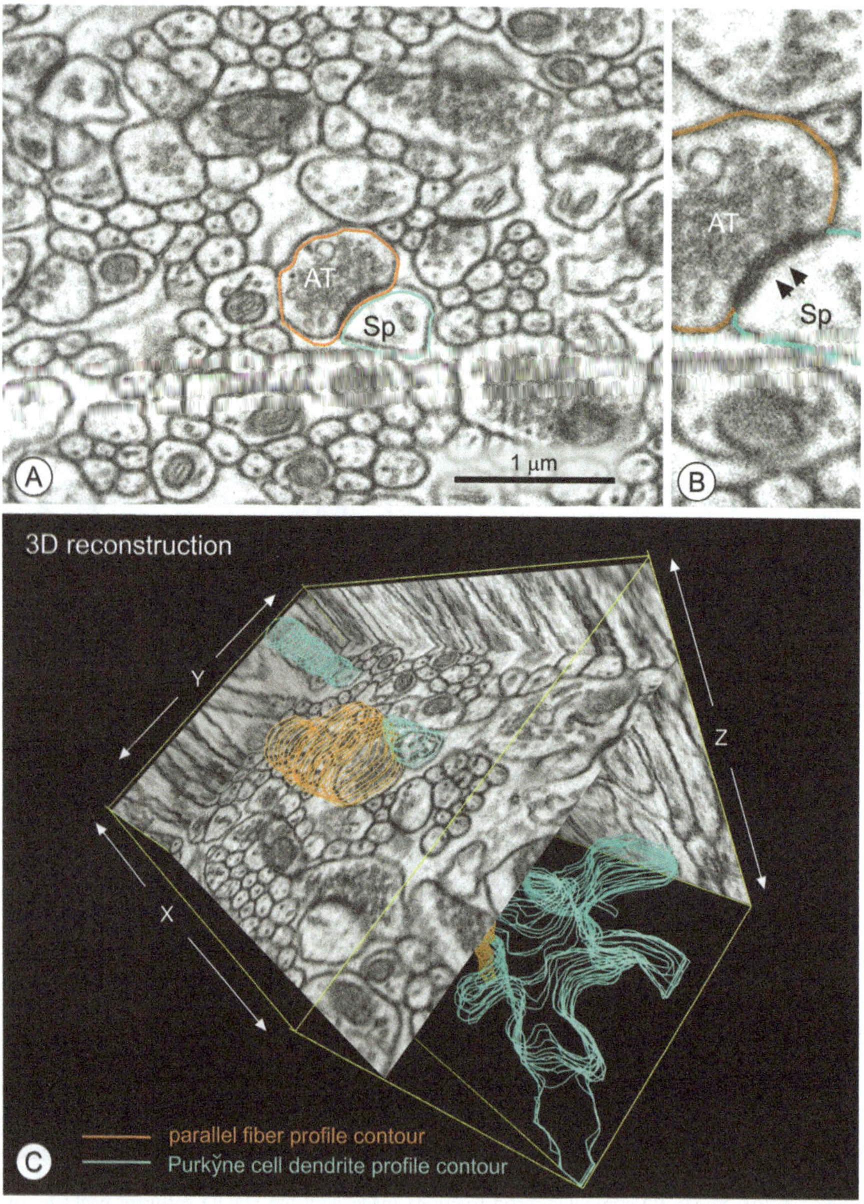

Fig. 2.3 Details of FIB-SEM 3D computer reconstruction of a Purkyně branchlet in the molecular layer of mouse cerebellum (with IMOD). Stack of 100 image frames, 25 nm abrasion depth. (**a**) Synapse between a parallel fiber axon terminal (orange outline) and dendritic spine head (blue outline). (**b**) With a high zoom factor, the quality of FIB-SEM can be assessed. Segmentation of the spine and its main dendritic trunk was done in all 100 *Z* planes of this dataset. Outlines are shown stacked in panel **c**. IMOD allows solid surface rendering such as represented in Fig. 2.2b, d, e

successive image planes much easier and much more reliable than in classical transmission electron microscopy while it allows fast and reliable computer 3D reconstruction of any object present in the dataset. An additional advantage of FIB-SEM is that the staining and embedding procedure for the material to be subjected to imaging does not deviate much from standard fixation and plastic embedding used in transmission electron microscopy. A main difference with standard EM staining is that block staining with heavy metals must be carefully performed to have high-contrast staining available necessary for successful scanning and imaging (review in Wanner et al. 2015).

FEI company, one of the manufacturers of FIB-SEM equipment, kindly supplied an image dataset that we imported in our 3D reconstruction program (IMOD) and played with to illustrate in Figs. 2.2 and 2.3 the versatility of FIB-SEM to visualize dendritic spines at very high magnification. The material "milled" in the FIB-SEM procedures was a sample of mouse cerebellum molecular layer that had been fixed and embedded with a procedure standard for transmission electron microscopy. The section had been block stained with uranyl acetate before final embedding in Durcupan resin followed by mounting in the FIB-SEM (sample provided to FEI by Graham Knott, EPFL, Switzerland).

All the FIB-SEM work presented in this chapter includes manual outlining of the profiles belonging to the Purkyně cell dendrite and selected parallel fibers. As manual segmentation is the most time-consuming and error-prone part of the job, algorithms have been developed that use artificial intelligence and machine learning to semi-automate the segmentation procedure (Wernitznig et al. 2015).

Recent examples of FIB-SEM volume work done on dendritic spines include studies by Merchán-Pérez et al. (2009) and Bosch et al. (2015). Merchán-Pérez et al. (2009) determined the actual numbers of synapses in a given volume of mouse brain tissue and compared the outcome with traditional TEM analysis. Bosch et al. (2015) labeled neurons in mouse brain via retrograde viral transfection tracing, immunostained the expressed GFP, and reconstructed via FIB-SEM and post-acquisition image processing all synaptic input on complete dendrites, specifically that on GFP containing dendritic spines. Kikuchi et al. (2020) published a 3D FIB-SEM reconstruction study of astrocytic processes in developing rat cerebral cortex. Anno Domini 2022 a website exists (MICrONS Explorer) where the general public can online browse through FIB-SEM image stacks sampled in mouse cerebral cortex: https://www.microns-explorer.org/cortical-mm3.

2.4 Light Microscopy: Golgi Metallic Silver Staining

As explained in the Introduction section, the original way of making dendritic spines visible is through Golgi metallic staining (also called "Golgi impregnation"). This is "old school" histology. The family of Golgi metallic impregnation methods encompasses two distinct subfamilies: those using mercury and those using silver. The mercury methods date back to the work of Cox (1891) and hence are called Golgi-Cox techniques. The silver-based methods are usually referred to as "Golgi-rapid" or plain "Golgi." The taxonomy of various Golgi impregnation techniques and modifications is covered by Facundo Valverde in his legendary Golgi monograph on mouse brain (Valverde 1998), while a recent review treats many of the details of mercury and silver Golgi procedures (Vints et al. 2019).

We discuss here one of the many varieties of silver Golgi impregnation, a Golgi-rapid class procedure called Golgi-Kopsch (developed nearly a hundred years ago by del Rio-Hortega (1928)). Other popular rapid Golgi impregnation procedures are those published by Blackstad, (1965) and Somogyi (1978; see Vints et al. 2019). The Golgi-Kopsch procedure is very well suited if combined light-electron microscopy is planned. Golgi Cox metallic staining is for light microscopy only.

2.4.1 Golgi-Kopsch Procedure

Fixation
For rats: perfusion fixation with iso-osmotic, 125–150 mM phosphate-buffered 4% formalin, pH 7.6, and 0.1% glutaraldehyde. The addition of glutaraldehyde was introduced by Colonnier (1964).

Golgi Silver Impregnation
Either whole brains or selected blocks are placed in a "mordant" consisting of 100 ml 3% potassium dichromate, 4% freshly depolymerized paraformaldehyde, and 1% chloral hydrate in distilled water. Just before placing the blocks in the mordant, they may be coated with gelatin or agar to prevent silver chromate crystal growth on their surfaces. The mordant is changed daily. Tissue blocks remain in mordant for 3–4 days in the dark and at room temperature. Then, they are transferred to a 0.75% silver nitrate solution in distilled water (dark; room temperature; 2–3 days). Slices 100 µm thick or with a thickness that fits the study are cut with a vibrating microtome and stored in a saturated silver chromate solution in distilled water.

Embedding and Coverslipping for Light Microscopy
It is advised to go fast through mounting, dehydration, and embedding-coverslipping. Counterstaining can be done at this stage using Anderson's (1954) alcoholic thionin method (Wouterlood 1986). Golgi silver staining can be combined with fluorescence counterstaining (Vints et al. 2019). Dehydration is in the standard series of ethanol through xylene. Slides can be parked for a while in 100% Xylene. We coverslip with a synthetic mounting medium such as Entellan®. After coverslipping, the slides are dried at room temperature in the dark. Storage is in a light-tight box at room temperature.

Silver-Gold Conversion, Embedding, and Coverslipping for Electron Microscopy Preparation for electron microscopy requires after the silver nitrate step a different continuation, notably a silver-gold conversion step, block staining, embedding in plastic, and curing (Fairén et al. 1977, 2005; Wouterlood and Mugnaini 1984; Wouterlood 1992). Thus, after rinsing three times in cold distilled water the blocks are deimpregnated during 10 minutes at room temperature and under constant gentle agitation in a "fixing" agent (4% sodium thiosulphate and 0.4% potassium-disulfite), and then treated during 30 minutes with 0.1% yellow gold chloride (hydrogen tetrachloroaurate (III), $HAuCl_4.H_2O$). This step is introduced to convert a fraction of the massive silver chromate deposit inside neurons into metallic gold particles and silver chloride. Afterward, the non-converted silver chromate and the silver chloride are removed by another stay in "fixing" agent (30 minutes, room temperature). During the gold chloride treatment, the sections have to be shielded from light to prevent a cascade of diffusion of silver ions outward from the neurons into the neuropil that subsequent catch chlorine ions to precipitate as silver chloride, followed by photoreduction, oxidation by gold chloride, and cycling through.

After these procedural steps, all sections are stored in cold 100 mM cacodylate buffer, pH 7.3; post-fixed during 1 hour in 2% OsO_4 in 100 mM cacodylate buffer, pH 7.3; block stained for 1 hour (dark, on melting ice) in 2% aqueous uranyl acetate; dehydrated through ascending grades of ethanol; and flat-embedded in epoxy resin (on slides and between two layers of polyethylene foil).

2.4.2 Notes on Golgi Metallic Staining

The black deposit that in Golgi silver-stained neurons so exquisitely outlines the details of all processes and their excrescences—with the exception of myelinated axons—has been identified as silver chromate (Fregerslev et al. 1971; Chan-Palay 1973). This salt is an unfavorable marker when it comes to the preparation of samples for transmission electron microscopy. To start with, it has a tendency to dissolve in acidic watery solutions and to fade during the embedding procedure in plastic resin. More important, silver chromate is so hard that it damages diamond knives during ultramicrotomy. It also tends to drop out of ultrathin sections, leaving holes. Finally, in the electron microscope beam, silver chromate absorbs so much energy relative to the surrounding section that the carrier film often collapses. These aspects fed the urge to replace for transmission electron microscopy purposes the "obnoxious" silver chromate with a more benign marker, especially if volume work followed by 3D reconstruction was in the planning. After several attempts with limited success, the procedure using gold chloride (discussed above) was developed by Fairén et al. (1977), and their technique has become the de facto standard (Fig. 2.4).

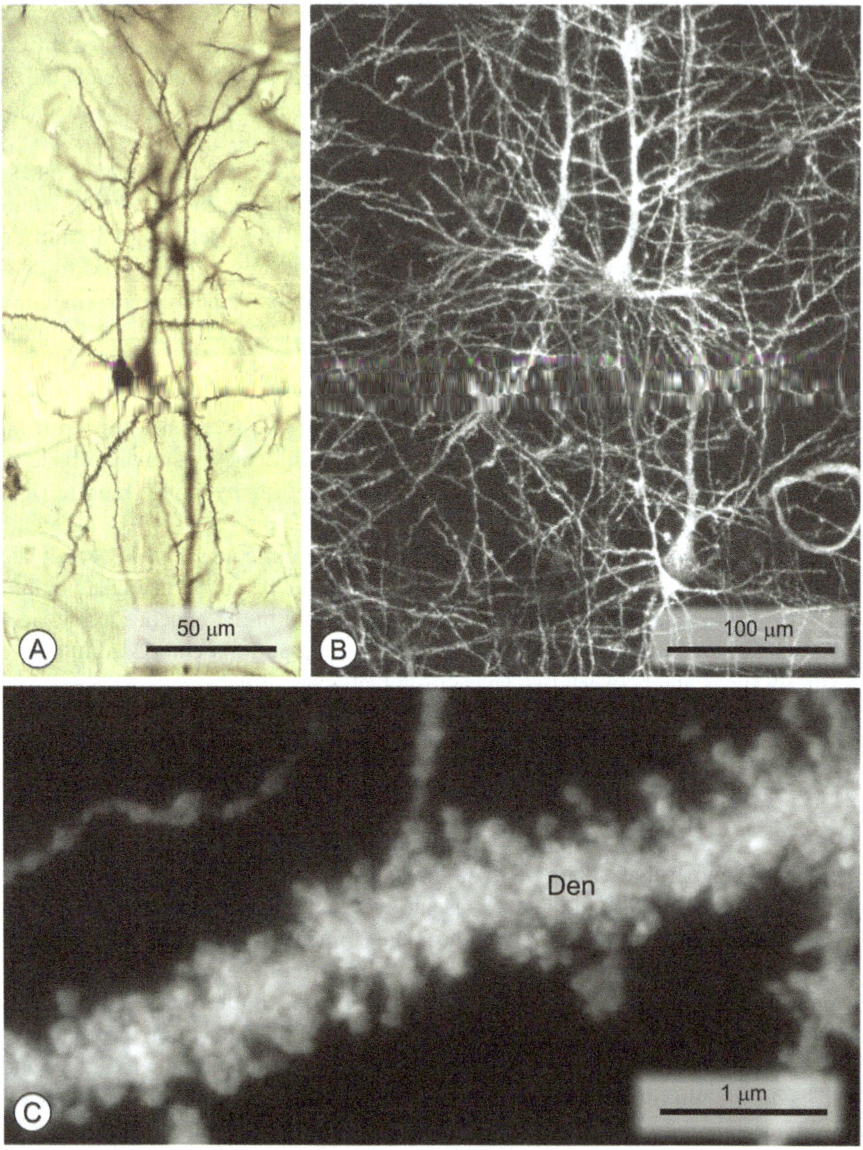

Fig. 2.4 Golgi silver staining. (**a**) Small pyramidal neuron (arrow) in rat cerebral cortex in conventional light microscopy. ax = initial axon segment. (**b**) Golgi silver-stained pyramidal neurons in human cerebral cortex, confocal laser scanning microscopy, reflective mode. (**c**) Thick spiny dendrite at high magnification in Golgi silver-stained human cerebral cortex, confocal laser scanning microscopy, reflective mode. sp = spine. (Material for **b** and **c** kindly provided by Dr. Harry Uijlings)

2.5 Dendritic Spine Visualization Through Intracellular Injection

Neuron visualization through one of the Golgi metallic staining methods, albeit sometimes spectacular, fails in being selective because with Golgi stains complete staining of discrete neurons is notoriously random. The histologist has a limited array of tools available to influence which type of neuron will be stained, how dense the population of stained neurons will be, and how completely stained individual neurons will be. Golgi metallic staining takes much effort to control (Rosoklija et al. 2014; Du 2019). This "capricious" aspect of Golgi stains is quite inconvenient if one aims to study specific neuron types. Taken "aiming" literally, a technique to impale a living neuron with a pipette-type electrode to record neurophysiological events and then complete the sequence with a nanoinjection of a fluorescent marker was introduced more than half a century ago by Stretton and Kravitz (1968) using the dye Procion yellow. Other, more convenient intracellular fluorescent markers appeared soon: Procion brown (Christensen 1973), and finally, Lucifer yellow (LY; Stewart 1978, 1981; Taghert et al. 1982; Katz et al. 1984). Horikawa and Armstrong (1988) introduced biocytin in intracellular recording-labeling, thus expanding the post-recording histology with avidin-biotin staining techniques. Note that all these intracellular recording-dye injection techniques involved neurophysiological preparations in in vitro brain slices or in situ.

The medium-size spiny striatal neuron with its formidable local axonal collateral projection system has been studied extensively through intracellular neurophysiology completed with injection of a marker (e.g., Kawaguchi et al. 1990). Perineuronal injection (Pinault 1996) and patch-clamp technology (Luebke et al. 2010) broadened the field. Current neurophysiological intracellular techniques (sharp pipette, juxtacellular, patch clamp) were recently reviewed by Cid and de la Prida (2019). One step aside from neurophysiology was made by Tauchi and Masland (1984, 1985) and Einstein (1988) who introduced a purely morphological technique: intracellular injection of fluorescent dye into neurons in slices of lightly fixed brain. From that moment on anatomists who wanted to completely label and study specific neurons of interest were relieved from the capriciousness of Golgi metallic staining as well as from the barrier of needing a neurophysiology setup. Even more exciting, Katz (1987), Rho and Sidman (1986), and Buhl and Lübke (1989) added to the lightly fixed brain method the pre-labeling of neurons of interest through retrograde neuronal transport. The latter preparation provides in the neuroanatomy lab a firm link to neurophysiological and cell biological experiments. The protocol introduced in 1989 by Buhl and Lübke has become a nearly universal application in neuroscience: neurons as diverse as dentate gyrus granule neurons in flying foxes (Buhl and Dann 1990) and cortical pyramidal cells in marmosets and macaque monkeys (Elston et al. 1996, 2011) have been intracellularly injected. Neurons in human material have received special attention (Benavides-Piccione et al. 2021). A review of what is possible through combination of intracellular labeling, neuroanatomical tracing, and immunohistochemistry is provided by Luo et al. (2001). In our lab,

Riichi Kajiwara intracellularly injected the fluorescent dye Alexa Fluor 555 (AF555) into hippocampal CA1 pyramidal neurons in slices of lightly fixed rat brain (Kajiwara et al. 2008). Some time prior to these injections, the rats had received a topical injection at two different locations with a suitable neuroanatomical tracer to differentially label axonal projections that might converge onto the apical dendrites of the intracellularly labeled CA1 neurons.

2.5.1 Intracellular Injection Procedure

Pre Labeling Through a Neuroanatomical Tracing Procedure

A young adult rat is anesthetized and introduced in a stereotaxic instrument. A focal, stereotaxically guided injection is made in a brain area of interest with a retrograde fluorescent neuroanatomical tracer. The purpose of this surgery is to label through retrograde axonal transport the cell bodies of the parent neurons of this particular axonal connectivity. Suitable retrograde fluorescent tracers are Fluoro-Gold and Fast Blue (Lanciego and Wouterlood 2011; Wouterlood et al. 2018).

Actual Intracellular Injection

1. After the post-surgery recovery period wherein the tracer is transported to the cell bodies of projection neurons, the experimental animal is sacrificed by an overdose of sodium pentobarbital and subsequently transcardially perfused with 100 ml of saline solution (0.85% NaCl, 0.025% KCl, 0.02% NaHCO3, pH 6.9, body temperature) followed by 500 ml of a mixture of freshly depolymerized 4% paraformaldehyde and 0.1% glutaraldehyde in 125 mM phosphate buffer (PB), pH 7.4, room temperature.
2. The brain is removed from the skull immediately after the perfusion fixation and post-fixed in the same fixative for 1 hour at 4 °C.
3. Alternating sections with a thickness of 400 μm ("thick") and 100 μm ("thin") are cut with a vibrating microtome. The "thin" sections serve as control sections to check the quality of the tissue as well as the presence of transported neuroanatomical tracers.
4. The "thick" slices are used for intracellular injection. They are transferred to a small Petri dish mounted in the intracellular injection unit: an epifluorescence microscope equipped with long-working distance objectives, a micromanipulator, and an iontophoretic nano-Ampère current injection device (detailed description by Buhl and Lübke 1989).
5. Approach, impalement, and subsequent intracellular filling of retrogradely fluorescent neurons now can begin. The pipette used for impalement is a 70–130 MΩ resistance glass pipette (microelectrode; tip diameter 0.5–0.8 μm outer diameter) containing 4% Lucifer yellow dilithium salt in distilled water or Alexa Fluor™ 555 hydrazide (AF555; Invitrogen-Molecular Probes; red-orange fluorescence, 10 mM in PB).

6. Impalement and injection. Under visual guidance, the pipette tip is gently positioned against a cell body that is visible because of the presence of fluorescent retrograde tracer. Most critical is the moment of penetration. This literally needs the master's touch: a single gentle tap on the stage may do the job. Successful impalement reports itself by "flaring up" of the neuron's cell body and initial portions of primary dendrites through filling with the fluorescent dye.

   After successful impalement, a pulsed negative current of 2 nA is applied to the micropipette (500 ms on, 50 ms off), for at least 10 minutes, to propel the fluorescent intracellular dye gently into the neuron's cytoplasm.
7. Filling of a neuron can be monitored by the operator. Whether filling is complete and successful is judged by the operator. We use the criterion of robust and homogeneous fluorescence inside primary and secondary dendrites, especially inside thin distal branches far away from the cell body. In pyramidal neurons, the appearance of fluorescence in dendritic spines and especially their thin necks is a good criterion. Conversely, unsuccessful filling results in an irregular spot of ejected intracellular fluorescent dye, very poor labeling of dendrites or just a blob of ejected fluorescent dye in the immediate vicinity of the site where impalement of a neuron had been attempted.

2.5.2 Processing After Intracellular Injection

Some post-injection histological processing is required because one cannot expect that imaging of intracellularly filled neurons lodged deep inside 400 μm thick rather opaque sections produces immediately picturesque results. In fluorescence microscopy and especially in confocal laser scanning, every extra photon that gets out of the section is favorable. Clearing is therefore a high priority issue. Histological processing may involve in toto clearing. A spread of techniques is available here, most of them developed for light sheet microscopy purposes, e.g., those of Becker et al. (2012), Lagerweij et al. (2017), Murakami et al. (2018), Jing et al. (2018) or Ueda et al. (2020).

As in toto clearing may have negative consequences for the histological quality of the tissue, an alternative is to resection the thick section into thinner subsections and to perform imaging on selected subsections. A big advantage of using subsections is that each of them can be subjected to incubation with a different set of antibodies, for instance to reveal neuroanatomical tracers or neuroactive substances. We usually implement the resectioning way of processing.

Resectioning Slices

Slices containing filled neurons are transferred to cold PB-buffered 4% formaldehyde, post-fixed overnight, and transferred to graded cryoprotective solutions with as final composition 20% glycerin and 2% dimethyl sulfoxide (DMSO, Merck, Germany) in PB (Rosene et al. 1986). Then, they are frozen on dry ice, parked in a freezer and later resectioned on a freezing-siding microtome at 40 μm, and stored in ice-cold PB at pH 7.6.

Imaging
Subsections are mounted on glass slides, dried, dehydrated, and cleared by a short dip in toluene, and coverslipped with one of the confocal laser scanning microscopy compliant media as mounting agent. Slides are introduced in a conventional confocal laser scanning microscope (CLSM) and imaged using the appropriate laser and channel settings (Fig. 2.5). First, structures of interest are imaged with the appropriate instrument settings at low magnification (Fig. 2.5a–d) after which a digital reconstruction of the intracellularly filled neuron can be made through superimposition (Fig. 2.5e). Using the digital reconstruction as a map regions of interest are scanned at high magnification (63× NA 1.3 glycerin immersion lens, 8× digital zoom). Filling, in case of Fig. 2.5, with AF555 was sufficient to reveal dendritic spines (Fig. 2.5f, arrowheads).

2.6 Lipophilic Carbocyanine Dye Staining

Hosokawa et al. (1995), working in a neurophysiology setting with dendritic spines of neurons in living slices, wanted to make these structures visible in a CLSM. For this purpose, they developed a method that consists of introducing microdroplets of oil saturated with mono-unsaturated DiI (1,1 '-dioleyl-3,3,3',3'-tetramethylindocarbocyanine methanesulfonate) into hippocampal slices. DiI and its spectral variations named DiO, DiC, DiD, and DiR are carbocyanine species, a family of dedicated lipophilic, intensely fluorescent molecules. Carbocyanine dyes fluoresce weakly in water while incorporated in lipid membranes, especially myelin, fluorescence picks up and becomes intense. DiI-labeled dendrites with their spines could confidently be imaged in the Hosokawa et al. (1995) study. DiI had several years before been introduced by Marcia Honig as a neuroanatomical tracing compound that fluoresces orange-red (Honig and Hume 1986). The dye is mostly being used in models that lend themselves with difficulty to standard experimental neuroanatomical tracing such as the embryonic brain environment (Godement et al. 1987), neuron lineage (Modrell et al. 2014), and postmortem human brain (Dall'Oglio et al. 2015; Sivukhina and Jirikowski 2021).

2.6.1 Application of Lipophilic Dye

Carbocyanine dye crystals may be deposited in myelinated fiber tracts or in brain parenchyma (e.g., Das et al. 2019); powdered DiI may be gently applied to slices of fixed brain (Rasia-Filho et al. 2010), whereas it is also applied dissolved 1% in methylene chloride through electrophoresis in a sharp pipette (Gan et al. 1999) or dissolved in 100% ethanol (1 mg/ml) through injection (Sivukhina and Jirikowski

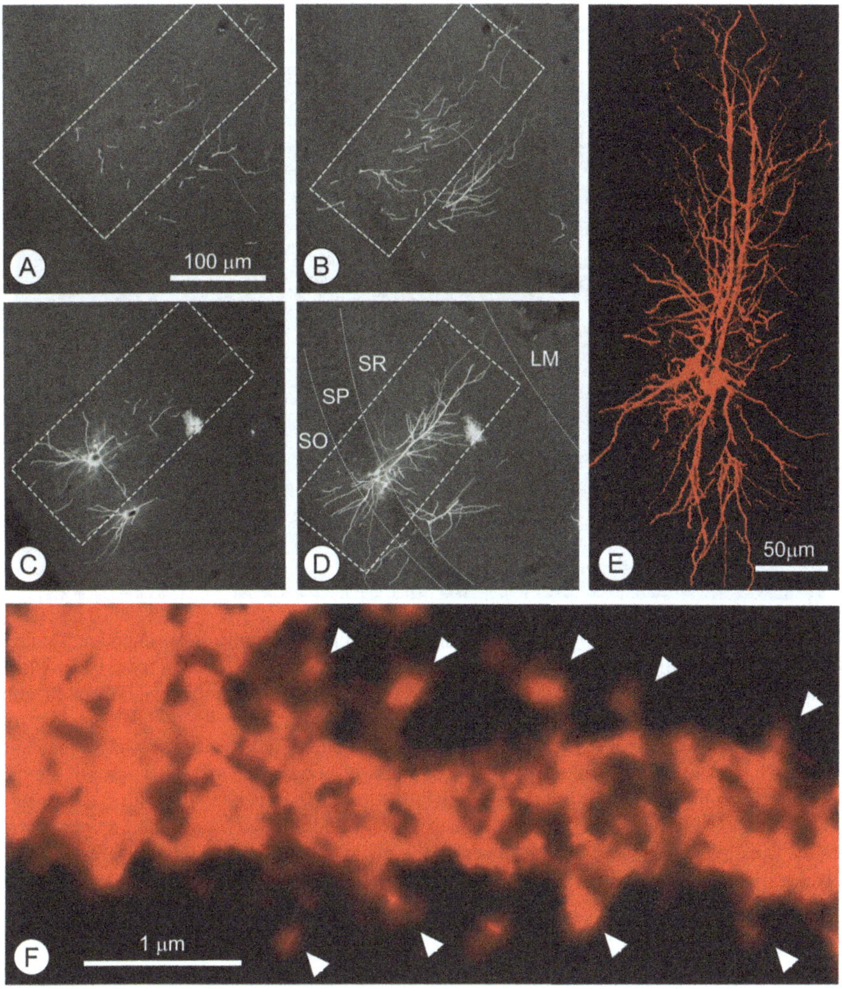

Fig. 2.5 Intracellularly AF555 injected rat hippocampal CA1 pyramidal neuron neuron. Injections in slices of lightly fixed rat brain. Images acquired in a CLSM at 543 nm laser excitation (absorption peak of AF555 is 555 nm. (**a–d**) Subsections used for mapping and reconstruction. Low magnification (20× objective lens). SO = stratum oriens, SP = stratum pyramidale, SR = stratum radiatum, LM = stratum lacunosum-moleculare. The most apical dendrites of this neuron extend all the way into stratum lacunosum-moleculare. Bar in **a** also valid in **b**, **c**, and **d**. (**e**) Quick reconstruction / neuron map made by superimposing images acquired from the subsections. (**f**) High magnification imaging (1.3 NA 63× glycerin immersion lens, 8× zoom). Dendritic spines are indicated with arrowheads. (Material prepared in collaboration with Dr. Riichi Kajiwara (cell #10-P2; Kajiwara et al. 2008))

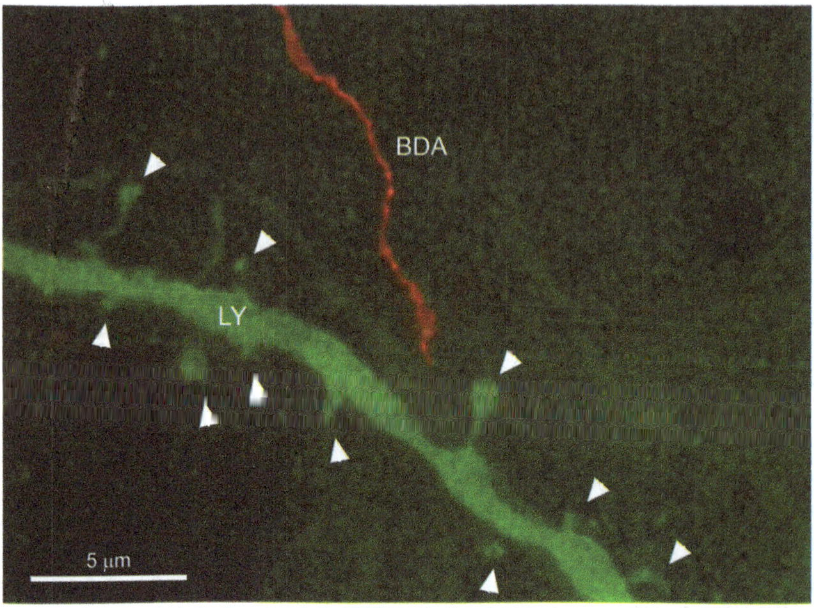

Fig. 2.6 Combination of retrograde neuroanatomical tracing and intracellular injection. Dendrite of a Lucifer yellow filled neuron in the substantia nigra pars reticulata (SNr) of a rat. Lightly fixed brain. Two-laser imaging in a CLSM. The green color of this dendrite is the result of stabilization of the Lucifer yellow via immunofluorescence with final marker Alexa 488. A fiber arriving from the striatum, labeled with the tracer biotinylated dextran amine (red, visualized with streptavidin-Alexa 546). This image was called by us "land of promise" because the (red) fiber abruptly stopped just before forming a synaptic-like contact with the dendrite. The Lucifer yellow filled cell belongs to a type of sparsely spiny SNr neurons. Actually, these dendrites carry quite some spines (arrowheads). Detailed description of the used neuroanatomical tracing—intracellular labeling—CLSM imaging method in Wouterlood et al. (2018)

2021). Dissolving DiI in oil microdroplets or an organic solvent like ethanol solves the most essential problem, notably the application of the carbocyanine dye to the target neurons. The essential characteristic of DiI is that the molecule consists of a hydrophilic head that lies above the lipid bilayer of a plasma cell membrane while two lipophilic alkyl chains anchor in the lipid bilayer (Honig and Hume 1989; Bruce et al. 1997). This anchoring forces a carbocyanine dye molecule to follow the dynamics of live lipid membrane bilayers. It diffuses hence within cell membranes in fixed tissue and particularly in myelin tracts. DiI is also stable: once inside myelin, it stays there and is hard to remove.

The spectral characteristics of the carbocyanine dyes hinge on the excitation-emission peaks: DiO: (excitation 483 nm/emission 501 nm), DiC (549/563), DiI (549/563), DiD (645/663), and DiR (754/778) (data: AAT Bioquest). Thus, these carbocyanine dyes combine well with conventional fluorescence microscopy (except of course the far infrared dye DiR) and confocal laser scanning microscopy. Fluorescence is robust and can be used as a neuronal marker in a CLSM (Lanciego and Wouterlood 2011, 2020).

Kim et al. (2007) reported that perfusion fixation with a low concentration of formalin (1.5%) gives better staining with DiI than in unfixed brain material. These observations were made in brains from experimental animals. At this point, more expertise is needed for the human setting. Das et al. (2019) found excellent labeling in 4% formalin post-fixed human hippocampal CA1 neurons with apical dendrites studded with spines that could be 3D reconstructed with high confidence in the CLSM.

Usually, the fixation of human brain tissue is block immersion in fixative. As human postmortem material typically is acquired several hours after death while the donor's medical condition and medication are barely documented, the quality of fixation depends on conditions beyond control.

Whereas the first observations with DiI as fluorescent marker were done with epifluorescence equipment (Honig and Hume 1989), a search to locate the first publication using confocal laser scanning microscopy led to a paper by Baker and Reese (1993) studying DiI-labeled neuronal projections. Not much later, Terasaki et al. (1994) reported their confocal laser scanning microscopy study of DiI-labeled living Purkyně cells in vitro slice preparations. Today, most of the work done with DiI is documented with CLSM images. It can safely be stated that DiI labeling technology has matured in combination with advances in confocal laser scanning microscopy. This holds in particular where human postmortem material is being studied, e.g., volume scanning of dendritic spines in human amygdaloid nuclei (Dall'Oglio et al. 2015).

2.7 DiOlistic Labeling of Neurons

The quest of bringing neurons in close contact with an extremely lipophilic compound has led to innovative solutions. One such a solution is ballistic delivery of lipophilic dye-coated particles with a gene gun (Gan et al. 2000; Grutzendler et al. 2003; Moolman et al. 2004; O'Brien and Lummis 2007; Hough and Brown 2017; Olive et al. 2018). This technique is nowadays known as "diolistics." Gan et al. (2000) show dendritic spines spectacularly labeled with seven combinations of carbocyanine dyes. O'Brien and Lummis (2007) managed to "shoot" DiI into a mouse Purkyně cell, with excellent visualization of the Purkyně cell's dendritic spines. Biolistics (the condensed generalized term of "biological ballistics," i.e., positioning plasmids into cells) can be used to deposit GFP coding plasmid-coated particles with physical means into neurons, forcing the latter to start metabolizing and accumulating GFP (Gan et al. 2000). The method to use a gene gun to shoot DNA into cells was invented 30 years ago by plant geneticists at Cornell (Klein et al. 1987).

Today, biolistics has widespread applications in agricultural research, genetics, and biomedicine. How biolistics works on rat hippocampal cells in vitro slices, together with a gene gun protocol, is explained by Woods and Zito (2008).

2.8 Genetic Engineering: GFP Expression and Accumulation in Neurons

With proper application of genetic engineering, a neuron can be forced into do-it-yourself mode. It then metabolizes a fluorescent marker that diffuses inside the neuron's cytoplasm into all axonal and dendritic compartments, including dendritic spines. Advances in genetics and molecular biology have delivered breeds of animals that carry foreign genes. The gene of interest here causes expression of GFP or a spectral variant like YFP, CFP, RFP, Cherry, Tomato, or a similar fluorescent protein. In standard fluorescence microscopy sections remain dark except for the neurons that express the foreign gene. These cells display without hesitation all their beauty in spectacular detail, no histology necessary nor painstaking intracellular neurophysiology with mechanical intracellular labeling, just that! Incredible!

Sauer (1998), Callahan et al. (1998), and Feng et al. (2000) were among the first who delivered on this idea. A superb example of mapping fiber tracts in eYFP expressing Thy1-eYFP-H transgenic mice is that of the corticospinal tract by Porrero et al. (2010). Here, the entire pyramidal tract can be followed in fluorescence, including the spiny parent pyramidal neurons and the entire trajectory followed by the pathway into the spinal cord.

In the first generation of genetically engineered mice, GFP was expressed spontaneously in neurons whose identity was uncontrolled. Our laboratory received in 2010 a gift consisting of brains of CRE-eYFP mice that for some mysterious reason expressed GFP in a few medium-size spiny neurons (MSNs) in the caudate-putamen (Chakravarthy et al. 2008). Several of these neurons were scanned in our CLSM at high magnification for demonstration purposes. Figure 2.7 is one example of an eYFP-labeled MSN.

This first wave of genetically engineered mice was followed up by the development at the Allen Institute for Brain Science of robust and universal Cre-responder lines where intense native eYFP accumulation occurs in neurons (Madisen et al. 2010). Particularly interesting is the Cre/loxP system, in which expression is conditionally triggered of the recombinase gene that is located next to a gene that determines the neurochemophenotype of the neuron. Triggering of the recombinase can occur through transfection (after injection) of a viral vector whose payload is an eYFP gene. Cre/loxP has been applied successfully with neurons that contain specific neuroactive substances markers, e.g., choline acetyltransferase (Bloem et al. 2014), somatostatin, tyrosine hydroxylase, NPY, calbindin, or parvalbumin. Cortical

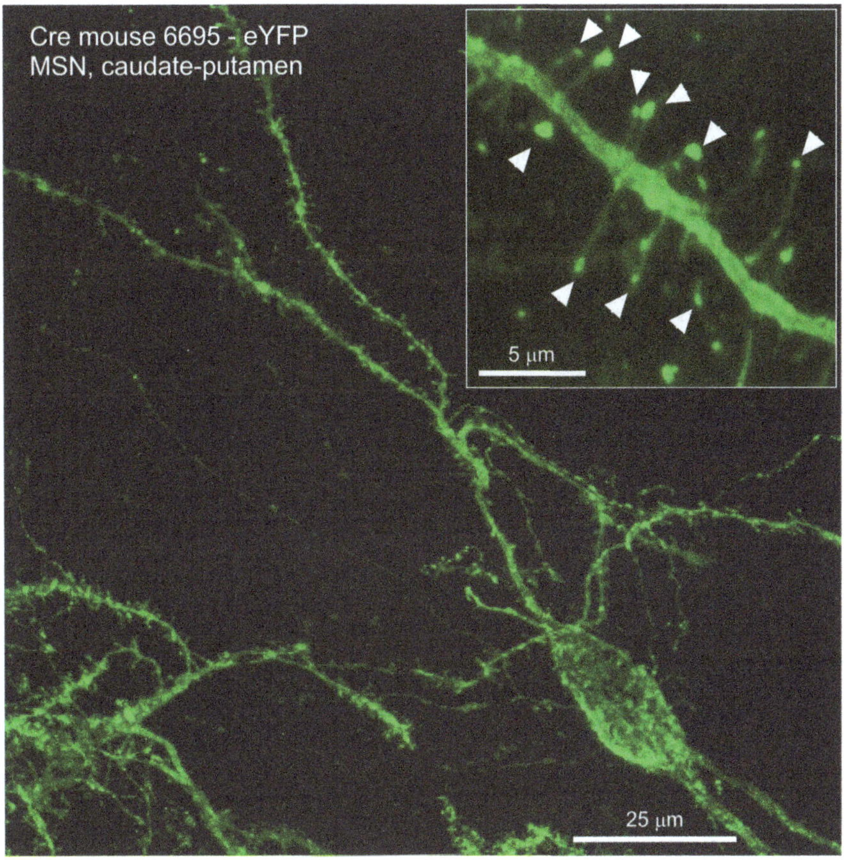

Fig. 2.7 Compound confocal laser scanning image of a medium-sized spiny neuron in the murine caudate-putamen. This mouse had a gene inserted that initializes metabolism of eYFP in some medium-size spiny neurons (MSNs) in the caudate-putamen and cerebral cortex. MSNs populate the caudate-putamen in high numbers making it in normal conditions difficult to distinguish individual specimens. The presence of the fluorescent marker in all the cell's components in the scattered individual MSNs makes this brain and those of the donor's siblings very attractive for the study of this type of neuron. A segment of one of the dendrites of this neuron was scanned at very high magnification (inset). The spines on this segment (arrowheads) display extravagantly long necks. Fluorescence here is native eYFP fluorescence illuminated with a 488 nm laser, images acquired by Irineu Bochaca; mouse material gift of Christiaan Levelt, Amsterdam Neuroscience—Cellular & Molecular Mechanisms, Vrije University, Amsterdam (Chakravarthy et al. 2008)

pyramidal neurons characterized by neurotransmitter, e.g., MSNs (GABA) and pyramidal neurons (glutamate), do not respond well to Cre/loxP tools. Today, a range of Cre-reporter mice is listed in a big catalog at JAX (https://mice.jax.org) and mice that are particularly interesting can simply be ordered by mail.

2.9 Transported Virus as Marker

Another approach to force a neuron to metabolize a fluorescent marker is to bring it in contact with a viral vector. The virus' payload is a gene that, once it has penetrated the neuron's outer membrane, is inserted in the genome and then forces the cell metabolism to produce the fluorescent protein. It is the accumulation of this protein that prepares the cell for imaging in a CLSM. The gene delivering virus, however, has a forerunner. Krister Kristensson, the founder of horseradish peroxidase transport (HRP)-based tracing (Kristensson and Olsson 1971) played at the same time as his landmark HRP publication with herpes simplex virus as an agent to travel through the brain's labyrinthine pathway organization with as destination parent cell bodies (Kristensson et al. 1971). Neurotropic virus tracing with another virus, alpha herpes virus, with the aim to use it as an amplifier in first- and second-order neurons wherein the viral capsid proteins acted as the marker, was introduced by Ugolini et al. (1987, 1989), see also Kuypers and Ugolini (1990) and Ugolini (2010) for review. Once it was discovered that rabies virus is a more potent "tracer" than herpes simplex (Loewy et al. 1991), this virus became a mainstream tracing agent, especially after the original rabbit species was replaced with rabies virus from swine (PRV) that is not dangerous for humans (Martin and Dolivo 1983). Amplification of the detection with antibodies directed against a soluble viral phosphoprotein (31G10, isolated by Raux et al. 1997), that spreads throughout the cytoplasm of RV-infected neurons produces a Golgi silver staining-like retrograde labeling of infected neurons (Salin et al. 2008, 2020).

2.9.1 Viruses Carrying a Fluorescent Protein-Coding Gene as Their Payload

Focal injection into cerebral cortex in rats of a Herpes simplex virus carrying a plasmid coding for green fluorescent protein (GFP), with the purpose of forcing pyramidal cortical neurons to start metabolizing GFP to such a level that dendritic spines become visible and their density can be counted, was achieved by Sánchez-González et al. (2021). Retrograde transport of pseudorabies virus carrying a gene for rCVS-N2c-P-mCherry (injected into the substantia nigra) produced in vivo red, Golgi like staining of medium size spiny neurons in the striatum (Salin et al. 2008, 2020).

While genetically modified swine rabies virus currently seems to be the most popular vector to transfect neurons whose dendrites carry spines there is competition. For instance, Hirano et al. (2013) introduced a "tamed" lentivirus based on human immunodeficiency virus type-1 (HIV-1), that is a pseudotyped lentiviral vector with a rabies virus fusion envelope glycoprotein, carrying a GFP gene payload. Marketed as "HiRet," this lentivirus been applied recently for both short-distance as well as long-distance retrograde tracing with GFP filling of parent cell bodies and dendrites in the corticospinal system (Sheikh et al. 2018). Indeed, after cervical spinal cord injection Sheikh et al. found labeled pyramidal cells in motor cortex (Sheikh et al. 2018—their Fig. 1C). Before Sheikh et al.'s publication, a group in Japan managed in mice and marmosets with an altered lentivirus, named "TET-Off vector" to effectively retrogradely program cortical pyramidal neurons projecting to the pons to metabolize GFP and red fluorescing protein (RFP). Accumulation of RFP in cortical pyramidal neurons appeared in such quantities that dendritic spines on the apical dendrites of the pyramidal neurons became confidentially labeled (Watakabe et al. 2012; their Figs. 2 and 3).

2.10 Light Microscopic Imaging of Spines Is Diffraction Sensitive

As a dendritic spine head's volume is in the 0.01–0.8 μm^3 range, the visualization of dendritic spines touches an essential issue. Dendritic spines belong to the category of "diffraction-sensitive" objects. This means that their dimensions hover around the bottom of the resolving power of a light microscope. Already Santiago Ramon y Cajal in his original research on dendritic "espinas" (Ramón y Cajal 1888) must have been painfully aware of this natural phenomenon. In the visible light range, the resolution limit of a good light microscope objective is at approximately 200 nm distance between two discrete points. Diffraction of light that occurs at this resolution limit makes small objects such as dendritic spines and especially their necks look fuzzy. It is a mistake to think that this fuzziness can be removed by using sharpening filters in free or commercial photomanipulation software. Fuzziness it is a real-life phenomenon dictated by the forces of nature. Ernst Abbe (1840–1905), the great German pioneer in optical lens theory, expressed diffraction in mathematical equations in 1873 (Abbe 1873). Lord Rayleigh (1842–1919) worded the essence of Abbe's work in 1891 in a definition that still holds today, the "Raleigh criterion": resolution is "the smallest distance between two point sources in which we are able to distinguish these as two points" (Rayleigh 1886; Rayleigh and Strutt 1891). This criterion holds for all optical instruments, including astronomy telescopes, field optics, microscopes, digital imaging equipment, and even electron microscopes.

For a microscope, Rayleigh's criterion lists as follows: r (resolution limit) = 0.61 λ/NA where λ is the wavelength of the used light and NA is the numeric aperture of the objective lens. The medium in between the lens and the object (water, glycerin, oil) is also a factor of importance. With the best immersion lens (NA 1.6) and green light emitted by GFP molecules (509 nm emission peak), r can be calculated: theoretically 194 nm. In the lab, a resolution of 200 nm is "normal" which means that two objects that are closer to each other than r will be visible as one fuzzy "thing." The electron micrographs of Figs. 2.1, 2.2, and 2.3 show dendritic spines with spine necks 100–200 nm in diameter and spine heads about half a micron diameter at their equator. This explains why in the light microscope we can see spine heads because they comply with Raleigh's criterion while spine necks do not comply and as a consequence are barely visible as fuzzy stalks or as "sometimes missing." With special tricks such as the use of monochromatic (laser) excitation light and post-acquisition image processing based on statistics theory, images can be improved approximately 20%, leading to a resolution (parameter "r") of approximately 120 nm. Images such as Figs. 2.5, 2.6, and 2.7 were produced with the aid of a standard confocal laser scanning microscope followed up with post-acquisition image processing. Compare these with Fig. 2.2a that were made with a photocamera mounted on a laboratory research microscope. This takes us to a small but necessary excursion into the world of laser scanning microscopy. After all, we are dealing in this chapter with methods and microscopes! The application of an essential patent making confocal imaging possible was submitted in 1957 by Marvin Minsky (1957) and granted in 1961. It can safely be argued that the introduction of the first reliable commercial instruments of this type around 1990 started a revolution in microscopy (Amos and White 2003).

2.10.1 *Imaging of Spines with a Conventional CLSM*

Imaging in a CLSM starts with an excitation laser that scans the region of interest (ROI) in *X* and *Y* with typically 512 * 512 or 1024 * 1024 pulses such that digital images can be constructed. A pinhole in front of the detector lets only the light emitted by the object in a thin focal plane to be detected. A photomultiplier and the electronics behind that sensor translate the detected light into a gray value, e.g., in 8-bit sampling in values on a scale between 0 and 255. The Z component in the imaging is introduced via a stepping motor that raises or lowers the stage with a small increment after every scan. Thus, a CLSM produces a series of images which can be compared with a stack of QR codes. Pixel dimensions in the CLSM used to acquire the image shown in Fig. 2.6 had dimensions of 55 * 55 nm. Stepping was 122 nm, which implies that voxels (volume elements) measure 55 * 55 * 122 nm. The consequence of this type of image acquisition is that the intuitive naive notion of a voxel being a cube is false: voxels resemble short sticks; their length (in *Z*) is twice their size in *X* and *Y*. The upper half of the voxel "sticks up" from the scanning plane while its lower half "sticks down," with the voxel's mathematical center of gravity exactly in the plane of scanning.

2.10.2 *Working with Voxels*

Inevitably the question arises: How many voxels in my confocal images correspond with a typical dendritic spine and its neck? In the previous section, we determined that efficient scanning produces at high-resolution images with in *XY* 55 * 55 nm pixels and in *Z* 122 nm due to stepping. Figure 2.8a–d may serve here as an illustration. A spine head 500 nm wide in *X*–*Y* means 9 pixels across. If a spine head were a sphere and one voxel takes 55 * 55 * 122 nm application of the classical sphere

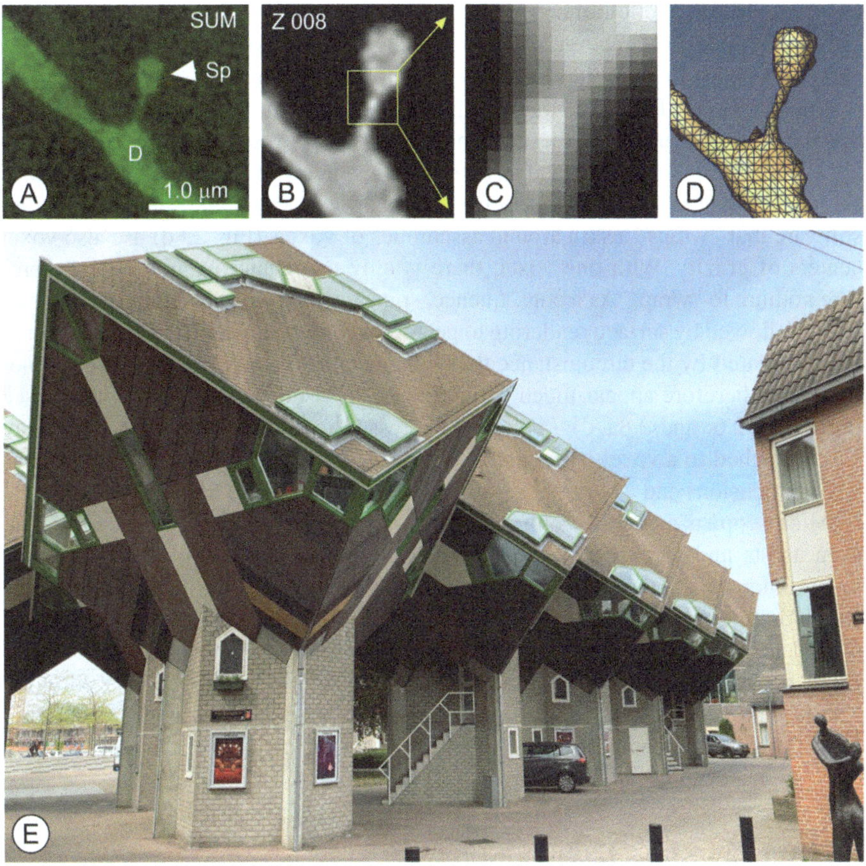

Fig. 2.8 Confocal laser scanning. (**a**) Spine (Sp) on a dendrite (D) imaged in the CLSM; scanned at 1024 × 1024 pixels. This is a summation image including 34 images of a Z-acquisition series. (**b**) frame 008 of the series. The rectangular area is magnified in **c**. (**c**) detail of B showing individual pixels. (**d**) 3D reconstruction of this spine with Amira software. (**e**) "Cube homes" located in Helmond, the Netherlands, resembling "voxelized" dendritic spines—architect Piet Blom. (Photo credit Niels Schoorlemmer, Eindhoven, the Netherlands)

volume formula (4/3 * π * r3) brings us at approximately 177 voxels. A spine neck is covered across by two or three voxels and, in the *Z* direction (length of the neck), by a variable number of voxels, say five. We ignore here the uncertainty introduced by a factor called the point spread function (PSF) of the microscope's detector pinhole. The PSF measured in the radial direction (*X*–*Y*) is different from that in the *Z* direction (*X*–*Z*) (see Pawley 2006).

Professional image processing software uses statistical algorithms to "deconvolute" images. Convolution—the opposite—is the diffraction process inherent in optical systems, from astronomic telescopes to binoculars and microscopes, wherein light rays statistically bend around small objects to produce a fuzzy image. Abbe has described diffraction in his 1873 paper. Deconvolution can be seen as an exercise in the opposite direction, sort of reversing Abbe's equations. In statistics, it is very hard if not impossible to reliably deconvolute images that contain structures two pixels across. Reason is that two pixels next to each other in an image means that the centers of gravity of these two pixels are only one pixel apart! Deconvolution algorithms use pixel centers and have a strong tendency to statistically remove pixels when they encounter structures two pixels or less across. 3D surface rendering software that "wraps" a skin around assemblies of voxels (Fig. 2.8d) use also voxel centers of gravity. With one voxel, there is only one center of gravity and therefore nothing to "wrap." As a consequence, spine necks have a tendency in deconvolution followed by surface rendering to get removed or be unseen! This phenomenon is exacerbated by the circumstance that spine necks do contain far less GFP molecules and therefore appear much "fainter" in CLSM images than spine heads (see Figs. 2.5f, 2.6, and 2.8a–d). "Faint" is expressed in CLSM images in a low gray value attached to a voxel (typically in a scale 0–255 or 8-bit). Thus, any voxel has fixed dimensions and a varying gray value: 0 is ink black and 255 is bright white.

Microscopists are not the only ones turning structures into assemblies of voxels, even artists and architects "voxelize" their creations. Sol LeWitt's created many three-dimensional cubic sculptures (https://en.wikipedia.org/wiki/Sol_LeWitt). The Dutch architect Piet Blom (1934–1999) designed "cube homes," i.e., cubes tilted while leaning on one corner on a pedestal (Fig. 2.8e) (https://en.wikipedia.org/wiki/Cube_house). In the cities of Helmond and Rotterdam, such homes can be admired. The resemblance of Piet's cubic homes with dendritic spines scanned and represented as voxels is stunning.

2.11 The Future Is Already Here: Super-Resolution Microscopy

The 2014 Nobel Prize in Chemistry was awarded to Stefan Hell (Max Planck Institutes of Heidelberg and Göttingen) because of his contribution to the development of stimulated emission depletion microscopy (STED). This award underscores the importance of this type of microscopy. Basically, a STED microscope is an advanced confocal laser scanning instrument that is equipped with two lasers: an illumination laser with a sharp beam and a depletion laser with a doughnut-shaped beam. The purpose of the depletion laser is to suppress fluorescence at the edges of the sharp beam. The doughnut laser beam wipes out diffraction blur produced by the sharp laser beam. The stunning result is nanopoint illumination wherein the factor "r" in Abbe's formula: r (resolution limit) = 0.61 λ/NA has become = 0.61 * $\sqrt{(1 + \sigma)}$ where σ is the ratio of the intensity of the sharp beam and the intensity divided by the saturation intensity. Microscopes designed around the STED principle achieve a resolution in X–Y in the range of 30–80 nm (Vicidomini et al. 2018). Thus, it can be said that Hell broke the resolution limit inherent in Abbe's equation. Abbe's "r" limit for light microscopes is 200 nm, for conventional CLSMs with post-acquisition deconvolution r equals 120 nm, while STED with post-acquisition deconvolution records 80 nm resolution. This is a very important development because STED brings dendritic spines out of the blurry, diffraction-sensitive zone into the spotlight of microscopy (Tønnesen and Nägerl 2016). The implications are immense. With STED dendritic spines can be so much better imaged than with conventional CLSM (Kashiwagi and Okabe 2021) that specific molecules can be pinpointed in the postsynaptic density area of the spine. STED and comparable microscopy techniques such as structured illumination microscopy and microscopes that reconstruct fluorochrome location: photoactivated localization microscopy (PALM) and stochastic optical reconstruction microscopy (STORM) truly fit the resolution gap that happened to exist between light and electron microscopy. Thus, understanding of dendritic spines has evolved from "suspect artifact" to neuronal appendages said to be true "molecular machines involved in synaptic transmission."

References

Abbe E (1873) Beiträge zur Theorie des Mikroskops und der mikroskopischen Wahrnehmung. Archiv für Mikroskopische Anatomie (in German). Verlag von Max Cohen & Sohn, Bonn 9(1):413–468. https://doi.org/10.1007/BF02956173

Amos WB, White JG (2003) How the confocal laser scanning microscope entered biological research. Biol Cell 95:335–342. https://doi.org/10.1016/s0248-4900(03)00078-9. PMID: 14519550

Anderson FD (1954) Alcoholic thionin counterstaining following Golgi dichromate-silver impregnation. Stain Technol 29:179–184. https://doi.org/10.3109/10520295409115465. PMID: 13178861

Baker GE, Reese BE (1993) Using confocal laser scanning microscopy to investigate the organization and development of neuronal projections labeled with DiI. Methods Cell Biol 38:325–344

Bang BH, Bang FB (1957) Graphic reconstruction of the third dimension from serial electron microphotographs. J Ultrastruct Res 1:138–139. https://doi.org/10.1016/s0022-5320(57)80002-1

Becker K, Jährling N, Saghafi S, Weiler R, Dodt H-U (2012) Chemical clearing and dehydration of GFP expressing mouse brains. PloS One 7(3):e33916. https://doi.org/10.1371/journal.pone.0033916. PMID: 22479475 PMCID: PMC3316521

Benavides-Piccione R, Rojo C, Kastanauskaite A, DeFelipe J (2021) Variation in pyramidal cell morphology across the human anterior temporal lobe. Cereb Cortex 31:3592–3609. https://doi.org/10.1093/cercor/bhab034

Blackstad TW (1965) Mapping of experimental axon degeneration by electron microscopy of Golgi preparations. Z Zellforsch Mikrosk Anat 67:819–834. https://doi.org/10.1007/BF00339303

Blackstad TW, Kjaerheim A (1961) Special axo-dendritic synapses in the hippocampal cortex: electron and light microscopic studies on the layer of mossy fibers. J Comp Neurol 117:133–159. https://doi.org/10.1002/cne.901170202

Bloem B, Schoppink L, Rotaru DC, Faiz A, Hendriks P, Mansvelder HD, van de Berg WD, Wouterlood FG (2014) Topographic mapping between basal forebrain cholinergic neurons and the medial prefrontal cortex in mice. J Neurosci 34:16234–11646. https://doi.org/10.1523/JNEUROSCI.3011-14.2014. PMID: 25471564

Bonhoeffer T, Yuste R (2002) Spine motility – phenomenology, mechanisms, and function. Neuron 35:1019–1027. https://doi.org/10.1016/S0896-6273(02)00906-6

Bosch C, Martínez A, Masachs N, Teixeira CM, Fernaud I, Ulloa F, Pérez-Martínez E, Lois C, Comella JX, DeFelipe J, Merchán-Pérez A, Soriano E (2015) FIB/SEM technology and high-throughput 3D reconstruction of dendritic spines and synapses in GFP-labeled adult-generated neurons. Front Neuroanat. https://doi.org/10.3389/fnana.2015.00060

Bruce LL, Christensen MA, Fritzsch B (1997) Electron microscopic differentiation of directly and transneuronally transported DiI and applications for studies of synaptogenesis. J Neurosci Methods 73:107–112. https://doi.org/10.1016/S0165-0270(96)02218-2

Buhl EH, Dann JF (1990) Basal dendrites are a regular feature of hippocampal granule cells in flying fox hippocampus. Neurosci Lett 116:263–268. https://doi.org/10.1016/0304-3940(90)90084-m. PMID: 2243603

Buhl EH, Lübke J (1989) Intracellular Lucifer Yellow injection in fixed brain slices combined with retrograde tracing, light and electron microscopy. Neuroscience 28:3–16. https://doi.org/10.1016/0306 4522(89)90227 3. PMID: 2668782

Callahan CA, Yoshikawa S, Thomas JB (1998) Tracing axons. Curr Opin Neurobiol 8:582–586 https://doi.org/10.1016/s0959 4388(98)80084 6. PMID: 9811618 Review

Chakravarthy S, Keck T, Roelandse M, Hartman R, Jeromin A, Perry S, Hofer SB, Mrsic-Flogel T, Levelt CN (2008) Cre-dependent expression of multiple transgenes in isolated neurons of the adult forebrain. PloS One 3:e3059. https://doi.org/10.1371/journal.pone.0003059. PMID: 18725976 PMCID: PMC2518110

Chan-Palay V (1973) A brief note on the chemical nature of the precipitate within nerve fibers after the rapid Golgi reaction: Selected area diffraction in high voltage electron microscopy. Z Anat Entwickl- Gesch 139:115–117. https://doi.org/10.1007/BF00523632. PMID: 4197941

Christensen B (1973) Procion brown: an intracellular dye for light and electron microscopy. Science 182:1255–1256. https://doi.org/10.1126/science.182.4118.1255. PMID: 4584668

Cid E, de la Prida LM (2019) Methods for single-cell recording and labeling in vivo. J Neurosci Methods 325:108354. https://doi.org/10.1016/j.jneumeth.2019.108354. PMID: 31302156

Colonnier M (1964) The tangential organization of the visual cortex. J Anat 98:327–344. PMID: 14193193

Cox W (1891) Imprägnation des Centralen Nervensystems mit Quecksilbersalzen. Arch Mikr Anat 98:327

Dall'Oglio A, Dutra ACL, Moreira JE, Rasia-Filho AA (2015) The human medial amygdala: structure, diversity, and complexity of dendritic spines. J Anat 227:440–459. PMID: 26218827

Das SC, Chen D, Callor WB, Christensen E, Coon H, Williams ME (2019) DiI-mediated analysis of pre- and postsynaptic structures in human postmortem brain tissue. J Comp Neurol 527:3087–3098. https://doi.org/10.1002/cne.24722. PMID: 31152449

DeFelipe J (2015) The dendritic spine story: an intriguing process of discovery. Front Neuroanat 9:14. https://doi.org/10.3389/fnana.2015.00014

del Rio-Hortega P (1928) Tercera aportación al conocimiento morfológico e interpretación de la oligodendroglia. Mem Real Soc Esp Hist Nat 14:1–128

Denk W, Horstmann H (2004) Serial block-face scanning electron microscopy to reconstruct three-dimensional tissue nanostructure. PLoS Biol 2:e329. https://doi.org/10.1371/journal.pbio.0020329. PMID: 15514700 PMCID: PMC524270

Du F (2019) Golgi-cox staining of neuronal dendrites and dendritic spines with FD rapid GolgiStain™ kit. Curr Protoc Neurosci 88:e69. https://doi.org/10.1002/cpns.69

Einstein G (1988) Intracellular injection of Lucifer yellow into cortical neurons in lightly fixed sections and its application to human autopsy material. J Neurosci Methods 26:95–103. https://doi.org/10.1016/0165-0270(88)90157-4. PMID: 3216684

Elston GN, Rosa MG, Calford MB (1996) Comparison of dendritic fields of layer III pyramidal neurons in striate and extrastriate visual areas of the marmoset: a Lucifer yellow intracellular injection. Cereb Cortex 6:807–813. https://doi.org/10.1093/cercor/6.6.807

Elston GN, OgaT OT, Fujita I (2011) Spinogenesis and pruning in the anterior ventral inferotemporal cortex of the macaque monkey: an intracellular injection study of layer III pyramidal cells. Front Neuroanat. https://doi.org/10.3389/fnana.2011.00042

Fairén A (2005) Pioneering a golden age of cerebral microcircuits: the births of the combined Golgi-electron microscope methods. Neuroscience 136:607–614. https://doi.org/10.1016/j.neuroscience.2005.08.011. PMID: 16344138

Fairén A, Peters A, Saldanha J (1977) A new procedure for examining Golgi impregnated neurons by light and electron microscopy. J Neurocytol 6:311–337. https://doi.org/10.1007/BF01175194. PMID: 71343

Feng G, Mellor RH, Bernstein M, Keller-Peck C, Nguyen QT, Wallace M, Nerbonne JM, Lichtman JW, Sanes JR (2000) Imaging neuronal subsets in transgenic mice expressing multiple spectral variants of GFP. Neuron 28:41–51. https://doi.org/10.1016/s0896-6273(00)00084-2. PMID: 11086982

Fregerslev A, Blackstad TW, Fredens K, Holm MJ (1971) Golgi potassium-dichromate silver-nitrate impregnation. Nature of the precipitate studied by X-ray powder diffraction methods. Histochemie 25:63–71. https://doi.org/10.1007/BF00303947. PMID: 4929087

Gan WB, Bishop DL, Turney SG, Lichtman JW (1999) Vital imaging and ultrastructural analysis of individual axon terminals labeled by iontophoretic application of lipophilic dye. J Neurosci Methods 93:13–20. https://doi.org/10.1016/s0165-0270(99)00096-5. MID: 10598860

Gan WB, Grutzendler J, Wong WT, Wong RO, Lichtman JW (2000) Multicolor "DiOlistic" labeling of the nervous system using lipophilic dye combinations. Neuron 27:219–225. https://doi.org/10.1016/s0896-6273(00)00031-3. PMID 10985343

Glauert AM, Lewis PR (2016) Biological specimen preparation for transmission electron microscopy. Princeton Legacy Library, 350 pp

Godement P, Vanselow J, Thanos S, Bonhoeffer F (1987) A study in developing visual systems with a new method of staining neurones and their processes in fixed tissue. Development 101:697–713. https://doi.org/10.1242/dev.101.4.697. PMID: 2460302

Gray EG (1959a) Axo-somatic and axo-dendritic synapses of the cerebral cortex: an electron microscope study. J Anat 93:420–433. PMID: 13829103 PMCID: PMC1244535

Gray EG (1959b) Electron microscopy of synaptic contacts on dendrite spines of the cerebral cortex. Nature 183(4675):1592–1593. https://doi.org/10.1038/1831592a0. PMID: 13666826

Grutzendler J, Kasthuri N, Gan WB (2002) Long-term dendritic spine stability in the adult cortex. Nature 420(6917):812–816. https://doi.org/10.1038/nature01276. PMID: 12490949

Grutzendler J, Tsai J, Gan WB (2003) Rapid labeling of neuronal populations by ballistic delivery of fluorescent dyes. Methods 30:79–85. PMID 12695105

Harris KM, Stevens JK (1988) Dendritic spines of rat cerebellar Purkinje cells: serial electron microscopy with reference to their biophysical characteristics. J Neurosci 8:4455–4469. https://doi.org/10.1523/JNEUROSCI.08-12-04455.1988

Harris KM, Perry E, Bourne J, Feinberg M, Ostroff L, Hurlburt J (2006) Uniform serial sectioning for transmission electron microscopy. J Neurosci 26:12101–12103. https://doi.org/10.1523/JNEUROSCI.3994-06.2006. PMID: 17122034 PMCID: PMC6675417

Hayworth K, Kasthuri N, Schalek R, Lichtman J (2006) Automating the collection of ultrathin serial sections for large volume TEM reconstructions. Microsc Microanal 12:86–87

Hirano M, Kato S, Kobayashi K, Okada T, Yaginuma H, Kobayashi K (2013) Highly efficient retrograde gene transfer into motor neurons by a lentiviral vector pseudotyped with fusion glycoprotein. PloS One 8(9):e75896. https://doi.org/10.1371/journal.pone.0075896. PMID: 24086660 PMCID: PMC3782444

Honig MG, Hume RI (1986) Fluorescent carbocyanine dyes allow living neurons of identified origin to be studied in long-term cultures. J Cell Biol 103:171–187. https://doi.org/10.1083/jcb.103.1.171. PMID: 2424918

Honig MG, Hume RI. (1989) Dil and diO: versatile fluorescent dyes for neuronal labelling and pathway tracing. Trends Neurosci 12:333–335, 340–1. PMID: 2480673

Horikawa K, Armstrong WE (1988) A versatile means of intracellular labeling: injection of biocytin and its detection with avidin conjugates. J Neurosci Methods 25:1–11. https://doi.org/10.1016/0165-0270(88)90114-8

Hosokawa T, Rusakov DA, Bliss TV, Fine A (1995) Repeated confocal imaging of individual dendritic spines in the living hippocampal slice: evidence for changes in length and orientation associated with chemically induced LTP. J Neurosci 15:5560–5573. https://doi.org/10.1523/JNEUROSCI.15-08-05560.1995. PMID 7643201

Hough LH, Brown ME (2017) Labeling of neuronal morphology using custom diolistic techniques. J Neurosci Methods 282:43–51. https://doi.org/10.1016/j.jneumeth.2017.03.004. PMID: 28274738

Jing D, Zhang S, Luo W, Gao X, Men Y, Ma C, Liu X, Yi Y, Bugde A, Zhou BO, Zhao Z, Yuan Q, Feng JQ, Gao L, Ge W-P, Zhao H (2018) Tissue clearing of both hard and soft tissue organs with the PEGASOS method. Cell Res 28:803–818. https://doi.org/10.1038/s41422-018-0049-z. PMCID: PMC6082844. PMID: 29844583

Kageyama GH, Meyer RL (1987) Dense HRP filling in pre-fixed brain tissue for light and electron microscopy. J Histochem Cytochem 35:1127–1136. https://doi.org/10.1177/35.10.3624853. PMID: 3624853

Kajiwara R, Wouterlood FG, Sah A, Boekel AJ, Baks te Bulte LTG, Witter MP (2008) Convergence of entorhinal and CA3 inputs onto pyramidal neurons and interneurons in hippocampal area CA1. An anatomical study in the rat. Hippocampus 18:266–280. https://doi.org/10.1002/hipo.20385. PMID: 18000818

Kang HW, Kim HK, Moon BH, Lee SJ, Lee SJ, Rhyu IJ (2017) Comprehensive review of Golgi staining methods for nervous tissue. Appl Microsc 47:63–69. https://doi.org/10.9729/AM.2017.47.2.63

Kashiwagi Y, Okabe S (2021) Imaging of spine synapses using super-resolution microscopy. Anat Sci Int 96:343–358. https://doi.org/10.1007/s12565-021-00603-0

Katz LC (1987) Local circuitry of identified projection neurons in cat visual cortex brain slices. J Neurosci 7:1223–1249. https://doi.org/10.1523/JNEUROSCI.07-04-01223.1987. PMID: 3553446 PMCID: PMC6569009

Katz LC, Burkhalter A, Dreyer WJ (1984) Fluorescent latex microspheres as a retrograde neuronal marker for in vivo and in vitro studies of visual cortex. Nature 310(5977):498-500. https://doi.org/10.1038/310498a0. PMID: 6205278

Kawaguchi Y, Wilson CJ, Emson PC (1990) Projection subtypes of rat neostriatal matrix cells revealed by intracellular injection of biocytin. J Neurosci 10:3421–3438. https://doi.org/10.1523/JNEUROSCI.10-10-03421.1990. PMCID: PMC6570194

Kikuchi T, Gonzalez-Soriano J, Kastanauskaite A, Benavides-Piccione R, Merchan-Perez A, DeFelipe J, Blazquez-Llorca L (2020) Volume electron microscopy study of the relationship between synapses and astrocytes in the developing rat somatosensory cortex. Cerebral Cortex 30:3800–3819. https://doi.org/10.1093/cercor/bhz343. Erratum in: Cerebral Cortex 30(2020):3427–3428. https://doi.org/10.1093/cercor/bhaa047

Kim BG, Dai H-N, McAtee M, Vicini S, Bregmana BS (2007) Labeling of dendritic spines with the carbocyanine dye DiI for confocal microscopic imaging in lightly fixed cortical slices. J Neurosci Methods 162:237–243. https://doi.org/10.1016/j.jneumeth.2007.01.016

Klein TM, Wolf ED, Wu R, Sanford JC (1987) High-velocity microprojectiles for delivering nucleic acids into living cells. Nature 327(6117):70–73. https://doi.org/10.1038/327070a0. S2CID 4265777

Knott G, Holtmaat A (2008) Dendritic spine plasticity—Current understanding from in vivo studies. Brain Res Rev 58:282–289. https://doi.org/10.1016/j.brainresrev.2008.01.002. PMID: 18353441

Knott G, Marchmann H, Wall D, Lich B (2008) Serial section scanning electron microscopy of adult brain tissue using focused ion beam milling. J Neurosci 28:2959–2964

Kremer JR, Mastronarde DN, McIntosh JR (1996) Computer visualization of three-dimensional image data using IMOD. J Struct Biol 116:71–76. https://doi.org/10.1006/jsbi.1996.0013. PMID: 8742726

Kristensson K, Olsson Y (1971) Uptake and retrograde axonal transport of peroxidase in hypoglossal neurons. Electron microscopical localization in the neuronal perikaryon. Acta Neuropathol 19(1):1–9. PMID: 4107378

Kristensson K, Lycke E, Sjöstrand J (1971) Spread of herpes simplex virus in peripheral nerves. Acta Neuropathol 17:44–53. https://doi.org/10.1007/BF00684740. PMID: 4100319

Kuypers HGJM, Ugolini G (1990) Viruses as transneuronal tracers. Trends Neurosci 13:71–75. PMID: 1690933

Lagerweij T, Dusoswa SA, Negrean A, Hendrikx EML, de Vries HE, Kole J, Garcia-Vallejo JJ, Mansvelder HD, Vandertop WP, Noske DP, Tannous BA, Musters RJP, van Kooyk Y, Wesseling P, Zhao XW, Wurdinger T (2017) Optical clearing and fluorescence deep-tissue imaging for 3D quantitative analysis of the brain tumor microenvironment. Angiogenesis 20:533–546. https://doi.org/10.1007/s10456-017-9565-6. Epub 2017 Jul 11. PMID: 28699046

Lanciego JL, Wouterlood FG (2011) A half century of experimental neuroanatomical tracing. J Chem Neuroanat 42:157–183. https://doi.org/10.1016/j.jchemneu.2011.07.001

Lanciego JL, Wouterlood FG (2020) Neuroanatomical tract-tracing techniques that did go viral. Brain Struct Funct 225:1193–1224. https://doi.org/10.1007/s00429-020-02041-6

Loewy AD, Bridgman PC, Mettenleiter TC (1991) beta-Galactosidase expressing recombinant pseudorabies virus for light and electron microscopic study of transneuronally labeled CNS neurons. Brain Res 555:346–352. https://doi.org/10.1016/0006-8993(91)90364-2. PMID: 1657302

Luebke JI, Weaver CM, Rocher AB, Rodriguez A, Crimins JL, Dickstein DL, Wearne SL, Hof PR (2010) Dendritic vulnerability in neurodegenerative disease: insights from analyses of cortical pyramidal neurons in transgenic mouse models. Brain Struct Funct 214:181–199. https://doi.org/10.1007/s00429-010-0244-2. PMID: 20177698

Luo P, Haines A, Dessem D (2001) Elucidation of neuronal circuitry: protocol(s) combining intracellular labeling, neuroanatomical tracing and immunocytochemical methodologies. Brain Res Brain Res Protoc 7:222–234. https://doi.org/10.1016/s1385-299x(01)00065-4. PMID: 11431123

Madisen L, Zwingman TA, Sunkin SM, Oh SW, Zariwala HA, Gu H, Ng LL, Palmiter RD, Hawrylycz MJ, Jones AR, Lein ES, Zeng H (2010) A robust and high-throughput Cre reporting and characterization system for the whole mouse brain. Nat Neurosci 13:133–140. https://doi.org/10.1038/nn.2467

Martin X, Dolivo M (1983) Neuronal and transneuronal tracing in the trigeminal system of the rat using the herpes virus suis. Brain Res 273:253–276. https://doi.org/10.1016/0006-8993(83)90850-8. PMID: 6311350

Merchán-Pérez A, Rodriguez J-R, Alonso-Nanclares L, Schertel A, DeFelipe J (2009) Counting synapses using FIB/SEM microscopy: a true revolution for ultrastructural volume reconstruction. Front Neuroanat. https://doi.org/10.3389/neuro.05.018.2009

Minsky M (1957) Microscopy Apparatus. US Patent 3.013.467, filed 7 November 1957, granted 19 December 1961

Modrell MS, Hockman D, Uy B, Buckley D, Sauka-Spengler T, Bronner ME, Baker CVH (2014) A fate-map for cranial sensory ganglia in the sea lamprey. Dev Biol 385:405–416. https://doi.org/10.1016/j.ydbio.2013.10.021. PMID: 24513489

Moolman DL, Vitolo OV, Vonsattel J-PG, Shelanski ML (2004) Dendrite and dendritic spine alterations in Alzheimer models. J Neurocytol 33:377–387. https://doi.org/10.1023/B:NEUR.0000044197.83514.64

Murakami TC, Mano T, Saikawa S, Horiguchi SA, Shigeta D, Baba K, Sekiya H, Shimizu Y, Tanaka KF, Kiyonari H, Iino M, Mochizuki H, Tainaka K, Ueda HR (2018) A three-dimensional single-cell-resolution whole-brain atlas using CUBIC-X expansion microscopy and tissue clearing. Nat Neurosci 21:625–637. https://doi.org/10.1038/s41593-018-0109-1. PMID: 29507408

Nimchinsky E, Sabatini B, Svoboda K (2002) Structure and function of dendritic spines. Annu Rev Physiol 64:313–353. https://doi.org/10.1146/annurev.physiol.64.081501.160008. PMID 11826272

O'Brien JA, Lummis SCR (2007) Diolistics: incorporating fluorescent dyes into biological samples using a gene gun. Trends Biotechnol 25:530–534. https://doi.org/10.1016/j.tibtech.2007.07.014. PMID: 17945370

Olive MF, Del Frabco AP, Gipson CD (2018) Diolistic labeling and analysis of dendritic spines. Methods Mol Biol 1727:179–200. https://doi.org/10.1007/978-1-4939-7571-6_14. PMID: 29222782 PMCID: PMC6112159

Parajuli LK, Koike M (2021) Three-dimensional structure of dendritic spines revealed by volume electron microscopy techniques. Front Neuroanat 15:627368. https://doi.org/10.3389/fnana.2021.627368

Pawley JB (ed) (2006) Handbook of biological confocal microscopy, 3rd edn. Springer, Berlin. ISBN 0-387-25921-X

Peddie CJ, Collinson LM (2014) Exploring the third dimension: volume electron microscopy comes of age. Micron 61:9–19. https://doi.org/10.1016/j.micron.2014.01.009. PMID: 24792442

Peter K (1906) Die Methoden der Rekonstruktion. Gustav Fischer Verlag, Jena, 150pp. Reprinted 2010 by Kessinger Publishing, ISBN-13?:? 978-1161237375

Peters A, Kaiserman Abramof IR (1970) The small pyramidal neuron of the rat cerebral cortex. The perikaryon, dendrites and spines. Am J Anat 127:321–355. https://doi.org/10.1002/aja.1001270402. PMID: 4905050

Peters A, Palay SL, Webster HdF (1991) The fine structure of the nervous system: neurons and their supporting cells, 3rd edn. Oxford University Press, 493 pp. ISBN 0-19-506571-9

Pinault D (1996) A novel single-cell staining procedure performed in vivo under electrophysiological control: morpho-functional features of juxtacellularly labeled thalamic cells and other central neurons with biocytin or Neurobiotin. J Neurosci Methods 65:113–136. https://doi.org/10.1016/0165-0270(95)00144-1

Porrero C, Rubio-Garrido P, Avendaño C, Clascá F (2010) Mapping of fluorescent protein-expressing neurons and axon pathways in adult and developing Thy1-eYFP-H transgenic mice. Brain Res 1345:59–72. https://doi.org/10.1016/j.brainres.2010.05.061

Ramón y Cajal S (1888) Estructura de los centros nerviosos de las aves. Rev Trim Histol Norm Pat 1:1–10

Rasia-Filho AA, Brusco J, Rocha LB, Moreire J (2010) Dendritic spines observed by extracellular DiI dye and immunolabeling under confocal microscopy. Nat Protoc/Protoc Exch. http://www.nature.com/protocolexchange/protocols/1890

Raux H, Iseni F, Lafay F, Blondel D (1997) Mapping of monoclonal antibody epitopes of the rabies virus P protein. J Gen Virol 78:119–124. https://doi.org/10.1099/0022-1317-78-1-119. PMID: 9010294

Rayleigh L (1886) On the theory of optical images, with special reference to the microscope. London, Edinburgh, and Dublin Philosophical Magazine and J Sci 42, Part XV:167–195. https://doi.org/10.1080/14786449608620902

Rayleigh L, Strutt JW (1891) On pin-hole photography. Philos Mag 11:87–99

Rho JH, Sidman RL (1986) Intracellular injection of lucifer yellow into lightly fixed cerebellar neurons. Neurosci Lett 72:21–24. https://doi.org/10.1016/0304-3940(86)90611-7. PMID: 2433645

Rosene DL, Roy NJ, Davis BJ (1986) A cryoprotection method that facilitates cutting frozen sections of whole monkey brains for histological and histochemical processing without freezing artifact. J Histochem Cytochem 34:1301–1315. https://doi.org/10.1177/34.10.3745909. PMID: 3745909

Rosoklija GB, Petrushevski VM, Stankov A, Dika A, Jakovski Z, Pavlovski G, Davcheva N, Lipkin R, Schnieder T, Scobie K, Duma A, Dwork AJ (2014) Reliable and durable Golgi staining of brain tissue from human autopsies and experimental animals. J Neurosci Methods 230:20–29. https://doi.org/10.1016/j.jneumeth.2014.04.006. PMID: 24747874 PMCID: PMC4080906

Salin P, Castle M, Kachidian P, Barroso-Chinea P, López IP, Rico AJ, Kerkerian-Le Goff L, Coulon P, Lanciego JL (2008) High-resolution neuroanatomical tract-tracing for the analysis of striatal microcircuits. Brain Res 1221:49–58. https://doi.org/10.1016/j.brainres.2008.05.011. PMID: 18561898

Salin P, Blondel D, Kerkerian-Le Goff L, Coulon P (2020) Golgi staining-like retrograde labeling of brain circuits using rabies virus: focus onto the striatonigral neurons. J Neurosci Methods 344:108872. https://doi.org/10.1016/j.jneumeth.2020.108872. PMID: 32693000. PMID: 32693000

Sánchez-González A, Thougaard E, Tapias-Espinosa C, Cañete T, Sampedro-Viana D, Saunders JM, Toneatti R, Tobeña A, Gónzalez-Maeso J, Aznar S, Fernández-Teruel A (2021) Increased thin-spine density in frontal cortex pyramidal neurons in a genetic rat model of schizophrenia-relevant features. Eur Neuropsychopharmacol 44:79–91. https://doi.org/10.1016/j.euroneuro.2021.01.006. PMID: 33485732. PMCID: PMC7902438

Sandell JH, Masland RH (1989) Photoconversion of some fluorescent markers to a diaminobenzidine product. J Histochem Cytochem 36:535–559. https://doi.org/10.1177/36.5.3356898. PMID: 3356898

Sauer B (1998) Inducible gene targeting in mice using the Cre/lox system. Methods 14:381–392. https://doi.org/10.1006/meth.1998.0593. PMID: 9608509 Review

Schindelin J, Arganda-Carreras I, Frise E, Kaynig V, Longair M, Pietzsch T, Preibisch S, Rueden C, Saalfeld S, Schmid B, Tinevez J-Y, White DJ, Hartenstein V, Eliceiri K, Tomancak P, Cardona A (2012) Fiji: an open-source platform for biological-image analysis. Nat Methods 9:676–682. https://doi.org/10.1038/nmeth.2019

Segal M (2010) Dendritic spines, synaptic plasticity and neuronal survival: activity shapes dendritic spines to enhance neuronal viability. Eur J Neurosci 31:2178–2184. https://doi.org/10.1111/j.1460-9568.2010.07270.x

Sheikh IS, Keefe KM, Sterling NA, Junker IP, Eneanya CI, Liu Y, Tang X-Q, Smith GM (2018) Retrogradely transportable lentivirus tracers for mapping spinal cord locomotor circuits. Front Neural Circuits 12:60. https://doi.org/10.3389/fncir.2018.00060. PMID: 30090059 PMCID: PMC6068242

Sivukhina EV, Jirikowski GF (2021) Oxytocin, but not arginine-vasopressin neurons project from the hypothalamus to amygdala in human: DiI-based tracing study in postmortem brain. J Chem Neuroanat 111:101882. https://doi.org/10.1016/j.jchemneu.2020.101882. Epub 2020 Nov 4. PMID: 33157259

Somogyi P (1978) The study of Golgi stained cells and of experimental degeneration under the electron microscope: a direct method for the identification in the visual cortex of three successive links in a neuron chain. Neuroscience 3:167–180. https://doi.org/10.1016/0306-4522(78)90099-4. PMID: 83552

Stewart WW (1978) Functional connections between cells as revealed by dye-coupling with a highly fluorescent naphthalimide tracer. Cell 14:741–759. https://doi.org/10.1016/0092-8674(78)90256-8. PMID: 688392

Stewart WW (1981) Lucifer dyes – highly fluorescent dyes for biological tracing. Nature 292:17–21. https://doi.org/10.1038/292017a0. PMID: 6168915

Stretton AOW, Kravitz EA (1968) Neuronal geometry: determination with a technique of intracellular dye injection. Science 162:132–134. https://doi.org/10.1126/science.162.3849.132. PMID: 4175300

Taghert PH, Bastiani MJ, Ho RK, Goodman CS (1982) Guidance of pioneer growth cones: filopodial contacts and coupling revealed with an antibody to Lucifer Yellow. Dev Biol 94:391–399. https://doi.org/10.1016/0012-1606(82)90356-6. PMID: 715211

Tauchi M, Masland RH (1984) The shape and arrangement of the cholinergic neurons in the rabbit retina. Proc R Soc Lond B Biol Sci 223(1230):101–119. https://doi.org/10.1098/rspb.1984.0085. PMID: 6151180

Tauchi M, Masland RH (1985) Local order among the dendrites of an amacrine cell population. J Neurosci 5(9):2494–2501. https://doi.org/10.1523/JNEUROSCI.05-09-02494.1985. PMID: 4032008

Terasaki M, Slater NT, Fein A, Schmidek A, Reese TS (1994) Continuous network of endoplasmic reticulum in cerebellar Purkinje neurons. Proc Natl Acad Sci U S A 91:7510–7514. https://doi.org/10.1073/pnas.91.16.7510

Tønnesen J, Nägerl UV (2016) Dendritic spines as tunable regulators of synaptic signals. Front Psychiatry 7:101. https://doi.org/10.3389/fpsyt.2016.00101

Ueda HR, Dodt HU, Osten P, Economo MN, Chandrashekar J, Keller PJ (2020) Whole-brain profiling of cells and circuits in mammals by tissue clearing and light-sheet microscopy. Neuron 106:369–387. https://doi.org/10.1016/j.neuron.2020.03.004. PMID: 32380050

Ugolini G (2010) Advances in viral transneuronal tracing. J Neurosci Methods 194:2–20. https://doi.org/10.1016/j.jneumeth.2009.12.001

Ugolini G, Kuypers HG, Simmons A (1987) Retrograde transneuronal transfer of herpes simplex virus type 1 (HSV 1) from motoneurones. Brain Res 422:242–256. https://doi.org/10.1016/0006-8993(87)90931-0. PMID: 2445438

Ugolini G, Kuypers HG, Strick PL (1989) Transneuronal transfer of herpes virus from peripheral nerves to cortex and brainstem. Science 243(4887):89–91. https://doi.org/10.1126/science.2536188. PMID: 2536188

Valverde F (1998) Golgi atlas of the postnatal mouse brain. Springer, pp 1–146. ISBN: 978-3-7091-6501-0

Van Hest JJHA, Agronskaia AV, Fokkema J, Montanarella F, Gregorio Puig A, De Mello Donega C, Meijerink A, Blab GA, Gerritsen HC (2019) Towards robust and versatile single nanoparticle fiducial markers for correlative light and electron microscopy. J Microsc 274:13–22. https://doi.org/10.1111/jmi.12778

Vicidomini G, Bianchini P, Diaspro A (2018) STED super-resolved microscopy. Nat Methods 15:173–182. https://doi.org/10.1038/nmeth.4593

Vints K, Vandael D, Baatsen P, Pavie B, Vernaillen F, Corthout N, Rybakin V, Munck S, Gounko NV (2019) Modernization of Golgi staining techniques for high-resolution, 3-dimensional imaging of individual neurons. Sci Rep 9:130. https://doi.org/10.1038/s41598-018-37377-x

Wanner AA, Kirschmann MA, Genoud C (2015) Challenges of microtome-based serial blockface scanning electron microscopy in neuroscience. J Microsc 259:137–142. https://doi.org/10.1111/jmi.12244

Watakabe A, Kato S, Kobayashi K, Takaji M, Nakagami Y, Sadakane O, Ohtsuka M, Hioki H, Kaneko T, Okuno H, Kawashima T, Bito H, Kitamura Y, Yamamori T (2012) Visualization of cortical projection neurons with retrograde TET-off lentiviral vector. PloS One 7:e46157. https://doi.org/10.1371/journal.pone.0046157

Wernitznig S, Sele M, Urschler M, Zankel A, Pölt P, Rind FC, Leitinger G (2015) Optimizing the 3D-reconstruction technique for serial block-face scanning electron microscopy. J Neurosci Methods 264:16–24. https://doi.org/10.1016/j.jneumeth.2016.02.019. PMID: 26928258

Westrum L (1970) Observations on initial segments of axons in the prepyriform cortex of the rat. J Comp Neurol 139:337–356. https://doi.org/10.1002/cne.901390306. PMID: 4914800

Wilson CJ, Groves PM, Kitai ST, Linder JC (1983) Three-dimensional structure of dendritic spines in the rat neostriatum. J Neurosci 3:383–398. https://doi.org/10.1523/JNEUROSCI.03-02-00383.1983. PMID: 6822869 PMCID: PMC6564480

Woods G, Zito K (2008) Preparation of gene gun bullets and biolistic transfection of neurons in slice culture. J Vis Exp 12:675. https://doi.org/10.3791/675. PMID: 19066564

Wouterlood FG (1986) Stabilization of silver chromate Golgi impregnation in rat central nervous system neurons using photographic developers. I. Light microscopy. Stain Technol 61:337–352. https://doi.org/10.3109/10520298609113582. PMID: 3798487

Wouterlood FG (1992) Techniques for converting Golgi precipitate in CNS neurons into stable electron microscopic markers. Microsc Res Tech 23:275–288. https://doi.org/10.1002/jemt.1070230404

Wouterlood FG, Mugnaini E (1984) Cartwheel neurons of the dorsal cochlear nucleus. A Golgi-electron microscopic study in the rat. J Comp Neurol 227:136–157. https://doi.org/10.1002/cne.902270114

Wouterlood FG, Engel A, Daal M, Houwen G, Meinderts A, Jordà Siquier T, Beliën JAM, van Dongen YC, Scheel-Krüger J, Thierry A-M, Groenewegen HJ, Deniau J-M (2018) Mesencephalic dopamine neurons interfacing the shell of nucleus accumbens and the dorsolateral striatum in the rat. J Neurosci Res 96:1518–1542. https://doi.org/10.1002/jnr.24242

Yuste R (2015) The discovery of dendritic spines by Cajal. Front Neuroanat. https://doi.org/10.3389/fnana.2015.00018

Chapter 3
Electrophysiology of Dendritic Spines: Information Processing, Dynamic Compartmentalization, and Synaptic Plasticity

Joseane Righes Marafiga and Maria Elisa Calcagnotto

Abstract For many years, synaptic transmission was considered as information transfer between presynaptic neuron and postsynaptic cell. At the synaptic level, it was thought that dendritic arbors were only receiving and integrating all information flow sent along to the soma, while axons were primarily responsible for point-to-point information transfer. However, it is important to highlight that dendritic spines play a crucial role as postsynaptic components in central nervous system (CNS) synapses, not only integrating and filtering signals to the soma but also facilitating diverse connections with axons from many different sources. The majority of excitatory connections from presynaptic axonal terminals occurs on postsynaptic spines, although a subset of GABAergic synapses also targets spine heads. Several studies have shown the vast heterogeneous morphological, biochemical, and functional features of dendritic spines related to synaptic processing. In this chapter (adding to the relevant data on the biophysics of spines described in Chap. 1 of this book), we address the up-to-date functional dendritic characteristics assessed through electrophysiological approaches, including backpropagating action potentials (bAPs) and synaptic potentials mediated in dendritic and

J. Righes Marafiga
Department of Neurological Surgery, University of California San Francisco, San Francisco, CA, USA

M. E. Calcagnotto (✉)
Department of Biochemistry, Universidade Federal do Rio Grande do Sul, Porto Alegre, RS, Brazil

Graduate Program in Neuroscience, Universidade Federal do Rio Grande do Sul, Porto Alegre, RS, Brazil

Graduate Program in Biochemistry, Universidade Federal do Rio Grande do Sul, Porto Alegre, RS, Brazil

Graduate Program in Psychiatry and Behavioral Science, Universidade Federal do Rio Grande do Sul, Porto Alegre, RS, Brazil
e-mail: elisa.calcagnotto@ufrgs.br

A. A. Rasia-Filho et al. (eds.), *Dendritic Spines*, Advances in Neurobiology 34,
https://doi.org/10.1007/978-3-031-36159-3_3

spine compartmentalization, as well as describing the temporal and spatial dynamics of glutamate receptors in the spines related to synaptic plasticity.

Keywords Spine electrophysiology · Compartmentalization · Synaptic plasticity

3.1 Dendritic Spine Electrophysiology

Dendritic spines, the subcellular formation characterized by tiny membrane protrusions (≲1 µm diameter) of dendrites, constitute the fundamental units of synaptic and information processing in the mammalian brain (Harris et al. 1992; Popovic et al. 2015b; Kwon et al. 2017). Most excitatory inputs on cortical pyramidal neurons are made on spines, rather than on dendritic shafts (Yuste 2010; Araya et al. 2014), but dendritic spines also modulate inhibitory synaptic inputs showing particularly high levels of plasticity (Chen et al. 2012; van Versendaal et al. 2012). Spines can individually detect temporal coincidence of synaptic activity, acting as basic functional units of neuronal integration (Yuste and Denk 1995; Sabatini et al. 2002; Noguchi et al. 2005; Yuste 2013). Additionally, spine shape and spine head size are correlated with changes in synaptic strength during synaptic plasticity (Yuste and Bonhoeffer 2001). While the complex and dynamic regional electrical properties of dendritic spines play an important role in determining the function and integration of neural circuits, the electrical behavior of spines is still not well understood and sometimes remains controversial.

For a long time, the limitations of available techniques prevented measuring voltage signals directly from the membrane of individual spines. Thus, the electrical properties of dendritic spines, including active conductances, were based on theoretical models (Harris and Stevens 1989; Koch and Zador 1993; Harris and Kater 1994; Johnston et al. 1996; Yuste and Tank 1996; Svoboda et al. 1996; Shepherd 1996; Yuste 2010) and based on channel densities and kinetics from recordings of dendritic patches. Whole-cell or cell-attached recordings of spine heads per se are still very difficult to perform in brain slices. However, advanced imaging techniques, including high voltage-sensitive dyes, low-affinity Ca^{2+} and Na^{+} indicators, two-photon Ca^{2+} imaging, and two-photon uncaging of glutamate, made it possible to detect membrane voltage signals from individual dendritic spines in brain slices (Holthoff et al. 2010; Popovic et al. 2015a, b; Miyazaki and Ross 2022). In addition, the intracellular voltage recordings from dendritic spines have been obtained by quantum-dot-coated nanopipettes in cells in culture and brain slices (Jayant et al. 2017), and patch-clamp recordings have been also performed in dissociated neurons and isolated spines (Priel et al. 2022). That is, the combination of advanced techniques in imaging and morphology with electrophysiology and spine models are providing valuable information about the spatiotemporal dynamics of biophysical properties of dendritic spines (Stuart et al. 1993; Stuart and Sakmann 1994; Denk et al. 1996; Mainen et al. 1999; Korngreen and Sakmann 2000; Oertner 2002; Nuriya et al. 2006; Araya et al. 2006).

3.1.1 Backpropagating Action Potentials and Synaptic Potentials in Dendritic Spine

Somatic action potentials (APs), after being generated in the axon initial segment, travel back through the dendritic tree as backpropagating action potentials (bAPs) (Häusser et al. 2000). Dendritic spines can regulate the synapses without attenuation of dendritic bAP (Popovic et al. 2015a; Cornejo et al. 2022), which can influence the processing of incoming synaptic inputs in the dendrites of most neurons (Stuart et al. 1997a, b; Xiong and Chen 2002). The coincidence and temporal sequence of bAP and excitatory postsynaptic potential (EPSP) within a dendritic spine determine the amount of Ca^{2+} entry through voltage-gated channels and/or ionotropic receptors, essential for synaptic plasticity (Yuste and Denk 1995; Schiller et al. 1998; Koester and Sakmann 1998; Yuste et al. 1999; Nevian and Sakmann 2004; Camiré and Topolnik 2014). The rapid time course of the bAP in spines may be a critical factor for the accurate regulation of spike timing-dependent synaptic plasticity within a very short time window, while the magnitude and time course of synaptically activated EPSP in dendritic spines determine how these synapses contribute to synaptic integration (Caporale and Dan 2008; Popovic et al. 2015a). Both synaptic integration and plasticity depend on spatiotemporal changes in the membrane voltage on the dendritic spine, and the ability to measure such modifications is improving.

Spine ionic conductance especially from voltage-gated sodium channels could play an important role in dynamically regulating bAP. Voltage-gated sodium channels have been found in spines (Caldwell et al. 2000; Araya et al. 2007; Bywalez et al. 2015; Miyazaki and Ross 2017). Calcium channel subunits (Caldwell et al. 2000; Bloodgood et al. 2009) and potassium channels (Kim et al. 2007; Kaufmann et al. 2013; Wang et al. 2014; Strobel et al. 2017; Tazerart et al. 2022) have also been localized in spines. These channels, together with functional N-methyl-*D*-aspartate (NMDA) and α-amino-3-hydroxy-5-methyl-4-isoxazolepropionic acid (AMPA) glutamate receptors (Yuste and Denk 1995; Schiller et al. 1998; Yuste et al. 1999; Kovalchuk et al. 2000; Makino and Malinow 2009; Park 2018), control synaptic potential. These relevant advances in detecting voltage changes in spine membrane are further discussed below.

Voltage-sensitive dye recordings based on wide-field laser-excitation fluorescence microscopy permit monitoring of subthreshold electrical signal integration between individual dendritic spines and adjacent dendritic shafts in acute brain slice preparations (Popovic et al. 2015b; Weng et al. 2023). It is, then, possible to measure membrane potential (V_m) transients from individual dendritic spines by using intracellular high-sensitivity voltage-sensing dyes and high-resolution wide-field epi-fluorescence microscopy applied at optical magnification 10 times higher than that used to record membrane voltage from other neuronal parts (Cohen and Salzberg 1978; Weng et al. 2023). These measurements have a sufficient spatial and temporal resolution (approximately 1 μm and submillisecond range) for recording neuronal AP and synaptic potential signals (Popovic et al. 2015a). With this approach, the AP at the soma can be recorded by patch-clamp technique, while voltage-imaging signals of bAP at dendritic spines can be recorded at a frame rate of 2 kHz at high optical magnification,

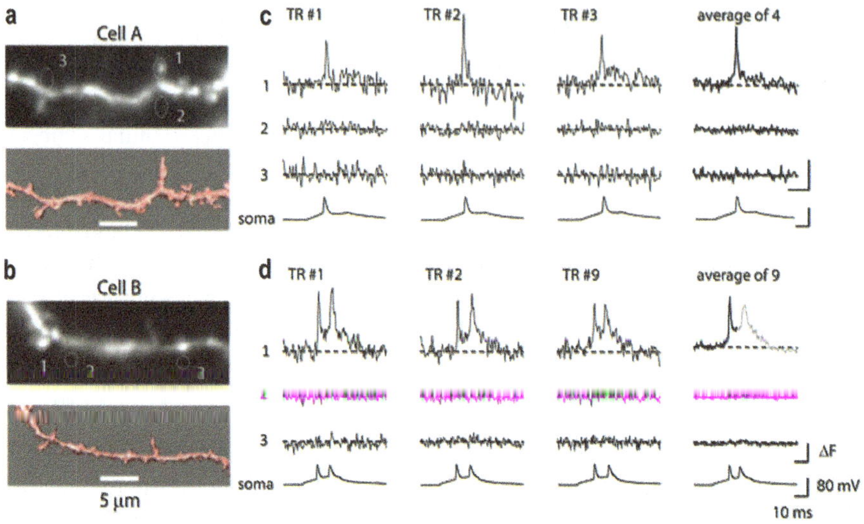

Fig. 3.1 Optical recordings of backpropagating action potentials (bAPs) signals from individual dendritic spines of two neurons. (**a**, **b**) Upper micrographs show fluorescence images of dendritic branches with spines obtained with a CCD camera for voltage imaging. Lower micrographs show spiny dendrites reconstructed after deconvoluting stacks of spinning-disk confocal images. (**c**, **d**) Fluorescence intensity traces from locations 1 to 3 outlined in CCD images. Single-trial recordings and temporal averages of 4 (cell A) and 9 trials (cell B) are shown. Bottom traces are electrode recordings from the soma. The bAP signals, clearly recorded in spines, are absent from regions without spines (locations 2 and 3). (Reproduced from Holthoff et al. (2010), Figure 4, https://doi.org/10.1113/jphysiol.2009.184960, under CCC RightsLink® license 5516071490678)

with a satisfactory signal-to-noise ratio (S/N) and insignificant interference of the scattered light or the photodynamic damage (Holthoff et al. 2010) (Fig. 3.1). It is also possible to detect the bAP invading spines with a rapid time course critical for regulating plasticity and very similar to the spikes recorded in the parent dendrite.

There is an important functional implication for these data. The magnitude of the synaptic depolarization (EPSP) at the dendritic spine, usually difficult to determine, could provide crucial information about the influence of synaptic inputs on local voltage-dependent processes and cellular responses (Acker et al. 2016). The EPSP signals present an amplitude fivefold to tenfold smaller than an AP. Therefore, the optical resolution used for monitoring the dendritic spines' bAP is insufficient for recording unitary subthreshold synaptic responses at the spatial scale of individual spines, considering an adequate sensitivity and spatiotemporal resolution with a suitable S/N to allow a precise quantitative analysis.

The combination of high-sensitivity voltage-sensing dyes with the 2-photon uncaging of glutamate could be a suitable strategy for recording dendritic spines (Popovic et al. 2015b; Weng et al. 2023). The increment of the sensitivity can be achieved by increasing excitation intensity from a laser at a required wavelength and reducing the photodynamic damage (Popovic et al. 2015b; Weng et al. 2023). Then, a subthreshold uncaging EPSP signal (uEPSP) can be evoked by stimulating one synapse on a dendritic spine using 2-photon uncaging of glutamate. For example, this procedure could be done by simultaneously recording optical signals along

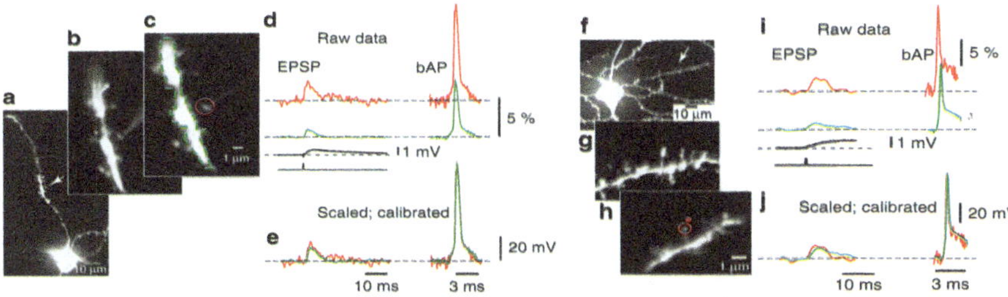

Fig. 3.2 Recording subthreshold electrical events from individual dendritic spines. (**a**) Low-magnification fluorescence image of a basal dendrite labeled with a voltage-sensitive dye; z-stack of confocal images. Arrow: recorded spine. (**b**) High magnification confocal image of the same spine. Tip of iontophoretic electrode (labeled with the fluorescent dye) in the immediate vicinity of the spine head. (**c**) Single frame image of a spine in recording position obtained with a CCD camera for voltage imaging. (**d**) Traces shown on the left are evoked subthreshold excitatory postsynaptic potential (EPSP) recordings from the spine head (red) and its parent dendrite (green). Average of 16 trials. Bottom black traces are somatic electrode recording and the uncaging command pulse. Traces on the right are backpropagating action potentials (bAPs) signals from the same locations. Average of nine trials. (**e**) Left traces showing superimposed evoked subthreshold EPSP signals from the spine head and its parent dendrite calibrated in terms of membrane potential. Right traces represent bAP signals corrected for recording sensitivity difference. (**f–j**) Two-photon uncaging of glutamate. Same information as shown in (**a–e**). Red dot in **h**: position and approximate size of uncaging light spot. The evoked subthreshold EPSP and bAP recordings are average of 8 and 4 trials, respectively. (Adapted from Popovic et al. (2015b), Figure 5a–j, https://doi.org/10.1038/ncomms9436, under CC BY license, originally published by Springer Nature)

a small segment of a basal dendrite from a cortical layer V pyramidal neuron at a frame rate of 2 kHz (Popovic et al. 2015a, b) (Fig. 3.2) and from cortical layers II/III, V, and VI pyramidal neurons at a frame rate of 5 kHz (Weng et al. 2023). The optical recordings of local uEPSP signals evoked by two-photon glutamate uncaging were followed by a bAP signal evoked by depolarizing pulse delivered by the somatic patch electrode (Popovic et al. 2015a, b; Weng et al. 2023).

Also, the combined voltage imaging and glutamate uncaging using patterned illumination based on computer-generated holography allowed one-photon uncaging of glutamate on multiple spines to be carried out in parallel with voltage imaging from the parent dendrite and neighboring spines. In this procedure, somatosensory cortex in brain slices was used to perform electrical and optical recordings from cortical layer V pyramidal neurons filled with a voltage-sensitive dye. Subthreshold uEPSP and bAP signals were recorded at a frame rate of 2 or 5 kHz. Optical signals of a subthreshold uEPSP onto one spine and bAP evoked by depolarizing current pulse were able to be recorded from spine head and parent dendrite and also in two spines simultaneously, while somatic patch-clamp recordings of the subthreshold responses related to glutamate uncaging onto one and two spines were obtained (Tanese et al. 2017).

These approaches improved the accuracy required for spatiotemporal resolution to simultaneously record and reconstruct individual uEPSP and AP signals from the spine head and the parent dendrite at near physiological temperature (Popovic et al. 2015a, b; Tanese et al. 2017; Weng et al. 2023).

Recently, direct electrical recording from dendritic spines in culture and brain slices was achieved by using quantum-dot-coated nanopipettes (Jayant et al. 2017).

The simultaneous electrical recordings on the soma and the spine of pyramidal cells from either primary hippocampal neuronal culture or neocortical slice preparations (Fig. 3.3) revealed that bAP invaded spines and was similar to a somatic AP. Also, the recorded spine bAP preceded the initial rise of EPSP measured in the soma (Fig. 3.4). Spontaneous EPSPs, with large amplitude (25–30 mV), fast rise time (1–2 ms), and half-width of approximately 5 ms were recorded as the nanopipette entered the spine head (Fig. 3.5) (Jayant et al. 2017). This possibility of direct measurement of the V_m at the spine level opens a way to explore the dynamics of local ionic conductances during synaptic integration and plasticity.

Furthermore, the changes in V_m of spines during EPSP are associated with variations of intracellular concentration of Na^+ and Ca^{2+} caused by changes in channels and receptor conductances. While the influx of Na^+ is important for driving the synaptic potential and bAP, the Ca^{2+} influx is essential for synaptic plasticity processing. Compared to Na^+ channels, Ca^{2+} channels are well described in terms of type, distribution, and function in dendritic spines (Bloodgood and Sabatini 2005).

Imaging studies of AP and synaptically evoked Ca^{2+} transients have revealed multiple classes of voltage-sensitive Ca^{2+} channels that contribute to Ca^{2+} influx in dendritic spines (Bloodgood and Sabatini 2007a). Ca^{2+} ion influx has been studied in some detail in a series of experiments, using two-photon Ca^{2+} imaging and two-photon uncaging of glutamate. For example, through a combination of conventional electrophysiological techniques with two-photon excitation laser scanning microscopy, Ca^{2+} imaging, and optogenetic, it was possible to investigate local Ca^{2+} transients in dendrites and spines (Nakamura et al. 1999; Camiré and Topolnik 2018). The bAPs in dendrites opened voltage-gated Ca^{2+} channels (Jaffe et al. 1992; Popovic et al. 2012) with consequent Ca^{2+} transients in dendrites and spines (Jaffe et al. 1992; Popovic et al. 2012). Moreover, using combined electrophysiology, two-photon Ca^{2+} imaging, and two-photon glutamate uncaging in layer II pyramidal cells from the rat medial entorhinal cortex in acute brain slices, it was possible to detect dendritic spine bAP-evoked Ca^{2+} transients by R- and T-type voltage-gated Ca^{2+} channel activation (Theis et al. 2018). These bAP-evoked Ca^{2+} transients did not affect the EPSP measured at the soma but induced a downscaling of NMDA receptor currents (Theis et al. 2018). Na^+ conductance on spines of mice neocortical pyramidal neurons has also been detected in experiments using two-photon uncaged glutamate associated with application of tetrodotoxin (TTX), a voltage-gated Na^+ channel blocker that significantly decreased the spine potentials (Araya et al. 2007). Interestingly, the Na^+ K^+-ATPase α3 isoform was found to be expressed in spines with excitatory synapses (Blom et al. 2011), which suggests the active maintenance of ionic gradients by the dendritic spine at the postsynaptic cell.

Recently, the time courses of synaptically induced changes in Ca^{2+} and Na^+ concentrations in the dendritic spines of hippocampal CA1 pyramidal cells were studied in brain slices of mice. Na^+ and Ca^{2+} transients in a spine were simultaneously measured in response to a minimal electrical stimulation, in order to generate normal neurotransmitter release (smaller than uncaging glutamate), combined with low-affinity linear Na^+ and Ca^{2+} indicators and single photon laser fluorescence stimulation with a high-speed, sensitive CCD camera (Miyazaki and Ross 2022). The intracellular Na^+ concentration in the spine increased following a single synaptic

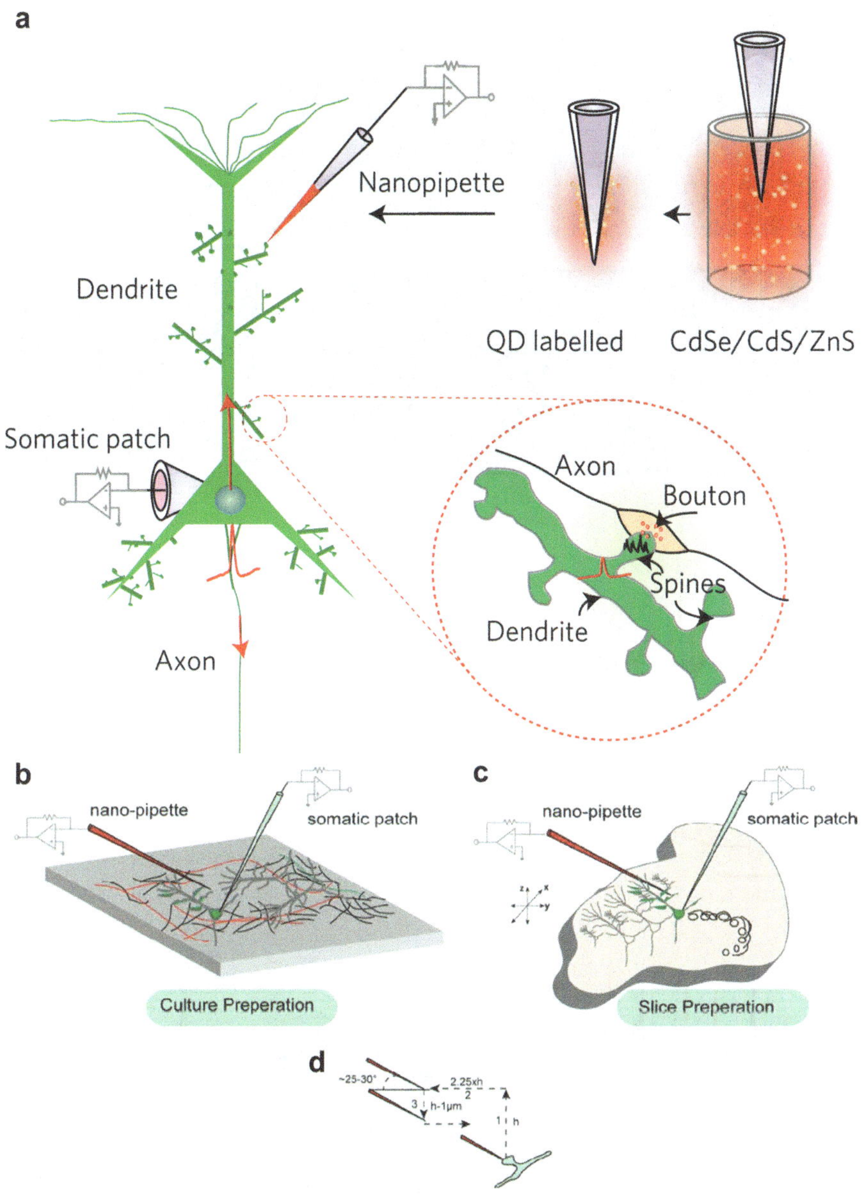

Fig. 3.3 Nanopipette dual somato-spine recordings. (**a**) Schematic describing the dual patch-nanopipette recording and associated signal transmission mechanisms to and from the spine. Action potentials (APs) originate in the axon hillock and propagate both forward (toward the postsynaptic cell) and backward into dendrites and spines. Backpropagating action potentials (bAPs) invade spines (inset) across the spine neck. Neurotransmitters released from the presynaptic axonal terminals activate receptors located on the spine head and cause spine head EPSPs (inset). Nanopipettes are labeled with QDs (CdSe/CdS/ZnS quantum dots) through adsorption (top right). (**b**) Schematic representation of dual somato-spine recordings in cultures and (**c**) slices. (**d**) Nanopipette "navigation" approach in slices toward a chosen target. (Adapted from Jayant et al. (2017), Figure 1a and Supplementary Figure 2 https://doi.org/10.1038/nnano.2016.268, under CCC RightsLink® license 5516540783944)

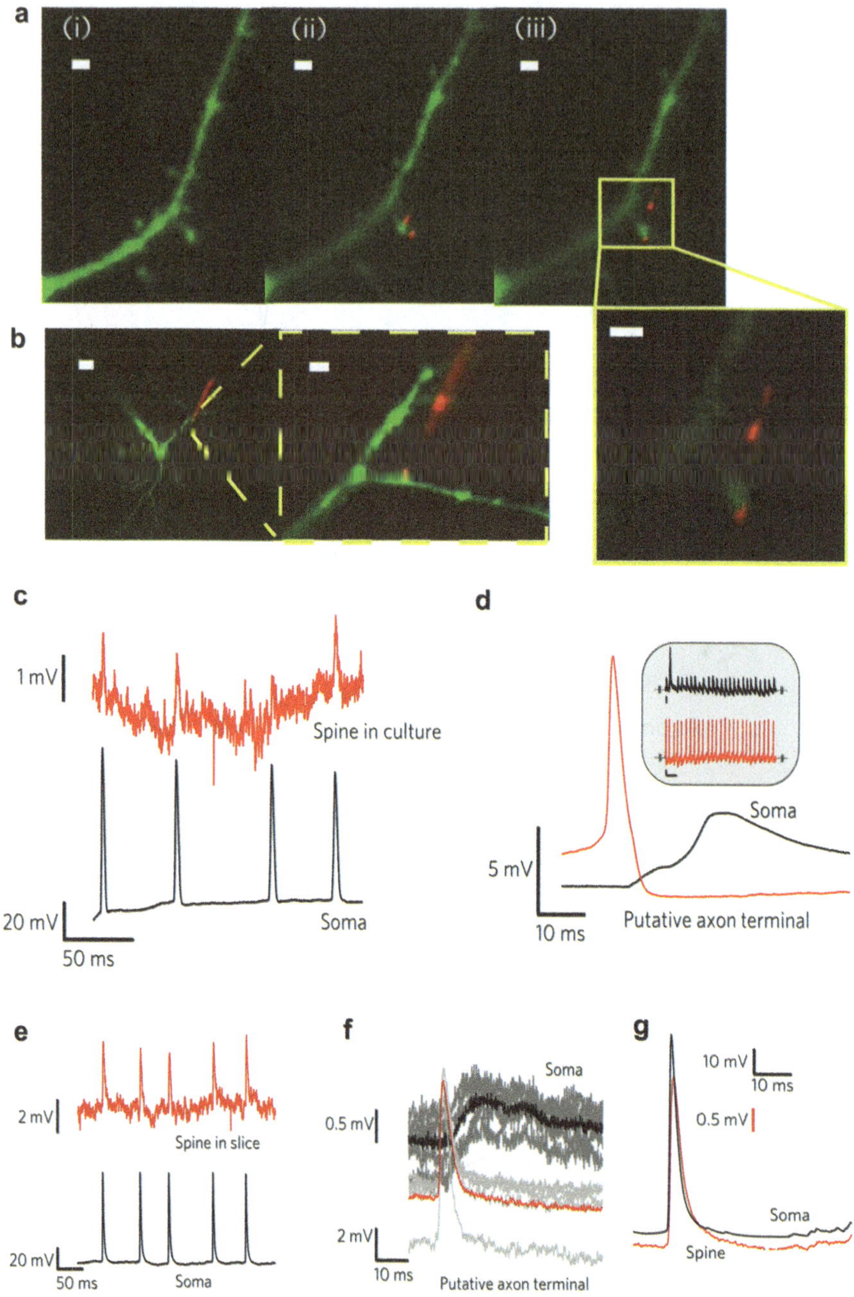

Fig. 3.4 Nanopipette dual somato-spine recordings. (**a**, **b**) Dual somato-spine recordings in cultures (**a**, scale bar, 1.5 μm, panels (i)–(iii) and callout) and slices (**b**, scale bar, 20 μm left, 3 μm, inset middle). Dendritic and spine nanopipette recordings were restricted to within 100 μm from the soma to avoid space clamping and backpropagating action potential (bAP) attenuation. (**c**, **f**) Spine and presynaptic axon terminal recordings in culture (**c**, **d**) and slices (**e**, **f**). (**c**) bAPs invasion into spines. Somatic AP (bottom) and bAPs in the spine head (top) reveal that the nanopipette is capable of registering bAPs without the need for averaging. (**d**) Nanopipette recordings in the presynaptic axon terminal and postsynaptic soma. Note the characteristic 2–4 ms time delay

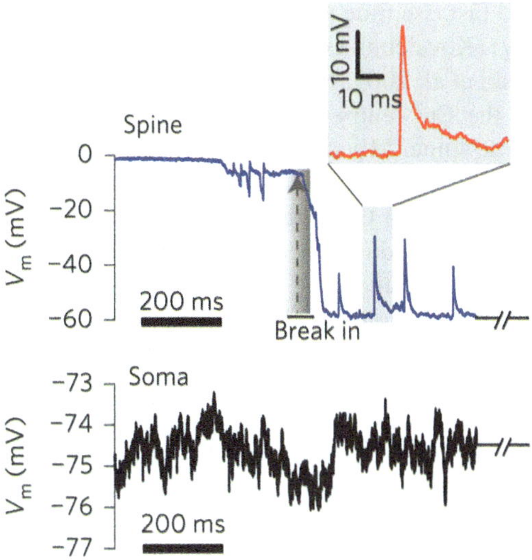

Fig. 3.5 Nanopipette synaptic potentials recorded in spine head. EPSPs in spines. Spontaneous break-in into the spine head (top) in cultures and corresponding somatic recordings (bottom). Note the large and fast-rising spontaneous EPSPs registered in the spine that are completely filtered from the soma. The gray-shaded region is zoomed in to show the typical rise and decay kinetics of the potential transient. (Adapted from Jayant et al. (2017), Figure 5j, https://doi.org/10.1038/nnano.2016.268, under CCC RightsLink® license 5516540783944)

stimulation. The kinetics of this synaptically activated increment of intracellular Na^+ concentration in the spines of hippocampal CA1 pyramidal cells of mice was consistent with previous measurements in rat pyramidal neuron spines (Miyazaki and Ross 2017, 2022). The rise time of the synaptically activated increment of intracellular Na^+ concentration was around 7.0 ms, about 2 ms slower than the evoked AP in the axon initial segment (Filipis and Canepari 2021; Miyazaki and Ross 2022). That is, a Na^+ current contributes to the fast EPSP in the spine (Palmer and Stuart 2009; Popovic et al. 2015b), that is slower than the duration of the Na^+ current. Almost all synaptically activated Na^+ currents were mediated by AMPA receptors with little Na^+ entry through either voltage-gated Na^+ channels or NMDA receptors (Grunditz et al. 2008; Miyazaki and Ross 2017, 2022). In fact, using a highly sensitive electron microscopic immunogold technique, no evidence was found for Na^+ channels in the dendritic spines of hippocampal CA1 pyramidal cells (Lorincz and Nusser 2010).

Similar to Na^+, the intracellular Ca^{2+} concentration in a spine, also, increased following a single synaptic stimulation. The synaptically activated Ca^{2+}

Fig. 3.4 (continued) between the presynaptic AP peak and onset of the EPSP (inset: raw data, scale bar = 2 mV, 500 ms). (**e**) bAPs registration in spines probed in slices. (**f**) Putative presynaptic terminal recordings in slice preparations. Note the very short delay between the putative presynaptic AP peak and the EPSP rise measured at the soma. (**g**) Raw bAPs and somatic APs. Note the close fit in rise time and half-width indicating the bAPs invade the spine with nearly no loss in temporal structure. (Adapted from Jayant et al. (2017), Figure 5b–e, g–i https://doi.org/10.1038/nnano.2016.268, under CCC RightsLink® license 5516540783944)

transients exhibited fast rise time and half-decay time kinetics (about 7.7 ms and 26 ms, respectively) (Kovalchuk et al. 2000; Sabatini et al. 2002; Bloodgood and Sabatini 2005; Enoki et al. 2009; Miyazaki and Ross 2022). Resembling the Na^+ transient, most of the Ca^{2+} entry was driven by the AMPA receptor-mediated EPSP in the dendritic spine (Miyazaki and Ross 2022). Therefore, processes that modulate AMPA receptor activation, such as desensitization or potentiation, might have a direct effect on Ca^{2+} signaling and synaptic potential in dendritic spines. In contrast, two-photon uncaging experiments (Hestrin et al. 1990; Spruston et al. 1995) and uncaging evoked signals by repetitive stimulation (Polsky et al. 2009) have demonstrated much larger Ca^{2+} entry into spines through NMDA receptors. However, the larger Ca^{2+} current through NMDA receptors may be attributed to the intense two photon uncaging activation compared with synaptically evoked responses by minimal electrical stimulation. In addition, bAP-induced spine depolarization opens NMDA receptors by lifting the Mg^{2+} block (Schiller et al. 1998; Yuste et al. 1999; Nevian and Sakmann 2004), further enhancing the spine V_m. This increase in spine V_m may be elicited either by additional activation of voltage-gated Na^+ channels during the coincidence of bAP and incoming excitatory synaptic input (Stuart and Häusser 2001) or by previous depolarization caused by AMPA receptors in the spine head (Grunditz et al. 2008). For example, using whole-cell recording in acute brain slices of mice coupled with two-photon Ca^{2+} imaging, Ca^{2+} transients in spines were driven mostly by the activity of NMDA receptors in GABAergic intercalated neurons, located between the basolateral (BLA, from which they receive glutamatergic inputs) and the central amygdala (CeA) (Royer et al. 1999; Strobel et al. 2015). Both bAP and evoked spine Ca^{2+} transients in this cell type attenuated rapidly with distance, a finding caused by voltage-gated K^+ channels present in the spine head (probably containing Kv4 channels) (Kaufmann et al. 2013; Strobel et al. 2017). Other techniques showed different results (Bloodgood and Sabatini 2005; Araya et al. 2007; Bywalez et al. 2015), which suggests that neurons and spines may exhibit distinct characteristics depending on the brain area studied, its connectivity and function, and the spine morphology (Bloodgood and Sabatini 2007b; Araya et al. 2007; Bywalez et al. 2015). Synaptically activated potentials can also involve distinct receptors and channel compositions that contribute to the variation in Ca^{2+} transients.

To date, there is no direct information on dendritic spine ionic conductance obtained from direct electrical activity measurement by whole-cell or cell-attached recordings of dendritic spines in slice preparation. However, electrical recordings of isolated and in situ dendritic spines in cultured hippocampal neurons of adult mice were performed using patch-clamping, detecting backpropagating dendritic electrical oscillations and NMDA-mediated electrical activity (Priel et al. 2022). First, dendritic spines were isolated from the dissected hippocampus and then reconstituted using a lipid bilayer membrane system. Ca^{2+}-dependent spontaneous current oscillations, with peak frequency at 1–2 Hz, and NMDA receptor single-channel currents were recorded. NMDA and Ca^{2+}-dependent oscillations at diverse frequencies and amplitudes were similarly recorded in the isolated dendritic spines using patch-clamp technique in both voltage and current-clamping configuration. Voltage-dependent conductances, consistent with both Na^+ and K^+ currents, and AP were also recorded in the isolated dendritic spine. The cell-attached patch-clamp recordings of in situ dendritic spines identified NMDA-dependent EPSP-like intrinsic oscillations at diverse frequencies and amplitudes. Double patch-clamp recordings of a dendritic spine and its parent dendrite detected a NMDA-dependent activity with diverse frequencies and waveforms coupled with and propagated to the parent dendrite. While synapses between dissociated cells do not necessarily exhibit the same properties as those in the intact brain tissue, they are more easily accessible for imaging. Valuable information can also be obtained about the biophysical properties of the dendritic spines and their interaction with the adjacent dendritic segments and the soma (Priel et al. 2022).

Although most spines receive an excitatory input, GABA-mediated inhibitory postsynaptic potentials (IPSPs) participate in the control of neuronal excitability. Experiments using cell type-specific optical stimulation in combination with two-photon Ca^{2+} imaging showed that the dendritic spines' heads of cortical pyramidal cells are targeted by somatostatin-expressing interneurons, compartmentalize GABAergic inhibition, limit both bAP and synaptically evoked Ca^{2+} entry, and regulate NMDAR-dependent synaptic integration. These inhibitory synapses on spines can participate in the control of electrical and biochemical signaling along dendritic segments, including Ca^{2+} signaling and synaptic plasticity (Chiu et al. 2013).

The conductance of different channels and receptors and the specific changes in the V_m of dendritic spines are now being revealed by the use of combined new technical approaches, evidencing spatiotemporal patterns of induced or spontaneous depolarizations of dendrites and spines during synaptic integration and plasticity.

3.2 Dendritic Spines as Dynamic Compartments

Since their discovery by Ramón y Cajal, it has been hypothesized that dendritic spines may be involved in the compartmentalization of synaptic signals (Yuste 2015). Such a compartmentalized microenvironment could modulate the functional properties of different synapses (Calabrese et al. 2006; Araya et al. 2014; Tønnesen and Nägerl 2016; Vallés and Barrantes 2021) affecting synaptic strength, cell information processing, circuits' organization, and behavior (Yuste and Denk 1995; Sabatini et al. 2002; Noguchi et al. 2005; Araya 2014).

3.2.1 Biochemical Compartmentalization

Dendritic spines, with a wide diversity of sizes and shapes, have organized membrane-bound enzymes, lipids, and proteins into delimited regions or subcompartments of varying extension and composition (Honigmann and Pralle 2016). Compartmentalization is generated by the spine neck, which relatively isolates the spine head from the parent dendrite (Yuste and Denk 1995; Nevian and Sakmann 2004; Honigmann and Pralle 2016).

A series of experiments, including those using two-photon excitation laser scanning microscopy in combination with Ca^{2+} imaging, has revealed that spines function as isolated biochemical compartments, not only in excitatory synapses, but also in GABAergic synapses that directly target spine heads (Chen et al. 2011; Chiu et al. 2013; Camiré and Topolnik 2018).

Spines act as biochemical and highly differentiated subcellular compartments. They have the ability to dynamically control the membrane area available for specific domains and clustering of neurotransmitter receptors as well as the diffusion of signaling and regulatory molecules, such as Ca^{2+}, kinases, second messengers, enzymes, scaffolding proteins, and cytoskeletal elements within a very small volume (See Fig. 1.13, Chap. 1 of this book). Thus, biochemical compartmentalization restricts and regulates biochemical processes in individual synapses to smaller regions, modulating synaptic output, integration, and plasticity (Yuste 2011; Tønnesen and Nägerl 2016). Various neurotransmitter receptors, including the cholinergic nicotinic receptor, for example, have been located in the raft-type lipid domains of the spine membrane (Brusés et al. 2001; Allen et al. 2007; Egawa et al. 2016). Also, spine membrane domains show temporal modifications that affect the timing of biochemical cascades following synaptic activation. These changes induce dynamic alterations in the spine head and neck width associated with the degree of the synaptic signals compartmentalization, thereby biochemically isolating synaptic inputs and enabling input-specific plasticity (Magee and Johnston 1997; Lee et al. 2009; Tønnesen et al. 2014). In addition, the dynamic changes of the spine neck length have been associated with the bidirectional modulation of the synaptic and the parent dendrite voltage, acting as an electrical compartment and modifying synaptic integration (Vanderklish and Edelman 2002; Araya et al. 2014; Cornejo et al. 2022). This spatiotemporal organization involving dendritic spines is crucial to coordinate the synaptic function in a variety of neurons throughout the CNS.

3.2.2 *Electrical Compartmentalization*

Electrical compartmentalization relates to the ability of dendritic spines to process synaptic inputs, modulating the amplitude, kinetics, and integration of synaptic potentials into the spine head (Tsay and Yuste 2004). Spine heads exhibit voltage-gated ion channels and neurotransmitter receptors (Kim et al. 2007; Araya et al. 2007; Chung et al. 2009), which modulate the voltage inputs. In addition, the geometry of the spine neck influences voltage propagation into the adjacent dendritic shaft, acting as a high resistance pathway. The process of compartmentalization is influenced by membrane passive electrotonic and active properties (Koch and Segev 2000; Sjöström et al. 2008). Then, the geometry and passive properties of dendritic spines, such as axial resistance, membrane capacitance, and diameter, are crucial factors to understand how a spine modulates synaptic potentials and alters synaptic strength (Spruston et al. 1994), input integration (Yuste 2010), and, consequently, neural circuits (Yuste 2010; Kwon et al. 2017). As opposed to the well-established biochemical compartments, the electrical compartmentalization of dendritic spines is a matter of debate.

The spine electrical compartmentalization has been investigated with various experimental techniques in vitro and in vivo. For example, the strategy of combining a high-sensitivity organic voltage-sensitivity dye with two-photon uncaging of glutamate (Popovic et al. 2015b) detected subthreshold uEPSP in dendritic spines of layer V pyramidal neurons (Popovic et al. 2015a, b). The subthreshold uEPSP presented no significant attenuation along the spine neck, which exhibited low electrical resistance relative to the input impedance of the parent dendrite (mean value of 27 ± 6 MΩ) (Popovic et al. 2015b). These data suggested that spine synapses were not electrically isolated from the parent dendrites and may exhibit the same electrical behavior as those contacts made directly on dendrites (Popovic et al. 2014, 2015b). The low resistance of the spine neck relative to the impedance of the parent dendrite is similar to previous theoretical predictions and diffusional resistance measurements (Wilson 1984; Koch and Zador 1993; Svoboda et al. 1996; Bloodgood and Sabatini 2005; Tønnesen et al. 2014; Takasaki and Sabatini 2014).

Experiments using simultaneous Na^+ and Ca^{2+} imaging with single-spine resolution in pyramidal neurons in rat hippocampal slices detected a rapid diffusion of Na^+ out of the spine head through the spine neck to the parent dendrite, also suggesting that the neck offers low resistance for ionic diffusion (Miyazaki and Ross 2017). Moreover, the amplitude of EPSP and Ca^{2+} transients evoked by 2-photon glutamate uncaging also showed no correlation with the neck geometry, supporting the lack of a functional impact of spine shape on electrical compartmentalization (Takasaki and Sabatini 2014; Bywalez et al. 2015). With that assumption in mind, constricted spine neck diameter would rather serve to isolate metabolic events by reducing diffusion of activated molecules to neighboring synapses, without significantly influencing the transfer of synaptic charge to the postsynaptic dendrite (Harris and Stevens 1988, 1989).

Alternatively, other experiments have demonstrated that dendritic spines may constitute fundamental electric compartments (Tsay and Yuste 2004; Harnett et al. 2012; Yuste 2013; Acker et al. 2016; Jayant et al. 2017; Kwon et al. 2017; Strobel et al. 2017; Cornejo et al. 2022). For example, experiments using combined dendritic patch electrophysiology, Ca^{2+} imaging, and glutamate uncaging in spines at apical dendrites of rat hippocampal CA1 pyramidal neurons demonstrated that, when spine neck resistance was high (mean value of 514 ± 44 MΩ) (Harnett et al. 2012), passive amplification of the spine EPSP up to 50-fold would occur compared to an unitary dendritic EPSP. This amplification was very much dependent on the spine and neck structure in different neurons (Harnett et al. 2012).

Another study used two-photon voltage-sensitive organic dye recording with two-photon glutamate uncaging to measure the amplitude and duration of an uEPSP in single spines from the basal dendrites of layer V pyramidal neurons in acute brain slices of mice (Acker et al. 2016). Data from this study showed high spine neck resistance with a mean value of 179 ± 25 MΩ (ranging from 23 to 420 MΩ), suggesting the existence of a substantial electrical compartmentalization in spines.

Study using a genetically encoded voltage indicator and two-photon glutamate uncaging to identify electrical properties of dendritic spines in cultured hippocampal neurons showed that the invading bAP was unchanged. However, the spine neck resistance had values ranging from 15 to 332 MΩ (mean value: 101 ± 95 MΩ) reducing the synaptic and intrinsic voltage-gated currents that reach the adjacent dendritic shaft. Thus, the spines could compartmentalize voltage and attenuate the synaptic inputs by about twofold as they drive through the spine neck to the parent dendrite. Spines with longer neck attenuated uEPSP by fourfold as they propagate to the parent dendrites and exhibited an estimated neck resistance of 332 MΩ (Kwon et al. 2017).

Intracellular recordings of dendritic spines with long and narrow necks from neurons in brain slices using quantum-dot-coated glass nanopipettes revealed neck resistances ranging from 250 to 536 MΩ (mean: 425 ± 102 MΩ) (Jayant et al. 2017). The EPSPs recorded were large in the spine head (mean 26 mV), but strongly attenuated at the soma (0.5–1 mV), and the estimated neck resistance was large enough to generate significant local voltage compartmentalization in spines (Jayant et al. 2017) (Fig. 3.6).

Two-photon Ca^{2+} imaging coupled with whole-cell recording in amygdaloid GABAergic intercalated neurons in acute brain slices showed that Ca^{2+} enters the spine head mediated by the activity of NMDA receptors and with little spread of the Ca^{2+} signal. That finding also suggested that a dendritic spine head can act as an isolated electrical compartment (Strobel et al. 2017). High spine neck resistances, associated with longer and thinner spine necks, would be expected to increase the voltage in the spine head during a given synaptic input, reducing the driving force for synaptic current flow (Koch and Zador 1993).

Interestingly, the size of the spine head has also been correlated with the area of the postsynaptic density (PSD), a characteristic specialization of the postsynaptic cell membrane (Harris and Stevens 1989; Trommald and Hulleberg 1997; Arellano 2007), and with the amplitude of the generated EPSP (Matsuzaki et al. 2001; Noguchi et al. 2011). Varied EPSPs' amplitudes have been observed in different experiments, which likely related to a spectrum of spine neck resistances correlated with distinct neck lengths, even in the same dendritic shaft and neuron type (Miyazaki and Ross 2022). Not only does EPSP modulation by spines have important repercussions for synaptic plasticity but also NMDA receptor activation and intra-spine Ca^{2+} levels activate a variety of signaling pathways that can modify actin filament assembly and contributing to activity-dependent alterations in spine morphology (Tada and Sheng 2006; Araya 2014; Yasuda 2017).

Finally, using in vivo two-photon microscopy and a genetically encoded voltage indicator, it was possible to measure the V_m of basal dendrites and their spines while performing simultaneous somatic whole-cell recordings in layers II/III pyramidal neurons of the somatosensory cortex of mice. Recordings were obtained during spontaneous activity and sensory stimulation (Cornejo et al. 2022) (Fig. 3.7). The estimated spine neck resistance was 226.6 ± 128.8 MΩ (mean ± SD), ranging from approximately 0 to 530.8 MΩ, similar to that seen in previous publications (Harnett et al. 2012; Tønnesen et al. 2014; Jayant et al. 2017). Spines and dendrites were depolarized together during AP, the dendritic potentials propagated into spines, and there was no attenuation of dendritic potentials or AP but synaptic potentials were

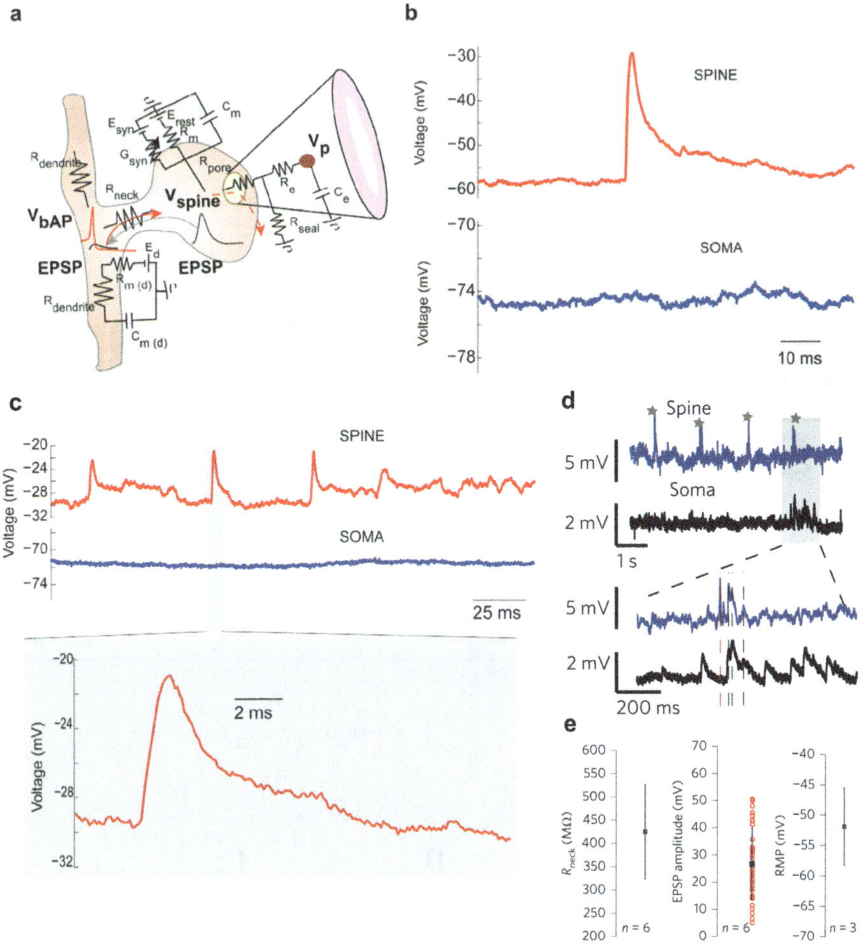

Fig. 3.6 Nanopipette recordings in spines reveal electrical compartmentalization. (**a**) Electrical equivalent circuit of the passive dendritic spine. E_{rest}, leak reversal potential; R_m, spine head passive membrane resistance; C_m, spine head passive membrane capacitance; E_{syn}, synaptic reversal potential; g_{syn}, synaptic conductance; R_{pore}, pore resistance; R_{seal}, seal resistance; R_{neck}, neck resistance; R_e, pipette resistance; C_e, pipette capacitance; $R_{dendrite}$, dendritic resistance; $R_{m(d)}$, dendrite passive membrane resistance; $C_{m(d)}$, dendrite passive membrane capacitance; E_d, dendrite reversal potential. (**b**, **c**) Typical rise and decay kinetics for spontaneous PSPs recorded in the spine head. Notice the millisecond resolution rise and decay indicative of an extremely small RC time constant. The data shown in (**b**) and (**c**) are not deconvolved. (**d**) Raw EPSPs in spines from slices (marked stars) are larger than somatic EPSPs (~0.5 mV flickers). Note the sudden registration of EPSPs in the soma (gray bar, bottom) that signify concomitant inputs from other spines. (**e**) Deconvolved R_{neck} (left), EPSP amplitudes (middle), and RMPs (right) in the spine head. The range of deconvolved EPSP amplitudes is in accordance with the spontaneous EPSP. The range for R_{neck} indicates that EPSPs that invade a low input resistance dendrite will be heavily attenuated. (Adapted from Jayant et al. (2017), (**a–c**) Supplementary Figure 12 and (**d**, **e**) Figure 5k, l, https://doi.org/10.1038/nnano.2016.268, under CCC RightsLink® license 5516540783944)

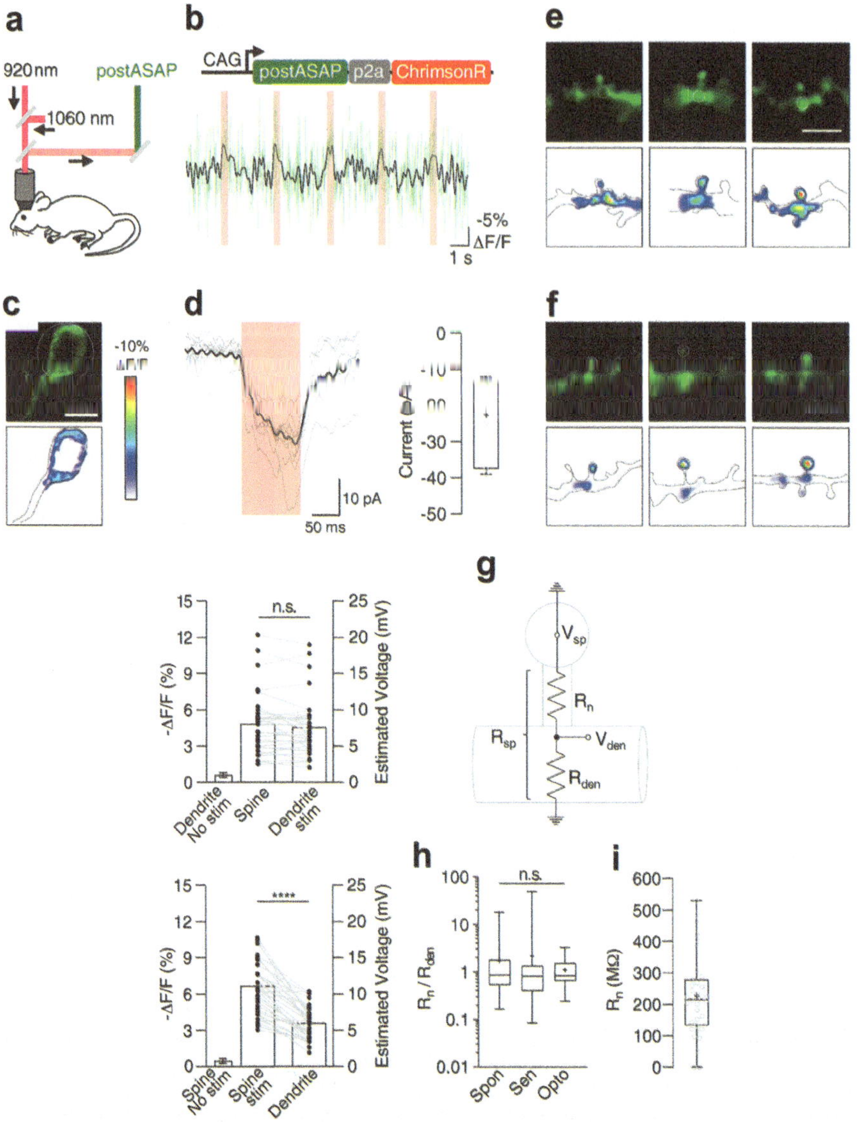

Fig. 3.7 Two-photon optogenetics and voltage imaging showing the voltage compartmentalization in dendritic spines in vivo. (**a**) Experimental design. (**b**) Construct and representative fluorescence changes in soma (light green, raw fluorescence; black, 10-Hz low-pass filtered) during 500-ms stimulation trials (red, 100 mW power). (**c**) Representative soma (top) and peak fluorescence response (bottom) during stimulation trials (×10, 500 ms, 100 mW). The dotted circle shows the stimulation area (scale bar, 10 μm). (**d**) Representative in vivo voltage-clamp recordings during optogenetic stimulation of proximal dendrites (100 mW, 100 ms) are shown on the left. Peak currents are shown on the right; −22.7 ± 11.3 pA (mean ± SD), 10 trials ($n = 7$ cells and 4 animals). (**e**) Representative peak fluorescence changes during optogenetic activation of dendritic shafts are shown on the left (stimulation ROIs are indicated by white dotted circles; 10 trials, 100 ms, 100 mW; color bar same as in (**c**); scale bar, 5 μm). Peak fluorescence changes in stimulated dendritic shaft (dendrite stim), adjacent dendritic spine (spine), and unstimulated dendritic shaft

significantly reduced. During subthreshold and resting potentials, spines could be activated independently, even in the absence of dendritic or somatic activity.

These findings, including the spine-independent depolarization during spontaneous and evoked activity, contribute to our understanding of dendritic spine biophysics, showing that dendritic spines serve not only for biochemical but also for electrical compartmentalization of synaptic inputs essential for synaptic integration, information processing, and plasticity (London and Häusser 2005; Yuste 2011; Cornejo et al. 2022).

3.3 Spatiotemporal Dynamics of Dendritic Spines Related to Synaptic Plasticity

Dendritic spines are known to be the site of long-term synaptic plasticity (Matsuzaki et al. 2004; Lai and Ip 2013). In fact, the activity-dependent morphological changes in the head and/or neck of spines (Matsuzaki et al. 2004; Tønnesen et al. 2014; Araya et al. 2014) have been correlated with modifications in synaptic strength in cortical pyramidal neurons involving biochemical and electrical mechanisms (Araya et al. 2006; Araya 2014).

Studies using signal imaging techniques based on fluorescence resonance energy transfer (FRET) in combination with photostimulation of single synapses revealed the temporal sequence of biochemical signaling cascades and their spatial spreading from stimulated spines (Nishiyama and Yasuda 2015). Dendritic spine plasticity directly modifies the synaptic strength of neuronal networks through changes in protein synthesis, spine cytoskeleton structure, spine morphology, and density of receptors at the synaptic membrane (Nakahata and Yasuda 2018). These modifications guide rearrangements of connections between presynaptic and postsynaptic elements and consequently promote a shift of synaptic efficacy, either potentiating or depressing synaptic transmission. Over the long term, two processes involving various molecular mechanisms promote synaptic strength or weakness, respectively: long-term potentiation (LTP) and long-term depression (LTD) (Citri and Malenka 2008). Together, spine synaptic plasticity allows the establishment and reorganization of connectivity within neuronal circuits during normal brain development to adulthood and aging. It also may be involved in degenerative processes and in a large spectrum of brain disorders that involve deficits in information processing, such as schizophrenia, Alzheimer's disease, autism spectrum disorder, and epilepsy (Penzes et al. 2011; Dickstein et al. 2013; Nishiyama 2019; Runge et al. 2020; Rossini et al. 2021; Zaccard et al. 2023).

Fig. 3.7 (continued) (dendrite no stim) are shown on the right (n = 34 dendrites and 9 animals). (**f**) Same as (**e**) during optogenetic activation of spines (n = 35 spines and 12 animals; scale bar, 5 μm). ****$p < 0.0001$. (**g**) Simplified electrical model. (**h**) Resistance ratio (R_n to R_{den}) of spine-only events during spontaneous (Spon, n = 116), sensory stimuli (Sen, n = 181), and optogenetic spine stimulation (Opto, n = 35). (**i**) Values of spine neck resistance for stimulated spines; median = 213.7 MΩ (n = 35 spines). In (**d**), (**h**), and (**i**), boxes and whiskers represent median (line), 25th to 75th percentiles (box), range (whiskers), and mean as a "+." *postASAP: postsynaptic accelerated sensor of APs. (Reproduced from Cornejo et al. (2022), Figure 4, https://doi.org/10.1126/science.abg0501, under CCC RightsLink® license 1338950-1)

Before describing the dynamics of receptors in the spines during LTP and LTD, it is important to consider a fundamental synaptic component that influences spine synaptic plasticity and the formation and stabilization of new dendritic spines, that is, the extracellular matrix.

The extracellular matrix, together with pre- and postsynaptic elements and glial processes, form the so-called tetrapartite synapse (Dityatev and Schachner 2003; Dityatev et al. 2006; Dityatev and Rusakov 2011; Song and Dityatev 2018). The fully developed extracellular matrix has the ability to stabilize neuronal connectivity, restraining morphological changes of dendritic spines and decreasing cortical plasticity in the adult brain. In fact, as the circuitry reaches its adult form, spine motility diminishes and connectivity stabilizes (Oray et al. 2004). A prominent component of the matrix, chondroitin sulfate proteoglycans (CSPGs), may play an important role in limiting cortical plasticity, restraining the structural remodeling of neurons, and inducing inhibition of synaptic transmission (Oray et al. 2004).

Degradation of CSPGs proved to affect ionic conductances, control receptors, and channel localization at the cell membrane (Gundelfinger et al. 2010; Dityatev et al. 2010), induce rapid structural changes of dendritic spines by increasing the size of the spine head, and enhance spine mobility thereby restoring plasticity (de Vivo et al. 2013). CSPGs modulate structural plasticity by preventing integrin activation and signaling at synaptic sites (Orlando et al. 2012).

Several other molecules of the extracellular matrix have been involved in the regulation of synaptic plasticity. For example, hyaluronan, the backbone of extracellular matrix, was found to regulate hippocampal synaptic plasticity by modulating postsynaptic L-type Ca^{2+} channels (Kochlamazashvili et al. 2010). Another important glycoprotein, Reelin, is secreted into the extracellular matrix by neurons, such as GABAergic cells. Reelin regulates the composition of NMDA receptors, the increment of AMPA responses by increasing the number of AMPA receptors on the postsynaptic membrane, and robustly enhances LTP (Qiu et al. 2006; Campo et al. 2009; Hellwig et al. 2011). Reelin strongly induces spine remodeling, regulating spine size and stability, and the number of synaptic contacts per spine (Beffert et al. 2006; Pujadas et al. 2010). Reelin interacts with integrins to regulate AMPA receptor internalization and cell membrane trafficking of NMDA receptor subunits (Dityatev et al. 2010; McGeachie et al. 2011), important for receptor dynamics during synaptic plasticity.

Furthermore, in mature neurons, integrins are involved in homeostatic synaptic scaling and interact with matrix metalloproteinases, promoting spine volume changes induced by long-term plasticity (Nagy et al. 2006; Wang et al. 2008). Cell adhesion molecules (CAMs) mediate extracellular matrix and PSD signaling that may influence the actin cytoskeleton network of dendritic spines, and consequently, spine shape (Huntley et al. 2002; Washbourne et al. 2004; Koleske 2013). Spine remodeling is, also, controlled by autocrine signaling of brain-derived neurotrophic factor (BDNF), crucial for LTP (Harward et al. 2016; Hedrick et al. 2016) and for proteolytic cleavage of extracellular matrix (Nakahata and Yasuda 2018). The increased intracellular Ca^{2+} concentration, mediated by NMDA receptors or by voltage-gated Ca^{2+} channels, induces lysosomal fusion with the plasma membrane, releasing the protease Cathepsin B to the extracellular matrix, which cleaves and inhibits the endogenous inhibitor of the matrix metalloprotease 9. The disinhibited matrix metalloprotease 9 cleaves the CAMs and the extracellular matrix, facilitating the structural remodeling of spines (Nakahata and Yasuda 2018). Thus, the extracellular matrix that fills perisynaptic spaces surrounding dendritic spines plays a crucial role in remodeling neural circuits, affecting long-term synaptic plasticity and receptor trafficking (Dityatev et al. 2010; Orlando et al. 2012).

3.3.1 Dynamics of Receptors in Spines During Long-Term Potentiation

The most studied form of spine structural plasticity is LTP (Nakahata and Yasuda 2018). The typical protocol for inducing LTP involves brief and high-frequency stimulation (for review, see Nicoll 2017), and paired pre- and postsynaptic stimulation, known as spike timing-dependent plasticity (STDP) (Markram et al. 1997; Debanne et al. 1998; Bi and Poo 1998; Sjöström et al. 2008; Feldman 2012). However, there are protocols applying lower-frequency stimulations to induce LTP mediated by metabotropic glutamate receptor (mGluR) and protein kinase A activation (Lanté et al. 2006). Importantly, LTP is input-specific at the level of individual synapses, that is, only stimulated spines may trigger LTP-induced synaptic modifications (Magee and Johnston 1997; Lee et al. 2009; Tønnesen et al. 2014). Input-specific LTP induction can be performed by two-photon glutamate uncaging. For example, using an uncaged glutamate-LTP protocol to study the visual cortex of anesthetized rats in vivo, it was possible to demonstrate the induction of individual spine enlargements (Noguchi et al. 2019) reinforcing the LTP-synaptic specificity (Harvey and Svoboda 2007). Although LTP can also be mediated by presynaptic mechanisms, here we focus on the NMDA receptor-dependent form of LTP (Luscher and Malenka 2012).

Multiple molecular and cellular mechanisms contribute to induction, expression, and maintenance of NMDA-dependent form of LTP. Ca^{2+} represents a key factor for long-term synaptic plasticity (Mateos-Aparicio and Rodríguez-Moreno 2020). During postsynaptic depolarization, Ca^{2+} entry through activated NMDA receptors results in a local increase in dendritic excitability and an increase in synaptic strength (Luscher and Malenka 2012). Activation of NMDA receptors with subsequent increase of Ca^{2+} influx is a well-known mechanism involved in the early phase of LTP, which lasts approximately 60 min (Becker and Tetzlaff 2021). Experimental approaches using chemical and selective activation of NMDA receptors were able to induce synaptic potentiation, that was blocked by a competitive antagonist, D-2-amino-5-phosphonopentanoate (AP5) (Lu et al. 2001). Following NMDA receptors agonist washout of cultured hippocampal neurons, synaptic potentiation was still enhanced, revealing the contribution of AMPA receptors in the maintenance of postsynaptic excitatory response potentiation (Lu et al. 2001). Additionally, NMDA-dependent LTP could be blocked by MK-801, a competitive NMDAR antagonist, and BAPTA, a Ca^{2+} chelator, highly supporting the involvement of NMDA and AMPA receptor activation and Ca^{2+} influx in LTP expression (Lu et al. 2001) (Fig. 3.8).

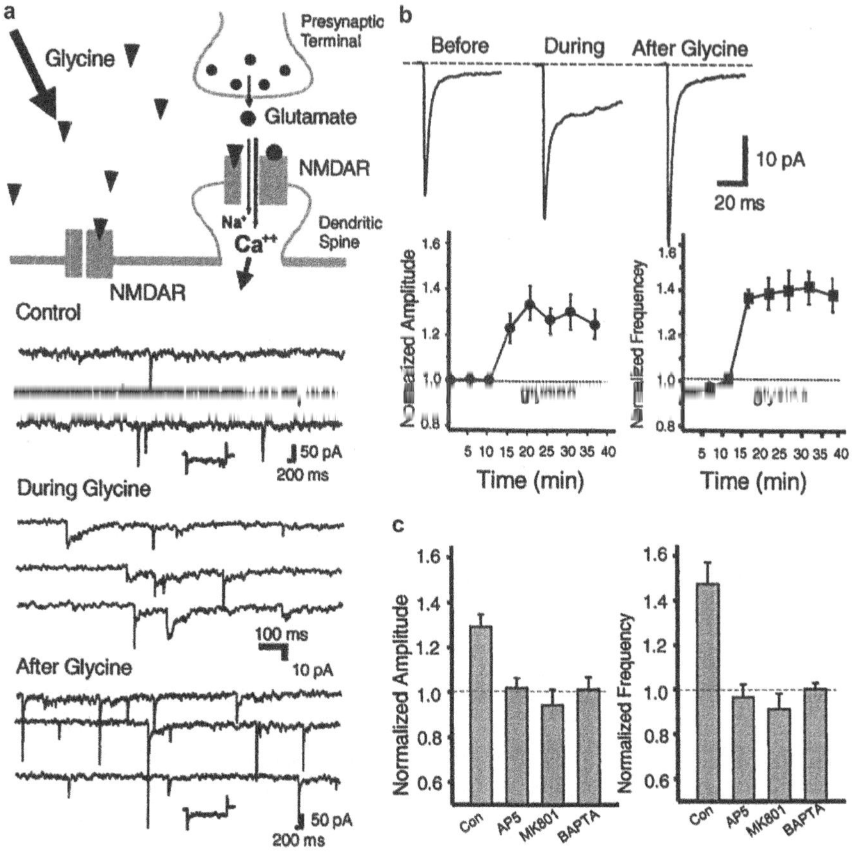

Fig. 3.8 LTP induction. (**a**) Up: The illustration demonstrates that brief stimulation of postsynaptic NMDA receptors with glycine may induce LTP. Bottom: Representative traces of mEPSC. Glycine-induced LTD increases mEPSC amplitude and frequency measured by whole-cell patch clamp in hippocampal neurons of primary dissociated cultures. (**b**) Averaged mEPSCs demonstrating before, during, and following the application of glycine. (**c**) Pretreatment with AP5, MK-801, and BAPTA blocked LTP, decreasing both mEPSC amplitude and frequency. (Adapted from Lu et al. (2001), https://doi.org/10.1016/S0896-6273(01)00194-5, (**a**) Figure 1A,D,F,G and (**b**, **c**) Figure 2E,F, under CCC RightsLink® license 5516571126317)

LTP induced by STDP at CA3-CA1 hippocampal synapses and some synapses of neocortical layers II/III pyramidal cells are dependent on NMDA receptors (Nishiyama et al. 2000; Froemke et al. 2005). NMDA receptor-mediated Ca^{2+} accumulation in a spine is also associated with actin polymerization-dependent shrinkage of the activated spine neck and an increase in synaptic strength, as demonstrated during the timing-dependent LTP triggered by two-photon glutamate uncaging at the presynaptic site paired with postsynaptic spikes in layer V pyramidal neurons from juvenile mice (Tazerart et al. 2020) (Fig. 3.9a, c).

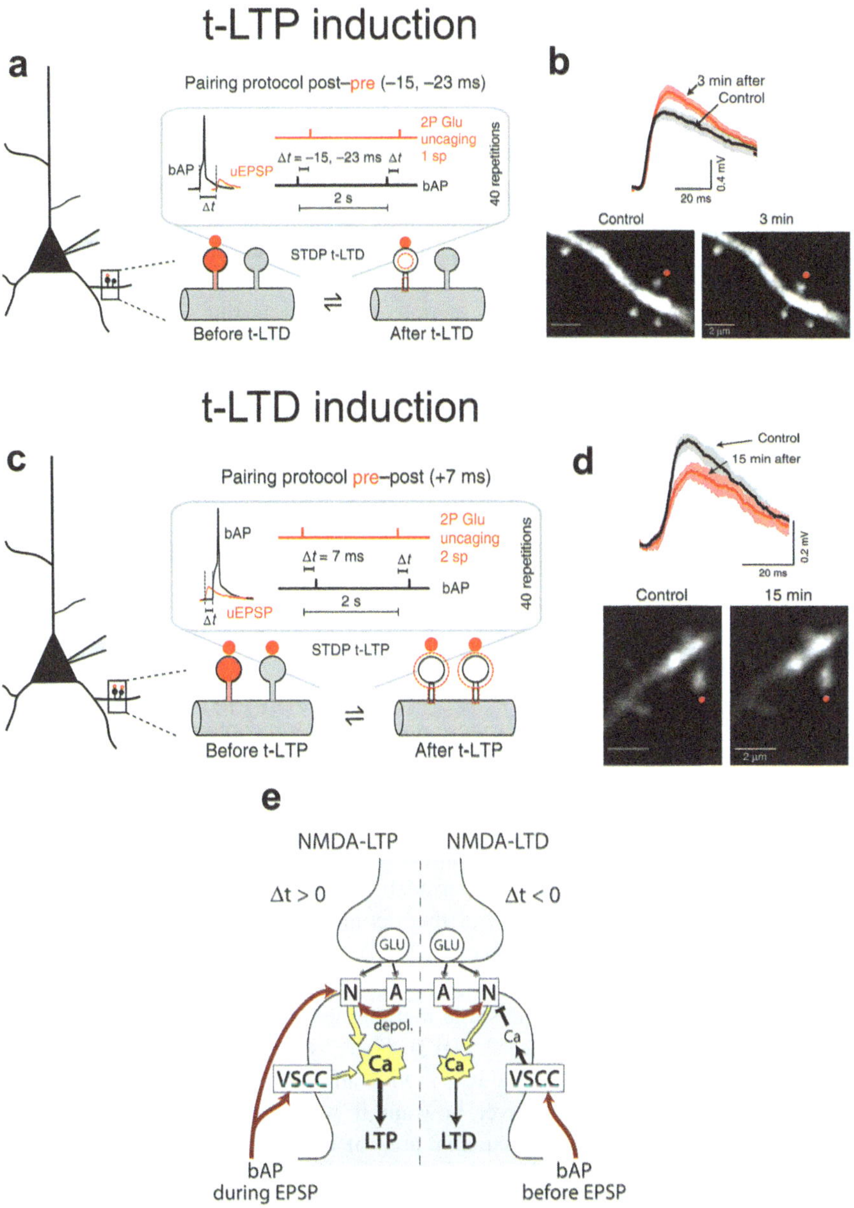

Fig. 3.9 Induction of t-LTP and t-LTD in single dendritic spines. (**a**) Experimental protocol for t-LTP induction in single dendritic spines (sp). (**b**) Representative experiment where a spine was activated with t-LTP pre- and post-pairing protocol of +13 ms. Traces correspond to average of ten uncaging EPSPs (uEPSPs) recorded in the soma and generated by 2P uncaging before (control, black trace) and after t-LTP induction (red trace) over the indicated spine (red dot). (**c**) Experimental protocol for t-LTD induction in single dendritic spines. (**d**) Representative experiment where a spine was activated with t-LTD post–pre pairing of −15 ms. Traces correspond to average of 10 uEPSPs recorded at the soma and generated by 2P uncaging before (control, black trace) and after **Fig. 3.9** (continued) t-LTD induction (red trace) over indicated spine (red dot). (**e**) Biochemical signaling pathways of the STDP induced by NMDA receptors. N and A, NMDA, and AMPA receptors, respectively. (Figure and legend adapted from (**a–d**) Tazerart et al. (2020), https://doi.org/10.1038/s41467-020-17861-7, Figures 1a,b and 2a,b, respectively, under CC BY license, originally published by Springer Nature; (**e**) Feldman (2012), https://doi.org/10.1016/j.neuron.2012.08.001, Figure 4A, under CCC RightsLink® license 5516580280377)

Furthermore, during synaptic potentiation, increased Ca^{2+} influx leads to activation of biochemical cascades through Ca^{2+}-calmodulin-dependent protein kinase II (CaMKII), followed by phosphorylation of AMPA receptors and insertion of new receptors in the postsynaptic membrane sites. CaMKII antagonists block the synaptic potentiation, preventing exocytosis of new receptors at the postsynaptic membrane (Matsuzaki et al. 2004; Lee et al. 2009; Makino and Malinow 2009; Lisman et al. 2012; Jaafari et al. 2013; Penn et al. 2017). In vivo studies also revealed that CaMKII is crucial for LTP, learning, memory, and long-lasting spine enlargements (Miller et al. 2002; Matsuzaki et al. 2004; Buard et al. 2010). Indeed, the enlargement of dendritic spines associated with increase in AMPA receptor-mediated currents (Matsuzaki et al. 2004) and the expression of new AMPA receptors in the spine membrane at the stimulated synapse (Tanaka and Hirano 2012; Vitureira and Goda 2013) contribute to the enhancement of synaptic transmission during LTP (Park 2018) and are associated with larger PSDs (Chater and Goda 2014). Those changes were dependent on NMDA receptors, Ca^{2+} entry, CaMKII, and actin polymerization (Matsuzaki et al. 2004; Lee et al. 2009; Makino and Malinow 2009; Lisman et al. 2012; Jaafari et al. 2013; Penn et al. 2017). Also, the enlargement of individual spines precedes the increase in AMPA receptors at the postsynaptic site at spines (Chater and Goda 2014).

A protocol of glycine-induced LTP in dissociated cortical and hippocampal neuronal cultures also showed an increase of AMPA receptors expression in the spine membrane by exocytosis from intracellular pools. It begins within a few minutes after the LTP-inducing stimulation and is associated with a long-lasting potentiation of AMPA receptor-mediated excitatory postsynaptic current (EPSC) (Lu et al. 2001; Jaafari et al. 2013; Vitureira and Goda 2013; Henley and Wilkinson 2016). These new AMPA receptors permeable to Ca^{2+} (CP-AMPA receptors) are GluA2-lacking receptors and most of them contain the GluA1 subunit (Lu et al. 2001; Clem and Barth 2006; Jaafari et al. 2013), whose reversible phosphorylation plays an important role in long-term plasticity (Lee et al. 2010). The reduction of AMPA receptors diffusion in the spine membrane directly impairs hippocampal-dependent fear learning (Penn et al. 2017). There is strong evidence that synaptic CP-AMPA receptors have a crucial role in vivo and in vitro, tuning neuronal excitability to induce plasticity. However, how and to what extent CP-AMPA receptors are incorporated during LTP remains an area of research (Henley and Wilkinson 2016) (Fig. 3.10a, b).

Besides AMPA receptors, NMDA receptors also exhibit gating and trafficking during LTP. This includes the rapid delivery of functional receptors and an increase in receptor clusters on the membrane of dendrites and dendritic spines thereby modulating neuronal excitability (Lan et al. 2001; Grosshans et al. 2002; Lau and Zukin 2007; Lau et al. 2010). The modifications in NMDA receptors are induced by

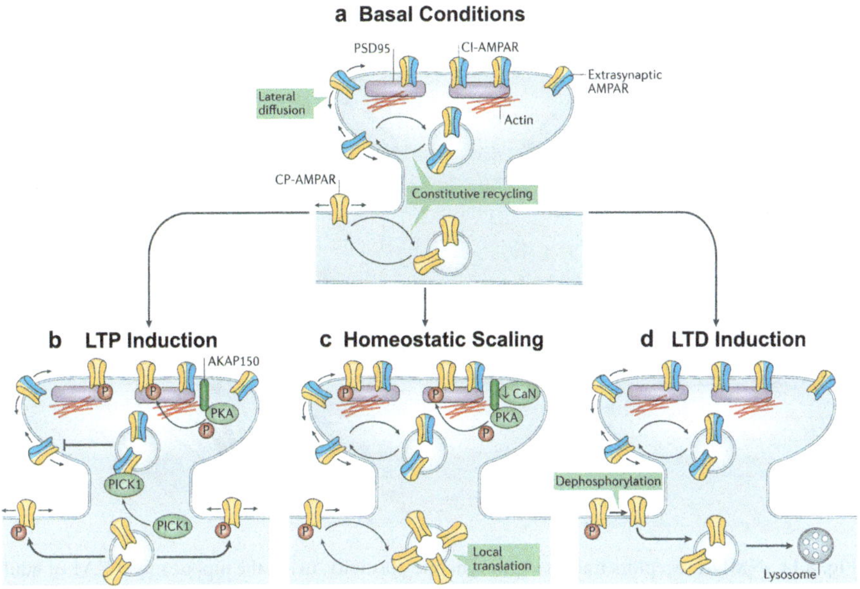

Fig. 3.10 AMPA receptors trafficking in synaptic plasticity. (**a**) Under basal conditions, synaptic AMPA receptors (AMPAR) are predominantly calcium impermeable and are anchored to postsynaptic density protein 95 (PSD95) and the actin cytoskeleton. They undergo both lateral diffusion in the membrane and rapid constitutive recycling. Furthermore, under certain conditions, Ca^{2+}-permeable AMPARs (CP-AMPARs) can be observed extrasynaptically. (**b**) During long-term potentiation (LTP) induction, activation of protein kinase A (PKA), which is anchored at synapses by the scaffold protein AKAP150, leads to phosphorylation (indicated by the P) of GluA1 at S845 in both spines and dendrites, promoting its surface expression and leading to the appearance of synaptic CP-AMPAR. In addition, enhanced binding of protein interacting with C kinase 1 (PICK1) to GluA2-containing Ca^{2+}-impermeable AMPARs (CI-AMPAR) in intracellular vesicles restricts their recycling to the cell surface, thus further enhancing the ratio of CP-AMPAR to CI-AMPAR. (**c**) During activity deprivation to induce synaptic homeostatic scaling, calcineurin (CaN) activity is reduced, favoring PKA-mediated phosphorylation of GluA1 subunits at S845, and promoting surface expression of GluA1-containing AMPAR and their synaptic incorporation. Furthermore, induction of homeostatic scaling via NMDA receptor (NMDAR) blockade promotes the local translation of GluA1 in dendrites, and an increase in surface expression of GluA1-containing CP-AMPAR. By contrast, (**d**) long-term depression (LTD) is associated with dephosphorylation of GluA1 at S845, reducing the extrasynaptic pool of GluA1-containing CP-AMPARs, which then undergo lysosomal degradation. (Reproduced from Henley and Wilkinson (2016), https://doi.org/10.1038/nrn.2016.37, Figure 4, under CCC RightsLink® license 5516911470380)

protein kinase C (PKC) and tyrosine kinase Src (Lan et al. 2001; Grosshans et al. 2002; Lau and Zukin 2007; Lau et al. 2010) (Fig. 3.11a). Therefore, both AMPA and NMDA receptors contribute to synaptic potentiation. Increase of both components of EPSP (AMPA and NMDA) with a prominent potentiation of AMPA over the NMDA was reported during early and late LTP (Dozmorov et al. 2006), as well as, for the delayed NMDA-dependent LTP (Watt et al. 2004) (Fig. 3.12).

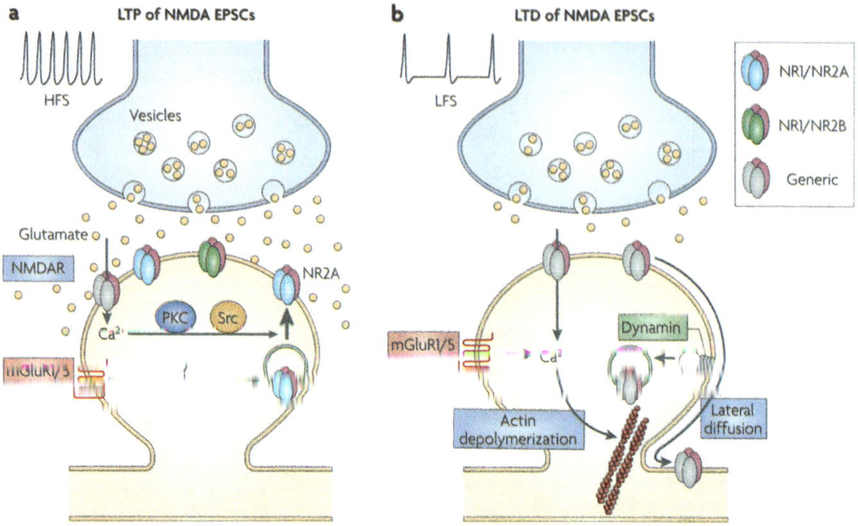

Fig. 3.11 NMDA receptors trafficking in synaptic plasticity. (**a**) In the hippocampal CA1 of adult animals, high-frequency stimulation (HFS) and long-term potentiation (LTP) promote rapid insertion of NR2A-containing NMDA receptors and an increase in NMDA field excitatory postsynaptic potentials (fEPSPs) at CA1 synapses through a protein kinase C (PKC)- and Src-dependent pathway (blue box). HFS-induced activation of mGluR5 can also induce LTP with NMDA receptors trafficking (red box). (**b**) Long-term depression (LTD) of synaptic NMDA EPSCs (LTD_{NMDAR}) occurs by retrieval of NMDA receptors from synaptic sites. Low-frequency stimulation (LFS) of Schaffer collaterals can induce metabotropic glutamate receptors (mGluR)-dependent LTD_{NMDAR} or NMDA-dependent LTD. (mGluR)-dependent LTD is associated with enhanced internalization of NMDA receptors (red box). NMDA-dependent LTD promotes actin depolymerization and lateral diffusion of NMDA receptors away from the synapse site (blue boxes). LFS of connected pairs of CA3 pyramidal neurons induces LTD_{NMDAR} by stimulating dynamin-dependent internalization of NMDA receptors (green box). (Adapted from Lau and Zukin (2007), https://doi.org/10.1038/nrn2153, Figure 5, under CCC RightsLink® license 5516591094919)

At later stages of LTP, structural modifications of spines are triggered to support synaptic strength enhancement. Spine structural modifications are essential to synaptic plasticity (Fig. 3.13) (Harvey and Svoboda 2007; Noguchi et al. 2019), since it is highly correlated with the molecular reorganization and functional expression of receptors, such as AMPA and NMDA in excitatory synapses (Cao and Harris 2012; Baltaci et al. 2019). Increased NMDA-mediated synaptic transmission not only can modify dendritic spine morphology but also can induce new spine formation, stabilizing recently formed spines and enlarging already-existing spines (Lai and Ip 2013). Indeed, an increase in synaptic strength by LTP triggers the pruning and replacement of nonactivated spines by new ones together with a selective stabilization of activated synapses (De Roo et al. 2008) and the rapid formation of new spines that can be blocked by NMDA receptors antagonists (Toni et al. 1999; Ruddy et al. 2015).

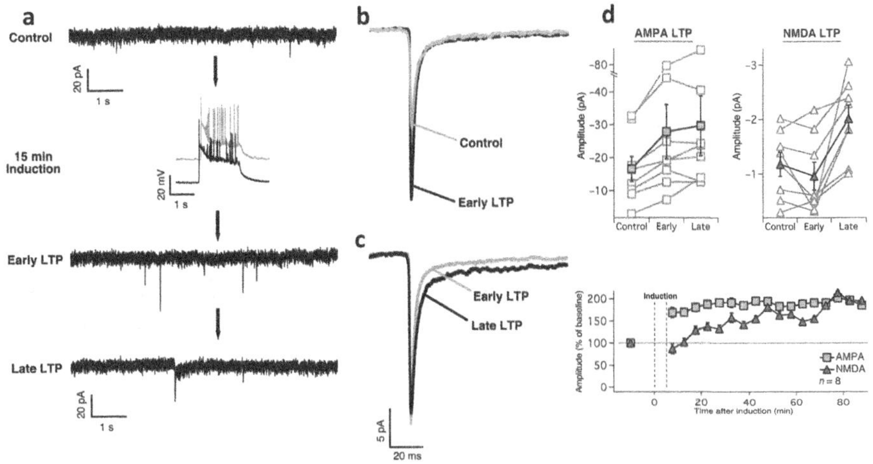

Fig. 3.12 AMPA and NMDA components of the LTP response. (**a**) Representative traces of mEPSCs before (control), immediately after LTP induction (early LTP), and >2 h after LTP induction (late LTP). At early LTP, mEPSC amplitude and frequency increase significantly. At times >2 h after LTP induction (late LTP), mEPSC frequency returned to control levels. (**b**) Average mEPSC waveforms from control (gray) and early LTP (black) showing that the rapid AMPA component of the mEPSC was potentiated (AMPA-LTP), whereas the slowly decaying NMDA component was unaffected. (**c**) Average mEPSC waveforms from early LTP (gray) and late LTP (black). During late LTP, potentiation of the NMDA component (NMDA-LTP) was now also evident. (**d**) AMPA and NMDA components of mEPSCs at early and late LTP. (Up) Increased AMPA component of mEPSCs at both periods, and a delayed potentiation of NMDA component of mEPSCs only at late LTP. (Bottom) Time course of AMPA (open boxes) and NMDA (filled triangles) mEPSC amplitudes, with a prominent contribution of AMPA component. (Adapted from Watt et al. (2004), https://doi.org/10.1038/nn1220, (**a–c**) Figures 1 and (**d**) 6a,b under CCC RightsLink® license 5516591392907)

NMDA receptor-dependent enlargement of spines can be reversed by low-frequency stimulation (LFS) via a phosphatase-dependent mechanism, a well-known mechanism of LTD (Luscher and Malenka 2012).

Therefore, spine dynamics may constitute the basis of long-term molecular reorganization correlated with long-term memory and cognition. Strengthening synaptic connections is believed to be a common mechanism of both processes, and abnormal spine profiles can be correlated with altered synaptic transmission in various neurological and neuropsychiatric conditions (Kasai et al. 2010; Lai and Ip 2013).

3.3.2 Dynamics of Receptors in Spines During Long-Term Depression

LTD has been considered as a parallel and opposite process of LTP, associated with a reduction of functional synapses and, consequently, reduction in reactivity to afferent stimulation (Luscher and Malenka 2012). There are several protocols for

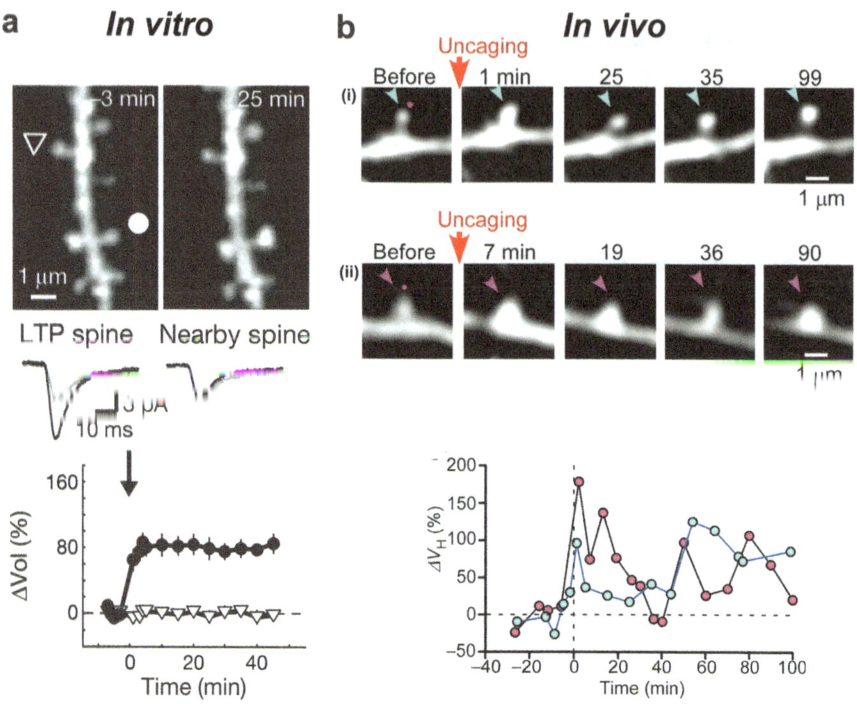

Fig. 3.13 In vitro and in vivo enlargement of spines under LTP. (**a**) (Up) LTP induction in vitro leads to rapid expansion of spine heads. Images before (23 min) and after (25 min) LTP induction. At time = 0, the LTP protocol (30 uncaging pulses at 0.5 Hz, 4-ms pulse duration, postsynaptic potential 0 mV) was applied to the spine marked by a circle (LTP spine). A triangle marks a tested nearby spine. (Middle) uEPSCs, averaged across all cells, in response to test stimuli before (23 min; gray) and after (40 min; black) the LTP protocol. (Bottom) Time course of the changes in uEPSC amplitude and spine volume at the LTP spine (filled circles; *n* 5 7) and at nearby spines (open triangles; uEPSC, *n* 5 7; Vol, *n* 5 31). The arrow marks the LTP protocol. (**b**) (Up) Induction of spine enlargement in the visual cortex in vivo using two-photon glutamate uncaging. Time-lapse images show an expansion of stimulated spines. (Bottom) Spine's volume increased after LTP stimulation. Spine (i) in cyan; and spine (ii) in magenta. (Adapted from (**a**) Harvey and Svoboda (2007), https://doi.org/10.1038/nature06416, Figure 1a,b under CCC RightsLink® license 5517380739551; and (**b**) from Noguchi et al. (2019), https://doi.org/10.1038/s41598-019-50445-0, Figure 1b,c under CC BY license, originally published by Springer Nature)

LTD induction, including a low-frequency stimulation (LFS: 0.5–3 Hz), chemical induction (He et al. 2011) and paired pre- and postsynaptic stimulation (STDP) (Markram et al. 1997; Debanne et al. 1998; Bi and Poo 1998; Sjöström et al. 2001; Feldman 2012) or mediated by metabotropic glutamate receptors (mGluR) (Hasegawa et al. 2015).

The NMDA receptor-mediated LTD involves the activation of NMDA receptors followed by a small but lasting elevation in postsynaptic Ca^{2+} concentration, and activation of a serine-threonine protein phosphatase cascade that act to modulate the spine actin cytoskeleton (Stein and Zito 2019). For example, NMDA receptor-mediated moderated intracellular Ca^{2+} signals in spines and parent dendrites were observed in the timing-dependent LTD triggered by two-photon glutamate uncaging using the post- and pre-protocol in a single spine of layer V pyramidal neurons in juvenile mice (Tazerart et al. 2020) (Fig. 3.9b, c). The LTD induced by STDP at CA3-CA1 hippocampal synapses and some synapses on neocortical layers II/III pyramidal cells is dependent on NMDA receptors (Nishiyama et al. 2000; Froemke et al. 2005; Feldman 2012) (Nishiyama et al. 2000; Froemke et al. 2005). The LTD induced by STDP at several synapses in layers II/III and V of somatosensory and visual cortex and in cortical synapses onto striatal medium spiny neurons is dependent on mGluR- and endocannabinoids receptors (CB1) (Nevian and Sakmann 2006; Feldman 2012).

Following Ca^{2+} entry through activated NMDA receptors, LTD requires the activation of protein phosphatase 1 (PP1), which is essential for the functional process of LTD but not for spine structural modifications. Cofilin, a family of actin-binding proteins that trigger a rapid depolymerization of actin microfilaments, mediates spine structural modifications, that is, spine shrinkage induced by LFS (Zhou et al. 2004). In addition, the activation of NMDA receptors with Ca^{2+} entry and subsequent activation of calcineurin and AP2 are required for both functional and structural processes of LTD (Zhou et al. 2004) and for triggering AMPA receptors endocytosis at the dendritic spine (Vitureira and Goda 2013). The NMDA receptor-induced internalization of AMPA receptors can be inhibited by preventing the increment of intracellular Ca^{2+} concentration or by specific calcineurin inhibitors (Carroll et al. 2001). The removal of extrasynaptic CP-AMPA receptors and replacement by Ca^{2+}-impermeable AMPA receptors (CI-AMPA receptors) is associated with LTD also in mGluR-induced LTD (Bellone and Lüscher 2005; Henley and Wilkinson 2016). During another form of plasticity, in which the neuron self-tunes its synapses, called the homeostatic synaptic scaling (Fig. 3.10a, c) (Turrigiano 2008), the recruitment of CP-AMPA receptors remains controversial and it may be dependent on the synapse, developmental stage and mode of induction. Overall, the activity-dependent removal of CP-AMPA receptors can define a time frame in which memories can be erased (Henley and Wilkinson 2016) (Fig. 3.10a, d).

NMDAR trafficking is also implicated in LTD following NMDA receptor activation (Carroll et al. 2001; Lau and Zukin 2007) (Fig 3.11b). In the hippocampus, for example, NMDAR trafficking requires actin depolymerization (Morishita et al. 2005) for the lateral diffusion of NMDA receptors between synaptic and extrasynaptic sites or dynamin-dependent internalization (Montgomery et al. 2005). Notably, activity-dependent alterations in NMDA synaptic strength influence metaplasticity, experience-dependent plasticity, and structural remodeling of local circuitry (Philpot et al. 2003; Lau and Zukin 2007).

Live imaging of dendritic spines after stimulation by low-frequency uncaging glutamate demonstrates that the stimulus of LTD leads to spine shrinkage specifically of the stimulated spine but not of neighboring spines. Like LTP-induced spine enlargement, spine shrinkage induced by LTD is also synapse-specific (Oh et al. 2013). Volume reduction was observed in both large and small spines after pharmacological manipulations in combination with two-photon glutamate uncaging. Shrinkage of small dendritic spines required activation of NMDA receptors, whereas shrinkage of large spines required signaling through both NMDA and metabotropic glutamate receptors mGluR and the second messenger inositol 1,4,5-trisphosphate receptors (IP3Rs) (Oh et al. 2013). LTD induction in dendritic spines followed or not by spine shrinkage is highly correlated with synaptic weakness (Fig. 3.14). However, these mechanisms also play a vital role in the neural circuit plasticity that underlies learning. Indeed, the formation and stabilization of new dendritic spines as new circuits are formed during learning is only possible due to the elimination of preexisting spines (Yang et al. 2008; De Roo et al. 2008) (see also Chap. 7 in this book).

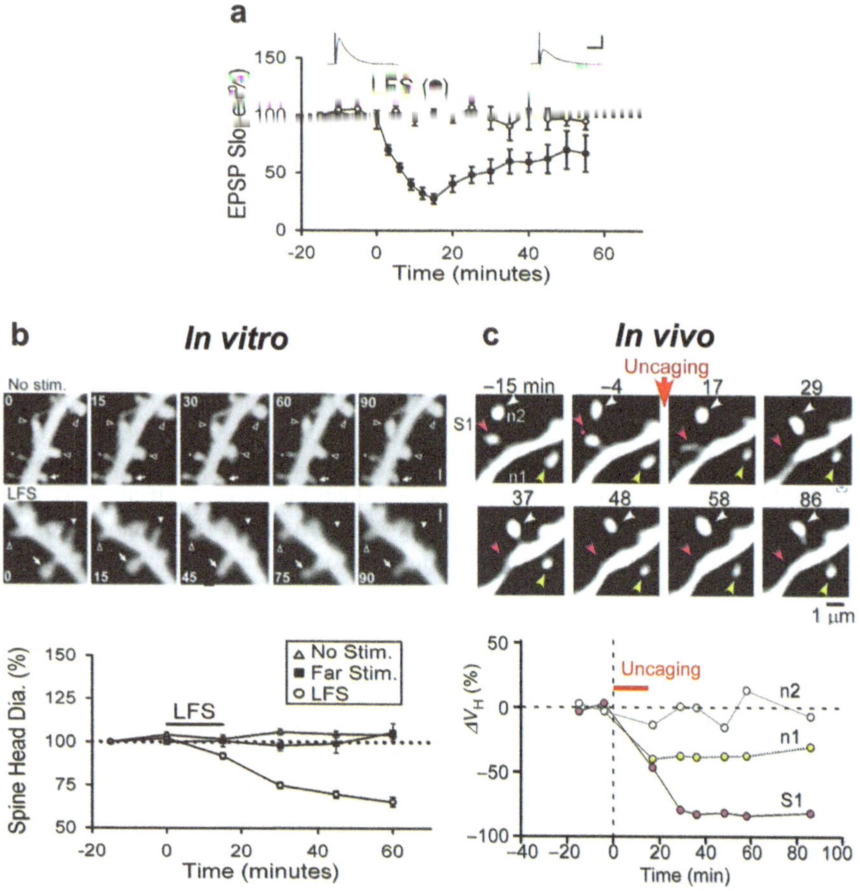

Fig. 3.14 LTD induction and structural plasticity of spines in vitro and in vivo. (**a**) Low-frequency stimulation (LFS) induces long-term depression. (**b**) (Up) In vitro shrinkage of spines at different time points before and after stimulation. (Bottom) Quantitative measure of spine head diameter demonstrating a decrease of spine volume after LTD induction. (**c**) (Up) In vivo shrinkage of spines after LTD induction, showing a decrease of spine volume in the stimulated spine. (Bottom) The magenta, yellow, and white circles indicate spine S1, n1, and n2 traces, respectively. (Adapted from (**a**) and (**b**): Zhou et al. (2004), https://doi.org/10.1016/j.neuron.2004.11.011, (**a**, **b** Bottom) Figure 2A,B and (**b** Up) Figure 1C,D, under CCC RightsLink® license 5518550626563; (**c**) from Noguchi et al. (2019), https://doi.org/10.1038/s41598-019-50445-0, Figure 3A,B under CC BY license, originally published by Springer Nature)

3.4 Concluding Remarks

For many years, the electrical properties of dendritic spines, such as membrane voltage and ionic conductances, were not able to be directly measured due to technical limitations usually difficult to overcome. However, several important biophysical properties of spines have been unveiled recently. Advancements result from the ongoing development of novel and integrated technologies, as described in this chapter.

We highlighted some important achievements for detecting the electrical properties of dendritic spines. First, we discussed the direct measurement of the V_m (bAPs and synaptic potentials) at the spine head, composed by a rich variety of ion channels and receptors and whose dynamics and trafficking are influenced by intracellular signaling and extracellular matrix during synaptic activation and plasticity. By comparing the modifications of the V_m of spines to the parent dendrites and soma, we can now have a better understanding about signal processing and integration. Second, we described the possibility to measure the spine neck resistance, which is associated with the electrical compartmentalization of spines. Spine neck resistance is crucial for tuning the amplitude and kinetics of synaptic signals along the dendrite and controls the spatiotemporal dynamics of information during the synaptic integration and plasticity. Although the electrical role of dendritic spines remains an open issue, these constant advances in technology are providing valuable evidence on the intrinsic properties of dendritic spines for the biochemical and electrical compartmentalization and associated molecular and structural modifications during synaptic plasticity that contribute actively to neuronal behavior.

References

Acker CD, Hoyos E, Loew LM (2016) EPSPs measured in proximal dendritic spines of cortical pyramidal neurons. eNeuro 3(2):ENEURO.0050-15.2016. https://doi.org/10.1523/ENEURO.0050-15.2016

Allen JA, Halverson-Tamboli RA, Rasenick MM (2007) Lipid raft microdomains and neurotransmitter signalling. Nat Rev Neurosci 8(2):128–140. https://doi.org/10.1038/nrn2059

Araya R (2014) Input transformation by dendritic spines of pyramidal neurons. Front Neuroanat 8:141. https://doi.org/10.3389/fnana.2014.00141

Araya R, Jiang J, Eisenthal KB, Yuste R (2006) The spine neck filters membrane potentials. Proc Natl Acad Sci U S A 103(47):17961–17966. https://doi.org/10.1073/pnas.0608755103

Araya R, Nikolenko V, Eisenthal KB, Yuste R (2007) Sodium channels amplify spine potentials. Proc Natl Acad Sci U S A 104(30):12347–12352. https://doi.org/10.1073/pnas.0705282104

Araya R, Vogels TP, Yuste R (2014) Activity-dependent dendritic spine neck changes are correlated with synaptic strength. Proc Natl Acad Sci U S A 111(28):E2895–E2904. https://doi.org/10.1073/pnas.1321869111

Arellano JI (2007) Ultrastructure of dendritic spines: correlation between synaptic and spine morphologies. Front Neurosci 1(1):131–143. https://doi.org/10.3389/neuro.01.1.1.010.2007

Baltaci SB, Mogulkoc R, Baltaci AK (2019) Molecular mechanisms of early and late LTP. Neurochem Res 44(2):281–296. https://doi.org/10.1007/s11064-018-2695-4

Becker MFP, Tetzlaff C (2021) The biophysical basis underlying the maintenance of early phase long-term potentiation. PLoS Comput Biol 17(3):e1008813. https://doi.org/10.1371/journal.pcbi.1008813

Beffert U, Durudas A, Weeber EJ, Stolt PC, Giehl KM, Sweatt JD, Hammer RE, Herz J (2006) Functional dissection of Reelin signaling by site-directed disruption of Disabled-1 adaptor binding to apolipoprotein E receptor 2: distinct roles in development and synaptic plasticity. J Neurosci 26(7):2041–2052. https://doi.org/10.1523/JNEUROSCI.4566-05.2006

Bellone C, Lüscher C (2005) mGluRs induce a long-term depression in the ventral tegmental area that involves a switch of the subunit composition of AMPA receptors. Eur J Neurosci 21(5):1280–1288. https://doi.org/10.1111/j.1460-9568.2005.03979.x

Bi G, Poo M (1998) Synaptic modifications in cultured hippocampal neurons: dependence on spike timing, synaptic strength, and postsynaptic cell type. J Neurosci 18(24):10464–10472. https://doi.org/10.1523/JNEUROSCI.18-24-10464.1998

Blom H, Rönnlund D, Scott L, Spicarova Z, Widengren J, Bondar A, Aperia A, Brismar H (2011) Spatial distribution of Na+/K+-ATPase in dendritic spines dissected by nanoscale superresolution STED microscopy. BMC Neurosci 12(1):16. https://doi.org/10.1186/1471-2202-12-16

Bloodgood BL, Sabatini BL (2005) Neuronal activity regulates diffusion across the neck of dendritic spines. Science 310(5749):866–869. https://doi.org/10.1126/science.1114816

Bloodgood BL, Sabatini BL (2007a) Ca2+ signaling in dendritic spines. Curr Opin Neurobiol 17(3):345–351. https://doi.org/10.1016/j.conb.2007.04.003

Bloodgood BL, Sabatini BL (2007b) Nonlinear regulation of unitary synaptic signals by CaV2.3 voltage-sensitive calcium channels located in dendritic spines. Neuron 53(2):249–260. https://doi.org/10.1016/j.neuron.2006.12.017

Bloodgood BL, Giessel AJ, Sabatini BL (2009) Biphasic synaptic Ca influx arising from compartmentalized electrical signals in dendritic spines. PLoS Biol 7(9):e1000190. https://doi.org/10.1371/journal.pbio.1000190

Brusés JL, Chauvet N, Rutishauser U (2001) Membrane lipid rafts are necessary for the maintenance of the α7 nicotinic acetylcholine receptor in somatic spines of ciliary neurons. J Neurosci 21(2):504–512. https://doi.org/10.1523/JNEUROSCI.21-02-00504.2001

Buard I, Coultrap SJ, Freund RK, Lee YS, Dell'Acqua ML, Silva AJ, Bayer KU (2010) CaMKII "autonomy" is required for initiating but not for maintaining neuronal long-term information storage. Neurosci 30(24):8214–8220. https://doi.org/10.1523/JNEUROSCI.1469-10.2010

Bywalez WG, Patirniche D, Rupprecht V, Stemmler M, Herz AVM, Pálfi D, Rózsa B, Egger V (2015) Local postsynaptic voltage-gated sodium channel activation in dendritic spines of olfactory bulb granule cells. Neuron 85(3):590–601. https://doi.org/10.1016/j.neuron.2014.12.051

Calabrese B, Wilson MS, Halpain S (2006) Development and regulation of dendritic spine synapses. Physiology 21(1):38–47. https://doi.org/10.1152/physiol.00042.2005

Caldwell JH, Schaller KL, Lasher RS, Peles E, Levinson SR (2000) Sodium channel Na(v)1.6 is localized at nodes of ranvier, dendrites, and synapses. Proc Natl Acad Sci U S A 97(10):5616–5620. https://doi.org/10.1073/pnas.090034797

Camiré O, Topolnik L (2014) Dendritic calcium nonlinearities switch the direction of synaptic plasticity in fast-spiking interneurons. J Neurosci 34(11):3864–3877. https://doi.org/10.1523/JNEUROSCI.2253-13.2014

Camiré O, Topolnik L (2018) Two-photon calcium imaging in neuronal dendrites in brain slices. J Vis Exp (133):56776. https://doi.org/10.3791/56776

Campo CG, Sinagra M, Verrier D, Manzoni OJ, Chavis P (2009) Reelin secreted by GABAergic neurons regulates glutamate receptor homeostasis. PLoS One 4(5):e5505. https://doi.org/10.1371/journal.pone.0005505

Cao G, Harris KM (2012) Developmental regulation of the late phase of long-term potentiation (L-LTP) and metaplasticity in hippocampal area CA1 of the rat. J Neurophysiol 107(3):902–912. https://doi.org/10.1152/jn.00780.2011

Caporale N, Dan Y (2008) Spike timing–dependent plasticity: a Hebbian learning rule. Annu Rev Neurosci 31(1):25–46. https://doi.org/10.1146/annurev.neuro.31.060407.125639

Carroll RC, Beattie EC, von Zastrow M, Malenka RC (2001) Role of AMPA receptor endocytosis in synaptic plasticity. Nat Rev Neurosci 2(5):315–324. https://doi.org/10.1038/35072500
Chater TE, Goda Y (2014) The role of AMPA receptors in postsynaptic mechanisms of synaptic plasticity. Front Cell Neurosci 8:401. https://doi.org/10.3389/fncel.2014.00401
Chen X, Leischner U, Rochefort NL, Nelken I, Konnerth A (2011) Functional mapping of single spines in cortical neurons in vivo. Nature 475(7357):501–505. https://doi.org/10.1038/nature10193
Chen JL, Villa KL, Cha JW, So PTC, Kubota Y, Nedivi E (2012) Clustered dynamics of inhibitory synapses and dendritic spines in the adult neocortex. Neuron 74(2):361–373. https://doi.org/10.1016/j.neuron.2012.02.030
Chiu CQ, Lur G, Morse TM, Carnevale NT, Ellis-Davies GCR, Higley MJ (2013) Compartmentalization of GABAergic inhibition by dendritic spines. Science 340(6133):759–762. https://doi.org/10.1126/science.1234274
Chung HJ, Qian X, Ehlers M, Jan YN, Jan LY (2009) Neuronal activity regulates phosphorylation-dependent surface delivery of G protein-activated inwardly rectifying potassium channels. Proc Natl Acad Sci U S A 106(2):629–634. https://doi.org/10.1073/pnas.0811615106
Citri A, Malenka RC (2008) Synaptic plasticity: multiple forms, functions, and mechanisms. Neuropsychopharmacology 33(1):18–41. https://doi.org/10.1038/sj.npp.1301559
Clem RL, Barth A (2006) Pathway-specific trafficking of native AMPARs by in vivo experience. Neuron 49(5):663–670. https://doi.org/10.1016/j.neuron.2006.01.019
Cohen LB, Salzberg BM (1978) Optical measurement of membrane potential. Rev Physiol Biochem Pharmacol 83:35–88. https://doi.org/10.1007/3-540-08907-1_2
Cornejo VH, Ofer N, Yuste R (2022) Voltage compartmentalization in dendritic spines in vivo. Science 375(6576):82–86. https://doi.org/10.1126/science.abg0501
De Roo M, Klauser P, Muller D (2008) LTP promotes a selective long-term stabilization and clustering of dendritic spines. PLoS Biol 6(9):e219. https://doi.org/10.1371/journal.pbio.0060219
de Vivo L, Landi S, Panniello M, Baroncelli L, Chierzi S, Mariotti L, Spolidoro M, Pizzorusso T, Maffei L, Ratto GM (2013) Extracellular matrix inhibits structural and functional plasticity of dendritic spines in the adult visual cortex. Nat Commun 4(1):1484. https://doi.org/10.1038/ncomms2491
Debanne D, Gähwiler BH, Thompson SM (1998) Long-term synaptic plasticity between pairs of individual CA3 pyramidal cells in rat hippocampal slice cultures. J Physiol 507(Pt 1):237–247. https://doi.org/10.1111/j.1469-7793.1998.237bu.x
Denk W, Yuste R, Svoboda K, Tank DW (1996) Imaging calcium dynamics in dendritic spines. Curr Opin Neurobiol 6(3):372–378. https://doi.org/10.1016/s0959-4388(96)80122-x
Dickstein DL, Weaver CM, Luebke JI, Hof PR (2013) Dendritic spine changes associated with normal aging. Neuroscience 251:21–32. https://doi.org/10.1016/j.neuroscience.2012.09.077
Dityatev A, Rusakov DA (2011) Molecular signals of plasticity at the tetrapartite synapse. Curr Opin Neurobiol 21(2):353–359. https://doi.org/10.1016/j.conb.2010.12.006
Dityatev A, Schachner M (2003) Extracellular matrix molecules and synaptic plasticity. Nat Rev Neurosci 4(6):456–468. https://doi.org/10.1038/nrn1115
Dityatev A, Frischknecht R, Seidenbecher CI (2006) Extracellular matrix and synaptic functions. In: Gundelfinger ED, Seidenbecher CI, Schraven B (eds) Cell communication in nervous and immune system. Springer, Berlin Heidelberg, pp 69–97
Dityatev A, Schachner M, Sonderegger P (2010) The dual role of the extracellular matrix in synaptic plasticity and homeostasis. Nat Rev Neurosci 11(11):735–746. https://doi.org/10.1038/nrn2898
Dozmorov M, Li R, Abbas A-K, Hellberg F, Farre C, Huang F-S, Jilderos B, Wigström H (2006) Contribution of AMPA and NMDA receptors to early and late phases of LTP in hippocampal slices. Neurosci Res 55(2):182–188. https://doi.org/10.1016/j.neures.2006.03.001
Egawa J, Pearn ML, Lemkuil BP, Patel PM, Head BP (2016) Membrane lipid rafts and neurobiology: age-related changes in membrane lipids and loss of neuronal function. J Physiol 594(16):4565–4579. https://doi.org/10.1113/JP270590

Enoki R, Hu Y, Hamilton D, Fine A (2009) Expression of long-term plasticity at individual synapses in hippocampus is graded, bidirectional, and mainly presynaptic: optical quantal analysis. Neuron 62(2):242–253. https://doi.org/10.1016/j.neuron.2009.02.026

Feldman DE (2012) The spike-timing dependence of plasticity. Neuron 75(4):556–571. https://doi.org/10.1016/j.neuron.2012.08.001

Filipis L, Canepari M (2021) Optical measurement of physiological sodium currents in the axon initial segment. J Physiol 599(1):49–66. https://doi.org/10.1113/JP280554

Froemke RC, Poo M-M, Dan Y (2005) Spike-timing-dependent synaptic plasticity depends on dendritic location. Nature 434(7030):221–225. https://doi.org/10.1038/nature03366

Grosshans DR, Clayton DA, Coultrap SJ, Browning MD (2002) LTP leads to rapid surface expression of NMDA but not AMPA receptors in adult rat CA1. Nat Neurosci 5(1):27–33. https://doi.org/10.1038/nn779

Grunditz Å, Holbro N, Tian L, Zuo Y, Oertner TG (2008) Spine neck plasticity controls postsynaptic calcium signals through electrical compartmentalization. J Neurosci 28(50):13457–13466. https://doi.org/10.1523/JNEUROSCI.2702-08.2008

Gundelfinger ED, Frischknecht R, Choquet D, Heine M (2010) Converting juvenile into adult plasticity: a role for the brain's extracellular matrix: converting juvenile into adult plasticity. Eur J Neurosci 31(12):2156–2165. https://doi.org/10.1111/j.1460-9568.2010.07253.x

Harnett MT, Makara JK, Spruston N, Kath WL, Magee JC (2012) Synaptic amplification by dendritic spines enhances input cooperativity. Nature 491(7425):599–602. https://doi.org/10.1038/nature11554

Harris KM, Kater SB (1994) Dendritic spines: cellular specializations imparting both stability and flexibility to synaptic function. Annu Rev Neurosci 17:341–371. https://doi.org/10.1146/annurev.ne.17.030194.002013

Harris K, Stevens J (1988) Dendritic spines of rat cerebellar Purkinje cells: serial electron microscopy with reference to their biophysical characteristics. J Neurosci 8(12):4455–4469. https://doi.org/10.1523/JNEUROSCI.08-12-04455.1988

Harris K, Stevens J (1989) Dendritic spines of CA 1 pyramidal cells in the rat hippocampus: serial electron microscopy with reference to their biophysical characteristics. J Neurosci 9(8):2982–2997. https://doi.org/10.1523/JNEUROSCI.09-08-02982.1989

Harris KM, Jensen FE, Tsao B (1992) Three-dimensional structure of dendritic spines and synapses in rat hippocampus (CA1) at postnatal day 15 and adult ages: implications for the maturation of synaptic physiology and long-term potentiation. J Neurosci 12(7):2685–2705. https://doi.org/10.1523/JNEUROSCI.12-07-02685.1992

Harvey CD, Svoboda K (2007) Locally dynamic synaptic learning rules in pyramidal neuron dendrites. Nature 450(7173):1195–1200. https://doi.org/10.1038/nature06416

Harward SC, Hedrick NG, Hall CE, Parra-Bueno P, Milner TA, Pan E, Laviv T, Hempstead BL, Yasuda R, McNamara JO (2016) Autocrine BDNF-TrkB signalling within a single dendritic spine. Nature 538(7623):99–103. https://doi.org/10.1038/nature19766

Hasegawa S, Sakuragi S, Tominaga-Yoshino K, Ogura A (2015) Dendritic spine dynamics leading to spine elimination after repeated inductions of LTD. Sci Rep 5:7707. https://doi.org/10.1038/srep07707

Häusser M, Spruston N, Stuart GJ (2000) Diversity and dynamics of dendritic signaling. Science 290(5492):739–744. https://doi.org/10.1126/science.290.5492.739

He K, Lee A, Song L, Kanold PO, Lee H-K (2011) AMPA receptor subunit GluR1 (GluA1) serine-845 site is involved in synaptic depression but not in spine shrinkage associated with chemical long-term depression. J Neurophysiol 105(4):1897–1907. https://doi.org/10.1152/jn.00913.2010

Hedrick NG, Harward SC, Hall CE, Murakoshi H, McNamara JO, Yasuda R (2016) Rho GTPase complementation underlies BDNF-dependent homo- and heterosynaptic plasticity. Nature 538(7623):104–108. https://doi.org/10.1038/nature19784

Hellwig S, Hack I, Kowalski J, Brunne B, Jarowyj J, Unger A, Bock HH, Junghans D, Frotscher M (2011) Role for Reelin in neurotransmitter release. J Neurosci 31(7):2352–2360. https://doi.org/10.1523/JNEUROSCI.3984-10.2011

Henley JM, Wilkinson KA (2016) Synaptic AMPA receptor composition in development, plasticity and disease. Nat Rev Neurosci 17(6):337–350. https://doi.org/10.1038/nrn.2016.37
Hestrin S, Nicoll RA, Perkel DJ, Sah P (1990) Analysis of excitatory synaptic action in pyramidal cells using whole-cell recording from rat hippocampal slices. J Physiol 422:203–225. https://doi.org/10.1113/jphysiol.1990.sp017980
Holthoff K, Zecevic D, Konnerth A (2010) Rapid time course of action potentials in spines and remote dendrites of mouse visual cortex neurons: optical recording of APs in dendritic spines. J Physiol 588(7):1085–1096. https://doi.org/10.1113/jphysiol.2009.184960
Honigmann A, Pralle A (2016) Compartmentalization of the cell membrane. J Mol Biol 428(24):4739–4748. https://doi.org/10.1016/j.jmb.2016.09.022
Huntley GW, Gil O, Bozdagi O (2002) The cadherin family of cell adhesion molecules: multiple roles in synaptic plasticity. Neuroscientist 8(3):221–233
Jaafari N, Konopacki FA, Owen TF, Kantamneni S, Rubin P, Craig TJ, Wilkinson KA, Henley JM (2013) SUMOylation is required for glycine-induced increases in AMPA receptor surface expression (ChemLTP) in hippocampal neurons. PLoS One 8(1):e52345. https://doi.org/10.1371/journal.pone.0052345
Jaffe DB, Johnston D, Lasser-Ross N, Lisman JE, Miyakawa H, Ross WN (1992) The spread of Na+ spikes determines the pattern of dendritic Ca2+ entry into hippocampal neurons. Nature 357(6375):244–246. https://doi.org/10.1038/357244a0
Jayant K, Hirtz JJ, Plante IJ-L, Tsai DM, De Boer WDAM, Semonche A, Peterka DS, Owen JS, Sahin O, Shepard KL, Yuste R (2017) Targeted intracellular voltage recordings from dendritic spines using quantum-dot-coated nanopipettes. Nat Nanotechnol 12(4):335–342. https://doi.org/10.1038/nnano.2016.268
Johnston D, Magee JC, Colbert CM, Christie BR (1996) Active properties of neuronal dendrites. Annu Rev Neurosci 19:165–186. https://doi.org/10.1146/annurev.ne.19.030196.001121
Kasai H, Fukuda M, Watanabe S, Hayashi-Takagi A, Noguchi J (2010) Structural dynamics of dendritic spines in memory and cognition. Trends Neurosci 33(3):121–129. https://doi.org/10.1016/j.tins.2010.01.001
Kaufmann WA, Matsui K, Jeromin A, Nerbonne JM, Ferraguti F (2013) Kv4.2 potassium channels segregate to extrasynaptic domains and influence intrasynaptic NMDA receptor NR2B subunit expression. Brain Struct Funct 218(5):1115–1132. https://doi.org/10.1007/s00429-012-0450-1
Kim J, Jung S-C, Clemens AM, Petralia RS, Hoffman DA (2007) Regulation of dendritic excitability by activity-dependent trafficking of the A-type K+ channel subunit Kv4.2 in hippocampal neurons. Neuron 54(6):933–947. https://doi.org/10.1016/j.neuron.2007.05.026
Koch C, Segev I (2000) The role of single neurons in information processing. Nat Neurosci 3(S11):1171–1177. https://doi.org/10.1038/81444
Koch C, Zador A (1993) The function of dendritic spines: devices subserving biochemical rather than electrical compartmentalization. J Neurosci 13(2):413–422. https://doi.org/10.1523/JNEUROSCI.13-02-00413.1993
Kochlamazashvili G, Henneberger C, Bukalo O, Dvoretskova E, Senkov O, Lievens PM-J, Westenbroek R, Engel AK, Catterall WA, Rusakov DA, Schachner M, Dityatev A (2010) The extracellular matrix molecule hyaluronic acid regulates hippocampal synaptic plasticity by modulating postsynaptic L-type Ca2+ channels. Neuron 67(1):116–128. https://doi.org/10.1016/j.neuron.2010.05.030
Koester HJ, Sakmann B (1998) Calcium dynamics in single spines during coincident pre- and postsynaptic activity depend on relative timing of back-propagating action potentials and subthreshold excitatory postsynaptic potentials. Proc Natl Acad Sci U S A 95(16):9596–9601. https://doi.org/10.1073/pnas.95.16.9596
Koleske AJ (2013) Molecular mechanisms of dendrite stability. Nat Rev Neurosci 14(8):536–550. https://doi.org/10.1038/nrn3486
Korngreen A, Sakmann B (2000) Voltage-gated K+ channels in layer 5 neocortical pyramidal neurones from young rats: subtypes and gradients. J Physiol 525(Pt 3):621–639. https://doi.org/10.1111/j.1469-7793.2000.00621.x

Kovalchuk Y, Eilers J, Lisman J, Konnerth A (2000) NMDA receptor-mediated subthreshold Ca^{2+} signals in spines of hippocampal neurons. J Neurosci 20(5):1791–1799. https://doi.org/10.1523/JNEUROSCI.20-05-01791.2000

Kwon T, Sakamoto M, Peterka DS, Yuste R (2017) Attenuation of synaptic potentials in dendritic spines. Cell Rep 20(5):1100–1110. https://doi.org/10.1016/j.celrep.2017.07.012

Lai K-O, Ip NY (2013) Structural plasticity of dendritic spines: the underlying mechanisms and its dysregulation in brain disorders. Biochim Biophys Acta (BBA) – Mol Basis Dis 1832(12):2257–2263. https://doi.org/10.1016/j.bbadis.2013.08.012

Lan J, Skeberdis VA, Jover T, Grooms SY, Lin Y, Araneda RC, Zheng X, Bennett MVL, Zukin RS (2001) Protein kinase C modulates NMDA receptor trafficking and gating. Nat Neurosci 4(4):382–390. https://doi.org/10.1038/86028

Lanté F, de Jésus Ferreira M-C, Guiramand J, Récasens M, Vignes M (2006) Low-frequency stimulation induces a new form of LTP, metabotropic glutamate (mGlu5) receptor- and PKA-dependent, in the CA1 area of the rat hippocampus. Hippocampus 16(1):[illegible]. [illegible]

Lau CG, Zukin RS (2007) NMDA receptor trafficking in synaptic plasticity and neuropsychiatric disorders. Nat Rev Neurosci 8(6):413–426. https://doi.org/10.1038/nrn2153

Lau CG, Takayasu Y, Rodenas-Ruano A, Paternain AV, Lerma J, Bennett MVL, Zukin RS (2010) SNAP-25 is a target of protein kinase C phosphorylation critical to NMDA receptor trafficking. J Neurosci 30(1):242–254. https://doi.org/10.1523/JNEUROSCI.4933-08.2010

Lee S-JR, Escobedo-Lozoya Y, Szatmari EM, Yasuda R (2009) Activation of CaMKII in single dendritic spines during long-term potentiation. Nature 458(7236):299–304. https://doi.org/10.1038/nature07842

Lee H-K, Takamiya K, He K, Song L, Huganir RL (2010) Specific roles of AMPA receptor subunit GluR1 (GluA1) phosphorylation sites in regulating synaptic plasticity in the CA1 region of hippocampus. J Neurophysiol 103(1):479–489. https://doi.org/10.1152/jn.00835.2009

Lisman J, Yasuda R, Raghavachari S (2012) Mechanisms of CaMKII action in long-term potentiation. Nat Rev Neurosci 13(3):169–182. https://doi.org/10.1038/nrn3192

London M, Häusser M (2005) Dendritic computation. Annu Rev Neurosci 28:503–532

Lorincz A, Nusser Z (2010) Molecular identity of dendritic voltage-gated sodium channels. Science 328(5980):906–909. https://doi.org/10.1126/science.1187958

Lu W-Y, Man H-Y, Ju W, Trimble WS, MacDonald JF, Wang YT (2001) Activation of synaptic NMDA receptors induces membrane insertion of new AMPA receptors and LTP in cultured hippocampal neurons. Neuron 29(1):243–254. https://doi.org/10.1016/S0896-6273(01)00194-5

Luscher C, Malenka RC (2012) NMDA receptor-dependent long-term potentiation and long-term depression (LTP/LTD). Cold Spring Harb Perspect Biol 4(6):a005710. https://doi.org/10.1101/cshperspect.a005710

Magee JC, Johnston D (1997) A synaptically controlled, associative signal for Hebbian plasticity in hippocampal neurons. Science 275(5297):209–213. https://doi.org/10.1126/science.275.5297.209

Mainen ZF, Maletic-Savatic M, Shi SH, Hayashi Y, Malinow R, Svoboda K (1999) Two-photon imaging in living brain slices. Methods 18(2):231–239. https://doi.org/10.1006/meth.1999.0776

Makino H, Malinow R (2009) AMPA receptor incorporation into synapses during LTP: the role of lateral movement and exocytosis. Neuron 64(3):381–390. https://doi.org/10.1016/j.neuron.2009.08.035

Markram H, Lübke J, Frotscher M, Sakmann B (1997) Regulation of synaptic efficacy by coincidence of postsynaptic APs and EPSPs. Science 275(5297):213–215. https://doi.org/10.1126/science.275.5297.213

Mateos-Aparicio P, Rodríguez-Moreno A (2020) Calcium dynamics and synaptic plasticity. Adv Exp Med Biol 1131:965–984. https://doi.org/10.1007/978-3-030-12457-1_38

Matsuzaki M, Ellis-Davies GCR, Nemoto T, Miyashita Y, Iino M, Kasai H (2001) Dendritic spine geometry is critical for AMPA receptor expression in hippocampal CA1 pyramidal neurons. Nat Neurosci 4(11):1086–1092. https://doi.org/10.1038/nn736

Matsuzaki M, Honkura N, Ellis-Davies GCR, Kasai H (2004) Structural basis of long-term potentiation in single dendritic spines. Nature 429(6993):761–766. https://doi.org/10.1038/nature02617
McGeachie AB, Cingolani LA, Goda Y (2011) A stabilising influence: integrins in regulation of synaptic plasticity. Neurosci Res 70(1):24–29. https://doi.org/10.1016/j.neures.2011.02.006
Miller S, Yasuda M, Coats JK, Jones Y, Martone ME, Mayford M (2002) Disruption of dendritic translation of CaMKIIalpha impairs stabilization of synaptic plasticity and memory consolidation Neuron 36(3):507–519. https://doi.org/10.1016/s0896-6273(02)00978-9
Miyazaki K, Ross WN (2017) Sodium dynamics in pyramidal neuron dendritic spines: synaptically evoked entry predominantly through AMPA receptors and removal by diffusion. J Neurosci 37(41):9964–9976. https://doi.org/10.1523/JNEUROSCI.1758-17.2017
Miyazaki K, Ross WN (2022) Fast synaptically activated calcium and sodium kinetics in hippocampal pyramidal neuron dendritic spines. eNeuro 9(6):ENEURO.0396-22.2022. https://doi.org/10.1523/ENEURO.0396-22.2022
Montgomery JM, Selcher JC, Hanson JE, Madison DV (2005) Dynamin-dependent NMDAR endocytosis during LTD and its dependence on synaptic state. BMC Neurosci 6:48. https://doi.org/10.1186/1471-2202-6-48
Morishita W, Marie H, Malenka RC (2005) Distinct triggering and expression mechanisms underlie LTD of AMPA and NMDA synaptic responses. Nat Neurosci 8(8):1043–1050. https://doi.org/10.1038/nn1506
Nagy V, Bozdagi O, Matynia A, Balcerzyk M, Okulski P, Dzwonek J, Costa RM, Silva AJ, Kaczmarek L, Huntley GW (2006) Matrix metalloproteinase-9 is required for hippocampal late-phase long-term potentiation and memory. J Neurosci 26(7):1923–1934. https://doi.org/10.1523/JNEUROSCI.4359-05.2006
Nakahata Y, Yasuda R (2018) Plasticity of spine structure: local signaling, translation and cytoskeletal reorganization. Front Synaptic Neurosci 10:29. https://doi.org/10.3389/fnsyn.2018.00029
Nakamura T, Barbara J-G, Nakamura K, Ross WN (1999) Synergistic release of Ca2+ from IP3-sensitive stores evoked by synaptic activation of mGluRs paired with backpropagating action potentials. Neuron 24(3):727–737. https://doi.org/10.1016/S0896-6273(00)81125-3
Nevian T, Sakmann B (2004) Single spine Ca^{2+} signals evoked by coincident EPSPs and backpropagating action potentials in spiny stellate cells of layer 4 in the juvenile rat somatosensory barrel cortex. J Neurosci 24(7):1689–1699. https://doi.org/10.1523/JNEUROSCI.3332-03.2004
Nevian T, Sakmann B (2006) Spine Ca^{2+} signaling in spike-timing-dependent plasticity. J Neurosci 26(43):11001–11013. https://doi.org/10.1523/JNEUROSCI.1749-06.2006
Nicoll RA (2017) A brief history of long-term potentiation. Neuron 93(2):281–290. https://doi.org/10.1016/j.neuron.2016.12.015
Nishiyama J (2019) Plasticity of dendritic spines: molecular function and dysfunction in neurodevelopmental disorders. Psychiatry Clin Neurosci 73(9):541–550. https://doi.org/10.1111/pcn.12899
Nishiyama J, Yasuda R (2015) Biochemical computation for spine structural plasticity. Neuron 87(1):63–75. https://doi.org/10.1016/j.neuron.2015.05.043
Nishiyama M, Hong K, Mikoshiba K, Poo MM, Kato K (2000) Calcium stores regulate the polarity and input specificity of synaptic modification. Nature 408(6812):584–588. https://doi.org/10.1038/35046067
Noguchi J, Matsuzaki M, Ellis-Davies GCR, Kasai H (2005) Spine-neck geometry determines NMDA receptor-dependent Ca2+ signaling in dendrites. Neuron 46(4):609–622. https://doi.org/10.1016/j.neuron.2005.03.015
Noguchi J, Nagaoka A, Watanabe S, Ellis-Davies GCR, Kitamura K, Kano M, Matsuzaki M, Kasai H (2011) In vivo two-photon uncaging of glutamate revealing the structure-function relationships of dendritic spines in the neocortex of adult mice. J Physiol 589(Pt 10):2447–2457. https://doi.org/10.1113/jphysiol.2011.207100
Noguchi J, Nagaoka A, Hayama T, Ucar H, Yagishita S, Takahashi N, Kasai H (2019) Bidirectional in vivo structural dendritic spine plasticity revealed by two-photon glutamate uncaging in the mouse neocortex. Sci Rep 9(1):13922. https://doi.org/10.1038/s41598-019-50445-0

Nuriya M, Jiang J, Nemet B, Eisenthal KB, Yuste R (2006) Imaging membrane potential in dendritic spines. Proc Natl Acad Sci U S A 103(3):786–790. https://doi.org/10.1073/pnas.0510092103

Oertner TG (2002) Functional imaging of single synapses in brain slices. Exp Physiol 87(6):733–736. https://doi.org/10.1113/eph8702482

Oh WC, Hill TC, Zito K (2013) Synapse-specific and size-dependent mechanisms of spine structural plasticity accompanying synaptic weakening. Proc Natl Acad Sci U S A 110(4):E305–E312. https://doi.org/10.1073/pnas.1214705110

Oray S, Majewska A, Sur M (2004) Dendritic spine dynamics are regulated by monocular deprivation and extracellular matrix degradation. Neuron 44(6):1021–1030. https://doi.org/10.1016/j.neuron.2004.12.001

Orlando C, Ster J, Gerber U, Fawcett JW, Raineteau O (2012) Perisynaptic chondroitin sulfate proteoglycans restrict structural plasticity in an integrin-dependent manner. J Neurosci 32(50):18009–18017. https://doi.org/10.1523/JNEUROSCI.2406-12.2012

Palmer LM, Stuart GJ (2009) Membrane potential changes in dendritic spines during action potentials and synaptic input. J Neurosci 29(21):6897–6903. https://doi.org/10.1523/JNEUROSCI.5847-08.2009

Park M (2018) AMPA receptor trafficking for postsynaptic potentiation. Front Cell Neurosci 12:361. https://doi.org/10.3389/fncel.2018.00361

Penn AC, Zhang CL, Georges F, Royer L, Breillat C, Hosy E, Petersen JD, Humeau Y, Choquet D (2017) Hippocampal LTP and contextual learning require surface diffusion of AMPA receptors. Nature 549(7672):384–388. https://doi.org/10.1038/nature23658

Penzes P, Cahill ME, Jones KA, VanLeeuwen J-E, Woolfrey KM (2011) Dendritic spine pathology in neuropsychiatric disorders. Nat Neurosci 14(3):285–293. https://doi.org/10.1038/nn.2741

Philpot BD, Espinosa JS, Bear MF (2003) Evidence for altered NMDA receptor function as a basis for metaplasticity in visual cortex. J Neurosci 23(13):5583–5588. https://doi.org/10.1523/JNEUROSCI.23-13-05583.2003

Polsky A, Mel B, Schiller J (2009) Encoding and decoding bursts by NMDA spikes in basal dendrites of layer 5 pyramidal neurons. J Neurosci 29(38):11891–11903. https://doi.org/10.1523/JNEUROSCI.5250-08.2009

Popovic M, Gao X, Zecevic D (2012) Voltage-sensitive dye recording from axons, dendrites and dendritic spines of individual neurons in brain slices. J Vis Exp (69):4261. https://doi.org/10.3791/4261

Popovic MA, Gao X, Carnevale NT, Zecevic D (2014) Cortical dendritic spine heads are not electrically isolated by the spine neck from membrane potential signals in parent dendrites. Cereb Cortex 24(2):385–395. https://doi.org/10.1093/cercor/bhs320

Popovic M, Vogt K, Holthoff K, Konnerth A, Salzberg BM, Grinvald A, Antic SD, Canepari M, Zecevic D (2015a) Imaging submillisecond membrane potential changes from individual regions of single axons, dendrites and spines. In: Canepari M, Zecevic D, Bernus O (eds) Membrane potential imaging in the nervous system and heart. Springer International Publishing, Cham, pp 57–101

Popovic MA, Carnevale N, Rozsa B, Zecevic D (2015b) Electrical behaviour of dendritic spines as revealed by voltage imaging. Nat Commun 6(1):8436. https://doi.org/10.1038/ncomms9436

Priel A, Dai X-Q, Chen X-Z, Scarinci N, Cantero MR, Cantiello HF (2022) Electrical recordings from dendritic spines of adult mouse hippocampus and effect of the actin cytoskeleton. Front Mol Neurosci 15:769725. https://doi.org/10.3389/fnmol.2022.769725

Pujadas L, Gruart A, Bosch C, Delgado L, Teixeira CM, Rossi D, de Lecea L, Martínez A, Delgado-García JM, Soriano E (2010) Reelin regulates postnatal neurogenesis and enhances spine hypertrophy and long-term potentiation. J Neurosci 30(13):4636–4649. https://doi.org/10.1523/JNEUROSCI.5284-09.2010

Qiu S, Zhao LF, Korwek KM, Weeber EJ (2006) Differential Reelin-induced enhancement of NMDA and AMPA receptor activity in the adult hippocampus. J Neurosci 26(50):12943–12955. https://doi.org/10.1523/JNEUROSCI.2561-06.2006

Rossini L, De Santis D, Mauceri RR, Tesoriero C, Bentivoglio M, Maderna E, Maiorana A, Deleo F, de Curtis M, Tringali G, Cossu M, Tumminelli G, Bramerio M, Spreafico R, Tassi L, Garbelli R (2021) Dendritic pathology, spine loss and synaptic reorganization in human cortex from epilepsy patients. Brain 144(1):251–265. https://doi.org/10.1093/brain/awaa387

Royer S, Martina M, Paré D (1999) An inhibitory interface gates impulse traffic between the input and output stations of the amygdala. J Neurosci 19(23):10575–10583. https://doi.org/10.1523/JNEUROSCI.19-23-10575.1999

Ruddy RM, Chen Y, Milenkovic M, Ramsey AJ (2015) Differential effects of NMDA receptor antagonism on spine density. Synapse 69(1):52–56. https://doi.org/10.1002/syn.21784

Runge K, Cardoso C, de Chevigny A (2020) Dendritic spine plasticity: function and mechanisms. Front Synaptic Neurosci 12:36. https://doi.org/10.3389/fnsyn.2020.00036

Sabatini BL, Oertner TG, Svoboda K (2002) The life cycle of Ca2+ ions in dendritic spines. Neuron 33(3):439–452. https://doi.org/10.1016/S0896-6273(02)00573-1

Schiller J, Schiller Y, Clapham DE (1998) NMDA receptors amplify calcium influx into dendritic spines during associative pre- and postsynaptic activation. Nat Neurosci 1(2):114–118. https://doi.org/10.1038/363

Shepherd GM (1996) The dendritic spine: a multifunctional integrative unit. J Neurophysiol 75(6):2197–2210. https://doi.org/10.1152/jn.1996.75.6.2197

Sjöström PJ, Turrigiano GG, Nelson SB (2001) Rate, timing, and cooperativity jointly determine cortical synaptic plasticity. Neuron 32(6):1149–1164. https://doi.org/10.1016/S0896-6273(01)00542-6

Sjöström PJ, Rancz EA, Roth A, Häusser M (2008) Dendritic excitability and synaptic plasticity. Physiol Rev 88(2):769–840. https://doi.org/10.1152/physrev.00016.2007

Song I, Dityatev A (2018) Crosstalk between glia, extracellular matrix and neurons. Brain Res Bull 136:101–108. https://doi.org/10.1016/j.brainresbull.2017.03.003

Spruston N, Jaffe DB, Johnston D (1994) Dendritic attenuation of synaptic potentials and currents: the role of passive membrane properties. Trends Neurosci 17(4):161–166. https://doi.org/10.1016/0166-2236(94)90094-9

Spruston N, Jonas P, Sakmann B (1995) Dendritic glutamate receptor channels in rat hippocampal CA3 and CA1 pyramidal neurons. J Physiol 482(2):325–352. https://doi.org/10.1113/jphysiol.1995.sp020521

Stein IS, Zito K (2019) Dendritic spine elimination: molecular mechanisms and implications. Neuroscientist 25(1):27–47. https://doi.org/10.1177/1073858418769644

Strobel C, Marek R, Gooch HM, Sullivan RKP, Sah P (2015) Prefrontal and auditory input to intercalated neurons of the amygdala. Cell Rep 10(9):1435–1442. https://doi.org/10.1016/j.celrep.2015.02.008

Strobel C, Sullivan RKP, Stratton P, Sah P (2017) Calcium signalling in medial intercalated cell dendrites and spines: dendritic properties of medial ITC neurons. J Physiol 595(16):5653–5669. https://doi.org/10.1113/JP274261

Stuart GJ, Häusser M (2001) Dendritic coincidence detection of EPSPs and action potentials. Nat Neurosci 4(1):63–71. https://doi.org/10.1038/82910

Stuart GJ, Sakmann B (1994) Active propagation of somatic action potentials into neocortical pyramidal cell dendrites. Nature 367(6458):69–72. https://doi.org/10.1038/367069a0

Stuart GJ, Dodt HU, Sakmann B (1993) Patch-clamp recordings from the soma and dendrites of neurons in brain slices using infrared video microscopy. Pflugers Arch 423(5–6):511–518. https://doi.org/10.1007/BF00374949

Stuart G, Schiller J, Sakmann B (1997a) Action potential initiation and propagation in rat neocortical pyramidal neurons. J Physiol 505(3):617–632. https://doi.org/10.1111/j.1469-7793.1997.617ba.x

Stuart G, Spruston N, Sakmann B, Häusser M (1997b) Action potential initiation and back-propagation in neurons of the mammalian CNS. Trends Neurosci 20(3):125–131. https://doi.org/10.1016/S0166-2236(96)10075-8

Svoboda K, Tank DW, Denk W (1996) Direct measurement of coupling between dendritic spines and shafts. Science 272(5262):716–719. https://doi.org/10.1126/science.272.5262.716

Tada T, Sheng M (2006) Molecular mechanisms of dendritic spine morphogenesis. Curr Opin Neurobiol 16(1):95–101. https://doi.org/10.1016/j.conb.2005.12.001

Takasaki K, Sabatini BL (2014) Super-resolution 2-photon microscopy reveals that the morphology of each dendritic spine correlates with diffusive but not synaptic properties. Front Neuroanat 8:29. https://doi.org/10.3389/fnana.2014.00029

Tanaka H, Hirano T (2012) Visualization of subunit-specific delivery of glutamate receptors to postsynaptic membrane during hippocampal long-term potentiation. Cell Rep 1(4):291–298. https://doi.org/10.1016/j.celrep.2012.02.004

Tanese D, Weng J-Y, Zampini V, De Sars V, Canepari M, Rozsa B, Emiliani V, Zecevic D (2017) Imaging membrane potential changes from dendritic spines using computer-generated holography. Neurophotonics 4(3):031211. https://doi.org/10.1117/1.NPh.4.3.031211

Tazerart S, Mitchell DE, Miranda-Rottmann S, Araya R (2020) A spike-timing-dependent plasticity rule for dendritic spines. Nat Commun 11(1):4276. https://doi.org/10.1038/s41467-020-17861-7

Tazerart S, Blanchard MG, Miranda-Rottmann S, Mitchell DE, Navea Pina B, Thomas CI, Kamasawa N, Araya R (2022) Selective activation of BK channels in small-headed dendritic spines suppresses excitatory postsynaptic potentials. J Physiol 600(9):2165–2187. https://doi.org/10.1113/JP282303

Theis A-K, Rózsa B, Katona G, Schmitz D, Johenning FW (2018) Voltage gated calcium channel activation by backpropagating action potentials downregulates NMDAR function. Front Cell Neurosci 12:109. https://doi.org/10.3389/fncel.2018.00109

Toni N, Buchs P-A, Nikonenko I, Bron CR, Muller D (1999) LTP promotes formation of multiple spine synapses between a single axon terminal and a dendrite. Nature 402(6760):421–425. https://doi.org/10.1038/46574

Tønnesen J, Nägerl UV (2016) Dendritic spines as tunable regulators of synaptic signals. Front Psych 7:101. https://doi.org/10.3389/fpsyt.2016.00101

Tønnesen J, Katona G, Rózsa B, Nägerl UV (2014) Spine neck plasticity regulates compartmentalization of synapses. Nat Neurosci 17(5):678–685. https://doi.org/10.1038/nn.3682

Trommald M, Hulleberg G (1997) Dimensions and density of dendritic spines from rat dentate granule cells based on reconstructions from serial electron micrographs. J Comp Neurol 377(1):15–28. https://doi.org/10.1002/(SICI)1096-9861(19970106)377:1<15::AID-CNE3>3.0.CO;2-M

Tsay D, Yuste R (2004) On the electrical function of dendritic spines. Trends Neurosci 27(2):77–83. https://doi.org/10.1016/j.tins.2003.11.008

Turrigiano GG (2008) The self-tuning neuron: synaptic scaling of excitatory synapses. Cell 135(3):422–435. https://doi.org/10.1016/j.cell.2008.10.008

Vallés AS, Barrantes FJ (2021) Nanoscale sub-compartmentalization of the dendritic spine compartment. Biomol Ther 11(11):1697. https://doi.org/10.3390/biom11111697

van Versendaal D, Rajendran R, Saiepour MH, Klooster J, Smit-Rigter L, Sommeijer J-P, De Zeeuw CI, Hofer SB, Heimel JA, Levelt CN (2012) Elimination of inhibitory synapses is a major component of adult ocular dominance plasticity. Neuron 74(2):374–383. https://doi.org/10.1016/j.neuron.2012.03.015

Vanderklish PW, Edelman GM (2002) Dendritic spines elongate after stimulation of group 1 metabotropic glutamate receptors in cultured hippocampal neurons. Proc Natl Acad Sci U S A 99(3):1639–1644. https://doi.org/10.1073/pnas.032681099

Vitureira N, Goda Y (2013) Cell biology in neuroscience: the interplay between Hebbian and homeostatic synaptic plasticity. J Cell Biol 203(2):175–186. https://doi.org/10.1083/jcb.201306030

Wang X, Bozdagi O, Nikitczuk JS, Zhai ZW, Zhou Q, Huntley GW (2008) Extracellular proteolysis by matrix metalloproteinase-9 drives dendritic spine enlargement and long-term potentiation coordinately. Proc Natl Acad Sci U S A 105(49):19520–19525. https://doi.org/10.1073/pnas.0807248105

Wang K, Lin MT, Adelman JP, Maylie J (2014) Distinct Ca2+ sources in dendritic spines of hippocampal CA1 neurons couple to SK and Kv4 channels. Neuron 81(2):379–387. https://doi.org/10.1016/j.neuron.2013.11.004

Washbourne P, Dityatev A, Scheiffele P, Biederer T, Weiner JA, Christopherson KS, El-Husseini A (2004) Cell adhesion molecules in synapse formation: figure 1. J Neurosci 24(42):9244–9249. https://doi.org/10.1523/JNEUROSCI.3339-04.2004

Watt AJ, Sjöström PJ, Häusser M, Nelson SB, Turrigiano GG (2004) A proportional but slower NMDA potentiation follows AMPA potentiation in LTP. Nat Neurosci 7(5):518–524. https://doi.org/10.1038/nn1220

Weng J-Y, Ceballos C, Zecevic D (2023) Contribution of individual excitatory synapses on dendritic spines to electrical signalling. bioRxiv:2022.02.15.480607. https://doi.org/10.1101/2022.02.15.480607

Wilson CJ (1984) Passive cable properties of dendritic spines and spiny neurons. J Neurosci 4(1):281–297. https://doi.org/10.1523/JNEUROSCI.04-01-00281.1984

Xiong W, Chen WR (2002) Dynamic gating of spike propagation in the mitral cell lateral dendrites. Neuron 34(1):115–126. https://doi.org/10.1016/S0896-6273(02)00628-1

Yang Y, Wang X, Frerking M, Zhou Q (2008) Spine expansion and stabilization associated with long-term potentiation. J Neurosci 28(22):5740–5751. https://doi.org/10.1523/JNEUROSCI.3998-07.2008

Yasuda R (2017) Biophysics of biochemical signaling in dendritic spines: implications in synaptic plasticity. Biophys J 113(10):2152–2159. https://doi.org/10.1016/j.bpj.2017.07.029

Yuste R (2010) Dendritic spines. MIT Press, Cambridge, MA

Yuste R (2011) Dendritic spines and distributed circuits. Neuron 71(5):772–781. https://doi.org/10.1016/j.neuron.2011.07.024

Yuste R (2013) Electrical compartmentalization in dendritic spines. Annu Rev Neurosci 36(1):429–449. https://doi.org/10.1146/annurev-neuro-062111-150455

Yuste R (2015) The discovery of dendritic spines by Cajal. Front Neuroanat 9:18. https://doi.org/10.3389/fnana.2015.00018

Yuste R, Bonhoeffer T (2001) Morphological changes in dendritic spines associated with long-term synaptic plasticity. Annu Rev Neurosci 24(1):1071–1089. https://doi.org/10.1146/annurev.neuro.24.1.1071

Yuste R, Denk W (1995) Dendritic spines as basic functional units of neuronal integration. Nature 375(6533):682–684. https://doi.org/10.1038/375682a0

Yuste R, Tank DW (1996) Dendritic integration in mammalian neurons, a century after Cajal. Neuron 16(4):701–716. https://doi.org/10.1016/S0896-6273(00)80091-4

Yuste R, Majewska A, Cash SS, Denk W (1999) Mechanisms of calcium influx into hippocampal spines: heterogeneity among spines, coincidence detection by NMDA receptors, and optical quantal analysis. J Neurosci 19(6):1976–1987. https://doi.org/10.1523/JNEUROSCI.19-06-01976.1999

Zaccard CR, Gippo I, Song A, Geula C, Penzes P (2023) Dendritic spinule-mediated structural synaptic plasticity: implications for development, aging, and psychiatric disease. Front Mol Neurosci 16:1059730. https://doi.org/10.3389/fnmol.2023.1059730

Zhou Q, Homma KJ, Poo M (2004) Shrinkage of dendritic spines associated with long-term depression of hippocampal synapses. Neuron 44(5):749–757. https://doi.org/10.1016/j.neuron.2004.11.011

Chapter 4
Dendritic Spines: Synaptogenesis and Synaptic Pruning for the Developmental Organization of Brain Circuits

Zdravko Petanjek, Ivan Banovac, Dora Sedmak, and Ana Hladnik

Abstract Synaptic overproduction and elimination is a regular developmental event in the mammalian brain. In the cerebral cortex, synaptic overproduction is almost exclusively correlated with glutamatergic synapses located on dendritic spines. Therefore, analysis of changes in spine density on different parts of the dendritic tree in identified classes of principal neurons could provide insight into developmental reorganization of specific microcircuits.

The activity-dependent stabilization and selective elimination of the initially overproduced synapses is a major mechanism for generating diversity of neural connections beyond their genetic determination. The largest number of overproduced synapses was found in the monkey and human cerebral cortex. The highest (exceeding adult values by two- to threefold) and most protracted overproduction (up to third decade of life) was described for associative layer IIIC pyramidal neurons in the human dorsolateral prefrontal cortex.

Therefore, the highest proportion and extraordinarily extended phase of synaptic spine overproduction is a hallmark of neural circuitry in human higher-order associative areas. This indicates that microcircuits processing the most complex human cognitive functions have the highest level of developmental plasticity. This finding is the backbone for understanding the effect of environmental impact on the development of the most complex, human-specific cognitive and emotional capacities, and on the late onset of human-specific neuropsychiatric disorders, such as autism and schizophrenia.

Z. Petanjek (✉) · I. Banovac · D. Sedmak · A. Hladnik
Department of Anatomy and Clinical Anatomy, School of Medicine, University of Zagreb, Zagreb, Croatia

Department of Neuroscience, Croatian Institute for Brain Research, School of Medicine, University of Zagreb, Zagreb, Croatia

Center of Excellence for Basic, Clinical and Translational Neuroscience, School of Medicine, University of Zagreb, Zagreb, Croatia
e-mail: zdravko.petanjek@mef.hr

A. A. Rasia-Filho et al. (eds.), *Dendritic Spines*, Advances in Neurobiology 34,
https://doi.org/10.1007/978-3-031-36159-3_4

Keywords Cerebral cortex · Development · Glutamate · Pyramidal neurons · Cognitive functions · Synapse overproduction

4.1 Introduction

Dendritic spines are small protrusions emerging from dendrites (Rasia-Filho et al. 2023a; Wouterlood 2023; Reberger et al. 2018). In the cerebral cortex, they are found on glutamatergic neurons, most of which are principal neurons (i.e., long projecting pyramidal and modified pyramidal neurons) (Rasia-Filho et al. 2021). Based on their morphological characteristics (their length and the size and shape of their head and neck), three major types of spines can be defined (stubby, mushroom, and thin). Mature dendritic spines are mainly mushroom-shaped, with a thin stalk and bulbous enlargements at their tips (spine heads). Dendritic spines have a specific molecular structure compared to the dendritic shaft (Rasia-Filho et al. 2023b; Tønnesen and Nägerl 2016). The head represents the postsynaptic part of a glutamatergic synapse, and most excitatory inputs on principal cortical neurons actually target dendritic spines (Marafiga and Calcagnotto 2023; Rasia-Filho 2022).

Cajal was the first to name dendritic spines as neuronal elements in 1888, using Golgi impregnation on Purkinje cells of the cerebellum (Cajal 1888; Yuste 2015). Later, Cajal concluded that dendritic spines serve as a site of contacts between axons and dendrites. Additionally, based on the notion that cells from *more highly evolved animals* have more spines, he argued that spines were probably associated with intelligence (Cajal 1899, 1904). Cajal proposed that changes in spine morphology could be related to neuronal functioning and learning, growing with activity, and retracting during inactivity. According to Cajal's interpretation, physical movements of dendritic spines could result in connecting or disconnecting neurons. This classical conclusion is somewhat in line with modern studies, which demonstrated that spines are not stable structures but are constantly morphologically changing (von Bohlen and Halbach 2023; Heck and Dos Santos 2023; Frankfurt et al. 2023; Bonilla-Quintana et al. 2020; Dunaevsky et al. 1999; Fischer et al. 1998; Gipson and Olive 2017; Kasai et al. 2021; Rangamani et al. 2016). Nowadays, it is generally accepted that spines are plastic, and their morphology depends on the functional stage of adjacent synapses (Rasia-Filho et al. 2023a,b; Renner and Rasia-Filho 2023; Shepherd 1996).

While describing dendritic spines, Cajal could not confirm the presence of anastomoses between axons and dendrites. Therefore, he proposed that neurons are independent units in the nervous system (Bock 2013). In the same publication, Cajal changed the principle of brain research with two fundamental observations: neurons are independent ("neuron doctrine") and covered with spines. The first ultrastructural analysis of synapses was performed in 1955 (Palay 1956; de Robertis and Bennett 1955), and a few years later, synapses were visualized on dendritic spines (Gray 1959). Additional research revealed that in cerebral cortex each dendritic spine contains one glutamatergic synapse (Alvarez and Sabatini 2007). Glutamatergic synapses change their strength by undergoing molecular modifications. The molecular changes in glutamatergic synapses are reflected on the morphology of corresponding dendritic spines. This is thought to be a key event in memory formation (Pan and Monje 2020). Modern techniques that allow imaging of individual spines in live animals demonstrate that spines are dynamic. Depending on neuronal activity, dendritic spines undergo continuous changes in shape, becoming stabilized or replaced (Runge et al. 2020).

Striking differences in spine densities were observed during development (Elston and Fujita 2014). In mammals, including humans, neurons in developing brains show a higher spine density than in adults (Rakic et al. 1994). In this chapter, we aim to review the data on the developmental pattern of synapto- and spinogenesis and synaptic and spine elimination in the human and monkey cerebral cortex. The focus is on the role of spontaneous and environmental-driven activity in these processes, particularly on the role of environment in the reshaping cortical circuitry to achieve the architecture able to sustain regular cognitive processing.

4.2 Overproduction and Elimination as a Regular Developmental Event During Formation of Neural Circuitry

4.2.1 Neuronal Death During Regular Development

The interaction of intrinsically determined genetic programs with a wide range of environmental stimuli determines the birth, death, and cellular characteristics of neurons and the formation and reorganization of their dendrites, axons, and synapses (Rakic 2009; Tau and Peterson 2010). Before synaptic overproduction was discovered as a regulating mechanism during the maturation of neural circuitry, extensive production of cells was described in many tissues and organs.

The overproduction of cells is a normal and widespread occurrence in the formation of the body, and the nervous system is no exception (Fricker et al. 2018; Ghose and Shaham 2020; Oppenheim 1991; Yamaguchi and Miura 2015). Coincidental with the processes of proliferation, migration, and differentiation, cell death is also a widespread phenomenon appearing during normal development of the central nervous system (Blaschke et al. 1998; Glucksmann 1951). Sympathetic neurons of the superior cervical ganglion of rat and mice have been extensively studied as a model of naturally occurring neuronal death. During development, one-third of these cells normally die by apoptosis during the first two postnatal weeks (Kristiansen and Ham 2014; Wright et al. 1983). Based on this finding, neuronal death was widely accepted as a mechanism involved in the formation of cortical circuitry, even in humans (Wong and Marín 2019).

In 1971, Dawkins proposed that neuronal death might serve as a mechanism for preserving connections, which would assemble into one of the brain's most complex functions—memory or information storage (Dawkins 1971). For confirmation of his theory, the author referred to two manuscripts that proposed massive neuronal death as a regular event in the human central nervous system throughout postnatal development (Brody 1955; Burns 1958). This became a widespread and persistent myth, even though the authors of these two manuscripts did not provide any evidence for the occurrence of massive neuronal death. More recent studies have not found evidence for widespread neuronal loss within the developing human brain (Giannaris and Rosene 2012; Leuba and Kraftsik 1994). They were able to identify only the sporadic presence of apoptotic cells within some of the fetal zones of the human telencephalon during a very limited time frame (Anlar et al. 2003; Chan et al. 2002).

In the rhesus monkey, only one study described massive developmental neuronal death (loss of 800,000 among 2,200,000 neurons) in the lateral geniculate nucleus (Williams and Rakic 1988). The neuron loss occurred between embryonic day (E) 50 and E100, before the establishment of connections with cortical neurons. Thus, neuronal death in the regularly developing mammalian telencephalon, particularly in the primate and human brain, is generally not an important developmental event in the formation of connections as might be the case in avians, amphibians, and reptiles (Pettmann and Henderson 1998; White and Barone 2001).

Indeed, in humans and monkeys apoptotic cell death was described only on some postmitotic neurons that establish transitory fetal circuits and on some of the remaining progenitors (Spreafico et al. 1999). The vast majority of transitory fetal neurons became partly incorporated into mature circuits, while others remain for a while without significant functional importance (Kostovic and Rakic 1980, 1984; Meyer and González-Gómez 2018; Sedmak and Judaš 2021). The majority of progenitors, which do not transform into neurons and glia, remain for a while as stem cells of subependymal zone. Moreover, there is no apoptotic cell death observed within the cortical plate, the fetal layer that will transform in neuronal layers (II–VI) of the cerebral cortex (Hansen et al. 2010; Pollen et al. 2015). Therefore, contrary to the widely held myth (Dawkins 1971), with few exceptions and excluding injury or pathology, in the human cortex most of the nerve cells we are born with are the same ones we die with (Peters and Rosene 2003). This shows that during human development there is absence of programed cell death within neuronal classes that will establish mature cortical circuitries. Thus, overproduction of neurons and programed cell death is not a mechanism that plays a role in formation and specification of cortical circuitries in humans.

4.2.2 *Dendritic and Axon Overgrowth*

Despite the absence of supernumerary neurons during development, it is possible that circuitry specialization is related to overgrowth of dendrites and axons (Luo and O'Leary 2005; Pease and Segal 2014; Riccomagno and Kolodkin 2015). However, overgrowth of dendrites during development is seen only in specific neuronal subpopulations, as are marginal zone neurons (Cajal Retzius neurons) and subplate neurons (Miškić et al. 2021; Mrzljak et al. 1988, 1992). These neurons have important roles in guiding development during the fetal and perinatal stages, but in the adult network, they usually have no important functions. Additionally, some subpopulations of modified pyramidal neurons, that have a typical pyramidal shape during development, change their shape and shrink parts of their dendritic arbor to adapt to their adult functional role. The most obvious example is spiny stellate neurons of the layer IV (Meyer et al. 1989). During development, they have a typical pyramidal shape with a characteristic apical dendrite and several basal dendrites. In later stages of development, they retract their apical dendrite, change in cell body shape, and undergo translocation of their basal dendrites, to achieve a final stellate morphology. Also, in layer V of the rat cortex, callosal projecting and subcortical projecting neurons undergo different types of dendritic differentiation (Koester and O'Leary 1992). Neurons of each class initially expand an apical dendrite into layer I. The apical dendrite of layer V callosal neurons is losing the segments of their apical dendrite superficial to layer IV later during development, generating their characteristic short pyramidal morphology.

Therefore, most neurons do not go through developmental dendritic arbor regression. For neurons that do, regression is a part of adjustment related to their functional role in the adult, and occurrs at early stages of synaptogenesis. This supports the notion that dendritic overgrowth is also not a solitary mechanism in specification of cortical connections.

Compared to adult stages, a higher number of axons are found in the commissures (Lamantia and Rakic 1990a, b, 1994) and the optic nerve during development (Provis et al. 1985; Rakic and Riley 1983a, b). Some of those are axon collaterals that will not establish connections with neurons, and some will make temporary projections to the contralateral area or areas in the same hemisphere that are usually not present in adults (Fig. 4.1a) (Chalupa and Killackey 1989; Dehay et al. 1988; Innocenti 1981; Innocenti et al. 2022). The functional significance of transitory axonal projections during regular development is still not clear (Innocenti and Price 2005). They might serve as a functional compensation in case of pre- or perinatal injury (Geden et al. 2019). The retraction of exuberant number of axons and transitory projections parallels the synaptogenesis, implying that synaptic overproduction and pruning are not related to an exuberant number of axon collaterals, but most likely includes changes within terminal ramifications (Fig. 4.1b).

4.2.3 Synaptic Elimination at the Neuromuscular Junction

During postnatal maturation, synaptic connections are overproduced and redundant (Purves and Lichtman 1980). Synaptic elimination during development, an event through which redundant synaptic connections are removed, is crucial to establish proper wiring and processing within a mature nervous system (Lohof et al. 1996). The neuromuscular junction in rodents was the first model on which the dynamics

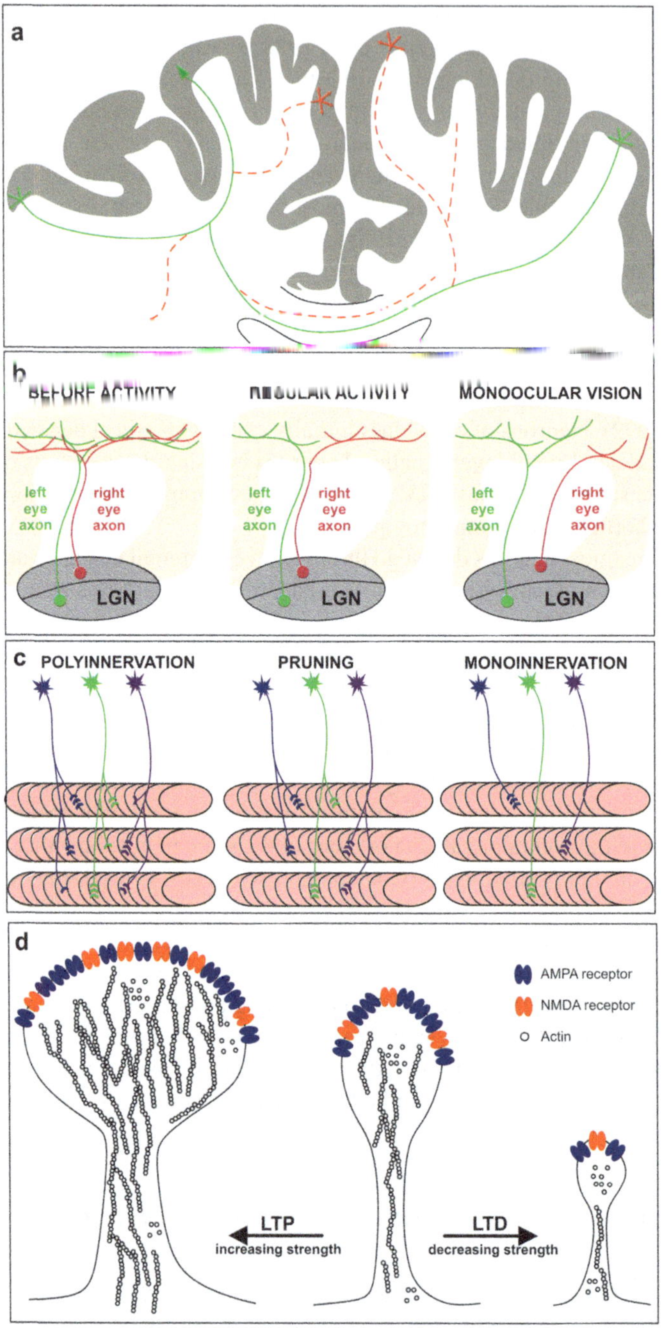

Fig. 4.1 Axon retraction during development of cortical connections (**a**), role of activity in formation of ocular dominant columns (**b**), axon organization of neuromuscular junction during development (**c**), and changes in spine size with LTP and LTD (**d**). (**a**) During development of the central

and mechanisms of synaptic formation and elimination were studied (Fig. 4.1c) (Lee 2020).

In adults, each mature muscle fiber receives input from a single motor neuron (Kugelberg 1976). During muscle contraction, the single (non-overlapping) innervation of individual muscle fiber is necessary for the proper selection of motor units (a particular motor neuron with muscle fibers it innervates) (Nagamori et al. 2021; Thompson et al. 1984). It was first assumed that the axon of a motor neuron seeks out and synapses with specific target muscle fibers. However, during development, a single muscle fiber typically receives innervation from two or more motor neurons (Bagust et al. 1973; Bennett and Pettigrew 1974; Brown et al. 1976; Redfern 1970).

Fig. 4.1 (continued) nervous system, neurons establish connections with areas that are not present in adult (red dotted lines with closed ending). There are some long axon collaterals that do not reach any target (red dotted lines with open ending), sometimes even crossing through the cortical commissures. During the normotypic formation of cortical circuits, excessive connections are removed. The branches that are not retracted (full green lines) increase the number of their terminal ramifications within the target cortical area, and this happens in parallel with synaptogenesis. Therefore, the synaptic elimination is not related to elimination of excessive projections, rather it is in correlation with the reorganization of terminal branches. In case of injury within the determined target region, axons can avoid retraction and retain established projections, which in normal situation will be retracted. It is even possible that new projections to regions that were not contacted could be established. (**b**) Data on development of ocular dominant columns reveal the role of activity in synaptic elimination and reorganization of terminal branches. During development, the terminal ramification of one axon extends to a far greater area of the primary visual cortex than in adults, overlapping with the terminal ramification of a neighboring axon projecting from the other eye. If activity is blocked during development, an ocular dominant column will not be formed and distribution of axon will remain as at the beginning of development (left schema). During normal development, axons will compete and retract branches, so that, by the end of development, an axon from each branch will occupy a similar territory of the visual cortex, without overlap (middle schema). If during the period of synaptic overproduction one eye is used significantly more than the other, more cortical synapses will be stabilized, and the corresponding oculo-dominant column will finally occupy a wider territory than the deprived eye (right schema). The reorganization of axon terminal branches appearing during formation of oculo-dominant columns shows similarity with the development of the neuromuscular junction. (**c**) Data on development of innervation of the striated muscles show that motor neurons establish supernumerary contacts with muscle fibers (polyinnervation). Thus, during development, a single muscle fiber is innervated by at least two motor neurons. Eventually, more efficient synapses will increase in strength, whereas the rest will decrease the number of acetylcholine receptors and finally retract (pruning), resulting in each muscle fiber being innervated by a single motor neuron by the end of the development (monoinnervation). (**d**) Long-term potentiation (LTP) is associated with an enlargement of the spine head, recruitment of more AMPA and NMDA receptors to the postsynaptic membrane and reorganization of the actin cytoskeleton. By contrast, long-term depression (LTD) results in shrinkage of spines. LTP and LTD either increase or decrease the number of receptors at the postsynaptic site. In addition, they affect the amount of neurotransmitters stored and released from the presynaptic site. LTP and LDP modulate the synaptic "strength," allowing changes in processing within neural network. Those changes are related to the formation of new memories without changing the network architecture (i.e., replacing synapses). During the period of synaptic overproduction, changes in synaptic strength and spine size, where LTP is acting as stabilizer, are dependent on activity. By this mechanism, it will be specified which synaptic spines will remain and which will be removed, and this is the way how environment affects formation of circuitry architecture

The axons of motor neurons are guided to innervate skeletal muscles (Rodríguez Cruz et al. 2020; Wu et al. 2010). Before innervation, acetylcholine receptor gene expression is enriched in the central region of embryonic skeletal muscles (Lin et al. 2005; Yang et al. 2001a). The area of acetylcholine receptor gene expression in muscles lacking motor axons is wider than usual, pointing that neural signals refine intrinsically guided muscle pre-patterning (Yang et al. 2001b). Subsequently, when the motor neurons innervate some of the pre-patterned acetylcholine receptor clusters, these become enlarged and stable, while aneural acetylcholine receptor clusters tend to disappear (Bai et al. 2022; Schaeffer et al. 2001). In rodents, during the first postnatal weeks, neuromuscular junctions undergo synaptic elimination, and axons gradually withdraw from muscle fibers innervated by multiple neurons, leaving a single innervating axon at each neuromuscular junction (Lichtman and Colman 2000). The developmental elimination of excessive motor neuron inputs from neuromuscular junctions is related to an active competition among the motor axons converging to individual endplates (Mayseless et al. 2023; Moron et al. 2021; Thompson and Jansen 1977; Walsh and Lichtman 2003).

The chronic and selective stimulation of a subset of motor units leads motor neurons to innervate more muscle fibers, suppressing the non-stimulated motor units that innervated the same muscle fiber (Balice-Gordon and Lichtman 1994; Buffelli et al. 2003; Engisch et al. 2022; Keller-Peck et al. 2001; O'Brien et al. 1978; Personius et al. 2016; Ridge and Betz 1984; Zempo et al. 2020). In contrary, a block of motor neuron impulse conduction and neuromuscular blockade decreases synapse elimination (Brown et al. 1981; Callaway and van Essen 1989; Garcia et al. 2022; Thompson et al. 1979; Tomàs et al. 2023). This clearly shows that neuromuscular synapse elimination is activity-driven, as a result of divergent motor neuron spike timing. It allows the muscle fibers to discriminate between a number of presynaptic inputs, strengthening and maintaining the inputs being able to produce coincident contractions in a Hebbian fashion (Lee 2020; Thompson 1983).

During development, the number of both motor neurons and target muscle fibers remains unchanged. This demonstrates that synaptic elimination is related to pruning of only local branches of motor axons (Balice-Gordon and Thompson 1988; Brown et al. 1976; Sharples and Miles 2021; Xu et al. 2022b). It is important to note that developmental neuromuscular synapse elimination should be discriminated from synapse elimination that occurs in adult muscles following injury-induced denervation, which is followed by reinnervation of motor endplates by multiple axons (Nazareth et al. 2021; Nguyen and Lichtman 1996; Wyatt and Balice-Gordon 2003).

Similar to the neuromuscular junction, in the rat submandibular ganglion each neuron is innervated by several (three to seven) preganglionic fibers at birth, while by 5 weeks of age, most of the neurons are innervated by a single axon (Lichtman 1977). This shows that the process of synapse elimination occurs in the autonomic nervous system as well.

Synaptic overproduction and elimination during development has also been described in the motor structures within the central nervous system (Conradi and Ronnevi 1975; Conradi and Skoglund 1969; Crepel et al. 1976). In the hypoglossal nucleus of the rat, the total number of synapses increased significantly from birth to day 20 and then decreased by 30% in adults (Kanjhan et al. 2016; O'Kusky 1998).

Synaptic overproduction and influence of the electrical activity on developmental synapse elimination within the peripheral nervous system and on lower motor neurons of the central nervous system might serve as a model on how neural activity affects synapse elimination within more complex structures of the central nervous system.

4.3 Role of Activity in the Formation and Maintenance of Synaptic Spines: Selective Synaptic Stabilization Hypothesis

4.3.1 Modifying Synaptic "Strength" Without Changing the Network Architecture

The architecture of a neural network, i.e., the way how neurons are interconnected, is defined by the number, proportion, and distribution of various synaptic types targeting different compartments of postsynaptic neurons (Heckman and Doe 2021; Hunnicutt and Krzywinski 2016; Seng et al. 2022).

The functional properties of neural circuits are dependent on microcircuitry architecture and include size and topology of dendritic structure and pattern of distribution for corresponding synapses. The architecture needs to be highly conserved throughout the lifespan to enable long-term memory and stability of personality (Chen et al. 2022; Pérez-Ortega et al. 2021; Petanjek et al. 2019; Yang et al. 2009). However, changes should also occur to allow the formation of new memories, i.e., to establish memory engrams (Gallinaro et al. 2022; Han et al. 2022; Josselyn and Tonegawa 2020; Lee et al. 2023; Poo et al. 2016; Rao-Ruiz et al. 2021). If the neural network maintains a stable architecture, adaptation to the environment (Pan and Monje 2020; Stampanoni Bassi et al. 2019) needs to be achieved by modifying the efficiency of individual synapses rather than by replacing them (Caroni et al. 2014; Holtmaat and Caroni 2016). This modification of synaptic efficiency is referred to as a change in synaptic strength. Synaptic strength is the molecular composition of a synapse that determines the efficacy of signalization between neurons. Changes in molecular composition can either strengthen or weaken the connection between two neurons. This theory was first proposed in 1949 by Donald Hebb (1949) and is often described as Hebb's rule. The rule assumes that when the axon of one neuron repeatedly activates another neuron, synapses adopt to increase the efficiency of the first neuron in activating the second neuron. As a result, the connection between those two neurons is strengthened. Hebbian theory can be summarized as "neurons that fire together, wire together" and attempts to explain how neuron interaction might form engrams (Chéreau et al. 2022; Ryan et al. 2021; Sweis et al. 2021). Modifying synaptic strength seems to be the dominant way how functional properties of the neural network change in response to neuronal activity in the mature cortex during one's lifetime (Citri and Malenka 2008; Cline et al. 2023; Ho et al. 2011; Katz and Shatz 1996).

Two forms of activity-dependent synaptic modification are: long-term potentiation (LTP)—the persistent enhancement of synaptic strength, and long-term depression (LTD)—the persistent reduction in synaptic strength. By these mechanisms, synaptic function can be modulated by changing the probability of vesicles being released or by adjusting the trafficking of neurotransmitter receptors to and from the postsynaptic membrane. The neuronal activity during environmental stimulation alters the amount of neurotransmitter release and synaptic receptor composition in behavior-relevant circuits. The activity can also change the size and shape of postsynaptic dendritic spines (Fig. 4.1d), which is related to molecular and functional changes on associated synapses (Cheetham et al. 2014; Chen et al. 2022; Hazan and Ziv 2020; Holtmaat et al. 2005, 2006; Kasai et al. 2021; Kruijssen and Wierenga 2019; Ma and Zuo 2022; Meyer et al. 2014; Ray et al. 2023; de Roo et al. 2008; Sheynikhovich et al. 2022; Sohn et al. 2022; Tong et al. 2020; Trachtenberg et al. 2002).

With the discovery of LTP, Hebb's rule was confirmed (Bliss and Gardner-Medwin 1973; Lomo 1968; Lømo 2018; Matsuzaki et al. 2004; Nicoll 2017), and many recent studies have advanced the understanding of activity-dependent synaptic plasticity (Abbott and Nelson 2000; Blundon and Zakharenko 2008; Caroni et al. 2012; Collingridge et al. 2010; Kasai et al. 2010; Lamprecht and LeDoux 2004; Magee and Grienberger 2020; Malenka and Bear 2004). This activity-dependent synaptic plasticity is today generally accepted as a central cellular correlate of learning and memory (Appelbaum et al. 2023; Cline et al. 2023; Ho et al. 2011; Katz and Shatz 1996; Pan and Monje 2020), showing that in the mature cortex there is no need for replacement or addition of synapses to establish new memories and keep a high level of memory capacity.

4.3.2 Structural Features and Lifelong Changes in Dendritic Spines

In the cerebral cortex, changes in synaptic strength are closely related to excitatory glutamatergic circuitries (Di Maio 2021). The majority of glutamatergic excitatory inputs are received by dendritic spines (Alvarez and Sabatini 2007). Dendritic spines are heterogeneous in shape. They represent a continuous spectrum of morphologies (Fig. 4.2) and often do not fall into clearly recognizable structural subtypes (Anton-Sanchez et al. 2017; Arellano et al. 2007; Benavides-Piccione et al. 2013; Fu and Zuo 2011; Luengo-Sanchez et al. 2018; Ofer et al. 2021, 2022; Pchitskaya and Bezprozvanny 2020). For spines that are very long and thin without a large bulb, the term "filopodia" is used. Filopodia are rare in the adult cortex and abundant during development (Vardalaki et al. 2022).

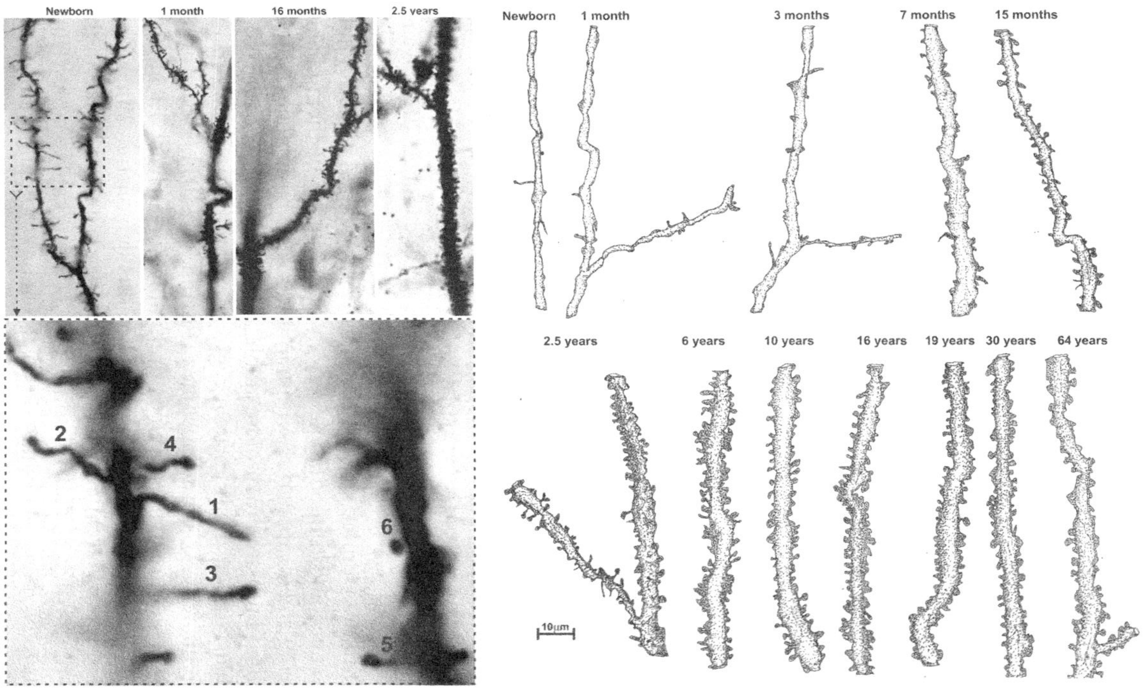

Fig. 4.2 Left side shows microphotographs of rapid Golgi-impregnated middle part of apical dendrite with side branches from layer III pyramidal neurons in prefrontal area 9 of newborn, 1-month-old, 16-month-old, and 2.5-year-old specimens. The photographs represent a composition of the same field taken at different depths. Note the transformation from filopodia, that dominate in newborn, through longer spines with clear head at 1 month to the presence of only typical mushroom type at 16 months and 2.5 years of age. The enlarged panel shows different morphology of dendritic spines representing developmental stages: very long and thin filopodia without a clear head (1–2) and with a small terminal enlargement (3), long spines with a small bulb (4), long mushroom spine (5), and short mushroom spine (6). Right side shows camera lucida drawing of the more proximal part of the apical dendrite of layer IIIC pyramidal neuron in specimens of the following ages: newborn, 1 month, 3 months, 7 months, 15 months, 2.5 years, 6 years, 10 years, 16 years, 19 years, 30 years, and 64 years

Dendritic filopodia have a high degree of motility and flexibility, with an average lifetime ranging from minutes to hours. They serve as an attractor for ingrowing axons, making repeatedly transient contacts with them. Thus, they appear to be particularly important as a precursor of dendritic spines during early development (Sheng et al. 2018; Wildenberg et al. 2023; Yoshihara et al. 2009; Yuste and Bonhoeffer 2004; Ziv and Smith 1996). Filopodia do not necessarily have functional synapses and frequently do not contain postsynaptic densities at their postsynaptic membranes (Fu and Zuo 2011; Ozcan 2017).

The process of transforming filopodia into mature and stable mushroom spines, which are characterized by large heads and short necks, is not very efficient. Imaging studies in living, young mice suggest that only 10–20% of filopodia may actually be transformed into stable spines. Only a small percentage of filopodia is accumulating synaptic proteins (Knott et al. 2006; Sheng et al. 2018). This is the first step in the process of morphological and functional transformation into classical mushroom-shaped spines. During the transitional stage between filopodia and mature stable mushroom spines, spines are thinner and elongated with a smaller bulb than in their mature form. Spines with such morphology are sometimes referred to as "learning" spines. Most of the "learning" spines also disappear within subsequent days, which is linked to lack of stabilization of adjacent synapses (Ehrlich et al. 2007; Yoshihara et al. 2009).

All of the aforementioned demonstrates that the stabilization of new synapses is a challenging process, and only a subset of initially formed contacts succeeded in reaching mature stage. This process is regulated by neuronal activity (Ehrlich et al. 2007). During the early phase of stabilization, when newly formed spines acquire a postsynaptic density, their spine head enlarges, a phenomenon that shows similarities with the head enlargement of mushroom spines during LTP induction (Bączyńska et al. 2021; Harvey and Svoboda 2007; Honkura et al. 2008; Kasai et al. 2010; Matsuzaki et al. 2004; Runge et al. 2020; Sun et al. 2021; Ucar et al. 2021; Weber et al. 2016).

Experimental studies in mice showed that during the first postnatal month, both "learning" spines and filopodia are very dynamic, many of them changing shape, appearing and disappearing within minutes. One month later, spines with a clear terminal bulb are much more stable. Up to 4 months of age, more spines are lost than formed in all cortical regions examined, resulting in the pruning of approximately 30% of total spines (Fu and Zuo 2011). After pruning, around 90% of mature-shaped spines are stable for at least 13 months (Grutzendler et al. 2002; Zuo et al. 2005a). All of this demonstrates that, during development, circuitry architecture is continually changing, unlike in the adult cerebral cortex, when circuitry architecture is remarkably stable (Hunt et al. 2023). It has to be mentioned that homeostatic condition might influence dendritic spine density and morphology even in adult. In the marmoset, fatherhood leads to increased spine density on layer III pyramidal neurons of the prefrontal cortex (Kozorovitskiy et al. 2006), and imaging studies suggest that such effects might appear in the humans (Horrell et al. 2021; Provenzi et al. 2021). In female rats during the estrus phase, a reduction in the dendritic spine density of hippocampal neurons was described, while an increase has been observed during metestrus. However, such changes could not be confirmed on neocortical neurons (Castillo-Fernández and Silva-Gómez 2022), still the susceptibility of cortical neurons to sensory-evoked structural plasticity may be dependent on the stage of the cycle, showing that dendritic spines are more plastic during proestrus and estrus cycle (Alexander et al. 2018). Developmental studies also suggest that pattern of neocortical pyramidal neurons spine development is not considerably influenced by gonadal hormones in mice (Boivin et al. 2018; Delevich et al. 2020) and monkeys (Anderson et al. 1995; Piekarski et al. 2017).

4.3.3 Role of Neuronal Activity in Formation of Neural Circuits and Experience-Dependent Synaptic Spine Plasticity

In adulthood, neuronal activity is crucial for preservation of circuitry architecture, and during development, neuronal activity governs circuit formation (Chandrasekaran et al. 2015; Fu and Zuo 2011; Hofer et al. 2009; Pan and Monje 2020; Tropea et al. 2010; Wilbrecht et al. 2010). The first and most numerous studies on the role of activity in circuit formation are studies on the development of the visual system. These have shown that formation of synaptic contacts starts as a diffuse event that is directed by neuronal activity (Katz and Shatz 1996; Shatz and Stryker 1988). By pharmacologically blocking (tetrodotoxin injection) voltage-gated sodium channels in prenatal cats and preventing action potentials of the retinal ganglion cells, the arborization of neurons and segregation of retinal synapses within the lateral geniculate nucleus were altered (Shatz and Stryker 1988; Sretavan et al. 1988). A similar effect in the visual cortex of fetal cats was observed by pharmacological blocking of activity within the thalamocortical visual pathway (Herrmann and Shatz 1995; Martini et al. 2021). The role of activity in the development of the visual cortex is reinforced by research done on ferrets, where inhibiting early spontaneous retinal activity leads to abnormal development of ocular dominance columns (Huberman et al. 2006). The fundamental role of spontaneous activity in neural circuit development (Hubel and Wiesel 1963; Sherk and Stryker 1976) has also been demonstrated in other developing brain regions (Pan and Monje 2020).

The abovementioned data indicate that spontaneous activity is required for early formation of appropriate connections. However, spontaneous activity is not sufficient to establish a fully mature circuit architecture, and appearance of environmentally driven neuronal activity is needed for proper circuitry development (Wiesel and Hubel 1963a, b). Monocular deprivation after birth leads to a dramatic increase in neuronal representation of the normal eye in the visual cortex (Hubel et al. 1977; Hubel and Wiesel 1965). In addition, it was shown that visual experience is essential for the reorganization of synaptic connections at thalamic axon terminals, which form the ocular dominance columns (Fig. 4.1c).

There is a specific time window during which experience-dependent neuronal activity regulates circuitry formation (Daw et al. 1992; Hubel and Wiesel 1970). In kittens, monocular deprivation during a specific period leads to atrophy of the lateral geniculate nucleus and reduced cortical response to the deprived eye. However, no such changes are observed in kittens exposed to monocular deprivation outside this period nor in adult cats that were exposed to monocular deprivation (Wiesel and Hubel 1963a, b).

After formation of synaptic contacts, rearrangement of cortical connections (creation of new and loss of present synaptic spines) is a structural substrate for experience-dependent plasticity (Fu et al. 2012; Hofer et al. 2009; Jensen et al. 2022; Kirchner and Gjorgjieva 2021; Ray et al. 2023; Xu et al. 2009). Most of the understanding about experience-dependent synaptic spine plasticity was extracted from sensory manipulation studies. During development, impoverished or enriched conditions were shown to influence dendritic spine production and stabilization. For example, mice raised in darkness since birth showed high spine motility and immature spine morphology in the visual cortex. A few days of light exposure during a specific time period decreased the spine motility and re-established the mature spine size and shape (Jenks et al. 2021; Tropea et al. 2010). The sensory deprivation induced by unilateral whiskers plucking during development reduces and delays spine pruning in the contralateral barrel cortex (Chen and Brumberg 2021; Zuo et al. 2005b). Enriched sensory stimulation increases spine turnover in the same brain region (Holtmaat et al. 2006; Trachtenberg et al. 2002; Yang et al. 2009). In rats, unilaterally trimming of all whiskers in a brief time window during development decreases both spine and filopodium motility in the contralateral barrel cortex (Fu and Zuo 2011).

In 1-month-old mice, training with a forelimb reaching task shows rapid formation of new spines on apical dendrites of layer V pyramidal neurons in the motor cortex of the other hemisphere (Xu et al. 2009). This is followed by enhanced spine elimination, and as a result, there is a similar overall spine density as observed in control. The degree of spine formation is associated with the degree of learning acquisition, whereas the survival of new spines correlates with the maintenance of motor skills (Fu and Zuo 2011; Jensen et al. 2022; Sohn et al. 2022; Wang et al. 2011). Focused retinal lesions lead to almost complete spine turnover within 2 months in the region where innervation was lost. Monocular deprivation was observed to increase spine formation on specific dendritic domains of layer V pyramids in the mouse visual cortex (Keck et al. 2008; Vasalauskaite et al. 2019).

The high levels of remodeling of network architecture that are present during development do not occur in the healthy adult brain (Fu and Zuo 2011; Petanjek et al. 2011). Therefore, the structural plasticity of cortical circuits is much higher during development, giving environmental stimulation opportunity to determine architecture of the mature cortical circuit (Kolb et al. 2012). In addition to the role in the formation of synaptic contacts, environmentally driven activity is necessary for synaptic elimination that is a regular event in development of cortical circuitry.

4.3.4 *Epigenesis During Neurodevelopment—Environmental Role in Shaping Cortical Circuits via the Selection and Stabilization of Synaptic Connections*

While the size and shape of synapses in the adult neural network are highly dynamic, the synapses themselves are highly consistent, making the entire network remarkably stable (Chidambaram et al. 2019; Di Maio 2021; Fu and Zuo 2011; Pan and Monje 2020; Segal 2017). The changes in size and shape of dendritic spines are dependent on LTP and LTD. Both phenomena take place during development when the synaptic number is much higher than in adults. LTP and LTD are modifying the synaptic strength and in fact predetermine the final developmental destiny of particular synapse (Faust et al. 2021; Midorikawa and Miyata 2021). Therefore, the efficiency of synapses will be proportional with their chances to endure developmentally induced elimination.

This concept is in line with the model of developmental synaptic specification proposed by Changeux and Danchin (1976). This model postulated that the genetic template, which determines molecular interactions between developing brain cells, directs the proper synaptic interaction between the main neuron categories, alongside neuronal activity. Both play an important role in further specification of connectivity. It proposes that during development, within a given category, several contacts formed at the same site are producing redundant connectivity. After synapses are formed, spontaneous or environmentally driven activity increases the accuracy of the system by reducing redundancy. Changeux and Danchin suggest that early synaptic contacts are highly plastic and appear in different states (Changeux and Danchin 1976). They propose that the growth process of synaptic spines might be observed as the formation of a labile state. With maturation, the synaptic spines will reach a stable state, or regress and finally disappear. An essential part of this theory is that the transition from a labile state to a stable state or regression is regulated in an "epigenetic" manner (Changeux et al. 1973).

It should be mentioned that the term "epigenesis" (or "epigenetics") is used in this context to describe a higher level of adaptation—the selection and stabilization of developing synaptic connections by neuronal activity. Today, the term "epigenesis" is almost exclusively used to describe molecular (nongenetic) mechanisms of gene expression, i.e., to describe the status of DNA methylation and histone modification in a particular genomic region. However, well before the term "epigenetics" was adopted by molecular biologists, the term was commonly used in developmental neurobiology (Changeux et al. 1973).

If the development of synaptic connections was mainly genetically determined, the adult circuitries would be almost identical in genetically identical individuals. However, this is not the case. Even in vertebrates that reproduce via parthenogenesis, dendrite and axonal morphology as well as the pattern of synaptic distribution show high levels of variation (Changeux 1983). Due to different environmental inputs during development, genetically identical individuals do not have identical connective patterns. Vice versa, the interindividual variability of the genome affects circuitry architecture. Because epigenetic variability is superimposed on genetic variability, the same learning input may not produce the same connective patterns in different individuals due to genetic variability (Hublin and Changeux 2022). This particularly accounts for variability in circuitry architecture found between human individuals.

At early childhood, each cortical area reaches the time point when synaptogenesis mainly diminishes, and synaptic (and dendritic spine) number reaches its peak. Later during childhood and adolescence, the plasticity of cortical circuits is related to synaptic overproduction (Jenks et al. 2021; Kourosh-Arami et al. 2021; Sakai 2020; Selemon 2013). The initially exuberant connections are refined by activity in the network as some synapses are reinforced and others are eliminated. Therefore, overproduction of synapses during development allows the environment to affect structure of the neural network by activity-dependent stabilization of synapses. In mice and rats, the maximum synaptic density is reached within a few weeks after birth (Alvarez and Sabatini 2007; Grutzendler et al. 2002; Holtmaat et al. 2005; Knott et al. 2006; Zuo et al. 2005a), whereas in the associative regions of the human cortex this peak is reached at the age of 3 years. Moreover, mice and rats show a relatively small loss of synapses (up to 30%) from maximum density reached, contrary to humans where in some associative cortical microcircuits the number of synapses decreases two to three times compared to the maximum number (Petanjek et al. 2011, 2019; Rakic et al. 1994).

In postnatal life, the vast majority of neural network activity results from inputs from the environment. The environmentally driven activity affects the organization of connections through the stabilization of some synapses, or leads to degeneration (pruning) of labile ones. Data about length and amount of synaptic overproduction showed that the role of environment in shaping circuitry architecture increases in parallel with the capabilities of microcircuits to process more complex functions (Marchetto et al. 2019). This is further confirmed with highest overproduction level in microcircuits that are processing the most complex cognitive functions (Petanjek et al. 2019). Therefore, the extension of the postnatal developmental period in humans might have been essential for the internalization of social and cultural attitudes and for acquiring individual emotional and cognitive attitudes (Changeux et al. 2021; Muthukrishna et al. 2018).

The dramatic differences in the structure and function between the brains of humans and other mammals are largely based on the high proportion of connections in the human brain established during postnatal life (Mashour et al. 2020). The highly protracted synaptic overproduction allows the continuation of changes to the postnatal environment made by epigenetic responses, increasing the role of environmentally driven activity in formation of circuitry architecture and producing a huge spectrum of interindividual differences.

It was shown that synaptic pruning can be altered by artificial modification of neuronal activity. Blocking of neuronal activity results in a higher number of connections, whereas increased neuronal activity enhances synaptic elimination (Benoit and Changeux 1975, 1978; Luo and O'Leary 2005; Pease and Segal 2014; Riccomagno and Kolodkin 2015). This process showed similarity with the development of motor neurons that innervate muscle cells (Sunesen and Changeux 2003).

In conclusion, the theory of epigenesis of neural networks through selective stabilization of synapses accounts for the critical and reciprocal developmental interactions that take place between the brain and its physical, social, and cultural environment. It allows the environment to determine the architecture of cortical microcircuits, particularly those processing the highest-level cognitive functions (Changeux 2017; Diniz and Crestani 2023; Hublin and Changeux 2022; Kolb et al. 2012; Volzhenin et al. 2022). The theory deals with two major features regarding the genetic evolution of the human brain. The first is the nonlinear increase in the organizational complexity of the brain despite a nearly constant number of genes (Changeux et al. 2021). The second is a long postnatal period of brain maturation in humans, which explains the variability in the brain's connectivity and behavior between individuals (Hodel 2018), associated with the variability of the developing environment (Leisman 2022; Ptak et al. 2021; Quartz and Sejnowski 1997).

4.4 Synaptogenesis in the Human Fetal Cerebral Cortex

The prenatal development of the human cerebral cortex is characterized by temporary neural organization blueprints (Kostović 2020; Ramos et al. 2022; Vasung et al. 2016). The vast majority of circuitries formed before the last trimester of gestation are transitory and include connections with postmigratory neurons located within the marginal and subplate zone. However, the number of synaptic contacts established by transitory circuits is much lower than the number of connections found in the adult cerebral cortex.

To reach the full complexity of mature human cortical circuitries, an enormous production of synaptic connections must occur. The exponential expansion in the number of synaptic contacts starts during the last trimester of gestation and continues in this manner during infancy and early childhood (Kostović et al. 2021; Kostović and Judaš 2015; Tau and Peterson 2010).

4.4.1 Early Appearance of Synapses in the Dorsal Telencephalon (Cortical Anlage)

The earliest neurons in the human embryonic forebrain have been found beneath the pial surface (Bystron et al. 2005, 2006) even before complete closure of the neural tube, at post-conceptional week (pcw) 4.5. These predecessor neurons most likely originate from the meso-diencephalic boundary and tangentially migrate into the dorsal telencephalon. Analyzing the period before the appearance of the cortical plate (post-conceptional weeks 5–7.5), it was found that specific changes in neural organization appear within the cortical anlage (Meyer et al. 2000; Supèr et al. 1998; Zecevic et al. 1999). During this time frame, several cell populations appear sequentially, expressing reelin, calretinin, and GABA or their synthetizing enzyme (GAD). Most of these cells form a thin layer (two to three cells thick) defined as the pioneer plate. Prospective principal glutamatergic neurons that will establish the cortical plate arrive by radial migration around post-conceptional week 7.5, settling within the pioneer plate and splitting it into an upper part and a lower part. The neurons of the upper part of the pioneer plate become the neuronal elements of the marginal zone, and the neurons from the lower part form the presubplate layer below the cortical plate. At post-conceptional week 8, the cortical plate is found in every part of the telencephalon. The same cellular composition in the stage between the embryonic and early fetal periods (E38–E55) was observed in the monkey cerebral cortex (Zecevic and Rakic 2001).

In the first study on synaptic development in the human fetal cortex (Molliver et al. 1973), synapses were found in all 24 fetuses analyzed, starting from post-conceptional week 8.5. The density of synapses in the youngest fetus analyzed was very low (5/mm^2). In more recent study, synapses were shown even in earlier stages (post-conceptional week 6.5), i.e., before the formation of the cortical plate (Zecevic 1998). In addition to the extremely low synaptic density during post-conceptional weeks 6–9, those synapses had a very immature morphology.

During the early fetal period, most of the cortical plate neurons are undifferentiated with dendrites that do not extend outside of the cortical plate (Eze et al. 2021; Kopić et al. 2023). In contrast, the neurons of the marginal zone and the primordial subplate already reach a certain level of differentiation. Synapses in the early fetal period are exclusively present in the marginal zone and the primordial subplate and not in the cortical plate (Fig. 4.3a). Electron microscopy (EM) studies performed on fetal human and monkey somatosensory and visual cortices show that the majority of early synapses below the cortical plate are located on the proximal dendritic shafts of subplate neurons (Kostovic and Rakic 1990).

The external sources of the fibers forming synapses in the marginal and subplate zones during the early fetal stage are mainly neurons located within the brainstem and medio-basal telencephalon. Monoaminergic afferents from the brainstem nuclei are first observed in the frontal part of the cerebral cortex around post-conceptional week 9 (Nobin and Björklund 1973; Verney et al. 1993; Verney 1999; Zecevic and Verney 1995). Dopaminergic fibers originating in the ventral tegmental area and serotoninergic fibers from the nuclei raphe run above and below the cortical plate, and their distribution thus corresponds to the bilaminar distribution of synapses. Around post-conceptional week 9, there is outgrowth of fibers from the nucleus basalis of Meynert, which contains cholinergic and GABAergic neurons. At the end of the early fetal period, they spread throughout the whole subplate zone of the developing frontal lobe (Kostović 1986).

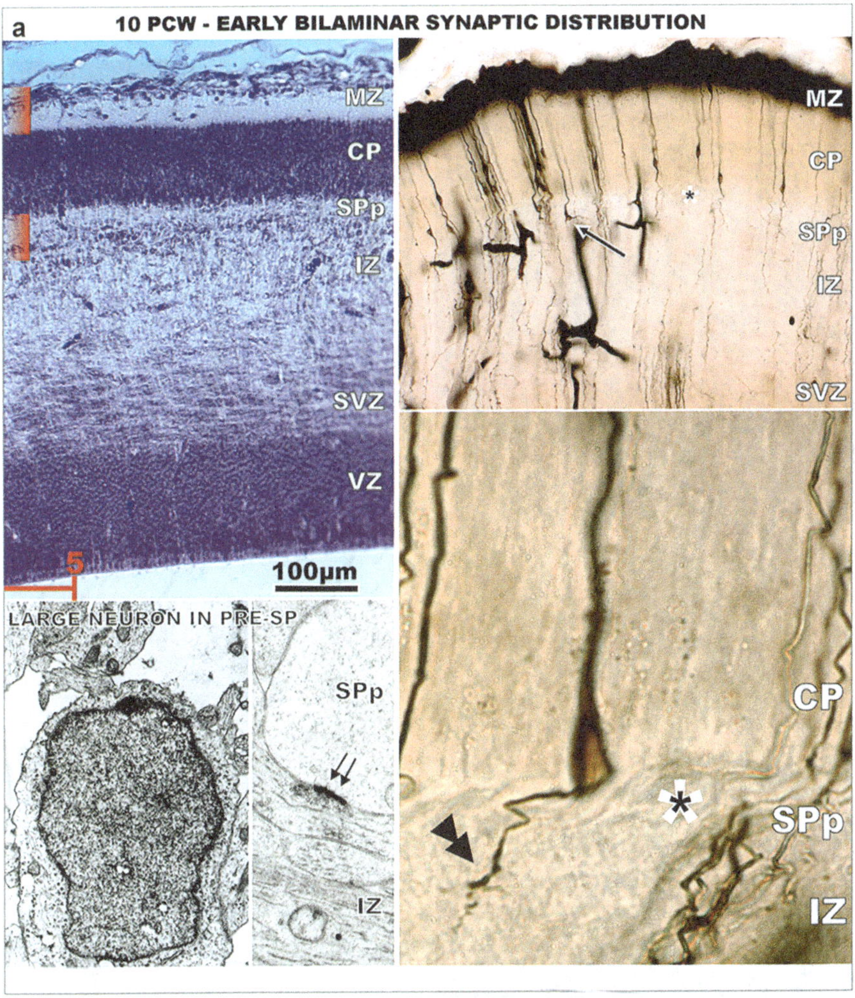

Fig. 4.3 Synaptic distribution in the fetal dorsal telencephalon, at post-conceptional week (pcw) 10 (composition **a**), post-conceptional week 13 (composition **b**), and post-conceptional week 15 (composition **c**). At post-conceptional week 10, (**a**) neurons of the cortical plate (CP) displays bipolar orientation of dendrites with arborizations (arrowheads) in the presubplate (SPp) below the CP (asterisk) and marginal zone (MZ) (above the CP). Thin presubplate zone contains scattered polymorphic neurons with well-developed dendrites (arrow). The electron microscopy photography (courtesy of Ivica Kostović, Zagreb neuroembryological collection) (Hrabač et al. 2018; Judaš et al. 2011) showed nucleus of large neuron in presubplate and synapse located on it (double arrow). The distribution of synapses is bilaminar: below and above the cortical plate (red). Other fetal zones (ventricular zone—VZ, subventricular zone—SVZ, intermediate zone—IZ) are clearly distinguishable on both Nissl and Golgi staining. Around post-conceptional week 13 (**b**), cells located deep within the cortical plate (CP) become more loosely arranged and lose their radial orientation, forming the so-called second cortical plate (Kostovic and Rakic 1990). Neurons in the second ("loose") part of cortical plate change their radial orientation (asterisk). The spread of the second cortical plate (CPs) cells leads to the development of a new, prominent lamina—the subplate zone (SP). The subplate zone becomes the thickest layer of the fetal cortex

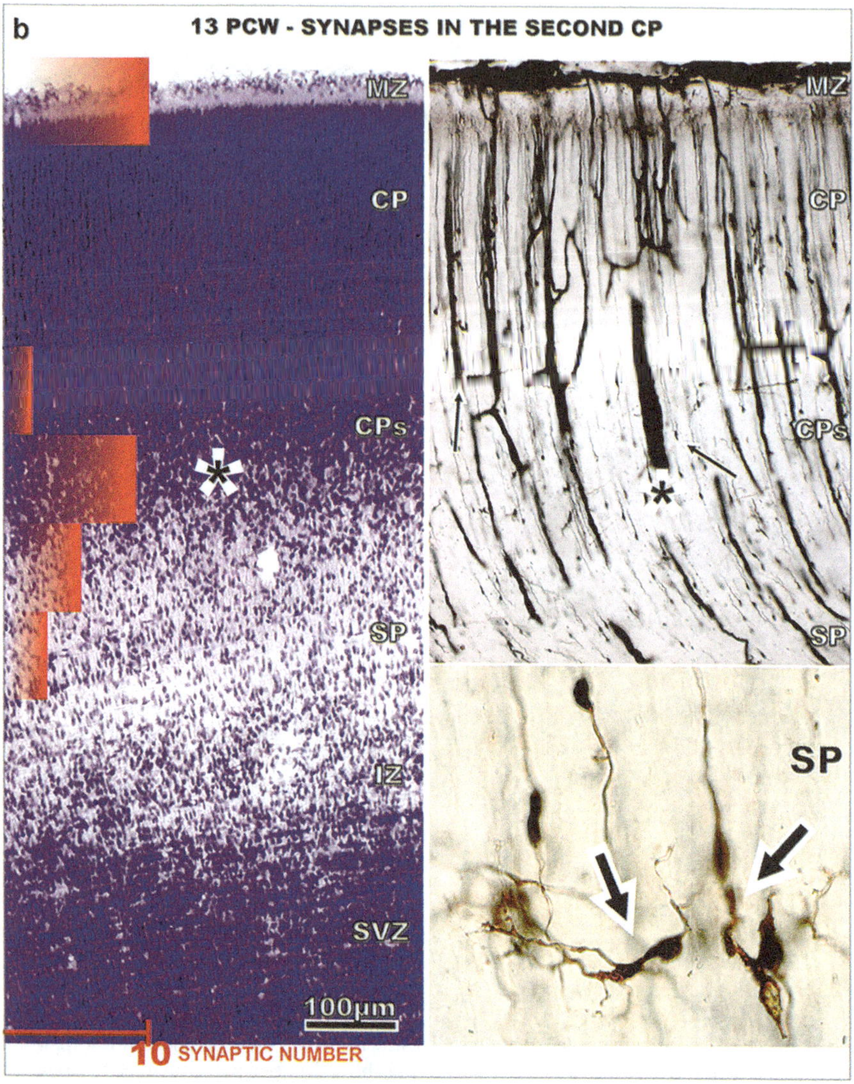

Fig. 4.3 (continued) containing most of the differentiated postmigratory neurons. Thus, the earliest born neurons, which formed the primordial subplate, represent, in humans, a very small proportion of subplate neurons. Most of the sublate neurons are actually those forming the cortical plate during the first 2 weeks of their persistence (post-conceptional weeks 8–10). A variety of neuronal shapes and orientations (arrow) appear in the transitional zone toward the developing deep subplate. Synapses are present throughout the second cortical plate and developing subplate (red). After post-conceptional week 15, (**c**) the cortical plate (CP) undergoes secondary consolidation and newly arrived postmigratory neurons show a radial orientation and immature pyramidal shapes (radial columns). This newly consolidated and dense cortical plate contains no synapses. In contrast, synapses are present throughout the entire subplate (SP). Subplate phenotypes are predominantly nonpyramidal, randomly oriented. First synapses within the cortical plate will be observed at post-conceptional weeks 18–19, and at post-conceptional week 24 synapses spread through whole cortical depth

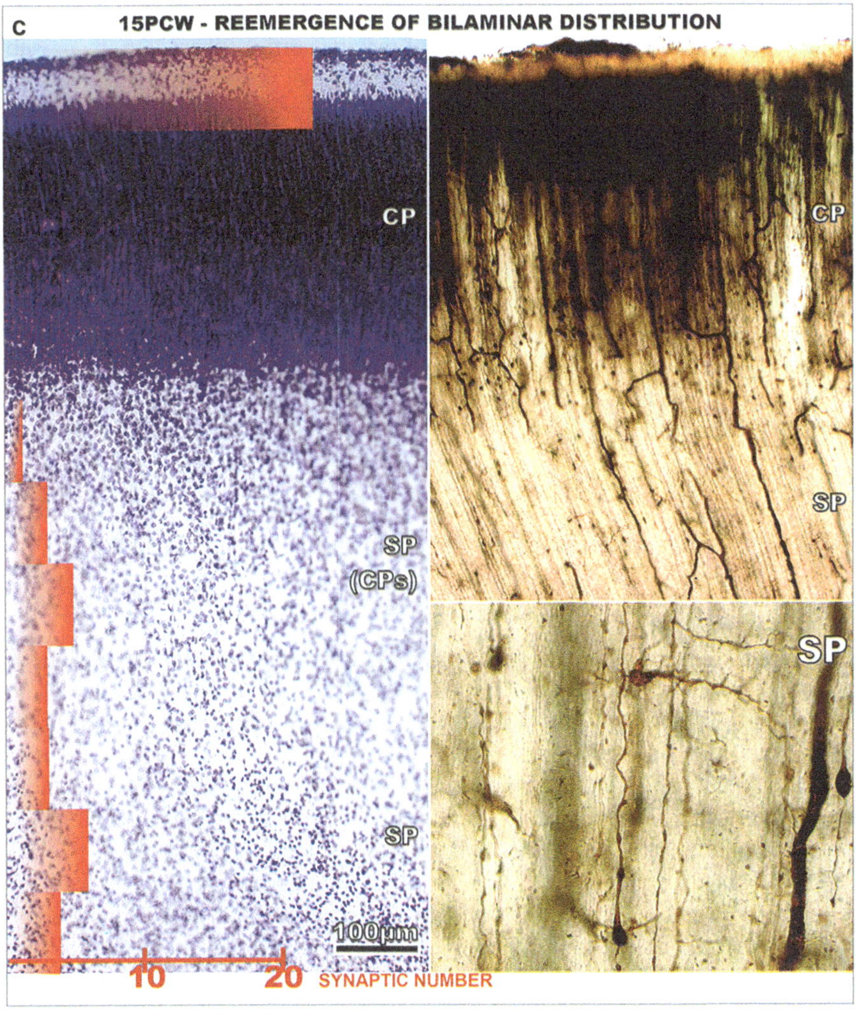

Fig. 4.3 (continued)

In parallel with the appearance of fibers from the brainstem and medio-basal telencephalon, there is also early growth of thalamocortical fibers into the presubplate and subplate zones (Alzu'bi et al. 2019; Kostovic and Rakic 1984; Kostović and Judas 2010; Krmpotić-Nemanić et al. 1983). However, the growth of projections from specific thalamic nuclei and their dense accumulation within the subplate start around mid-gestation (Krsnik et al. 2017).

Data from humans and rodents have shown that neurons at early stages predominantly communicate via gap junctions ("electrical synapses") and form the initial network using correlated firing (Elias and Kriegstein 2008; Luhmann and Khazipov 2018; Moore et al. 2014; Mukherjee and Kanold 2022). This stage may correspond to stage when the first synchronous assemblies were described in experimental rodents (Luhmann et al. 2018), and human stage around post-conceptional week 10.5 when spontaneous electroencephalography (EEG) activity is recorded (Borkowski and Bernstine 1955).

4.4.2 Laminar and Circuitry Organization During the Middle Trimester of Gestation

To understand circuitry changes during the middle trimester of gestation, it is important to understand changes in laminar organization (Ding et al. 2022). Up to post-conceptional week 18, synapses were found above (in the marginal zone) and below (presublate and sublate zone) the cortical plate (prospective layers II–VI), but never within it. At the end of the first trimester of gestation (Fig. 4.3b), a special laminar event starts—the formation of the true subplate zone (SP) (Duque et al. 2016; Kostovic and Rakic 1990).

The subplate zone serves as a "waiting" compartment for transient cellular interactions and as a substrate for competition, segregation, and growth of afferents originating sequentially from the brainstem, basal forebrain, thalamus, and the ipsi- and contralateral cerebral cortex. The subplate is not a vestige of the phylogenetically old network but a transient embryonic structure that expanded during evolution (Fig. 4.3c) to support the increasing number of cortical connections (Kostović 2020). The most protracted presence of the subplate and marginal zone neurons was found in the human prefrontal cortex higher-order associative areas with profuse cortico-cortical connections (Mrzljak et al. 1992).

During the second trimester, thalamocortical fibers start to densely accumulate within the subplate (Krsnik et al. 2017). The major developmental accumulation of thalamocortical fibers below the cortical plate occurs around the middle of gestation (post-conceptional weeks 20–22). The emergence of the earliest thalamocortical synapses in layer IV of the visual cortex occurs around post-conceptional week 19 (Hevner 2000). Thalamocortical fibers grow more intensively into the cortical plate between post-conceptional weeks 22 and 28, and form synapses with subplate neurons (Kostovic and Goldman-Rakic 1983; Kostović and Judas 2002). These connections represent a major cortical input during the middle trimester of gestation.

The interaction of thalamocortical fibers with neurons of the subplate zone is both morphogenetic and functional (Herrmann et al. 1994; Luhmann et al. 2018; Martini et al. 2021; Mukherjee and Kanold 2022; Ohtaka-Maruyama 2020). Glutamatergic thalamocortical afferents establish synapses with subplate neurons and may have an inductive role. During mid-gestation, there is a gradual increase in the density of synapses on subplate neurons. After post-conceptional week 24, in some cortical regions, such as the somatosensory, synapses are distributed through the whole depth of the cortical plate, and numerous synapses are still present in the subplate zone. The establishment of thalamocortical connections within the cortical plate coincides with the appearance of evoked and event-related potentials (Kostović 2020).

Although chemical synapses in the human cortex are found at the embryonic stage, mature membrane properties were not observed until post-conceptional week 14 (Cadwell et al. 2019). Neurons capable of firing repetitive action potentials are not observed until post-conceptional week 18 (Moore et al. 2009), when spontaneous activity is first observed (Moore et al. 2011). Spontaneous activity precedes the appearance of the thalamocortical synapses in layer IV of the primary visual cortex (Hevner 2000). Data from the monkey hippocampus revealed that the first synaptic currents can be observed around the middle of gestation and are related to GABAergic synapses (Khazipov et al. 2001).

The onset of synaptogenesis in the cortical plate and the formation of connections between ingrowing thalamocortical fibers and neurons of the cortical plate are crucial for the new stage of circuitry development. This event may be sensory-driven because thalamic fibers are able to transmit impulses from the sensory periphery to the cortex, and consequently, intracortical synapses might be activated (Zecevic 1998).

4.4.3 Synaptogenesis in the Last Trimester of Gestation and Appearance of Dendritic Spines

Growth of dendrites and formation of new synapses accelerate during the third trimester (Fig. 4.4). Post-conceptional week 32 marks the beginning of the peak period of synaptogenesis, during which millions of new synapses are formed every second, and intensive synaptogenesis continues into early postnatal life (Huttenlocher and Dabholkar 1997; Rakic et al. 1994; Tau and Peterson 2010). The massive synaptogenesis in the human cortical plate is first a result of thalamocortical fiber ingrowth and afterward a result of cortico-cortical fibers growth (Kostović 2020). The ingrowth of cortico-cortical fiber is particularly responsible for the acceleration of synaptogenesis, because the number of cortico-cortical projections exceeds the number of thalamocortical fibers by more than 10 times (Rosen and Halgren 2022).

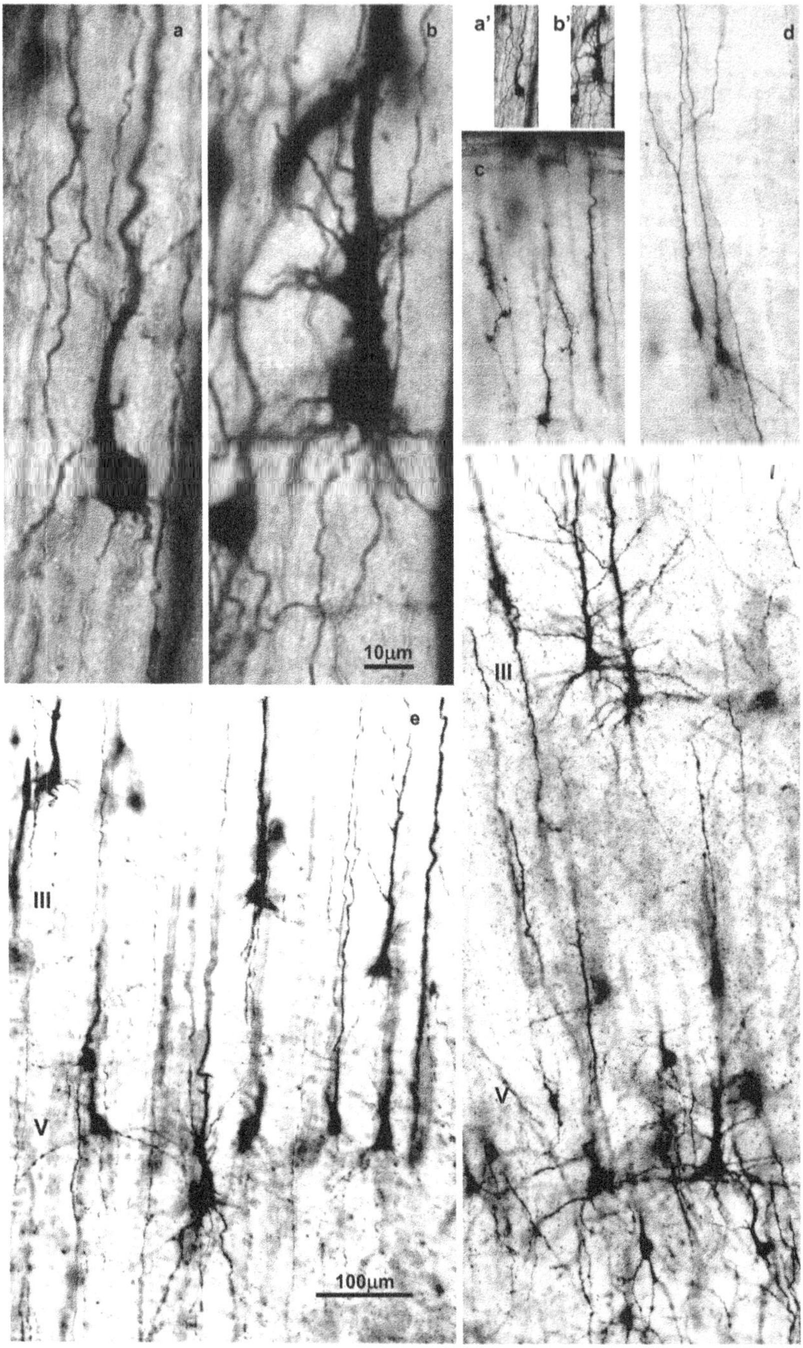

Fig. 4.4 Changes in dendritic morphology of rapid Golgi-impregnated pyramidal neurons in the dorsolateral part of the prefrontal cortex during the second half of gestation (**a**–**d**) and first postnatal month (**e**, **f**). Microphotograph of rapid Golgi-impregnated sections in the human fetal prospective dorsolateral prefrontal cortex at post-conceptional week 21 (**a**, **b**), and post-conceptional week 32 (**c**, **d**), newborn (**e**), and 1-month-old infant (**f**). Scale bar: 10 µm (**a**, **b**) and 100 µm (**c**–**f**). Small panels (**a'**, **b'**) are neurons from panels **a** and **b** shown at the same magnification as panel **c**–**f**. (Reproduced from Petanjek et al. (2019), fig. 3 without change, used under CC-BY 4.0 International license and copyright)

In human associative regions, such as the prefrontal cortex, where cortico-cortical afferents are most abundant, the subplate and marginal zone neurons continue to grow up to the end of gestation and even during the first postnatal month (Kostovic and Judas 2006; Kostović et al. 2014; Mrzljak et al. 1992). Note that around 80% of synapses originate from excitatory glutamatergic neurons located within the same area, and growth of this fibers is expanding their contribution in synaptogenesis during infancy and childhood (Petanjek et al. 2019).

Experiments in primates provide additional understanding of the process of synaptogenesis (Rakic et al. 1994). The most rapid accumulation of synapses in the macaque monkey starts at the beginning of the third trimester (E112) when synapses are found to be distributed throughout the whole depth of the cortical plate, albeit in low density. To illustrate, in the visual cortex (Bourgeois and Rakic 1993, 1996), the exponential growth of synapses occurs approximately 2 weeks after the end of the neurogenesis and just after the completion of neuronal migration. Between 2 months before and 2 months after birth, the synaptic density in the striate cortex increases ten times faster (17-fold) than the increase in the total volume of this area (1.7-fold). The rate of accumulation of synapses during this exponential phase is strikingly high. It is estimated that around 10,000–40,000 synapses are formed every second in each striate cortex of the macaque monkey. This high rate continues through infancy and begins to subside only around the third postnatal month.

In humans, synapses are found throughout the whole depth of the somatosensory cortex at post-conceptional week 24, when the rapid accumulation of synapses within the cortical plate started. This corresponds to E112 in monkey, the time point when the vast majority of neurons has attained their final position and when basic laminar organization is established (Kostovic and Rakic 1990).

A Golgi study of neuronal development in the human prefrontal cortex showed that the first dendritic spines appear on cortical plate neurons in parallel with ingrowth of thalamic and cortical afferents (Mrzljak et al. 1988, 1990, 1992). Due to the high increase in synaptic density during the last trimester of gestation, a significant proportion of the neonate cortical pyramidal neurons already show a considerable number of dendritic spines (Petanjek et al. 2008, 2011). The most frequent spine type was hair-like, 5–8 μm long with a small terminal expansion, which can be classified as filopodia (Fig. 4.2). Stubby and mushroom spines were less frequently observed. Long, hair-like spines were also observed on nonpyramidal neurons.

Spine-like protrusions appeared on subplate neurons at the end of the middle trimester of gestation. They were not very frequent and there is no further increase in their density up to the end of gestation. This is in line with electron microscopy data, which show that in the subplate and the marginal zones the dendritic shaft is the main synaptic location (Kostovic and Rakic 1990). This suggests that even the glutamatergic cells located in these zones do not possess a large number of spines.

4.4.4 Synaptic-Driven Telencephalic Activity During Gestation

Due to the role of activity in synaptogenesis, it is worth to elaborate which type of intrinsically or environmentally driven activity is present during particular developmental stage (Luhmann et al. 2022). During the middle trimester of gestation, functional chemical synapses are readily found throughout the subplate and marginal zone, but not within the cortical plate (Kostović 2020). It was indicated that cholinergic input propagating network oscillations in the beta frequency range may be related to classical chemical synapses within transitory circuits. Physiological studies suggest that these connections play an increasing role in spontaneous cortical activity found at that time (Kostović and Judas 2010; Luhmann and Khazipov 2018). Between post-conceptional weeks 22 and 24, evoked cortical responses may be elicited upon peripheral stimulation.

The subplate neurons potentially respond to sensory expectant thalamocortically generated excitation by the end of the second trimester. At this time, there is a massive penetration of thalamic axons to the cortical plate, and these fibers establish synapses with cortical plate neurons. During this phase, thalamic axons may activate subplate and cortical neurons simultaneously—this is described as disynaptic cortical activation (Friauf et al. 1990). The existence of spontaneous transient electrical phenomena, such as delta brushes, was observed starting from post-conceptional week 24 (Dreyfus-Brisac and Larroche 1971; Milh et al. 2007), simultaneously with a highly discontinuous temporal organization of electroencephalography (de Asis-Cruz et al. 2021; Derbyshire and Bockmann 2020; Fabrizi et al. 2011; Fitzgerald 2005; Neumane et al. 2022; Tolonen et al. 2007).

Based on functional magnetic resonance imaging (MRI), it seems plausible to propose that subplate neurons are engaged in physiological networks during the transition from fetal spontaneous (intrinsic) activity via sensory expectant activity (by the beginning of the last trimester), to sensory-driven activity later on (Arichi et al. 2017; Fulford et al. 2003; Huang et al. 2020; Jardri et al. 2008; Lee et al. 2012; Mencía et al. 2022; Polese et al. 2022).

There are numerous classical and ongoing clinical physiological studies indicating that peripheral stimulation in preterm infants initiates cortical-evoked responses and event-related cortical potentials (Cheour-Luhtanen et al. 1996; Kostović 2020). It is proposed that transient subplate circuitry participates in cortical activity even after the onset of permanent circuitry (Kostović and Judas 2010).

The question remains whether spontaneous motor behavioral phenomena described in human fetuses and infants (Dall'Orso et al. 2022; Fagard et al. 2018; Hadders-Algra 2007, 2018b; Ji et al. 2023; Kanazawa et al. 2023; Prechtl and Hopkins 1986) have a functional relationship with the transient circuitry of the subplate (Molnár et al. 2020; Wess et al. 2017), or if these activities are more related with already established connections within the cortical plate (Katušić et al. 2021; Kostović et al. 2021; Peyton et al. 2020).

4.5 Synaptic Overproduction During Development of the Cerebral Cortex

4.5.1 *Synaptic Development in the Human*

The first study showing synaptic overproduction during development in humans (Fig. 4.5a) was performed by Huttenlocher (1979). The study was undertaken to assess the usefulness of a phosphotungstic acid histochemical method for demonstration of synapses by using electron microscopy analysis on human material and to obtain normal values on synaptic density in the normotypical human cerebral cortex. Synaptic density in layer III from the middle frontal gyrus was analyzed in histological samples of 21 specimens ranging from newborn to 90 years. It was shown that values are stable in the post-puberty period (16 years onward), but between 1 and 7 years of age, synaptic density was 50% higher than in adults. In newborns, the synaptic density is around adult values, and by age of 1 year, the synaptic density is doubled. In parallel, the neuron density dropped six times, indicating that the first postnatal year is the period when the vast majority of synapses are formed.

The significantly greater synaptic density compared to adult values that were observed during late infancy and childhood was at that time a somewhat unexpected finding. Studies on synaptic development in animals performed before the Huttenlocher study have shown a similar trend, but not as significant. In the molecular layer of the rat parietal cortex between postnatal day 12 and the 3rd postnatal month, Aghajanian and Bloom found that the synaptic density at the age of 35 days was slightly (10%) higher than the synaptic density at the age of 3 months (Aghajanian and Bloom 1967). By studying the visual cortex of the kitten (Cragg 1975), it was found that the synaptic density at the age of 36 days was about 50% above the adult level. The synaptic density in the kitten cerebral cortex continued to be above the adult value by the age of 15 weeks (the time when the kitten approaches maturity). Thus, it took a while after Huttenlocher's first study to recognize synaptic overproduction as an important developmental event.

Using the same method as in the first study, lifespan changes in synaptic density were analyzed in the human striate cortex (Huttenlocher et al. 1982; Huttenlocher and de Courten 1987; Nelson 1994). Synaptogenesis in the human primary visual cortex was found to be most rapid between the ages of 2 and 4 months. Synapse elimination occurred subsequently with a loss of about 40% of synapses between the ages of 8 months and 11 years.

In their last study (Huttenlocher and Dabholkar 1997) about the formation of synaptic contacts in the human cerebral cortex, the auditory cortex (Heschl's gyrus) was compared with the prefrontal cortex (middle frontal gyrus) in specimens ranging from post-conceptional week 27 to age of 59 years. Synaptic density increases more rapidly in the auditory cortex, where the maximum is reached at the postnatal age of 3 months. In parallel with increase in synaptic density during first year, the neuron density decreased up to age of 1 year and remains around stable values

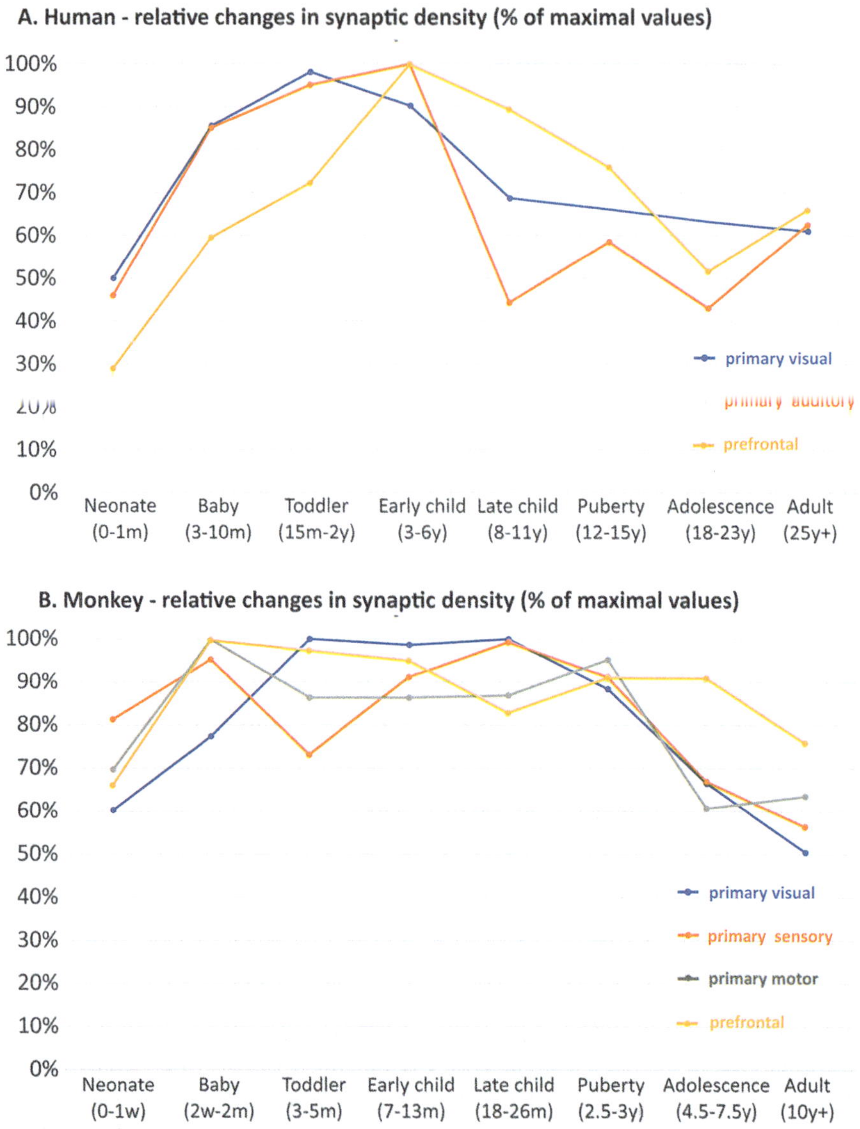

Fig. 4.5 Graphs are showing pattern of relative changes in synaptic density in humans (**a**) and macaque monkey (**b**) at different stages of postnatal development in several cortical regions revealed by electron microscopy analysis. Data are calculated as average values from specimens within particular age group published by Huttenlocher and Dabholkar (1997) (**a**) and Rakic et al. (1986, 1994) (**b**). Data are presented as percentage to maximum values, which are estimated as 100% (w—week, m—month, y—year). Both the human and monkey data show tendency that changes through primary regions appear concurrently, whereas the curve in the associative prefrontal cortex suggested delayed pruning in both of species and delayed synaptogenesis in human. Graphs **c**–**f** are presenting macaque monkey data about dendritic spine number and developmental changes form basal dendrites of layer III pyramidal neurons intracellularly injected with lucifer yellow (Elston et al. 2009, 2010a, b, 2011b). Data are presented taking into account the functional hierarchy (from left to right) of cortical areas, from primary auditory (A1) and primary visual area

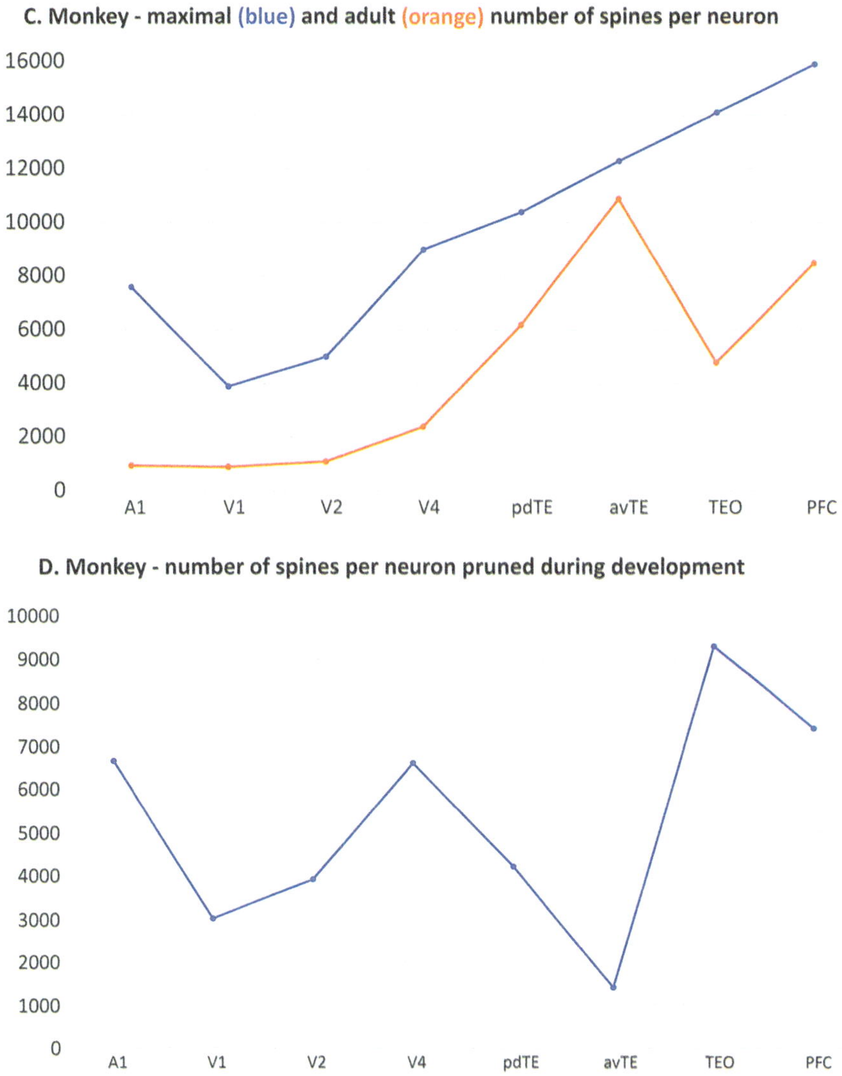

Fig. 4.5 (continued) (V1), through secondary (V2) and tertiary visual area (V4) to highest-order visual areas (TE and TEO), and finally ventrolateral multimodal associative prefrontal area 12 (PFC). Graph **c** is showing maximal (blue) and final (orange) number of dendritic spines per neuron on basal dendritic tree, and graph **d** is showing net number of spines per neuron lost during development. Note the significantly higher number of spines per neuron in adult associative versus lower-order areas. The number of spines pruned also increases across functional hierarchy, but interestingly, the layer III neurons in primary auditory area (A1) pruned two times more spines than neurons in primary and secondary visual areas (V1 and V2). Graphs **e** and **f** show percentage (compared to maximum—100%) remaining in the adult (**e**) and removed during development (**f**). Lower-order areas show relatively the highest pruning rate (around 80% of spines are pruned), whereas in higher-order associative areas the proportion of pruned spines is below 50% (except TEO where it is 65%). The most interesting differences are between the anterior-ventral (av) and posterior-dorsal (pd) parts of area TE, with only 10% of spines pruned in avTE, whereas in the pdTE this proportion is four times higher. Therefore, the layer III neurons in avTE have adult values at 90% of the maximum values, whereas in pdTE the adult values are at 60% of the maximum values

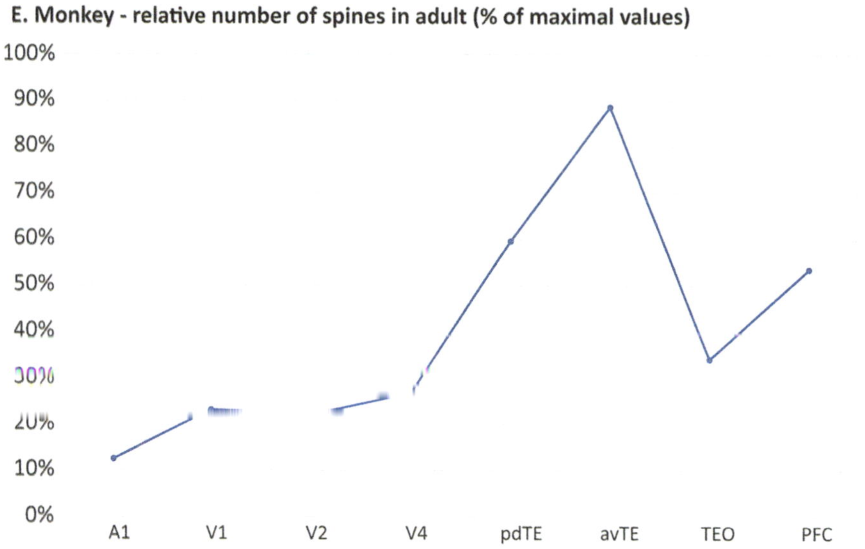

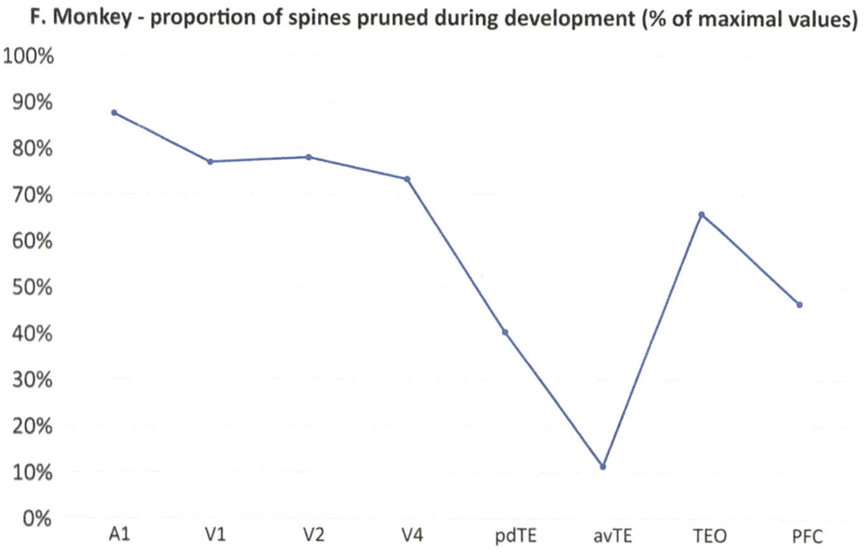

Fig. 4.5 (continued)

through rest of life. So, it can be concluded that the maximal synaptic number is reached around age of 1 year, similar to visual cortex. The maximum synaptic density in the middle frontal gyrus was reached later than in primary regions, between 15 months and 3.5 years. This period can be considered as a stage with a maximal number of synapses.

In the auditory cortex, there is a plateau in synaptic density lasting up to the 3rd year with significant decrease happening already during childhood. The adult level of synaptic density was reached by the beginning of adolescence. Data from primary visual cortex suggest that the peak might be reached slightly earlier than in the auditory cortex, during the first year, and that a significant drop could be observed already by the age of 3 years. However, small sampling size does not allow conclusion if subtle differences in timing of synaptic development occur between primary visual and primary auditory cortex.

In the prefrontal cortex, there was a slight drop in synaptic density (10–20%) observed in three specimens aged 12–15 years when compared to peak at 3.5 years (Huttenlocher and Dabholkar 1997). It seems that during childhood and early adolescence synaptic number within middle frontal gyrus remains stable and significantly higher than in adult. Therefore, in addition to delayed synaptogenesis, data also suggest more protracted maturation of neural network in the prefrontal than in the auditory and visual cortices. Based on the mentioned electron microscopy data, it seems that circuitry development in the human cerebral cortex follows an asynchronous and hierarchical pattern.

4.5.2 Synaptic Development in the Monkey

The most systematic quantitative electron microscopy analysis was done in the macaque monkey. Regarding pattern, the data are opposite to human and more in line with concurrent, i.e., parallel synaptic development through neocortical areas (Rakic et al. 1986). The analysis was performed starting from E50 (early fetal stage) up to 20 years of postnatal age in four major cortical areas—the primary visual (Bourgeois and Rakic 1993), somatosensory (Zecevic and Rakic 1991) and motor cortex (Zecevic et al. 1989), and in association prefrontal cortex (Bourgeois et al. 1994). The most rapid accumulation of synapses starts at E112 in all analyzed areas, after which exponential synaptic growth occurred. In the visual cortex, this happens approximately 2–3 weeks after the end of neurogenesis and not long after the completion of neuronal migration. This high rate of synaptic production continues during infancy (Fig. 4.5b) and begins to subside around the third postnatal month (Bourgeois and Rakic 1993). It is noteworthy that the course of synaptogenesis does not seem to change significantly at birth as a consequence of exposure to visual stimulation (Bourgeois et al. 1989; Bourgeois and Rakic 1996).

The rapid phase of synapse formation in the monkey cerebral cortex subsides around 2–3 months postnatally. Thereafter, synaptic density remains higher than in adults for several years (Missler et al. 1993; O'Kusky and Colonnier 1982; Zielinski and Hendrickson 1992). Between 3 months and 3 years of age (Bourgeois et al. 1994), synaptic numbers decrease gradually, and after this period, there is a sharp decrease. Note that cognitive abilities in 3-month-old macaque monkey can be compared with a stage of 1- to 2-year-old human infant. The 3-year-old monkey can be roughly compared to the stage of a 15-year-old human adolescent.

It should be mentioned that a somewhat different time course exists in the prefrontal cortex, where the highest level of synaptic density appears to be constant from the 2nd postnatal month until the 3rd year after birth, and then, the values are declining very gradually to the adult level (Rakic et al. 1994).

In all cortical areas studied, the decrease in synaptic density is mainly due to the elimination of asymmetrical junctions situated on dendritic spines, while symmetric synapses situated on shafts remain relatively constant (Bourgeois et al. 1994). Since asymmetric synapses are considered to be excitatory glutamatergic, their massive loss is likely to be reflected in the overall excitability of the cerebral hemispheres. In addition, in both the human and monkey, there is a sharp decline in cortical metabolic activity during the same time period (Chugani et al. 1987).

The total number of synapses eliminated has been calculated for the monkey primary visual cortex (Bourgeois and Rakic 1993) showing that as many as 1.9×10^{11} synapses may be lost between adolescence and adulthood in area 17 of each hemisphere. This means that, on average, 2450 synapses may be lost per second in each striate cortex between 2.5 and 5 years of age (Bourgeois and Rakic 1993). The same was reported in other macaque species and in marmoset monkey (Peters 1987). This means that there is almost 50% loss of synapses per neuron during puberty in area 17 of the monkey cortex (Bourgeois and Rakic 1993; O'Kusky and Colonnier 1982). It can be estimated that, in the entire cerebral cortex of both monkey hemispheres, as much as 30,000 synapses may be lost per second during the period of sexual maturation.

4.5.3 Synchronous Versus Hierarchical Synaptic Development

It can be concluded that in the visual, motor, somatosensory, and prefrontal cortex (Rakic et al. 1994) of the macaque monkey, synaptogenesis occurs synchronously. Even synaptogenesis in the limbic cortex seemed to follow the same pattern as in the neocortex (Granger et al. 1995). This is not in line with human electron microscopy data, which suggest a delay in the rapid phase of synaptogenesis and delay in the phase of synaptic elimination in the associative prefrontal cortex compared to that in the primary visual and primary auditory area (Huttenlocher and Dabholkar 1997). Technical and procedural factors might explain this discrepancy. For example, the studies in humans were done using phosphotungstic acid staining of synaptic junctions, a method that does not allow clear identification of synaptic structures. In addition, different sampling procedures were performed in different studies, and the number of cases was rather small. Due to the high interindividual differences that are expected in humans, there should be more cases to reach a higher sample size being able to clearly identify differences in pattern of synaptic development between different areas. Even in the macaque monkey, where lower interindividual differences are expected and sample error was reduced because tissue was dissected from the same set of animal specimens, it is difficult to identify subtle differences in pattern of synaptic development between different areas.

Rakic et al. normalized and replotted data on synaptogenesis obtained from the human prefrontal and visual cortex on a semi-logarithmic scale (Rakic et al. 1994). They found both curves to overlap, as they do in the macaque monkey. Therefore, the course of synaptogenesis in the human and monkey could be remarkably similar across the entire cortex. The study performed in developing chimpanzees by counting synaptophysin puncta in the primary somatosensory (area 3b), primary motor (area 4), prestriate visual (area 18), and prefrontal (area 10) cortices found synchronous synaptic development within the analyzed areas (Bianchi et al. 2013). Here, we organized the data about synaptic development in the monkey (Rakic et al. 1994) and humans (Huttenlocher and Dabholkar 1997) cerebral cortex in relation to maximum values through equivalent developmental stages (Fig. 4.5a, b). With respect to sample size, data support the view that, in both species, the general temporal pattern of synaptic development occurred synchronously with indications of delayed pruning in the associative prefrontal cortex compared to primary areas.

The synchronous time course of synaptogenesis in different cortical areas corresponds well with the changes occurring simultaneously in cerebral metabolism in both, the human and monkey cerebral cortexes. In humans, the use of fluorodeoxyglucose, which indicates the level of metabolic activity in positron-emission tomography (PET), reveals that, after birth, metabolic activity also increases concurrently in the prefrontal, motor, somatosensory, and visual cortex (Chugani et al. 1987).

The glucose utilization is mainly used for maintenance of membrane potential. So, in sedated children after the age of 2 years, the positron-emission tomography activity mainly correlates with changes in synaptic number (Chugani and Phelps 1986; Koehler 2021). During the first 3 years, there was increase in glucose utilization, and adult level is reached around the age of 1 year for all regions. Between 3 and 8 years of age, glucose utilization for all regions was twice as high as in adults, and between 9 and 15 years, it was 60–70% higher than in adults. Developmental positron-emission tomography studies support the idea that maturation of diverse cortical areas in both, monkeys (Jacobs et al. 1995; Moore et al. 2000) and humans (Chugani et al. 1987; Muzik et al. 1999), occurs simultaneously rather than in a markedly sequential order.

Further, the synchrony in synaptogenesis observed in nonhuman primates is in tune with biochemical and functional data on cortical maturation. Biochemical studies suggest that the concentrations of dopamine, noradrenaline, and serotonin increase rapidly in the cortex of the macaque over the first 2 months and approach adult levels by the 5th postnatal month (Goldman-Rakic and Brown 1982). Additional studies on the accumulation of major neurotransmitter receptor sites in different cortical areas show that their maximum density is also reached between 2 and 4 months after birth (Lidow et al. 1991; Lidow and Rakic 1992). The curves of the increase in receptor density are very similar to those of synaptogenesis in all the examined areas. However, the phase of decline of receptor density appears to slightly precede the phase of synaptic decline. These observations from several divergent cortical areas suggest that the initial formation and maintenance of synapses and their biochemical maturation may be determined by intrinsic signals, which are common for the entire cortical mantle.

However, it should be taken into account that the electron microscopy data from both monkey and human studies suggest a slightly different and prolonged period of synaptic development for the prefrontal cortex. In addition, despite concurrent changes in functional activity, the local cerebral glucose utilization obtained during positron-emission tomography showed delayed decrease and the highest and most protracted overactivity in the frontal cortical regions. The extended synaptic maturation in the human prefrontal cortex was supported by data on changes in synaptophysin and postsynaptic density protein 95 in the human prefrontal cortex from mid-gestation to young adulthood (Glantz et al. 2007), and also by analyzing human-specific gene expression changes during postnatal brain development in the prefrontal cortex of humans, chimpanzees, and rhesus macaques (Bakken et al. 2016; Liu et al. 2012).

Laminar changes in synaptic numbers in monkeys showed a tendency for higher and slightly protracted development of synapses in layers II–III, layers that are predominantly recipients (and sources) of cortico-cortical projections (Rakic et al. 1994). It should be emphasized that both methods (electron microscopy and positron-emission tomography) estimate the total number of synapses and measure the overall functional activity without the possibility to assess changes within a particular microcircuit. These data are a summation of changes within all microcircuits and might mask the changes in synaptic numbers that may appear in individual microcircuits. Electron microscopy studies showed that asymmetric synapses located on dendritic spines are those that are overproduced during development (Rakic et al. 1994). Thus, analyzing changes in spine density on identified classes of neurons, even on different parts of their dendritic tree (apical versus basal), might provide a better insight about changes within different microcircuits.

4.5.4 Stages of Cortical Synaptic Development

After evaluating present data about synaptic development in the human and monkey, here we present a modified staging (Table 4.1) according to previously described phases of synaptic development (Bourgeois 1997).

Stage 1 includes slow synaptogenesis that leads to a relatively low synaptic density in the marginal and subplate zones. These events are characteristics of the early fetal stage of human cortical development (post-conceptional weeks 8–12, E46–E60 in monkey), starting even during embryonic stage (post-conceptional week 6, E40).

Table 4.1 Stages of synaptic development in the cerebral cortex of human and macaque monkey

	Title	Main feature	Timing (H—humans, M—monkey)	Description
Stage 1	Initial synaptogenesis	Present only in the transitory zones (marginal zone and presubplate)	H: 6–12 pcw M: E40–E60	Very low number of synapses that have immature morphology
Stage 2	Transitory fetal circuits	Well-developed circuits within the transitory zones (marginal zone and subplate)	H: 14–24 pcw M: E65–E110	Subplate is dominant layer containing most differentiated neurons
Sub-stage 2B	Initial synaptogenesis in cortical plate	First synapses found at 18 pcw (E85) and at 24 pcw (E110) synapses present throughout entire cortical plate	H: 18–24 pcw M: E85–E110	Accumulation of thalamocortical fibers within subplate and first penetration into cortical plate
Stage 3	Intensive cortical synaptogenesis (part 1)	Rapid accumulation of synapses within cortical plate, while transitory circuits are still highly functional without massive regression	H: 26 pcw–3 months M: E110–3 weeks	At the beginning induced by thalamocortical ingrowth and afterward by cortico-cortical fiber ingrowth
Stage 4	Intensive cortical synaptogenesis (part 2)	Rapid accumulation of synapses within cortical layers with functional regression of transitory circuits of marginal zone and subplate	H: 3 months–2 years M: 1–3 months	Up to one million of synapses in humans and 100,000 in monkey produced per second
Stage 5	Synaptic overproduction	Period of synaptic fine-tuning modulated by activity	H: 1–20 years M: 3 months–5 years	High regional differences in duration and amount of overproduction—regional differences in environmental role to shape neural network architecture
Sub-stage 5B	Synaptic pruning	Synapses with low activity that have not reached sufficient stability during period of overproduction are eliminated	H: 3–25 years M: 6 months–7 years	Elimination trend could show different patterns between different microcircuits
Adult	Stable values	High structural stability of network architecture with stable synaptic number through lifespan	Beginning H: 15–25 years M: 5–7 years	Changes include modulation of synaptic strength without synaptic turnover

Timing of prenatal period is presented in post-conceptional weeks (pcw) for humans (H) and embryonic days (E) for monkey (M). Staging presented here is modified according to Bourgeois (1997).

Stage 2 occurs during the middle trimester of gestation (post-conceptional weeks 14–24, E65–E110) when the synaptic density increases in the marginal zone and the subplate, as a result of ingrowing fibers accumulation. Stages 1 and 2 represent the stages of transitory fetal circuits. In the second half of stage 2, the first synapses appear in the cortical plate (the prospective cerebral cortex). The synapses appear first in the lower part (prospective infragranular layers), and at the end of this stage, they are present across the entire depth of the cortical plate.

Stage 3 occurs during the last trimester of gestation and lasts up to the 1st and the 3rd postnatal months. This stage is characterized by a very rapid accumulation of synapses within the cortical plate, where the highest increase in synaptic number is due to the ingrowth of cortico-cortical fibers. Therefore, this stage is more extended in the regions with a higher amount of cortico-cortical connections, such as associative areas. During this stage, transitory circuits are still highly functional without significant morphological regression (Hadders-Algra 2018a). This feature distinguishes stage 3 from stage 4, which is also characterized by a very rapid accumulation of synapses.

Stage 4 corresponds to the human infant stage, the first year, and in some regions continuing into the second postnatal year. The transitory circuits lose their functional role, and only a small part of them is incorporated into the mature circuits. Rapid increase in synaptic number during stage 4 is not induced by important ingrowth of new afferents, but mainly by the increase in terminal ramification of already grown fibers and growth of intracortical excitatory connections. The rapid accumulation of synapses in the cerebral cortex is in continuity with stage 3. By the end of stage 4, the peak in synaptic number is reached, depending on the region, occurs around the second half of the 1st and during the 2nd year. Despite an abrupt slowdown in the tempo of synaptogenesis, in some of regions there might still be an increase in synaptic number during the rest of childhood.

It is important to mention that in the classification given by Bouergeois (1997), stages 3 and 4 were not recognized as separate stages but as a single stage, phase 3. During stages of intensive synaptogenesis (stages 3 and 4), it can be estimated that in the rhesus monkey at some point more than 100,000 synapses are formed in the cerebral cortex every second, and in humans, this can reach several millions.

However, the synaptic development does not end with stage 4 and there is an additional stage of synaptic development. During stage 5, the synaptic number is higher than it is adulthood. This stage is a dominant developmental stage in primates and particularly characterizes humans. In some of the microcircuits in the human cerebral cortex, the peak values in synaptic number could be up to three times higher than in adult. Also, the period of synaptic overproduction could last up to the third decade of life, i.e., synaptic overproduction might last for a period of more than 20 years. After developmental stage 5, the synaptic number dropped to values that will be stable during the rest of the lifespan.

It is difficult to postulate a distinct demarcation between the stages proposed here. This especially applies to stage 5 and the transition to the adult stage. Data suggest that for some areas (i.e., layers or circuits), the peak values in synaptic number are preserved without any important decrease during a longer period of time, after a sharp and fast drop occurs to reach the final adult values. In other areas, the reduction in synaptic number can occur immediately and may show different tempo and patter. The synaptic number can be reduced slowly and continuously, but it can also show changes in tempo and amount of reduction. So, there might be one or two time points with sharp decrease, after which a new plateau is reached with synaptic number still above the adult level. It should be mentioned that during stage 5, when the synaptic number is above adult values, new synaptogenesis is most likely not a frequent event (Fu and Zuo 2011). Thus, during this stage, synapses disappear without replacement, or will became preserved during the rest of the lifespan, suggesting that the vast majority of synapses found in adults are already present at the beginning of stage 5.

4.6 Changes in Dendritic Spine Number on Cortical Neurons in Human and Monkey

In the studies mentioned in the previous chapter (Huttenlocher and Dabholkar 1997; Rakic et al. 1994), an estimate of the number of synapses per neuron was made at different ages by dividing the density of synapses in the neuropil with the number of neurons. These calculations do not give information about the changes for different classes of neurons nor between different compartments of the dendritic tree (e.g., apical vs. basal dendrites). Different classes of principal neurons could be targeted from different sources, and even different compartments of dendritic tree within particular neuron class can be targeted by different afferent system (Bitanihirwe and Woo 2021). By studying changes in synaptic number on different classes of principal neurons or on specific parts of the dendritic tree, it is possible to presume the pattern of development for particular microcircuits. As stated before, an individual dendritic spine receives one excitatory glutamatergic input (Cano-Astorga et al. 2021; Rasia-Filho 2022) and dendritic spine synapses are overproduced during development (Lübke and Albus 1989; Markus and Petit 1987). Therefore, it is possible to estimate the total number of excitatory synapses of a particular neuron class (and even dendritic compartment) by quantifying the number of dendritic spines.

Numerous Golgi studies have been performed to analyze the changes in dendritic spine density of the developing human cerebral cortex (Garey 1984; Garey and de Courten 1983; Michel and Garey 1984; Mrzljak et al. 1988, 1990, 1992; Petanjek et al. 2008, 2011; Purpura 1975; Takashima et al. 1980). The first comprehensive study was performed by Michel and Garey who measured the density of spines on 50 μm long segment of apical dendrites from layer III pyramidal neurons in the primary visual cortex of specimens aged from post-conceptional week 33 to age of

30 years (Michel and Garey 1984). In the fetus, around 30 spines were found on average per segment, in the neonate about 50 spines, and at 5 months there were approximately 80 spines per segment. This was the maximum number of spines reached, after which the spine number decreased, reaching a level of about 50 per segments again by 2 years of age, a value that is maintained into adulthood.

Golgi studies about dendritic spine development were also performed in monkeys (Anderson et al. 1995; Boothe et al. 1979; Fritschy and Garey 1986; Garey and de Courten 1983; Lund et al. 1977, 1991; Lund and Holbach 1991; Mates and Lund 1983). Quantitative analysis of Golgi-impregnated mid-layer III pyramidal neurons in area 9 and 46 of the rhesus monkey prefrontal cortex showed that spine density on both the apical and basal dendrites increased during the first two postnatal months (Anderson et al. 1995). Spine density remained at a plateau until 1.5 years of age, and then decreased during the peripubertal period until stable levels were achieved and kept for the rest of the lifespan. This corresponds to the data on synaptogenesis in the monkey prefrontal cortex (Rakic et al. 1994). Interestingly, the density of parvalbumin-immunoreactive axon terminals belonging to the chandelier class of local circuit neurons exhibited a temporal pattern of change that exactly paralleled the changes in dendritic spine density. This was somewhat surprising, because electron microscopy studies did not find overproduction for symmetric, presumably inhibitory synapses. The decrease in density of parvalbumin-immunoreactive chandelier axon cartridges during adolescence in monkeys was confirmed by a subsequent study (Chung et al. 2017; Fish et al. 2013). At the same time, the density of parvalbumin reactive basket interneurons puncta increased during adolescence. This suggests that some GABAergic circuitries might undergo synaptic overproduction and pruning through childhood and adolescence, similar to glutamatergic principal cells. At the same time, other GABAergic circuitry underwent continuous production without elimination, which can be extended up to the end of adolescence. This clearly shows that counting the total synaptic number is merely a summation of changes within all microcircuitries, and changes in an individual microcircuit could potentially exhibit completely different patterns.

The quantitative analysis of synaptic spine development of Golgi-impregnated neurons was also performed in the monkey visual cortex from E145 to adult (Boothe et al. 1979). Different patterns of changes in dendritic spine number were found between different neuron classes. The analysis was performed on spiny stellate neurons from laminae IVC-alpha and IVC-beta and pyramidal neurons of lamina IIIB and upper lamina VI. All neurons showed a gradual increase in spine number up to the 2nd postnatal month. The pattern of synaptic spine elimination differs between different classes of neurons and even between different dendritic compartments within the same neuronal population. However, for all neuronal populations, the spine number decreases between 9 months and 5 years. In a later study (Lund and Holbach 1991), it was shown that the alpha neurons peak in spine number at 5–8 weeks postnatal, whereas the beta neurons peak in spine number between 8 and 24 weeks postnatal. At all ages prior to 30 weeks, the two sets of neurons have different total spine numbers. Close to 30 weeks of postnatal age, the total spine density falls in both sets of neurons and becomes identical in both alpha and beta neurons. Data from mentioned studies strongly support the opinion that there is different patterns of synaptic development between microcircuitries, particularly referring to the level of overproduction and timing of pruning.

In addition to Golgi studies, spine development in monkeys was analyzed on neurons stained using Lucifer Yellow during electrophysiological recordings (Duan et al. 2003; Elston et al. 2011a; Elston and Fujita 2014; Oga et al. 2013, 2017; Sasaki et al. 2015). It was shown that pyramidal cells in functionally different cortical regions of macaque monkey have different rates of axon and dendritic growth and vary in tempo of spine formation as well as in the degree of pruning (Elston et al. 2009, 2010a, b, 2011b). The overall temporal pattern of dendritic spine development is strikingly similar on the basal dendritic tree of layer III pyramidal neurons across various cortical areas. However, there are large differences in the number and degree of spines per pyramidal neurons present in adult or overproduced during development.

The changes in spine density and number were studied in macaque monkeys at the ages of 2 days, 3 weeks, 3.5 months, 7 months, 18 months, and 4.5 years (Fig. 4.5c–f). The basal dendrites were analyzed in visual areas through functional hierarchy. This includes in occipital lobe primary visual area V1, dorsal stream secondary visual area V2, and ventral stream tertiary visual area V4. In the temporal lobe, it includes area TE, the highest-order associative area of the ventral visual stream, where two parts (anterior-ventral and posterior-dorsal) were analyzed separately. It includes the area TEO (located at the temporo-occipital border), which is a high-order visual associative area responsible for pattern perception and selective visual attention. The analysis was also performed in the primary auditory cortex (A1) and in ventrolateral area 12, corresponding to orbital executive prefrontal cortex.

The total number of spines per neuron within basal dendritic tree varied considerably in adult (Fig. 4.5c) as wel as the number of spines overproduced (Fig. 4.5d). The lowest number of spines in adult was found in primary regions (A1 and V1) that were around 900 spines per neuron. The rather low number of spines was found in secondary (V2) and tertiary (V3) visual areas (1,100 and 2,400 spines per neuron), whereas in associative areas the number of spines in adult was several times higher, with highest number reached in prefrontal area 12 (8,500 spines) and anterior-ventral area TE (10,900 spines). It should be mentioned that anterior-ventral area TE has the lowest number of spines overproduced (Fig. 4.5d), 1,400 spines, i.e., only 10% of spines were pruned (Fig. 4.5e). On the other side, in primary and secondary visual areas (V1 and V2) almost 80% of spines were pruned during development, and in primary auditory cortex (A1), 90% of spines were eliminated. However, the highest number of spines pruned per neuron was in higher associative areas.

In the occipital visual cortex (V1, V2, and V4), adult spine number and number of spines pruned, were increasing with functional hierarchy (Fig. 4.5c). Interestingly, despite similar number of spines found in adult primary areas analyzed, in the auditory A1 area number of spines overproduced was two times higher than in visual area V1 (6,650 versus 3,000). The number of spines overproduced in A1 was around number overproduced in tertiary visual area V4. The highest number of spines overproduced was in highest-order associative areas, prefrontal area 12 (7,400), and visual area TEO (9,300). Interesting result was found for associative area TE, where in the anterior-ventral part only slightly more than 10% of spines were eliminated (1,400 spines), whereas in the posterior-dorsal part 40% of spines were pruned (4,200 spines). So, despite TE being highest-order visual area, the total number of spines eliminated per neuron is below some of lower-order visual areas. It should be emphasized that despite the highest number of spines eliminated per neuron is largest in high-order associative areas (PFC and TEO), relatively more spines per neuron are overproduced in areas of lower functional order (Fig. 4.5f).

Nevertheless, present data showed that different neuron classes, i.e., different neural circuitries, have different complexity of connectivity in adult and different amount of synapses overproduced. Regarding the temporal pattern of changes in synaptic number, in most of the regions the highest synaptic number is found at 3.5 months. At the age of 4.5 years, the values dropped to final level. However, the curve of decrease based on values in specimens 7 and 18 months old varied slightly between areas. The same was in period from birth up to 3.5 months. So, in 2-day-old monkey, both relative and absolute level of spine number varies significantly between areas, as in the 3-week-old specimens. Therefore, despite a general temporal pattern of synaptic production and elimination across areas was similar, there might be differences in pattern of the both ascending and descending curve of synaptic development. The most pronounced deviation in developmental curve was seen in basal dendrites of neurons in area V4. These neurons were the most spinous at 2 days among all areas analyzed, reaching approximately 9,000 spines per neuron, that is almost at maximal level found at the 3.5 months. Another example of significant deviation from curve is layer V pyramidal cells in V1, because they exhibited peak spine number earlier, reaching greatest values at 3 postnatal weeks.

Mentioned data lead us to conclusion that the total number of spines in adult and the total number of spines overproduced increases with the functional hierarchy of cortical area. As the functional hierarchy of area increases, there is also a higher chance for spinogenesis to be delayed, for pruning to occur more slowly and the period of overproduction to be extended. However, this rule could not be strictly applied, and there seem to be many area (i.e., circuit)-specific features in the pattern of postnatal synaptic development.

A particularly interesting pattern of spine development was described for basal dendrites of layer III neurons within area V4. Despite reaching almost maximal values at 2nd postnatal day, in coming weeks layer III neurons appeared to prune spines, after they undergoes considerable spinogenesis to the age of 3.5 months. These data suggest that even if the maximal number of synaptic spines is reached, during the period of net decrease in the number of spines there might be additional synaptogenesis and turnover of synapses. However, experimental data in rodents have found that from synapses found in adult brain, at least 90% were present at time point when the synaptic number has reached maximal values (Alvarez and Sabatini 2007; Grutzendler et al. 2002; Holtmaat et al. 2005; Knott et al. 2006; Zuo et al. 2005a).

The synaptic spine counting performed on particular classes of neurons offers clues that are, indeed, more precise in understanding the developmental pattern of different microcircuits than data about total number of synapses. At the same time, there is no information about synaptic turnover, removal of present and formation of new synapses. The morphological features of dendritic spines and corresponding synapses in humans and monkey during period of synaptic overproduction, as well as experimental data from rodents, favor the option that huge turnover during period of overproduction is not present. However, there is no direct proof of this assumption.

4.7 Dendritic Spine Development of Principal Neurons in the Human Prefrontal Cortex

A detailed analysis studying dendrites and dendritic spine development in humans was performed on basal and apical oblique dendrites of large deep layer III (layer IIIC) and large layer V pyramidal neurons within area 9 ranging from post-conceptional week 12 to age of 91 years (Koenderink et al. 1994; Koenderink and Uylings 1995; Mrzljak et al. 1988, 1990, 1992; Petanjek et al. 2008, 2011, 2019; Sedmak et al. 2018). Large pyramidal neurons of layers IIIC and V represent the key elements of the dorsolateral prefrontal cortical circuitry. Layer V pyramidal neurons project to the basal ganglia, as principal cells of the associative cortex basal ganglia circuit (Borra et al. 2021, 2022; Finn et al. 2019; Groenewegen et al. 1997; Korponay et al. 2020; McGuire et al. 1991; Uylings and de Brabander 2002; Yeterian and Pandya 1994). Pyramidal neurons of the layer IIIC have long ipsi- and contralateral cortico-cortical projections (Innocenti et al. 2022; Schwartz and Goldman-Rakic 1984; Xu et al. 2022a). Individual layer IIIC neurons establish projections to several different areas suggesting a major role in inter-areal integration that grants them the title “associative” neurons (Goldman-Rakic 1999; Petanjek et al. 2019; Schwartz and Goldman-Rakic 1982).

Quantitative spine analysis was preceded by the studies analyzing quantitatively pattern of lifespan dendritic changes in these classes of neurons (Mrzljak et al. 1992; Petanjek et al. 2008; Sedmak et al. 2018). Dendritic growth begins immediately after the arrival of the neuron in the cortical plate (Fig. 4.4), which is between post-conceptional weeks 12 and 20 in the human prefrontal cortex. This phase is characterized by the protrusion of primary dendrites, which includes the growth of primary basal dendrites and the apical dendrite with oblique dendrites. It is also characterized by intensive axon growth (Mrzljak et al. 1988, 1990).

After initial growth, when basic dendritic structure is established, dendritic growth intensifies. This phase begins with a massive outgrowth of new dendritic segments, followed by large elongation without significant formation of new dendrites. This coincides with the ingrowth of afferent fibers into the cortical plate (Kostovic and Goldman-Rakic 1983; Kostović 2020; Kostović et al. 2021; Kostović and Jovanov-Milosević 2006; Krsnik et al. 2017; Vasung et al. 2017). The intensive growth of dendrites of layer V pyramidal cells begins around post-conceptional week 24, when thalamocortical afferents penetrate the cortical plate, and around post-conceptional week 32 for layer IIIC pyramidal cells, when cortico-cortical axons invade the cortical plate. Since both thalamocortical and cortico-cortical afferents are glutamatergic, this finding is in agreement with experimental data showing that an intensive dendritic differentiation starts with the appearance of N-methyl-D-aspartate (NMDA) receptors on neurons, making a functional path for activity to influence dendritic growth (Cline et al. 2023; Khazipov et al. 2001; Sin et al. 2002).

After fast and intensive growth through last trimester of gestation (He et al. 2020; Petanjek et al. 2019) and first few months of postnatal life (Fig. 4.6), the dendritic growth slows down. This last, third phase of dendritic growth, is longer and lasts for

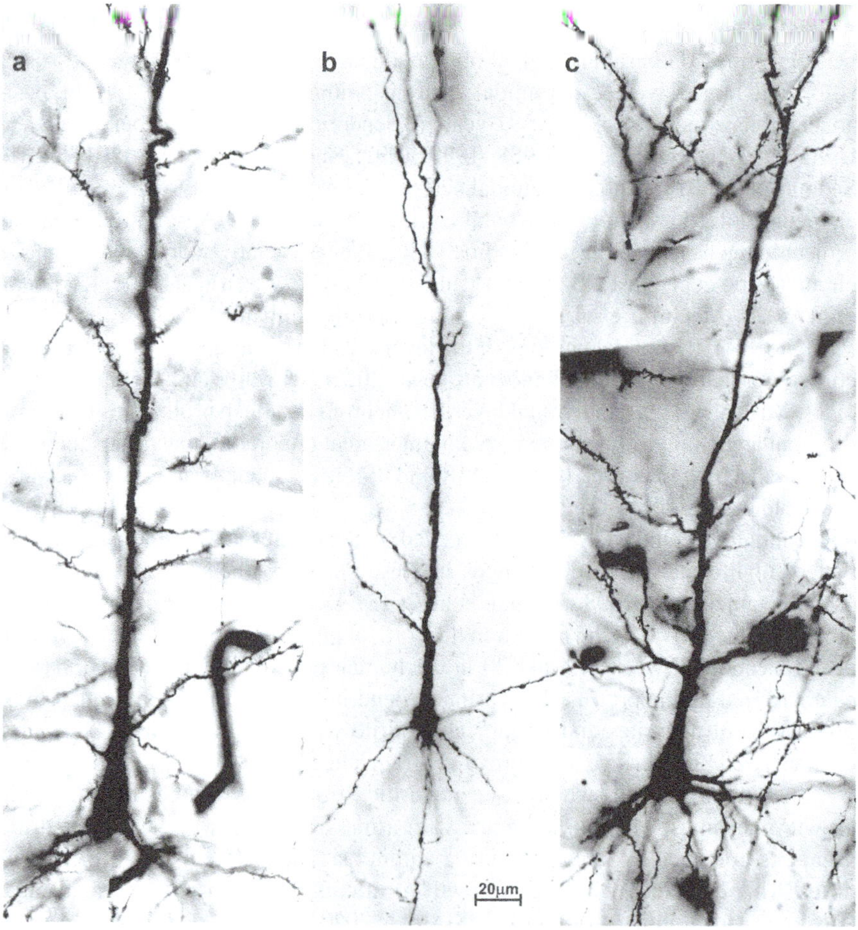

Fig. 4.6 Golgi microphotographs of layer V (**a**) and layer IIIC (**b**, **c**) pyramidal neurons in newborn (**a**, **b**) and 1-month-old infant (**c**). Note the immaturity of the layer IIIC pyramidal neuron in newborn compared to the layer V neuron, which in newborn is at the level of the layer IIIC neuron in a 1-month-old infant

about 1–1.5 years. Usually, up to 20% of dendritic length is established during that period (Fig. 4.7). During the third phase of dendritic growth, when the growth is slow, there is a large increase in dendritic spine density and number of synapses (Anderson et al. 1995; Bourgeois et al. 1994; Huttenlocher and Dabholkar 1997; Rakic et al. 1994). The morphological and experimental studies described three-phasic dendritic growth as a typical model of dendritic tree formation (McAllister 2000; Okabe 2009; Uylings et al. 1994; van Pelt and Uylings 2002; Whitford et al. 2002). The large layer V pyramidal neurons of the human prefrontal cortex follow a typical pattern of dendritic growth through 3 phases, but layer IIIC does not (Fig. 4.7c, d, e). For large layer IIIC pyramidal neurons, the period that lasts between 2.5 and 16 months and should correspond to the last phase of growth, is a dormant period regarding dendritic growth. During this period, large layer V neurons enter the last, third phase, and by the 16 postnatal month, layer V neurons reach values very close to adult (Figs. 4.7a, b and 4.8).

For large layer IIIC pyramidal neurons, a last phase of dendritic development occurs after 1-year-long dormant growth period (Fig. 4.8a). So, in period between 16 months and 2.5 years there is an increase in dendritic length for about 50% in comparison with values that are present between 2.5 and 16 months. It means that layer IIIC neurons have two stages of massive dendritic growth. First, massive growth occurs during phase 2 (between 32nd post-conceptional week and 3rd postnatal month), when around 50% of dendritic tree is established. The second stage of large growth occurs during phase 4 (16 months–2.5 years), when close to 40% of dendritic tree grows. Such biphasic growth was not described in previous studies of dendritic development in humans, neither in other primates during equivalent postnatal ages (Petanjek et al. 2019; Sedmak et al. 2018). For most subpopulations of principal neurons in the prefrontal cortex (Jacobs et al. 1997; Koenderink et al. 1994; Koenderink and Uylings 1995; Schade and van Groenigen 1961), as well as other regions of the human cerebral cortex (Becker et al. 1984; Bianchi et al. 2013; Jacobs and Scheibel 1993; Purpura 1975; Takashima et al. 1980), major postnatal dendritic growth occurs during the first year. So, the large elongation of basal and oblique dendrites by the age of 2 years is unique feature of associative layer III neurons, most likely related to their unique role for integrating processing through cortical network (Lagercrantz and Changeux 2010).

First postnatal years are characterized by intensive synaptogenesis (Chugani 1998; Huttenlocher and Dabholkar 1997; Rakic et al. 1994; Somel et al. 2009; Yin et al. 2019), and this is also reflected on spinogenesis (Petanjek et al. 2011). The vast majority of spines (about 90% after the neonatal period) belong to the mushroom type characterized by a neck 1.5–3.5 μm long and up to 0.5 μm thick that expands into a terminal bulb (head) with a radius of 1–2 μm. The morphology of the head and neck of mushroom spines remained relatively constant except in the late adolescent stage (age 16–20 years), when spine heads become obviously larger. During the first postnatal month, numerous hair-like thin spines (usually around 5 μm long, but can extend up to 8 μm), which lack an obvious bulb and have terminal expansions slightly wider than the neck, were observed, and their percentage diminished afterward (Fig. 4.2).

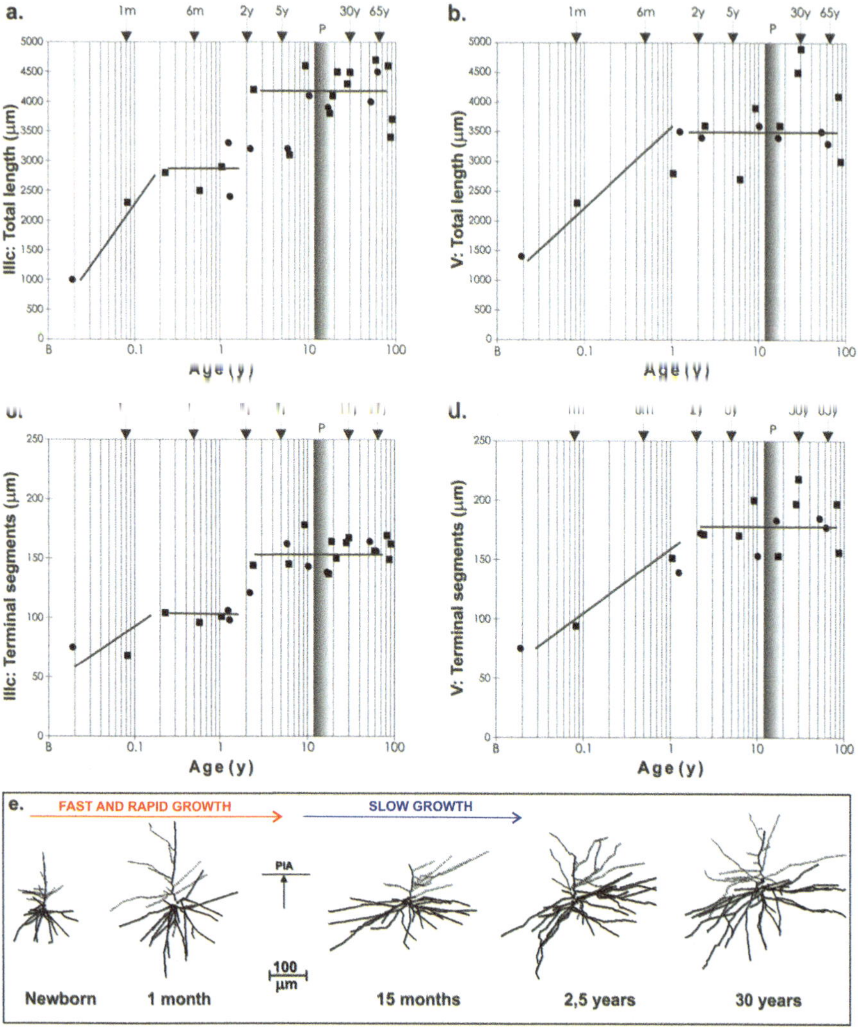

Fig. 4.7 The total length of the basal dendritic tree per neuron (**a**, **b**) and the mean length (**c**, **d**) of terminal segments for layer IIIC (**a**, **c**) and layer V pyramidal cells (**b**, **d**). Observe that the most rapid and intensive dendritic growth occurred, for both classes of neurons, between birth and 2.5 months of age. However, dendrites of layer V pyramidal cells continue to grow slowly until the ages of 16 months, whereas dendrites of layer IIIC pyramidal neurons did not display significant dendritic growth between 2.5 and 16 months. They displayed a second growth spurt later, between 16 months and 2.5 years. This is in line with data on the individual terminal segment length. The age is presented in postnatal years on a logarithmic scale in order to fit the entire human life span onto a single graph. The period of puberty is marked by a shaded bar. Squares represent males, and circles females. P, puberty; B, birth (this actually corresponds to 4th postnatal day); m, months; y, years. Panel **e** shows three-dimensional reconstructions of basal and apical dendritic trees of rapid Golgi-impregnated layer V pyramidal cells in the prefrontal cortex projected onto the coronal plane. The orientation toward the pia is indicated by an arrow. Oblique dendrites originate from the apical dendrite and are represented by dashed lines. All pyramidal cells are represented at the same magnification (bar scale = 100 μm). Fast and rapid growth occurred during first postnatal month,

On basal and apical oblique dendrites of both layer IIIC and layer V pyramidal neurons (Fig. 4.9), the dendritic spine density increased significantly during infancy. It reached its peak earlier during childhood, when the density was, on average, more than two times higher than in adults. In the newborn, there were less than five spines per 50 μm long part of dendrite on all measured segments. In the 2.5-month-old infant, the number of spines per 50 μm segment increases up to 3 times. The length of the dendritic tree increased during the same period with a similar rate, so it can be estimated that the total number of dendritic spines per neuron increases at least five times during the first 3 months of postnatal life.

Between the 3rd postnatal month and time point when maximal values were measured later during childhood, spine density increases for 4–5 times on layer IIIC neurons and 2–3 times on layer V neurons (Fig. 4.9d). The maximum density observed for distal oblique dendrites was more than 1 spine counted per 1 μm dendritic length on neurons in both layers. The density of proximal oblique and basal was ~0.7 spines per 1 μm dendritic length on layer IIIC and ~0.6 spines for layer V neurons. It has to be mentioned that in parallel with the increase in spine density, there was further increase in dendritic length, about 20–30% on layer V neurons, and by about 50% on layer IIIC neurons.

In adulthood, the spine density for distal oblique dendrites was slightly below 0.5 spines per μm of dendritic segment and around 0.25 per 1 μm spine per μm for proximal oblique and basal dendrites. For all segments, the spine density of layer IIIC neurons was ~20% higher than on layer V neurons (Petanjek et al. 2008, 2011). It should be noted that both basal and apical oblique dendrites of layer III pyramidal neurons are ~10% longer than those of layer V neurons (Petanjek et al. 2008; Sedmak et al. 2018). Thus, the total number of dendritic spines on basal and oblique dendrites of layer IIIC is greater than in layer V pyramidal neurons, in both the adult prefrontal cortex and during development. This finding contrasts with the finding in high-order unimodal areas where layer V pyramidal neurons demonstrate a greater degree of dendritic complexity and a higher density of dendritic spines (Benavides-Piccione et al. 2021).

The highest density of spines for all dendritic segments in layer IIIC pyramidal neurons was reached by the age of 2.5–7 years. Note that for layer V there was a slightly smaller sample and higher interindividual variability. The presented data showed that the dendritic spine density on all segments of layer V pyramids displayed the highest values between the ages of 2.5 and 9 years, before beginning to decline. Although on most of the measured segments dendritic spine density diminished gradually later during childhood and adolescence (9–22 years), it remained

Fig. 4.7 (continued) whereas during the rest of the first and during the second year growth is slow. Around 20% of dendritic length is established after third postnatal month. Note that there are no obvious differences for layer V pyramidal cells (**e**) of the 15-month-old infant in comparison with all subsequent ages (compare with layer IIIC on fig. 4.8A). (Reproduced from Petanjek et al. (2008), compilation of part of figs. 4.1 and 4.5 with modification, used under RightLinks license #5502690568405, originally published by Oxford University Press)

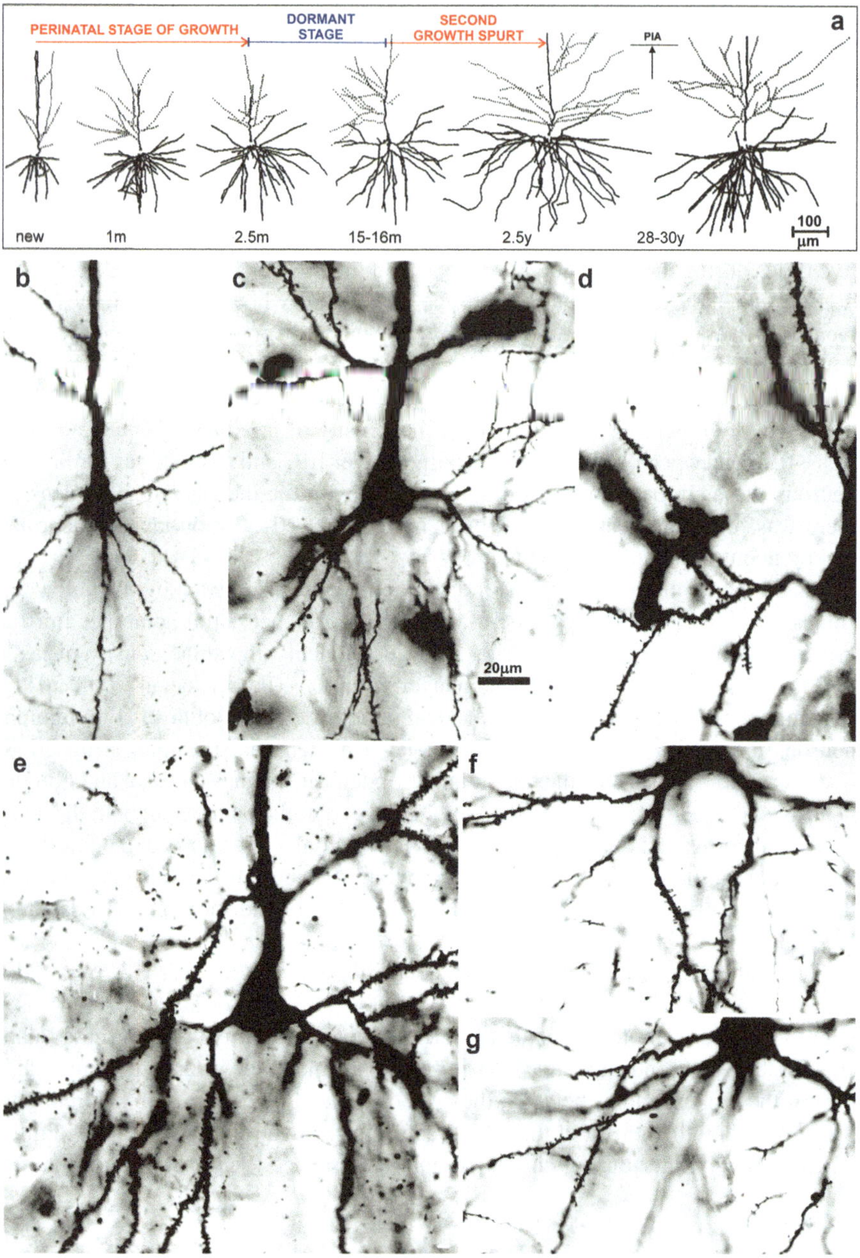

Fig. 4.8 Postnatal development of rapid Golgi-impregnated large layer IIIC pyramidal neurons in the magnopyramidal area 9 of the human prefrontal cortex. Three-dimensional reconstructions of basal and apical dendritic trees of rapid Golgi-impregnated pyramidal cells in layer IIIC, projected onto the coronal plane (**a**). Orientation toward the pia is indicated by the arrow. Oblique dendrites originate from the apical dendrite and are represented by dashed lines. All layer IIIC pyramidal

significantly higher throughout this period than in the specimens older than 30 years. In almost all subjects older than 30 years, dendritic spine density for all segments was significantly lower than in subjects between 15 months and 22 years of age.

Regression analysis of changes in spine density showed that the overall developmental course of changes was similar in both layer IIIC and layer V pyramidal neurons (Fig. 4.10). The overproduction extends up to the third decade of life, and the density on stable adult value appeared around the age of 30 years. The regression analysis confirmed that in adulthood there was 10–20% higher spine density on segments of layer IIIC neurons. During developmental peak in spine number, these differences rose and layer IIIC neurons have 20–40% higher spine density when compared to the corresponding segments of layer V neurons. This showed that the pruning of supernumerary dendritic spines is more pronounced in layer IIIC (cortico–cortical projecting neurons) than in comparable segments of layer V (subcortically projecting neurons). The lowest overproduction was found on segments targeted mainly by thalamocortical afferents, i.e., on basal dendrites of both classes of neurons and proximal oblique dendrites of layer V neurons (Garcia-Marin et al. 2013; Garey and Powell 1971; Miyashita 2022; Winfield et al. 1982). The highest overproduction with the most gradual reduction in density was observed for distal oblique dendrites of layer IIIC neurons that are the main target for inter-areal projections, but also for local intra-areal excitatory projections (Petanjek et al. 2019).

The long projections of large layer IIIC pyramidal neuron modulate inter-areal processing, while their local axonal collaterals are in control of the whole prefrontal cortico-cortical output (Fig. 4.11). The connectivity pattern of the human prefrontal cortex is characterized by massive reciprocal projections to both multimodal and unimodal parasensory associative areas (Barbas 2015; Fuster 2008; Groenewegen and Uylings 2000). Such a pattern of connectivity is responsible for regulation coordinated and synchronous activity among different areas of cerebral cortex (Friedman and Robbins 2022; Preuss and Wise 2022; Teffer and Semendeferi 2012). In this way, due to rich interconnections between the prefrontal cortex and other associative areas, prefrontal layer IIIC pyramids enable high efficiency in network

Fig. 4.8 (continued) cells are represented at the same magnification (scale bar 100 μm) and at the following ages: newborn, 1-month-old, 2.5-month-old, 15-month-old infants, 2.5-year-old child, and 28-year-old adult. Dendritic trees of layer IIIC pyramidal cells showed a marked increase in first 3 months and between 16 months and 2.5 years of age (compare with the fig. 4.7e). There are no obvious differences between layer IIIC pyramidal cells of 2.5-month-old and 16-month-old infants (dormant stage), as well as 2.5-year-old and 28-year-old subjects. Microphotographs showing changes in morphology of rapid Golgi-impregnated layer IIIC pyramidal cells of the Brodmann area 9 between newborn (**b**), 1-month-old-infant (**c**), 16-month-old infant (**d**), 2.5-year-old child (**e**), 19-year-old (**f**), and 73-year-old (**g**) adults (the magnification is same for all microphotographs; scale bar 20 μm). Even in these high-power microphotographs, the increase in dendritic complexity (an outgrowth of new segments) between newborn (**b**) and 1-month-old infant (**c**) is obvious. (Reproduced from Petanjek et al. (2008), compilation of part of fig. 1 with modification, merged with full fig. 2, used under RightLinks license #5502690568405, originally published by Oxford University Press)

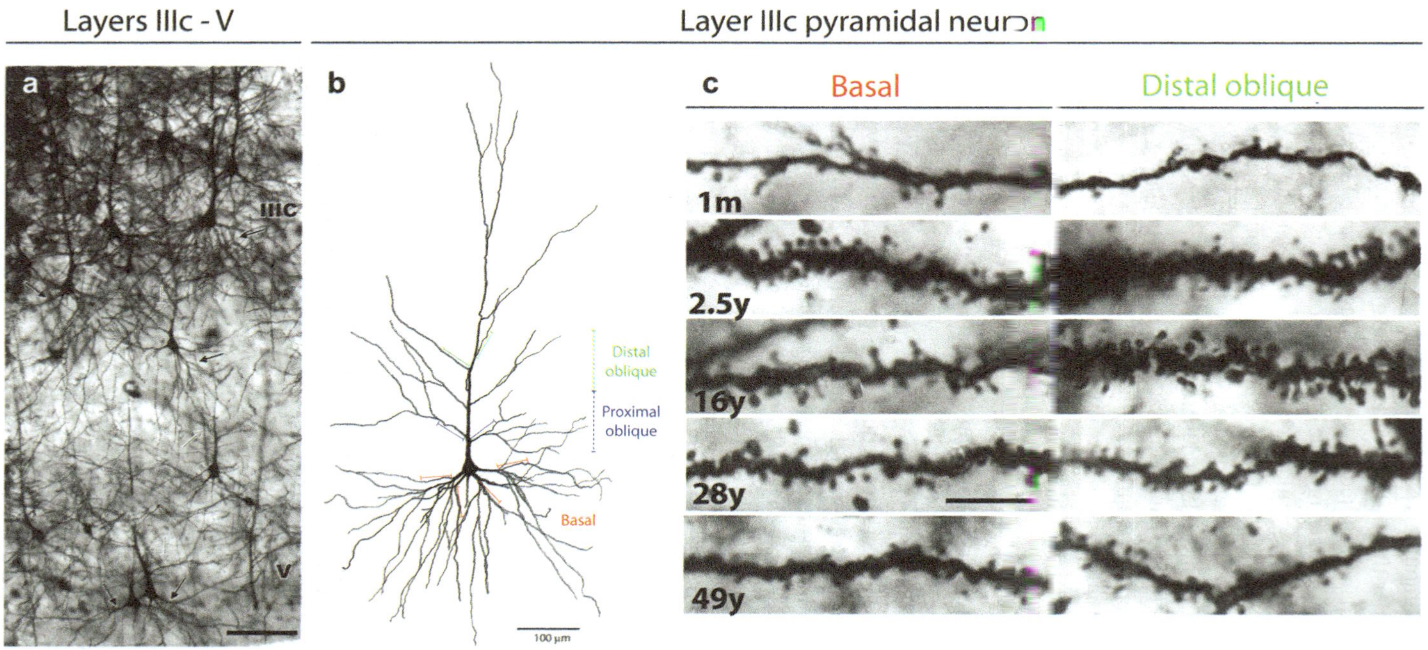

Fig. 4.9 Changes in dendritic synaptic spine density on rapid Golgi-impregnated large layer IIIC and layer V pyramidal neurons in magnopyramidal area 9 of the human prefrontal cortex from birth to 86 years. (**a**) Representative low magnification photographs of the rapid Golgi-impregnated layer IIIC and layer V pyramidal cells in the dorsolateral prefrontal cortex of a 16-year-old subject. (**b**) Neurolucida reconstruction of layer IIIC pyramidal neuron of a 49-year-old subject, illustrating dendrites selected for counting spines. (**c**) Representative high-power magnification image of layer IIIC pyramidal neuron dendrites. (**d**) Graphs represent changes in spine numbers per 50 μm of dendritic length. The age is presented in postnatal years on a logarithmic scale. Puberty is marked by a shaded bar. Squares represent males; circles, females. P, puberty; B, birth (fourth postnatal day); m, months; y, years. (Reproduced from Petanjek et al. (2011), Fig. 1 without change, used under CC-BY 4.0 International license and copyright)

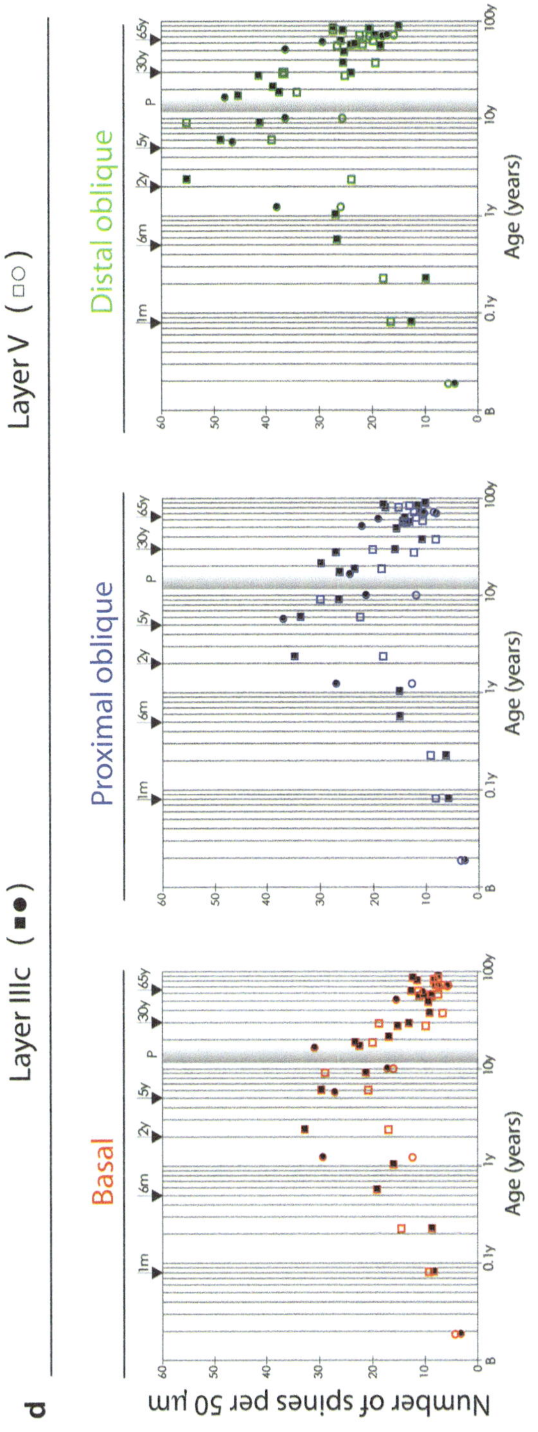

Fig. 4.9 (continued)

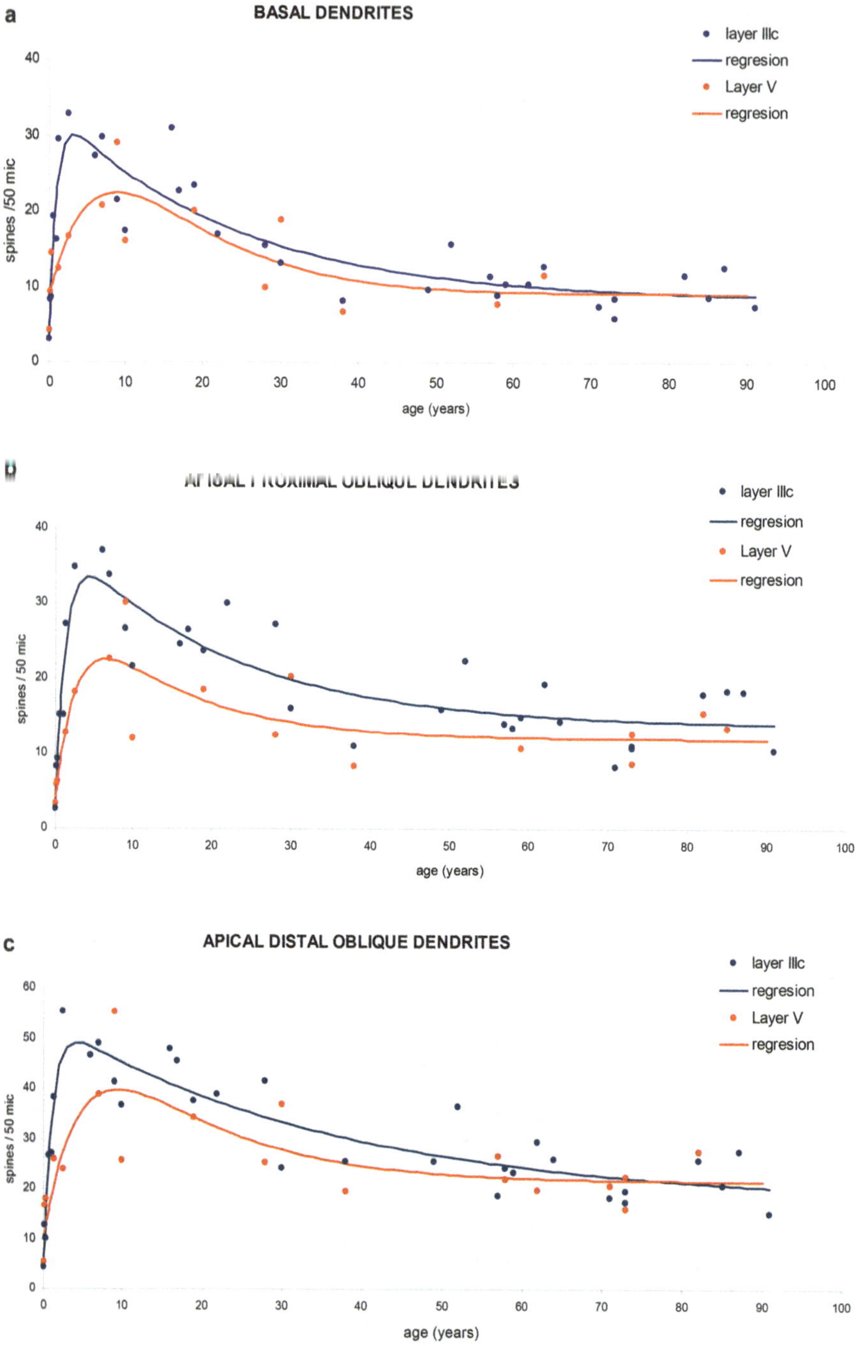

Fig. 4.10 The dendritic spine density, as defined in fig. 4.9, plotted at the linear scale to illustrate the dynamics of changes occurring during the human lifespan. Regression curves fit the distribution of data from the basal dendrites (**a**), apical proximal oblique dendrites (**b**), and apical distal oblique dendrites (**c**) of pyramidal cells from layer IIIC and V. (Reproduced from Petanjek et al. (2011), fig. 2 without change, used under CC-BY 4.0 International license and copyright)

integration throughout the human cerebral cortex. So, changes in their connectivity patterns that appear during development or in pathological condition will be reflected on global cortical functioning.

Developmental changes during the preschool period, i.e., the second growth spurt of dendrites occurring in parallel with a high level of synaptic spine overproduction on associative layer IIIC neurons, are necessary for the appearance of complex cognitive abilities. These abilities are crucial to develop mental understanding of higher levels of inter-personal interaction that is a basic step for socioemotional maturity (Petanjek et al. 2019). It is possible that selective alterations of associative layer III neurons have a pivotal role in the "dis-connectivity" of prefrontal cortex found in autism spectrum disorder (Carroll et al. 2021), but also in other prefrontal cortex-associated disorders, like schizophrenia (Arion et al. 2015; Banovac et al. 2020; Batiuk et al. 2022; Datta and Arnsten 2018; Dienel et al. 2017, 2022; Glausier and Lewis 2018; Hoftman et al. 2017; Zikopoulos and Barbas 2013).

Structural changes through the cortical network are not finished by the age of 5–6 years, and circuitry reorganization continues throughout the rest of childhood and adolescence (Amlien et al. 2016; Averbeck 2022; Blakemore and Mills 2014; Charvet and Finlay 2012; Larsen and Luna 2018; Somel et al. 2013; Sousa et al. 2018; Sydnor et al. 2021; Tamnes et al. 2017; Tousignant et al. 2018; Werchan and Amso 2017). During the period when synaptic spine numbers exceed adult values, molecular tuning of synaptic strength occurred. This is proposed to be a major mechanism how the environment is influencing circuitry reorganization (Brant et al. 2013; Uylings 2006). Synaptic spine development of principal neurons in the human prefrontal cortex showed that the stage of developmental plasticity of cortical circuitry extends even up to the third decade of life. Concomitantly, there is a prolonged peak in expression of genes regulating neuronal development, including those associated with schizophrenia (Bilecki and Maćkowiak 2023; Dienel et al. 2022; Guardiola-Ripoll and Fatjó-Vilas 2023; Harris et al. 2009; Nagy et al. 2020; Selemon and Zecevic 2015). The comparative analysis of mRNA expression in the prefrontal cortex shows that in the human brain, relative to nonhuman primates, the dramatic changes in transcriptome profiles are delayed (Berg et al. 2021; Berto and Nowick 2018; Beveridge et al. 2014; Breen et al. 2018; Enwright Iii et al. 2022; Ganapathee and Gunz 2023; He et al. 2017; Hoftman et al. 2021; Huuki-Myers et al. 2023; Islam et al. 2021; Kroeze et al. 2018; Liu et al. 2012; Ma et al. 2022; Marchetto et al. 2019; Maynard et al. 2021; Pollen et al. 2023; Salamon and Rasin 2021; Shapiro et al. 2017; Somel et al. 2009; Vanderhaeghen and Polleux 2023; Won et al. 2019).

Thus, extraordinarily protracted circuitry reorganization seems to be a specific feature of human higher-order associative areas. Particularly, the microcircuits that are processing the highest cognitive functions, such as social abilities, are subject to the large-scale developmental remodeling induced by psycho-social and emotional environment (Chailangkarn et al. 2016; Galakhova et al. 2022; Pendl et al. 2017; Semendeferi et al. 2011; Zimmermann et al. 2019). Quantitative analysis of dendritic and synaptic spine development has identified the associative layer IIIC neurons as main neuronal candidate within microcircuit processing the most

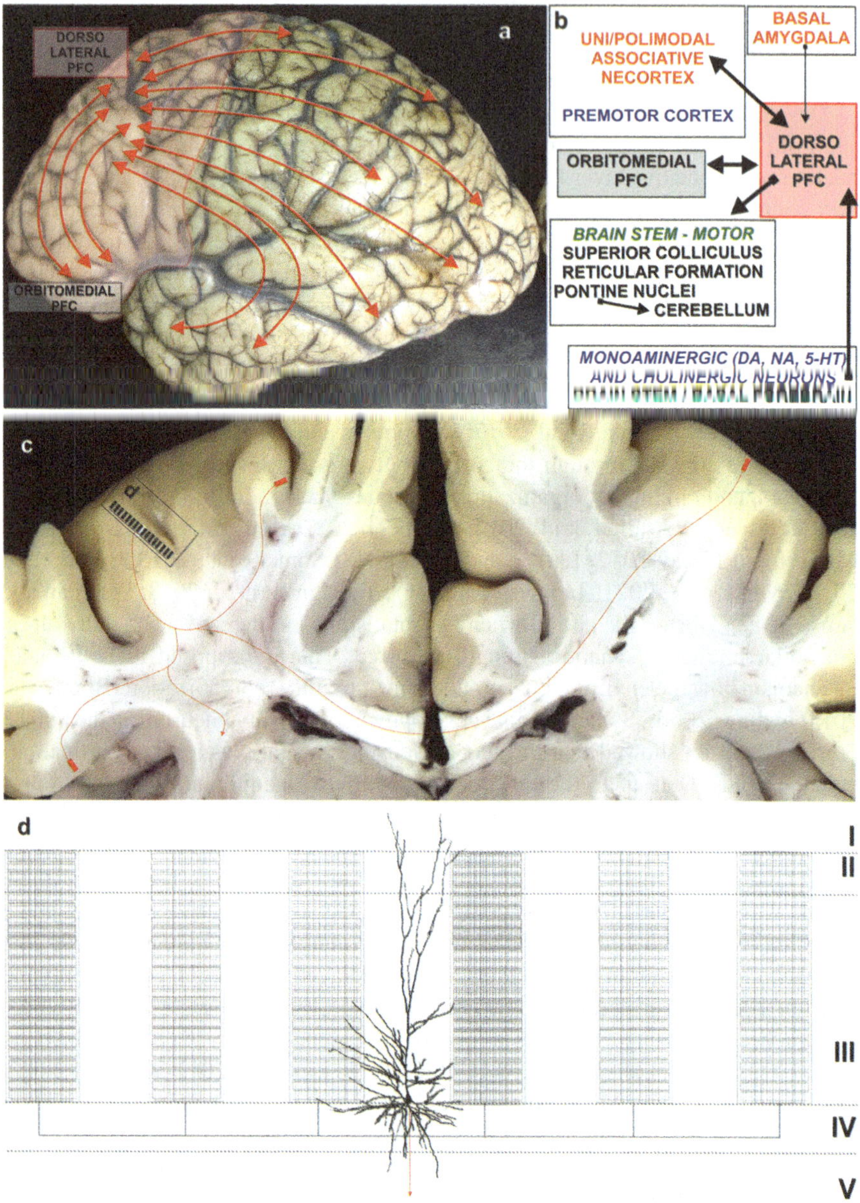

Fig. 4.11 Connectivity of prefrontal cortex (**a**, **b**) and schematic organization of extrinsic and intrinsic cortical projections of associative layer IIIC neurons (**c**, **d**). In the adult human brain (**a**), the frontal granular cortex occupies 80% of the frontal lobe and almost one third (red) of the total cortical surface. The two main subdivisions of the prefrontal cortex are the orbitomedial and dorsolateral part. The dorsolateral prefrontal cortex establishes rich interconnections with all neocortical areas (**b**, and red lines with arrows on both sides extending out of the prefrontal cortex, panel **a**), except primary regions. Note that the dorsolateral and orbitomedial prefrontal cortex are

Fig. 4.11 (continued) densely interconnected. The orbitomedial prefrontal cortex is mainly connected with the hippocampus and cortical regions processing visceral information. Panel **b** is based on Groenewegen and Uylings (2000). DA, dopamine; NA, noradrenaline; 5HT, serotonin. Large layer IIIC neurons are considered to be associative neurons, connecting several higher-order areas in the ipsi- and contralateral hemisphere (**c**), with the columnar pattern of axon ramification. However, around 80% of synapses are established within the area of origin. Local axon branches are forming numerous terminal ramifications, which have columnar distribution through layers II and III (**d**) and extend several millimeters around the cell body. (Reproduced from Petanjek et al. (2019), compilation of figs. 4.1 and 4.2 with modifications, used under CC-BY 4.0 International license and copyright)

sophisticated cognitive functions. Their unique developmental features—second growth spurt of dendritic development and high synaptic overproduction during preschool period, as well as heavily protracted period of synaptic spine tuning with massive postadolescent pruning—allow the highest developmental plasticity. Therefore, plasticity of associative neurons and corresponding microcircuitries might lay a foundation for a further increase in cognitive capacity later throughout childhood and adolescence (Hrvoj-Mihic and Semendeferi 2019). It is also advanced for acquisition of the highest level for most complex brain functions in humans, including affective modulation of emotional cues, self-conceptualization, mentalization, cognitive flexibility, and working memory. However, it provides also extended window of opportunity for pathological events to disrupt normal formation of circuits processing these complex cognitive functions.

References

Abbott LF, Nelson SB (2000) Synaptic plasticity: taming the beast. Nat Neurosci 3(Suppl):1178–1183. https://doi.org/10.1038/81453

Aghajanian GK, Bloom FE (1967) The formation of synaptic junctions in developing rat brain: a quantitative electron microscopic study. Brain Res 6:716–727. https://doi.org/10.1016/0006-8993(67)90128-x

Alexander BH, Barnes HM, Trimmer E, Davidson AM, Ogola BO, Lindsey SH, Mostany R (2018) Stable density and dynamics of dendritic spines of cortical neurons across the estrous cycle while expressing differential levels of sensory-evoked plasticity. Front Mol Neurosci 11:83. https://doi.org/10.3389/fnmol.2018.00083

Alvarez VA, Sabatini BL (2007) Anatomical and physiological plasticity of dendritic spines. Annu Rev Neurosci 30:79–97. https://doi.org/10.1146/annurev.neuro.30.051606.094222

Alzu'bi A, Homman-Ludiye J, Bourne JA, Clowry GJ (2019) Thalamocortical afferents innervate the cortical subplate much earlier in development in primate than in rodent. Cereb Cortex 29:1706–1718. https://doi.org/10.1093/cercor/bhy327

Amlien IK, Fjell AM, Tamnes CK, Grydeland H, Krogsrud SK, Chaplin TA, Rosa MGP, Walhovd KB (2016) Organizing principles of human cortical development--thickness and area from 4 to 30 years: insights from comparative primate neuroanatomy. Cereb Cortex 26:257–267. https://doi.org/10.1093/cercor/bhu214

Anderson SA, Classey JD, Condé F, Lund JS, Lewis DA (1995) Synchronous development of pyramidal neuron dendritic spines and parvalbumin-immunoreactive chandelier neuron axon terminals in layer III of monkey prefrontal cortex. Neuroscience 67:7–22. https://doi.org/10.1016/0306-4522(95)00051-J

Anlar B, Atilla P, Cakar N, Tombakoglu M, Bulun A (2003) Apoptosis in the developing human brain: a preliminary study of the frontal region. Early Hum Dev 71:53–60. https://doi.org/10.1016/s0378-3782(02)00116-0

Anton-Sanchez L, Larrañaga P, Benavides-Piccione R, Fernaud-Espinosa I, DeFelipe J, Bielza C (2017) Three-dimensional spatial modeling of spines along dendritic networks in human cortical pyramidal neurons. PLoS One 12:e0180400. https://doi.org/10.1371/journal.pone.0180400

Appelbaum LG, Shenasa MA, Stolz L, Daskalakis Z (2023) Synaptic plasticity and mental health: methods, challenges and opportunities. Neuropsychopharmacology 48:113–120. https://doi.org/10.1038/s41386-022-01370-w

Arellano JI, Benavides-Piccione R, DeFelipe J, Yuste R (2007) Ultrastructure of dendritic spines: correlation between synaptic and spine morphologies. Front Neurosci 1:131–143. https://doi.org/10.3389/neuro.01.1.1.010.2007

Arichi T, Whitehead K, Barone G, Pressler R, Padormo F, Edwards AD, Fabrizi L (2017) Localization of spontaneous bursting neuronal activity in the preterm human brain with simultaneous EEG-fMRI. Elife 6. https://doi.org/10.7554/eLife.27814

Arion D, Corradi JP, Tang S, Datta D, Boothe F, He A, Cacace AM, Zaczek R, Albright CF, Tseng G, Lewis DA (2015) Distinctive transcriptome alterations of prefrontal pyramidal neurons in schizophrenia and schizoaffective disorder. Mol Psychiatry 20:1397–1405. https://doi.org/10.1038/mp.2014.171

Averbeck BB (2022) Pruning recurrent neural networks replicates adolescent changes in working memory and reinforcement learning. Proc Natl Acad Sci U S A 119:e2121331119. https://doi.org/10.1073/pnas.2121331119

Bączyńska E, Pels KK, Basu S, Włodarczyk J, Ruszczycki B (2021) Quantification of dendritic spines remodeling under physiological stimuli and in pathological conditions. Int J Mol Sci 22. https://doi.org/10.3390/ijms22084053

Bagust J, Lewis DM, Westerman RA (1973) Polyneuronal innervation of kitten skeletal muscle. J Physiol 229:241–255. https://doi.org/10.1113/jphysiol.1973.sp010136

Bai L, Tu WY, Xiao Y, Zhang K, Shen C (2022) Motoneurons innervation determines the distinct gene expressions in multinucleated myofibers. Cell Biosci 12:140. https://doi.org/10.1186/s13578-022-00876-6

Bakken TE, Miller JA, Ding S-L, Sunkin SM, Smith KA, Ng L, Szafer A, Dalley RA, Royall JJ, Lemon T, Shapouri S, Aiona K, Arnold J, Bennett JL, Bertagnolli D, Bickley K, Boe A, Brouner K, Butler S, Byrnes E, Caldejon S, Carey A, Cate S, Chapin M, Chen J, Dee N, Desta T, Dolbeare TA, Dotson N, Ebbert A, Fulfs E, Gee G, Gilbert TL, Goldy J, Gourley L, Gregor B, Gu G, Hall J, Haradon Z, Haynor DR, Hejazinia N, Hoerder-Suabedissen A, Howard R, Jochim J, Kinnunen M, Kriedberg A, Kuan CL, Lau C, Lee C-K, Lee F, Luong L, Mastan N, May R, Melchor J, Mosqueda N, Mott E, Ngo K, Nyhus J, Oldre A, Olson E, Parente J, Parker PD, Parry S, Pendergraft J, Potekhina L, Reding M, Riley ZL, Roberts T, Rogers B, Roll K, Rosen D, Sandman D, Sarreal M, Shapovalova N, Shi S, Sjoquist N, Sodt AJ, Townsend R, Velasquez L, Wagley U, Wakeman WB, White C, Bennett C, Wu J, Young R, Youngstrom BL, Wohnoutka P, Gibbs RA, Rogers J, Hohmann JG, Hawrylycz MJ, Hevner RF, Molnár Z, Phillips JW, Dang C, Jones AR, Amaral DG, Bernard A, Lein ES (2016) A comprehensive transcriptional map of primate brain development. Nature 535:367–375. https://doi.org/10.1038/nature18637

Balice-Gordon RJ, Lichtman JW (1994) Long-term synapse loss induced by focal blockade of postsynaptic receptors. Nature 372:519–524. https://doi.org/10.1038/372519a0

Balice-Gordon RJ, Thompson WJ (1988) Synaptic rearrangements and alterations in motor unit properties in neonatal rat extensor digitorum longus muscle. J Physiol 398:191–210. https://doi.org/10.1113/jphysiol.1988.sp017038

Banovac I, Sedmak D, Rojnić Kuzman M, Hladnik A, Petanjek Z (2020) Axon morphology of rapid Golgi-stained pyramidal neurons in the prefrontal cortex in schizophrenia. Croat Med J 61:354–365. https://doi.org/10.3325/cmj.2020.61.354

Barbas H (2015) General cortical and special prefrontal connections: principles from structure to function. Annu Rev Neurosci 38:269–289. https://doi.org/10.1146/annurev-neuro-071714-033936

Batiuk MY, Tyler T, Dragicevic K, Mei S, Rydbirk R, Petukhov V, Deviatiiarov R, Sedmak D, Frank E, Feher V, Habek N, Hu Q, Igolkina A, Roszik L, Pfisterer U, Garcia-Gonzalez D, Petanjek Z, Adorjan I, Kharchenko PV, Khodosevich K (2022) Upper cortical layer-driven network impairment in schizophrenia. Sci Adv 8:eabn8367. https://doi.org/10.1126/sciadv.abn8367

Becker LE, Armstrong DL, Chan F, Wood MM (1984) Dendritic development in human occipital cortical neurons. Brain Res 315:117–124. https://doi.org/10.1016/0165-3806(84)90083-X

Benavides-Piccione R, Fernaud-Espinosa I, Robles V, Yuste R, DeFelipe J (2013) Age-based comparison of human dendritic spine structure using complete three-dimensional reconstructions. Cereb Cortex 23:1798–1810. https://doi.org/10.1093/cercor/bhs154

Benavides-Piccione R, Rojo C, Kastanauskaite A, DeFelipe J (2021) Variation in pyramidal cell morphology across the human anterior temporal lobe. Cereb Cortex 31:3592–3609. https://doi.org/10.1093/cercor/bhab034

Bennett MR, Pettigrew AG (1974) The formation of synapses in striated muscle during development. J Physiol 241:515–545. https://doi.org/10.1113/jphysiol.1974.sp010670

Benoit P, Changeux JP (1975) Consequences of tenotomy on the evolution of multi-innervation in developing rat soleus muscle. Brain Res 99:354–358. https://doi.org/10.1016/0006-8993(75)90036-0

Benoit P, Changeux JP (1978) Consequences of blocking the nerve with a local anaesthetic on the evolution of multiinnervation at the regenerating neuromuscular junction of the rat. Brain Res 149:89–96. https://doi.org/10.1016/0006-8993(78)90589-9

Berg J, Sorensen SA, Ting JT, Miller JA, Chartrand T, Buchin A, Bakken TE, Budzillo A, Dee N, Ding S-L, Gouwens NW, Hodge RD, Kalmbach B, Lee C, Lee BR, Alfiler L, Baker K, Barkan E, Beller A, Berry K, Bertagnolli D, Bickley K, Bomben J, Braun T, Brouner K, Casper T, Chong P, Crichton K, Dalley R, de Frates R, Desta T, Lee SD, D'Orazi F, Dotson N, Egdorf T, Enstrom R, Farrell C, Feng D, Fong O, Furdan S, Galakhova AA, Gamlin C, Gary A, Glandon A, Goldy J, Gorham M, Goriounova NA, Gratiy S, Graybuck L, Gu H, Hadley K, Hansen N, Heistek TS, Henry AM, Heyer DB, Hill D, Hill C, Hupp M, Jarsky T, Kebede S, Keene L, Kim L, Kim M-H, Kroll M, Latimer C, Levi BP, Link KE, Mallory M, Mann R, Marshall D, Maxwell M, McGraw M, McMillen D, Melief E, Mertens EJ, Mezei L, Mihut N, Mok S, Molnar G, Mukora A, Ng L, Ngo K, Nicovich PR, Nyhus J, Olah G, Oldre A, Omstead V, Ozsvar A, Park D, Peng H, Pham T, Pom CA, Potekhina L, Rajanbabu R, Ransford S, Reid D, Rimorin C, Ruiz A, Sandman D, Sulc J, Sunkin SM, Szafer A, Szemenyei V, Thomsen ER, Tieu M, Torkelson A, Trinh J, Tung H, Wakeman W, Waleboer F, Ward K, Wilbers R, Williams G, Yao Z, Yoon J-G, Anastassiou C, Arkhipov A, Barzo P, Bernard A, Cobbs C, de Witt Hamer PC, Ellenbogen RG, Esposito L, Ferreira M, Gwinn RP, Hawrylycz MJ, Hof PR, Idema S, Jones AR, Keene CD, Ko AL, Murphy GJ, Ng L, Ojemann JG, Patel AP, Phillips JW, Silbergeld DL, Smith K, Tasic B, Yuste R, Segev I, de Kock CPJ, Mansvelder HD, Tamas G, Zeng H, Koch C, Lein ES (2021) Human neocortical expansion involves glutamatergic neuron diversification. Nature 598:151–158. https://doi.org/10.1038/s41586-021-03813-8

Berto S, Nowick K (2018) Species-specific changes in a primate transcription factor network provide insights into the molecular evolution of the primate prefrontal cortex. Genome Biol Evol 10:2023–2036. https://doi.org/10.1093/gbe/evy149

Beveridge NJ, Santarelli DM, Wang X, Tooney PA, Webster MJ, Weickert CS, Cairns MJ (2014) Maturation of the human dorsolateral prefrontal cortex coincides with a dynamic shift in microRNA expression. Schizophr Bull 40:399–409. https://doi.org/10.1093/schbul/sbs198

Bianchi S, Stimpson CD, Duka T, Larsen MD, Janssen WGM, Collins Z, Bauernfeind AL, Schapiro SJ, Baze WB, McArthur MJ, Hopkins WD, Wildman DE, Lipovich L, Kuzawa CW, Jacobs B, Hof PR, Sherwood CC (2013) Synaptogenesis and development of pyramidal neuron dendritic morphology in the chimpanzee neocortex resembles humans. Proc Natl Acad Sci U S A 110(Suppl 2):10395–10401. https://doi.org/10.1073/pnas.1301224110

Bilecki W, Maćkowiak M (2023) Gene expression and epigenetic regulation in the prefrontal cortex of schizophrenia. Genes (Basel) 14. https://doi.org/10.3390/genes14020243

Bitanihirwe BK, Woo T-UW (2021) Pyramidal neurons. In: Factors affecting neurodevelopment. Elsevier, San Diego, pp 433–445

Blakemore SJ, Mills KL (2014) Is adolescence a sensitive period for sociocultural processing? Annu Rev Psychol 65:187–207. https://doi.org/10.1146/annurev-psych-010213-115202

Blaschke AJ, Weiner JA, Chun J (1998) Programmed cell death is a universal feature of embryonic and postnatal neuroproliferative regions throughout the central nervous system. J Comp Neurol 396:39–50. https://doi.org/10.1002/(sici)1096-9861(19980622)396:1<39:aid-cne4>3.0.co;2-j

Bliss TV, Gardner-Medwin AR (1973) Long-lasting potentiation of synaptic transmission in the dentate area of the unanaestetized rabbit following stimulation of the perforant path. J Physiol 232:357–374. https://doi.org/10.1113/jphysiol.1973.sp010274

Blundon JA, Zakharenko SS (2008) Dissecting the components of long-term potentiation. Neuroscientist 14:598–608. https://doi.org/10.1177/1073858408320643

Bock O (2013) Cajal, Golgi, Nansen, Schäfer and the neuron doctrine. Endeavour 37:228–234. https://doi.org/10.1016/j.endeavour.2013.06.006

Boivin JR, Piekarski DJ, Thomas AW, Wilbrecht L (2018) Adolescent pruning and stabilization of dendritic spines on cortical layer 5 pyramidal neurons do not depend on gonadal hormones. Dev Cogn Neurosci 30:100–107. https://doi.org/10.1016/j.dcn.2018.01.007

Bonilla-Quintana M, Wörgötter F, Tetzlaff C, Fauth M (2020) Modeling the shape of synaptic spines by their actin dynamics. Front Synaptic Neurosci 12:9. https://doi.org/10.3389/fnsyn.2020.00009

Boothe RG, Greenough WT, Lund JS, Wrege K (1979) A quantitative investigation of spine and dendrite development of neurons in visual cortex (area 17) of Macaca nemestrina monkeys. J Comp Neurol 186:473–489. https://doi.org/10.1002/cne.901860310

Borkowski WJ, Bernstine RL (1955) Electroencephalography of the fetus. Neurology 5:362–365. https://doi.org/10.1212/wnl.5.5.362

Borra E, Rizzo M, Gerbella M, Rozzi S, Luppino G (2021) Laminar origin of corticostriatal projections to the motor putamen in the macaque brain. J Neurosci 41:1455–1469. https://doi.org/10.1523/JNEUROSCI.1475-20.2020

Borra E, Biancheri D, Rizzo M, Leonardi F, Luppino G (2022) Crossed corticostriatal projections in the macaque brain. J Neurosci 42:7060–7076. https://doi.org/10.1523/JNEUROSCI.0071-22.2022

Bourgeois JP (1997) Synaptogenesis, heterochrony and epigenesis in the mammalian neocortex. Acta Paediatr Suppl 422:27–33. https://doi.org/10.1111/j.1651-2227.1997.tb18340.x

Bourgeois JP, Rakic P (1993) Changes of synaptic density in the primary visual cortex of the macaque monkey from fetal to adult stage. J Neurosci 13:2801–2820. https://doi.org/10.1523/JNEUROSCI.13-07-02801.1993

Bourgeois JP, Rakic P (1996) Synaptogenesis in the occipital cortex of macaque monkey devoid of retinal input from early embryonic stages. Eur J Neurosci 8:942–950. https://doi.org/10.1111/j.1460-9568.1996.tb01581.x

Bourgeois JP, Jastreboff PJ, Rakic P (1989) Synaptogenesis in visual cortex of normal and preterm monkeys: evidence for intrinsic regulation of synaptic overproduction. Proc Natl Acad Sci U S A 86:4297–4301. https://doi.org/10.1073/pnas.86.11.4297

Bourgeois JP, Goldman-Rakic PS, Rakic P (1994) Synaptogenesis in the prefrontal cortex of rhesus monkeys. Cereb Cortex 4:78–96. https://doi.org/10.1093/cercor/4.1.78

Brant AM, Munakata Y, Boomsma DI, Defries JC, Haworth CMA, Keller MC, Martin NG, McGue M, Petrill SA, Plomin R, Wadsworth SJ, Wright MJ, Hewitt JK (2013) The nature and nurture of high IQ: an extended sensitive period for intellectual development. Psychol Sci 24:1487–1495. https://doi.org/10.1177/0956797612473119

Breen MS, Ozcan S, Ramsey JM, Wang Z, Ma'ayan A, Rustogi N, Gottschalk MG, Webster MJ, Weickert CS, Buxbaum JD, Bahn S (2018) Temporal proteomic profiling of postnatal human cortical development. Transl Psychiatry 8:267. https://doi.org/10.1038/s41398-018-0306-4

Brody H (1955) Organization of the cerebral cortex. III. A study of aging in the human cerebral cortex. J Comp Neurol 102:511–516. https://doi.org/10.1002/cne.901020206
Brown MC, Jansen JK, van Essen D (1976) Polyneuronal innervation of skeletal muscle in newborn rats and its elimination during maturation. J Physiol 261:387–422. https://doi.org/10.1113/jphysiol.1976.sp011565
Brown MC, Holland RL, Hopkins WG (1981) Restoration of focal multiple innervation in rat muscles by transmission block during a critical stage of development. J Physiol 318:355–364. https://doi.org/10.1113/jphysiol.1981.sp013869
Buffelli M, Burgess RW, Feng G, Lobe CG, Lichtman JW, Sanes JR (2003) Genetic evidence that relative synaptic efficacy biases the outcome of synaptic competition. Nature 424:430–434. https://doi.org/10.1038/nature01844
Burns BD (1958) The mammalian cerebral cortex. Arnold, London
Bystron I, Molnár Z, Otellin V, Blakemore C (2005) Tangential networks of precocious neurons and early axonal outgrowth in the embryonic human forebrain. J Neurosci 25:2781–2792. https://doi.org/10.1523/JNEUROSCI.4770-04.2005
Bystron I, Rakic P, Molnár Z, Blakemore C (2006) The first neurons of the human cerebral cortex. Nat Neurosci 9:880–886. https://doi.org/10.1038/nn1726
Cadwell CR, Bhaduri A, Mostajo-Radji MA, Keefe MG, Nowakowski TJ (2019) Development and arealization of the cerebral cortex. Neuron 103:980–1004. https://doi.org/10.1016/j.neuron.2019.07.009
Cajal SR (1888) Estructura de los centros nerviosos de las aves. Rev Trim Histol Norm Pat:1–10
Cajal SR (1899) Estudios sobre la cortexa cerebral humana. Estructura de la cortex motriz del hombre y mamiferos. Rev Trim Microgr 4, 117–200
Cajal SR (1904) La Textura del Sistema Nerviosa del Hombre y los Vertebrados. Moya, Madrid
Callaway EM, van Essen DC (1989) Slowing of synapse elimination by alpha-bungarotoxin superfusion of the neonatal rabbit soleus muscle. Dev Biol 131:356–365. https://doi.org/10.1016/s0012-1606(89)80009-0
Cano-Astorga N, DeFelipe J, Alonso-Nanclares L (2021) Three-dimensional synaptic organization of layer III of the human temporal neocortex. Cereb Cortex 31:4742–4764. https://doi.org/10.1093/cercor/bhab120
Caroni P, Donato F, Muller D (2012) Structural plasticity upon learning: regulation and functions. Nat Rev Neurosci 13:478–490. https://doi.org/10.1038/nrn3258
Caroni P, Chowdhury A, Lahr M (2014) Synapse rearrangements upon learning: from divergent-sparse connectivity to dedicated sub-circuits. Trends Neurosci 37:604–614. https://doi.org/10.1016/j.tins.2014.08.011
Carroll L, Braeutigam S, Dawes JM, Krsnik Z, Kostovic I, Coutinho E, Dewing JM, Horton CA, Gomez-Nicola D, Menassa DA (2021) Autism spectrum disorders: multiple routes to, and multiple consequences of, abnormal synaptic function and connectivity. Neuroscientist 27:10–29. https://doi.org/10.1177/1073858420921378
Castillo-Fernández S, Silva-Gómez AB (2022) Changes in dendritic arborization related to the estrous cycle in pyramidal neurons of layer V of the motor cortex. J Chem Neuroanat 119:102042. https://doi.org/10.1016/j.jchemneu.2021.102042
Chailangkarn T, Trujillo CA, Freitas BC, Hrvoj-Mihic B, Herai RH, Yu DX, Brown TT, Marchetto MC, Bardy C, McHenry L, Stefanacci L, Järvinen A, Searcy YM, DeWitt M, Wong W, Lai P, Ard MC, Hanson KL, Romero S, Jacobs B, Dale AM, Dai L, Korenberg JR, Gage FH, Bellugi U, Halgren E, Semendeferi K, Muotri AR (2016) A human neurodevelopmental model for Williams syndrome. Nature 536:338–343. https://doi.org/10.1038/nature19067
Chalupa LM, Killackey HP (1989) Process elimination underlies ontogenetic change in the distribution of callosal projection neurons in the postcentral gyrus of the fetal rhesus monkey. Proc Natl Acad Sci U S A 86:1076–1079. https://doi.org/10.1073/pnas.86.3.1076
Chan WY, Lorke DE, Tiu SC, Yew DT (2002) Proliferation and apoptosis in the developing human neocortex. Anat Rec 267:261–276. https://doi.org/10.1002/ar.10100

Chandrasekaran S, Navlakha S, Audette NJ, McCreary DD, Suhan J, Bar-Joseph Z, Barth AL (2015) Unbiased, high-throughput electron microscopy analysis of experience-dependent synaptic changes in the neocortex. J Neurosci 35:16450–16462. https://doi.org/10.1523/JNEUROSCI.1573-15.2015

Changeux JP (1983) L'homme neuronal. Le Temps des sciences, Fayard, Paris

Changeux JP (2017) Climbing brain levels of organisation from genes to consciousness. Trends Cogn Sci 21:168–181. https://doi.org/10.1016/j.tics.2017.01.004

Changeux JP, Danchin A (1976) Selective stabilisation of developing synapses as a mechanism for the specification of neuronal networks. Nature 264:705–712. https://doi.org/10.1038/264705a0

Changeux JP, Courrège P, Danchin A (1973) A theory of the epigenesis of neuronal networks by selective stabilization of synapses. Proc Natl Acad Sci U S A 70:2974–2978. https://doi.org/10.1073/pnas.70.10.2974

Changeux JP, Goulas A, Hilgetag CC (2021) A connectomic hypothesis for the hominization of the brain. Cereb Cortex 31:2425–2449. https://doi.org/10.1093/cercor/bhaa365

Charvet CJ, Finlay BL (2012) Embracing covariation in brain evolution: large brains, extended development, and flexible primate social systems. Prog Brain Res 195:71–87. https://doi.org/10.1016/B978-0-444-53860-4.00004-0

Cheetham CEJ, Barnes SJ, Albieri G, Knott GW, Finnerty GT (2014) Pansynaptic enlargement at adult cortical connections strengthened by experience. Cereb Cortex 24:521–531. https://doi.org/10.1093/cercor/bhs334

Chen CC, Brumberg JC (2021) Sensory experience as a regulator of structural plasticity in the developing Whisker-to-Barrel system. Front Cell Neurosci 15:770453. https://doi.org/10.3389/fncel.2021.770453

Chen M, Qi J, Poo M, Yang Y (2022) Stability and dynamics of dendritic spines in macaque prefrontal cortex. Natl Sci Rev 9:nwac125. https://doi.org/10.1093/nsr/nwac125

Cheour-Luhtanen M, Alho K, Sainio K, Rinne T, Reinikainen K, Pohjavuori M, Renlund M, Aaltonen O, Eerola O, Näätänen R (1996) The ontogenetically earliest discriminative response of the human brain. Psychophysiology 33:478–481. https://doi.org/10.1111/j.1469-8986.1996.tb01074.x

Chéreau R, Williams LE, Bawa T, Holtmaat A (2022) Circuit mechanisms for cortical plasticity and learning. Semin Cell Dev Biol 125:68–75. https://doi.org/10.1016/j.semcdb.2021.07.012

Chidambaram SB, Rathipriya AG, Bolla SR, Bhat A, Ray B, Mahalakshmi AM, Manivasagam T, Thenmozhi AJ, Essa MM, Guillemin GJ, Chandra R, Sakharkar MK (2019) Dendritic spines: revisiting the physiological role. Prog Neuro-Psychopharmacol Biol Psychiatry 92:161–193. https://doi.org/10.1016/j.pnpbp.2019.01.005

Chugani HT (1998) A critical period of brain development: studies of cerebral glucose utilization with PET. Prev Med 27:184–188. https://doi.org/10.1006/pmed.1998.0274

Chugani HT, Phelps ME (1986) Maturational changes in cerebral function in infants determined by 18FDG positron emission tomography. Science 231:840–843. https://doi.org/10.1126/science.3945811

Chugani HT, Phelps ME, Mazziotta JC (1987) Positron emission tomography study of human brain functional development. Ann Neurol 22:487–497. https://doi.org/10.1002/ana.410220408

Chung DW, Wills ZP, Fish KN, Lewis DA (2017) Developmental pruning of excitatory synaptic inputs to parvalbumin interneurons in monkey prefrontal cortex. Proc Natl Acad Sci U S A 114:E629–E637. https://doi.org/10.1073/pnas.1610077114

Citri A, Malenka RC (2008) Synaptic plasticity: multiple forms, functions, and mechanisms. Neuropsychopharmacology 33:18–41. https://doi.org/10.1038/sj.npp.1301559

Cline HT, Lau M, Hiramoto M (2023) Activity-dependent organization of topographic neural circuits. Neuroscience 508:3–18. https://doi.org/10.1016/j.neuroscience.2022.11.032

Collingridge GL, Peineau S, Howland JG, Wang YT (2010) Long-term depression in the CNS. Nat Rev Neurosci 11:459–473. https://doi.org/10.1038/nrn2867

Conradi S, Ronnevi LO (1975) Spontaneous elimination of synapses on cat spinal motoneurons after birth: do half of the synapses on the cell bodies disappear? Brain Res 92:505–510. https://doi.org/10.1016/0006-8993(75)90338-8

Conradi S, Skoglund S (1969) Observations on the ultrastructure of the initial motor axon segment and dorsal root boutons on the motoneurons in the lumbosacral spinal cord of the cat during postnatal development. Acta Physiol Scand Suppl 333:53–76
Cragg BG (1975) The development of synapses in the visual system of the cat. J Comp Neurol 160:147–166. https://doi.org/10.1002/cne.901600202
Crepel F, Mariani J, Delhaye-Bouchaud N (1976) Evidence for a multiple innervation of Purkinje cells by climbing fibers in the immature rat cerebellum. J Neurobiol 7:567–578. https://doi.org/10.1002/neu.480070609
Dall'Orso S, Arichi T, Fitzgibbon SP, Edwards AD, Burdet E, Muceli S (2022) Development of functional organization within the sensorimotor network across the perinatal period. Hum Brain Mapp 43:2249–2261. https://doi.org/10.1002/hbm.25785
Datta D, Arnsten AFT (2018) Unique molecular regulation of higher-order prefrontal cortical circuits: insights into the neurobiology of schizophrenia. ACS Chem Neurosci 9:2127–2145. https://doi.org/10.1021/acschemneuro.7b00505
Daw NW, Fox K, Sato H, Czepita D (1992) Critical period for monocular deprivation in the cat visual cortex. J Neurophysiol 67:197–202. https://doi.org/10.1152/jn.1992.67.1.197
Dawkins R (1971) Selective neurone death as a possible memory mechanism. Nature 229:118–119. https://doi.org/10.1038/229118a0
de Asis-Cruz J, Barnett SD, Kim J-H, Limperopoulos C (2021) Functional connectivity-derived optimal gestational-age cut points for fetal brain network maturity. Brain Sci 11. https://doi.org/10.3390/brainsci11070921
de Robertis ED, Bennett HS (1955) Some features of the submicroscopic morphology of synapses in frog and earthworm. J Biophys Biochem Cytol 1:47–58. https://doi.org/10.1083/jcb.1.1.47
de Roo M, Klauser P, Muller D (2008) LTP promotes a selective long-term stabilization and clustering of dendritic spines. PLoS Biol 6:e219. https://doi.org/10.1371/journal.pbio.0060219
Dehay C, Kennedy H, Bullier J, Berland M (1988) Absence of interhemispheric connections of area 17 during development in the monkey. Nature 331:348–350. https://doi.org/10.1038/331348a0
Delevich K, Okada NJ, Rahane A, Zhang Z, Hall CD, Wilbrecht L (2020) Sex and pubertal status influence dendritic spine density on frontal corticostriatal projection neurons in mice. Cereb Cortex 30:3543–3557. https://doi.org/10.1093/cercor/bhz325
Derbyshire SW, Bockmann JC (2020) Reconsidering fetal pain. J Med Ethics 46:3–6. https://doi.org/10.1136/medethics-2019-105701
Di Maio V (2021) The glutamatergic synapse: a complex machinery for information processing. Cogn Neurodyn 15:757–781. https://doi.org/10.1007/s11571-021-09679-w
Dienel SJ, Bazmi HH, Lewis DA (2017) Development of transcripts regulating dendritic spines in layer 3 pyramidal cells of the monkey prefrontal cortex: implications for the pathogenesis of schizophrenia. Neurobiol Dis 105:132–141. https://doi.org/10.1016/j.nbd.2017.05.016
Dienel SJ, Schoonover KE, Lewis DA (2022) Cognitive dysfunction and prefrontal cortical circuit alterations in schizophrenia: developmental trajectories. Biol Psychiatry 92:450–459. https://doi.org/10.1016/j.biopsych.2022.03.002
Ding SL, Royall JJ, Lesnar P, Facer BAC, Smith KA, Wei Y, Brouner K, Dalley RA, Dee N, Dolbeare TA, Ebbert A, Glass IA, Keller NH, Lee F, Lemon TA, Nyhus J, Pendergraft J, Reid R, Sarreal M, Shapovalova NV, Szafer A, Phillips JW, Sunkin SM, Hohmann JG, Jones AR, Hawrylycz MJ, Hof PR, Ng L, Bernard A, Lein ES (2022) Cellular resolution anatomical and molecular atlases for prenatal human brains. J Comp Neurol 530:6–503. https://doi.org/10.1002/cne.25243
Diniz CRAF, Crestani AP (2023) The times they are a-changin': a proposal on how brain flexibility goes beyond the obvious to include the concepts of "upward" and "downward" to neuroplasticity. Mol Psychiatry 28:977–992. https://doi.org/10.1038/s41380-022-01931-x
Dreyfus-Brisac C, Larroche JC (1971) Electroencéphalogrammes discontinus du nouveau-né prématuré et à terme. Corrélations électro-anatoma-cliniques (Discontinuous electroencephalograms in the premature newborn and at term. Electro-anatomo-clinical correlations). Rev Electroencephalogr Neurophysiol Clin 1:95–99. https://doi.org/10.1016/s0370-4475(71)80022-9

Duan H, Wearne SL, Rocher AB, Macedo A, Morrison JH, Hof PR (2003) Age-related dendritic and spine changes in corticocortically projecting neurons in macaque monkeys. Cereb Cortex 13:950–961. https://doi.org/10.1093/cercor/13.9.950

Dunaevsky A, Tashiro A, Majewska A, Mason C, Yuste R (1999) Developmental regulation of spine motility in the mammalian central nervous system. Proc Natl Acad Sci U S A 96:13438–13443. https://doi.org/10.1073/pnas.96.23.13438

Duque A, Krsnik Z, Kostović I, Rakic P (2016) Secondary expansion of the transient subplate zone in the developing cerebrum of human and nonhuman primates. Proc Natl Acad Sci U S A 113:9892–9897. https://doi.org/10.1073/pnas.1610078113

Ehrlich I, Klein M, Rumpel S, Malinow R (2007) PSD-95 is required for activity-driven synapse stabilization. Proc Natl Acad Sci U S A 104:4176–4181. https://doi.org/10.1073/pnas.0609307104

Elias LAB, Kriegstein AR (2008) Gap junctions: multifaceted regulators of embryonic cortical development. Trends Neurosci 31:243–250. https://doi.org/10.1016/j.tins.2008.02.007

Elston GN, Fujita I (2014) Pyramidal cell development: postnatal spinogenesis, dendritic growth, axon growth, and electrophysiology. Front Neuroanat 8:78. https://doi.org/10.3389/fnana.2014.00078

Elston GN, Oga T, Fujita I (2009) Spinogenesis and pruning scales across functional hierarchies. J Neurosci 29:3271–3275. https://doi.org/10.1523/JNEUROSCI.5216-08.2009

Elston GN, Oga T, Okamoto T, Fujita I (2010a) Spinogenesis and pruning from early visual onset to adulthood: an intracellular injection study of layer III pyramidal cells in the ventral visual cortical pathway of the macaque monkey. Cereb Cortex 20:1398–1408. https://doi.org/10.1093/cercor/bhp203

Elston GN, Okamoto T, Oga T, Dornan D, Fujita I (2010b) Spinogenesis and pruning in the primary auditory cortex of the macaque monkey (Macaca fascicularis): an intracellular injection study of layer III pyramidal cells. Brain Res 1316:35–42. https://doi.org/10.1016/j.brainres.2009.12.056

Elston GN, Benavides-Piccione R, Elston A, Manger PR, DeFelipe J (2011a) Pyramidal cells in prefrontal cortex of primates: marked differences in neuronal structure among species. Front Neuroanat 5:2. https://doi.org/10.3389/fnana.2011.00002

Elston GN, Oga T, Okamoto T, Fujita I (2011b) Spinogenesis and pruning in the anterior ventral inferotemporal cortex of the macaque monkey: an intracellular injection study of layer III pyramidal cells. Front Neuroanat 5:42. https://doi.org/10.3389/fnana.2011.00042

Engisch KL, Wang X, Rich MM (2022) Homeostatic plasticity of the mammalian neuromuscular junction. Adv Neurobiol 28:111–130. https://doi.org/10.1007/978-3-031-07167-6_5

Enwright Iii JF, Arion D, MacDonald WA, Elbakri R, Pan Y, Vyas G, Berndt A, Lewis DA (2022) Differential gene expression in layer 3 pyramidal neurons across 3 regions of the human cortical visual spatial working memory network. Cereb Cortex 32:5216–5229. https://doi.org/10.1093/cercor/bhac009

Eze UC, Bhaduri A, Haeussler M, Nowakowski TJ, Kriegstein AR (2021) Single-cell atlas of early human brain development highlights heterogeneity of human neuroepithelial cells and early radial glia. Nat Neurosci 24:584–594. https://doi.org/10.1038/s41593-020-00794-1

Fabrizi L, Slater R, Worley A, Meek J, Boyd S, Olhede S, Fitzgerald M (2011) A shift in sensory processing that enables the developing human brain to discriminate touch from pain. Curr Biol 21:1552–1558. https://doi.org/10.1016/j.cub.2011.08.010

Fagard J, Esseily R, Jacquey L, O'Regan K, Somogyi E (2018) Fetal origin of sensorimotor behavior. Front Neurorobot 12:23. https://doi.org/10.3389/fnbot.2018.00023

Faust TE, Gunner G, Schafer DP (2021) Mechanisms governing activity-dependent synaptic pruning in the developing mammalian CNS. Nat Rev Neurosci 22:657–673. https://doi.org/10.1038/s41583-021-00507-y

Finn ES, Huber L, Jangraw DC, Molfese PJ, Bandettini PA (2019) Layer-dependent activity in human prefrontal cortex during working memory. Nat Neurosci 22:1687–1695. https://doi.org/10.1038/s41593-019-0487-z

Fischer M, Kaech S, Knutti D, Matus A (1998) Rapid actin-based plasticity in dendritic spines. Neuron 20:847–854. https://doi.org/10.1016/S0896-6273(00)80467-5

Fish KN, Hoftman GD, Sheikh W, Kitchens M, Lewis DA (2013) Parvalbumin-containing chandelier and basket cell boutons have distinctive modes of maturation in monkey prefrontal cortex. J Neurosci 33:8352–8358. https://doi.org/10.1523/JNEUROSCI.0306-13.2013

Fitzgerald M (2005) The development of nociceptive circuits. Nat Rev Neurosci 6:507–520. https://doi.org/10.1038/nrn1701

Friauf E, McConnell SK, Shatz CJ (1990) Functional synaptic circuits in the subplate during fetal and early postnatal development of cat visual cortex. J Neurosci 10:2601–2613. https://doi.org/10.1523/JNEUROSCI.10-08-02601.1990

Frankfurt M, Nassrallah Z, Luine V (2023) Steroid hormone interaction with dendritic spines: implications for neuropsychiatric disease. In: Rasia-Filho AA et al (eds) Dendritic spines. Springer, Cham

Fricker M, Tolkovsky AM, Borutaite V, Coleman M, Brown GC (2018) Neuronal cell death. Physiol Rev 98:813–880. https://doi.org/10.1152/physrev.00011.2017

Friedman NP, Robbins TW (2022) The role of prefrontal cortex in cognitive control and executive function. Neuropsychopharmacology 47:72–89. https://doi.org/10.1038/s41386-021-01132-0

Fritschy JM, Garey LJ (1986) Quantitative changes in morphological parameters in the developing visual cortex of the marmoset monkey. Brain Res 394:173–188. https://doi.org/10.1016/0165-3806(86)90093-3

Fu M, Zuo Y (2011) Experience-dependent structural plasticity in the cortex. Trends Neurosci 34:177–187. https://doi.org/10.1016/j.tins.2011.02.001

Fu M, Yu X, Lu J, Zuo Y (2012) Repetitive motor learning induces coordinated formation of clustered dendritic spines in vivo. Nature 483:92–95. https://doi.org/10.1038/nature10844

Fulford J, Vadeyar SH, Dodampahala SH, Moore RJ, Young P, Baker PN, James DK, Gowland PA (2003) Fetal brain activity in response to a visual stimulus. Hum Brain Mapp 20:239–245. https://doi.org/10.1002/hbm.10139

Fuster JM (2008) Anatomy of the prefrontal cortex. In: The prefrontal cortex. Elsevier, Amsterdam, pp 7–58

Galakhova AA, Hunt S, Wilbers R, Heyer DB, de Kock CPJ, Mansvelder HD, Goriounova NA (2022) Evolution of cortical neurons supporting human cognition. Trends Cogn Sci 26:909–922. https://doi.org/10.1016/j.tics.2022.08.012

Gallinaro JV, Gašparović N, Rotter S (2022) Homeostatic control of synaptic rewiring in recurrent networks induces the formation of stable memory engrams. PLoS Comput Biol 18:e1009836. https://doi.org/10.1371/journal.pcbi.1009836

Ganapathee DS, Gunz P (2023) Insights into brain evolution through the genotype-phenotype connection. Prog Brain Res 275:73–92. https://doi.org/10.1016/bs.pbr.2022.12.013

Garcia N, Hernández P, Lanuza MA, Tomàs M, Cilleros-Mañé V, Just-Borràs L, Duran-Vigara M, Polishchuk A, Balanyà-Segura M, Tomàs J (2022) Involvement of the voltage-gated calcium channels L- P/Q- and N-types in synapse elimination during neuromuscular junction development. Mol Neurobiol 59:4044–4064. https://doi.org/10.1007/s12035-022-02818-2

Garcia-Marin V, Ahmed TH, Afzal YC, Hawken MJ (2013) Distribution of vesicular glutamate transporter 2 (VGluT2) in the primary visual cortex of the macaque and human. J Comp Neurol 521:130–151. https://doi.org/10.1002/cne.23165

Garey LJ (1984) Structural development of the visual system of man. Hum Neurobiol 3:75–80

Garey LJ, de Courten C (1983) Structural development of the lateral geniculate nucleus and visual cortex in monkey and man. Behav Brain Res 10:3–13. https://doi.org/10.1016/0166-4328(83)90145-6

Garey LJ, Powell TP (1971) An experimental study of the termination of the lateral geniculo-cortical pathway in the cat and monkey. Proc R Soc Lond B Biol Sci 179:41–63. https://doi.org/10.1098/rspb.1971.0080

Geden MJ, Romero SE, Deshmukh M (2019) Apoptosis versus axon pruning: molecular intersection of two distinct pathways for axon degeneration. Neurosci Res 139:3–8. https://doi.org/10.1016/j.neures.2018.11.007

Ghose P, Shaham S (2020) Cell death in animal development. Development 147. https://doi.org/10.1242/dev.191882

Giannaris EL, Rosene DL (2012) A stereological study of the numbers of neurons and glia in the primary visual cortex across the lifespan of male and female rhesus monkeys. J Comp Neurol 520:3492–3508. https://doi.org/10.1002/cne.23101

Gipson CD, Olive MF (2017) Structural and functional plasticity of dendritic spines – root or result of behavior? Genes Brain Behav 16:101–117. https://doi.org/10.1111/gbb.12324

Glantz LA, Gilmore JH, Hamer RM, Lieberman JA, Jarskog LF (2007) Synaptophysin and post-synaptic density protein 95 in the human prefrontal cortex from mid-gestation into early adulthood. Neuroscience 149:582–591. https://doi.org/10.1016/j.neuroscience.2007.06.036

Glausier JR, Lewis DA (2018) Mapping pathologic circuitry in schizophrenia. Handb Clin Neurol 150:389–417. https://doi.org/10.1016/B978-0-444-63639-3.00025-6

Glucksmann A (1951) Cell deaths in normal vertebrate ontogeny. Biol Rev Camb Philos Soc 26:59–86. https://doi.org/10.1111/j.1469-185x.1951.tb00774.x

Goldman-Rakic PS (1999) The "psychic" neuron of the cerebral cortex. Ann N Y Acad Sci 868:13–26. https://doi.org/10.1111/j.1749-6632.1999.tb11270.x

Goldman-Rakic PS, Brown RM (1982) Postnatal development of monoamine content and synthesis in the cerebral cortex of rhesus monkeys. Brain Res 256:339–349. https://doi.org/10.1016/0165-3806(82)90146-8

Granger B, Tekaia F, Le Sourd AM, Rakic P, Bourgeois JP (1995) Tempo of neurogenesis and synaptogenesis in the primate cingulate mesocortex: comparison with the neocortex. J Comp Neurol 360:363–376. https://doi.org/10.1002/cne.903600212

Gray EG (1959) Electron microscopy of synaptic contacts on dendrite spines of the cerebral cortex. Nature 183:1592–1593. https://doi.org/10.1038/1831592a0

Groenewegen HJ, Uylings HB (2000) The prefrontal cortex and the integration of sensory, limbic and autonomic information. Prog Brain Res 126:3–28. https://doi.org/10.1016/S0079-6123(00)26003-2

Groenewegen HJ, Wright CI, Uylings HB (1997) The anatomical relationships of the prefrontal cortex with limbic structures and the basal ganglia. J Psychopharmacol 11:99–106. https://doi.org/10.1177/026988119701100202

Grutzendler J, Kasthuri N, Gan W-B (2002) Long-term dendritic spine stability in the adult cortex. Nature 420:812–816. https://doi.org/10.1038/nature01276

Guardiola-Ripoll M, Fatjó-Vilas M (2023) A systematic review of the human accelerated regions in schizophrenia and related disorders: where the evolutionary and neurodevelopmental hypotheses converge. Int J Mol Sci 24. https://doi.org/10.3390/ijms24043597

Hadders-Algra M (2007) Putative neural substrate of normal and abnormal general movements. Neurosci Biobehav Rev 31:1181–1190. https://doi.org/10.1016/j.neubiorev.2007.04.009

Hadders-Algra M (2018a) Early human brain development: starring the subplate. Neurosci Biobehav Rev 92:276–290. https://doi.org/10.1016/j.neubiorev.2018.06.017

Hadders-Algra M (2018b) Neural substrate and clinical significance of general movements: an update. Dev Med Child Neurol 60:39–46. https://doi.org/10.1111/dmcn.13540

Han DH, Park P, Choi DI, Bliss TVP, Kaang B-K (2022) The essence of the engram: cellular or synaptic? Semin Cell Dev Biol 125:122–135. https://doi.org/10.1016/j.semcdb.2021.05.033

Hansen DV, Lui JH, Parker PRL, Kriegstein AR (2010) Neurogenic radial glia in the outer subventricular zone of human neocortex. Nature 464:554–561. https://doi.org/10.1038/nature08845

Harris LW, Lockstone HE, Khaitovich P, Weickert CS, Webster MJ, Bahn S (2009) Gene expression in the prefrontal cortex during adolescence: implications for the onset of schizophrenia. BMC Med Genet 2:28. https://doi.org/10.1186/1755-8794-2-28

Harvey CD, Svoboda K (2007) Locally dynamic synaptic learning rules in pyramidal neuron dendrites. Nature 450:1195–1200. https://doi.org/10.1038/nature06416

Hazan L, Ziv NE (2020) Activity dependent and independent determinants of synaptic size diversity. J Neurosci 40:2828–2848. https://doi.org/10.1523/JNEUROSCI.2181-19.2020

He Z, Han D, Efimova O, Guijarro P, Yu Q, Oleksiak A, Jiang S, Anokhin K, Velichkovsky B, Grünewald S, Khaitovich P (2017) Comprehensive transcriptome analysis of neocortical layers in humans, chimpanzees and macaques. Nat Neurosci 20:886–895. https://doi.org/10.1038/nn.4548

He LX, Wan L, Xiang W, Li JM, Pan AH, Lu DH (2020) Synaptic development of layer V pyramidal neurons in the prenatal human prefrontal neocortex: a Neurolucida-aided Golgi study. Neural Regen Res 15:1490–1495. https://doi.org/10.4103/1673-5374.274345

Hebb DO (1949) The organization of behavior; a neuropsychological theory. The organization of behavior, a neuropsychological theory. Wiley, Oxford

Heck N, Santos MD (2023) Dendritic spines in learning and memory: from first discoveries to current insights. In: Rasia-Filho AA et al (eds) Dendritic spines. Springer, Cham

Heckman EL, Doe CQ (2021) Establishment and maintenance of neural circuit architecture. J Neurosci 41:1119–1129. https://doi.org/10.1523/JNEUROSCI.1143-20.2020

Herrmann K, Shatz CJ (1995) Blockade of action potential activity alters initial arborization of thalamic axons within cortical layer 4. Proc Natl Acad Sci U S A 92:11244–11248. https://doi.org/10.1073/pnas.92.24.11244

Herrmann K, Antonini A, Shatz CJ (1994) Ultrastructural evidence for synaptic interactions between thalamocortical axons and subplate neurons. Eur J Neurosci 6:1729–1742. https://doi.org/10.1111/j.1460-9568.1994.tb00565.x

Hevner RF (2000) Development of connections in the human visual system during fetal midgestation: a DiI-tracing study. J Neuropathol Exp Neurol 59:385–392. https://doi.org/10.1093/jnen/59.5.385

Ho VM, Lee JA, Martin KC (2011) The cell biology of synaptic plasticity. Science 334:623–628. https://doi.org/10.1126/science.1209236

Hodel AS (2018) Rapid infant prefrontal cortex development and sensitivity to early environmental experience. Dev Rev 48:113–144. https://doi.org/10.1016/j.dr.2018.02.003

Hofer SB, Mrsic-Flogel TD, Bonhoeffer T, Hübener M (2009) Experience leaves a lasting structural trace in cortical circuits. Nature 457:313–317. https://doi.org/10.1038/nature07487

Hoftman GD, Datta D, Lewis DA (2017) Layer 3 excitatory and inhibitory circuitry in the prefrontal cortex: developmental trajectories and alterations in schizophrenia. Biol Psychiatry 81:862–873. https://doi.org/10.1016/j.biopsych.2016.05.022

Hoftman GD, Bazmi HH, Ciesielski AJ, Dinka LA, Chen K, Lewis DA (2021) Postnatal development of glutamate and GABA transcript expression in monkey visual, parietal, and prefrontal cortices. Cereb Cortex 31:2026–2037. https://doi.org/10.1093/cercor/bhaa342

Holtmaat A, Caroni P (2016) Functional and structural underpinnings of neuronal assembly formation in learning. Nat Neurosci 19:1553–1562. https://doi.org/10.1038/nn.4418

Holtmaat AJGD, Trachtenberg JT, Wilbrecht L, Shepherd GM, Zhang X, Knott GW, Svoboda K (2005) Transient and persistent dendritic spines in the neocortex in vivo. Neuron 45:279–291. https://doi.org/10.1016/j.neuron.2005.01.003

Holtmaat A, Wilbrecht L, Knott GW, Welker E, Svoboda K (2006) Experience-dependent and cell-type-specific spine growth in the neocortex. Nature 441:979–983. https://doi.org/10.1038/nature04783

Honkura N, Matsuzaki M, Noguchi J, Ellis-Davies GC, Kasai H (2008) The subspine organization of actin fibers regulates the structure and plasticity of dendritic spines. Neuron 57:719–729. https://doi.org/10.1016/j.neuron.2008.01.013

Horrell ND, Acosta MC, Saltzman W (2021) Plasticity of the paternal brain: effects of fatherhood on neural structure and function. Dev Psychobiol 63:1499–1520. https://doi.org/10.1002/dev.22097

Hrabač P, Bosak A, Vukšić M, Judaš M, Kostović I, Krsnik Ž (2018) The Zagreb collection of human brains: entering the virtual world. Croat Med J 59:283–287. https://doi.org/10.3325/10.3325/cmj.2018.59.283

Hrvoj-Mihic B, Semendeferi K (2019) Neurodevelopmental disorders of the prefrontal cortex in an evolutionary context. Prog Brain Res 250:109–127. https://doi.org/10.1016/bs.pbr.2019.05.003

Huang Z, Wang Q, Zhou S, Tang C, Yi F, Nie J (2020) Exploring functional brain activity in neonates: a resting-state fMRI study. Dev Cogn Neurosci 45:100850. https://doi.org/10.1016/j.dcn.2020.100850

Hubel DH, Wiesel TN (1963) Receptive fields of cells in striate cortex of very young, visually inexperienced kittens. J Neurophysiol 26:994–1002. https://doi.org/10.1152/jn.1963.26.6.994

Hubel DH, Wiesel TN (1965) Binocular interaction in striate cortex of kittens reared with artificial squint. J Neurophysiol 28:1041–1059. https://doi.org/10.1152/jn.1965.28.6.1041

Hubel DH, Wiesel TN (1970) The period of susceptibility to the physiological effects of unilateral eye closure in kittens. J Physiol 206:419–436. https://doi.org/10.1113/jphysiol.1970.sp009022

Hubel DH, Wiesel TN, LeVay S (1977) Plasticity of ocular dominance columns in monkey striate cortex. Philos Trans R Soc Lond Ser B Biol Sci 278:377–409. https://doi.org/10.1098/rstb.1977.0050

Huberman AD, Speer CM, Chapman B (2006) Spontaneous retinal activity mediates development of ocular dominance columns and binocular receptive fields in v1. Neuron 52:247–254. https://doi.org/10.1016/j.neuron.2006.07.028

Hublin J-J, Changeux J-P (2022) Paleoanthropology of cognition: an overview on hominins brain evolution. C R Biol 345:57–75. https://doi.org/10.5802/crbiol.92

Hunnicutt BJ, Krzywinski M (2016) Neural circuit diagrams. Nat Methods 13:189. https://doi.org/10.1038/nmeth.3777

Hunt S, Leibner Y, Mertens EJ, Barros-Zulaica N, Kanari L, Heistek TS, Karnani MM, Aardse R, Wilbers R, Heyer DB, Goriounova NA, Verhoog MB, Testa-Silva G, Obermayer J, Versluis T, Benavides-Piccione R, de Witt-Hamer P, Idema S, Noske DP, Baayen JC, Lein ES, DeFelipe J, Markram H, Mansvelder HD, Schürmann F, Segev I, de Kock CPJ (2023) Strong and reliable synaptic communication between pyramidal neurons in adult human cerebral cortex. Cereb Cortex 33:2857–2878. https://doi.org/10.1093/cercor/bhac246

Huttenlocher PR (1979) Synaptic density in human frontal cortex – developmental changes and effects of aging. Brain Res 163:195–205. https://doi.org/10.1016/0006-8993(79)90349-4

Huttenlocher PR, Dabholkar AS (1997) Regional differences in synaptogenesis in human cerebral cortex. J Comp Neurol 387:167–178. https://doi.org/10.1002/(SICI)1096-9861(19971020)387:2<167:AID-CNE1>3.0.CO;2-Z

Huttenlocher PR, de Courten C (1987) The development of synapses in striate cortex of man. Hum Neurobiol 6:1–9

Huttenlocher PR, de Courten C, Garey LJ, van der Loos H (1982) Synaptogenesis in human visual cortex--evidence for synapse elimination during normal development. Neurosci Lett 33:247–252. https://doi.org/10.1016/0304-3940(82)90379-2

Huuki-Myers L, Spangler A, Eagles N, Montgomery KD, Kwon SH, Guo B, Grant-Peters M, Divecha HR, Tippani M, Sriworarat C, Nguyen AB, Ravichandran P, Tran MN, Seyedian A, Hyde TM, Kleinman JE, Battle A, Page SC, Ryten M, Hicks SC, Martinowich K, Collado-Torres L, Maynard KR (2023) Integrated single cell and unsupervised spatial transcriptomic analysis defines molecular anatomy of the human dorsolateral prefrontal cortex. bioRxiv. https://doi.org/10.1101/2023.02.15.528722

Innocenti GM (1981) Growth and reshaping of axons in the establishment of visual callosal connections. Science 212:824–827. https://doi.org/10.1126/science.7221566

Innocenti GM, Price DJ (2005) Exuberance in the development of cortical networks. Nat Rev Neurosci 6:955–965. https://doi.org/10.1038/nrn1790

Innocenti GM, Schmidt K, Milleret C, Fabri M, Knyazeva MG, Battaglia-Mayer A, Aboitiz F, Ptito M, Caleo M, Marzi CA, Barakovic M, Lepore F, Caminiti R (2022) The functional characterization of callosal connections. Prog Neurobiol 208:102186. https://doi.org/10.1016/j.pneurobio.2021.102186

Islam KUS, Meli N, Blaess S (2021) The development of the mesoprefrontal dopaminergic system in health and disease. Front Neural Circuits 15:746582. https://doi.org/10.3389/fncir.2021.746582

Jacobs B, Scheibel AB (1993) A quantitative dendritic analysis of Wernicke's area in humans. I. Lifespan changes. J Comp Neurol 327:83–96. https://doi.org/10.1002/cne.903270107

Jacobs B, Chugani HT, Allada V, Chen S, Phelps ME, Pollack DB, Raleigh MJ (1995) Developmental changes in brain metabolism in sedated rhesus macaques and vervet monkeys revealed by positron emission tomography. Cereb Cortex 5:222–233. https://doi.org/10.1093/cercor/5.3.222

Jacobs B, Driscoll L, Schall M (1997) Life-span dendritic and spine changes in areas 10 and 18 of human cortex: a quantitative Golgi study. J Comp Neurol 386:661–680. https://doi.org/10.1002/(SICI)1096-9861(19971006)386:4<661:AID-CNE11>3.0.CO;2-N

Jardri R, Pins D, Houfflin-Debarge V, Chaffiotte C, Rocourt N, Pruvo J-P, Steinling M, Delion P, Thomas P (2008) Fetal cortical activation to sound at 33 weeks of gestation: a functional MRI study. NeuroImage 42:10–18. https://doi.org/10.1016/j.neuroimage.2008.04.247

Jenks KR, Tsimring K, Ip JPK, Zepeda JC, Sur M (2021) Heterosynaptic plasticity and the experience-dependent refinement of developing neuronal circuits. Front Neural Circuits 15:803401. https://doi.org/10.3389/fncir.2021.803401

Jensen KT, Kadmon Harpaz N, Dhawale AK, Wolff SBE, Ölveczky BP (2022) Long-term stability of single neuron activity in the motor system. Nat Neurosci 25:1664–1674. https://doi.org/10.1038/s41593-022-01194-3

Ji L, Majbri A, Hendrix CL, Thomason ME (2023) Fetal behavior during MRI changes with age and relates to network dynamics. Hum Brain Mapp 44:1683–1694. https://doi.org/10.1002/hbm.26167

Josselyn SA, Tonegawa S (2020) Memory engrams: recalling the past and imagining the future. Science 367. https://doi.org/10.1126/science.aaw4325

Judaš M, Šimić G, Petanjek Z, Jovanov-Milošević N, Pletikos M, Vasung L, Vukšić M, Kostović I (2011) The Zagreb collection of human brains: a unique, versatile, but underexploited resource for the neuroscience community. Ann N Y Acad Sci 1225(Suppl 1):E105–E130. https://doi.org/10.1111/j.1749-6632.2011.05993.x

Kanazawa H, Yamada Y, Tanaka K, Kawai M, Niwa F, Iwanaga K, Kuniyoshi Y (2023) Open-ended movements structure sensorimotor information in early human development. Proc Natl Acad Sci U S A 120:e2209953120. https://doi.org/10.1073/pnas.2209953120

Kanjhan R, Fogarty MJ, Noakes PG, Bellingham MC (2016) Developmental changes in the morphology of mouse hypoglossal motor neurons. Brain Struct Funct 221:3755–3786. https://doi.org/10.1007/s00429-015-1130-8

Kasai H, Hayama T, Ishikawa M, Watanabe S, Yagishita S, Noguchi J (2010) Learning rules and persistence of dendritic spines. Eur J Neurosci 32:241–249. https://doi.org/10.1111/j.1460-9568.2010.07344.x

Kasai H, Ziv NE, Okazaki H, Yagishita S, Toyoizumi T (2021) Spine dynamics in the brain, mental disorders and artificial neural networks. Nat Rev Neurosci 22:407–422. https://doi.org/10.1038/s41583-021-00467-3

Katušić A, Žunić Išasegi I, Radoš M, Raguž M, Grizelj R, Ferrari F, Kostović I (2021) Transient structural MRI patterns correlate with the motor functions in preterm infants. Brain and Development 43:363–371. https://doi.org/10.1016/j.braindev.2020.11.002

Katz LC, Shatz CJ (1996) Synaptic activity and the construction of cortical circuits. Science 274:1133–1138. https://doi.org/10.1126/science.274.5290.1133

Keck T, Mrsic-Flogel TD, Vaz Afonso M, Eysel UT, Bonhoeffer T, Hübener M (2008) Massive restructuring of neuronal circuits during functional reorganization of adult visual cortex. Nat Neurosci 11:1162–1167. https://doi.org/10.1038/nn.2181

Keller-Peck CR, Walsh MK, Gan WB, Feng G, Sanes JR, Lichtman JW (2001) Asynchronous synapse elimination in neonatal motor units: studies using GFP transgenic mice. Neuron 31:381–394. https://doi.org/10.1016/s0896-6273(01)00383-x

Khazipov R, Esclapez M, Caillard O, Bernard C, Khalilov I, Tyzio R, Hirsch J, Dzhala V, Berger B, Ben-Ari Y (2001) Early development of neuronal activity in the primate hippocampus in utero. J Neurosci 21:9770–9781. https://doi.org/10.1523/JNEUROSCI.21-24-09770.2001

Kirchner JH, Gjorgjieva J (2021) Emergence of local and global synaptic organization on cortical dendrites. Nat Commun 12:4005. https://doi.org/10.1038/s41467-021-23557-3

Knott GW, Holtmaat A, Wilbrecht L, Welker E, Svoboda K (2006) Spine growth precedes synapse formation in the adult neocortex in vivo. Nat Neurosci 9:1117–1124. https://doi.org/10.1038/nn1747

Koehler RC (2021) Regulation of the cerebral circulation during development. Compr Physiol 11:2371–2432. https://doi.org/10.1002/cphy.c200028

Koenderink MJ, Uylings HB (1995) Postnatal maturation of layer V pyramidal neurons in the human prefrontal cortex. A quantitative Golgi analysis. Brain Res 678:233–243. https://doi.org/10.1016/0006-8993(95)00206-6

Koenderink MJ, Uylings HB, Mrzljak L (1994) Postnatal maturation of the layer III pyramidal neurons in the human prefrontal cortex: a quantitative Golgi analysis. Brain Res 653:173–182. https://doi.org/10.1016/0006-8993(94)90387-5

Koester SE, O'Leary DD (1992) Functional classes of cortical projection neurons develop dendritic distinctions by class-specific sculpting of an early common pattern. J Neurosci 12:1382–1393. https://doi.org/10.1523/JNEUROSCI.12-04-01382.1992

Kolb B, Mychasiuk R, Muhammad A, Li Y, Frost DO, Gibb R (2012) Experience and the developing prefrontal cortex. Proc Natl Acad Sci U S A 109(Suppl 2):17186–17193. https://doi.org/10.1073/pnas.1121251109

Kopić J, Junaković A, Salamon I, Rasin M-R, Kostović I, Krsnik Ž (2023) Early regional patterning in the human prefrontal cortex revealed by laminar dynamics of deep projection neuron markers. Cell 12. https://doi.org/10.3390/cells12020231

Korponay C, Choi EY, Haber SN (2020) Corticostriatal projections of macaque area 44. Cereb Cortex Commun 1:tgaa079. https://doi.org/10.1093/texcom/tgaa079

Kostović I (1986) Prenatal development of nucleus basalis complex and related fiber systems in man: a histochemical study. Neuroscience 17:1047–1077. https://doi.org/10.1016/0306-4522(86)90077-1

Kostović I (2020) The enigmatic fetal subplate compartment forms an early tangential cortical nexus and provides the framework for construction of cortical connectivity. Prog Neurobiol 194:101883. https://doi.org/10.1016/j.pneurobio.2020.101883

Kostovic I, Goldman-Rakic PS (1983) Transient cholinesterase staining in the mediodorsal nucleus of the thalamus and its connections in the developing human and monkey brain. J Comp Neurol 219:431–447. https://doi.org/10.1002/cne.902190405

Kostović I, Jovanov-Milosević N (2006) The development of cerebral connections during the first 20–45 weeks' gestation. Semin Fetal Neonatal Med 11:415–422. https://doi.org/10.1016/j.siny.2006.07.001

Kostović I, Judas M (2002) Correlation between the sequential ingrowth of afferents and transient patterns of cortical lamination in preterm infants. Anat Rec 267:1–6. https://doi.org/10.1002/ar.10069

Kostovic I, Judas M (2006) Prolonged coexistence of transient and permanent circuitry elements in the developing cerebral cortex of fetuses and preterm infants. Dev Med Child Neurol 48:388–393. https://doi.org/10.1017/S0012162206000831

Kostović I, Judas M (2010) The development of the subplate and thalamocortical connections in the human foetal brain. Acta Paediatr 99:1119–1127. https://doi.org/10.1111/j.1651-2227.2010.01811.x

Kostović I, Judaš M (2015) Embryonic and fetal development of the human cerebral cortex. In: Brain mapping. Elsevier, Amsterdam, pp 167–175

Kostovic I, Rakic P (1980) Cytology and time of origin of interstitial neurons in the white matter in infant and adult human and monkey telencephalon. J Neurocytol 9:219–242. https://doi.org/10.1007/BF01205159

Kostovic I, Rakic P (1984) Development of prestriate visual projections in the monkey and human fetal cerebrum revealed by transient cholinesterase staining. J Neurosci 4:25–42. https://doi.org/10.1523/JNEUROSCI.04-01-00025.1984

Kostovic I, Rakic P (1990) Developmental history of the transient subplate zone in the visual and somatosensory cortex of the macaque monkey and human brain. J Comp Neurol 297:441–470. https://doi.org/10.1002/cne.902970309

Kostović I, Jovanov-Milošević N, Radoš M, Sedmak G, Benjak V, Kostović-Srzentić M, Vasung L, Čuljat M, Radoš M, Hüppi P, Judaš M (2014) Perinatal and early postnatal reorganization of the subplate and related cellular compartments in the human cerebral wall as revealed by histological and MRI approaches. Brain Struct Funct 219:231–253. https://doi.org/10.1007/s00429-012-0496-0

Kostović I, Radoš M, Kostović-Srzentić M, Krsnik Ž (2021) Fundamentals of the development of connectivity in the human fetal brain in late gestation: from 24 weeks gestational age to term. J Neuropathol Exp Neurol 80:393–414. https://doi.org/10.1093/jnen/nlab024

Kourosh-Arami M, Hosseini N, Komaki A (2021) Brain is modulated by neuronal plasticity during postnatal development. J Physiol Sci 71:34. https://doi.org/10.1186/s12576-021-00819-9

Kozorovitskiy Y, Hughes M, Lee K, Gould E (2006) Fatherhood affects dendritic spines and vasopressin V1a receptors in the primate prefrontal cortex. Nat Neurosci 9:1094–1095. https://doi.org/10.1038/nn1753

Kristiansen M, Ham J (2014) Programmed cell death during neuronal development: the sympathetic neuron model. Cell Death Differ 21:1025–1035. https://doi.org/10.1038/cdd.2014.47

Krmpotić-Nemanić J, Kostović I, Kelović Z, Nemanić D, Mrzljak L (1983) Development of the human fetal auditory cortex: growth of afferent fibres. Acta Anat (Basel) 116:69–73

Kroeze Y, Oti M, van Beusekom E, Cooijmans RHM, van Bokhoven H, Kolk SM, Homberg JR, Zhou H (2018) Transcriptome analysis identifies multifaceted regulatory mechanisms dictating a genetic switch from neuronal network establishment to maintenance during postnatal prefrontal cortex development. Cereb Cortex 28:833–851. https://doi.org/10.1093/cercor/bhw407

Krsnik Ž, Majić V, Vasung L, Huang H, Kostović I (2017) Growth of thalamocortical fibers to the somatosensory cortex in the human fetal brain. Front Neurosci 11:233. https://doi.org/10.3389/fnins.2017.00233

Kruijssen DLH, Wierenga CJ (2019) Single synapse LTP: a matter of context? Front Cell Neurosci 13:496. https://doi.org/10.3389/fncel.2019.00496

Kugelberg E (1976) Adaptive transformation of rat soleus motor units during growth. J Neurol Sci 27:269–289. https://doi.org/10.1016/0022-510x(76)90001-0

Lagercrantz H, Changeux J-P (2010) Basic consciousness of the newborn. Semin Perinatol 34:201–206. https://doi.org/10.1053/j.semperi.2010.02.004

Lamantia AS, Rakic P (1990a) Axon overproduction and elimination in the corpus callosum of the developing rhesus monkey. J Neurosci 10:2156–2175. https://doi.org/10.1523/JNEUROSCI.10-07-02156.1990

Lamantia AS, Rakic P (1990b) Cytological and quantitative characteristics of four cerebral commissures in the rhesus monkey. J Comp Neurol 291:520–537. https://doi.org/10.1002/cne.902910404

Lamantia AS, Rakic P (1994) Axon overproduction and elimination in the anterior commissure of the developing rhesus monkey. J Comp Neurol 340:328–336. https://doi.org/10.1002/cne.903400304

Lamprecht R, LeDoux J (2004) Structural plasticity and memory. Nat Rev Neurosci 5:45–54. https://doi.org/10.1038/nrn1301

Larsen B, Luna B (2018) Adolescence as a neurobiological critical period for the development of higher-order cognition. Neurosci Biobehav Rev 94:179–195. https://doi.org/10.1016/j.neubiorev.2018.09.005

Lee YI (2020) Developmental neuromuscular synapse elimination: activity-dependence and potential downstream effector mechanisms. Neurosci Lett 718:134724. https://doi.org/10.1016/j.neulet.2019.134724

Lee W, Donner EJ, Nossin-Manor R, Whyte HEA, Sled JG, Taylor MJ (2012) Visual functional magnetic resonance imaging of preterm infants. Dev Med Child Neurol 54:724–729. https://doi.org/10.1111/j.1469-8749.2012.04342.x

Lee JH, Kim WB, Park EH, Cho JH (2023) Neocortical synaptic engrams for remote contextual memories. Nat Neurosci 26:259–273. https://doi.org/10.1038/s41593-022-01223-1

Leisman G (2022) On the application of developmental cognitive neuroscience in educational environments. Brain Sci 12. https://doi.org/10.3390/brainsci12111501

Leuba G, Kraftsik R (1994) Changes in volume, surface estimate, three-dimensional shape and total number of neurons of the human primary visual cortex from midgestation until old age. Anat Embryol (Berl) 190:351–366. https://doi.org/10.1007/BF00187293

Lichtman JW (1977) The reorganization of synaptic connexions in the rat submandibular ganglion during post-natal development. J Physiol 273:155–177. https://doi.org/10.1113/jphysiol.1977.sp012087

Lichtman JW, Colman H (2000) Synapse elimination and indelible memory. Neuron 25:269–278. https://doi.org/10.1016/S0896-6273(00)80893-4

Lidow MS, Rakic P (1992) Scheduling of monoaminergic neurotransmitter receptor expression in the primate neocortex during postnatal development. Cereb Cortex 2:401–416. https://doi.org/10.1093/cercor/2.5.401

Lidow MS, Goldman-Rakic PS, Rakic P (1991) Synchronized overproduction of neurotransmitter receptors in diverse regions of the primate cerebral cortex. Proc Natl Acad Sci U S A 88:10218–10221. https://doi.org/10.1073/pnas.88.22.10218

Lin W, Dominguez B, Yang J, Aryal P, Brandon EP, Gage FH, Lee KF (2005) Neurotransmitter acetylcholine negatively regulates neuromuscular synapse formation by a Cdk5-dependent mechanism. Neuron 46:569–579. https://doi.org/10.1016/j.neuron.2005.04.002

Liu X, Somel M, Tang L, Yan Z, Jiang X, Guo S, Yuan Y, He L, Oleksiak A, Zhang Y, Li N, Hu Y, Chen W, Qiu Z, Pääbo S, Khaitovich P (2012) Extension of cortical synaptic development distinguishes humans from chimpanzees and macaques. Genome Res 22:611–622. https://doi.org/10.1101/gr.127324.111

Lohof AM, Delhaye-Bouchaud N, Mariani J (1996) Synapse elimination in the central nervous system: functional significance and cellular mechanisms. Rev Neurosci 7:85–101. https://doi.org/10.1515/revneuro.1996.7.2.85

Lomo T (1968) Potentiation of monosynaptic EPSPs in cortical cells by single and repetitive afferent volleys. J Physiol 194:84–5P

Lømo T (2018) Discovering long-term potentiation (LTP) – recollections and reflections on what came after. Acta Physiol (Oxf) 222. https://doi.org/10.1111/apha.12921

Lübke J, Albus K (1989) The postnatal development of layer VI pyramidal neurons in the cat's striate cortex, as visualized by intracellular Lucifer yellow injections in aldehyde-fixed tissue. Brain Res Dev Brain Res 45:29–38. https://doi.org/10.1016/0165-3806(89)90004-7

Luengo-Sanchez S, Fernaud-Espinosa I, Bielza C, Benavides-Piccione R, Larrañaga P, DeFelipe J (2018) 3D morphology-based clustering and simulation of human pyramidal cell dendritic spines. PLoS Comput Biol 14:e1006221. https://doi.org/10.1371/journal.pcbi.1006221

Luhmann HJ, Khazipov R (2018) Neuronal activity patterns in the developing barrel cortex. Neuroscience 368:256–267. https://doi.org/10.1016/j.neuroscience.2017.05.025

Luhmann HJ, Kirischuk S, Kilb W (2018) The superior function of the subplate in early neocortical development. Front Neuroanat 12:97. https://doi.org/10.3389/fnana.2018.00097

Luhmann HJ, Kanold PO, Molnár Z, Vanhatalo S (2022) Early brain activity: translations between bedside and laboratory. Prog Neurobiol 213:102268. https://doi.org/10.1016/j.pneurobio.2022.102268

Lund JS, Holbach SM (1991) Postnatal development of thalamic recipient neurons in the monkey striate cortex: I. Comparison of spine acquisition and dendritic growth of layer 4C alpha and beta spiny stellate neurons. J Comp Neurol 309:115–128. https://doi.org/10.1002/cne.903090108

Lund JS, Boothe RG, Lund RD (1977) Development of neurons in the visual cortex (area 17) of the monkey (Macaca nemestrina): a Golgi study from fetal day 127 to postnatal maturity. J Comp Neurol 176:149–188. https://doi.org/10.1002/cne.901760203

Lund JS, Holbach SM, Chung WW (1991) Postnatal development of thalamic recipient neurons in the monkey striate cortex: II. Influence of afferent driving on spine acquisition and dendritic growth of layer 4C spiny stellate neurons. J Comp Neurol 309:129–140. https://doi.org/10.1002/cne.903090109

Luo L, O'Leary DDM (2005) Axon retraction and degeneration in development and disease. Annu Rev Neurosci 28:127–156. https://doi.org/10.1146/annurev.neuro.28.061604.135632

Ma S, Zuo Y (2022) Synaptic modifications in learning and memory – a dendritic spine story. Semin Cell Dev Biol 125:84–90. https://doi.org/10.1016/j.semcdb.2021.05.015

Ma S, Skarica M, Li Q, Xu C, Risgaard RD, Tebbenkamp ATN, Mato-Blanco X, Kovner R, Krsnik Ž, de Martin X, Luria V, Martí-Pérez X, Liang D, Karger A, Schmidt DK, Gomez-Sanchez

Z, Qi C, Gobeske KT, Pochareddy S, Debnath A, Hottman CJ, Spurrier J, Teo L, Boghdadi AG, Homman-Ludiye J, Ely JJ, Daadi EW, Mi D, Daadi M, Marín O, Hof PR, Rasin M-R, Bourne J, Sherwood CC, Santpere G, Girgenti MJ, Strittmatter SM, Sousa AMM, Sestan N (2022) Molecular and cellular evolution of the primate dorsolateral prefrontal cortex. Science 377:eabo7257. https://doi.org/10.1126/science.abo7257

Magee JC, Grienberger C (2020) Synaptic plasticity forms and functions. Annu Rev Neurosci 43:95–117. https://doi.org/10.1146/annurev-neuro-090919-022842

Malenka RC, Bear MF (2004) LTP and LTD: an embarrassment of riches. Neuron 44:5–21. https://doi.org/10.1016/j.neuron.2004.09.012

Marafiga JR, Calcagnotto ME (2023) Electrophysiology of dendritic spines: information processing, dynamic compartmentalization, and synaptic plasticity. In: Rasia-Filho AA et al (eds) Dendritic spines. Springer, Cham

Marchetto MC, Hrvoj-Mihic B, Kerman BE, Yu DX, Vadodaria KC, Linker SB, Narvaiza I, Santos R, Denli AM, Mendes AP, Oefner R, Cook J, McHenry L, Grasmick JM, Heard K, Fredlender C, Randolph-Moore L, Kshirsagar R, Xenitopoulos R, Chou G, Hah N, Muotri AR, Padmanabhan K, Semendeferi K, Gage FH (2019) Species-specific maturation profiles of human, chimpanzee and bonobo neural cells. Elife 8. https://doi.org/10.7554/eLife.37527

Markus EJ, Petit TL (1987) Neocortical synaptogenesis, aging, and behavior: lifespan development in the motor-sensory system of the rat. Exp Neurol 96:262–278. https://doi.org/10.1016/0014-4886(87)90045-8

Martini FJ, Guillamón-Vivancos T, Moreno-Juan V, Valdeolmillos M, López-Bendito G (2021) Spontaneous activity in developing thalamic and cortical sensory networks. Neuron 109:2519–2534. https://doi.org/10.1016/j.neuron.2021.06.026

Mashour GA, Roelfsema P, Changeux J-P, Dehaene S (2020) Conscious processing and the global neuronal workspace hypothesis. Neuron 105:776–798. https://doi.org/10.1016/j.neuron.2020.01.026

Mates SL, Lund JS (1983) Neuronal composition and development in lamina 4C of monkey striate cortex. J Comp Neurol 221:60–90. https://doi.org/10.1002/cne.902210106

Matsuzaki M, Honkura N, Ellis-Davies GCR, Kasai H (2004) Structural basis of long-term potentiation in single dendritic spines. Nature 429:761–766. https://doi.org/10.1038/nature02617

Maynard KR, Collado-Torres L, Weber LM, Uytingco C, Barry BK, Williams SR, Catallini JL, Tran MN, Besich Z, Tippani M, Chew J, Yin Y, Kleinman JE, Hyde TM, Rao N, Hicks SC, Martinowich K, Jaffe AE (2021) Transcriptome-scale spatial gene expression in the human dorsolateral prefrontal cortex. Nat Neurosci 24:425–436. https://doi.org/10.1038/s41593-020-00787-0

Mayseless O, Shapira G, Rachad EY, Fiala A, Schuldiner O (2023) Neuronal excitability as a regulator of circuit remodeling. Curr Biol 33:981–989.e3. https://doi.org/10.1016/j.cub.2023.01.032

McAllister AK (2000) Cellular and molecular mechanisms of dendrite growth. Cereb Cortex 10:963–973. https://doi.org/10.1093/cercor/10.10.963

McGuire PK, Bates JF, Goldman-Rakic PS (1991) Interhemispheric integration: II. Symmetry and convergence of the corticostriatal projections of the left and the right principal sulcus (PS) and the left and the right supplementary motor area (SMA) of the rhesus monkey. Cereb Cortex 1:408–417. https://doi.org/10.1093/cercor/1.5.408

Mencía S, Alonso C, Pallás-Alonso C, López-Herce J, Maternal ACHADNS, II (2022) Evaluation and treatment of pain in fetuses, neonates and children. Children (Basel) 9. https://doi.org/10.3390/children9111688

Meyer G, González-Gómez M (2018) The heterogeneity of human Cajal-Retzius neurons. Semin Cell Dev Biol 76:101–111. https://doi.org/10.1016/j.semcdb.2017.08.059

Meyer G, González-Hernández TH, Ferres-Torres R (1989) The spiny stellate neurons in layer IV of the human auditory cortex. A Golgi study. Neuroscience 33:489–498. https://doi.org/10.1016/0306-4522(89)90401-6

Meyer G, Schaaps JP, Moreau L, Goffinet AM (2000) Embryonic and early fetal development of the human neocortex. J Neurosci 20:1858–1868. https://doi.org/10.1523/JNEUROSCI.20-05-01858.2000

Meyer D, Bonhoeffer T, Scheuss V (2014) Balance and stability of synaptic structures during synaptic plasticity. Neuron 82:430–443. https://doi.org/10.1016/j.neuron.2014.02.031

Michel AE, Garey LJ (1984) The development of dendritic spines in the human visual cortex. Hum Neurobiol 3:223–227

Midorikawa M, Miyata M (2021) Distinct functional developments of surviving and eliminated presynaptic terminals. Proc Natl Acad Sci U S A 118. https://doi.org/10.1073/pnas.2022423118

Milh M, Kaminska A, Huon C, Lapillonne A, Ben-Ari Y, Khazipov R (2007) Rapid cortical oscillations and early motor activity in premature human neonate. Cereb Cortex 17:1582–1594. https://doi.org/10.1093/cercor/bhl069

Miškić T, Kostović I, Rašin M-R, Krsnik Ž (2021) Adult upper cortical layer specific transcription factor CUX2 is expressed in transient subplate and marginal zone neurons of the developing human brain. Cells 10. https://doi.org/10.3390/cells10020415

Missler M, Wolff A, Merker HJ, Wolff JR (1993) Pre- and postnatal development of the primary visual cortex of the common marmoset. II. Formation, remodelling, and elimination of synapses as overlapping processes. J Comp Neurol 333:53–67. https://doi.org/10.1002/cne.903330105

Miyashita Y (2022) Operating principles of the cerebral cortex as a six layered network in primates: beyond the classic canonical circuit model. Proc Jpn Acad Ser B Phys Biol Sci 98:93–111. https://doi.org/10.2183/pjab.98.007

Molliver ME, Kostović I, van der Loos H (1973) The development of synapses in cerebral cortex of the human fetus. Brain Res 50:403–407. https://doi.org/10.1016/0006-8993(73)90741-5

Molnár Z, Luhmann HJ, Kanold PO (2020) Transient cortical circuits match spontaneous and sensory-driven activity during development. Science 370. https://doi.org/10.1126/science.abb2153

Moore AH, Hovda DA, Cherry SR, Villablanca JP, Pollack DB, Phelps ME (2000) Dynamic changes in cerebral glucose metabolism in conscious infant monkeys during the first year of life as measured by positron emission tomography. Brain Res Dev Brain Res 120:141–150. https://doi.org/10.1016/s0165-3806(00)00005-5

Moore AR, Filipovic R, Mo Z, Rasband MN, Zecevic N, Antic SD (2009) Electrical excitability of early neurons in the human cerebral cortex during the second trimester of gestation. Cereb Cortex 19:1795–1805. https://doi.org/10.1093/cercor/bhn206

Moore AR, Zhou W-L, Jakovcevski I, Zecevic N, Antic SD (2011) Spontaneous electrical activity in the human fetal cortex in vitro. J Neurosci 31:2391–2398. https://doi.org/10.1523/JNEUROSCI.3886-10.2011

Moore AR, Zhou W-L, Sirois CL, Belinsky GS, Zecevic N, Antic SD (2014) Connexin hemichannels contribute to spontaneous electrical activity in the human fetal cortex. Proc Natl Acad Sci U S A 111:E3919–E3928. https://doi.org/10.1073/pnas.1405253111

Moron H, Gagnard-Landra C, Guiraud D, Dupeyron A (2021) Contribution of single-fiber evaluation on monitoring outcomes following injection of botulinum toxin-A: a narrative review of the literature. Toxins (Basel) 13. https://doi.org/10.3390/toxins13050356

Mrzljak L, Uylings HB, Kostovic I, van Eden CG (1988) Prenatal development of neurons in the human prefrontal cortex: I. A qualitative Golgi study. J Comp Neurol 271:355–386. https://doi.org/10.1002/cne.902710306

Mrzljak L, Uylings HB, van Eden CG, Judás M (1990) Neuronal development in human prefrontal cortex in prenatal and postnatal stages. Prog Brain Res 85:185–222. https://doi.org/10.1016/s0079-6123(08)62681-3

Mrzljak L, Uylings HB, Kostovic I, van Eden CG (1992) Prenatal development of neurons in the human prefrontal cortex. II. A quantitative Golgi study. J Comp Neurol 316:485–496. https://doi.org/10.1002/cne.903160408

Mukherjee D, Kanold PO (2022) Changing subplate circuits: early activity dependent circuit plasticity. Front Cell Neurosci 16:1067365. https://doi.org/10.3389/fncel.2022.1067365

Muthukrishna M, Doebeli M, Chudek M, Henrich J (2018) The cultural brain hypothesis: how culture drives brain expansion, sociality, and life history. PLoS Comput Biol 14:e1006504. https://doi.org/10.1371/journal.pcbi.1006504

Muzik O, Janisse J, Ager J, Shen C, Chugani DC, Chugani HT (1999) A mathematical model for the analysis of cross-sectional brain glucose metabolism data in children. Prog Neuro-Psychopharmacol Biol Psychiatry 23:589–600. https://doi.org/10.1016/s0278-5846(99)00018-4

Nagamori A, Laine CM, Loeb GE, Valero-Cuevas FJ (2021) Force variability is mostly not motor noise: theoretical implications for motor control. PLoS Comput Biol 17:e1008707. https://doi.org/10.1371/journal.pcbi.1008707

Nagy C, Maitra M, Tanti A, Suderman M, Théroux J-F, Davoli MA, Perlman K, Yerko V, Wang YC, Tripathy SJ, Pavlidis P, Mechawar N, Ragoussis J, Turecki G (2020) Single-nucleus transcriptomics of the prefrontal cortex in major depressive disorder implicates oligodendrocyte precursor cells and excitatory neurons. Nat Neurosci 23:771–781. https://doi.org/10.1038/s41593-020-0621-y

Nazareth L, St John J, Murtaza M, Ekberg J (2021) Phagocytosis by peripheral glia: importance for nervous system functions and implications in injury and disease. Front Cell Dev Biol 9:660259. https://doi.org/10.3389/fcell.2021.660259

Nelson CA (1994) Threats to optimal development: integrating biological, psychological, and social risk factors. Minnesota symposia on child psychology, vol 27. L. Erlbaum Associates, Hillsdale

Neumane S, Gondova A, Leprince Y, Hertz-Pannier L, Arichi T, Dubois J (2022) Early structural connectivity within the sensorimotor network: deviations related to prematurity and association to neurodevelopmental outcome. Front Neurosci 16:932386. https://doi.org/10.3389/fnins.2022.932386

Nguyen QT, Lichtman JW (1996) Mechanism of synapse disassembly at the developing neuromuscular junction. Curr Opin Neurobiol 6:104–112. https://doi.org/10.1016/s0959-4388(96)80015-8

Nicoll RA (2017) A brief history of long-term potentiation. Neuron 93:281–290. https://doi.org/10.1016/j.neuron.2016.12.015

Nobin A, Björklund A (1973) Topography of the monoamine neuron systems in the human brain as revealed in fetuses. Acta Physiol Scand Suppl 388:1–40

O'Brien RA, Ostberg AJ, Vrbová G (1978) Observations on the elimination of polyneuronal innervation in developing mammalian skeletal muscle. J Physiol 282:571–582. https://doi.org/10.1113/jphysiol.1978.sp012482

O'Kusky JR (1998) Postnatal changes in the numerical density and total number of asymmetric and symmetric synapses in the hypoglossal nucleus of the rat. Brain Res Dev Brain Res 108:179–191. https://doi.org/10.1016/s0165-3806(98)00048-0

O'Kusky J, Colonnier M (1982) Postnatal changes in the number of neurons and synapses in the visual cortex (area 17) of the macaque monkey: a stereological analysis in normal and monocularly deprived animals. J Comp Neurol 210:291–306. https://doi.org/10.1002/cne.902100308

Ofer N, Berger DR, Kasthuri N, Lichtman JW, Yuste R (2021) Ultrastructural analysis of dendritic spine necks reveals a continuum of spine morphologies. Dev Neurobiol 81:746–757. https://doi.org/10.1002/dneu.22829

Ofer N, Benavides-Piccione R, DeFelipe J, Yuste R (2022) Structural analysis of human and mouse dendritic spines reveals a morphological continuum and differences across ages and species. eNeuro 9. https://doi.org/10.1523/ENEURO.0039-22.2022

Oga T, Aoi H, Sasaki T, Fujita I, Ichinohe N (2013) Postnatal development of layer III pyramidal cells in the primary visual, inferior temporal, and prefrontal cortices of the marmoset. Front Neural Circuits 7:31. https://doi.org/10.3389/fncir.2013.00031

Oga T, Elston GN, Fujita I (2017) Postnatal dendritic growth and spinogenesis of layer-V pyramidal cells differ between visual, inferotemporal, and prefrontal cortex of the macaque monkey. Front Neurosci 11:118. https://doi.org/10.3389/fnins.2017.00118

Ohtaka-Maruyama C (2020) Subplate neurons as an organizer of mammalian neocortical development. Front Neuroanat 14:8. https://doi.org/10.3389/fnana.2020.00008

Okabe S (2009) Dendritic growth. In: Binder MD, Hirokawa N, Windhorst U (eds) Encyclopedia of neuroscience. Springer Berlin Heidelberg, Berlin, Heidelberg, pp 937–942

Oppenheim RW (1991) Cell death during development of the nervous system. Annu Rev Neurosci 14:453–501. https://doi.org/10.1146/annurev.ne.14.030191.002321

Ozcan AS (2017) Filopodia: a rapid structural plasticity substrate for fast learning. Front Synaptic Neurosci 9:12. https://doi.org/10.3389/fnsyn.2017.00012

Palay SL (1956) Synapses in the central nervous system. J Biophys Biochem Cytol 2:193–202. https://doi.org/10.1083/jcb.2.4.193

Pan Y, Monje M (2020) Activity shapes neural circuit form and function: a historical perspective. J Neurosci 40:944–954. https://doi.org/10.1523/JNEUROSCI.0740-19.2019

Pchitskaya E, Bezprozvanny I (2020) Dendritic spines shape analysis-classification or clusterization? Perspective. Front Synaptic Neurosci 12:31. https://doi.org/10.3389/fnsyn.2020.00031

Pease SE, Segal RA (2014) Preserve and protect: maintaining axons within functional circuits. Trends Neurosci 37:572–582. https://doi.org/10.1016/j.tins.2014.07.007

Pendl SL, Salzwedel AP, Goldman BD, Barrett LF, Lin W, Gilmore JH, Gao W (2017) Emergence of a hierarchical brain during infancy reflected by stepwise functional connectivity. Hum Brain Mapp 38:2666–2682. https://doi.org/10.1002/hbm.23552

Pérez-Ortega J, Alejandre-García T, Yuste R (2021) Long-term stability of cortical ensembles. Elife 10. https://doi.org/10.7554/eLife.64449

Personius KE, Slusher BS, Udin SB (2016) Neuromuscular NMDA receptors modulate developmental synapse elimination. J Neurosci 36:8783–8789. https://doi.org/10.1523/JNEUROSCI.1181-16.2016

Petanjek Z, Judas M, Kostović I, Uylings HBM (2008) Lifespan alterations of basal dendritic trees of pyramidal neurons in the human prefrontal cortex: a layer-specific pattern. Cereb Cortex 18:915–929. https://doi.org/10.1093/cercor/bhm124

Petanjek Z, Judaš M, Šimic G, Rasin MR, Uylings HBM, Rakic P, Kostovic I (2011) Extraordinary neoteny of synaptic spines in the human prefrontal cortex. Proc Natl Acad Sci U S A 108:13281–13286. https://doi.org/10.1073/pnas.1105108108

Petanjek Z, Sedmak D, Džaja D, Hladnik A, Rašin MR, Jovanov-Milosevic N (2019) The protracted maturation of associative layer IIIC pyramidal neurons in the human prefrontal cortex during childhood: a major role in cognitive development and selective alteration in autism. Front Psychiatry 10:122. https://doi.org/10.3389/fpsyt.2019.00122

Peters A (1987) Number of neurons and synapses in primary visual cortex. In: Jones EG, Peters A (eds) Cerebral cortex, vol 6. Springer US, Boston, pp 267–294

Peters A, Rosene DL (2003) In aging, is it gray or white? J Comp Neurol 462:139–143. https://doi.org/10.1002/cne.10715

Pettmann B, Henderson CE (1998) Neuronal cell death. Neuron 20:633–647. https://doi.org/10.1016/s0896-6273(00)81004-1

Peyton C, Einspieler C, Fjørtoft T, Adde L, Schreiber MD, Drobyshevsky A, Marks JD (2020) Correlates of normal and abnormal general movements in infancy and long-term neurodevelopment of preterm infants: insights from functional connectivity studies at term equivalence. J Clin Med 9. https://doi.org/10.3390/jcm9030834

Piekarski DJ, Johnson CM, Boivin JR, Thomas AW, Lin WC, Delevich K, Galarce EM, Wilbrecht L (2017) Does puberty mark a transition in sensitive periods for plasticity in the associative neocortex? Brain Res 1654:123–144. https://doi.org/10.1016/j.brainres.2016.08.042

Polese D, Riccio ML, Fagioli M, Mazzetta A, Fagioli F, Parisi P, Fagioli M (2022) The newborn's reaction to light as the determinant of the brain's activation at human birth. Front Integr Neurosci 16:933426. https://doi.org/10.3389/fnint.2022.933426

Pollen AA, Nowakowski TJ, Chen J, Retallack H, Sandoval-Espinosa C, Nicholas CR, Shuga J, Liu SJ, Oldham MC, Diaz A, Lim DA, Leyrat AA, West JA, Kriegstein AR (2015) Molecular identity of human outer radial glia during cortical development. Cell 163:55–67. https://doi.org/10.1016/j.cell.2015.09.004

Pollen AA, Kilik U, Lowe CB, Camp JG (2023) Human-specific genetics: new tools to explore the molecular and cellular basis of human evolution. Nat Rev Genet:1–25. https://doi.org/10.1038/s41576-022-00568-4

Poo MM, Pignatelli M, Ryan TJ, Tonegawa S, Bonhoeffer T, Martin KC, Rudenko A, Tsai LH, Tsien RW, Fishell G, Mullins C, Gonçalves JT, Shtrahman M, Johnston ST, Gage FH, Dan Y, Long J, Buzsáki G, Stevens C (2016) What is memory? The present state of the engram. BMC Biol 14:40. https://doi.org/10.1186/s12915-016-0261-6

Prechtl HF, Hopkins B (1986) Developmental transformations of spontaneous movements in early infancy. Early Hum Dev 14:233–238. https://doi.org/10.1016/0378-3782(86)90184-2

Preuss TM, Wise SP (2022) Evolution of prefrontal cortex. Neuropsychopharmacology 47:3–19. https://doi.org/10.1038/s41386-021-01076-5

Provenzi L, Lindstedt J, de Coen K, Gasparini L, Peruzzo D, Grumi S, Arrigoni F, Ahlqvist-Björkroth S (2021) The paternal brain in action: a review of human fathers' fMRI brain responses to child-related stimuli. Brain Sci 11. https://doi.org/10.3390/brainsci11060816

Provis JM, van Driel D, Billson FA, Russell P (1985) Human fetal optic nerve: overproduction and elimination of retinal axons during development. J Comp Neurol 238:92–100. https://doi.org/10.1002/cne.902380108

Ptak R, Doganci N, Bourgeois A (2021) From action to cognition: neural reuse, network theory and the emergence of higher cognitive functions. Brain Sci 11. https://doi.org/10.3390/brainsci11121652

Purpura DP (1975) Normal and aberrant neuronal development in the cerebral cortex of human fetus and young infant. UCLA Forum Med Sci:141–169. https://doi.org/10.1016/b978-0-12-139050-1.50014-8

Purves D, Lichtman JW (1980) Elimination of synapses in the developing nervous system. Science 210:153–157. https://doi.org/10.1126/science.7414326

Quartz SR, Sejnowski TJ (1997) The neural basis of cognitive development: a constructivist manifesto. Behav Brain Sci 20:537–556. https://doi.org/10.1017/s0140525x97001581. discussion 556–96

Rakic P (2009) Evolution of the neocortex: a perspective from developmental biology. Nat Rev Neurosci 10:724–735. https://doi.org/10.1038/nrn2719

Rakic P, Riley KP (1983a) Overproduction and elimination of retinal axons in the fetal rhesus monkey. Science 219:1441–1444. https://doi.org/10.1126/science.6828871

Rakic P, Riley KP (1983b) Regulation of axon number in primate optic nerve by prenatal binocular competition. Nature 305:135–137. https://doi.org/10.1038/305135a0

Rakic P, Bourgeois JP, Eckenhoff MF, Zecevic N, Goldman-Rakic PS (1986) Concurrent overproduction of synapses in diverse regions of the primate cerebral cortex. Science 232:232–235. https://doi.org/10.1126/science.3952506

Rakic P, Bourgeois JP, Goldman-Rakic PS (1994) Synaptic development of the cerebral cortex: implications for learning, memory, and mental illness. Prog Brain Res 102:227–243. https://doi.org/10.1016/S0079-6123(08)60543-9

Ramos SI, Mussa ZM, Falk EN, Pai B, Giotti B, Allette K, Cai P, Dekio F, Sebra R, Beaumont KG, Tsankov AM, Tsankova NM (2022) An atlas of late prenatal human neurodevelopment resolved by single-nucleus transcriptomics. Nat Commun 13:7671. https://doi.org/10.1038/s41467-022-34975-2

Rangamani P, Levy MG, Khan S, Oster G (2016) Paradoxical signaling regulates structural plasticity in dendritic spines. Proc Natl Acad Sci U S A 113:E5298–E5307. https://doi.org/10.1073/pnas.1610391113

Rao-Ruiz P, Visser E, Mitrić M, Smit AB, van den Oever MC (2021) A synaptic framework for the persistence of memory engrams. Front Synaptic Neurosci 13:661476. https://doi.org/10.3389/fnsyn.2021.661476

Rasia-Filho AA (2022) Unraveling brain microcircuits, dendritic spines, and synaptic processing using multiple complementary approaches. Front Physiol 13:831568. https://doi.org/10.3389/fphys.2022.831568

Rasia-Filho AA, Guerra KTK, Vásquez CE, Dall'Oglio A, Reberger R, Jung CR, Calcagnotto ME (2021) The subcortical-allocortical- neocortical continuum for the emergence and morphological heterogeneity of pyramidal neurons in the human brain. Front Synaptic Neurosci 13:616607. https://doi.org/10.3389/fnsyn.2021.616607

Rasia-Filho AA, Calcagnotto ME, von Bohlen und Halbach O (2023a) Introduction: what are dendritic spines? In: Rasia-Filho AA et al (eds) Dendritic spines. Springer, Cham

Rasia-Filho AA, Calcagnotto ME, von Bohlen und Halbach O (2023b) Glial cell modulation of dendritic spine structure and synaptic function. In: Rasia-Filho AA et al (eds) Dendritic spines. Springer, Cham

Ray A, Christian JA, Mosso MB, Park E, Wegner W, Willig KI, Barth AL (2023) Quantitative fluorescence analysis reveals dendrite-specific thalamocortical plasticity in L5 pyramidal neurons during learning. J Neurosci 43:584–600. https://doi.org/10.1523/JNEUROSCI.1372-22.2022

Reberger R, Dall'Oglio A, Jung CR, Rasia-Filho AA (2018) Structure and diversity of human dendritic spines evidenced by a new three-dimensional reconstruction procedure for Golgi staining and light microscopy. J Neurosci Methods 293:27–36. https://doi.org/10.1016/j.jneumeth.2017.09.001

Redfern PA (1970) Neuromuscular transmission in new-born rats. J Physiol 209:701–709. https://doi.org/10.1113/jphysiol.1970.sp009187

Renner L, Rasia-Filho AA (2023) Morphological features of human dendritic spines. In: Rasia-Filho AA et al (eds) Dendritic spines. Springer, Cham

Riccomagno MM, Kolodkin AL (2015) Sculpting neural circuits by axon and dendrite pruning. Annu Rev Cell Dev Biol 31:779–805. https://doi.org/10.1146/annurev-cellbio-100913-013038

Ridge RM, Betz WJ (1984) The effect of selective, chronic stimulation on motor unit size in developing rat muscle. J Neurosci 4:2614–2620. https://doi.org/10.1523/JNEUROSCI.04-10-02614.1984

Rodríguez Cruz PM, Cossins J, Beeson D, Vincent A (2020) The neuromuscular junction in health and disease: molecular mechanisms governing synaptic formation and homeostasis. Front Mol Neurosci 13:610964. https://doi.org/10.3389/fnmol.2020.610964

Rosen BQ, Halgren E (2022) An estimation of the absolute number of axons indicates that human cortical areas are sparsely connected. PLoS Biol 20:e3001575. https://doi.org/10.1371/journal.pbio.3001575

Runge K, Cardoso C, de Chevigny A (2020) Dendritic spine plasticity: function and mechanisms. Front Synaptic Neurosci 12:36. https://doi.org/10.3389/fnsyn.2020.00036

Ryan TJ, Ortega-de San Luis C, Pezzoli M, Sen S (2021) Engram cell connectivity: an evolving substrate for information storage. Curr Opin Neurobiol 67:215–225. https://doi.org/10.1016/j.conb.2021.01.006

Sakai J (2020) Core concept: how synaptic pruning shapes neural wiring during development and, possibly, in disease. Proc Natl Acad Sci U S A 117:16096–16099. https://doi.org/10.1073/pnas.2010281117

Salamon I, Rasin M-R (2021) Evolution of the neocortex through RNA-binding proteins and post-transcriptional regulation. Front Neurosci 15:803107. https://doi.org/10.3389/fnins.2021.803107

Sasaki T, Aoi H, Oga T, Fujita I, Ichinohe N (2015) Postnatal development of dendritic structure of layer III pyramidal neurons in the medial prefrontal cortex of marmoset. Brain Struct Funct 220:3245–3258. https://doi.org/10.1007/s00429-014-0853-2

Schade JP, van Groenigen W (1961) Structural organization of the human cerebral cortex. 1. Maturation of the middle frontal gyrus. Acta Anat (Basel) 47:74–111. https://doi.org/10.1159/000141802

Schaeffer L, de Kerchove d'Exaerde A, Changeux JP (2001) Targeting transcription to the neuromuscular synapse. Neuron 31:15–22. https://doi.org/10.1016/S0896-6273(01)00353-1

Schwartz ML, Goldman-Rakic PS (1982) Single cortical neurones have axon collaterals to ipsilateral and contralateral cortex in fetal and adult primates. Nature 299:154–155. https://doi.org/10.1038/299154a0

Schwartz ML, Goldman-Rakic PS (1984) Callosal and intrahemispheric connectivity of the prefrontal association cortex in rhesus monkey: relation between intraparietal and principal sulcal cortex. J Comp Neurol 226:403–420. https://doi.org/10.1002/cne.902260309

Sedmak G, Judaš M (2021) White matter interstitial neurons in the adult human brain: 3% of cortical neurons in quest for recognition. Cell 10. https://doi.org/10.3390/cells10010190

Sedmak D, Hrvoj-Mihić B, Džaja D, Habek N, Uylings HB, Petanjek Z (2018) Biphasic dendritic growth of dorsolateral prefrontal cortex associative neurons and early cognitive development. Croat Med J 59:189–202. https://doi.org/10.3325/cmj.2018.59.189

Segal M (2017) Dendritic spines: morphological building blocks of memory. Neurobiol Learn Mem 138:3–9. https://doi.org/10.1016/j.nlm.2016.06.007

Selemon LD (2013) A role for synaptic plasticity in the adolescent development of executive function. Transl Psychiatry 3:e238. https://doi.org/10.1038/tp.2013.7

Selemon LD, Zecevic N (2015) Schizophrenia: a tale of two critical periods for prefrontal cortical development. Transl Psychiatry 5:e623. https://doi.org/10.1038/tp.2015.115

Semendeferi K, Teffer K, Buxhoeveden DP, Park MS, Bludau S, Amunts K, Travis K, Buckwalter J (2011) Spatial organization of neurons in the frontal pole sets humans apart from great apes. Cereb Cortex 21:1485–1497. https://doi.org/10.1093/cercor/bhq191

Seng C, Luo W, Földy C (2022) Circuit formation in the adult brain. Eur J Neurosci 56:4187–4213. https://doi.org/10.1111/ejn.15742

Shapiro LP, Parsons RG, Koleske AJ, Gourley SL (2017) Differential expression of cytoskeletal regulatory factors in the adolescent prefrontal cortex: implications for cortical development. J Neurosci Res 95:1123–1143. https://doi.org/10.1002/jnr.23960

Sharples SA, Miles GB (2021) Maturation of persistent and hyperpolarization-activated inward currents shapes the differential activation of motoneuron subtypes during postnatal development. Elife 10. https://doi.org/10.7554/eLife.71385

Shatz CJ, Stryker MP (1988) Prenatal tetrodotoxin infusion blocks segregation of retinogeniculate afferents. Science 242:87–89. https://doi.org/10.1126/science.3175636

Sheng C, Javed U, Gibbs M, Long C, Yin J, Qin B, Yuan Q (2018) Experience-dependent structural plasticity targets dynamic filopodia in regulating dendrite maturation and synaptogenesis. Nat Commun 9:3362. https://doi.org/10.1038/s41467-018-05871-5

Shepherd GM (1996) The dendritic spine: a multifunctional integrative unit. J Neurophysiol 75:2197–2210. https://doi.org/10.1152/jn.1996.75.6.2197

Sherk H, Stryker MP (1976) Quantitative study of cortical orientation selectivity in visually inexperienced kitten. J Neurophysiol 39:63–70. https://doi.org/10.1152/jn.1976.39.1.63

Sheynikhovich D, Otani S, Bai J, Arleo A (2022) Long-term memory, synaptic plasticity and dopamine in rodent medial prefrontal cortex: role in executive functions. Front Behav Neurosci 16:1068271. https://doi.org/10.3389/fnbeh.2022.1068271

Sin WC, Haas K, Ruthazer ES, Cline HT (2002) Dendrite growth increased by visual activity requires NMDA receptor and Rho GTPases. Nature 419:475–480. https://doi.org/10.1038/nature00987

Sohn J, Suzuki M, Youssef M, Hatada S, Larkum ME, Kawaguchi Y, Kubota Y (2022) Presynaptic supervision of cortical spine dynamics in motor learning. Sci Adv 8:eabm0531. https://doi.org/10.1126/sciadv.abm0531

Somel M, Franz H, Yan Z, Lorenc A, Guo S, Giger T, Kelso J, Nickel B, Dannemann M, Bahn S, Webster MJ, Weickert CS, Lachmann M, Pääbo S, Khaitovich P (2009) Transcriptional neoteny in the human brain. Proc Natl Acad Sci U S A 106:5743–5748. https://doi.org/10.1073/pnas.0900544106

Somel M, Liu X, Khaitovich P (2013) Human brain evolution: transcripts, metabolites and their regulators. Nat Rev Neurosci 14:112–127. https://doi.org/10.1038/nrn3372

Sousa SS, Amaro E, Crego A, Gonçalves ÓF, Sampaio A (2018) Developmental trajectory of the prefrontal cortex: a systematic review of diffusion tensor imaging studies. Brain Imaging Behav 12:1197–1210. https://doi.org/10.1007/s11682-017-9761-4

Spreafico R, Arcelli P, Frassoni C, Canetti P, Giaccone G, Rizzuti T, Mastrangelo M, Bentivoglio M (1999) Development of layer I of the human cerebral cortex after midgestation: architectonic findings, immunocytochemical identification of neurons and glia, and in situ labeling of apoptotic cells. J Comp Neurol 410:126–142. https://doi.org/10.1002/(sici)1096-9861(19990719)410:1<126:aid-cne11>3.0.co;2-5

Sretavan DW, Shatz CJ, Stryker MP (1988) Modification of retinal ganglion cell axon morphology by prenatal infusion of tetrodotoxin. Nature 336:468–471. https://doi.org/10.1038/336468a0

Stampanoni Bassi M, Iezzi E, Gilio L, Centonze D, Buttari F (2019) Synaptic plasticity shapes brain connectivity: implications for network topology. Int J Mol Sci 20. https://doi.org/10.3390/ijms20246193

Sun Y, Smirnov M, Kamasawa N, Yasuda R (2021) Rapid ultrastructural changes in the PSD and surrounding membrane after induction of structural LTP in single dendritic spines. J Neurosci 41:7003–7014. https://doi.org/10.1523/JNEUROSCI.1964-20.2021

Sunesen M, Changeux JP (2003) Transcription in neuromuscular junction formation: who turns on whom? J Neurocytol 32:677–684. https://doi.org/10.1023/B:NEUR.0000020616.53664.80

Supèr H, Soriano E, Uylings HB (1998) The functions of the preplate in development and evolution of the neocortex and hippocampus. Brain Res Brain Res Rev 27:40–64. https://doi.org/10.1016/s0165-0173(98)00005-8

Sweis BM, Mau W, Rabinowitz S, Cai DJ (2021) Dynamic and heterogeneous neural ensembles contribute to a memory engram. Curr Opin Neurobiol 67:199–206. https://doi.org/10.1016/j.conb.2020.11.017

Sydnor VJ, Larsen B, Bassett DS, Alexander-Bloch A, Fair DA, Liston C, Mackey AP, Milham MP, Pines A, Roalf DR, Seidlitz J, Xu T, Raznahan A, Satterthwaite TD (2021) Neurodevelopment of the association cortices: patterns, mechanisms, and implications for psychopathology. Neuron 109:2820–2846. https://doi.org/10.1016/j.neuron.2021.06.016

Takashima S, Chan F, Becker LE, Armstrong DL (1980) Morphology of the developing visual cortex of the human infant: a quantitative and qualitative Golgi study. J Neuropathol Exp Neurol 39:487–501. https://doi.org/10.1097/00005072-198007000-00007

Tamnes CK, Herting MM, Goddings A-L, Meuwese R, Blakemore S-J, Dahl RE, Güroğlu B, Raznahan A, Sowell ER, Crone EA, Mills KL (2017) Development of the cerebral cortex across adolescence: a multisample study of inter-related longitudinal changes in cortical volume, surface area, and thickness. J Neurosci 37:3402–3412. https://doi.org/10.1523/JNEUROSCI.3302-16.2017

Tau GZ, Peterson BS (2010) Normal development of brain circuits. Neuropsychopharmacology 35:147–168. https://doi.org/10.1038/npp.2009.115

Teffer K, Semendeferi K (2012) Human prefrontal cortex: evolution, development, and pathology. Prog Brain Res 195:191–218. https://doi.org/10.1016/B978-0-444-53860-4.00009-X

Thompson W (1983) Synapse elimination in neonatal rat muscle is sensitive to pattern of muscle use. Nature 302:614–616. https://doi.org/10.1038/302614a0

Thompson W, Jansen JK (1977) The extent of sprouting of remaining motor units in partly denervated immature and adult rat soleus muscle. Neuroscience 2:523–535. https://doi.org/10.1016/0306-4522(77)90049-5

Thompson W, Kuffler DP, Jansen JK (1979) The effect of prolonged, reversible block of nerve impulses on the elimination of polyneuronal innervation of new-born rat skeletal muscle fibers. Neuroscience 4:271–281. https://doi.org/10.1016/0306-4522(79)90088-5

Thompson WJ, Sutton LA, Riley DA (1984) Fibre type composition of single motor units during synapse elimination in neonatal rat soleus muscle. Nature 309:709–711. https://doi.org/10.1038/309709a0

Tolonen M, Palva JM, Andersson S, Vanhatalo S (2007) Development of the spontaneous activity transients and ongoing cortical activity in human preterm babies. Neuroscience 145:997–1006. https://doi.org/10.1016/j.neuroscience.2006.12.070

Tomàs J, Lanuza MA, Santafé MM, Cilleros-Mañé V, Just-Borràs L, Balanyà-Segura M, Polishchuk A, Nadal L, Tomàs M, Garcia N (2023) Muscarinic receptors in developmental axonal competition at the neuromuscular junction. Mol Neurobiol 60:1580–1593. https://doi.org/10.1007/s12035-022-03154-1

Tong R, Emptage NJ, Padamsey Z (2020) A two-compartment model of synaptic computation and plasticity. Mol Brain 13:79. https://doi.org/10.1186/s13041-020-00617-1

Tønnesen J, Nägerl UV (2016) Dendritic spines as tunable regulators of synaptic signals. Front Psychiatry 7:101. https://doi.org/10.3389/fpsyt.2016.00101

Tousignant B, Eugène F, Sirois K, Jackson PL (2018) Difference in neural response to social exclusion observation and subsequent altruism between adolescents and adults. Neuropsychologia 116:15–25. https://doi.org/10.1016/j.neuropsychologia.2017.04.017

Trachtenberg JT, Chen BE, Knott GW, Feng G, Sanes JR, Welker E, Svoboda K (2002) Long-term in vivo imaging of experience-dependent synaptic plasticity in adult cortex. Nature 420:788–794. https://doi.org/10.1038/nature01273

Tropea D, Majewska AK, Garcia R, Sur M (2010) Structural dynamics of synapses in vivo correlate with functional changes during experience-dependent plasticity in visual cortex. J Neurosci 30:11086–11095. https://doi.org/10.1523/JNEUROSCI.1661-10.2010

Ucar H, Watanabe S, Noguchi J, Morimoto Y, Iino Y, Yagishita S, Takahashi N, Kasai H (2021) Mechanical actions of dendritic-spine enlargement on presynaptic exocytosis. Nature 600:686–689. https://doi.org/10.1038/s41586-021-04125-7

Uylings HBM (2006) Development of the human cortex and the concept of "critical" or "sensitive" periods. Lang Learn 56:59–90. https://doi.org/10.1111/j.1467-9922.2006.00355.x

Uylings HBM, de Brabander JM (2002) Neuronal changes in normal human aging and Alzheimer's disease. Brain Cogn 49:268–276. https://doi.org/10.1006/brcg.2001.1500

Uylings HBM, van Pelt J, Parnavelas JG, Ruiz-Marcos A (1994) Geometrical and topological characteristics in the dendritic development of cortical pyramidal and non-pyramidal neurons. Prog Brain Res 102:109–123. https://doi.org/10.1016/s0079-6123(08)60535-x

van Pelt J, Uylings HBM (2002) Branching rates and growth functions in the outgrowth of dendritic branching patterns. Network 13:261–281. https://doi.org/10.1088/0954-898x/13/3/302

Vanderhaeghen P, Polleux F (2023) Developmental mechanisms underlying the evolution of human cortical circuits. Nat Rev Neurosci. https://doi.org/10.1038/s41583-023-00675-z

Vardalaki D, Chung K, Harnett MT (2022) Filopodia are a structural substrate for silent synapses in adult neocortex. Nature 612:323–327. https://doi.org/10.1038/s41586-022-05483-6

Vasalauskaite A, Morgan JE, Sengpiel F (2019) Plasticity in adult mouse visual cortex following optic nerve injury. Cereb Cortex 29:1767–1777. https://doi.org/10.1093/cercor/bhy347

Vasung L, Lepage C, Radoš M, Pletikos M, Goldman JS, Richiardi J, Raguž M, Fischi-Gómez E, Karama S, Huppi PS, Evans AC, Kostovic I (2016) Quantitative and qualitative analysis of transient fetal compartments during prenatal human brain development. Front Neuroanat 10:11. https://doi.org/10.3389/fnana.2016.00011

Vasung L, Raguz M, Kostovic I, Takahashi E (2017) Spatiotemporal relationship of brain pathways during human fetal development using high-angular resolution diffusion MR imaging and histology. Front Neurosci 11:348. https://doi.org/10.3389/fnins.2017.00348

Verney C (1999) Distribution of the catecholaminergic neurons in the central nervous system of human embryos and fetuses. Microsc Res Tech 46:24–47. https://doi.org/10.1002/(SICI)1097-0029(19990701)46:1%3C24:AID-JEMT3%3E3.0.CO;2-E

Verney C, Milosevic A, Alvarez C, Berger B (1993) Immunocytochemical evidence of well-developed dopaminergic and noradrenergic innervations in the frontal cerebral cortex of human fetuses at midgestation. J Comp Neurol 336:331–344. https://doi.org/10.1002/cne.903360303

Volzhenin K, Changeux JP, Dumas G (2022) Multilevel development of cognitive abilities in an artificial neural network. Proc Natl Acad Sci U S A 119:e2201304119. https://doi.org/10.1073/pnas.2201304119

von Bohlen und Halbach O (2023) Neurotrophic factors and dendritic spines. In: Rasia-Filho AA et al (eds) Dendritic spines. Springer, Cham

Walsh MK, Lichtman JW (2003) In vivo time-lapse imaging of synaptic takeover associated with naturally occurring synapse elimination. Neuron 37:67–73. https://doi.org/10.1016/s0896-6273(02)01142-x

Wang L, Conner JM, Rickert J, Tuszynski MH (2011) Structural plasticity within highly specific neuronal populations identifies a unique parcellation of motor learning in the adult brain. Proc Natl Acad Sci U S A 108:2545–2550. https://doi.org/10.1073/pnas.1014335108

Weber JP, Andrásfalvy BK, Polito M, Magó Á, Ujfalussy BB, Makara JK (2016) Location-dependent synaptic plasticity rules by dendritic spine cooperativity. Nat Commun 7:11380. https://doi.org/10.1038/ncomms11380

Werchan DM, Amso D (2017) A novel ecological account of prefrontal cortex functional development. Psychol Rev 124:720–739. https://doi.org/10.1037/rev0000078

Wess JM, Isaiah A, Watkins PV, Kanold PO (2017) Subplate neurons are the first cortical neurons to respond to sensory stimuli. Proc Natl Acad Sci U S A 114:12602–12607. https://doi.org/10.1073/pnas.1710793114

White LD, Barone S (2001) Qualitative and quantitative estimates of apoptosis from birth to senescence in the rat brain. Cell Death Differ 8:345–356. https://doi.org/10.1038/sj.cdd.4400816

Whitford KL, Dijkhuizen P, Polleux F, Ghosh A (2002) Molecular control of cortical dendrite development. Annu Rev Neurosci 25:127–149. https://doi.org/10.1146/annurev.neuro.25.112701.142932

Wiesel TN, Hubel DH (1963a) Effects of visual deprivation on morphology and physiology of cells in the cats lateral geniculate body. J Neurophysiol 26:978–993. https://doi.org/10.1152/jn.1963.26.6.978

Wiesel TN, Hubel DH (1963b) Single-cell responses in striate cortex of kittens deprived of vision in one eye. J Neurophysiol 26:1003–1017. https://doi.org/10.1152/jn.1963.26.6.1003

Wilbrecht L, Holtmaat A, Wright N, Fox K, Svoboda K (2010) Structural plasticity underlies experience-dependent functional plasticity of cortical circuits. J Neurosci 30:4927–4932. https://doi.org/10.1523/JNEUROSCI.6403-09.2010

Wildenberg G, Li H, Kasthuri N (2023) The development of synapses in mouse and macaque primary sensory cortices. bioRxiv. https://doi.org/10.1101/2023.02.15.528564

Williams RW, Rakic P (1988) Elimination of neurons from the rhesus monkey's lateral geniculate nucleus during development. J Comp Neurol 272:424–436. https://doi.org/10.1002/cne.902720310

Winfield DA, Rivera-Dominguez M, Powell TP (1982) The termination of geniculocortical fibres in area 17 of the visual cortex in the macaque monkey. Brain Res 231:19–32. https://doi.org/10.1016/0006-8993(82)90004-x

Won H, Huang J, Opland CK, Hartl CL, Geschwind DH (2019) Human evolved regulatory elements modulate genes involved in cortical expansion and neurodevelopmental disease susceptibility. Nat Commun 10:2396. https://doi.org/10.1038/s41467-019-10248-3

Wong FK, Marín O (2019) Developmental cell death in the cerebral cortex. Annu Rev Cell Dev Biol 35:523–542. https://doi.org/10.1146/annurev-cellbio-100818-125204

Wouterlood FG (2023) Techniques to render dendritic spines visible in the microscope. In: Rasia-Filho AA et al (eds) Dendritic spines. Springer, Cham

Wright LL, Cunningham TJ, Smolen AJ (1983) Developmental neuron death in the rat superior cervical sympathetic ganglion: cell counts and ultrastructure. J Neurocytol 12:727–738. https://doi.org/10.1007/BF01258147

Wu H, Xiong WC, Mei L (2010) To build a synapse: signaling pathways in neuromuscular junction assembly. Development 137:1017–1033. https://doi.org/10.1242/dev.038711

Wyatt RM, Balice-Gordon RJ (2003) Activity-dependent elimination of neuromuscular synapses. J Neurocytol 32:777–794. https://doi.org/10.1023/B:NEUR.0000020623.62043.33

Xu T, Yu X, Perlik AJ, Tobin WF, Zweig JA, Tennant K, Jones T, Zuo Y (2009) Rapid formation and selective stabilization of synapses for enduring motor memories. Nature 462:915–919. https://doi.org/10.1038/nature08389

Xu R, Bichot NP, Takahashi A, Desimone R (2022a) The cortical connectome of primate lateral prefrontal cortex. Neuron 110:312–327.e7. https://doi.org/10.1016/j.neuron.2021.10.018

Xu J, Zhu J, Li Y, Yao Y, Xuan A, Li D, Yu T, Zhu D (2022b) Three-dimensional mapping reveals heterochronic development of the neuromuscular system in postnatal mouse skeletal muscles. Commun Biol 5:1200. https://doi.org/10.1038/s42003-022-04159-1

Yamaguchi Y, Miura M (2015) Programmed cell death in neurodevelopment. Dev Cell 32:478–490. https://doi.org/10.1016/j.devcel.2015.01.019

Yang X, Arber S, William C, Li L, Tanabe Y, Jessell TM, Birchmeier C, Burden SJ (2001a) Patterning of muscle acetylcholine receptor gene expression in the absence of motor innervation. Neuron 30:399–410. https://doi.org/10.1016/S0896-6273(01)00287-2

Yang JF, Cao G, Koirala S, Reddy LV, Ko CP (2001b) Schwann cells express active agrin and enhance aggregation of acetylcholine receptors on muscle fibers. J Neurosci 21:9572–9584. https://doi.org/10.1523/jneurosci.21-24-09572.2001

Yang G, Pan F, Gan W-B (2009) Stably maintained dendritic spines are associated with lifelong memories. Nature 462:920–924. https://doi.org/10.1038/nature08577

Yeterian EH, Pandya DN (1994) Laminar origin of striatal and thalamic projections of the prefrontal cortex in rhesus monkeys. Exp Brain Res 99:383–398. https://doi.org/10.1007/BF00228975

Yin W, Chen M-H, Hung S-C, Baluyot KR, Li T, Lin W (2019) Brain functional development separates into three distinct time periods in the first two years of life. Neuroimage 189:715–726. https://doi.org/10.1016/j.neuroimage.2019.01.025

Yoshihara Y, de Roo M, Muller D (2009) Dendritic spine formation and stabilization. Curr Opin Neurobiol 19:146–153. https://doi.org/10.1016/j.conb.2009.05.013

Yuste R (2015) The discovery of dendritic spines by Cajal. Front Neuroanat 9:18. https://doi.org/10.3389/fnana.2015.00018

Yuste R, Bonhoeffer T (2004) Genesis of dendritic spines: insights from ultrastructural and imaging studies. Nat Rev Neurosci 5:24–34. https://doi.org/10.1038/nrn1300

Zecevic N (1998) Synaptogenesis in layer I of the human cerebral cortex in the first half of gestation. Cereb Cortex 8:245–252. https://doi.org/10.1093/cercor/8.3.245

Zecevic N, Rakic P (1991) Synaptogenesis in monkey somatosensory cortex. Cereb Cortex 1:510–523. https://doi.org/10.1093/cercor/1.6.510

Zecevic N, Rakic P (2001) Development of layer I neurons in the primate cerebral cortex. J Neurosci 21:5607–5619. https://doi.org/10.1523/JNEUROSCI.21-15-05607.2001

Zecevic N, Verney C (1995) Development of the catecholamine neurons in human embryos and fetuses, with special emphasis on the innervation of the cerebral cortex. J Comp Neurol 351:509–535. https://doi.org/10.1002/cne.903510404

Zecevic N, Bourgeois JP, Rakic P (1989) Changes in synaptic density in motor cortex of rhesus monkey during fetal and postnatal life. Brain Res Dev Brain Res 50:11–32. https://doi.org/10.1016/0165-3806(89)90124-7

Zecevic N, Milosevic A, Rakic S, Marín-Padilla M (1999) Early development and composition of the human primordial plexiform layer: an immunohistochemical study. J Comp Neurol 412:241–254

Zempo B, Yamamoto Y, Williams T, Ono F (2020) Synaptic silencing of fast muscle is compensated by rewired innervation of slow muscle. Sci Adv 6:eaax8382. https://doi.org/10.1126/sciadv.aax8382

Zielinski BS, Hendrickson AE (1992) Development of synapses in macaque monkey striate cortex. Vis Neurosci 8:491–504. https://doi.org/10.1017/s0952523800005599

Zikopoulos B, Barbas H (2013) Altered neural connectivity in excitatory and inhibitory cortical circuits in autism. Front Hum Neurosci 7:609. https://doi.org/10.3389/fnhum.2013.00609

Zimmermann KS, Richardson R, Baker KD (2019) Maturational changes in prefrontal and amygdala circuits in adolescence: implications for understanding fear inhibition during a vulnerable period of development. Brain Sci 9. https://doi.org/10.3390/brainsci9030065

Ziv NE, Smith SJ (1996) Evidence for a role of dendritic filopodia in synaptogenesis and spine formation. Neuron 17:91–102. https://doi.org/10.1016/s0896-6273(00)80283-4

Zuo Y, Lin A, Chang P, Gan W-B (2005a) Development of long-term dendritic spine stability in diverse regions of cerebral cortex. Neuron 46:181–189. https://doi.org/10.1016/j.neuron.2005.04.001

Zuo Y, Yang G, Kwon E, Gan W-B (2005b) Long-term sensory deprivation prevents dendritic spine loss in primary somatosensory cortex. Nature 436:261–265. https://doi.org/10.1038/nature03715

Chapter 5
Neurotrophic Factors and Dendritic Spines

Oliver von Bohlen und Halbach

Abstract Dendritic spines are highly dynamic structures that play important roles in neuronal plasticity. The morphologies and the numbers of dendritic spines are highly variable, and this diversity is correlated with the different morphological and physiological features of this neuronal compartment. Dendritic spines can change their morphology and number rapidly, allowing them to adapt to plastic changes. Neurotrophic factors play important roles in the brain during development. However, these factors are also necessary for a variety of processes in the postnatal brain. Neurotrophic factors, especially members of the neurotrophin family and the ephrin family, are involved in the modulation of long-lasting effects induced by neuronal plasticity by acting on dendritic spines, either directly or indirectly. Thereby, the neurotrophic factors play important roles in processes attributed, for example, to learning and memory.

Keywords Neurotrophic factor · Neurotrophin · Ephrin · Neuronal plasticity · Synaptic plasticity

5.1 Dendritic Spines

More than 130 years ago, the Spanish anatomist and Nobel laureate Santiago Ramón y Cajal discovered the structure of dendritic spines by using the Golgi silver-impregnation method (Garcia-Lopez et al. 2007) and different types of microscopes (see Chap. 1, Fig. 1.1)—a necessary setup as nineteenth-century microscopes had only one objective and were illuminated by reflected light via a mirror (Fig. 5.1). Nevertheless, Cajal hypothesized that dendritic spines increase the surface area of dendrites and thus serve as site of contacts between dendrites and axons. Moreover, he argued that dendritic spines were probably related to intelligence, since he noted

O. von Bohlen und Halbach (✉)
Institut für Anatomie und Zellbiologie, Universitätsmedizin Greifswald, Greifswald, Germany
e-mail: oliver.vonbohlen@uni-greifswald.de

A. A. Rasia-Filho et al. (eds.), *Dendritic Spines*, Advances in Neurobiology 34,
https://doi.org/10.1007/978-3-031-36159-3_5

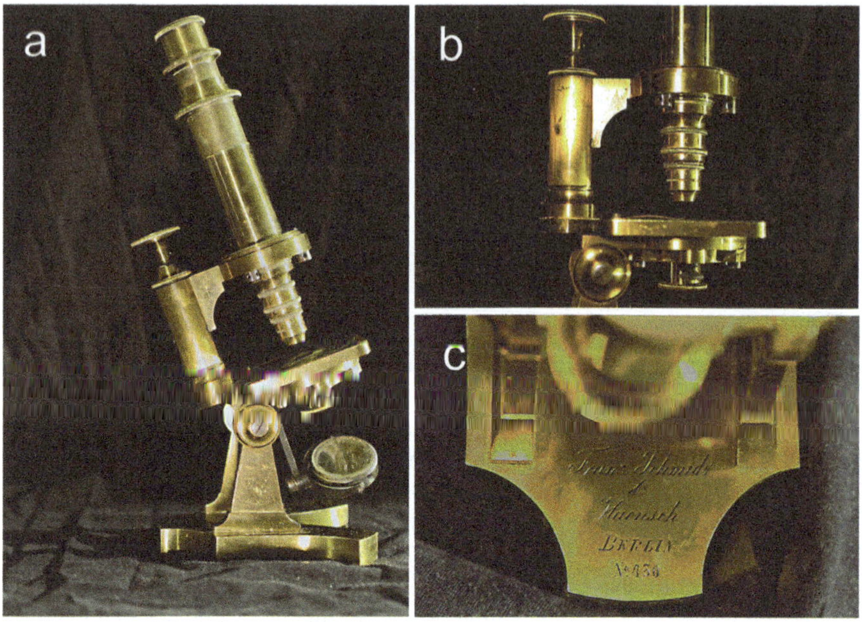

Fig. 5.1 A microscope from the nineteenth century. (**a**) The microscope has been constructed and manufactured around 1880. The design of that microscope is comparable to those S. Ramón y Cajal uses for his studies (see Chap. 1, Fig. 1.1). The microscope consists of fenced and blackened brass and blued steel. (**b**) The optics supplied with this microscope consists of an objective that is a three-part divisible objective. (**c**) The signature of the manufacturer is: Franz Schmidt & Haensch BERLIN. No 430

that cells from more highly evolved animals have more dendritic spines (see for details: Yuste (2015)). Nowadays, based on numerous results using different scientific approaches and—as compared to the time of Cajal—modern microscopes, it is widely accepted that the number of dendritic spines as well as their morphology is highly variable and that this diversity is correlated with the diverse morphological and physiological features of this neuronal compartment.

A classical dendritic spine consists of a head that is connected to the dendritic shaft by a narrow neck (Risher et al. 2014). Depending on their shape, the dendritic spines can be subdivided in different categories (Fig. 5.2). Dendritic spines were initially classified into three categories based on the relative sizes of the dendritic spine head and neck (Peters and Kaiserman-Abramof 1970):

(i) Mushroom spines are composed of a large head and a narrow neck.
(ii) Thin spines have a smaller head and a narrow neck.
(iii) The so-called stubby spines that display no obvious constriction between the head and the attachment to the shaft.

A further group of dendritic spines are the so-called branched spines. These dendritic spines have multiple heads that emerge from a shared origin. These heads can

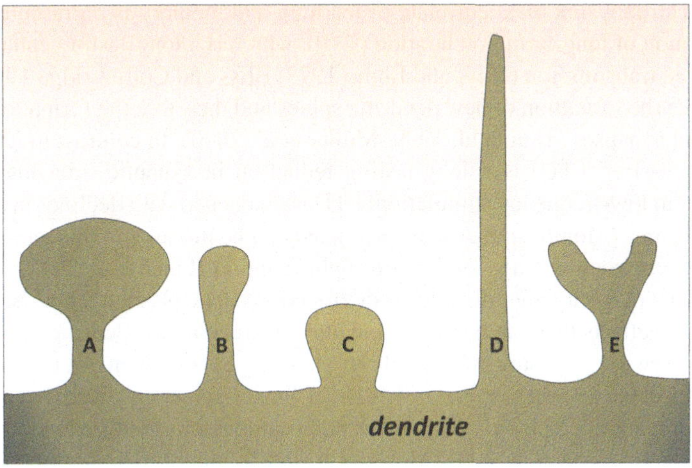

Fig. 5.2 Depending on their shape, dendritic spines can be subdivided into different categories. (A) Mushroom spines have a large head and a narrow neck, whereas (B) thin spines have a smaller head and a narrow neck. The so-called stubby spines (C) display no obvious subdivision in head and neck. Filopodia (D) are long and thin without an obvious head. Branched spines (E) have multiple heads that emerge from a shared origin. (Adapted from von Bohlen und Halbach (2009), Yuste and Bonhoeffer (2004), and Risher et al. (2014))

establish contacts with different presynaptic axons, while some heads have no presynaptic partners (Sorra et al. 1998). Furthermore, immature dendritic spines have been described as well. These spines possess a long filopodia-like morphology (Skoff and Hamburger 1974). The filopodia are often short-lived and seemed to be capable of choosing between potential synaptic partners before a mature synapse is established. During development, some filopodia become dendritic spines with synapses, while others withdraw back into the dendrite to form synapses on the dendritic shaft (Fiala et al. 1998; Bourne and Harris 2008; Lohmann and Bonhoeffer 2008).

Concerning the classification of dendritic spines by their shape, it should be kept in mind that this classification is based on their morphology in the moment the tissue was obtained. However, and in contrast to what has been believed before, the shape of dendritic spines is not stable and, thus, can vary within a short time-period. Therefore, the morphological plasticity of dendritic spines indicates that these categories do not represent different populations of dendritic spines; rather, they reflect temporal snapshots of a dynamic phenomenon (Parnass et al. 2000; von Bohlen und Halbach 2009). These dynamics of dendritic spines are, among others, associated with neuronal plasticity. Thus, dendritic spines play fundamental roles in neuronal plasticity. Several forms of learning have been shown to increase dendritic spine densities, for example, in the hippocampus (Moser et al. 1994; Geinisman 2000; Yuste and Bonhoeffer 2001; Nimchinsky et al. 2002; Leuner et al. 2003), whereas during aging, a decrease in the number of hippocampal dendritic spines has been observed (von Bohlen und Halbach et al. 2006b).

An electrophysiological correlate of learning and memory is represented by the phenomenon of long-term potentiation (LTP), which is a long-lasting enhancement of synaptic transmission (Bliss and Lomo 1973; Bliss and Collingridge 1993). LTP can induce the formation of new dendritic spines and therefore the formation of new functional synapses (Toni et al. 1999; Muller et al. 2000). In contrast to LTP, long-term depression (LTD) is a long-lasting reduction in synaptic transmission that results from low-frequency stimulation; LTD is associated with declines in dendritic spine densities (Monfils and Teskey 2004) and can be modulated by factors including stress, environment, age, and neurotrophic support (Pinar et al. 2017).

Neuronal plasticity not only affects short-term synaptic plasticity, but also induces long-lasting effects that can be obvious in altered structure and densities of dendritic spines. Likewise, exposure of animals to enriched environment leads not only to improvements in several learning tasks (Bruel Jungerman et al. 2005), but also to an increase in dendritic spine density in the hippocampus of rodents as well as of primates (Kozorovitskiy et al. 2005; Moser et al. 1997). Since the hippocampus is critically involved in spatial learning, an enriched environment might have an impact on neuronal plasticity; it is thought that changes in the number or shape of dendritic spines represent morphological correlates of neuronal plasticity, learning, and memory.

Changes in dendritic spines can also be observed in a number of neurological diseases (e.g., intellectual disability or dementia) as well as under certain psychopathological conditions, for example, major depression. The mechanisms of major depression as well as the neurobiological basis of antidepressant therapy are still enigmatic, but it is likely that the pathogenesis as well as the treatment of major depression involve changes in neuronal plasticity, that, among others, are accompanied by changes in dendritic spines in the hippocampus (Serafini 2012; Jiang and Salton 2013).

Neuronal plasticity is thought to depend on the availability of neurotrophic factors that are, on the one hand, capable of stabilizing and maintaining the brain architecture and, on the other hand, able to support the remodeling of existent pathways. Neurotrophic factors may therefore act as mediators of neuronal plasticity or factors contributing to neuronal plasticity. Neurotrophic factors might be involved in LTP and/or LTD as well as in changes in dendritic spines, pointing toward a role of these factors in learning and memory. Defects or deficits in the supply or in the signaling of neurotrophic factors may thus contribute to various neurological and psychiatric disorders.

5.2 Neurotrophic Factors

Neurotrophic factors are neuropeptides that have an impact upon growth, differentiation, and survival of cells within both the developing nervous system, as well as the postnatal nervous system. In the postnatal nervous system, especially in the peripheral nervous system (PNS), neurotrophic factors are involved in repair mechanisms, for example, in the regrowth of damaged neuronal processes (More et al. 2012). Neurotrophic factors not only play specific roles in the maintenance of neuronal populations, but also in plastic changes, including synaptic and neuronal plasticity, which can have an impact on processes attributed to learning and memory.

Furthermore, neurotrophic factors have both a direct as well as an indirect impact on brain activity since they can influence energy balance—that is, the relationship between energy intake and energy expenditure (Xu and Xie 2016). Most neurotrophic factors belong to different families that share specific properties, for example, the neurotrophins, or the family of fibroblast growth factors. Several of these factors are involved in neuronal plasticity and the signaling through their cognate receptors can lead to a remodeling of dendritic spines.

5.2.1 Neurotrophins

In 1951, Rita Levi-Montalcini (Fig. 5.3) and Viktor Hamburg published that a tumor released a diffusible factor that promoted neurogenesis (Levi-Montalcini and Hamburger 1951). Together with Stanley Cohen, they were able to purify a growth-promoting factor (Cohen et al. 1954), which later was termed nerve growth factor (NGF). NGF represents the progenitor of the neurotrophin family of growth factors.

A second family member of the neurotrophins, the brain-derived neurotrophic factor (BDNF), was originally identified in 1982 as a factor that promoted survival of cultured embryonic chick sensory neurons (Barde et al. 1982). Since then, a variety of further members of the neurotrophin family have been discovered.

The family of neurotrophins consist of BDNF, NGF, and the neurotrophins 3 (NT-3) and 4 (NT-4). Neurotrophins bind to specific receptors that belong to the family of tyrosine protein kinase receptors (NGF mainly binds to trkA; BDNF, like NT-4, mainly to trkB (Fig. 5.4) and NT-3 mainly to trkC).

Binding of BDNF to the trkB receptor induces ligand–receptor dimerization and autophosphorylation of tyrosine residues. Three main intracellular signaling cascades are activated by the trkB receptor (Minichiello 2009): (i) the Ras pathway, (ii) the PI3K-Akt pathway, and (iii) the PLCgamma–Ca^{2+} pathway (Fig. 5.4). In addition, all neurotrophins can signal through the low-affinity p75 neurotrophin receptor p75NTR (Barbacid 1994; Barker 1998).

Neurotrophins play important roles during development. Lack of functional NGF, BDNF, or NT3 genes results in severe neuronal deficits and in early postnatal death (Conover and Yancopoulos 1997). For example, mice lacking BDNF usually die during the second postnatal week (Ernfors et al. 1994). Therefore, other mouse models have been developed to analyze the roles of neurotrophins in the postnatal brain. The analysis of mice with conditional or heterozygous deletion of the neurotrophins or their cognate tyrosine kinase receptors (as well as homozygous p75NTR mice) revealed that neurotrophin signaling is essential for the postnatal brain and processes that involve synaptic and neuronal plasticity. Another approach for investigating the effects of BDNF deficiency, reduced availability or signaling in the central nervous system (CNS) is the use of brain or neuron-specific conditional knockout mice. Concerning the analysis of BDNF functions in the postnatal hippocampus, calcium-calmodulin-dependent protein kinase II (CaMKII)-Cre transgenic mice or glial fibrillary acid protein (GFAP)-Cre transgenic mice have been created (Monteggia et al. 2007). Moreover, conditional trkB knockout mice have

Fig. 5.3 Rita Levi-Montalcini (22 April 1909–30 December 2012) was awarded the Nobel Prize in Physiology or Medicine in 1986 together with Stanley Cohen (15 November 1922–5 February 2020) for the discovery and research of NGF. In the same year they also received the Albert Lasker Award for Basic Medical Research
The photograph has been generously provided by the European Brain Research Institute (EBRI) "Rita Levi-Montalcini." The EBRI was founded 2005 by Rita Levi-Montalcini. The Institute was established in response to the need in Italy for a center dedicated to the study of the brain and diseases that affect its function. For further information see: https://www.ebri.it/en/

been generated, for example, CaMKII-Cre transgenic mice (Minichiello et al. 1999) or mice with a targeted mutation in the Shc- and PLCgamma-docking sites of trkB (Medina et al. 2004) or mice with a conditional ablation of trkB in parvalbumin-expressing interneurons (Zheng et al. 2011; Xenos et al. 2017).

In humans, polymorphisms in the BDNF gene are associated with major depression (Frodl et al. 2007; Gonul et al. 2011) and reduced levels of BDNF can contribute to major depression. Moreover, it has been reported that low plasma BDNF levels are associated with suicidal behavior in people living with major depression (Kim et al. 2007).

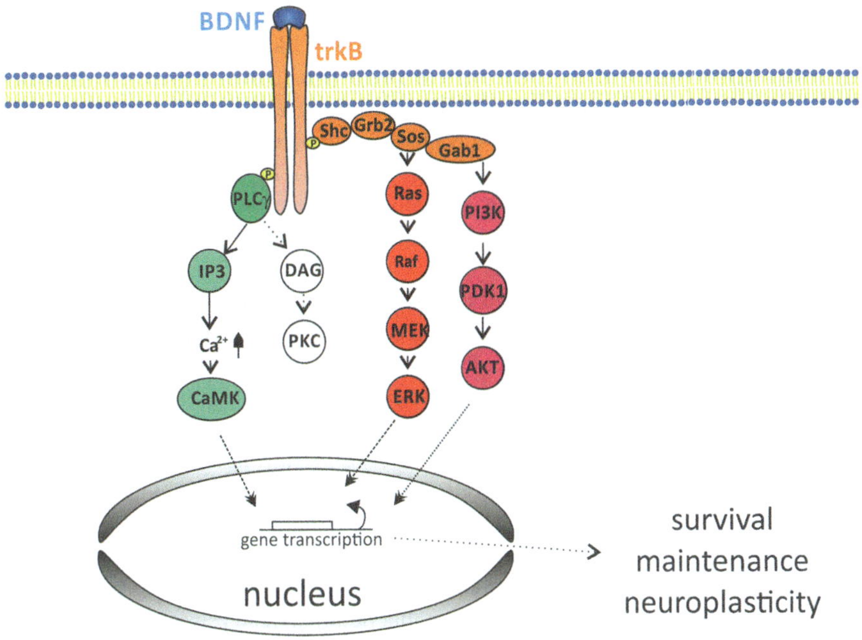

Fig. 5.4 BDNF binds to trkB and induces receptor dimerization and activation of tyrosine kinase domains, resulting in autophosphorylation, allowing the interaction with adapter proteins that induces intracellular signal cascades, such as (i) the Ras–microtubule-associated protein kinase (MAPK), (ii) the phosphatidyl inositol-3 kinase (PI3K)/serine threonine kinase (Akt), and (iii) phospholipase C (PLC)-γ pathways. Phosphorylation of TrkB tyrosine residue 515 leads to recruitment of the Src homology 2 domain-containing (Shc) adaptor protein that is followed by the recruitment of growth factor receptor-bound protein 2 (Grb2) and son of sevenless (SOS).
This can lead to the activation of the Ras (rat sarcoma) pathway (red). In addition, the recruitment of Grb2-associated binder-1 (GAB1) and activation of the PI3K–Akt pathway (magenta) can be induced. Phosphorylation of the TrkB tyrosine residue 816 leads to the recruitment of phospholipase C γ (PLCγ), which leads to the formation of inositol triphosphate (IP3) that upregulates intracellular Ca^{2+}, which further activates Ca^{2+}/calmodulin-dependent protein kinase (CaMK, green). Furthermore, diacylglycerol (DAG) is stimulated, leading to the activation of protein kinase C (PKC). The Ras pathway, the PI3K-Akt pathway, and the PLCgamma–Ca^{2+} pathway can induce gene transcription affecting not only survival and growth but also neuroplasticity.
Further abbreviations: ERK: extracellular signal-regulated kinase; MEK: MAP/ERK kinase; MKP1: MAP kinase phosphatase 1; PDK1: 3-phosphoinositide-dependent protein kinase 1; Raf: rapidly accelerated fibrosarcoma

Antidepressants are capable of increasing peripheral BDNF levels (Zhou et al. 2017) and may also stimulate the production of neurotrophins within the CNS and improve neurotrophin signaling, which in turn, improve the recovery from major depression (Castren and Rantamaki 2008). In rodents, treatment with the antidepressant fluoxetine completely blocks the BDNF decrease induced by psychological stress and enhances the gradual increase in the expression of BDNF mRNA and protein in the brain (Li et al. 2017). Thus, there are at least some substantial hints pointing toward a role for BDNF in the context of major depression. Indeed, diverse antidepressants have been shown to activate receptor trkB and induce PLCgamma signaling pathways, at least in the mouse brain (Rantamaki et al. 2007). Major depression can cause hippocampal volume shrinkage (Mervaala et al. 2000; Xia et al. 2004), yet this volume reduction is not due to major neuronal cell loss (Muller et al. 2001), but seems to be due to structural changes of hippocampal neurons, possibly in retraction of dendritic spines and dendrites. These results thus offer further evidence for a potential role of neurotrophins in contributing to structural and physiological neuronal plasticity.

As mentioned above, an enriched environment has an impact on neuronal plasticity and can induce adult hippocampal neurogenesis in the dentate gyrus (Fabel et al. 2009) as well as an increase in dendritic spine densities in area CA1 (Rampon and Tsien 2000). Moreover, an enriched environment can increase expression of BDNF in the hippocampus (Šolvsten et al. 2018), as well as dendritic spine densities within the hippocampus (Stranahan et al. 2009). This might be seen as offering additional evidence for a critical role of BDNF in plasticity-related induced changes in dendritic spines. Correlates of altered synaptic or neuronal plasticity on the electrophysiological level are represented by LTP and LTD (see above). During the establishing phase of LTP, the expression of several genes, including BDNF, is induced (Hall et al. 2000) and it is known that de novo gene transcription and protein synthesis enables the formation of dendritic spines (Engert and Bonhoeffer 1999). Thus, LTP can lead to increased levels of BDNF and an increase in dendritic spine densities (Muller et al. 2000). This does not prove that BDNF is responsible for the increase in dendritic spine densities, but it shows that LTP is able to induce both BDNF expression and formation of dendritic spines.

Since BDNF and NGF are the best analyzed neurotrophins and animal models of BDNF and NGF signaling are mainly based on rodents with altered expression of BDNF or NGF or the receptors trkA, trkB or p75NTR, the main focus of the following paragraphs will be on these factors. NT-3 stimulates dendritic growth and leads to more complex dendritic arborization; however, as in case of BDNF, this effect seems to be driven by the pan-neurotrophin receptor p75NTR (Gascon et al. 2005). At least in the murine hippocampus, trkC deficiency does not affect dendritic spine densities (von Bohlen und Halbach et al. 2008), though it affects axonal arborization (Martinez et al. 1998; Otal et al. 2005).

5.2.1.1 Dendritic Spines and BDNF Signaling

BDNF seems to be involved in the modulation of long-lasting effects induced by neuronal plasticity and may have an impact on learning and memory. BDNF seems to contribute to neuronal plasticity since it is, for instance, selectively induced during hippocampus-dependent contextual learning (Hall et al. 2000). Likewise, after learning a spatial memory task in the Morris water maze (MWM), rodents show elevated expression of BDNF (Vaynman et al. 2004) and its receptor trkB (Gomez-Pinilla et al. 2001). Furthermore, MWM-training can lead to a transient increase in dendritic spine densities (O'Malley et al. 2000). Comparable results have been described for the beneficial effects of enriched environment. Given that BDNF/trkB signaling has an essential role in these effects on neuronal plasticity and dendritic spine formation, reduced availability of deletion of trkB should also affect neuronal plasticity and the morphology or density of dendritic spines. Conditional deletion of trkB in mice affects hippocampus-dependent learning and memory and reduces hippocampal LTP (Minichiello et al. 1999). Moreover, the trkB-deficient mice display alterations in the length and densities of hippocampal dendritic spines (von Bohlen und Halbach et al. 2006a). The mechanisms underlying the effects of BDNF on dendritic spines are not fully understood, since the actions of BDNF can be diverse and may not only be regulated by the classic signaling of BDNF via its receptors. Thus, by using specific knockout mice lacking BDNF production from exon 1, 2, 4, or 6 splice variants, it was found that loss of BDNF from different mRNA variants differentially affected dendritic complexity and spine morphology in the hippocampus (Maynard et al. 2017). These splice variants may differentially active the trkB receptors or activate them for diverging durations of times. It seems likely that BDNF could either enhance basal transmission or facilitate LTP, depending on the BDNF-TrkB signaling kinetics determined by the speed with which the brain slices are exposed to BDNF (Ji et al. 2010). Thus, low exposure to BDNF facilitates LTP, whereas acute exposure rapidly increases basal synaptic transmission (Guo et al. 2018). Moreover, transient BDNF/trkB activation promotes dendritic elongation, whereas sustained BDNF/trkB activation is associated with dendritic branching (Guo et al. 2018). Thus, trkB signaling can be regulated at various levels. These features allow the BDNF/trkB signaling pathway to contribute to the plastic functioning of dendritic spines since (i) synthesis and secretion of the short-lived molecule BDNF as well as the insertion of trkB into synapses are regulated by neuronal activity and (ii) since BDNF and TrkB mRNA are transported into dendrites in an activity-dependent manner, conferring an excellent mechanism for local modulation of synapses (Guo et al. 2018). Moreover, unlike other growth factors, whose signaling is terminated by endocytosis of the receptor–ligand complex, activity-induced BDNF-TrkB endocytosis is a critical step in BDNF/CREB-dependent gene expression (Guo et al. 2018), enabling BDNF/trkB to regulate dendritic spines and synapses in an activity-dependent manner. The BDNF mRNA localized in dendrites plays an important role in the maturation and pruning of dendritic spines. Neuronal activity stimulates both the translation of dendritic BDNF mRNA and its secretion, mainly as proBDNF. ProBDNF can subsequently promote maturation and pruning of dendritic spines, an effect that seems to be mediated by p75NTR (Orefice et al. 2016). On the other hand, somatically synthesized BDNF has a stimulatory effect on dendritic spine formation (Orefice et al. 2016). Moreover, the effects of BDNF seem not only to depend on the source and availability of BDNF, but also on the neuron itself. The effect of BDNF on the dendritic architecture of hippocampal neurons depends on the neuron's maturation stage, since in mature primary hippocampal neurons BDNF is specifically required for the activity-dependent maintenance of the mature spine phenotype (Kellner et al. 2014).

Currently, it is thought that binding of BDNF to trkB promotes dendritic spine formation and maturation via the Shc-binding site that activates Erk1/2 to control gene regulation as well of the PI-3 kinase and of the PLCγ site that promotes the release of Ca^{2+} from the internal stores (Zagrebelsky et al. 2020). Dendritic spine head enlargement seems to be regulated in a different way: binding of BDNF to trkB induces the polymerization of actin by promoting the activity of Rac1 and Cdc42 within dendritic spines. In addition, the activity-dependent activation of NMDA receptors results in an increase in local protein synthesis in a CaMKII-dependent manner, possibly also providing the BDNF required to activate trkB at dendritic spines in an autocrine way (Zagrebelsky et al. 2020). The maturation of a dendritic spine leads to an enlargement of the postsynaptic density (PSD)—this may directly depend on BDNF/trkB signaling, since signaling via trkB increases PSD-95 localization at dendritic spines by prolonging the average microtubule dwell time in spines (Hu et al. 2011). BDNF signaling may modulate the activation of actin-binding proteins, since exogenous application of BDNF promotes actin polymerization, resulting in an increase in the number of dendritic spines containing F-actin (Rex et al. 2007).

5.2.1.2 Dendritic Spines and NGF Signaling

During development, NGF and BDNF are involved in the genesis, differentiation, growth, and maintenance of specific neuronal populations in the PNS and CNS. Within the adult brain, NGF is still expressed, albeit in the soma and in the processes of specific neuronal populations.

Exogenously provided NGF enhances cognitive performance in impaired rodents and humans (Frielingsdorf et al. 2007). This NGF-dependent cognitive improvement may be associated with changes on the morphological level within cortical and/or hippocampal areas. Treatment of rats with NGF increases hippocampal cholinergic activity (Frielingsdorf et al. 2007). In this context, it is important to mention that, within the cortex and hippocampus, trkA immunoreactivity is not seen in the soma of neurons, but in fibers that mainly arise from the cholinergic system (Sobreviela et al. 1994)—somewhat comparable to the expression of p75NTR in the adult brain (see below). In the postnatal brain, NGF has positive effects on adult hippocampal neurogenesis (Frielingsdorf et al. 2007) and it is likely that the NGF-promoting effect on adult neurogenesis is mediated through the increased cholinergic tone (Frielingsdorf et al. 2007). Thus, despite the fact that NGF is one of the well-known neurotrophins with very specific properties that may hint toward a role of NGF in plastic changes in the CNS (and presumably on dendritic spines), the current available data seem not to uniformly support this view.

By looking on the distribution of high-affinity receptor for NGF, trkA, in the adult brain, it seems not surprisingly that the actions of NGF are restricted to certain brain areas and that NGF-trkA signaling may not play an important and direct role for dendritic spines in the brain areas involved in neuronal plasticity, learning, and memory, for example, the amygdala, cortex, or hippocampal formation. TrkA expression seems to be restricted to the basal forebrain, striatum, and the nucleus reuniens of the thalamus (Barbacid 1994). Comparable to the effects described for NGF in the adult brain, a knockout of trkA affects mainly the cholinergic system and the cholinergic innervation pattern of cortical brain areas (Fagan et al. 1997).

5.2.1.3 Dendritic Spines and Signaling via the Pan-Neurotrophin p75NTR Receptor

The p75NTR receptor is not only capable of binding the mature neurotrophins but is also capable of binding the proneurotrophins as well. In addition, p75NTR can interact with trk receptors as well as with sortilin. Proneurotrophin signaling via the p75NTR/sortilin receptor complex seems to mediate neuronal cell death (Nykjaer et al. 2004; Teng et al. 2005). However, the signaling mechanisms of p75NTR are even more complex and can therefore contribute to a variety of biological actions. P75NTR, among others, can also interact with ephrin A (another neurotrophic factor that is important for dendritic spines (see below)), lingo-1, or neurophilin-1 (see for details: Teng et al. (2010)). Great advances in the understanding of the roles of p75NTR in the CNS were achieved by the generation of mice deficient for p75NTR (Lee et al. 1992; von Schack et al. 2001). These mice, in contrast to homozygous BDNF or trkB knockout mice, survive into adulthood. P75NTR-deficient mice display deficits in the PNS as well as in the CNS, including changes in the basal forebrain cholinergic system accompanied by altered choline acetyltransferase activity and cholinergic target innervation (Yeo et al. 1997), leading to increases in the densities of cholinergic fibers within the hippocampus (Dokter et al. 2015; Poser et al. 2015) and cortex (von Bohlen und Halbach and von Bohlen und Halbach 2016). Since the cholinergic axonal system is increased in the cortical areas, this might affect the wiring of the brain and, among others, lead to increased synaptogenesis. Experiments using organotypic slice cultures derived from p75NTR-deficient mice at an age of 5 or 6 days postnatally (P5–P6) revealed increased dendritic spine densities of CA1 pyramidal neurons (Zagrebelsky et al. 2005). Likewise, in organotypic hippocampal cultures derived from Sprague–Dawley rats from 7 to 10 postnatal days (Chapleau et al. 2008; Chapleau and Pozzo-Miller 2012), BDNF seems to require p75NTR to increase dendritic spine density and modulate dendritic spine morphology (Chapleau and Pozzo-Miller 2012). Similarly, organotypic hippocampal cultures prepared from mice 5 (P5) to 6 days (P6) postnatally express p75NTR mRNA in the pyramidal layer within the layers CA1–CA3, comparable to the murine hippocampus at postnatal day P21 (Zagrebelsky et al. 2005). However, it should be kept in mind that P75NTR is not expressed by hippocampal or cortical neurons in the adult brain (Lee et al. 1998; Roux et al. 1999), but mainly by cholinergic fibers innervating these areas (Mrzljak and Goldman-Rakic 1993).

In adult p75NTR knockout mice, current data is somewhat conflicting as one paper reported a small but significant increase in the number of dendritic spines in the dentate gyrus (by analyzing p75NTR Exon III knockout mice (Dokter et al. 2015)), whereas another report observed no change as compared to age-matched controls (by analyzing p75NTR Exon IV knockout mice (Poser et al. 2015)). Based on the fact that p75NTR is not expressed in the adult brain, the slight effect on dendritic spines seen in the p75NTR ExonIII knockout mice might be related to systems that under normal conditions express p75NTR in adulthood and project into the hippocampal formation, for example, the cholinergic system.

5.2.2 Ephrins

Eph (erythropoietin-producing hepatocellular) receptor tyrosine kinases and their ligands are membrane-associated proteins that typically function at sites where cells make contact with one another (Henderson and Dalva 2018). The mammalian Eph receptors family can be divided into two subgroups based on their structure and their linkage to the cell membrane: the EphA receptors composed of nine members (EphA1–EphA8 and EphA10) and the EphB receptors composed of five members (EphB1–EphB4 and EphB6). EphA receptors bind most or all types of ephrinA ligands (ephrin A 1–5), whereas EphB receptors bind most or all ephrinB ligands (ephrin B 1–3). However, EphA4 can also bind to ephrin B 2 and ephrin B 3, whereas EphB2 can bind to ephrin A 5 (Hruska and Dalva 2012; Dines and Lamprecht 2016; Henderson and Dalva 2018).

Ephrins are proteins localized in the cell membrane. Ephrin A is attached to the cell membrane by a glycosylphosphatidylinositol anchor (GPI anchor), whereas ephrin B has a transmembrane domain that is followed by a short cytoplasmic region. Since ephrins and Eph receptors are both membrane-bound proteins, binding and activation of Eph/ephrin intracellular signaling pathways can only occur via direct cell–cell interactions. Interestingly, ephrins and Eph receptors can act as receptors as well as ligands, allowing unidirectional or bidirectional signaling, which in turn can lead to parallel and antiparallel signaling modes between two neighboring cells (Fig. 5.5). Moreover, Ephs and ephrins can also interact on the surface of the same cell attenuating Eph signaling, possibly by inhibiting the formation of Eph clusters (Kania and Klein 2016).

Eph receptors are expressed in various brain regions that have been shown to be involved in memory formation such as the hippocampus, amygdala, and cortex both cerebellar and cerebral (Liebl et al. 2003; Dines and Lamprecht 2016). For example, concerning the postnatal hippocampus, EphA4 has been described to be highly expressed in the hippocampal fields CA1–CA3 as well as in the dentate gyrus (Liebl et al. 2003). EphB1, EphB2, and EphB3 are expressed at different levels within the hippocampal formation, whereby the highest expression was mapped to EphB3 in area CA1 (Liebl et al. 2003). In addition, all three ephrin Bs are also expressed in hippocampus. Ephrin B1 is expressed at low levels in the hippocampal areas

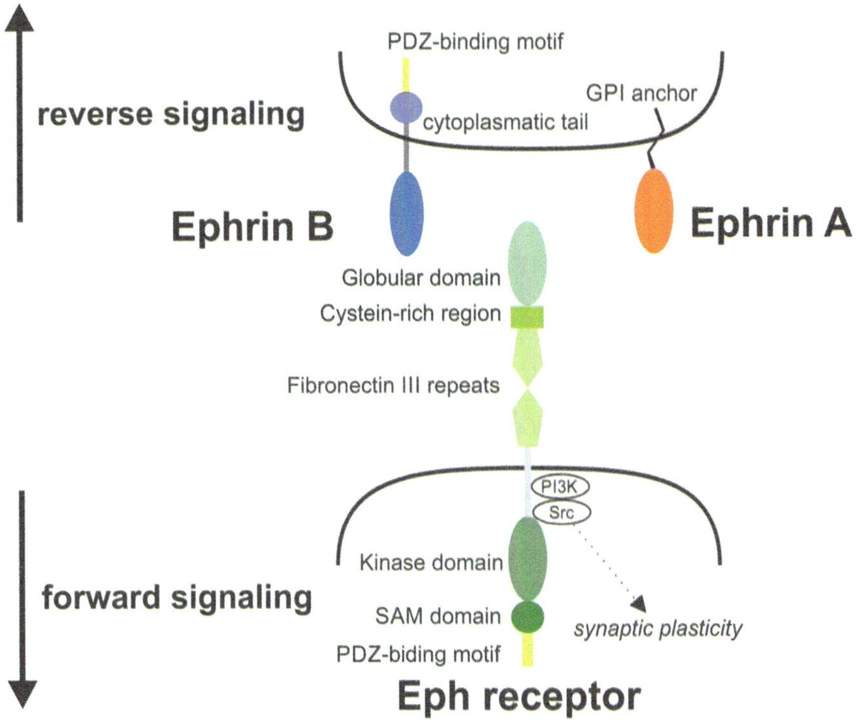

Fig. 5.5 Structure of ephrins and Eph receptors. Ephrin A and ephrin B proteins are structurally different. Ephrin A binds the cell membrane by a glycosylphosphatidylinositol anchor (GPI anchor), and ephrin B contains a transmembrane domain that is followed by a short cytoplasmic domain. Activation of Eph receptors induces multiple different pathways, which among others lead to altered synaptic plasticity by activation of pathways involving phosphoinositol-3-kinase (PI3K) or effector molecules such as Src. In addition, ephrin/Eph interaction allows the possibility for bidirectional signaling. (Adapted from Goldshmit et al. (2006) and Dines and Lamprecht (2016))

CA1–CA3 and the dentate gyrus, while ephrin B2 is enriched in the CA1 region of the hippocampus. Ephrin B3 is highly expressed in the dentate gyrus and in the hippocampal CA1 region (Liebl et al. 2003; Henderson and Dalva 2018).

Based on their expression in the postnatal brain, it is not surprising that ephrins not only play crucial roles during development of the CNS, but also have specific effects in the postnatal brain. Both EphA and EphB receptors play prominent roles in the formation and function of excitatory synapses and participate in activity-induced changes in synaptic strength, such as LTP (Klein 2009). For example, EphB2 mediates long-lasting LTP and LTD in a kinase-independent fashion (Grunwald et al. 2001). Moreover, EphA4 was shown to be required for the early stages of LTP at the CA3–CA1 synapse in a kinase-independent fashion (Grunwald et al. 2004). Along this line, EphA4 receptor is also required for synaptic plasticity in the amygdala (Deininger et al. 2008) and ephrin A4-binding sites in the lateral nucleus of the amygdala are essential for long-term fear memory formation (Dines and Lamprecht 2014). EphBs, for example, are expressed and localized to pre- and postsynaptic terminals in the postnatal brain. When ephrin Bs are localized presynaptically, they can serve as ligands for postsynaptic EphBs and mediate development and plasticity of presynaptic terminals, and when ephrin Bs are localized postsynaptically they are involved in plasticity, learning, and memory by linking neuronal activity and synapse formation (Henderson and Dalva 2018).

5.2.2.1 Ephrin A Receptors (EphA)

EphA receptors are important regulators of dendritic spine morphology in the hippocampus (Murai et al. 2003). However, as compared to EphB receptors (see below), the roles of EphA ligands in synapse formation and plasticity are less well understood. The development of synapses and their function is also regulated by the communication of neurons and glia (Eroglu and Barres 2010). EphA3-expressing hippocampal astrocytes might regulate synapse function through ephrin-mediated interactions (Murai et al. 2003). EphA3-expressing astrocytes not only regulate morphogenesis of dendritic spines, but they are also involved in glutamate transport and hippocampal LTP interacting with EphA4 receptors on dendrites of CA1 pyramidal neurons (Carmona et al. 2009; Filosa et al. 2009). It is assumed that astrocytes receive a signal from dendritic EphA4 receptors through ephrin A3 at their membrane, which prevents them from upregulating glial glutamate transporter expression and thus regulating the glutamate concentration at the synapse (Filosa et al. 2009). Along this line, it has been shown that EphA4-mediated ephrin A3 reverse signaling in astrocytes controls glial glutamate transporters and thereby protects rat hippocampal neurons from glutamate excitotoxicity, for example, seen under ischemic conditions (Yang et al. 2014). Altogether, EphA activation in neurons and in glial cells seems to be involved in the control of synaptic maturation, function, and plasticity.

5.2.2.2 Ephrin B Receptors (EphB)

EphBs are involved in the regulation of presynaptic development and are required for dendritic filopodia motility, spine synapse formation, and for the recruitment and clustering of glutamate receptors to synapses (Kayser et al. 2008; Hruska and Dalva 2012) and synapse density through diverse molecular mechanisms that involve both ephrin B signaling and protein–protein interactions (Klein 2009; Hruska and Dalva 2012). All the different EphB receptor tyrosine kinases (EphB1, EphB2, and EphB3) are important for dendritic spine development and all of these receptors, although to varying degrees, are involved in dendritic spine morphogenesis and synapse formation in the hippocampus. Hippocampal neurons lacking EphB expression fail to form dendritic spines in vitro and they develop abnormal dendritic spines in vivo (Henkemeyer et al. 2003). By using cultured hippocampal neurons, it was further demonstrated that ephrin Bs regulate maturation of dendritic spines and synapses by interacting with presynaptic EphB receptors, and ephrin Bs seem to control the transition from the filopodia to dendritic spines by recruiting G-protein-coupled receptor kinase-interacting protein 1 (GIT1) to sites of ephrin B signaling (Segura et al. 2007; Hruska and Dalva 2012). In developing dendrites of mouse hippocampal neurons, ephrin B3 seems to act postsynaptically through ephrin B reverse signaling to transduce reverse signals by acting through three distinct pathways that mediate the maturation of postsynaptic neurons: (i) tyrosine phosphorylation-dependent GRB4 SH2/SH3 adaptor-mediated signals and (ii) postsynaptic density-95/Discs large/zona occludens-1 (PDZ) domain-dependent signals are required for inhibition of dendrite branching, whereas (iii) PDZ interactions are necessary for dendritic spine formation and excitatory synaptic function (Xu et al. 2011). The regulation of dendritic spine morphology is mediated by the actin cytoskeleton, which is enriched in dendritic spines and Ephs control the signaling molecules in dendritic spines that are key regulators of the actin cytoskeleton (Dines and Lamprecht 2014). Thus, filopodial motility and synapse formation need EphB activation of PAK, a serine/threonine kinase, which regulates actin dynamics (Kayser et al. 2008). EphBs also signal to phosphorylate guanine exchange factors, which catalyze the Rho-family GTPases Rac1 and Cdc42 into the active state. These GTPases are needed for actin cytoskeleton reorganization and dendritic spine morphogenesis (Irie and Yamaguchi 2002; Penzes et al. 2003; Tolias et al. 2007).

In addition, EphB2 has been identified to control AMPA-type glutamate receptor localization through PDZ-binding domain interactions and to trigger presynaptic differentiation via its ephrin-binding domain (Kayser et al. 2008). Synapse development as well as function is, as mentioned above, regulated by neuron-glia communication.

Altering the availability of membrane-bound ephrin B1 in astrocytes differentially regulates development of excitatory and inhibitory circuits in the hippocampus during early postnatal development (Nguyen et al. 2020b) and high levels of astrocytic ephrinB1 resulted in reduced formation of new dendritic spines rather than in dendritic spine maturation in CA1 hippocampal neurons (Nguyen et al. 2020a). Thus, ephrins play important roles in the formation, stabilization, and maturation of synapses. A specific feature of the ephrins is their role as a "bridge" between neurons and glia cells. However, this "bridge" is not static but highly dynamic, allowing proper functioning of existent synapses while also contributing to synaptic plasticity.

5.2.3 *Epidermal Growth Factor Family*

Epidermal growth factor (EGF) is the founding member of the EGF family. Members of EGF family have highly structural and functional similarities. Family members of the EFG family include among others: amphiregulin, betacellulin, epigen, epiregulin, and heparin-binding EGF-like growth factor (HB-EGF) (Singh et al. 2016) as well as neuregulins (Ou et al. 2021) and CALEB (Schumacher et al. 1997). Some of the EGF family members seem to affect dendritic spine formation and maturation.

CALEB (chicken acidic leucine-rich EGF-like domain containing brain protein), for example, is highly expressed in the brain and especially upregulated in times of dendritic differentiation (Schumacher et al. 1997). Overexpression of CALEB enhances dendritic branching and increases the complexity of dendritic spines and filopodia, whereas inactivation of CALEB impairs dendritic branching and dendritic spine formation. In particular, the EGF-like domain of CALEB drives both dendritic branching and spine morphogenesis. The phosphatidylinositide 3-kinase (PI3K)/Akt/mammalian target of rapamycin (mTOR) signaling pathway and protein kinase C (PKC) are important for CALEB-induced stimulation of dendritic branching, whereas CALEB induced dendritic spine morphogenesis is independent of PI3K, but depends on PKC (Brandt et al. 2007). Thus, CALEB can act as a mediator of dendritic spine formation and dendritic arborization.

Aside from CALEB, neuregulin-1, which can bind to and signal through ErbB4 receptors, has an impact upon dendritic spines. Thus, it has been shown that synaptic activity leads to the activation and recruitment of ErbB4 into the synapse and that overexpressed ErbB4 selectively enhances AMPA synaptic currents and increases dendritic spine size. To the contrary, blocking NRG1/ErbB4 signaling destabilizes synaptic AMPA receptors and leads to loss of synaptic NMDA currents and dendritic spines (Li et al. 2007). These results indicate that ErbB4 signaling may play an important role in glutamatergic synapse formation. These results are strengthened by data showing that inactivation of ErbB signaling exclusively in the CNS impairs dendritic spine maturation and perturbs interactions of postsynaptic scaffold proteins with glutamate receptors. These morphological changes were accompanied by altered behavior (increased aggression) in these knockout mice (Barros et al. 2009). In another study was shown that long-term incubation of cortical pyramidal neurons with NRG1 increases dendritic spine size and that this effect was mediated by the Rac-GEF kalirin (Cahill et al. 2013). Somewhat comparable to this, it was demonstrated that ErbB4 conditional whole-brain knockout mice display reduced dendritic spine densities in the dorsomedial prefrontal cortex and that selective ablation of ErbB4 from excitatory neurons leads to a decrease in the proportion of mature dendritic spines and an overall reduction in dendritic spine densities in the prefrontal cortex (Cooper and Koleske 2014).

5.2.4 *Fibroblast Growth Factors*

The members of the fibroblast growth factor (FGF) family can signal through different receptors. Four FGF receptors (FGFR 1–4) represent receptor tyrosine kinases, and a fifth FGFR, FGFRL1, lacks the tyrosine kinase domain and is thought to represent a coreceptor for FGFR1 (Regeenes et al. 2018). The FGF family comprises of more than 20 family members that are highly conserved in gene structure and amino acid sequence (Reuss and von Bohlen und Halbach 2003), but not all of them are expressed in the adult brain (Klimaschewski and Claus 2021). FGFs are especially necessary during brain development and are involved in neuroprotection and lesion repair in the adult brain. Dysregulation of the FGF system in major depression involves the FGF receptors FGFR2 and FGFR3, as well as several FGFs, for example, FGF-1, FGF-2, FGF-9, and FGF-12 (Evans et al. 2004). In addition, there is evidence to suggest that the FGF system might play a role in the action of antidepressants (Maragnoli et al. 2004). FGFs seemed to play roles in mechanisms related to learning and memory and different members of the FGF family are capable of altering LTP-induction and/or maintenance of LTP. Aside from FGF-21 (Sa-Nguanmoo et al. 2016) and FGF-22 (Hu et al. 2016), FGF-2 seems to play an important role in the plastic development of dendritic spines and the maintenance of dendritic spines in the postnatal brain. With regard to the formation of dendritic spines, endogenous FGF-2 seems not to be essential for spinogenesis within the hippocampus, but lack of FGF-2 affects the length of individual hippocampal dendritic spines (Zechel et al. 2009). Within the postnatal cortex, FGF-2 seems to play a different role. Within the motor sensory cortex of FGF-2-deficient mice, dendritic spine densities were increased and the length of dendritic spines was reduced and these changes were accompanied by a downregulation of Arhgef6 (a gene that in its mutated form can cause X-linked intellectual disability) mRNA and protein (Baum et al. 2016). In animal models of depression, FGF-2 applied intraventricularly generated antidepressant-like effects (Jarosik et al. 2011). Moreover, different substances, for example, liquiritin (Chen et al. 2020) or peoniflorin (Cheng et al. 2021) produce antidepressant-like effect in rodents, accompanied by increases in hippocampal FGF-2 levels and in densities of hippocampal dendritic spines. However, the importance of FGF-2 for dendritic spines is far from being completely understood in detail. Likewise, whether other members of the FGF family have an impact on dendritic spines, in relation to maintenance or plasticity is not completely understood at the moment.

5.2.5 Ghrelin and Insulin

Hormones that have initially been identified in the periphery and are known for their effects in regulating food intake and appetite have also been shown to regulate different brain functions. One might expect that these effects are restricted to the hypothalamus, but it turned out that receptors for insulin and ghrelin are expressed in a variety of brain areas. Insulin as well as ghrelin contribute to a variety of different functions in the brain and, among others, are able to modulate learning and memory (Carlini et al. 2004).

Ghrelin administration is capable of increasing dendritic spine densities within the hippocampal area CA1, whereas ghrelin-knockout mice display impaired novel object memory as well reduced dendritic spine densities. The latter effect could partially be restored by ghrelin administration (Diano et al. 2006). Likewise, it has been shown that ghrelin promotes the reorganization of dendritic spines in cultured rat hippocampal slices (Berrout and Isokawa 2012). Moreover, ghrelin application has been found to increase mainly the densities of mushroom-type spines on secondary and tertiary extensions in hippocampal pyramidal neurons; an effect that was accompanied by an increase in the expression of BDNF mRNA (Perea Vega et al. 2021). In contrast to the effects of ghrelin-deficiency on dendritic spines of the pyramidal CA1 neurons, no effect on dendritic spine densities was observed in the dentate gyrus of ghrelin-deficient mice (Cahill et al. 2014). Nevertheless, ghrelin has an impact on neuronal plasticity in the dentate gyrus, not on the level of alterations in dendritic spines, but on adult hippocampal neurogenesis (Moon et al. 2009; Li et al. 2013). Thus, the effects of ghrelin on dendritic spine formation seem to be restricted to the pyramidal cells and it might be possible that these effects are mediated or accompanied by changes in BDNF levels.

Insulin is thought to represent a modulator of hippocampal function, since in contrast to the majority of brain regions, the hippocampus expresses both insulin receptors as well as insulin-regulated glucose transporters GluT4 and GluTX (McNay 2007). Interestingly, the capability of insulin to enhance memory appears to be restricted to the hippocampus, as administration to other memory-processing areas (like the amygdala and the striatum) did neither alter memory performance nor affect motor activity or anxiolysis, for instance (McNay 2007). Insulin promotes dendritic spine formation in primary cultures of rat hippocampal neurons. In these cultures, the increase in dendritic spine densities involves the phosphatidylinositol 3-kinase (PI3K)/Akt/mTOR signaling pathway and the GTPase Rac1, since pharmacological blockade of them prevented the insulin-induced increase in dendritic spine densities (Lee et al. 2011).

5.2.6 Glial Cell Line-Derived Neurotrophic Factor

The members of the glial cell line-derived neurotrophic factors (GDNF) family of ligands (GFL) consist of GDNF, neurturin, artemin, and persephin. GFLs belong to the transforming growth factor-β (TGF-β) superfamily, containing seven cysteine residues with the same relative spacing as other members of this family. All GFLs can signal through Ret (receptor tyrosine kinase). Ret is only activated when the GFL first binds to GFRα (GDNF-family receptor α) receptors, which are linked to the plasma membrane by a GPI anchor. Several GFRα receptors have been identified and it has been shown that GDNF binds to GFRα1, neurturin to GFRα2, artemin to GFRα3, and persephin to GFRα4 (see for more details: Airaksinen and Saarma (2002)). Neurturin and especially GDNF are important survival factors for midbrain dopaminergic neurons. Since GDNF was found to prevent neurotoxin-induced death of dopaminergic neurons and to promote functional recovery in various animal models of Parkinson's disease (PD), GDNF was thought to represent a promising candidate in the treatment of PD. Unfortunately, different series of clinical trials in PD patients have failed so far (Kordower et al. 1999; Barker et al. 2020).

Although GDNF and its receptor GFRα1 are expressed in the forebrain, the role of this system in the postnatal forebrain is largely unclear. Both GDNF and its cognate receptor are highly expressed in the postnatal hippocampus (Bonafina et al. 2019). Running seems to increase the expression of GDNF and promotes GFRα1-dependent CREB (cAMP response element-binding protein) activation and maturation of dendrites (Bonafina et al. 2019). This hints for a role of GDNF in neuronal plasticity. However, it should be kept in mind that aside from modulations in dendritic spines, adult hippocampal neurogenesis—which takes place in the postnatal dentate gyrus—is a further morphological hallmark of neuronal plasticity. Indeed, Bonafina and coworkers showed in 2019 that GDNF, acting through its GFRα1 receptor, controls dendritic structure and dendritic spine density of adult-born granule cells, which reveals that GFRα1 is required for their integration into preexisting circuits. In addition, GFRα1 is expressed on dendrites and dendritic spines of CA1 and CA3 hippocampal neurons. Irala and coworkers demonstrated that downregulation of GFRα1 reduces hippocampal dendrite complexity in vivo. Moreover, the authors could show that in the presence of GDNF, GFRα1 promotes dendritic spine formation and postsynaptic differentiation of dissociated hippocampal neurons. Furthermore, they showed that Ret is not detectable in the hippocampus and instead of signaling through Ret, it is thought that the effects are mediated through NCAM180, that is located mainly at postsynaptic sites within the hippocampus (Irala et al. 2016).

When patients suffering from PD were treated for influenza with amantadine, unexpected motor symptom improvements were observed (Schwab et al. 1969). Nowadays, amantadine is commonly used in combination with levodopamine (L-DOPA) to reduce the motor disorders of PD patients (Rascol et al. 2021) and there is evidence that that amantadine may also exert neuroprotective effects in several neurological disorders. Concerning PD, it has been shown that amantadine can protect dopaminergic neurons by reducing the release of proinflammatory factors and by increasing the expression of GNDF in astroglia (Ossola et al. 2011). Moreover, amantadine has been shown to attenuate postoperative learning and memory dysfunction. In a 2020 study, it could be shown that surgery and anesthesia impair learning and memory in aged rodents, accompanied by reduced expression of BDNF and GDNF and a decline in dendritic spine density in the hippocampus. These effects were attenuated by amantadine application (Zhong et al. 2020). The results of that study indicate that GDNF alone or in combination with BDNF might be beneficial in attenuating postoperative learning and memory dysfunctions and reductions in dendritic spine densities in the hippocampus. Indeed, a 2022 study showed that these described postoperative effects could be attenuated by GDNF injection (Xie et al. 2022). Thus, exogenously applied GDNF might play a role in repair mechanisms on the level of dendritic spines.

Whether or not the other members of the GDNF family have an impact on dendritic spines is currently not investigated in detail.

5.2.7 *Insulin-Like Growth Factor*

Within the cerebellum, insulin-like growth factor 1 (IGF-1) as well as IGF-1 receptors are expressed. In 1997, Nieto-Bona and colleagues showed that decreasing IGF-1 levels in the cerebellum of rats resulted in a reduction in the size of dendritic spines on Purkinje cells accompanied by a reduction in the numerical density of dendritic spines (Nieto-Bona et al. 1997). In subsequent years, researchers accumulated evidence for IGF-1 effects on dendritic spines which are not restricted to Purkinje cells in the cerebellum. For instance, IGF-1-deficient mice display abnormalities in the composition of dendrites in the frontoparietal cortex, including a reduction in dendritic spine densities (Cheng et al. 2003). Aside from IGF-1, IGF-2 also seems to play a role in neuronal plasticity. As a result of an overexpression of IGF-2 in the hippocampus, an increase in dendritic spine formation could be observed accompanied by improved hippocampus-dependent memory. Furthermore, increasing IGF-2 or IGF-1 levels in the hippocampus of APP mice (an animal model of Alzheimer's disease (AD)) rescued behavioral deficits and promoted dendritic spine formation accompanied by a significant reduction in amyloid levels (Pascual-Lucas et al. 2014). IGFs seemed to have beneficial effects on dendritic spines not only in neurodegenerative disorders, like AD, but also in neurodevelopmental disorders. For example, there are reports that dendritic spine instability is rescued by IGF-1 in a mouse model of the CDKL5 (cyclin-dependent kinase-like 5) disorder (Della Sala et al. 2016). A more well-known neurodevelopmental disorder is the Rett-Syndrome which is linked to mutations in the X-linked gene methyl-CpG-binding protein 2 (MeCP2). Mecp2-deficient mice have reduced IGF-1 levels; treatment with IGF1 restored dendritic spine densities (Castro et al. 2014). Taken together, these reports seem to point toward a potential therapeutic role of IGFs in treating a host of neurodegenerative and neurodevelopmental disorders linked to dendritic spine abnormalities.

5.2.8 *Leptin*

Leptin is a product of the obese (ob) gene and, following synthesis and secretion, can bind to its cognate receptor, the leptin receptor (LEP-R). The hormone leptin is mainly known for its regulatory effects on food intake and body mass. Leptin regulates appetite and metabolism by inhibiting the synthesis and release of neuropeptide Y in the arcuate nucleus of the hypothalamus. Leptin can inhibit neural pathways activated by appetite stimulants (orexigenic) to reduce energy intake and activate pathways targeted by anorexigenic to suppress appetite (Obradovic et al. 2021). Thus, leptin deficiency (due to mutations in the ob gene), can be causal for obesity (see, e.g., Finger et al. (2010)). The ob/ob mice model, first described by Ingalls and colleagues in 1950, represents one well-established animal model of obesity (Ingalls

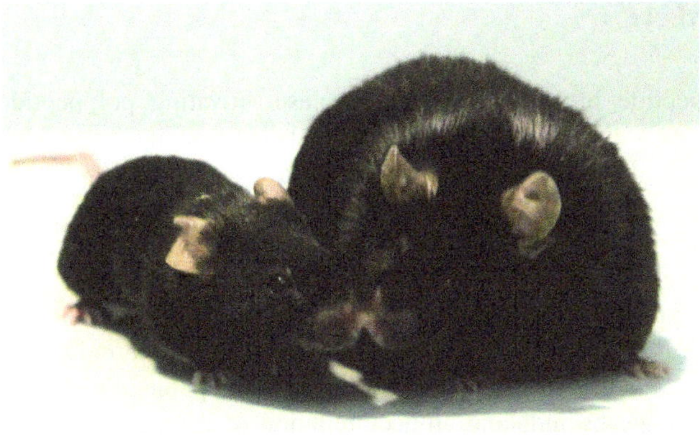

Fig. 5.6 An adult leptin-deficient (ob/ob) mouse (right) in comparison to an age-matched control mouse (left). (Figure reused from: Bracke et al. (2019), with permission from Sage Publishing, Right and Permission RP-7633)

et al. 1950). During the first 4 month after birth weight nearly doubles in ob/ob mice (Bracke et al. 2019) and the mice get obese (Fig. 5.6).

Moreover, the ob/ob mice display deficits in hippocampal synaptic plasticity. Hippocampal LTP is completely abolished in the hippocampal area CA1 of ob/ob mice (Porter et al. 2013); adult hippocampal neurogenesis is impaired but dendritic spine densities are not altered in the hippocampus (Bracke et al. 2019). LTP in the amygdala is, however, increased, but no major effect on dendritic spine densities has been observed (Schepers et al. 2020). In contrast to this, mice deficient for the leptin receptor (db/db mice) display reduced number of hippocampal dendritic spines in area CA1 as well as in area CA3 (Dhar et al. 2014), whereas hippocampal LTP was found to be impaired, but not abolished, in db/db mice (Li et al. 2002).

As the ob/ob mice represent an animal model of obesity and the db/db mice a model of diabetes, it seems possible that the effects on dendritic spines are not directly a consequence of reduced availability of leptin or its receptor but a consequence of getting obese or diabetic—this seems particularly possible considering that high-fat diets lead to dendritic remodeling in the amygdala and hippocampus (Janthakhin et al. 2017) and to changes in the dendritic spine densities in the medial prefrontal cortex (Dingess et al. 2017), a part of the brain reward system. Likewise, it has been shown that prolonged consumption of sucrose in a binge-like manner affects dendritic complexity and dendritic spine densities in the nucleus accumbens (Klenowski et al. 2016), a key structure in the processing of reward. Therefore, it might be possible that the effects on dendritic spines seen in the ob/ob and db/db mice are not directly related to leptin and its receptor, but to motivational changes in the reward system and the limbic system.

5.2.9 PACAP

The neuropeptide pituitary adenylate cyclase-activating polypeptide (PACAP), which is not only expressed in the pituitary, but also in many brain areas (e.g., hypothalamus, amygdala, and hippocampus), was discovered in 1989. Two substances were isolated that exhibit adenylate cyclase-activating properties; one protein consisting of 38 amino acids (PACAP38) while the other represents a 27 amino acid long fragment of the former. PACAP not only acts as a neurotrophic factor but also as a neuromodulator in the brain (for details see, e.g., Dermietzel and von Bohlen und Halbach (2006)). PACAP38 has been implicated in the induction of synaptic plasticity at excitatory glutamatergic synapse and is capable of modulating dendritic spine morphology by inducing an accumulation of ADAM10 at the postsynaptic membrane, at least in primary hippocampal neurons (Marcello et al. 2012). Moreover, it has been reported that PACAP modifies the density and morphology of PSD-95-positive dendritic spines in primary cultured hippocampal neurons. Likewise, the number of PSD-95-positive dendritic spines were found to be reduced in PACAP-deficient (PACAP (–/–)) mice. Golgi staining of hippocampal CA1 neurons revealed reduced dendritic spine densities and atypical morphologies in male PACAP (–/–) mice (Hayata-Takano et al. 2019). However, whether PACAP is necessary for dendritic spines is far from being completely understood. In the context of neuronal plasticity, PACAP seems to be capable of specific modulations. Genetic deletion of PACAP in animals produces mild to severe impairments in certain forms of hippocampus-dependent learning, pointing toward a potential PACAP role in learning, memory, and different forms of behavior, including aversive memory or its extinction (Gilmartin and Ferrara 2021). Furthermore, hippocampal LTP is impaired in mice lacking PACAP (Matsuyama et al. 2003). Whether these effects are related to alterations in adult neurogenesis in the dentate gyrus or to changes in the morphology of individual mature neurons and their connections is still a subject of scientific debate.

5.2.10 TGF-ß Superfamily

The transforming growth factor-ß (TGF-ß) superfamily is a large family of related growth and differentiation factors, including transforming growth factor-ß as well as bone morphogenetic proteins (BMPs), growth/differentiation factors (GDFs), inhibin, and activin (Wrana 1998; Unsicker and Krieglstein 2002; Hinck 2012). Concerning the latter, it is known that activin is expressed in the postnatal hippocampus (Hasegawa et al. 2014) where it plays a role in the maintenance of long-term memory and in late phases of hippocampal LTP (Ageta et al. 2010). In a 2014 study, Hasegawa and colleagues showed that incubation of hippocampal acute slices with activin for 2 h is capable of altering the density and morphology of dendritic spines in CA1 pyramidal neurons—dendritic spine densities increased by 1.2-fold upon activin treatments, to give but one example (Hasegawa et al. 2014).

Using cultured hippocampal neurons, researchers proved that activin increases the average number of presynaptic contacts on dendritic spines and induces dendritic spine lengthening resulting in longer mushroom-shaped spines (Shoji-Kasai et al. 2007). However, the number of dendritic spines and their head size were not significantly affected by activin treatment in these cultured neurons. The activin-mediated effects on dendritic spine length and on contact formation could be blocked by inhibition of cytoskeletal actin dynamics or of the mitogen-activated protein (MAP) kinase pathway (Shoji-Kasai et al. 2007). These data seem to indicate that activin plays a very dynamic role in the regulation of neuronal plasticity, both on the physiological as well as on the morphological level. Activin may represent one of the neurotrophic factors that is not involved in the maintenance of dendritic spines, but in the dynamic regulation of dendritic spine morphology and function in relation to neuronal plasticity, learning, and memory. It would be interesting to see whether activin-deficiency in the brain might inhibit dynamic dendritic spine changes and affect certain forms of learning and memory.

5.3 Dendritic Spines and Neurotrophic Factors

Dendritic spines can serve as multifunctional integrative units and they are capable of adapting rapidly to new demands. Dendritic spines can undergo plastic changes in shape and size; they appear and disappear over time. Interestingly enough, some dendritic spines change over time, while other dendritic spines remain stable. Thus, factors are needed that control the maintenance of the required and available connections. In addition, factors are required that help to adapt dendritic spines to new demands. Neurotrophic factors have been shown to play fundamental roles in the maintenance of preexisting connections, in the spatial modulation of single dendritic spines, as well as in the generation of new and functional new spines that integrate into the network. For that purpose, regulators are necessary to fine-tune the spines' dynamics. The various neurotrophic factors that are still expressed in the postnatal brain may have such regulatory functions; however, many more regulators of dendritic spines have been identified—and an unknown number of further regulators remain potentially to be identified. In addition, many questions remain regarding the cytoskeletal basis of dendritic spine formation, as well as the growth, retraction, and stabilization of these tiny structures. It seems likely that the above-mentioned diverse effects of various neurotrophic factors on dendritic spines form only a partial picture of the entire neuronal complexity. Further studies will hopefully help to gather further insights. Moreover, although a number of single players involved in these processes have been identified, it remains still unknown how these different regulators (e.g., neurotrophic factors) act together either (i) to preserve the shape and functioning of a dendritic spine or (ii) to alter the shape and efficacy of a dendritic spine in response to altered requirements. To understand both processes in more detail may help to unravel the similarities and differences of memory preservation and the capacity of learning on the level of dendritic spines.

References

Ageta H, Ikegami S, Miura M, Masuda M, Migishima R, Hino T, Takashima N, Murayama A, Sugino H, Setou M, Kida S, Yokoyama M, Hasegawa Y, Tsuchida K, Aosaki T, Inokuchi K (2010) Activin plays a key role in the maintenance of long-term memory and late-LTP. Learn Mem 17:176–185

Airaksinen MS, Saarma M (2002) The GDNF family: signalling, biological functions and therapeutic value. Nat Rev Neurosci 3:383–394

Barbacid M (1994) The Trk family of neurotrophin receptors. J Neurobiol 25:1386–1403

Barde YA, Edgar D, Thoenen H (1982) Purification of a new neurotrophic factor from mammalian brain. EMBO J 1:549–553

Barker PA (1998) p75NTR: a study in contrasts. Cell Death Differ 5:346–356

Barker RA, Bjorklund A, Gash DM, Whone A, Van Laar A, Kordower JH, Bankiewicz K, Kieburtz K, Saarma M, Booms S, Huttunen HJ, Kells AP, Fiandaca MS, Stoessl AJ, Eidelberg D, Federoff H, Voutilainen MH, Dexter DT, Eberling J, Brundin P, Isaacs L, Mursaleen L, Bresolin E, Carroll C, Coles A, Fiske B, Matthews H, Lungu C, Wyse RK, Stott S, Lang AE (2020) GDNF and Parkinson's disease: where next? A summary from a recent workshop. J Parkinsons Dis 10:875–891

Barros CS, Calabrese B, Chamero P, Roberts AJ, Korzus E, Lloyd K, Stowers L, Mayford M, Halpain S, Muller U (2009) Impaired maturation of dendritic spines without disorganization of cortical cell layers in mice lacking NRG1/ErbB signaling in the central nervous system. Proc Natl Acad Sci U S A 106:4507–4512

Baum P, Vogt MA, Gass P, Unsicker K, von Bohlen und Halbach O (2016) FGF-2 deficiency causes dysregulation of Arhgef6 and downstream targets in the cerebral cortex accompanied by altered neurite outgrowth and dendritic spine morphology. Int J Dev Neurosci 50:55–64

Berrout L, Isokawa M (2012) Ghrelin promotes reorganization of dendritic spines in cultured rat hippocampal slices. Neurosci Lett 516:280–284

Bliss TV, Collingridge GL (1993) A synaptic model of memory: long-term potentiation in the hippocampus. Nature 361:31–39

Bliss TV, Lomo T (1973) Long-lasting potentiation of synaptic transmission in the dentate area of the anaesthetized rabbit following stimulation of the perforant path. J Physiol 232:331–356

Bonafina A, Trinchero MF, Rios AS, Bekinschtein P, Schinder AF, Paratcha G, Ledda F (2019) GDNF and GFRalpha1 are required for proper integration of adult-born hippocampal neurons. Cell Rep 29:4308–4319.e4304

Bourne JN, Harris KM (2008) Balancing structure and function at hippocampal dendritic spines. Annu Rev Neurosci 31:47–67

Bracke A, Domanska G, Bracke K, Harzsch S, van den Brandt J, Broker B, von Bohlen und Halbach O (2019) Obesity impairs mobility and adult hippocampal neurogenesis. J Exp Neurosci 13:1179069519883588

Brandt N, Franke K, Rasin MR, Baumgart J, Vogt J, Khrulev S, Hassel B, Pohl EE, Sestan N, Nitsch R, Schumacher S (2007) The neural EGF family member CALEB/NGC mediates dendritic tree and spine complexity. EMBO J 26:2371–2386

Bruel-Jungerman E, Laroche S, Rampon C (2005) New neurons in the dentate gyrus are involved in the expression of enhanced long-term memory following environmental enrichment. Eur J Neurosci 21:513–521

Cahill ME, Remmers C, Jones KA, Xie Z, Sweet RA, Penzes P (2013) Neuregulin1 signaling promotes dendritic spine growth through kalirin. J Neurochem 126:625–635

Cahill SP, Hatchard T, Abizaid A, Holahan MR (2014) An examination of early neural and cognitive alterations in hippocampal-spatial function of ghrelin receptor-deficient rats. Behav Brain Res 264:105–115

Carlini VP, Varas MM, Cragnolini AB, Schioth HB, Scimonelli TN, de Barioglio SR (2004) Differential role of the hippocampus, amygdala, and dorsal raphe nucleus in regulating feeding, memory, and anxiety-like behavioral responses to ghrelin. Biochem Biophys Res Commun 313:635–641

Carmona MA, Murai KK, Wang L, Roberts AJ, Pasquale EB (2009) Glial ephrin-A3 regulates hippocampal dendritic spine morphology and glutamate transport. Proc Natl Acad Sci U S A 106:12524–12529

Castren E, Rantamaki T (2008) Neurotrophins in depression and antidepressant effects. Novartis Found Symp 289:43–52; discussion 53–49, 87–93

Castro J, Garcia RI, Kwok S, Banerjee A, Petravicz J, Woodson J, Mellios N, Tropea D, Sur M (2014) Functional recovery with recombinant human IGF1 treatment in a mouse model of Rett syndrome. Proc Natl Acad Sci U S A 111:9941–9946

Chapleau CA, Pozzo-Miller L (2012) Divergent roles of p75(NTR) and Trk receptors in BDNF's effects on dendritic spine density and morphology. Neural Plast 2012:578057

Chapleau CA, Carlo ME, Larimore JL, Pozzo-Miller L (2008) The actions of BDNF on dendritic spine density and morphology in organotypic slice cultures depend on the presence of serum in culture media. J Neurosci Methods 169:182–190

Chen M, Zhang QP, Zhu JX, Cheng J, Liu Q, Xu GH, Li CF, Yi LT (2020) Involvement of FGF-2 modulation in the antidepressant-like effects of liquiritin in mice. Eur J Pharmacol 881:173297

Cheng CM, Mervis RF, Niu SL, Salem N Jr, Witters LA, Tseng V, Reinhardt R, Bondy CA (2003) Insulin-like growth factor 1 is essential for normal dendritic growth. J Neurosci Res 73:1–9

Cheng J, Chen M, Wan HQ, Chen XQ, Li CF, Zhu JX, Liu Q, Xu GH, Yi LT (2021) Paeoniflorin exerts antidepressant-like effects through enhancing neuronal FGF-2 by microglial inactivation. J Ethnopharmacol 274:114046

Cohen S, Levi-Montalcini R, Hamburger V (1954) A nerve growth-stimulating factor isolated from sarcom as 37 and 180. Proc Natl Acad Sci U S A 40:1014–1018

Conover JC, Yancopoulos GD (1997) Neurotrophin regulation of the developing nervous system: analyses of knockout mice. Rev Neurosci 8:13–27

Cooper MA, Koleske AJ (2014) Ablation of ErbB4 from excitatory neurons leads to reduced dendritic spine density in mouse prefrontal cortex. J Comp Neurol 522:3351–3362

Deininger K, Eder M, Kramer ER, Zieglgansberger W, Dodt HU, Dornmair K, Colicelli J, Klein R (2008) The Rab5 guanylate exchange factor Rin1 regulates endocytosis of the EphA4 receptor in mature excitatory neurons. Proc Natl Acad Sci U S A 105:12539–12544

Della Sala G, Putignano E, Chelini G, Melani R, Calcagno E, Michele Ratto G, Amendola E, Gross CT, Giustetto M, Pizzorusso T (2016) Dendritic spine instability in a mouse model of CDKL5 disorder is rescued by insulin-like growth factor 1. Biol Psychiatry 80:302–311

Dermietzel R, von Bohlen und Halbach O (2006) Neurotransmitters and neuromodulators. Wiley-VCH, Weinheim

Dhar M, Zhu M, Impey S, Lambert TJ, Bland T, Karatsoreos IN, Nakazawa T, Appleyard SM, Wayman GA (2014) Leptin induces hippocampal synaptogenesis via CREB-regulated microRNA-132 suppression of p250GAP. Mol Endocrinol 28:1073–1087

Diano S, Farr SA, Benoit SC, McNay EC, da Silva I, Horvath B, Gaskin FS, Nonaka N, Jaeger LB, Banks WA, Morley JE, Pinto S, Sherwin RS, Xu L, Yamada KA, Sleeman MW, Tschop MH, Horvath TL (2006) Ghrelin controls hippocampal spine synapse density and memory performance. Nat Neurosci 9:381–388

Dines M, Lamprecht R (2014) Ephrina4 mimetic peptide targeted to EphA binding site impairs the formation of long-term fear memory in lateral amygdala. Transl Psychiatry 4:e450

Dines M, Lamprecht R (2016) The role of Ephs and ephrins in memory formation. Int J Neuropsychopharmacol 19:pyv106

Dingess PM, Darling RA, Kurt Dolence E, Culver BW, Brown TE (2017) Exposure to a diet high in fat attenuates dendritic spine density in the medial prefrontal cortex. Brain Struct Funct 222:1077–1085

Dokter M, Busch R, Poser R, Vogt MA, von Bohlen und Halbach V, Gass P, Unsicker K, von Bohlen und Halbach O (2015) Implications of p75NTR for dentate gyrus morphology and hippocampus-related behavior revisited. Brain Struct Funct 220:1449–1462

Engert F, Bonhoeffer T (1999) Dendritic spine changes associated with hippocampal long-term synaptic plasticity. Nature 399:66–70

Ernfors P, Lee KF, Jaenisch R (1994) Mice lacking brain-derived neurotrophic factor develop with sensory deficits. Nature 368:147–150

Eroglu C, Barres BA (2010) Regulation of synaptic connectivity by glia. Nature 468:223–231

Evans SJ, Choudary PV, Neal CR, Li JZ, Vawter MP, Tomita H, Lopez JF, Thompson RC, Meng F, Stead JD, Walsh DM, Myers RM, Bunney WE, Watson SJ, Jones EG, Akil H (2004) Dysregulation of the fibroblast growth factor system in major depression. Proc Natl Acad Sci U S A 101:15506–15511

Fabel K, Wolf SA, Ehninger D, Babu H, Leal-Galicia P, Kempermann G (2009) Additive effects of physical exercise and environmental enrichment on adult hippocampal neurogenesis in mice. Front Neurosci 3:50

Fagan AM, Garber M, Barbacid M, Silos-Santiago I, Holtzman DM (1997) A role for TrkA during maturation of striatal and basal forebrain cholinergic neurons in vivo. J Neurosci 17:7644–7654

Fiala JC, Feinberg M, Popov V, Harris KM (1998) Synaptogenesis via dendritic filopodia in developing hippocampal area CA1. J Neurosci 18:8900–8911

Filosa A, Paixao S, Honsek SD, Carmona MA, Becker L, Feddersen B, Gaitanos L, Rudhard Y, Schoepfer R, Klopstock T, Kullander K, Rose CR, Pasquale EB, Klein R (2009) Neuron-glia communication via EphA4/ephrin-A3 modulates LTP through glial glutamate transport. Nat Neurosci 12:1285–1292

Finger BC, Dinan TG, Cryan JF (2010) Leptin-deficient mice retain normal appetitive spatial learning yet exhibit marked increases in anxiety-related behaviours. Psychopharmacology 210:559–568

Frielingsdorf H, Simpson DR, Thal LJ, Pizzo DP (2007) Nerve growth factor promotes survival of new neurons in the adult hippocampus. Neurobiol Dis 26:47–55

Frodl T, Schule C, Schmitt G, Born C, Baghai T, Zill P, Bottlender R, Rupprecht R, Bondy B, Reiser M, Moller HJ, Meisenzahl EM (2007) Association of the brain-derived neurotrophic factor Val66Met polymorphism with reduced hippocampal volumes in major depression. Arch Gen Psychiatry 64:410–416

Garcia-Lopez P, Garcia-Marin V, Freire M (2007) The discovery of dendritic spines by Cajal in 1888 and its relevance in the present neuroscience. Prog Neurobiol 83:110–130

Gardoni F, Saraceno C, Malinverno M, Marcello E, Verpelli C, Sala C, Di Luca M (2012) The neuropeptide PACAP38 induces dendritic spine remodeling through ADAM10-n-cadherin signaling pathway. J Cell Sci 125:1401–1406

Gascon E, Vutskits L, Zhang H, Barral-Moran MJ, Kiss PJ, Mas C, Kiss JZ (2005) Sequential activation of p75 and TrkB is involved in dendritic development of subventricular zone-derived neuronal progenitors in vitro. Eur J Neurosci 21:69–80

Geinisman Y (2000) Structural synaptic modifications associated with hippocampal LTP and behavioral learning. Cereb Cortex 10:952–962

Gilmartin MR, Ferrara NC (2021) Pituitary adenylate cyclase-activating polypeptide in learning and memory. Front Cell Neurosci 15:663418

Goldshmit Y, McLenachan S, Turnley A (2006) Roles of Eph receptors and ephrins in the normal and damaged adult CNS. Brain Res Rev 52:327–345

Gomez-Pinilla F, So V, Kesslak JP (2001) Spatial learning induces neurotrophin receptor and synapsin I in the hippocampus. Brain Res 904:13–19

Gonul AS, Kitis O, Eker MC, Eker OD, Ozan E, Coburn K (2011) Association of the brain-derived neurotrophic factor Val66Met polymorphism with hippocampus volumes in drug-free depressed patients. World J Biol Psychiatry 12:110–118

Grunwald IC, Korte M, Wolfer D, Wilkinson GA, Unsicker K, Lipp HP, Bonhoeffer T, Klein R (2001) Kinase-independent requirement of EphB2 receptors in hippocampal synaptic plasticity. Neuron 32:1027–1040

Grunwald IC, Korte M, Adelmann G, Plueck A, Kullander K, Adams RH, Frotscher M, Bonhoeffer T, Klein R (2004) Hippocampal plasticity requires postsynaptic ephrinBs. Nat Neurosci 7:33–40

Guo W, Nagappan G, Lu B (2018) Differential effects of transient and sustained activation of BDNF-TrkB signaling. Dev Neurobiol 78:647–659
Hall J, Thomas KL, Everitt BJ (2000) Rapid and selective induction of BDNF expression in the hippocampus during contextual learning. Nat Neurosci 3:533–535
Hasegawa Y, Mukai H, Asashima M, Hojo Y, Ikeda M, Komatsuzaki Y, Ooishi Y, Kawato S (2014) Acute modulation of synaptic plasticity of pyramidal neurons by activin in adult hippocampus. Front Neural Circuits 8:56
Hayata-Takano A, Kamo T, Kijima H, Seiriki K, Ogata K, Ago Y, Nakazawa T, Shintani Y, Higashino K, Nagayasu K, Shintani N, Kasai A, Waschek JA, Hashimoto H (2019) Pituitary adenylate cyclase-activating polypeptide modulates dendritic spine maturation and morphogenesis via microRNA-132 upregulation. J Neurosci 39:4208–4220
Henderson NT, Dalva MB (2018) EphBs and ephrin-Bs: trans-synaptic organizers of synapse development and function. Mol Cell Neurosci 91:108–121
Henkemeyer M, Itkis OS, Ngo M, Hickmott PW, Ethell IM (2003) Multiple EphB receptor tyrosine kinases shape dendritic spines in the hippocampus. J Cell Biol 163:1313–1326
Hinck AP (2012) Structural studies of the TGF-betas and their receptors – insights into evolution of the TGF-beta superfamily. FEBS Lett 586:1860–1870
Hruska M, Dalva MB (2012) Ephrin regulation of synapse formation, function and plasticity. Mol Cell Neurosci 50:35–44
Hu X, Ballo L, Pietila L, Viesselmann C, Ballweg J, Lumbard D, Stevenson M, Merriam E, Dent EW (2011) BDNF-induced increase of PSD-95 in dendritic spines requires dynamic microtubule invasions. J Neurosci 31:15597–15603
Hu HT, Umemori H, Hsueh YP (2016) Postsynaptic SDC2 induces transsynaptic signaling via FGF22 for bidirectional synaptic formation. Sci Rep 6:33592
Ingalls AM, Dickie MM, Snell GD (1950) Obese, a new mutation in the house mouse. J Hered 41:317–318
Irala D, Bonafina A, Fontanet PA, Alsina FC, Paratcha G, Ledda F (2016) The GDNF-GFRalpha1 complex promotes the development of hippocampal dendritic arbors and spines via NCAM. Development 143:4224–4235
Irie F, Yamaguchi Y (2002) EphB receptors regulate dendritic spine development via intersectin, Cdc42 and N-WASP. Nat Neurosci 5:1117–1118
Janthakhin Y, Rincel M, Costa AM, Darnaudery M, Ferreira G (2017) Maternal high-fat diet leads to hippocampal and amygdala dendritic remodeling in adult male offspring. Psychoneuroendocrinology 83:49–57
Jarosik J, Legutko B, Werner S, Unsicker K, von Bohlen und Halbach O (2011) Roles of exogenous and endogenous FGF-2 in animal models of depression. Restor Neurol Neurosci 29:153–165
Ji Y, Lu Y, Yang F, Shen W, Tang TT, Feng L, Duan S, Lu B (2010) Acute and gradual increases in BDNF concentration elicit distinct signaling and functions in neurons. Nat Neurosci 13:302–309
Jiang C, Salton SR (2013) The role of neurotrophins in major depressive disorder. Transl Neurosci 4:46–58
Kania A, Klein R (2016) Mechanisms of ephrin-Eph signalling in development, physiology and disease. Nat Rev Mol Cell Biol 17:240–256
Kayser MS, Nolt MJ, Dalva MB (2008) EphB receptors couple dendritic filopodia motility to synapse formation. Neuron 59:56–69
Kellner Y, Godecke N, Dierkes T, Thieme N, Zagrebelsky M, Korte M (2014) The BDNF effects on dendritic spines of mature hippocampal neurons depend on neuronal activity. Front Synaptic Neurosci 6:5
Kim YK, Lee HP, Won SD, Park EY, Lee HY, Lee BH, Lee SW, Yoon D, Han C, Kim DJ, Choi SH (2007) Low plasma BDNF is associated with suicidal behavior in major depression. Prog Neuro-Psychopharmacol Biol Psychiatry 31:78–85
Klein R (2009) Bidirectional modulation of synaptic functions by Eph/ephrin signaling. Nat Neurosci 12:15–20

Klenowski PM, Shariff MR, Belmer A, Fogarty MJ, Mu EW, Bellingham MC, Bartlett SE (2016) Prolonged consumption of sucrose in a binge-like manner, alters the morphology of medium spiny neurons in the nucleus accumbens shell. Front Behav Neurosci 10:54

Klimaschewski L, Claus P (2021) Fibroblast growth factor signalling in the diseased nervous system. Mol Neurobiol 58:3884–3902

Kordower JH, Palfi S, Chen EY, Ma SY, Sendera T, Cochran EJ, Mufson EJ, Penn R, Goetz CG, Comella CD (1999) Clinicopathological findings following intraventricular glial-derived neurotrophic factor treatment in a patient with Parkinson's disease. Ann Neurol 46:419–424

Kozorovitskiy Y, Gross CG, Kopil C, Battaglia L, McBreen M, Stranahan AM, Gould E (2005) Experience induces structural and biochemical changes in the adult primate brain. Proc Natl Acad Sci U S A 102:17478–17482

Lee KF, Li E, Huber LJ, Landis SC, Sharpe AH, Chao MV, Jaenisch R (1992) Targeted mutation of the gene encoding the low affinity NGF receptor p75 leads to deficits in the peripheral sensory nervous system. Cell 69:737–749

Lee TH, Kato H, Pan LH, Ryu JH, Kogure K, Itoyama Y (1998) Localization of nerve growth factor, TrkA and p75 immunoreactivity in the hippocampal formation and basal forebrain of adult rats. Neuroscience 83:335–349

Lee CC, Huang CC, Hsu KS (2011) Insulin promotes dendritic spine and synapse formation by the PI3K/Akt/mTOR and Rac1 signaling pathways. Neuropharmacology 61:867–879

Leuner B, Falduto J, Shors TJ (2003) Associative memory formation increases the observation of dendritic spines in the hippocampus. J Neurosci 23:659–665

Levi-Montalcini R, Hamburger V (1951) Selective growth stimulating effects of mouse sarcoma on the sensory and sympathetic nervous system of the chick embryo. J Exp Zool 116:321–361

Li XL, Aou S, Oomura Y, Hori N, Fukunaga K, Hori T (2002) Impairment of long-term potentiation and spatial memory in leptin receptor-deficient rodents. Neuroscience 113:607–615

Li B, Woo RS, Mei L, Malinow R (2007) The neuregulin-1 receptor ErbB4 controls glutamatergic synapse maturation and plasticity. Neuron 54:583–597

Li E, Chung H, Kim Y, Kim DH, Ryu JH, Sato T, Kojima M, Park S (2013) Ghrelin directly stimulates adult hippocampal neurogenesis: implications for learning and memory. Endocr J 60:781–789

Li G, Jing P, Liu Z, Li Z, Ma H, Tu W, Zhang W, Zhuo C (2017) Beneficial effect of fluoxetine treatment against psychological stress is mediated by increasing BDNF expression in selected brain areas. Oncotarget 8:69527–69537

Liebl DJ, Morris CJ, Henkemeyer M, Parada LF (2003) Mrna expression of ephrins and Eph receptor tyrosine kinases in the neonatal and adult mouse central nervous system. J Neurosci Res 71:7–22

Lohmann C, Bonhoeffer T (2008) A role for local calcium signaling in rapid synaptic partner selection by dendritic filopodia. Neuron 59:253–260

Maragnoli ME, Fumagalli F, Gennarelli M, Racagni G, Riva MA (2004) Fluoxetine and olanzapine have synergistic effects in the modulation of fibroblast growth factor 2 expression within the rat brain. Biol Psychiatry 55:1095–1102

Martinez A, Alcantara S, Borrell V, Del Rio JA, Blasi J, Otal R, Campos N, Boronat A, Barbacid M, Silos-Santiago I, Soriano E (1998) TrkB and TrkC signaling are required for maturation and synaptogenesis of hippocampal connections. J Neurosci 18:7336–7350

Matsuyama S, Matsumoto A, Hashimoto H, Shintani N, Baba A (2003) Impaired long-term potentiation in vivo in the dentate gyrus of pituitary adenylate cyclase-activating polypeptide (PACAP) or PACAP type 1 receptor-mutant mice. Neuroreport 14:2095–2098

Maynard KR, Hobbs JW, Sukumar M, Kardian AS, Jimenez DV, Schloesser RJ, Martinowich K (2017) BDNF mRNA splice variants differentially impact CA1 and CA3 dendrite complexity and spine morphology in the hippocampus. Brain Struct Funct 222:3295

McNay EC (2007) Insulin and ghrelin: peripheral hormones modulating memory and hippocampal function. Curr Opin Pharmacol 7:628–632

Medina DL, Sciarretta C, Calella AM, von Bohlen und Halbach O, Unsicker K, Minichiello L (2004) TrkB regulates neocortex formation through the Shc/PLCgamma-mediated control of neuronal migration. EMBO J 23:3803–3814

Mervaala E, Fohr J, Kononen M, Valkonen-Korhonen M, Vainio P, Partanen K, Partanen J, Tiihonen J, Viinamaki H, Karjalainen AK, Lehtonen J (2000) Quantitative MRI of the hippocampus and amygdala in severe depression. Psychol Med 30:117–125

Minichiello L (2009) TrkB signalling pathways in LTP and learning. Nat Rev Neurosci 10:850–860

Minichiello L, Korte M, Wolfer D, Kuhn R, Unsicker K, Cestari V, Rossi-Arnaud C, Lipp HP, Bonhoeffer T, Klein R (1999) Essential role for TrkB receptors in hippocampus-mediated learning. Neuron 24:401–414

Monfils MH, Teskey GC (2004) Induction of long-term depression is associated with decreased dendritic length and spine density in layers III and V of sensorimotor neocortex. Synapse 53:114–121

Monteggia LM, Luikart B, Barrot M, Theobold D, Malkovska I, Nef S, Parada LF, Nestler EJ (2007) Brain-derived neurotrophic factor conditional knockouts show gender differences in depression-related behaviors. Biol Psychiatry 61:187–197

Moon M, Kim S, Hwang L, Park S (2009) Ghrelin regulates hippocampal neurogenesis in adult mice. Endocr J 56:525–531

More SV, Koppula S, Kim IS, Kumar H, Kim BW, Choi DK (2012) The role of bioactive compounds on the promotion of neurite outgrowth. Molecules 17:6728–6753

Moser MB, Trommald M, Andersen P (1994) An increase in dendritic spine density on hippocampal CA1 pyramidal cells following spatial learning in adult rats suggests the formation of new synapses. Proc Natl Acad Sci U S A 91:12673–12675

Moser MB, Trommald M, Egeland T, Andersen P (1997) Spatial training in a complex environment and isolation alter the spine distribution differently in rat CA1 pyramidal cells. J Comp Neurol 380:373–381

Mrzljak L, Goldman-Rakic PS (1993) Low-affinity nerve growth factor receptor (p75NGFR)- and choline acetyltransferase (ChAT)-immunoreactive axons in the cerebral cortex and hippocampus of adult macaque monkeys and humans. Cereb Cortex 3:133–147

Muller D, Toni N, Buchs PA (2000) Spine changes associated with long-term potentiation. Hippocampus 10:596–604

Muller MB, Lucassen PJ, Yassouridis A, Hoogendijk WJ, Holsboer F, Swaab DF (2001) Neither major depression nor glucocorticoid treatment affects the cellular integrity of the human hippocampus. Eur J Neurosci 14:1603–1612

Murai KK, Nguyen LN, Irie F, Yamaguchi Y, Pasquale EB (2003) Control of hippocampal dendritic spine morphology through ephrin-A3/EphA4 signaling. Nat Neurosci 6:153–160

Nguyen AQ, Koeppen J, Woodruff S, Mina K, Figueroa Z, Ethell IM (2020a) Astrocytic ephrin-B1 controls synapse formation in the hippocampus during learning and memory. Front Synaptic Neurosci 12:10

Nguyen AQ, Sutley S, Koeppen J, Mina K, Woodruff S, Hanna S, Vengala A, Hickmott PW, Obenaus A, Ethell IM (2020b) Astrocytic ephrin-B1 controls excitatory-inhibitory balance in developing hippocampus. J Neurosci 40:6854–6871

Nieto-Bona MP, Garcia-Segura LM, Torres-Aleman I (1997) Transynaptic modulation by insulin-like growth factor I of dendritic spines in Purkinje cells. Int J Dev Neurosci 15:749–754

Nimchinsky EA, Sabatini BL, Svoboda K (2002) Structure and function of dendritic spines. Annu Rev Physiol 64:313–353

Nykjaer A, Lee R, Teng KK, Jansen P, Madsen P, Nielsen MS, Jacobsen C, Kliemannel M, Schwarz E, Willnow TE, Hempstead BL, Petersen CM (2004) Sortilin is essential for proNGF-induced neuronal cell death. Nature 427:843–848

O'Malley A, O'Connell C, Murphy KJ, Regan CM (2000) Transient spine density increases in the mid-molecular layer of hippocampal dentate gyrus accompany consolidation of a spatial learning task in the rodent. Neuroscience 99:229–232

Obradovic M, Sudar-Milovanovic E, Soskic S, Essack M, Arya S, Stewart AJ, Gojobori T, Isenovic ER (2021) Leptin and obesity: role and clinical implication. Front Endocrinol (Lausanne) 12:585887

Orefice LL, Shih CC, Xu H, Waterhouse EG, Xu B (2016) Control of spine maturation and pruning through proBDNF synthesized and released in dendrites. Mol Cell Neurosci 71:66–79

Ossola B, Schendzielorz N, Chen SH, Bird GS, Tuominen RK, Mannisto PT, Hong JS (2011) Amantadine protects dopamine neurons by a dual action: reducing activation of microglia and inducing expression of GDNF in astroglia [corrected]. Neuropharmacology 61:574–582

Otal R, Martinez A, Soriano E (2005) Lack of TrkB and TrkC signaling alters the synaptogenesis and maturation of mossy fiber terminals in the hippocampus. Cell Tissue Res 319:349–358

Ou GY, Lin WW, Zhao WJ (2021) Neuregulins in neurodegenerative diseases. Front Aging Neurosci 13:662474

Parnass Z, Tashiro A, Yuste R (2000) Analysis of spine morphological plasticity in developing hippocampal pyramidal neurons. Hippocampus 10:561–568

Pascual Lucas M, Viana da Silva S, Di Scala M, Garcia Barroso C, Gonzalez Aseguinolaza G, Mulle C, Alberini CM, Cuadrado Tejedor M, Garcia Osta A (2014) Insulin like growth factor 2 reverses memory and synaptic deficits in APP transgenic mice. EMBO Mol Med 6:1246–1262

Penzes P, Beeser A, Chernoff J, Schiller MR, Eipper BA, Mains RE, Huganir RL (2003) Rapid induction of dendritic spine morphogenesis by trans-synaptic ephrinB-EphB receptor activation of the Rho-GEF kalirin. Neuron 37:263–274

Perea Vega ML, Sanchez MS, Fernandez G, Paglini MG, Martin M, de Barioglio SR (2021) Ghrelin treatment leads to dendritic spine remodeling in hippocampal neurons and increases the expression of specific BDNF-mRNA species. Neurobiol Learn Mem 179:107409

Peters A, Kaiserman-Abramof IR (1970) The small pyramidal neuron of the rat cerebral cortex. The perikaryon, dendrites and spines. Am J Anat 127:321–355

Pinar C, Fontaine CJ, Trivino-Paredes J, Lottenberg CP, Gil-Mohapel J, Christie BR (2017) Revisiting the flip side: long-term depression of synaptic efficacy in the hippocampus. Neurosci Biobehav Rev 80:394–413

Porter WD, Flatt PR, Holscher C, Gault VA (2013) Liraglutide improves hippocampal synaptic plasticity associated with increased expression of Mash1 in ob/ob mice. Int J Obes 37:678–684

Poser R, Dokter M, von Bohlen und Halbach V, Berger SM, Busch R, Baldus M, Unsicker K, von Bohlen und Halbach O (2015) Impact of a deletion of the full-length and short isoform of p75NTR on cholinergic innervation and the population of postmitotic doublecortin positive cells in the dentate gyrus. Front Neuroanat 9:63

Rampon C, Tsien JZ (2000) Genetic analysis of learning behavior-induced structural plasticity. Hippocampus 10:605–609

Rantamaki T, Hendolin P, Kankaanpaa A, Mijatovic J, Piepponen P, Domenici E, Chao MV, Mannisto PT, Castren E (2007) Pharmacologically diverse antidepressants rapidly activate brain-derived neurotrophic factor receptor TrkB and induce phospholipase-cgamma signaling pathways in mouse brain. Neuropsychopharmacology 32:2152–2162

Rascol O, Fabbri M, Poewe W (2021) Amantadine in the treatment of Parkinson's disease and other movement disorders. Lancet Neurol 20:1048–1056

Regeenes R, Silva PN, Chang HH, Arany EJ, Shukalyuk AI, Audet J, Kilkenny DM, Rocheleau JV (2018) Fibroblast growth factor receptor 5 (FGFR5) is a co-receptor for FGFR1 that is up-regulated in beta-cells by cytokine-induced inflammation. J Biol Chem 293:17218–17228

Reuss B, von Bohlen und Halbach O (2003) Fibroblast growth factors and their receptors in the central nervous system. Cell Tissue Res 313:139–157

Rex CS, Lin CY, Kramar EA, Chen LY, Gall CM, Lynch G (2007) Brain-derived neurotrophic factor promotes long-term potentiation-related cytoskeletal changes in adult hippocampus. J Neurosci 27:3017–3029

Risher WC, Ustunkaya T, Singh Alvarado J, Eroglu C (2014) Rapid golgi analysis method for efficient and unbiased classification of dendritic spines. PLoS One 9:e107591

Roux PP, Colicos MA, Barker PA, Kennedy TE (1999) P75 neurotrophin receptor expression is induced in apoptotic neurons after seizure. J Neurosci 19:6887–6896

Sa-Nguanmoo P, Tanajak P, Kerdphoo S, Satjaritanun P, Wang X, Liang G, Li X, Jiang C, Pratchayasakul W, Chattipakorn N, Chattipakorn SC (2016) FGF21 improves cognition by restored synaptic plasticity, dendritic spine density, brain mitochondrial function and cell apoptosis in obese-insulin resistant male rats. Horm Behav 85:86–95

Schepers J, Gebhardt C, Bracke A, Eiffler I, von Bohlen und Halbach O (2020) Structural and functional consequences in the amygdala of leptin-deficient mice. Cell Tissue Res 382:421–426

Schumacher S, Volkmer H, Buck F, Otto A, Tarnok A, Roth S, Rathjen FG (1997) Chicken acidic leucine-rich EGF-like domain containing brain protein (CALEB), a neural member of the EGF family of differentiation factors, is implicated in neurite formation. J Cell Biol 136:895–906

Schwab RS, England AC Jr, Poskanzer DC, Young RR (1969) Amantadine in the treatment of Parkinson's disease. JAMA 208:1168–1170

Segura I, Essmann CL, Weinges S, Acker-Palmer A (2007) Grb4 and GIT1 transduce ephrinB reverse signals modulating spine morphogenesis and synapse formation. Nat Neurosci 10:301–310

Serafini G (2012) Neuroplasticity and major depression, the role of modern antidepressant drugs. World J Psychiatry 2:49–57

Shoji-Kasai Y, Ageta H, Hasegawa Y, Tsuchida K, Sugino H, Inokuchi K (2007) Activin increases the number of synaptic contacts and the length of dendritic spine necks by modulating spinal actin dynamics. J Cell Sci 120:3830–3837

Singh B, Carpenter G, Coffey RJ (2016) EGF receptor ligands: recent advances. F1000Res 5:F1000 Faculty Rev-2270

Skoff RP, Hamburger V (1974) Fine structure of dendritic and axonal growth cones in embryonic chick spinal cord. J Comp Neurol 153:107–147

Sobreviela T, Clary DO, Reichardt LF, Brandabur MM, Kordower JH, Mufson EJ (1994) TrkA-immunoreactive profiles in the central nervous system: colocalization with neurons containing p75 nerve growth factor receptor, choline acetyltransferase, and serotonin. J Comp Neurol 350:587–611

Solvsten CA, de Paoli F, Christensen JH, Nielsen AL (2018) Voluntary physical exercise induces expression and epigenetic remodeling of VegfA in the rat hippocampus. Mol Neurobiol 55:567–582

Sorra KE, Fiala JC, Harris KM (1998) Critical assessment of the involvement of perforations, spinules, and spine branching in hippocampal synapse formation. J Comp Neurol 398:225–240

Stranahan AM, Lee K, Martin B, Maudsley S, Golden E, Cutler RG, Mattson MP (2009) Voluntary exercise and caloric restriction enhance hippocampal dendritic spine density and BDNF levels in diabetic mice. Hippocampus 19:951–961

Teng HK, Teng KK, Lee R, Wright S, Tevar S, Almeida RD, Kermani P, Torkin R, Chen ZY, Lee FS, Kraemer RT, Nykjaer A, Hempstead BL (2005) ProBDNF induces neuronal apoptosis via activation of a receptor complex of p75NTR and sortilin. J Neurosci 25:5455–5463

Teng KK, Felice S, Kim T, Hempstead BL (2010) Understanding proneurotrophin actions: recent advances and challenges. Dev Neurobiol 70:350–359

Tolias KF, Bikoff JB, Kane CG, Tolias CS, Hu L, Greenberg ME (2007) The Rac1 guanine nucleotide exchange factor Tiam1 mediates EphB receptor-dependent dendritic spine development. Proc Natl Acad Sci U S A 104:7265–7270

Toni N, Buchs PA, Nikonenko I, Bron CR, Muller D (1999) LTP promotes formation of multiple spine synapses between a single axon terminal and a dendrite. Nature 402:421–425

Unsicker K, Krieglstein K (2002) TGF-betas and their roles in the regulation of neuron survival. Adv Exp Med Biol 513:353–374

von Bohlen und Halbach O (2009) Structure and function of dendritic spines within the hippocampus. Ann Anat 191(6):518–531

von Bohlen und Halbach V, von Bohlen und Halbach O (2016) Deletion of p75NTR enhances the cholinergic innervation pattern of the visual cortex. Vis Neurosci 33:1–6 von Bohlen und Halbach O, Krause S, Medina D, Sciarretta C, Minichiello L, Unsicker K (2006a) Regional- and age-dependent reduction in TrkB receptor expression in the hippocampus is associated with altered spine morphologies. Biol Psychiatry 59:793–800

von Bohlen und Halbach O, Zacher C, Gass P, Unsicker K (2006b) Age-related alterations in hippocampal spines and deficiencies in spatial memory in mice. J Neurosci Res 83:525–531

von Bohlen und Halbach O, Minichiello L, Unsicker K (2008) TrkB but not TrkC receptors are necessary for postnatal maintenance of hippocampal spines. Neurobiol Aging 29:1247–1255

von Schack D, Casademunt E, Schweigreiter R, Meyer M, Bibel M, Dechant G (2001) Complete ablation of the neurotrophin receptor p75NTR causes defects both in the nervous and the vascular system. Nat Neurosci 4:977–978

Vaynman S, Ying Z, Gomez-Pinilla F (2004) Hippocampal BDNF mediates the efficacy of exercise on synaptic plasticity and cognition. Eur J Neurosci 20:2580–2590

Wrana JL (1998) TGF-beta receptors and signalling mechanisms. Miner Electrolyte Metab 24:120–130

Xenos D, Kamceva M, Tomasi S, Cardin JA, Schwartz ML, Vaccarino FM (2017) Loss of TrkB signaling in parvalbumin-expressing basket cells results in network activity disruption and abnormal behavior. Cereb Cortex 28(10):3399–3413

Xia J, Chen J, Zhou Y, Zhang J, Yang B, Xia L, Wang C (2004) Volumetric MRI analysis of the amygdala and hippocampus in subjects with major depression. J Huazhong Univ Sci Technolog Med Sci 24:500–502, 506

Xie Y, Zhao W, Zuo Z (2022) Glial cell-derived neurotrophic factor decrease may mediate learning, memory and behavior impairments in rats after neonatal surgery. Brain Res Bull 178:9–16

Xu B, Xie X (2016) Neurotrophic factor control of satiety and body weight. Nat Rev Neurosci 17:282–292

Xu NJ, Sun S, Gibson JR, Henkemeyer M (2011) A dual shaping mechanism for postsynaptic ephrin-B3 as a receptor that sculpts dendrites and synapses. Nat Neurosci 14:1421–1429

Yang J, Luo X, Huang X, Ning Q, Xie M, Wang W (2014) Ephrin-A3 reverse signaling regulates hippocampal neuronal damage and astrocytic glutamate transport after transient global ischemia. J Neurochem 131:383–394

Yeo TT, Chua-Couzens J, Butcher LL, Bredesen DE, Cooper JD, Valletta JS, Mobley WC, Longo FM (1997) Absence of p75NTR causes increased basal forebrain cholinergic neuron size, choline acetyltransferase activity, and target innervation. J Neurosci 17:7594–7605

Yuste R (2015) The discovery of dendritic spines by Cajal. Front Neuroanat 9:18

Yuste R, Bonhoeffer T (2001) Morphological changes in dendritic spines associated with long-term synaptic plasticity. Annu Rev Neurosci 24:1071–1089

Yuste R, Bonhoeffer T (2004) Genesis of dendritic spines: insights from ultrastructural and imaging studies. Nat Rev Neurosci 5:24–34

Zagrebelsky M, Holz A, Dechant G, Barde YA, Bonhoeffer T, Korte M (2005) The p75 neurotrophin receptor negatively modulates dendrite complexity and spine density in hippocampal neurons. J Neurosci 25:9989–9999

Zagrebelsky M, Tacke C, Korte M (2020) BDNF signaling during the lifetime of dendritic spines. Cell Tissue Res 382:185–199

Zechel S, Unsicker K, von Bohlen und Halbach O (2009) Fibroblast growth factor-2 deficiency affects hippocampal spine morphology, but not hippocampal catecholaminergic or cholinergic innervation. Dev Dyn 238:343–350

Zheng K, An JJ, Yang F, Xu W, Xu ZQ, Wu J, Hokfelt TG, Fisahn A, Xu B, Lu B (2011) TrkB signaling in parvalbumin-positive interneurons is critical for gamma-band network synchronization in hippocampus. Proc Natl Acad Sci U S A 108:17201–17206

Zhong J, Li J, Ni C, Zuo Z (2020) Amantadine alleviates postoperative cognitive dysfunction possibly by preserving neurotrophic factor expression and dendritic arborization in the hippocampus of old rodents. Front Aging Neurosci 12:605330

Zhou C, Zhong J, Zou B, Fang L, Chen J, Deng X, Zhang L, Zhao X, Qu Z, Lei Y, Lei T (2017) Meta-analyses of comparative efficacy of antidepressant medications on peripheral BDNF concentration in patients with depression. PLoS One 12:e0172270

Chapter 6
Glial Cell Modulation of Dendritic Spine Structure and Synaptic Function

Alberto A. Rasia-Filho, Maria Elisa Calcagnotto, and Oliver von Bohlen und Halbach

Abstract Glia comprise a heterogeneous group of cells involved in the structure and function of the central and peripheral nervous system. Glial cells are found from invertebrates to humans with morphological specializations related to the neural circuits in which they are embedded. Glial cells modulate neuronal functions, brain wiring and myelination, and information processing. For example, astrocytes send processes to the synaptic cleft, actively participate in the metabolism of neurotransmitters, and release gliotransmitters, whose multiple effects depend on the targeting cells. Human astrocytes are larger and more complex than their mice and rats counterparts. Astrocytes and microglia participate in the development and plasticity of neural circuits by modulating dendritic spines. Spines enhance neuronal connectivity, integrate most postsynaptic excitatory potentials, and balance the strength of each input. Not all central synapses are engulfed by astrocytic processes. When that relationship occurs, a different pattern for thin and large spines reflects an activity-dependent remodeling of motile astrocytic processes around presynaptic and postsynaptic elements. Microglia are equally relevant for synaptic processing, and both

A. A. Rasia-Filho (✉)
Department of Basic Sciences/Physiology and Graduate Program in Biosciences, Universidade Federal de Ciências da Saúde de Porto Alegre, Porto Alegre, RS, Brazil

Graduate Program in Neuroscience, Universidade Federal do Rio Grande do Sul, Porto Alegre, RS, Brazil

M. E. Calcagnotto
Graduate Program in Neuroscience, Universidade Federal do Rio Grande do Sul, Porto Alegre, RS, Brazil

Department of Biochemistry, Universidade Federal do Rio Grande do Sul, Porto Alegre, RS, Brazil

Graduate Program in Biochemistry, Universidade Federal do Rio Grande do Sul, Porto Alegre, RS, Brazil

Graduate Program in Psychiatry and Behavioral Science, Universidade Federal do Rio Grande do Sul, Porto Alegre, RS, Brazil

O. von Bohlen und Halbach
Institut für Anatomie und Zellbiologie, Universitätsmedizin Greifswald, Greifswald, Germany

A. A. Rasia-Filho et al. (eds.), *Dendritic Spines*, Advances in Neurobiology 34,
https://doi.org/10.1007/978-3-031-36159-3_6

glial cells modulate the switch of neuroendocrine secretion and behavioral display needed for reproduction. In this chapter, we provide an overview of the structure, function, and plasticity of glial cells and relate them to synaptic maturation and modulation, also involving neurotrophic factors. Together, neurons and glia coordinate synaptic transmission in both normal and abnormal conditions. Neglected over decades, this exciting research field can unravel the complexity of species-specific neural cytoarchitecture as well as the dynamic region-specific functional interactions between diverse neurons and glial subtypes.

Keywords Astrocytes · Microglia · Tripartite synapses · Tetrapartite synapses · Synaptic plasticity · Neurotrophic factors · Neural circuits · Behavior

6.1 Relevance of Glial Cells to Neural Cytoarchitecture and Function

For decades, the study of glial cells was second to neuronal morphology, electrophysiology, and functional properties in brain wiring (Kettenmann and Ransom 1995). "Neuroglia" ("Nervenkitt") was the original name for "glia," which derives from the Greek "glue." Over time, it was considered that highly diverse glial cells throughout the central (CNS) and peripheral nervous system (PNS) served as a structural support for neurons. Long glial processes were initially compared to collagen fibers in connective tissue (mentioned in Cajal 1909–1911). This scenario changed dramatically; glial cells are crucial elements completely integrated with nervous structure and function. Currently, it is estimated that the human brain has approximately 86 billion neurons and other 85 billion nonneuronal cells (e.g., mainly NeuN-negative glial cell types and few ependymal and endothelial cells; Azevedo et al. 2009). Glial cells are found with evolutionarily conserved molecular mechanisms from flies to mammals (Losada-Perez (2018), see evolution of astroglia in Verkhratsky and Nedergaard (2018)), and these cells did not vary in size and density as mammalian brains evolved (Herculano-Houzel 2011). On the other hand, the cellular structure and volume of cortical astrocytes are higher in humans than in other animals (Oberheim et al. 2009). That phenomenon indicates the need to maintain similar, conserved features for glial functions along evolution (Herculano-Houzel 2011; Losada-Perez 2018), while at the same time an increased morphological complexity amplified the functional implications for each cell in each nervous region.

Original descriptions and interpretations for glial cells were provided by Deiters and Virchow (Deiters and Guillery 2013; Hirbec et al. 2020), and further expanded by Cajal and collaborators (Sala y Pons, Terrazas, Tello, Achúcarro, Río-Hortega, and De Castro, for example) among others distinguished histologists, namely, Ranvier, Andriezen, Greppin, Lugaro, Weigert, von Lenhossék, Azoulay, Retzius, Kölliker, Golgi, and His (Cajal 1909–1911; Garcia-Segura 2002; Matyash and Kettenmann 2010; Navarrete and Araque 2014; Tremblay et al. 2015; Verkhratsky and Nedergaard 2018; Figs. 6.1 and 6.2). For example, Terrazas described the morphology of glial cells in the cerebellum, including "cells with a panache" (the

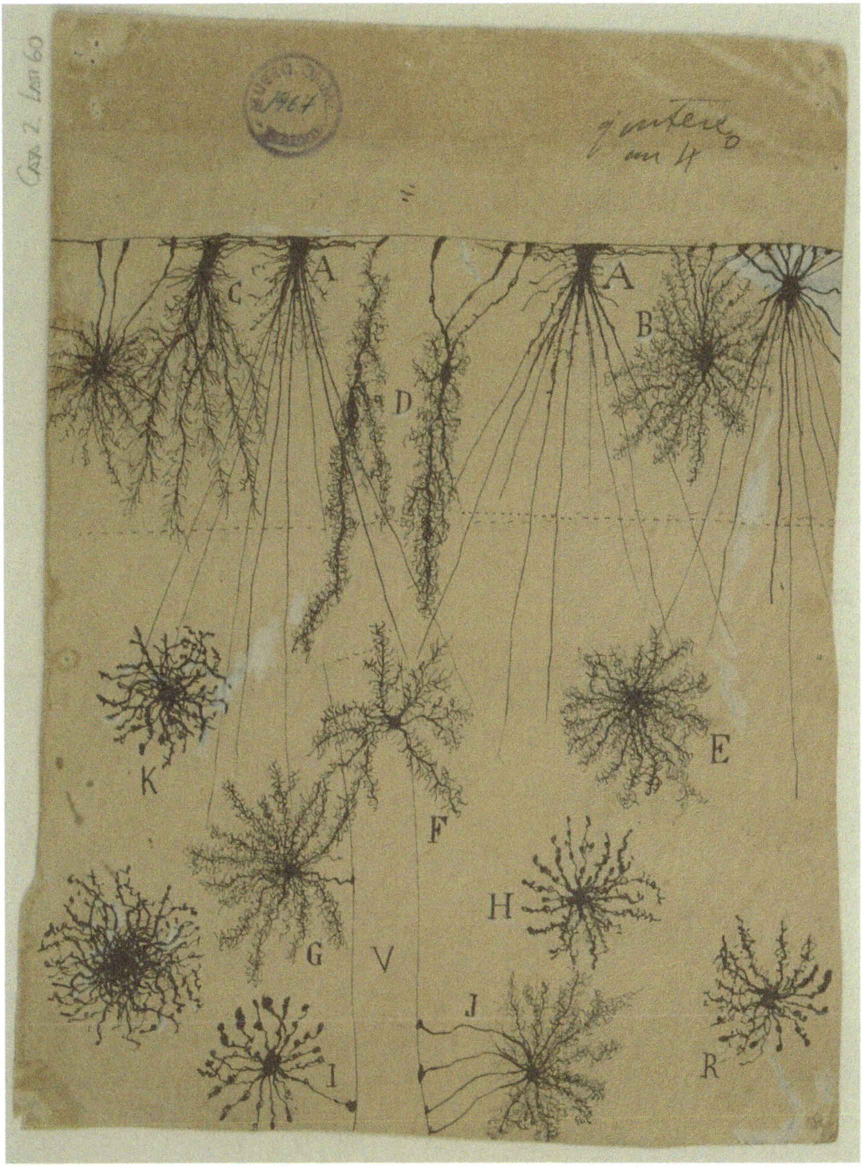

Fig. 6.1 "Neuroglia in superficial layers of the cerebral cortex of a child (male)" originally drawn by Santiago Ramón y Cajal in 1904. Note the morphological heterogeneity of astrocytes (A–K and R) in the gray matter. Some types have unique morphological features in humans, as those located in the superficial cortical layer I near the pial surface. Long processes would communicate cells across different layers. Endfeet specializations of cells G, I, and J are represented at the wall of a blood vessel (V). (Courtesy of the Cajal Legacy, Cajal Institute (CSIC), Madrid, Spain)

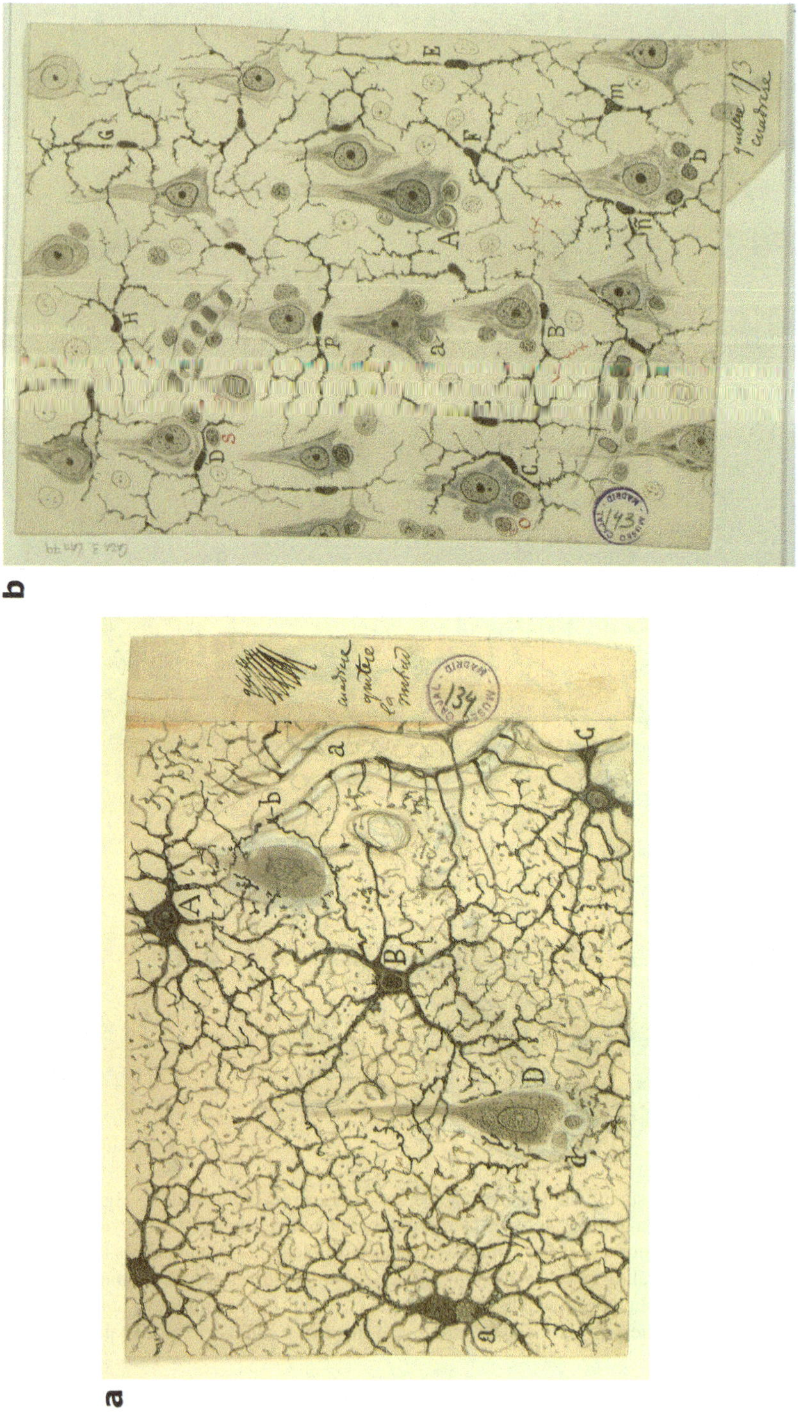

Fig. 6.2 (*caption on p. 259*)

Bergmann glial cells parallel to Purkinje cells) and cells with long processes from the white matter to the upper molecular layer (Cajal 1909–1911). Likewise, more than a century ago, Lugaro proposed that, besides an insulating function and close apposition to blood vessels, glial cells also provide "elasticity to neural tissue, inactivate chemically the metabolic by-products of neurons, and serve as guides for developing axons by providing chemotactic cues, and so on" (Cajal (1909–1911), see also the nicely illustrated review by García-Marín et al. (2007)).

Glia represent embryologically, morphologically, and functionally distinct subtypes of cells in both CNS and PNS (Verkhratsky and Nedergaard 2018). Despite existing heterogeneous morphology, gene expression profile, and responses to stimuli for each subpopulation even within the same region, glial cells are classified according to shape and lineage in microglia and macroglia (Zhang and Barres 2010; Losada-Perez 2018; Miller 2018; Seguella and Gulbransen 2021; Figs. 6.1, 6.2, and 6.3). Microglia can act as immune cells of the CNS and they have a role comparable to macrophages outside the nervous system. Macroglia is a collective term for glia cells different from microglia cells. Macroglia include protoplasmic and fibrous astrocytes and subpopulations, radial glia, ependymal cells of various types (e.g., tanycytes), transition zone glia (e.g., neural crest-derived boundary cap cells and olfactory ensheathing cells), soma-associated (satellite) glial cells, NG2-glia and oligodendrocytes, Schwann cells, blood–brain and blood–nerve barrier glia (e.g., astrocytes and pericyte-associated microglia in the CNS or perineurial glia in the PNS), and heterogeneous enteric glial cells (Dimou and Simons 2017; Radomska and Topilko 2017; Verkhratsky and Nedergaard 2018; Zeisel et al. 2018; Lago-Baldaia et al. 2020; MacDonald et al. 2021; Seguella and Gulbransen 2021; Morris et al. 2023, see a discussion for astrocyte diversity in Matyash and Kettenmann (2010) and for oligodendrocytes in Foerster et al. (2019)).

Fig. 6.2 (**a**) "Astrocytes in gray matter of the cerebral cortex" originally drawn by Santiago Ramón y Cajal in 1913. Note the identification of astrocytes (A–B) near pyramidal neurons (D), which have a visible cell body and primary apical dendritic shaft, and a pair of small perineuronal (satellite) glial cells immediately adjacent to the basal part of the neuronal soma (d, resembling oligodendrocytes). Note the high astrocytic branching pattern and the spreading of multiple sinuous thin processes of different diameters coursing in all directions in the neuropil, giving the idea of close appositions between processes of neighbor cells. An astrocyte (C) with a thick shaft directed to a local blood vessel and showing a "swelling" at the end of this pedicle (bottom right). Endfeet specializations (a, b) are seen contacting the blood vessel or embracing a transversally sectioned vessel (close to the letter "a"). Also note the presence of some appendages in astrocytic processes and an astrocyte with two nuclei attached (a), likely representing the mitotic capacity of these cells in the adult brain (but see Sosunov et al. 2020). A similar image was drawn and named "twin astrocytes" in the cerebral cortex of the dog (Swanson et al. 2017). Morphological descriptions based on Cajal (1909–1911) and García-Marín et al. (2007). (**b**) "Microglia in cerebral cortex of a normal man" originally drawn by Santiago Ramón y Cajal in 1920. Note the distribution of various microglia cells (e.g., B–G) with shafts, branching pattern, and thin processes close to adjacent neurons (A). Microglia "are characterized by their small, dark nucleus enveloped by scant protoplasm and its long, tortuous, ramified expansions adorned with lateral spines" (Río-Hortega 1919; Tremblay et al. 2015). Groups of small perineuronal glial cells close to basal dendrites and the axon hillock are also identified (a, b). (Courtesy of the Cajal Legacy, Cajal Institute (CSIC), Madrid, Spain)

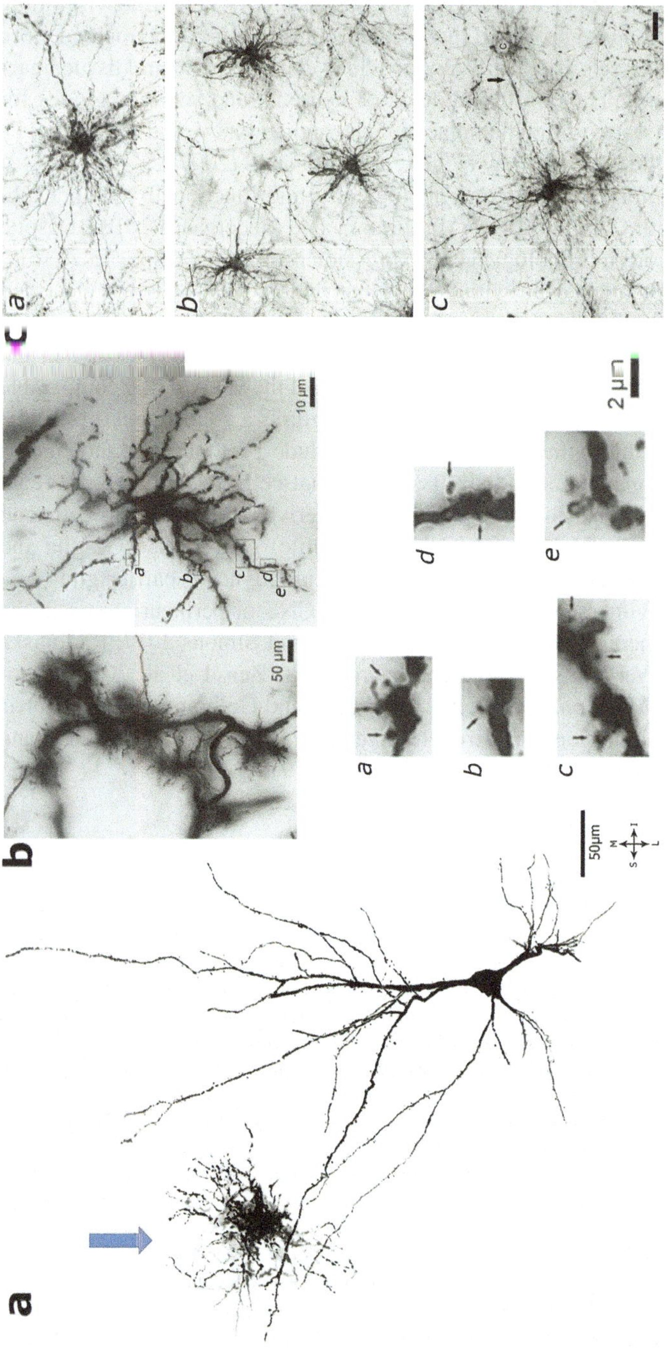

Fig. 6.3 (*caption on p. 261*)

Glial cells in the CNS and PNS bring about multiple activities, such as the following:

1. Regulate neuroblast proliferation, migration, neurogenesis, axon outgrowth, and connectional development
2. Selectively control the cellular components at transition zones between the CNS and PNS
3. Promote the ensheathment and myelination of axons
4. Induce synaptogenesis secreting extracellular matrix components (ECM) and modulate the synaptic transmission controlling the availability, turnover, and spillover of neurotransmitters
5. Modulate spine density (maturation or shrinkage) and show plastic, activity-dependent structural changes when covering dendritic spines
6. Secrete gliotransmitters and extracellular vesicles
7. Adjust neuroendocrine secretion and various behavioral displays, circadian rhythms, sleep, and pain sensation
8. Participate in the blood–brain barrier maintenance and selectivity, surround blood vessels with a specialized endfeet, release vasoactive substances, and modulate the local microcirculation and functional hyperemia
9. Promote ion buffering, water and osmotic homeostasis, regulate the extracellular pH, allow movements of inorganic and organic molecules
10. Synthesize glycogen, provide a metabolic/energetic support in the form of lactate to neurons, and participate in systemic energy balance
11. Serve as chemosensors of oxygen, CO_2, and pH to regulate respiration
12. Generate immune and inflammatory responses, have phagocytic properties, remove neuronal debris, respond to insults to nervous tissue, repair damages, and generate regenerative responses

Fig. 6.3 (**a**) Digitized and reconstructed light microscopy image of Golgi-impregnated cells from the human cortical nucleus of the amygdala. The arrow points to an astrocyte with a bushy appearance and radiating thin processes reconstructed at the left side of a pyramidal-like neuron. (Figure reproduced from Vásquez et al. (2018) under CCC RightsLink® license # 5527711164904, originally published by John Wiley & Sons, Inc). (**b**, **c**) Golgi-impregnated fibrous (**b**) and protoplasmic (**c**) astrocytes from the human medial nucleus of the amygdala. Note the close and intermingled relationship of astrocytes with local blood vessels in (**b**), and the multiple spine-like structures with heterogeneous sizes and shapes (exemplified in boxed regions in (**c**), shown in higher magnification and indicated by arrows in the inserts *a–e*). These small protrusions are found along branched processes radiating in all directions from the protoplasmic astrocyte cell body. (Figure reprinted from Dall'Oglio et al. (2015) under CCC RightsLink® license # 5527711419997, originally published by John Wiley & Sons, Inc). (**c**) Digitized photomicrographs of glial fibrillary acidic protein (GFAP)-immunoreactive cells in the human medial nucleus of the amygdala. Note the cell bodies and multiple thick and thin branches with variable lengths of protoplasmic astrocytes found in isolation (*a*), in clusters (*b*), and in the proximity of a capillary (c) blood vessel (*c*). Background contrast and brightness were adjusted in Adobe Photoshop 7.0 (USA). Scale bar = 30 μm. (Legend and figure reprinted from Dall'Oglio et al. (2013) under CCC RightsLink® license # 5527720258534, originally published by John Wiley & Sons, Inc)

13. Are the crucial elements of the CNS antioxidative system for the homeostasis of reactive oxygen species
14. Regulate axon pruning, synapse elimination, and programmed cell death
15. Form the glial cell-dependent paravascular network for interstitial waste solutes clearance, named the glymphatic system[1]

The strategic position of astrocytes with extended processes engulfing synaptic sites, the ability to uptake and respond to neurotransmitters, and the release of neuroactive substances provide a capacity for some of these cells to coordinate activity and modulate information processing within an "active milieu" and networks (Perea et al. 2014; Semyanov and Verkhratsky 2021, see further data below). The following sections provide an overview of the structure, function and plasticity of glial cells associated with dendritic spines and synaptic processing. Importantly, species-specific, subregion-specific, and layer-specific features were reported for the morphology of glial cells and their dynamic roles in neural circuits. This is an exciting research field with many unrevealed aspects about the glial function in network development and function, synaptic plasticity, neuronal output, network functioning, behavior, and diseases (see also the reviews and vast complementary data in Jäkel and Dimou (2017), Losada-Perez (2018), Mederos et al. (2018), Miller (2018), Verkhratsky and Nedergaard (2018), Hirbec et al. (2020), Lago-Baldaia et al. (2020), Pistono et al. (2021), Benfey et al. (2022), Lyon and Allen (2022) and references therein), including the gastrointestinal tract (Seguella and Gulbransen 2021).

6.2 Glial Cell Features in Neural Circuits

Neuronal geometry is adapted to nervous area availability, the density of elements within it, and the course of extrinsic and local, intrinsic afferent pathways, and connectional organization. So are the diverse adjacent glial cells, which are, also, functionally specialized, developed (including changes related to age), and adapted to the cytoarchitecture and circuits in which they are embedded (De Zeeuw and Hoogland 2015; MacDonald et al. 2021; Endo et al. 2022; see further commentary and data in Ascoli 2023). The preferred spatial orientation of processes seen in Bergmann glial cells of the cerebellum and Müller glia of the retina illustrate that observation. Other examples are oligodendrocytes (in the CNS) and Schwann cells (in the PNS) associated with and isolating each single axon in myelinated bundles, and astrocytes following the contours of dendrites and axons in the surrounding neuropil in the cerebral cortex (Peters et al. 1991).

[1] Based on Kettenmann and Ransom (1995), Magistretti et al. (1999), Garcia-Segura and McCarthy (2004), Kato et al. (2011), Bernardinelli et al. (2014), Ostroff et al. (2014), Ishikawa et al. (2018), Jinnou et al. (2018), Mederos et al. (2018), Miller (2018), Losada-Perez (2018), Verkhratsky and Nedergaard (2018), Lago-Baldaia et al. (2020), Reddy and van der Werf (2020), Fan and Huo (2021), and Pistono et al. (2021). Additional data on the integrated CNS immune surveillance were recently published (Proulx and Engelhardt 2023) as well as new findings for pathological conditions involving glia and vice versa (Chowen and Garcia-Segura 2021; Endo et al. 2022, see also the abovementioned references).

Glial cells can form dense arrays or be sparse in the neuropil, vary in shape and distribution in the gray and white matter, show short or long processes, thin or thick fibers with different radial branching pattern, and display "irregular outgrowths, which occasionally form lamellae, or with short, varicose, more or less branched spines or processes" (Cajal 1909–1911; Figs. 6.1, 6.2, and 6.3). For example, astrocytes comprise a high morphologically heterogeneous group (at least nine types proposed), whose array of shape, density, and proliferation rate would delimitate discrete regions in the brain (Emsley and Macklis 2006). The density of astrocytes differs across brain areas, ranging from a very low number in the core of the nucleus accumbens to a very high number in the subventricular zone (Matyash and Kettenmann 2010), or differ even in sub-regions of the same complex structure (e.g., the thalamic nuclei; Emsley and Macklis 2006). Therefore, the diversity of astrocytes likely indicates functional differences for these cells at the same time that it reflects each particular subregion architecture of cells and connections. In addition, astrocytes have a notable variety of membrane currents, expression, and effects of transmitters (e.g., glutamate, γ-aminobutyric acid (GABA), glycine, dopamine, serotonin, as well as beta-adrenergic, purinergic, neuropeptide, and complement receptors among others), glutamate transporter expression (e.g., distinct between astrocytes in brain and spinal cord), gap junction coupling (affected by age), and spatiotemporal organization of spontaneous or synaptically induced intracellular Ca^{2+} wave signaling (Matyash and Kettenmann (2010), for an extensive review see Verkhratsky and Nedergaard (2018), for different astrocytic roles on brain following CNS insults see Fan and Huo (2021), and for differences in inflammation and glia in PNS and CNS responses see Mietto et al. (2015)).

6.2.1 Morphological Features and Functional Implications for Complex Astrocytes (Including Human Glial Cells)

There is a bidirectional neuron-glia and glia-neuron communication relevant for neural network dynamic operations (Araque et al. 2014; Perea et al. 2014; Lago-Baldaia et al. 2020). Both classic staining techniques and advanced microscopic and cutting edge technologies (including the human/mouse chimeric model) demonstrate the shape and functional role of glial cells in specific brain areas (Hirbec et al. 2020) also related to dysfunctions observed in neurodevelopmental and neurodegenerative diseases (Lago-Baldaia et al. 2020; Chowen and Garcia-Segura 2021). For example, optogenetics and chemogenetics in combination with electrophysiology and behavior, single-nucleus RNA-seq and spatial transcriptomics, as well as noninvasive brain imaging methods in preclinical models and in humans are significant advancements to the field and for future research (Hirbec et al. 2020 and references therein).

However, the study of the function of glial subtypes is not a simple task because they are heterogeneous in form and function, do not have the same synaptic responses as recorded in neurons, and differ in their effects according to the brain regions studied (Hirbec et al. 2020). Astrocytic processes can be organelle-containing branches or organelle-free leaflets, which show higher motility, represent a part of the cell with high surface and low volume, and usually contact multiple and more synapses than branches

(Semyanov and Verkhratsky 2021). Some additional particular morphological features of astrocytes have important structural and functional implications. It is estimated that a single Lucifer Yellow-filled protoplasmic astrocyte in the rat hippocampal CA1 *stratum radiatum* occupies on average neuropilar volume of 66,000 μm^3 and would oversee approximately 140,000 synapses (Bushong et al. 2002). In the mice cerebral cortex in vivo, a single astrocyte occupies on average neuropilar volume of 21,000–24,000 μm^3, envelops on average four (maximum of eight) neuronal somata, 300–600 dendrites, and 36 (or more) spines per dendrite to form "functional islands of synapses" (Halassa et al. 2007; Fig. 6.4). In this regard, cell-to-cell communication can be altered by a single released gliotransmitter and by paracrine glia-derived extracellular vesicles, which elicit diverse responses depending on the target cells either locally or over relatively longer distances (Araque et al. 2014; Perea et al. 2014; Verkhratsky and Nedergaard 2018; Pistono et al. 2021). There are multiple signaling neuroactive molecules released by glial cells, such as glutamate, ATP, D-serine, and hormone metabolites (Araque et al. 2014; Chowen and Garcia-Segura 2021). For example, astrocytes activated by endocannabinoids released from neurons can secrete glutamate (Navarrete and Araque 2010). This and other gliotransmitters can alter neuronal function and synaptic efficacy from seconds to minutes, as well as provide spatial integration and coordination of networks (as "bridges for intersynaptic communication") several tens of micrometers away from the originally activated synaptic site (Araque et al. 2014).

Interastrocytic and astrocyte-oligodendrocyte are coupled as a functional syncytia by gap junction proteins connexin 30 and 43 (Gosejacob et al. 2011; Verkhratsky

Fig. 6.4 (continued) enhanced green fluorescent protein (EGFP) labeling of astrocytes, (right) structures with a volume value within 90% of the first Gaussian function are shown in green, representing single astrocytes, and structures with volumes equivalent to two astrocytes are shown in violet. (**b**) Distribution of the dendrite length (in μm) covered by a single astrocyte. From four neurons, the average length of basal dendrites (within 200 μm from the cell body) that extends across a single astrocyte is approximately 50 μm. (**c**) Two examples (left) of biocytin-filled cortical neurons stained with Alexa-conjugated streptavidin, and (right) higher-magnification images to evidence their dendritic spines. (**d**) Spine density was calculated from nine neurons, 105 basal dendrite segments of variable length (9–72 μm), excluding spines positioned above and underneath the dendritic shafts. The average linear spine is 0.7 ± 0.1 spine/μm and, on average, a single astrocyte has an estimated potential to contact at least 36 spines per dendrite. Up to 600 dendrites are associated with a single astrocyte (Halassa et al. 2007). (**e**) Ultrastructure of a presynaptic bouton (PrB) contacting a dendritic spine (Sp, purple) with a postsynaptic density (PSD) visible (arrow). Note that the synaptic site is completely surrounded by an astrocyte (Ast, yellow) at this focal plane. Scale bar = 0.5 μm. (**f, g**) Schematic representation of "functional synaptic islands" including neurons and astrocytes. In (**f**), one cell body of a neuron is represented among three adjacent astrocytes (within tissue volume colored in gray). Note that the dendrite can extend to another neighbor domain, indicating that different dendritic segments might be modulated by different astrocytes. Furthermore, (**g**) several spiny dendrites from different neurons can be enwrapped by a single astrocyte. "Synapses localized within the territory of this astrocyte have the potential to be modulated in a coordinated manner by gliotransmitter(s) released from this glial cell" (Halassa et al. 2007). (Legends adapted and figures (**a**)–(**d**), (**f**), and (**g**) reprinted from Halassa et al. (2007), https://doi.org/10.1523/JNEUROSCI.1419-07.2007, Copyright [2007] Society for Neuroscience; legend and figure (**e**) reprinted from the chapter published by Stewart et al. (2014), http://dx.doi.org/10.1016/B978-0-12-418675-0.00001-8, under license #(1349447) [230414-027626], Copyright Academic Press/Elsevier (2014))

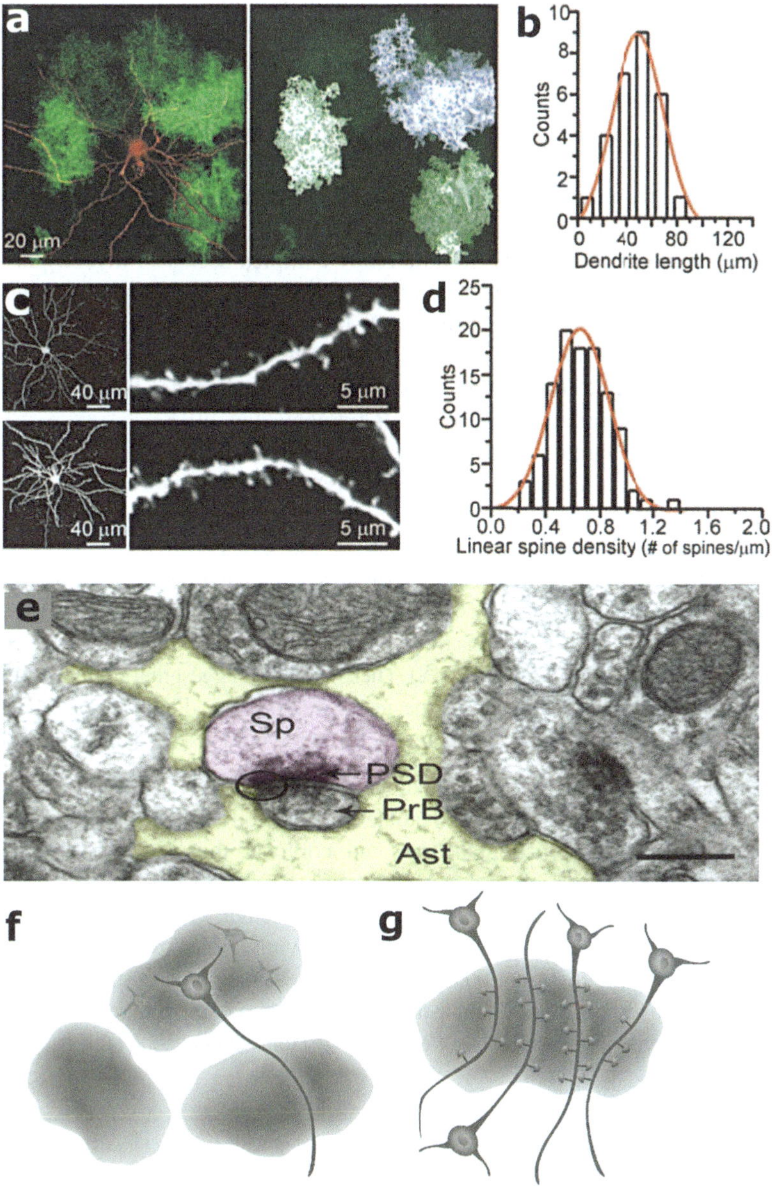

Fig. 6.4 Morphological relationship of astrocytes and neurons in cortical layers II/III of dnSNARE transgenic mice. Up to 600 dendrites are associated with a single astrocyte. (**a**) Top-view reconstruction (left) of a biocytin-filled cortical neuron (red) in a slice showing adjacent astrocytes enwrapping different dendrites of the same neuron. Based on the analysis of volume distribution of

and Nedergaard 2018). This feature is involved in the control of hippocampal excitatory synaptic transmission, the synaptic location of astroglial processes, proper astroglial extracellular glutamate and potassium removal, short- and long-term synaptic plasticity, and sensorimotor and spatial memory tasks (Pannasch et al. 2014; Lutz et al. 2009, reviewed in Perea et al. 2014). Some neighboring astrocytes are uncoupled, but close pairs of astrocytes in the gray matter have on average 230 gap junctions for intercellular diffusion of multiple molecules, and one single astrocyte can be connected with approximately 50–100 adjacent astroglial cells (Verkhratsky and Nedergaard 2018). Adding to an already complex scenario, myelin plasticity can serve strategically to adjust action potentials conduction velocity and to regulate the precise timing for integration of signals into neuronal circuits (Chorghay et al. 2018). This plasticity involves neuronal activity-dependent changes in the genesis, differentiation, and proliferation of the oligodendrocyte lineage and the glutamatergic signaling in "axomyelinic synapses" (Chorghay et al. 2018).

Similar to neural circuits, neurons, dendritic spines, and synapses higher complexity in diverse mammalian species, including humans (see Chap. 9 in this book and van den Heuvel et al. 2015), the diversity of astrocytes also shows species-specific differences. For example, cortical interlaminar and polarized astrocytes of higher primates are not found in other studied animals, cortical varicose projection astrocytes exist only in humans, and cortical astrocytes are larger and show a higher microanatomic complexity and phenotypic diversity in humans than in other primates or rodents (Oberheim et al. 2009; Verkhratsky and Nedergaard 2018). These findings suggest that some functions of local astrocytes may be unique in our species as well (see an additional commentary in Hirbec et al. 2020). Relevant data on human astrocytes demonstrated that:

1. Protoplasmic astrocytes in the human temporal neocortex can be visualized using immunofluorescence for the intermediate cytoskeleton filament glial fibrillary acid protein (GFAP), DiOlistic labeling, and confocal microscopy. These highly branched, bushy cells are organized in domains as in rodents, where astrocytes occupy their own anatomical neuropilar space together with neuronal cell bodies, dendrites and spines, synapses, and vessels, showing limited overlap between adjacent cells (Oberheim et al. 2009; Fig. 6.5a). However, human astrocytic domains show more overlaps than rodent astrocytes and, when compared to the rat allocortex and mice neocortex (Bushong et al. 2002; Halassa et al. 2007), it is estimated

Fig. 6.5 (continued) *f*: Human astrocyte diolistically labeled with a characteristic highly complicated network of fine process. Scale bar = 20 μm. Inset demonstrates the colocalization of GFAP immunolabeling (green) in a human protoplasmic astrocyte diolistically labeled. Scale bar = 20 μm. (**b**) Cortical fibrous astrocytes are also significantly larger in human than in rodent. *a*: Typical mouse fibrous astrocyte in white matter. GFAP (white) and sytox (blue) labeling. Scale bar = 10 μm. *b*: Typical human fibrous astrocytes in white matter. Scale bar = 10 μm. *c*: Human fibrous astrocytes are ~2.14-fold larger in diameter than the rodent counterpart. $^{*}p < 0.0001$. *d:* Human fibrous astrocyte labeled with DiI revealing the full structure of the cell. DiI (red) and sytox (blue). Scale bar = 10 μm. *e*: Mouse fibrous astrocyte labeled with DiI. Scale bar = 10 μm. (Legends with slight changes and figures reproduced from Oberheim et al. (2009), https://doi.org/10.1523/JNEUROSCI.4707-08.2009, Copyright [2009] Society for Neuroscience)

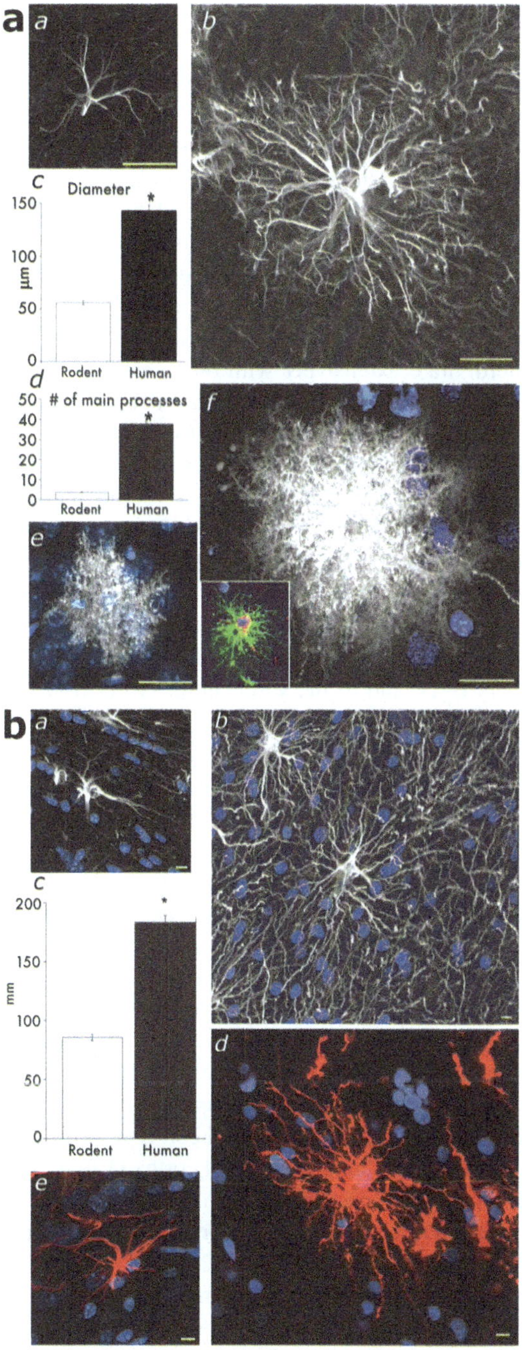

Fig. 6.5 (**a**) Human cortical protoplasmic astrocytes are larger and more complex than the rodent counterpart. *a*: Typical mouse protoplasmic astrocyte. GFAP labeling (white). Scale bar = 20 μm. *b*: Typical human protoplasmic astrocyte in the same scale. Scale bar = 20 μm. *c* and *d*: Human protoplasmic astrocytes are 2.55-fold larger and have tenfold more main GFAP processes than mouse astrocytes (mean ± SEM; $*p < 0.005$). *e*: Mouse protoplasmic astrocyte diolistically labeled with DiI (white) and sytox (blue) with numerous astrocytic fine processes. Scale bar = 20 μm.

that a single human large neocortical protoplasmic astrocyte with its fine bulbous processes covers from 270,000 to 2 million synapses within a single domain (Oberheim et al. 2009). This suggests that glioneuronal units in the human cortex integrate and process more inputs and more complex information than other species, implying that the computational power of human neural processing also increases with the size of its astrocytic domains and disproportionately relative to brain size alone (Oberheim et al. 2009, see also Herculano-Houzel 2012).

Likewise, temporal neocortex fibrous astrocytes found in the white matter of humans are larger than those in mice and rats, show relatively unbranched processes that interdigitate extensively with other neighboring fibrous astrocytes, and contact axons tracts and the local vasculature (Oberheim et al. 2009; Fig. 6.5b). These cells are larger than protoplasmic astrocytes, do not form domains, and display less fine bulbous processes likely related to the paucity of synaptic contacts in the surrounding neuropil (Oberheim et al. 2009).

Cortical layer-specific features of human GFAP-immunoreactive (GFAP-ir) astrocytes differ in terms of cellular complexity from those of chimpanzees, rhesus macaque, squirrel monkeys, rats, and mice. Temporal gray matter layer I has densely interlaminar astrocytes with both short processes, which form a thick network of fibers that contribute to the pial glial limitans, and long tortuous processes directed to layers II–IV and to the vasculature; layers II–VI show protoplasmic astrocytes; and, layers V and VI display sparse astrocytes with short radial "spiny" processes and one to five randomly radiating long (up to 1 mm in length) and straight processes with prominent varicosities, which are present only in higher-order primates although smaller and less complex in chimpanzee cortex than in humans (Oberheim et al. 2009, see also Fig. 6.1 from Cajal). These latter cells in humans might be related to the functioning of other cells along the cortical layers, transfering information between gray and white matter, and coordinating functional hyperemia (Oberheim et al. 2009). In addition to these morphological findings, human protoplasmic astrocytes also propagate Ca^{2+} signals to other cells several fold faster than rodents do, with functional implications for short- or long-distance communication, and responding by increasing intracellular Ca^{2+} to both ATP and metabotropic glutamate receptors stimulation (Oberheim et al. 2009).

2. Likely reflecting a high interconnectivity between cortical areas in humans, oligodendrocytes were the most common glial cell type, with a glia/neuron ratio of 2:1, within a sample of the human temporal cortex (1 mm^3, gray and white matter) imaged using a high-speed multibeam scanning electron microscopy and three-dimensional (3D) reconstruction (approximately 133 million synapses in a 1.4-petabyte volume; Shapson-Coe et al. 2021). Oligodendrocytes were more numerous in the white matter and found a minimum in cortical layer II. On the other hand, the astrocyte density divided cortical layer I into a superficial dense region and a deep region, which together with the adjacent layer II, showed fewer astrocytes than the upper layer I part or when compared to layers III–VI and the white matter (Shapson-Coe et al. 2021).
3. Many genes of CNS astrocytes relate to metabolism, cholesterol, and neurotransmitter synthesis and uptake (Endo et al. 2022). However, using single-nucleus RNA-sequencing analysis of the human middle temporal gyrus, nonneuronal types

showed 3643 genes expression with more than tenfold difference than in the same region of the mouse cerebral cortex (Hodge et al. 2019). This divergence of non-neuronal expression patterns suggests that the transcriptomic organization of cell classes and subclasses can be conserved along evolution, but species-specific and subregion-specific variations might reside at "the finest level of cell-type distinction" (Hodge et al. 2019, see relevant data in Xu et al. 2018; Hawrylycz et al. 2015; Miller et al. 2019), such as in the number and extension of astrocytic processes (Oberheim et al. 2009) and relation to multiple spine synapses.

The most-divergent gene expressions in the human neocortex include cell-adhesion molecules, ion channels neurotransmitter receptors, and ECM elements (Hodge et al. 2019). This finding suggests that the neuropil (including astrocytic processes) might have developed with a higher connectivity repertoire throughout mammalian evolution of the cerebral cortex (Oberheim et al. 2009; DeFelipe 2011; Navarrete and Araque 2014; Mederos et al. 2018; Rasia-Filho et al. 2021; Shapson-Coe et al. 2021, and relevant commentaries on cell type and species-specific genes in Xu et al. (2018)).

Indeed, it has been demonstrated that neurons, dendritic spines, and synapses show particularities in humans (see Chap. 9 in this book), which might also reflect the structure and function of glial cell counterparts. Humans have unique astrocytes of specific morphological types at the same time that these and other glial cells show morphological variants depending on region and developmental stage along the CNS and PNS (Oberheim et al. 2009; Matyash and Kettenmann 2010; Seguella and Gulbransen 2021). Reduction of hippocampal astrocyte morphology affects neural circuit function and impairs the performance of mice in a cognitive test (Endo et al. 2022). That is, astrocytic changes in ionic channels, neurotransmitters receptor and transporter expression, gliotransmitters, and extension of astrocytic domains covering various synaptic sites would enhance or decrease synaptic strength and circuitry functioning, which, in turn, would regulate the operational capabilities of neural networks and the elaboration of higher functions of the human brain (Mederos et al. 2018). "Therefore, the major difference in brain volume between humans and primates is not only due to a higher development of the neuronal neuropil but also to a higher number and complexity of astrocytes. Perhaps what makes us human is in part due to astrocytes" and additional glial cells (Navarrete and Araque (2014), see a complementary comment in Verkhratsky and Nedergaard (2018)).

6.3 Structural and Functional Relationships Between Glial Cells, Dendritic Spines, and Synaptic Plasticity

As mentioned above, the diversity in the phenotypic and molecular architecture and in the neuropil microenvironment of each area is evident for glial cells in the CNS, PNS, and enteric nervous system. Astrocyte heterogeneity is expected because of the intricate interactions with other cells whose shape and function are plastic in the CNS (Miller 2018). The morphology of astrocytes is non-uniform even within the same brain subregion (e.g., the CA1 *stratum radiatum* of rats; Bushong et al. 2002), which shows a notable plasticity related to the formation of new memories and learning based on activity-dependent circuit structural and synaptic modifications (Toni et al. 1999; Neves et al. 2008). Astroglial gene expression can be upregulated by neuronal activity and show GFAP mRNA levels increased at sites of stimulation (Steward et al. 1991). In other words, glial cells within and between subtypes can be molecularly and phenotypically plastic, adapting their morphology (more or less extensively) and functions in response to cues and demands from cellular and network microenvironments to maintain local homeostasis (see relevant comments in Seguella and Gulbransen 2021). That may also be related to synaptic plasticity. For example, the motility of hippocampal synapse-associated astrocytic processes was enhanced in fewer than 5 min after long-term potentiation (LTP), an effect dependent on presynaptic activity and on G-protein-mediated Ca^{2+} elevations in these glial cells (Perez-Alvarez et al. 2014). Environmental enrichment stimulation increases synaptic and spine densities in the mice hippocampal *stratum radiatum* CA1 region and, although not altering astrocyte number or the density of GFAP-ir, increased the ramification of local astrocytic processes (Viola et al. 2009).

Some astrocytic thin terminal processes are observed in close structural association with synaptic elements, as evidenced by transmission electron microscopy and 3D reconstructions of serial sections or 3D-Stimulated Emission Depletion (STED) microscopy in living organotypic brain slices (Bernardinelli et al. 2014; Stewart et al. 2014; Arizono et al. 2020; Fig. 6.6a–c). This functional assemblage would provide

Fig. 6.6 (**a**–**c**) Ultrastructure and 3D reconstructions illustrating glial coverage of excitatory spine synapses in the adult mouse cerebral cortex. (**a**) Electron microscopy micrograph of the cortical neuropil showing dendritic spine synapses (sp) surrounded by astrocytic processes (indicated by asterisks in blue). Note mitochondria (m) in the branching point of astrocytic processes and perisynaptic astrocytic processes (PAPs) contacting axon-spine interface (ASI, arrows). (**b**) 3D reconstruction of the PAP near spine synapses with simple PSDs in adult rat hippocampus, evidencing a dendritic segment with two spines (yellow), PSDs (red), and PAP (blue). (**c**) Similar 3D reconstruction for a large spine with complex PSD. Arrows point an ASI covered by the PAP. Note that in (**b**) and (**c**), a large fraction of ASI is devoid of PAPs and the pattern of small astrocytic processes surrounding the base, neck, and head of a dendritic spine and the adjacent dendritic shaft. Scale bar = 1 μm. (Legends slightly adapted and figures reprinted from Bernardinelli et al. (2014), https://doi.org/10.1155/2014/232105, Copyright © 2014 Yann Bernardinelli et al. under the Creative Commons Attribution License, originally published by Hindawi). This glial coverage of dendritic spines can show dynamic structural interplay or form stable astrocyte–spine interactions. For example, (**d**) in the order of few minutes, an astrocytic process (green) extends toward a dendritic spine (red). (**e**) Otherwise, interaction can be lost between a dendritic spine (red) and a retracting astrocytic process (green). Insets in (**d**) and (**e**) show 90° or top-down views of the astrocyte–spine interactions over time, respectively. Scale bars = 1 μm. (Legend slightly adapted and figure reprinted from Haber et al. (2006), https://doi.org/10.1523/JNEUROSCI.1302-06.2006, Copyright [2009] Society for Neuroscience). Astrocytic processes contact both immature protrusions with no postsynaptic specializations or those that have already formed synaptic contacts. (**f**–**j**) Dendritic protrusions (red) persist longer and stabilize by astrocytic contacts (green, **f**, **h**, **j**, arrows) or retract if they do not contact astrocytic processes (**g**, arrow). New dendritic protrusions (**h**, **j**) can have either prolonged (**h**) or transient (**j**, arrowheads) astrocytic contacts to subsequently stabilize. The same dendritic protrusion is presented with different colors in (**h**) (red) and (**i**) (white). (Legend slightly adapted and figure reprinted from Nishida and Okabe (2007), https://doi.org/10.1523/JNEUROSCI.4466-06.2007, Copyright [2007] Society for Neuroscience)

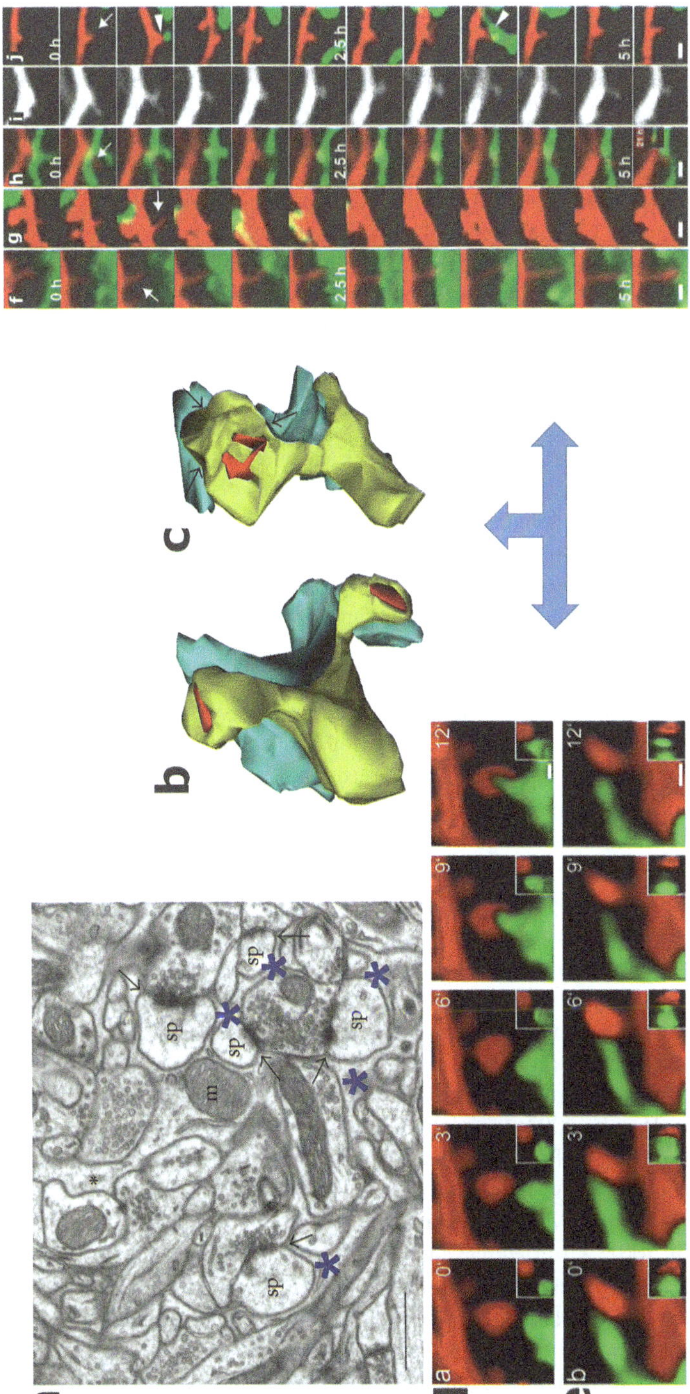

Fig. 6.6 (*caption on p. 270*)

fine-tuning of transmission at the same time that incorporates astrocytes as an active component of the synapses. This view of the synapse-glial association led to the term "tripartite synapse." This relationship may involve presynaptic and postsynaptic elements with perisynaptic Schwann cells or astrocytes (Araque et al. 1999, 2014).

Dendritic spines are specialized postsynaptic integrative units (see Chap. 1 in this book). Some synapses can have tight contacts as evidenced by the presence of *puncta adhaerens* at the edges of synapses on one-third of dendritic spines in the rat *stratum radiatum* of hippocampal CA1 area (Spacek and Harris 1998). In these same samples, approximately two-thirds of these *puncta adhaerens* occurred between the nonsynaptic surface of dendritic spines and adjacent astrocytic processes (Spacek and Harris 1998).

All brain regions have perisynaptic astrocytic process (PAP), but the difference resides in the proportion of synapses having PAPs and the extent of synaptic coverage by glial processes (Bernardinelli et al. 2014). That is, not every synapse is embraced by PAPs (Lavialle et al. 2011). When this relationship occurs, there are various patterns for the astrocytic engulfment of the synaptic elements (Fiala and Harris 1999; Fig. 6.6b–e). As reported in rats and mice, axospinous synapses vary in the surface/volume extent surrounded by glial processes, ranging from a complete glial covering of spines in most cerebellar Purkinje cells to approximately 29% in pyramidal cells of the visual cortex (Spacek 1985); half of synapses with an adjacent glial process in the somatosensory cortex (Kasthuri et al. 2015); and near 40% of the synapses in the CA1 *stratum radiatum* show no astroglial ensheathment (Ventura and Harris 1999). The nonuniform distribution of PAPs is also reflected in some CA1 synapses with no PAPs (thus lacking glial isolation) to almost all big spine synapses with a complex postsynaptic density (PSD) displaying a partial astrocytic terminal coverage following the induction of synaptic potentiation (Bernardinelli et al. 2014, see also Lushnikova et al. 2009). Accordingly, synaptic potentiation induced an enlargement of the spine head volume, PSD area, presynaptic terminals, and glial coverage of both pre- and postsynaptic structures (prevented by a NMDA receptor antagonist), indicating an activity-mediated structural synaptic remodeling in the hippocampus in vitro (Lushnikova et al. 2009). In this same area, thin and mushroom dendritic spines were associated with plastic properties for learning or synaptic stability and intraspine composition (Bourne and Harris 2007).

Another example of the astrocytic engulfment function in synapses involves the cerebellar function. In cerebellar mossy fiber glomeruli and granule cell synapses, glia do not form a physical barrier and insulate most release sites from each other; thus, desensitization and short-term depression occur after paired-pulse stimulation (Xu-Friedman and Regehr 2003). The same situation does not occur in cerebellar climbing fibers, where 87% of the perimeter of the synaptic cleft was surrounded and segregated by glia, suggesting that a glia-free area with glutamate spillover from a single site can desensitize AMPA receptors at many neighboring postsynaptic sites (Xu-Friedman et al. 2001; Xu-Friedman and Regehr 2003). In hippocampal synapses, bigger spines have less coverage by astroglial processes than small spines and, then, spine size can inversely correlate with the efficacy of local glutamate uptake or possibility of spillover (Herde et al. 2020). Transmitters leaving the synaptic cleft can serve to extra- and heterosynaptic signaling and synaptic cross talk, as well as let astrocytes with possibilities to affect the composition and diffusional flow of bioactive molecules and volume in the extracellular space (Bernardinelli et al. 2014 and references therein, see a recent critical discussion in Rusakov and Stewart 2021).

6.3.1 *Motility of Perisynaptic Astrocytic Processes Toward Dendritic Spines*

Thin astrocytic terminal processes are notably dynamic, can rapidly extend and retract fine processes, and change their shape and position in the neuropil with an average motility higher than dendritic spine counterparts (Haber et al. 2006; Nishida and Okabe 2007; Fig. 6.6d–j). The rapid remodeling of astrocytic processes modifies the coverage of the synaptic elements on a spine-by-spine basis. In hippocampal slice cultures, astrocytic processes contact both immature protrusions with no postsynaptic specializations or those that have already formed synaptic contacts at their base or neck. Then, protrusions retract if they do not have a contact with astrocytic processes or dendritic protrusions persist longer, stabilize, and have a higher tendency to form spines (Nishida and Okabe 2007, see also Sultan et al. 2015; Fig. 6.6f–j).

PAPs can extend to or withdraw from synaptic sites in an activity-dependent manner by involving the actin-binding protein ezrin and the metabotropic glutamate receptors 3 and 5 compartmentalized to astrocytic terminals, but not to the GFAP-containing branches (Lavialle et al. 2011). In addition, during synaptic development in the mouse cortex, the normal thalamocortical synaptic connectivity and refinement on specific dendritic spines involve the astrocyte-secreted protein hevin (Risher et al. 2014). Also secreted from astrocytes, small extracellular vesicles containing fibulin-2 increase the formation of spines and promote synapse formation involving the transforming growth factor β signaling (Patel and Weaver 2021). Astrocytic contacts can also occur on dendritic shaft synapses and perikaryon (Ostroff et al. 2014; Zancan et al. 2015; Semyanov and Verkhratsky 2021). For example, astrocytes can modulate spine occurrence (including somatic spines), axosomatic and axospinous synapses, and electrophysiological properties of cerebellar Purkinje cells (Seil 2001).

Interestingly, network activity can modulate spine and glial structure. For example, the coverage of spines and synaptic cleft by astrocyte terminals changes in the rat lateral amygdala during consolidation of fear conditioning or conditioned inhibition into long-term memory (Ostroff et al. 2014). The former procedure induces a transient increase in the density of synapses with no astrocytic contact and a net increase in astrocyte-free perimeter on enlarged spines with polyribosomes and a spine apparatus (Ostroff et al. 2014). Additional experimental findings indicate that synaptic perimeter and spine size may be inversely proportional to the fraction of the spine membrane covered with PAPs (see an important discussion in Bernardinelli et al. 2014). In dentate granule cells of rats, PAPs selectively approach synapses on thin dendritic spines (Medvedev et al. 2014), but in the hippocampal CA1 *stratum radiatum,* a similar astrocytic coverage of both small and larger spines was observed in glutamatergic synapses (Gavrilov et al. 2018). In synapses of the hippocampal *stratum radiatum* and in cerebellar parallel fibers, the occurrence of glia postsynaptically is three- to fourfold higher than presynaptically (Lehre and Rusakov 2002). Such asymmetry can restrict glutamate spillover, reuptake, and action outside the synaptic cleft, and differently affect presynaptic terminals and postsynaptic responses (Lehre and Rusakov 2002).

Microglia are also very active and "sculpt" dendritic spine density in the following ways: by phagocytosing spines not receiving synaptic inputs; modulating signaling cascades that promote the formation of dendritic spines; regulating synaptogenesis

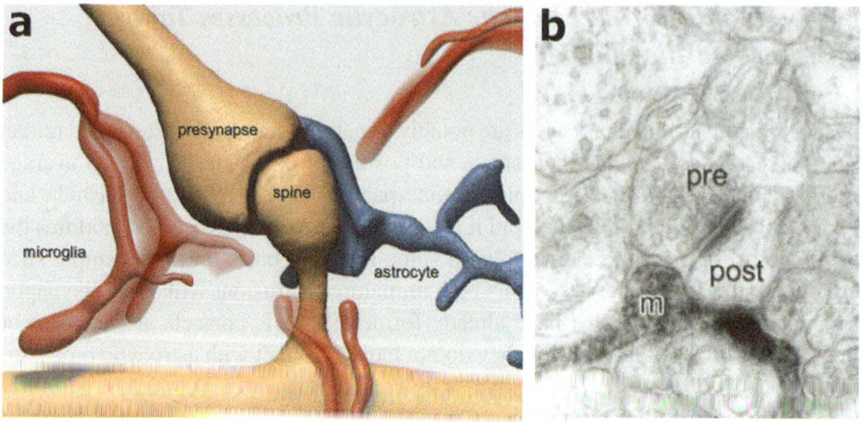

Fig. 6.7 Microglia participation in a tripartite synapse. (**a**) Microglial processes (in red) contact the presynaptic axonal terminal and the postsynaptic dendritic spine (both in brown) as well as the enwrapping perisynaptic astroglial process (in blue). (**b**) Ultrastructural identification of a microglial process (m) with a direct contact at the synaptic site of presynaptic (pre) and postsynaptic (post) elements. (Legend adapted and figure reprinted from Kettenmann et al. (2013) under CCC RightsLink® license # 5527280846542, originally published by Elsevier, and (**b**) from Wake et al. (2009), https://doi.org/10.1523/JNEUROSCI.4363-08.2009, Copyright [2009] Society for Neuroscience)

and synaptic pruning; interacting with small and transiently growing dendritic spines or large ones depending on sensorial stimulation and neuronal activity; eliminating dysfunctional synapses; and contributing to brain wiring, maturation, and function (Kettenmann et al. 2013; Squarzoni et al. 2015; Colonna and Butovsky 2017; Fig. 6.7). An interplay among neurons, astrocytes, and microglia can, also, occur in oxidative damage and inflammation related to aging, a process that also decreases dendritic spines density and synaptic plasticity in cortical areas (Aguilar-Hernández et al. (2023), see beneficial effects of exercise and hippocampal microglia function in Wang et al. (2023) and aging effects in microglia and cortical spines in Chan et al. (2018)).

6.3.2 Glial Cells Modulate Function from Synapses to Neural Circuits and Behavior

Glial cells are sensitive to a variety of substances related to synaptic activity within their microenvironment and by levels of circulating hormones (Chowen and Garcia-Segura 2021). For example, T_3 regulates the differentiation of oligodendrocyte precursor cells and Schwann cell function and, like testosterone and androgen metabolites or progesterone and progesterone derivatives, promotes the formation of central and peripheral myelin (Garcia-Segura and McCarthy 2004 and references therein, see sex differences in hormonal actions on glial cells in Chowen and Garcia-Segura 2021).

Gonadal hormones-induced morphological and functional changes in neurons, astrocytes, and synapses include, but are not limited to, the hypothalamus and the amygdaloid complex. These are two areas relevant for the timely neuroendocrine secretion and coordination of the reproductive behavior display in rats (Garcia-Segura et al. 1995; Rasia-Filho et al. 2002, 2012). In the rat hypothalamic arcuate nucleus, gonadal steroids affect the architecture and the formation of synapses on neurons that secrete gonadotrophin-releasing hormone (GnRH). This area modulates the negative and positive feedback actions of ovarian hormones by dynamically changing the coverage of the perikaryal membrane available for synapses (Fernandez-Galaz et al. 1997). The switch in GnRH secretion leads to a peak of luteinizing hormone (LH) and follicle-stimulating hormone (FSH) in circulation and, then, ovulation. During the preovulatory (proestrus) and ovulatory (estrus) phases of the ovarian cycle, steroid-responsive local glia promote a transient disconnection of the tonic inhibitory synaptic inputs reaching the cell body of arcuate neurons, releasing the activity and secretion of GnRH neurons. This precise, phasic synaptic remodeling is induced by estradiol, dependent on insulin-like growth factor I (IGF-I), and is blocked by progesterone. It involves the extension of astroglial process to ensheath and restricts the amount of neuronal membrane available for the GABAergic synaptic inputs during some specific hours of the 4–5 days ovarian cycle in females (Garcia-Segura et al. 1994; Fernandez-Galaz et al. 1997, 1999). Males show another astroglial pattern, with a higher amount of neuronal surface covered by astrocytic processes and a reduced number of synaptic contacts on the neuronal cell body compared to females (Garcia-Segura et al. 1995). Males also have more GFAP-ir stellate cells with longer processes than females in the developing hypothalamic arcuate nucleus (Mong et al. 1996). The hormonally mediated neuronal/glial cross talk for synaptic modulation also occurs in the developing preoptic area in the anterior hypothalamus, another area involved with reproduction. Local astrocytes promote dendritic spine synapse formation by local release of glutamate, which, in turn, depends on the action of estradiol-induced prostaglandin E_2 likely synthesized in neurons and, involving histamine from mast cells, by local microglia as well (Amateau and McCarthy 2002; McCarthy 2023). Therefore, bidirectional neuroglial signaling regulates phasic GnRH secretion by integrating estrogen and progesterone actions, plastic astrocytic structural rearrangements, and direct cell-cell communication also including the local release and actions of several growth factors (Ojeda et al. 2000, see also Neal-Perry et al. 2014).

At the same time that structural transformations mediated by sex steroids remodel neuroglial circuitry in the hypothalamus, other brain areas also show phasic changes in astrocytic processes (Verkhratsky and Nedergaard 2018). Although noted in only a fraction of the astrocytic structure, the expression of GFAP was found to be sexually dimorphic and affected by ovarian steroid fluctuation in the hippocampal dentate gyrus, CA1 and CA3 regions (Arias et al. 2009), and in the posterodorsal medial amygdala (MePD) of adult rats (Rasia-Filho et al. 2002; Martinez et al. 2006). GFAP-ir was higher in proestrus females in these three hippocampal regions and, in the dentate gyrus, the lowest GFAP values were found in males (Arias et al. 2009). Proestrus females also showed rounded astrocytes with various short processes, whereas in astrocytes with stellate shape, few and long processes were observed in males and diestrus females (Arias et al. 2009). In ovariectomized females, GFAP-ir in the dentate gyrus significantly increased by 5 h after a single injection of 17β estradiol and by 1 h after the injection of progesterone (Luquin et al. 1993).

The MePD is part of a plastic network that modulates sensorial processing and the sexual behavior of males (intromission and ejaculation) and females (disinhibition of sexual display and facilitation of sexual receptivity) and is connected with hypothalamic nuclei for GnRH secretion; changing with emotional/social processing and learning (Rasia-Filho et al. 2012; Zancan et al. 2015). The specific function of local glial cells remains unknown, but astrocytic remodeling can occur in physiological and pathological conditions in rats (Rasia-Filho et al. 2002; Martinez et al. 2006; Salazar et al. 2014).

The right McPD contains more astrocytes than the left MePD in rats, and males have a higher astrocyte number in the right and more complex astrocytes in the left MePD than females (Johnson et al. 2008). Additional findings in the MePD of both hemispheres demonstrated a sex difference in dendritic spines temporally coincident with phasic changes in GFAP-ir across the estrous cycle. GFAP-ir was higher in females in proestrus than in the other phases of the estrous cycle (Martinez et al. 2006). At the same time, the number and shape complexity of somatic spines increase (Zancan et al. 2015), the density of proximal dendritic spines significantly decreases by approximately 30% (Rasia-Filho et al. 2004), and a higher proportion of shaft synapses with a switch to mainly inhibitory inputs occur in proestrus, indicating a local dynamic remodeling of synapses (Brusco et al. 2014; Dalpian et al. 2019). The reduction in dendritic spine density during proestrus is coincident with less synapsin reactivity at the same time that an increase in local GFAP-ir is evident (Martinez et al. 2006; Oberlander and Erskine 2008). In another set of experiments, estradiol alone or estradiol plus progesterone increased GFAP-ir in the MePD of adult ovariectomized females (Martinez et al. 2006). In conjunction, these data indicate naturally morphological and electrophysiological changes in synaptic and spine plasticity paralleled by astrocytic remodeling in the MePD.

A causal relationship between astrocyte structural plasticity and neuroendocrine secretion or behavioral display remains to be demonstrated for various brain areas. This possibility opens up another avenue for further research. Sex differences in brain wiring may also involve microglial function. During the first postnatal weeks, hippocampal microglia volume and phagocytic capacity show precocious development and a high density of dendritic spines connected by functional, synaptophysin-containing boutons in females than male mice (Weinhard et al. 2018b), for example.

6.4 Synaptic Modulation by Glial Cells: Integrating Multiple Functions

Synapses in the CNS, sites of electrochemical signaling between dynamic circuits, exhibit morphological reorganizations during development, aging, and behavior (Grutzendler et al. 2002; Trachtenberg et al. 2002; Holtmaat et al. 2005; Zuo et al.

2005). The appropriate control of synaptic structure may be essential for neural network processing and information in the brain. Importantly, glial cells have a crucial role in synapse development and physiology (Castonguay et al. 2001; Haydon 2001; Volterra and Meldolesi 2005). In particular, dendritic spines undergo geometric changes that can provide rapid adjustments to postsynaptic ionic balance (Volfovsky et al. 1999), protein trafficking (Richards et al. 2004), and signaling events, regulating synaptic efficacy (Segal 2005) related to long-term plasticity (Matsuzaki et al. 2004; Nägerl et al. 2004; Zhou and Sutherland 2004; Zhou et al. 2004). The cellular compartments of the synapse (glia and neuronal pre- and postsynaptic compartments), together with a noncellular element, the functional scaffold ECM (see Chap. 3 in this book), named tetrapartite synapse (Chelini et al. 2018), are morphologically and functionally plastic (Haber et al. 2006; Theodosis et al. 2008; Reichenbach et al. 2010; Semyanov and Verkhratsky 2021). The glial component of synapses consists of (i) astrocytes; (ii) NG2 cells; and (iii) microglia (Chelini et al. 2018).

Astrocytes, with their different and specialized morphologies (e.g., protoplasmic, fibrous, perivascular, and Bergman glia), control neuronal activity and synaptic plasticity (Matyash and Kettenmann 2010; Farmer and Murai 2017). NG2 cells (also termed "polydendrocytes"), or oligodendrocytes, represent a distinct glial cell type with unique properties that play a significant role in the regulation of synaptic functions and plasticity (Wang et al. 2004; Sakry et al. 2011; Hill and Nishiyama 2014; Dimou and Gallo 2015; Eugenín-von Bernhardi and Dimou 2016; Larson et al. 2016). Microglia, in addition to their role as specialized macrophages of the CNS, interact with neurons to guide axonal growth, dendritic sprouting, and synapse spine remodeling (Rodríguez-Iglesias et al. 2019; De Luca et al. 2020). Oligodendrocytes, even without direct contact with dendritic spine, have the potential role to act as signal transducers and control the extracellular environment (Wang et al. 2004; Fig. 6.8).

6.4.1 *Astrocyte–Dendritic Spine Interaction*

Astrocytes, although electrically "silent", buffer the extracellular environment, respond to presynaptic activation with G-protein-mediated Ca^{2+} signals, and propagate Ca^{2+} signals (Haydon 2001), triggering the release of glutamate, ATP, D-serine, and GABA, which modulate local synaptic transmission (Charles et al. 1991; Perea and Araque 2007; Halassa and Haydon 2010; Parpura and Zorec 2010; Butt 2011; Araque et al. 2014; Harada et al. 2015). Also, astrocytes rapidly control extracellular glutamate and GABA levels through their specific uptake transporters to restrict synaptic transmission (Danbolt 2001) and modulate synaptic function (Volterra and Meldolesi 2005; Panatier and Oliet 2006; De Luca et al. 2020). For example, a decrease in the astrocytic glutamate transporter 1 (GLT1) can induce an increase in dendritic spine density in the basal dendrites of layer II-III excitatory neurons (Sipe et al. 2021) and increased paired-pulse facilitation in adult-born neurons of the

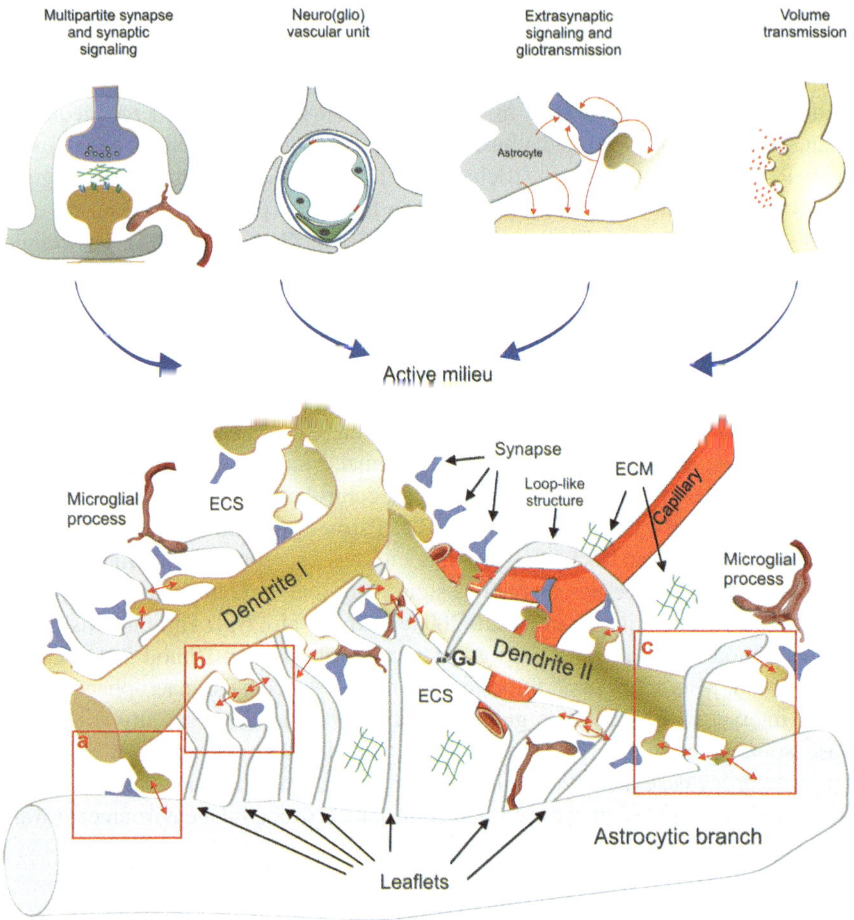

Fig. 6.8 Astrocyte-neuron communication in the active milieu. The active milieu concept integrates multiple theories that address different aspects of local functional organization of the brain: multipartite synapse, neuro(glio)vascular unit, extrasynaptic signaling, and volume transmission. The active milieu is formed through dynamic interactions between neuronal elements (somata, axons, dendrites, and spines), nonneuronal cell elements (astrocytic and microglial processes), vasculature (capillary), extracellular space (ECS), and extracellular matrix (ECM). In an active milieu, synapses can contact, signal, and be homeostatically controlled by astrocyte branches (a), by single or by several leaflets (b). A single astrocytic branch or leaflet may be contacted by several synapses (c). Dynamic changes in the morphology of astrocytic processes affect diffusional barriers, neurotransmitter clearance, K^+ dynamics, and the supply of glutamine or energy substrates, thus regulating neuronal plasticity. Astrocytic processes form loop-like structures through reciprocal gap junctions (GJs). (Reproduced from Semyanov and Verkhratsky (2021), https://doi.org/10.1016/j.tins.2021.07.006, Figure 1 under CCC RightsLink® license #5525900218857)

hippocampal dentate gyrus from mice (Krzisch et al. 2015), suggesting that perisynaptic processes modulate synaptic transmission in these cells.

In cortical and hippocampal synapses, astrocytic processes are localized closer to dendritic spines than to presynaptic terminals (Lehre and Rusakov 2002; Krzisch et al. 2015; Arizono et al. 2020; Fig. 6.9). They show cooperative and dynamic motility with spines, indicating a complex mutual interaction (Haber et al. 2006) involving several cell adhesion molecules (CAMs) and ECM factors to modify synaptic connectivity and remodeling. This astrocyte–synapse interaction adds another level of complexity to the synapse, mainly for long-term plasticity and circuit maintenance (Saint-Martin and Goda 2022; Fig. 6.10). Numerous findings revealed the effects of astrocyte-derived CAMs and ECM factors on the formation, maturation, maintenance, and function of the tripartite synapse (Hillen et al. 2018). In addition to neurons, astrocytes are connected to the vascular system through the ECM, particularly during sleep, by the glymphatic system (Plog and Nedergaard 2018). This is a possible CNS lymphatic-like system mediated by astrocytes that is responsible for ECM clearance of accumulated molecules and control of metabolic supplies (Lundgaard et al. 2017). Cell adhesion proteins of the ECM form bridges between neurons (presynaptic and postsynaptic compartments) and the PAPs at the synapse making that presynaptic neuron–astrocyte, postsynaptic neuron–astrocyte, and astrocyte–astrocyte interactions may dynamically affect spine morphology, synapse formation, efficacy, and function.

The Ephrin (Eph) system (tyrosine kinase receptors (Eph receptors) and associated ephrins ligands), for example, although not strictly classified as CAMs, is involved in the synaptic maturation of neuronal and astrocytic synaptic elements (Saint-Martin and Goda 2022). The interaction of ephrin-A3 on perisynaptic astrocytic processes with

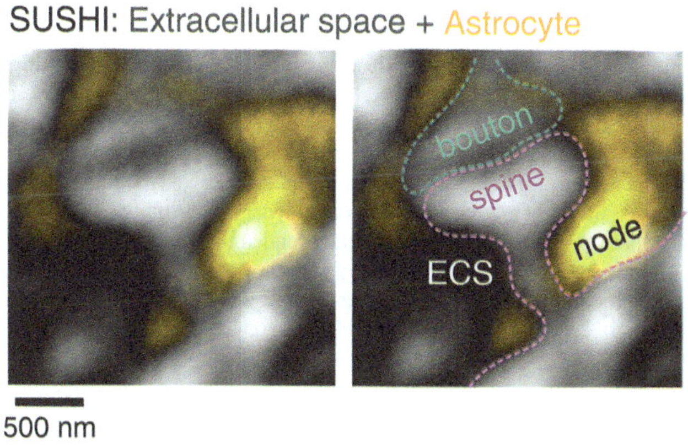

Fig. 6.9 Super-resolution shadow imaging (SUSHI) technique showing image of extracellular space (black) surrounding a synapse (gray) and astrocyte (yellow). Right image indicates the putative identity of the synaptic components. (Reproduced from Arizono et al. (2020), https://doi.org/10.1038/s41467-020-15648-4, Figure 3d under CC BY license, originally published by Springer Nature)

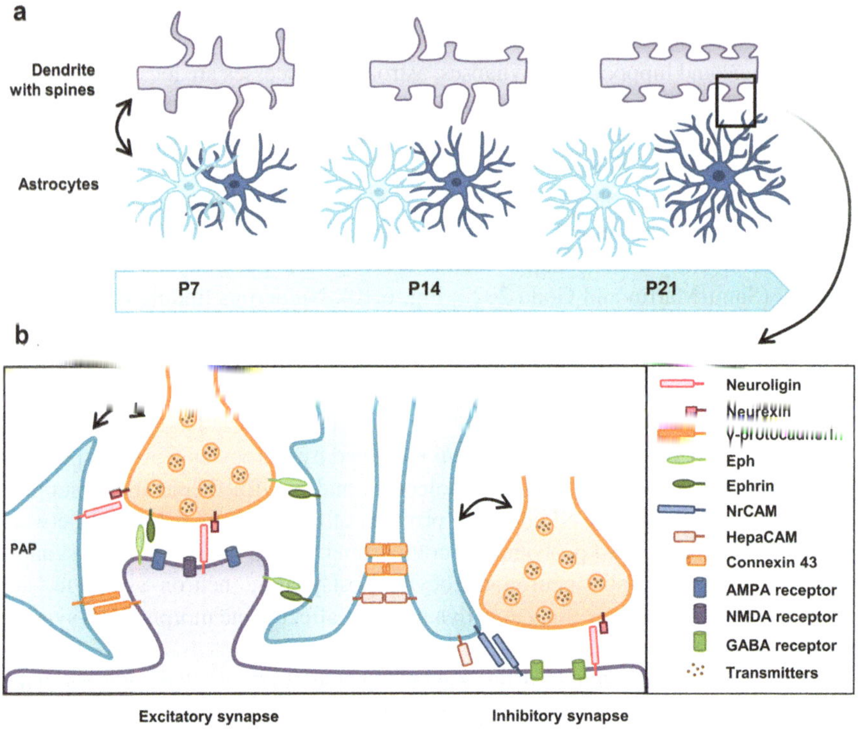

Fig. 6.10 Astrocyte–neuron cross talk. (**a**) During development, astrocyte–neuron crosstalk regulates synaptogenesis and the morphological elaboration of astrocytes that occur in the same time frame. Synapse formation starts during the first postnatal week, peaks around postnatal day (P) 14, and stabilizes around P21. In parallel, astrocytes at P7 extend long processes with little ramification that overlap with processes of neighboring astrocytes. At P14, astrocytic processes become increasingly ramified, and between P21 and P28, astrocytes reach a mature densely ramified morphology. During the course of maturation, astrocytes segregate into nonoverlapping territories. (**b**) Astrocyte–synapse crosstalk is mediated by cell adhesion proteins expressed in perisynaptic astrocytic processes (PAPs, in blue) such as neuroligins, γ-protocadherin, ephrins or NrCAM. Astrocyte cell adhesion proteins interact with proteins of the presynaptic compartment (orange) and the postsynaptic compartment (purple) to modulate diverse functions such as astrocyte morphology elaboration, excitatory and inhibitory synapse formation or synaptic plasticity. The cell adhesion protein hepaCAM, found both at astrocyte-astrocyte contacts and at the synapse, modulates astrocyte morphogenesis and segregation as well as inhibitory synapse formation. The double-headed arrows in (**a**, **b**) indicate reciprocal interactions between astrocytes and synapses. (Reproduced from Saint-Martin and Goda (2022), https://doi.org/10.1111/febs.16540, Figure 1 under CCC RightsLink® license 5525910759781)

postsynaptic EphA4 tyrosine kinase receptors contributes to spine morphology (Murai et al. 2003; Carmona et al. 2009). It also controls glial glutamate transporter levels in adult mice through posttranscriptional regulation (Carmona et al. 2009), affecting glutamate levels in the synaptic cleft and long-term plasticity (Filosa et al. 2009; Tanasic et al. 2016). Another ephrin ligand, ephrin-B1, also plays a role in astrocyte–neuron interactions. The astrocytic ephrin-B1 functions as a negative regulator of synapse formation in the hippocampus, where it modulates new excitatory synapse formation (Koeppen et al. 2018; Nguyen et al. 2020), probably by binding to the EphB receptor on the synaptic membrane (Koeppen et al. 2018). Astrocytes also participate in the secretion of proneurotrophins and their processing in the ECM (De Luca et al. 2018) and rapidly internalize proneurotrophins into perineuronal astrocytes via p75NTR-mediated endocytosis and recycling (Papa et al. 2014). Other ECM molecules important for synapse formation and plasticity are thrombospondin (TSP) and hevin. TSP is secreted by astrocytes (Liauw et al. 2008). TSP regulates synaptogenesis through binding of EGF (epidermal growth factor)-like TSP-2 domains to the VWF-A (von Willebrand factor) domain of postsynaptic $\alpha 2\delta$-1 subunits of L-type calcium channels (Eroglu et al. 2009). A deficit in TSP-1 protein expression in astrocytes from patients with Down syndrome was detected in human astrocyte/rat hippocampal neuron coculture, leading to dendritic spine malformation and reduced synaptic density (Garcia et al. 2010). Hevin, a member of the secreted protein acidic and rich in cysteine (SPARC) family, is expressed both by neurons and by astrocytes and has been localized in the PAP (Jones et al. 2011; Kucukdereli et al. 2011; Buosi et al. 2018). The astrocytic hevin interacts with the neurexin/neuroligin system to ensure spine contact and synaptic cleft stability (Singh et al. 2016; De Luca et al. 2020). Also, astrocytic SPARC expression is dynamically regulated and plays a critical role in determining the level of synaptic AMPA receptors (Jones et al. 2011). A number of other CAMs and ECM factors are involved in astrocyte–(glial)–neuron interaction for remodeling spines and controlling synaptic efficacy (for detailed information on individual astrocyte–synapse adhesion proteins, see Hillen et al. 2018).

The PAPs are highly dynamic to respond to synaptic signals, presenting increased mobility during synaptic activity associated with dynamic changes in the correspondent dendritic spines and reduced motility related to stable dendritic spines (Haber et al. 2006). Although the impact of synaptic activity on PAP motility is not fully understood, this mobility involves Na-K-Cl cotransporter-1 (NKCC1) and the actin-binding protein cofilin (Henneberger et al. 2020). The PSP reversibly can detach from synapses without compromising the structural integrity of the synapses. There are several possible ways in which astrocyte–synapse contact-mediated CAMs could contribute to the detachment of the PAP. For example, by competing for protein binding amongst CAMs and their components, by protein cleavage resulting in physical destruction of adhesion, and/or by degradation, leading to the loss of interacting protein substrates (Saint-Martin and Goda 2022; Fig. 6.11). The importance of understanding astrocytic function in the synapse and interaction with ECM has a great impact on a number of synaptic pathologies (Chung et al. 2015; Kim et al. 2017), such as schizophrenia, autism spectrum disorder, and epilepsy.

The PAPs in the cortex and hippocampus come into close physical contact with many synapses, appearing to wrap around them in electron microscopy images (Witcher et al. 2007). Neurotransmitters released from presynaptic neurons not only bind to postsynaptic receptors but are also detected by astrocytes, leading to an increase in intracellular Ca^{2+} concentration. Astrocytes thus release glutamate, D-serine, and ATP (Araque et al. 2014) that can influence synaptic function,

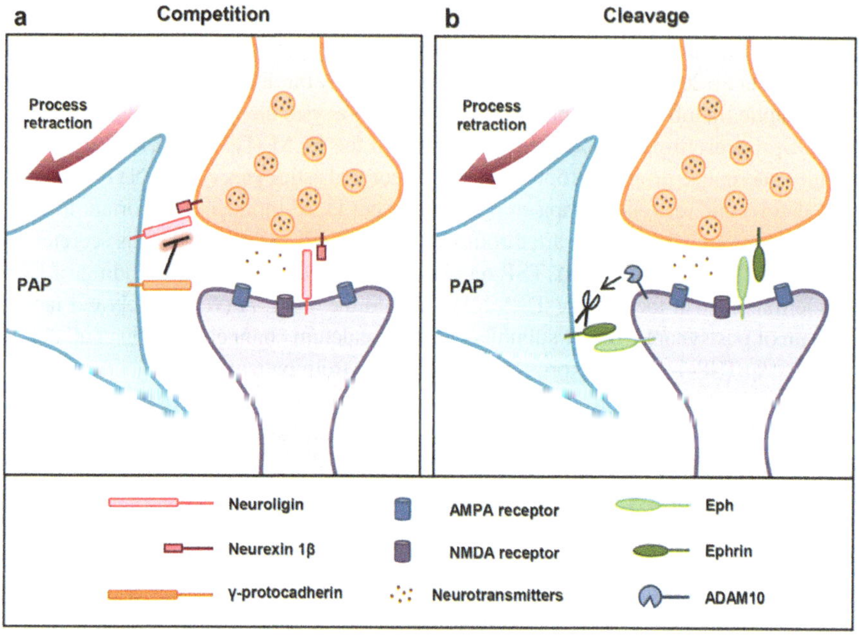

Fig. 6.11 Potential mechanisms for PAP motility. Perisynaptic astrocytic processes (PAPs, in blue) interact with the presynaptic compartment (orange) and the postsynaptic compartment (purple) via direct cell–cell adhesion. The motility of PAPs increases during synaptic activity. The present hypothetical mechanisms would drive the retraction of a PAP from a synapse, such as by competing for protein binding (**a**) or protein cleavage (**b**). In an example of the first scenario, the sequestering of astrocytic neuroligin by its binding to astrocytic γ-protocadherin in cis depletes the pool of astrocytic neuroligins that are available to interact with the presynaptic neurexin in trans (**a**). In an example of the second scenario, ADAM10 cleaves astrocytic ephrin upon its binding to Eph that severs the ephrin–Eph interaction (**b**). (Reproduced from Saint-Martin and Goda (2022), https://doi.org/10.1111/febs.16540, Figure 2 under CCC RightsLink® license 5525910759781)

including basal synaptic transmission (Panatier and Oliet 2006; Di Castro et al. 2011) and long-term synaptic plasticity (Henneberger et al. 2010; Papouin et al. 2017; Robin et al. 2018). The PAPs exhibit fast local Ca^{2+} signals in microdomains, similar to neurons (Panatier and Oliet 2006; Di Castro et al. 2011), supporting the fact that astrocytes have the ability to regulate single synapses. To study the morphological details of astrocytes and the astrocyte-dendritic spine contact, live-cell 3D-STED microscopy was performed in living brain tissue (Arizono et al. 2020). With this strategy, it was possible to observe that in the hippocampus and barrel cortex, the majority of dendritic spines were contacted by nodes, which were grouped along thin astrocytic processes or shafts and had the size correlated with the spines (Fig. 6.12). In addition, fluorescence recovery after photobleaching (FRAP) experiments and confocal Ca^{2+} imaging showed that nodes were isolated biochemical compartments that hosted highly localized spontaneous Ca^{2+} transients in microdomains (Fig. 6.13). Mapping the Ca^{2+} transients onto STED images of nodes and dendrites revealed that these localized Ca^{2+} signals occurred in nodes that were in contact with dendritic spines, showing that they were associated with individual synapses. Astrocytic element are nodes and these nodes may be a primary site for Ca^{2+} signaling at excitatory tripartite synapses, providing the structural framework and molecular machinery for synapse-specific communication between astrocytes and neurons (Arizono et al. 2020; Fig. 6.14).

Another study using a combination of serial-section immunoelectron microscopy, confocal microscopy, and electrophysiology on adult-born neurons in the hippocampal dentate gyrus from mice found that the afferent and efferent synapses of newborn neurons have PAPs, regardless of the age of the neurons or the size of their synapses. Interestingly, preexisting astrocytes remodel their processes to actively participate in the synapses of adult-born neurons, suggesting that astrocytes play a functional and structural role in the integration of these cells into the hippocampal network (Krzisch et al. 2015; Fig. 6.15).

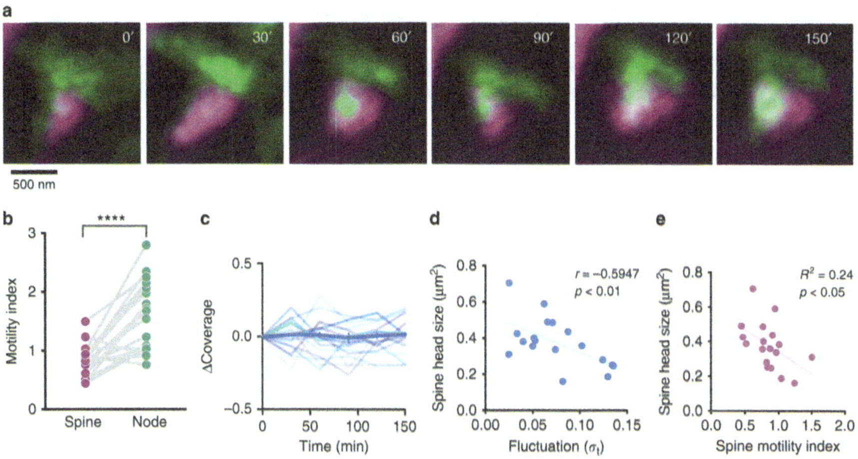

Fig. 6.12 Node-spine contacts are largely stable. (**a**) Time-lapse STED imaging of a spine (magenta) and a node (green) over 150 min acquired every 30 min. While contacting node can move on the spine surface, the overall contact is maintained. (**b**) Comparison between spine motility index and contacting nodes. Motility index of the node was significantly higher than that of spines ($n = 18$ structures from five slices; paired t-test, ****p (two-tailed) < 0.0001). (**c**) Changes of coverage state over time normalized to the coverage state at 0 min ($n = 18$ spines from five slices; Spearman r = −0.5947, p (two-tailed) = 0.0092). (**d**) Correlation between fluctuation (the temporal variance of coverage fraction) and spine head area ($n = 18$ spines from five slices; Spearman $r = -0.5947$, p (two-tailed) = 0.0092). Large node/large spine contact was more stable. (**e**) Correlation between spine head size and spine motility index. Large spines were less motile than small spines ($n = 18$ spines from five slices; Pearson R2 = 0.24, p (two-tailed) = 0.039). (Reproduced from Arizono et al. (2020), https://doi.org/10.1038/s41467-020-15648-4, Figure 4 under CC BY license, originally published by Springer Nature)

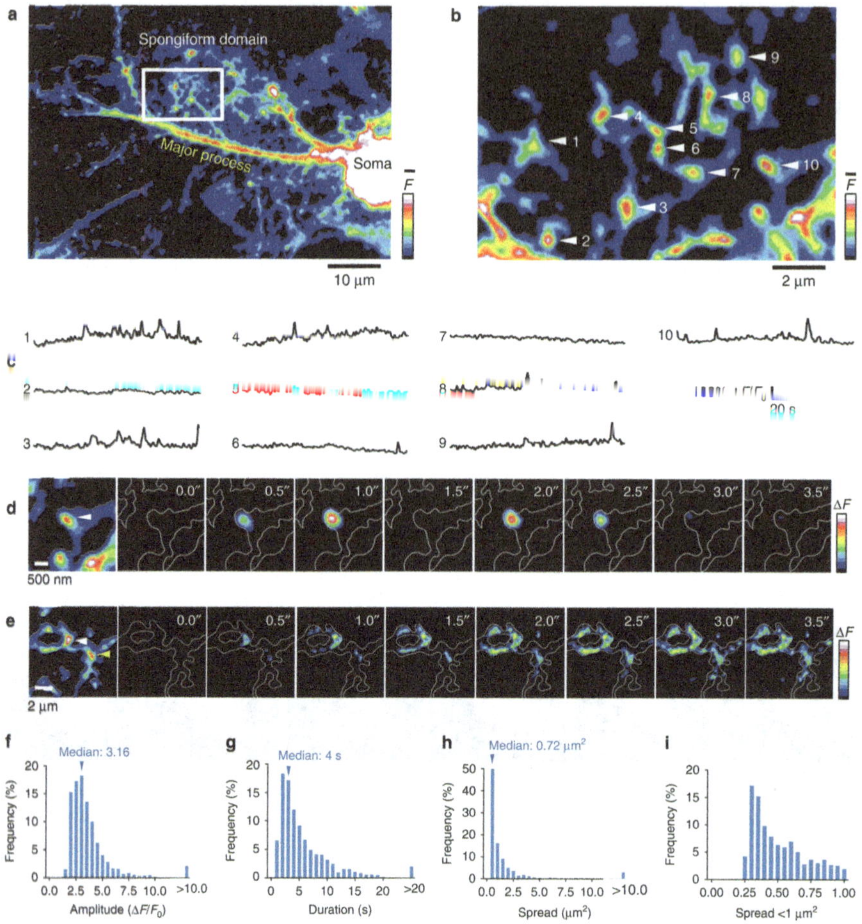

Fig. 6.13 Astrocytic nodes exhibit microdomain Ca^{2+} signals. (**a**) A confocal overview image of a Ca^{2+} sensor GCaMP6s-expressing astrocyte. (**b**) A zoom image of a box in (**a**). ROIs were manually placed on typical nodes. The general anatomical layout of shafts and nodes is still recognizable in confocal images. Nodes typically appear as bright spots at branch points. (**c**) Spontaneous Ca^{2+} traces from ROIs shown in b, showing that each node exhibits unique pattern of Ca^{2+} signals. (**d**) Time-lapse images of Ca^{2+} events, which were spatially restricted to a single node (right), and the corresponding confocal image of the structure where node is indicated with a white triangle (left). (**e**) Time-lapse images of a Ca^{2+} event, which originated at a node and spread via the connecting shafts to a neighboring node (right), and the corresponding confocal image of the structure where the originating node is indicated with a white triangle and neighbor node is indicated with a yellow triangle (left). (**f**) Frequency distribution of amplitude of Ca^{2+} events (n = 1583 events from five slices). (**g**) Frequency distribution of duration of Ca^{2+} events (n = 1583 events from five slices). (**h**) Frequency distribution of spread of Ca^{2+} events. In 59% of the events, the spread was less than 1 μm^2 (n = 1583 events from five slices). (**i**) Frequency distribution of spread of Ca^{2+} events under 1 μm^2, suggesting that the majority of Ca^{2+} events in the spongiform domain were confined to local domains corresponding to single nodes (n = 941 events from five slices). (Reproduced from Arizono et al. (2020), https://doi.org/10.1038/s41467-020-15648-4, Figure 6 under CC BY license, originally published by Springer Nature)

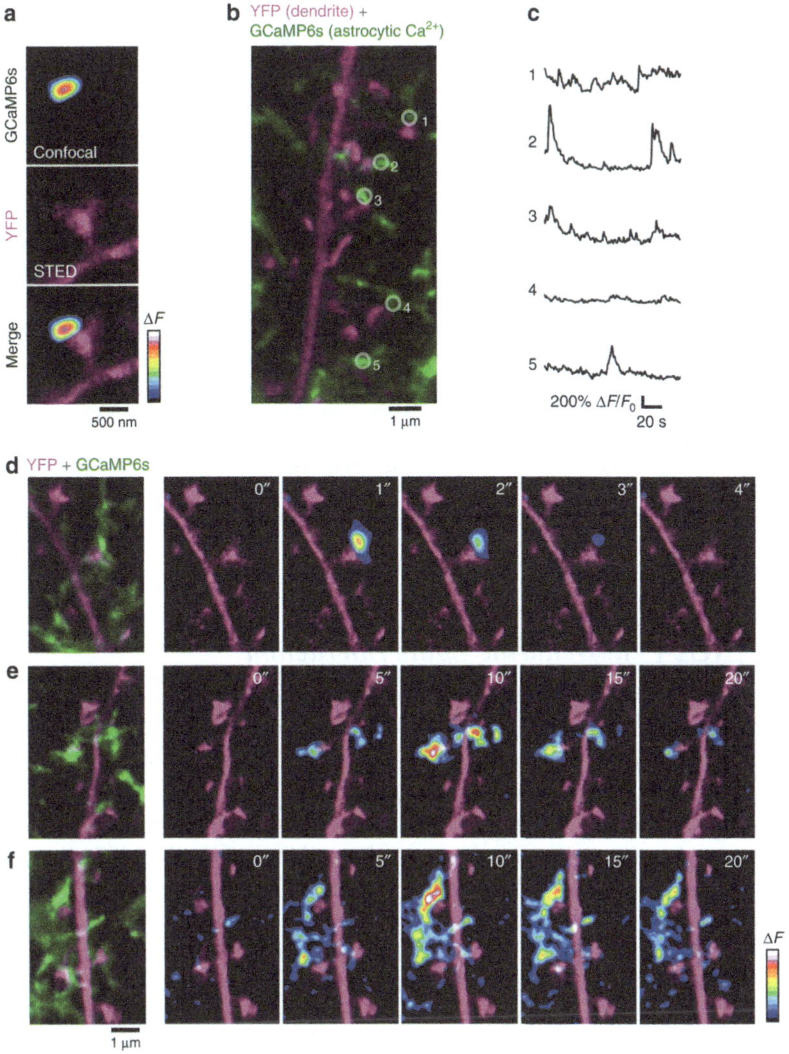

Fig. 6.14 Astrocytic nodes are likely functional component of excitatory tripartite synapses. (**a**) Confocal Ca^{2+} signal in node mapped onto STED image of spine morphology. (**b**) A STED image of YFP-labeled neuron and GCaMP6s-expressing astrocyte. (**c**) Spontaneous Ca^{2+} traces from ROIs shown in b. Note that nodes exhibit unique activation patterns. (**d–f**) Left: STED images of YFP-labeled neurons and GCaMP6s-expressing astrocytes. Right: Ca^{2+} activity mapped onto dendritic morphology, where the Ca^{2+} event remained confined to a single node (**d**), where two perisynaptic nodes were coactive (**e**) or where a Ca^{2+} wave appeared to propagate within the local astrocytic network (**f**).

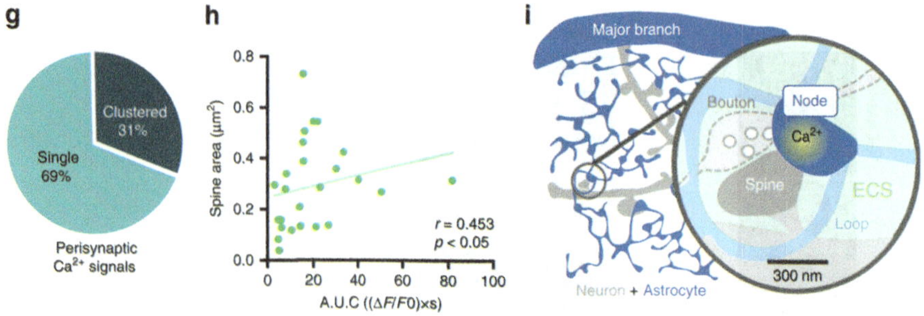

Fig. 6.14 (continued) (**g**) Percentage of perisynaptic Ca^{2+} events confined to single nodes or involving multiple nodes, showing that majority of Ca^{2+} events were confined ($n = 61$ events from 30 slices). (**h**) Correlation between spine size and area under curve (AUC) of the node Ca^{2+} event. Large spines were associated with large Ca^{2+} events in nodes ($n = 26$ structures from 24 slices; Spearman $r = 0.453$, *p (two-tailed) = 0.0201). (**i**) Summary schematic of the tripartite synapse microenvironment consisting of a presynaptic bouton, dendritic spine, astrocytic node, shaft, loop, and extracellular space (ECS). Nodes appear as the functional astrocytic component of excitatory tripartite synapses, which may provide the anatomical basis for synapse-specific communication. (Reproduced from Arizono et al. (2020), https://doi.org/10.1038/s41467-020-15648-4, Figure 8 under CC BY license, originally published by Springer Nature)

6.4.2 NG2 Cell–Dendritic Spine Interaction

NG2 cells are glial cells expressing Chondroitin sulfate proteoglycan NG2 platelet-derived growth factor a receptors (PDGFaR), O4 sulfatide (O4) (Reynolds and Hardy 1997), and represent the main source of mature oligodendrocytes (Dawson et al. 2000). Although the morphological features are similar to mature astrocytes, they have a distinct molecular signature (Reynolds and Hardy 1997; Butt et al. 2005). NG2 cells express voltage-gated ion channels and ligand-gated ionotropic neurotransmitter receptors, which enable them to have a distinctive electrophysiological profile (Larson et al. 2016). They are the only glial cell type capable of receiving synaptic GABAergic and glutamatergic contacts from neurons and responding to excitatory inputs with excitatory postsynaptic currents and

Fig. 6.15 (continued) timepoint, 85 neurons in total). (**d**) Confocal micrographs of newborn neurons (red) in GFAP-GFP mice (left panels) and Aldh1l1-GFP mice (express GFP under the control of the aldehyde dehydrogenase 1 (ALDH1L1) promoter) (right panels), illustrating one neuron intersecting the territory of several astrocytes (upper panels, arrows) and several neurons intersecting the territory of the same astrocyte (in white squares, lower panels). Scale bars = 20 μm. (**e**) Electron micrographs of multiple synapse boutons (stars) formed by the spine of a newborn neuron (dark immunolabeling) and embraced by a perisynaptic process (false-colored in green). Arrowheads point to dendritic spines. (**f**) 3D reconstruction of the multiple synapse boutons illustrated in the lower panel of (**e**). Scale bars in (**e**) and (**f**) = 0.5 μm. (**g**) Electron micrographs of mossy fiber terminal from new neurons (dark immunolabeling, false-colored in red) synapsing with the dendrite or thorny excrescence of a CA3 pyramidal cell (not colorized) that synapse with another nonlabeled mossy fiber terminal (blue). The perisynaptic processes are colored in green. (**h**) 3D reconstruction of a perisynaptic process (green) embracing a mossy fiber terminal from a new neuron (red) and from a nonlabeled neuron (blue). Scale bars in (**g**) and (**h**) = 1 μm. (Reproduced by Krzisch et al. (2015), https://doi.org/10.1007/s00429-014-0768-y, Figure 3 under CC BY license, originally published by Springer Nature)

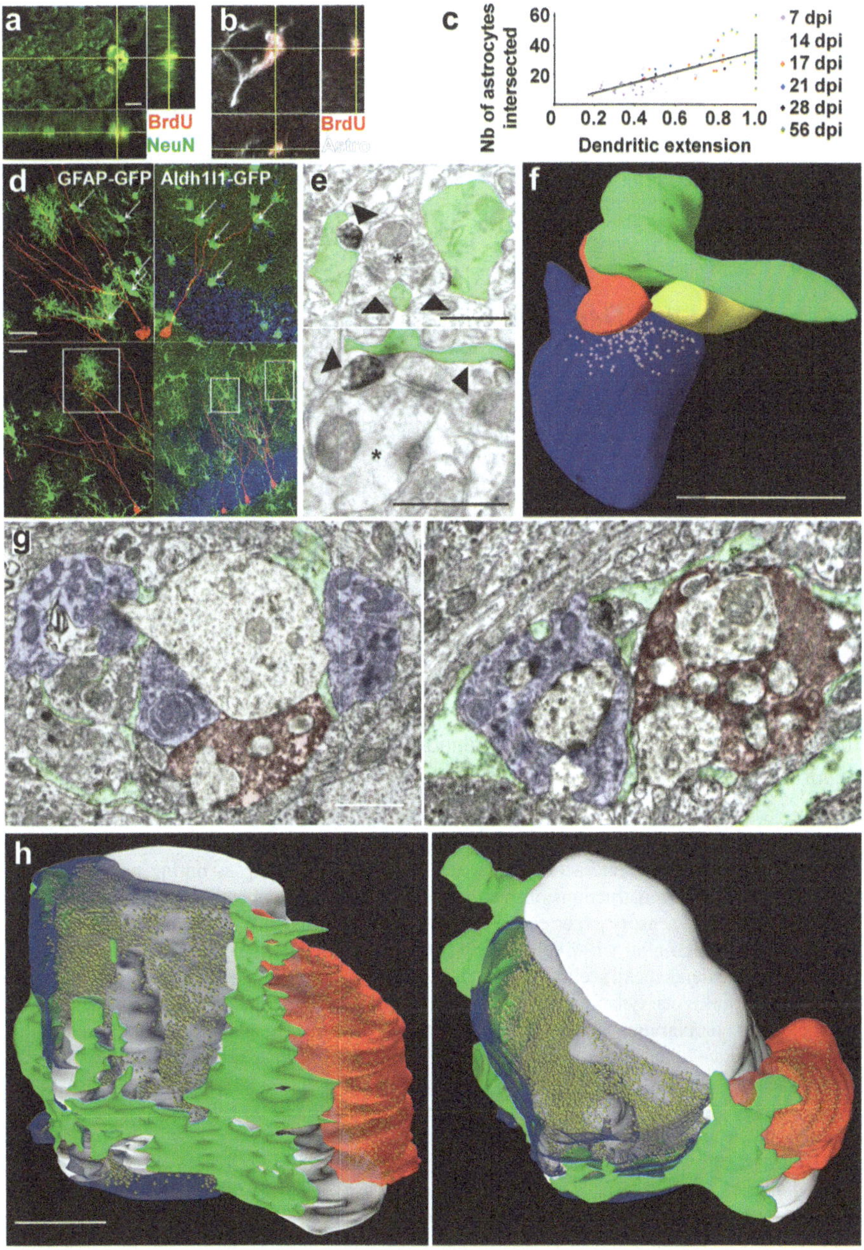

Fig. 6.15 Preexisting astrocytes embrace synapses from adult-born neurons. (**a**, **b**) Confocal micrographs and orthogonal projections of hippocampal sections immunostained for BrdU and NeuN (**a**) and BrdU and GFAP/S100b (Astro). (**b**). Scale bar = 5 µm. (**c**) Scatter plot showing the number of astrocytic territories intersected by newborn neurons, as a function of their normalized dendritic extension (Spearman's rank correlation test, $p = 0.0001$, R2 = 0.64, n = 6–26 neurons per

activity-dependent modifications equivalent to LTP at excitatory synapses (Bergles et al. 2000; Lin and Bergles 2004; Ge et al. 2006). Electron microscopy studies demonstrated that NG2 cell processes often encapsulate neuronal cell bodies and contact synapses (Butt et al. 2002). Downregulation of NG2 expression results in altered subunit composition of AMPA receptors and a marked reduction of NMDA-dependent LTP (Sakry et al. 2014), suggesting an important influence of these cells on synaptic transmission.

6.4.3 Microglia–Dendritic Spine Interaction

Microglia make frequent contact with synapses wrapping around dendritic spines, playing important roles in synaptic pruning, removing inactive synapses, remodeling dendritic spines, and maintaining plasticity during brain development and adulthood (Paolicelli et al. 2011; Schafer et al. 2012). Microglia regulate neuronal synapses and/or spine formation either directly or indirectly (Parkhurst et al. 2013; Miyamoto et al. 2016; Reshef et al. 2017). Studies have shown that almost all of the microglial processes (approximately 94%) form adjacent contact with axonal terminals, dendritic spines, PAPs, or synaptic clefts and about 68% of all microglial processes contact more than one synaptic element simultaneously (Tremblay et al. 2010; Šišková and Tremblay 2013; Fig. 6.16).

The interposition of microglial cells in the synaptic cleft (Chen et al. 2014) and the ability of their filopodia to sustain dendritic spine formation and relocation on the dendritic shaft are important to guide synaptic plasticity (Weinhard et al. 2018a; De

Fig. 6.16 Ultrastructural interactions between microglia and synapses during normal sensory experience and 3D reconstructions of the ultrastructural interactions between microglia and synapses during normal sensory experience. (**a–c**) Electron microscopy (EM) images showing IBA1-immunostained microglial (m+) cell bodies (**a**), as well as large (**b**) and small (**c**) processes, surrounded by extended extracellular space (asterisks) and contacting axon terminals (blue), dendritic spines (pink), perisynaptic astrocytes (green), and synaptic clefts (arrowheads). d, dendrite; N, nucleus; p, perikaryon. Scale bars = 250 nm. (**d**) EM image showing extended microglia-associated extracellular spaces (asterisks) after glutaraldehyde instead of acrolein fixation. The unlabeled microglial process (m) makes direct contacts with dendritic spines (s) and axon terminals (t), and displays an inclusion (in), as well as a clathrin-coated pit (black arrow) at the site of contact with a spine. Scale bar = 250 nm. (**e**) Extracellular space areas with or without contact with IBA1-positive microglial process (mean ± SEM). $^{**}p = 0.01$. (**f**) Correlation between the areas of microglial processes and associated extracellular space ($r = 0.48$; $p = 0.0001$). (**g**) Partial 3D reconstruction of the microglial proximal process (P; taupe) cut in transverse. Purple indicates the inclusion. Both processes simultaneously contact multiple axon terminals (blue), dendritic spines (red), and perisynaptic astrocytic processes (green), and are distinctively surrounded by extracellular space pockets of various size and shape (white; black arrows). (**h**) Additional view showing only microglia, dendritic spines, and extracellular space. (**i, j**) Insets illuminating the 3D geometry of the distal protrusion and its structural relationships with one axon terminal (t1), two dendritic spines (s1 and s2; postsynaptic density in dark red), and a pocket of extracellular space (white), which are partially reconstructed. For clarity, an astrocytic process was removed from the scene. Scale bars = 250 nm. (Reproduced from Tremblay et al. (2010), https://doi.org/10.1371/journal.pbio.1000527, (**a–f**) from Figure 2 and (**g–i**) from Figure 3 under CC BY license originally published by PloS)

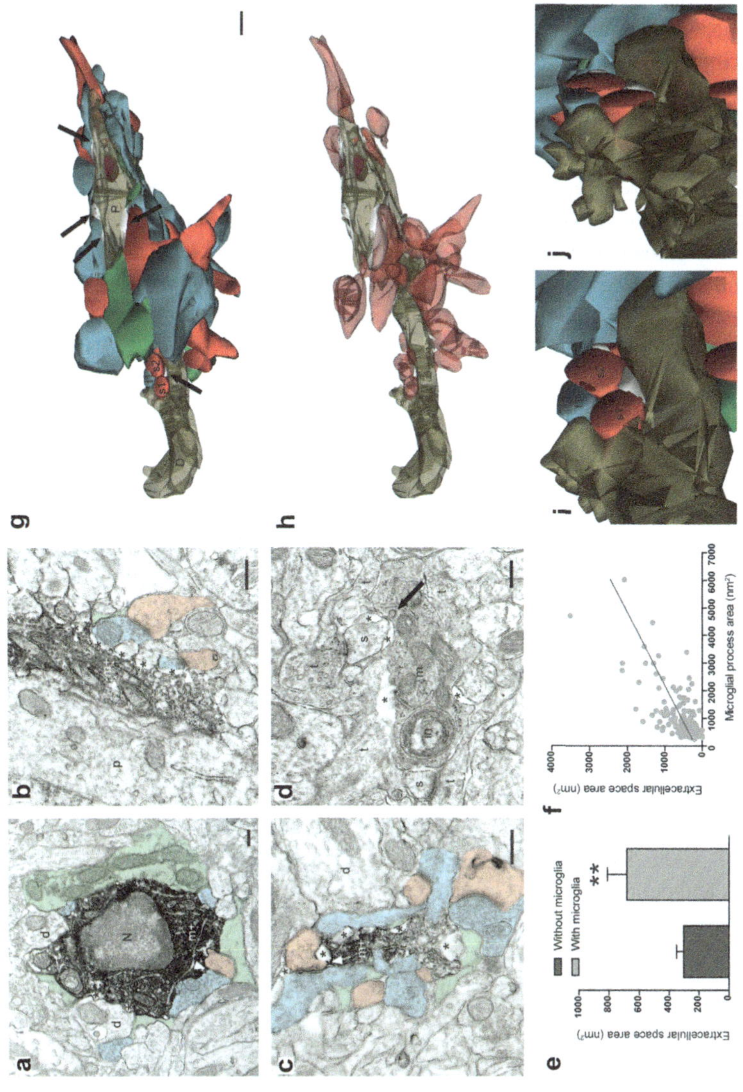

Fig. 6.16 *(caption on p. 288)*

Luca et al. 2020). Experiments using in vivo two-photon microscopy to evaluate microglia–synapse interactions in the living brain of mice (Wake et al. 2009) demonstrated that microglial processes make direct and brief (about 5 min/h) contacts with synapses located on dendritic spines in an activity-dependent manner. Later, the combined approach of in vivo high-resolution imaging of physiologic and activated microglia with imaging of cortical layer V neuronal activity in awake mice, at both the single synapse level and across neural circuits, was used to evaluate the interaction between microglia and neuronal synapses and how it can modulate neural circuit activity (Akiyoshi et al. 2018). With this approach, it was possible to detect and quantify localized Ca^{2+} transients within single dendritic spines. The frequency of Ca^{2+} transients and associated back-propagating action potentials (bAPs) increased during the microglial contact period (Akiyoshi et al. 2018). Spines without any apparent microglial contact had no significant changes in the frequency of Ca^{2+} transients, as well as in the parent dendrite before microglia contact. Also, the consistent increase in spine activity was reversible with microglial process withdrawal from the spine (Akiyoshi et al. 2018). Thus, physiologic microglial contact with the spine can selectively enhance neuronal activity (Akiyoshi et al. 2018; Fig. 6.17).

Notably, microglia are not only attracted by neurons but are also able to affect neural activities. The microglia-neuron contact results in an increase in the synchronization of activity across local neuronal populations. This increase in synchrony is inhibited by microglial activation, proving a possible mechanism by which immune status may affect neural circuit function (Akiyoshi et al. 2018) and suggesting that microglia could potentially play a primary role in cognitive dysfunction (Fig. 6.18).

Microglia can also modulate synaptic plasticity by the production and secretion of molecules, such as ECM components, neurotrophic factors (e.g., brain-derived neurotrophic factor (BDNF)) essential for learning-induced synaptic formation (Parkhurst et al. 2013), and endocannabinoids, cytokines (e.g., tumor necrosis factor α (TNFα)), microribonucleic acids by surface-anchored release, or the formation of extracellular vesicles (Szepesi et al. 2018). The expression of a variety of receptors also allows microglia to mediate neural-glia communications and regulate synaptic structural plasticity (Pocock and Kettenmann 2007; Wu et al. 2015; Liu et al. 2016; Colonna and Butovsky 2017), such as the receptor for chemokine CX3CR1 (Tarozzo et al. 2003; Reshef et al. 2017; Szepesi et al. 2018), receptors for neurotransmitters and neuropeptides (Colonna and Butovsky 2017), glutamate, GABA, ATP-P2Y12, dopamine, catecholamines, and acetylcholine.

In response to neuroinflammation, activated microglia lose their original ability to maintain synapses and reduce frequency of contact with synapses, disturbing the proper synchronized activity of neurons and the plasticity of neural circuits (Akiyoshi et al. 2018). TNFα, for example, released by microglia and astrocyte/microglia cross talk in maladaptive plasticity or neuroinflammation, modulates expression of NMDA and AMPA receptors at the synapse (Pickering et al. 2005). Moreover, the absence of microglial CX3CR1 receptors for CX3C chemokine ligand 1 (CX3CL1) released by neurons in development leads to an increase in PSD95 signal and spine density with reduced PSD95 inclusion, suggesting that synapses were not phagocytosed by microglia (Tarozzo et al. 2003; Reshef et al. 2017; Szepesi et al. 2018; Fig. 6.19).

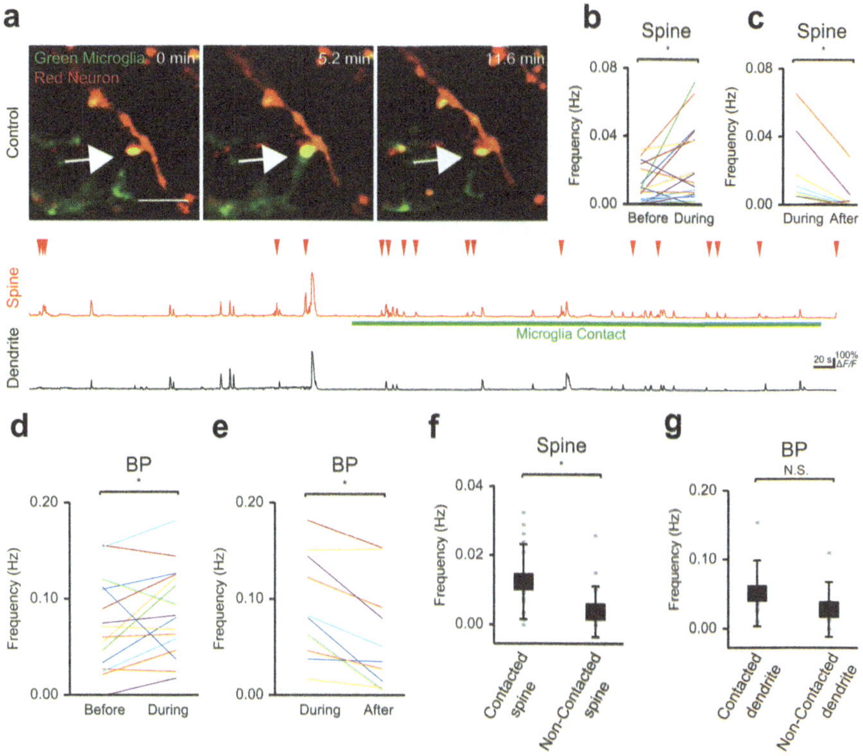

Fig. 6.17 Enhanced synapse activity follows microglia contact. (**a**) Top: Fluorescent image of a dendrite in layers II/III of the primary motor cortex showing the simultaneous imaging of spines and their parent dendrite (tdTomato; red) and microglia (EGFP, green). Scale bar = 5 μm. The image panels show microglial process contacting a spine (center). Lower traces, activity in the neuronal compartments was measured with Ca^{2+} sensor GCaMP6f and quantified as $\Delta F/F$. Sample Ca^{2+} traces from a spine and its parent dendrite before, during, and after microglia contact are shown. Putative back-propagating action potentials cause synchronous responses in both dendrite and spine, whereas presumed local synaptic activity results in spine transients only. Note the enhanced local (spine) synaptic activity with microglial contact. The white arrow indicate the spine contacted by microglia process. (**b**, **c**) The frequency of local spine Ca^{2+} transients was increased during microglia contact periods relative to those before (**b**) or after (**c**) contact. (**d**, **e**) The frequency of back-propagating action potentials reflected as dendritic Ca^{2+} transients was increased during the period of microglia-synapse contact, relative to that before (**d**) or after (**e**) contact. (**f**) Microglia were attracted to synapses with a higher basal (precontact) frequency of local Ca^{2+} responses. (**g**) In contrast, contact frequency did not depend on the basal frequency of back-propagating action potentials. Data are presented as the mean ± SEM; $^*p < 0.05$. (Reproduced from Akiyoshi et al. (2018), https://doi.org/10.1523/ENEURO.0088-18.2018, Figure 1 under CC 4.0 BY license)

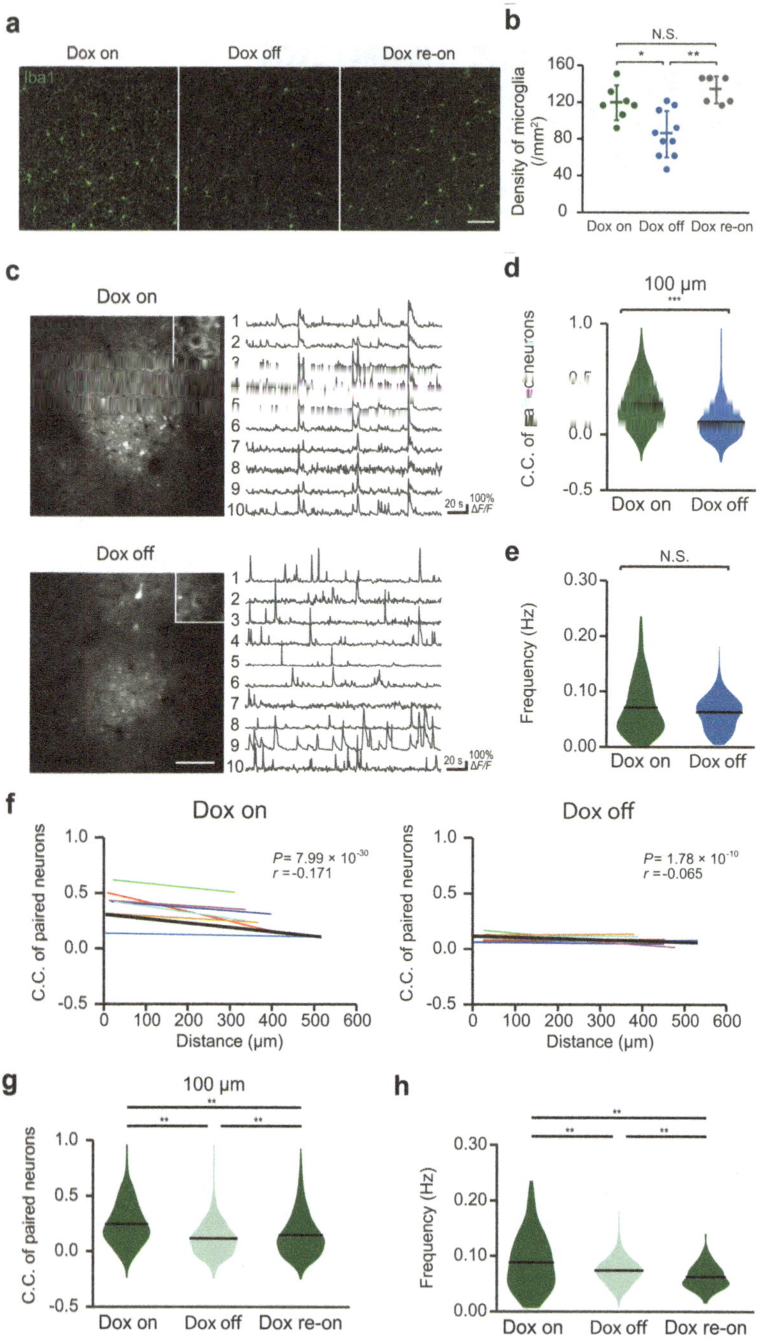

Fig. 6.18 Decreased synchronization of evoked neural activity in microglia ablated mice. (**a**) Fluorescent (immunohistochemistry) images of Iba1-positive microglia in cortical sections of Dox-on (with doxycycline) (left). Dox-off (withdrawal of doxycycline) (center). Re-Dox on mice showing reduced density of Iba1 (microglia)-positive cells in Iba1-tTA::tetO-DTA transgenic mice

Both glutamate and GABA affect resting microglia by binding to their respective receptors. Glutamate enhanced microglial process motility, while GABA caused the reverse (Fontainhas et al. 2011), indicating that microglia discriminate between the excitatory and inhibitory activity of synapses. Moreover, the blockage of the ATP-P2Y12 receptor pathway can prevent microglia activation (Sipe et al. 2016). The abundant expression of various neurotransmitter receptors on microglia and differences in the consequences of their activation highlight the wide spectrum of participation of microglia as a sensor and modulator of neural networks (Ikegami et al. 2019; Fig. 6.20).

Some mechanisms are involved in the ability of microglia to remove inactive synapses during brain development and adulthood. Cortical microglia secrete proteases, such as cathepsins, matrix metalloproteinases (MMPs), and tissue plasminogen activation, to reduce the spine density through proteolytic modulation of extracellular matrix molecules (Mataga et al. 2002, 2004; Tremblay et al. 2010). The microglial protease cathepsin S (CatS) and P2Y12 receptors, for example, have their expression regulated by the microglial molecular circadian clock. Both had a peak of expression 2 h after the beginning of the dark phase and regulate the maturation and elimination of dendritic spines of cortical neurons through modification of the perineuronal nets (Takayama et al. 2017; Nonaka and Nakanishi 2019), reducing synaptic activity (Hayashi et al. 2013) and spine density toward the beginning of

Fig. 6.18 (continued) following 7 days of Dox withdrawal as compared with Iba1-tTA::tetO-DTA transgenic mice with Dox maintained in the diet and in Dox-off mice after Dox had been returned to the diet for a week (Dox re-on). Scale bar in right panel = 50 m. (**b**) Genetic ablation of microglia was verified by quantification of microglia density. Dox-off mice showed a significant reduction in the density of microglia. The density of microglia recovered to control levels following 1 week of Dox return to the diet (**c**) left, typical fields of view of L2/3 primary motor cortex neurons expressing Ca^{2+} sensor GCaMP6f in Dox-on mice (upper) and Dox-off mice (lower). Right, representative $\Delta F/F$ Ca^{2+} traces from individual GCaMP6f-positive neurons from Dox-on (upper) and Dox-off mice (lower). (**d**) Synchronous neuronal firing was estimated by measuring the proportion of neuron pairs in which Ca^{2+} transients were seen at the same time, and quantifying this as a correlation coefficient (C.C.). The mean C.C. of paired neurons within 100 μm of each other was lower in Dox-off mice as compared with that in Dox-on mice. (**e**) There was no difference in the frequency of Ca^{2+} responses between Dox-on and Dox-off mice. (**f**) The C.C.s in each pair of neurons were negatively correlated with the distance separating each pair of neurons in control (Dox-on) mice (left). However, this negative correlation was absent in Dox-off mice (right). Individual slope correlations from each mouse are shown by the thin colored regression lines, while the thicker black line shows the correlation obtained by fitting all the data simultaneously. The value of the slope of this group correlation and the p value obtained from the Pearson's correlation test are shown in the top right of each graph. (**g**, **h**) In an additional cohort of mice, the C.C.s of neurons located within 100 μm of each other (**g**) and the neuron-firing frequencies (**h**) were measured during the doc on diet (left), following 7 days of Dox-off diet (center) and again 42 days after returning to dox diet (Dox re-on, right). Dox-off again reduced the CC which partially recovered after returning to dox. Data in panel (**b**) are presented as the mean ± SEM, while data in panels (**d**), (**e**), (**g**), (**h**) show means and distributions of each data point; $*p < 0.05$, $***p < 0.001$, n.s.: nonsignificant. (Reproduced from Akiyoshi et al. (2018), https://doi.org/10.1523/ENEURO.0088-18.2018, Figure 3 under CC 4.0 BY license)

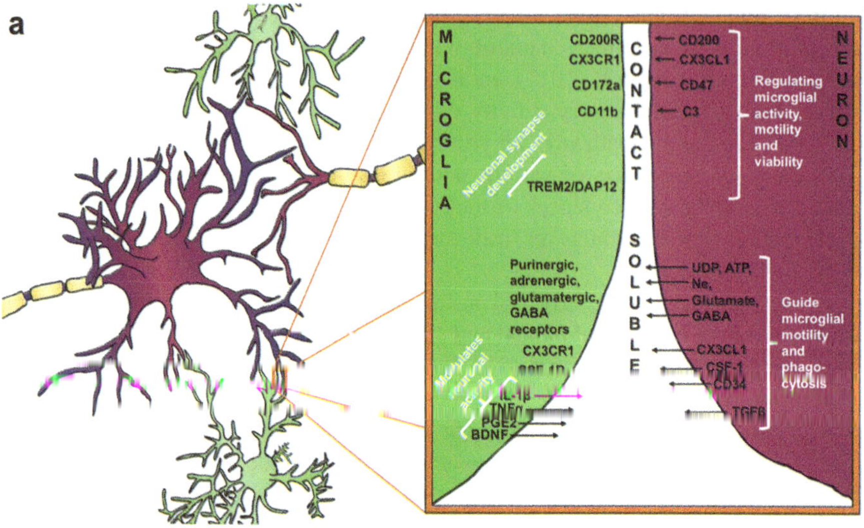

Fig. 6.19 (**a**) Bidirectional signaling between microglia (green) and neuron (purple). Neuron–microglia communication is mediated by receptor–ligand interactions as well as by various soluble factors. Microglia are equipped with a group of surface receptors, which trigger signals and regulate specific microglia function like phagocytosis, motility, and viability. Many of the receptor ligands, such as CX3CR1, CD200R, and CD172a are released or expressed on the surface of neurons. Receptor–ligand interactions represent a classical contact-dependent communication between microglia and neurons. Microglia and neurons reciprocally release soluble factors that can modulate cell functions and promote tissue homeostasis. (**b**) Fractalkine (CX3CR1–CX3CL1) signaling between microglia (green) and neuron (purple) in normal (green square) and pathologic (red square) states. CX3CL1 is either membrane-bound or cleaved by metalloproteinases to become soluble. Membrane-bound CX3CL1 is an OFF-signal, that keeps microglia in a surveying state, and soluble CX3CL1 is believed to act as a chemoattractant, stimulating migration of inflammatory cells. Fractalkine signaling has many roles in both healthy and pathologic conditions. Black triangles indicate increase (N), decrease (H) or not specified. (**c**) A summary of the effects of microglial depletion and Cx3cr1 deficiency on the formation, elimination, head size, and density of spines on adult-born granule cells. Under normal conditions (middle panel), microglia (green cells) play an important role in both facilitation of spine formation and elimination of nonfunctional spines, leading to normal maturation and plasticity of adult-born granule cells in the olfactory bulb. These effects are at least partly dependent on the microglial expression of CX3CR1, which mediates neuronal-microglial communication. Microglial depletion in PLX5622-treated mice (left panel) reduced the formation and elimination of spines on adult-born granule cells. Because, in control mice, spine formation exceeds spine elimination during the initial developmental period of adult-born granule cells, the reduced synaptic turnover in microglia-depleted mice resulted in lower spine density. Abrogation of normal microglia–neuronal interactions in Cx3cr1$^{-/-}$ mice (right panel) induced a reduction in spine formation, possibly due to the lower number of contacts between microglial processes and dendritic shafts, as well as lower spine head size, possibly due to the increased proportion of spines that were contacted by microglia. Spine elimination was also reduced, suggesting that CX3CL1-CX3CR1-mediated communication between specific spines on adult-born granule cells and adjacent microglial processes plays an important role in spine pruning. Similarly, to the effects of microglia depletion, spine density was also reduced in the Cx3cr1$^{-/-}$ mice. (Reproduced from (**a**, **b**) Szepesi et al. (2018), https://doi.org/10.3389/fncel.2018.00323, Figures 1 and 2 under CC BY license originally published by Frontiers; (**c**) from Reshef et al. (2017), https://doi.org/10.7554/eLife.30809, Figure 10 under CC 4.0 BY license)

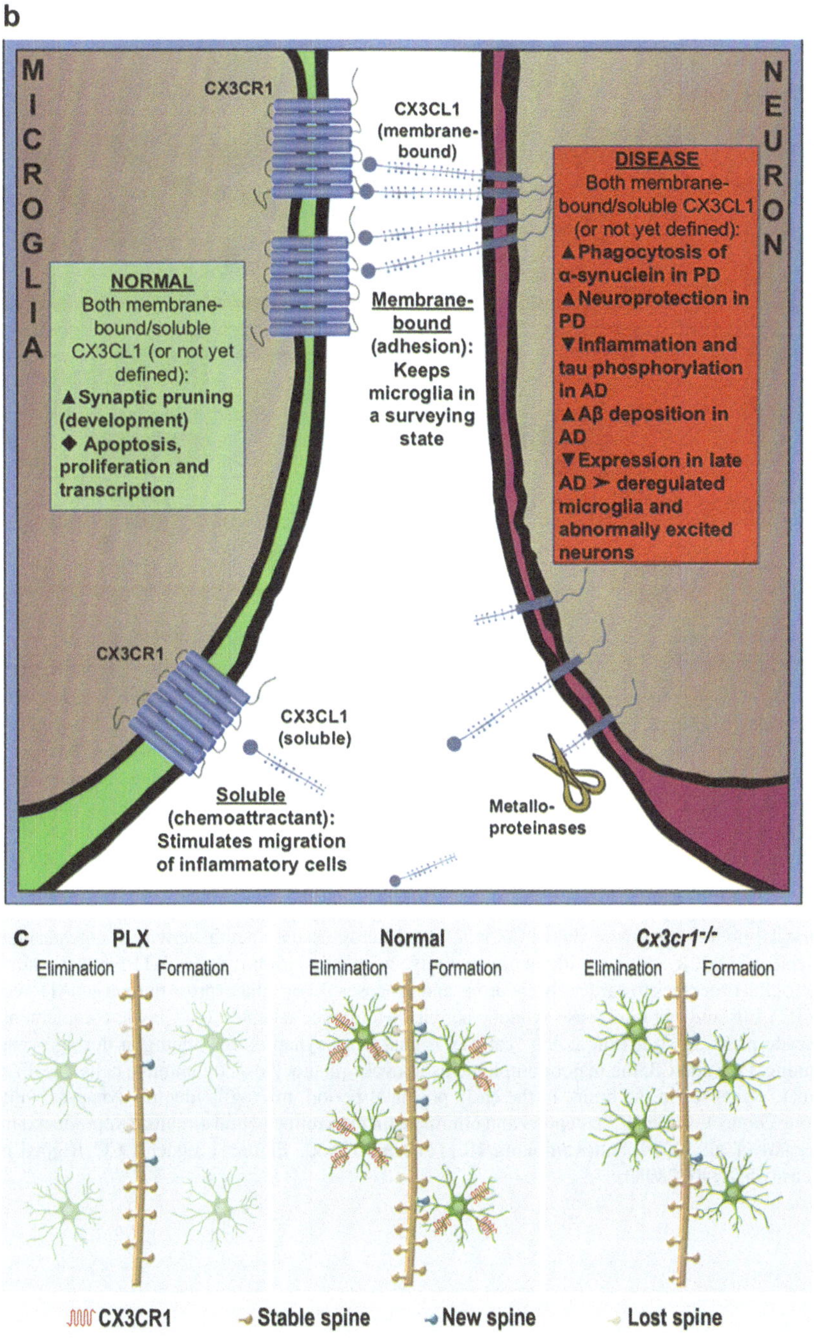

Fig. 6.19 (continued)

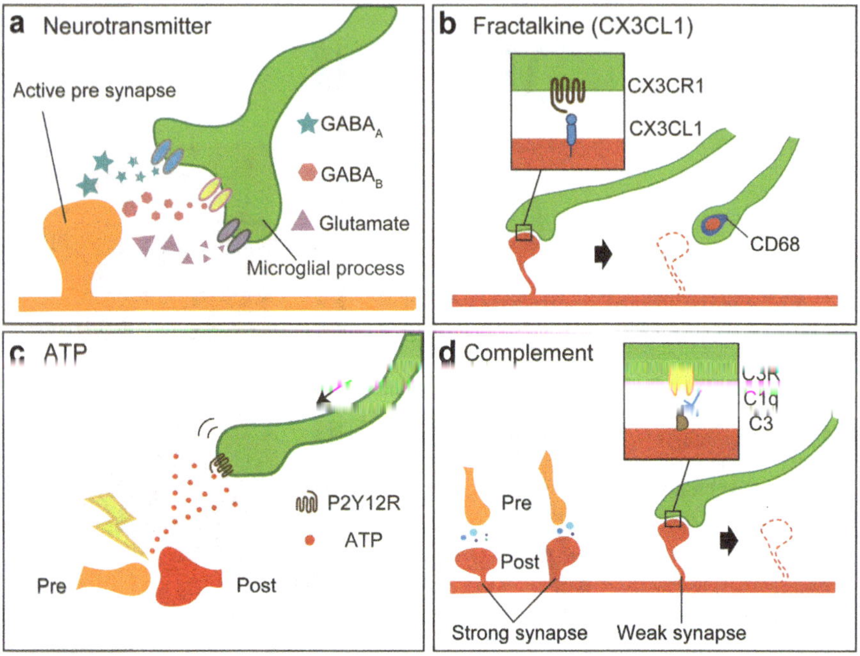

Fig. 6.20 Microglial machinery to wire synapses. In mature, intact brains, microglia are in "resting" states with highly branched processes surveying local neurons and synapses. They not only monitor them, but also wire neural circuits in the developmental stages. (**a**) Neurotransmitter receptors on microglia. Microglia have various neurotransmitter receptors and can sense neural signaling. Several members of the glutamate receptor family such as $GABA_A$ and $GABA_B$ receptors appear on microglia. Microglia respond to these signals in various ways, both neurotoxic and neuroprotective. (**b**) Fractalkine-CX3CR1 signaling between microglia and neurons is essential for the establishment of mature neural circuits. This immune pathway is reported to drive appropriate synaptic pruning by microglia. Microglia find and phagocytose synapses that express chemokine CX3CL1 on its cell surface via the CX3CR1. (**c**) Microglia sense neural activity via the purinergic receptor P2Y12R. Along with neural activity, adenosine triphosphate (ATP) release attracts microglial processes toward active neurons and synapses. Microglia express purinergic ATP receptor P2Y12R and can thus sense synaptic activity. Membrane receptor. (**d**) Classic complementary cascade pathways function as the "eat me" signal from synapses for microglia during synaptic pruning. C1q and C3, the major complements to evoke phagocytosis by immune cells, localize on "weak" synapses in the brain. In the early postnatal period, microglia identify synapses without strong connections via C3 receptors and eliminate them to refine neural circuits. (Reproduced from Ikegami et al. (2019), https://doi.org/10.1111/neup.12560, Figure 1 under CCC RightsLink® license 5525940728983)

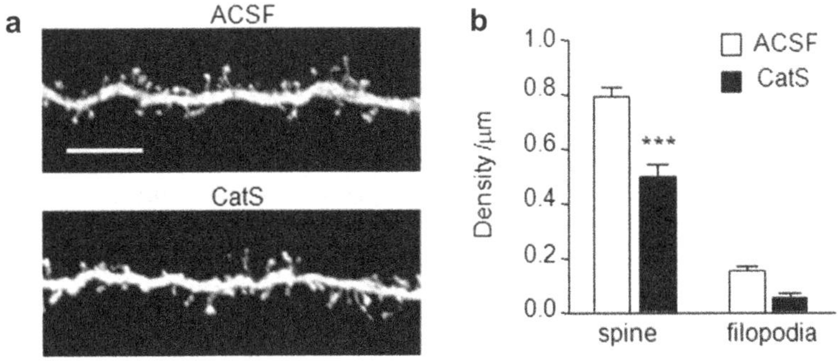

Fig. 6.21 Decline in the synaptic density of cortical neurons following treatment with human recombinant Cathepsin S (CatS) and the secretion of CatS from cultured microglia. (**a**) Confocal laser scanning microscope images of dendrites were captured from Lucifer Yellow-injected cortical neurons obtained from wild-type mice following treatment with ACSF or 50 nM human recombinant CatS. Scale bar = 10 μm. (**b**) The mean spine density of Lucifer Yellow-injected cortical neurons following treatment with artificial cerebrospinal fluid (ACSF) or 50 nM human recombinant CatS for 2 h. The data are the means ± SEM. ***$p < 0.001$; two-way factorial ANOVA, interaction (ACSF vs. CatS). (Reproduced from Takayama et al. (2017), https://doi.org/10.1016/j.bbrc.2017.06.061, Figure 2 under CCC RightsLink® license 5525970669210)

sleep (Nakanishi et al. 2021; Fig. 6.21). Microglia eliminate synapses during sleep via a mechanism other than phagocytosis (Andoh and Koyama 2021). Similarly, cortical microglia exhibit a ramified shape during sleep, while they have a hyper-ramified shape during wakefulness, which is characterized by their longer processes with increased branching points (Nakanishi et al. 2021). Therefore, the cortical microglia regulation of spines is related to the circadian clock. It is worth noting that the downscaling of both spine density and synaptic strength of cortical neurons during sleep could improve the signal-to-noise ratio, which would benefit memory consolidation or allow for new learning to occur during subsequent waking (Tononi and Cirelli 2006; Diekelmann and Born 2010). The overall bidirectional controlled microglia–neuron interaction (Akiyoshi et al. 2018; Szepesi et al. 2018) may underlie brain homeostasis, cognition, learning, and memory.

6.5 Glial Cells, Dendritic Spines, and Neurotrophic Factors

Neurotrophic factors are essential for dendritic spines. They help maintain existing spines and regulate plastic changes in the morphology and functionality of dendritic spines. Neurotrophic factors are secreted either by neurons, glia cells or by both. Glia cells, via the release of neurotrophic factors, have a great impact on dendritic spines as well. Mainly two types of glia cells provide essential neurotrophic support for neurons, namely astrocytes and oligodendrocytes.

The family of neurotrophins consists of nerve growth factor (NGF), BDNF, neurotrophin 3 (NT-3), and neurotrophin 4 (NT-4). Neurotrophins are, among others, important for processes related to neuronal and synaptic plasticity (see Chap. 5 in this book). Astrocytes from several brain areas are capable of synthesizing the neurotrophins NGF and NT-3. Furthermore, BDNF-expressing astroglia cells have been detected in the brain (Ivanova and Beyer 2001). Interestingly, treatment of astrocytes with the antidepressants, like fluoxetine, induces an upregulation of BDNF in astrocytes (Allaman et al. 2011). This increase in BDNF derived from astrocytes may support the regrowth of dendrites and dendritic spines seen in animal models of depression and treatements with antidepressants (Jiang and Salton 2013). Aside from astrocytes, oligodendrocytes are known for expressing NGF, BDNF, NT-3, and NT-4 mRNA. Neurotrophins seem to have a direct impact on the functions of oligodendrocytes, since they also express neurotrophin receptors (trkA, trkB, trkC, as well as p75NTR), which suggests possible autocrine signaling (Du and Dreyfus 2002). BDNF, for example, seems to play a role in demyelinating lesions by regulating the number of progenitors. Moreover, BDNF seems to have an impact on the expression of myelin proteins (VonDran et al. 2011). BDNF can also be produced by microglia cells. The microglia-derived BDNF contributes to self-renewal and density of microglia in the brain whereas ablation of BDNF in microglia stimulates neural precursor proliferation and alters working memory performance (Harley et al. 2021). However, ablation of BDNF in microglia does not alter the dendritic morphology of adult-born neurons within the dentate gyrus (Harley et al. 2021). Whether the microglia-derived BDNF has an impact on dendritic spines of postmitotic adult pyramidal neurons has not been analyzed yet.

Astrocytes play a fundamental role in synapse formation, pruning, and plasticity, which are associated with learning and memory, and it is thought that ephrins are involved in these processes. The ephrin-A3 ligand, which is located in the perisynaptic processes of astrocytes is essential for maintaining EphA4 activation and normal spine morphology (Carmona et al. 2009). Ephrin-A3/EphA4–mediated crosstalk between protoplasmic astrocytes and CA1 pyramidal neurons seems to be critical for maintaining the organization of dendritic spines during synaptic remodeling (Murai et al. 2003). Aside from ephrin-A3, other ephrins seemed also to be involved in the crosstalk of astrocytes and neurons, especially in the context of neuronal plasticity. For example, astrocyte-specific ephrin-B1 knockout enhances contextual memory recall following fear conditioning and increases the overall spine densities in the hippocampal area CA1 (Nguyen et al. 2020).

The members of the fibroblast growth factors (FGFs) as well as their receptors (FGFRs) are widely expressed in the CNS. FGFR-1 and FGFR-4 mRNAs are preferentially expressed in neurons, whereas FGFR-2 and FGFR-3 mRNAs are preferentially expressed in oligodendrocytes and in astrocytes (Asai et al. 1993; Miyake et al. 1996). The main sources of FGF-2 are astrocytes, although FGF-2 can also be secreted by microglia (Frautschy et al. 1991) and oligodendrocytes. Both, astrocytes and oligodendrocytes, express FGF-9 as well (Nakamura et al. 1999). Since FGF-2 is mainly derived from astrocytes, the effects of FGF-2 on neuronal plasticity and dendritic spine morphology and function (as described in Chap. 5) seemed to depend on the release of astrocytic FGF-2.

Glial cell line-derived neurotrophic factor (GDNF) is expressed by astrocytes (Du and Dreyfus 2002). The GDNF-receptor GFRα1 is expressed on dendrites and dendritic spines (Irala et al. 2016) that may hint for a crosstalk between astrocytes and dendritic compartments. In the presence of GDNF, GFRα1 promotes dendritic growth and complexity. GFRα1 seems to be required for proper synapse formation as well (Irala et al. 2016). The events that trigger the astrocytes to release GDNF in order to affect neuronal plasticity are largely enigmatic. Likewise, it is not clear whether the GDNF is released nearby a single synapse, which would allow a specific modulation of single synapses or spines, or effects are more globally. The TGF-β family consists of a group of at least 30 members of structurally related polypeptide growth factors (Diniz et al. 2019; von Bohlen und Halbach and Unsicker 2012). TGF-ßs are expressed by astrocytes, and astrocytes also respond to TGF-ß (Krieglstein 2006). Moreover, TGF-ßs were expressed by oligodendrocytes (Du and Dreyfus 2002).

6.6 Conclusion

Glia, especially astrocytes and oligodendrocytes, can secrete neurotrophic factors as another piece of the complex properties executed by these cells. These neurotrophic factors play important roles in the postnatal brain by affecting the dendritic composition of a neuron, the number and individual morphology of dendritic spines and, thereby, can have an impact on structural as well as synaptic plasticity. Astrocytes are involved in the control of neuronal activity and synaptic plasticity and oligodendrocytes, also, play significant roles in the regulation of neuronal functions and plasticity. It is evident that glial cells are more than just "Nervenkitt" or "resting" cells (for microglia, Tremblay et al. 2015). Astrocytes were described as playing a supportive role for neurons, but they have recently been recognized as active participants in the development and plasticity of dendritic spines and synapses. By the release of neurotrophic factors, they not only modulate synaptic transmission, but also can affect spinogenesis and structural remodeling of existing spines and synapses. Astrocytes induce synapse formation, and regulate neurotransmission and plasticity through a variety of direct and indirect signals. Astrocytic processes can dynamically associate with synapses, functionally interacting with dendritic spines and synaptic terminals. These astrocytic contacts regulate maturation of dendritic spines and their plastic structural remodeling. Microglia do the same. Oligodendrocytes modulate circuit function, and so on for the other evolved heterogeneous and locally specialized neuroglia. Therefore, glial cells are key components of the CNS and PNS (and gastrointestinal tract) that play more roles in neuronal processing than has previously been thought several decades ago, including in our species. "The human cortex differs from that of other animals not only in the huge amount of glandular cells [astrocytes] that it contains, but in their smallness and the wealth of the interstitial glial plexus [astrocyte appendages]" (Cajal 1913; Swanson et al. 2017). Dendrites, spines, and synapses are included in this network as well.

References

Aguilar-Hernández L, Alejandre R, Morales-Medina JC et al (2023) Cellular mechanisms in brain aging: focus on physiological and pathological aging. J Chem Neuroanat 128:102210. https://doi.org/10.1016/j.jchemneu.2022.102210

Akiyoshi R, Wake H, Kato D et al (2018) Microglia enhance synapse activity to promote local network synchronization. eNeuro 5:ENEURO.0088-18.2018. https://doi.org/10.1523/ENEURO.0088-18.2018

Allaman I, Fiumelli H, Magistretti PJ et al (2011) Fluoxetine regulates the expression of neurotrophic/growth factors and glucose metabolism in astrocytes. Psychopharmacology 216:75–84

Amateau SK, McCarthy MM (2002) A novel mechanism of dendritic spine plasticity involving estradiol induction of prostaglandin-E_2. J Neurosci 22:8586–8596

Andoh M, Koyama R (2021) Microglia regulate synaptic development and plasticity. Dev Neurobiol 81:568–590

Araque A, Parpura V, Sanzgiri RP et al (1999) Tripartite synapses: glia, the unacknowledged partner. Trends Neurosci 22:208–215

Araque A, Carmignoto G, Haydon PG et al (2014) Gliotransmitters travel in time and space. Neuron 81:728–739

Arias C, Zepeda A, Hernández-Ortega K et al (2009) Sex and estrous cycle-dependent differences in glial fibrillary acidic protein immunoreactivity in the adult rat hippocampus. Horm Behav 55:257–263

Arizono M, Inavalli VVGK, Panatier A et al (2020) Structural basis of astrocytic Ca2+ signals at tripartite synapses. Nat Commun 11:1906. https://doi.org/10.1038/s41467-020-15648-4

Asai T, Wanaka A, Kato H et al (1993) Differential expression of two members of FGF receptor gene family, FGFR-1 and FGFR-2 mRNA, in the adult rat central nervous system. Brain Res Mol Brain Res 17:174–178

Ascoli GA (2023) Cell morphologies in the nervous system: Glia steal the limelight. J Comp Neurol 531:338–343

Azevedo FA, Carvalho LR, Grinberg LT et al (2009) Equal numbers of neuronal and nonneuronal cells make the human brain an isometrically scaled-up primate brain. J Comp Neurol 513:532–541

Benfey N, Foubert D, Ruthazer ES (2022) Glia regulate the development, function, and plasticity of the visual system from retina to cortex. Front Neural Circuits 16:826664. https://doi.org/10.3389/fncir.2022.826664

Bergles DE, Roberts JD, Somogyi P et al (2000) Glutamatergic synapses on oligodendrocyte precursor cells in the hippocampus. Nature 405:187–191

Bernardinelli Y, Muller D, Nikonenko I (2014) Astrocyte-synapse structural plasticity. Neural Plast 2014:232105. https://doi.org/10.1155/2014/232105

Bourne JN, Harris KM (2007) Do thin spines learn to be mushroom spines that remember? Curr Opin Neurobiol 17:381–386

Brusco J, Merlo S, Ikeda ÉT et al (2014) Inhibitory and multisynaptic spines, and hemispherical synaptic specialization in the posterodorsal medial amygdala of male and female rats. J Comp Neurol 522:2075–2088

Buosi AS, Matias I, Araujo APB et al (2018) Heterogeneity in synaptogenic profile of astrocytes from different brain regions. Mol Neurobiol 55:751–762

Bushong EA, Martone ME, Jones YZ et al (2002) Protoplasmic astrocytes in CA1 stratum radiatum occupy separate anatomical domains. J Neurosci 22:183–192
Butt AM (2011) ATP: a ubiquitous gliotransmitter integrating neuron-glial networks. Semin Cell Dev Biol 22:205–213
Butt AM, Kiff J, Hubbard P et al (2002) Synantocytes: new functions for novel NG2 expressing glia. J Neurocytol 31:551–565
Butt AM, Hamilton N, Hubbard P et al (2005) Synantocytes: the fifth element. J Anat 207:695–706
Cajal SR (1909–1911) Histologie du système nerveux de l'homme et des vertébrés. Maloine, Paris. English edition: Cajal SR (1995) Histology of the nervous system of man and vertebrates (trans: Swanson N, Swanson LW). Oxford University Press, New York
Cajal SR (1913) Contribución al conocimiento de la neuroglia del cerebro humano. Trab Lab Invest Biol XI:255–314
Carmona MA, Murai KK, Wang L et al (2009) Glial ephrin-A3 regulates hippocampal dendritic spine morphology and glutamate transport. Proc Natl Acad Sci U S A 106:12524–12529
Castonguay A, Lévesque S, Robitaille R (2001) Glial cells as active partners in synaptic functions. Prog Brain Res 132:227–240
Chan TE, Grossman YS, Bloss EB et al (2018) Cell-type specific changes in glial morphology and glucocorticoid expression during stress and aging in the medial prefrontal cortex. Front Aging Neurosci 10:146. https://doi.org/10.3389/fnagi.2018.00146
Charles AC, Merrill JE, Dirksen ER et al (1991) Intercellular signaling in glial cells: calcium waves and oscillations in response to mechanical stimulation and glutamate. Neuron 6:983–992
Chelini G, Pantazopoulos H, Durning P et al (2018) The tetrapartite synapse: a key concept in the pathophysiology of schizophrenia. Eur Psychiatry 50:60–69
Chen Z, Jalabi W, Hu W et al (2014) Microglial displacement of inhibitory synapses provides neuroprotection in the adult brain. Nat Commun 5:4486. https://doi.org/10.1038/ncomms5486
Chorghay Z, Káradóttir RT, Ruthazer ES (2018) White matter plasticity keeps the brain in tune: axons conduct while glia wrap. Front Cell Neurosci 12:428. https://doi.org/10.3389/fncel.2018.00428
Chowen JA, Garcia-Segura LM (2021) Role of glial cells in the generation of sex differences in neurodegenerative diseases and brain aging. Mech Ageing Dev 196:111473. https://doi.org/10.1016/j.mad.2021.111473
Chung W-S, Welsh CA, Barres BA et al (2015) Do glia drive synaptic and cognitive impairment in disease? Nat Neurosci 18:1539–1545
Colonna M, Butovsky O (2017) Microglia function in the central nervous system during health and neurodegeneration. Annu Rev Immunol 35:441–468
Dall'Oglio A, Xavier LL, Hilbig A et al (2013) Cellular components of the human medial amygdaloid nucleus. J Comp Neurol 521:589–611
Dall'Oglio A, Dutra ACL, Moreira JE et al (2015) The human medial amygdala: structure, diversity, and complexity of dendritic spines. J Anat 227:440–459
Dalpian F, Rasia-Filho AA, Calcagnotto ME (2019) Sexual dimorphism, estrous cycle and laterality determine the intrinsic and synaptic properties of medial amygdala neurons in rat. J Cell Sci 132(9):jcs227793. https://doi.org/10.1242/jcs.227793
Danbolt NC (2001) Glutamate uptake. Prog Neurobiol 65:1–105
Dawson MR, Levine JM, Reynolds R (2000) NG2-expressing cells in the central nervous system: are they oligodendroglial progenitors? J Neurosci Res 61:471–479
De Luca C, Colangelo AM, Alberghina L et al (2018) Neuro-immune hemostasis: homeostasis and diseases in the central nervous system. Front Cell Neurosci 12:459. https://doi.org/10.3389/fncel.2018.00459
De Luca C, Colangelo AM, Virtuoso A et al (2020) Neurons, glia, extracellular matrix and neurovascular unit: a systems biology approach to the complexity of synaptic plasticity in health and disease. Int J Mol Sci 21:1539. https://doi.org/10.3390/ijms21041539
De Zeeuw CI, Hoogland TM (2015) Reappraisal of Bergmann glial cells as modulators of cerebellar circuit function. Front Cell Neurosci 9:246. https://doi.org/10.3389/fncel.2015.00246

DeFelipe J (2011) The evolution of the brain, the human nature of cortical circuits, and intellectual creativity. Front Neuroanat 5:29. https://doi.org/10.3389/fnana.2011.00029
Deiters VS, Guillery RW (2013) Otto Friedrich Karl Deiters (1834-1863). J Comp Neurol 521:1929–1953
Di Castro MA, Chuquet J, Liaudet N et al (2011) Local Ca^{2+} detection and modulation of synaptic release by astrocytes. Nat Neurosci 14:1276–1284
Diekelmann S, Born J (2010) The memory function of sleep. Nat Rev Neurosci 11:114–126
Dimou L, Gallo V (2015) NG2-glia and their functions in the central nervous system. Glia 63:1429–1451
Dimou L, Simons M (2017) Diversity of oligodendrocytes and their progenitors. Curr Opin Neurobiol 47:73–79
Diniz LP, Matias I, Siqueira M et al (2019) Astrocytes and the TGF-beta1 pathway in the healthy and diseased brain: a double-edged sword. Mol Neurobiol 56:4653–4679
Du Y, Dreyfus CF (2002) Oligodendrocytes as providers of growth factors. J Neurosci Res 68:647–654
Emsley JG, Macklis JD (2006) Astroglial heterogeneity closely reflects the neuronal-defined anatomy of the adult murine CNS. Neuron Glia Biol 2:175–186
Endo F, Kasai A, Soto JS et al (2022) Molecular basis of astrocyte diversity and morphology across the CNS in health and disease. Science 378:6619. https://doi.org/10.1126/science.adc9020
Eroglu C, Allen NJ, Susman MW et al (2009) Gabapentin receptor alpha2delta-1 is a neuronal thrombospondin receptor responsible for excitatory CNS synaptogenesis. Cell 139:380–392
Eugenín-von Bernhardi J, Dimou L (2016) NG2-glia, more than progenitor cells. Adv Exp Med Biol 949:27–45
Fan YY, Huo J (2021) A1/A2 astrocytes in central nervous system injuries and diseases: angels or devils? Neurochem Int 148:105080. https://doi.org/10.1016/j.neuint.2021.105080
Farmer WT, Murai K (2017) Resolving astrocyte heterogeneity in the CNS. Front Cell Neurosci 11:300. https://doi.org/10.3389/fncel.2017.00300
Fernandez-Galaz MC, Morschl E, Chowen JA et al (1997) Role of astroglia and insulin-like growth factor-I in gonadal hormone-dependent synaptic plasticity. Brain Res Bull 44:525–531
Fernandez-Galaz MC, Naftolin F, Garcia-Segura LM (1999) Phasic synaptic remodeling of the rat arcuate nucleus during the estrous cycle depends on insulin-like growth factor-I receptor activation. J Neurosci Res 55:286–292
Fiala JC, Harris KM (1999) Dendrite structure. In: Stuart G, Spruston N, Häusser M (eds) Dendrites. Oxford University Press, New York, pp 1–34
Filosa A, Paixão S, Honsek SD et al (2009) Neuron-glia communication via EphA4/ephrin-A3 modulates LTP through glial glutamate transport. Nat Neurosci 12:1285–1292. https://doi.org/10.1038/nn.2394
Foerster S, Hill MFE, Franklin RJM (2019) Diversity in the oligodendrocyte lineage: plasticity or heterogeneity? Glia 67:1797–1805
Fontainhas AM, Wang M, Liang KJ et al (2011) Microglial morphology and dynamic behavior is regulated by ionotropic glutamatergic and GABAergic neurotransmission. PLoS One 6:e15973. https://doi.org/10.1371/journal.pone.0015973
Frautschy SA, Walicke PA, Baird A (1991) Localization of basic fibroblast growth factor and its mRNA after CNS injury. Brain Res 553:291–299
Garcia O, Torres M, Helguera P et al (2010) A role for thrombospondin-1 deficits in astrocyte-mediated spine and synaptic pathology in Down's syndrome. PLoS One 5:e14200. https://doi.org/10.1371/journal.pone.0014200
García-Marín V, García-López P, Freire M (2007) Cajal's contributions to glia research. Trends Neurosci 30:479–487
Garcia-Segura LM (2002) Chapter 19 – Cajal and glial cells. Prog Brain Res 136:255–260
Garcia-Segura LM, McCarthy MM (2004) Minireview: role of glia in neuroendocrine function. Endocrinology 145:1082–1086
Garcia-Segura LM, Chowen JA, Dueñas M et al (1994) Gonadal steroids as promoters of neuro-glial plasticity. Psychoneuroendocrinology 19:445–453

Garcia-Segura LM, Dueñas M, Busiguina S et al (1995) Gonadal hormone regulation of neuronal-glial interactions in the developing neuroendocrine hypothalamus. J Steroid Biochem Mol Biol 53:293–298

Gavrilov N, Golyagina I, Brazhe A et al (2018) Astrocytic coverage of dendritic spines, dendritic shafts, and axonal boutons in hippocampal neuropil. Front Cell Neurosci 12:248. https://doi.org/10.3389/fncel.2018.00248

Ge W-P, Yang X-J, Zhang Z et al (2006) Long-term potentiation of neuron-glia synapses mediated by Ca^{2+}-permeable AMPA receptors. Science 312:1533–1537

Gosejacob D, Dublin P, Bedner P et al (2011) Role of astroglial connexin30 in hippocampal gap junction coupling. Glia 59:511–519

Grutzendler J, Kasthuri N, Gan W-B (2002) Long-term dendritic spine stability in the adult cortex. Nature 420:812–816

Haber M, Zhou L, Murai KK (2006) Cooperative astrocyte and dendritic spine dynamics at hippocampal excitatory synapses. J Neurosci 26:8881–8889

Halassa MM, Haydon PG (2010) Integrated brain circuits: astrocytic networks modulate neuronal activity and behavior. Annu Rev Physiol 72:335–355. https://doi.org/10.1146/annurev-physiol-021909-135843

Halassa MM, Fellin T, Takano H et al (2007) Synaptic islands defined by the territory of a single astrocyte. J Neurosci 27:6473–6477

Harada K, Kamiya T, Tsuboi T (2015) Gliotransmitter release from astrocytes: functional, developmental, and pathological implications in the brain. Front Neurosci 9:499. https://doi.org/10.3389/fnins.2015.00499

Harley SBR, Willis EF, Shaikh SN et al (2021) Selective ablation of BDNF from microglia reveals novel roles in self-renewal and hippocampal neurogenesis. J Neurosci 41:4172–4186

Hawrylycz M, Miller JA, Menon V et al (2015) Canonical genetic signatures of the adult human brain. Nat Neurosci 18:1832–1844

Hayashi Y, Koyanagi S, Kusunose N et al (2013) The intrinsic microglial molecular clock controls synaptic strength via the circadian expression of cathepsin S. Sci Rep 3:2744. https://doi.org/10.1038/srep02744

Haydon PG (2001) GLIA: listening and talking to the synapse. Nat Rev Neurosci 2:185–193

Henneberger C, Papouin T, Oliet SHR et al (2010) Long-term potentiation depends on release of D-serine from astrocytes. Nature 463:232–236

Henneberger C, Bard L, Panatier A et al (2020) LTP induction boosts glutamate spillover by driving withdrawal of perisynaptic astroglia. Neuron 108:919–936.e11. https://doi.org/10.1016/j.neuron.2020.08.030

Herculano-Houzel S (2011) Not all brains are made the same: new views on brain scaling in evolution. Brain Behav Evol 78:22–36

Herculano-Houzel S (2012) The remarkable, yet not extraordinary, human brain as a scaled-up primate brain and its associated cost. Proc Natl Acad Sci U S A 109(Suppl 1):10661–10668

Herde MK, Bohmbach K, Domingos C et al (2020) Local efficacy of glutamate uptake decreases with synapse size. Cell Rep 32(12):108182. https://doi.org/10.1016/j.celrep.2020.108182

Hill RA, Nishiyama A (2014) NG2 cells (polydendrocytes): listeners to the neural network with diverse properties. Glia 62:1195–1210

Hillen AEJ, Burbach JPH, Hol EM (2018) Cell adhesion and matricellular support by astrocytes of the tripartite synapse. Prog Neurobiol 165–167:66–86

Hirbec H, Déglon N, Foo LC et al (2020) Emerging technologies to study glial cells. Glia 68:1692–1728. Erratum in Glia 69:507

Hodge RD, Bakken TE, Miller JA et al (2019) Conserved cell types with divergent features in human versus mouse cortex. Nature 573:61–68

Holtmaat AJGD, Trachtenberg JT, Wilbrecht L et al (2005) Transient and persistent dendritic spines in the neocortex in vivo. Neuron 45:279–291

Ikegami A, Haruwaka K, Wake H (2019) Microglia: lifelong modulator of neural circuits. Neuropathology 39:173–180

Irala D, Bonafina A, Fontanet PA et al (2016) The GDNF-GFRalpha1 complex promotes the development of hippocampal dendritic arbors and spines via NCAM. Development 143:4224–4235

Ishikawa T, Eto K, Kim SK et al (2018) Cortical astrocytes prime the induction of spine plasticity and mirror image pain. Pain 159:1592–1606

Ivanova T, Beyer C (2001) Pre- and postnatal expression of brain-derived neurotrophic factor mRNA/protein and tyrosine protein kinase receptor B mRNA in the mouse hippocampus. Neurosci Lett 307:21–24

Jäkel S, Dimou L (2017) Glial cells and their function in the adult brain: a journey through the history of their ablation. Front Cell Neurosci 11:24. https://doi.org/10.3389/fncel.2017.00024

Jiang C, Salton SR (2013) The role of neurotrophins in major depressive disorder. Transl Neurosci 4:46–58

Jinnou H, Sawada M, Kawase K et al (2018) Radial glial fibers promote neuronal migration and functional recovery after neonatal brain injury. Cell Stem Cell 22:128–137.e9

Johnson RT, Breedlove SM, Jordan CL (2008) Sex differences and laterality in astrocyte number and complexity in the adult rat medial amygdala. J Comp Neurol 511:599–609

Jones EV, Bernardinelli Y, Tse YC et al (2011) Astrocytes control glutamate receptor levels at developing synapses through SPARC beta integrin interactions. J Neurosci 31:4154–4165

Kasthuri N, Hayworth KJ, Berger DR et al (2015) Saturated reconstruction of a volume of neocortex. Cell 162:648–661

Kato K, Forero MG, Fenton JC et al (2011) The glial regenerative response to central nervous system injury is enabled by pros-notch and pros-NFκB feedback. PLoS Biol 9(8):e1001133. https://doi.org/10.1371/journal.pbio.1001133

Kettenmann H, Ransom BR (1995) Neuroglia. Oxford University Press, New York

Kettenmann H, Kirchhoff F, Verkhratsky A (2013) Microglia: new roles for the synaptic stripper. Neuron 77:10–18

Kim SK, Nabekura J, Koizumi S (2017) Astrocyte-mediated synapse remodeling in the pathological brain. Glia 65:1719–1727

Koeppen J, Nguyen AQ, Nikolakopoulou AM et al (2018) Functional consequences of synapse remodeling following astrocyte-specific regulation of Ephrin-B1 in the adult hippocampus. J Neurosci 38:5710–5726

Krieglstein K (2006) Transforming growth factor-ßs in the brain. In: Lajtha A, Lim R (eds) Handbook of neurochemistry and molecular neurobiology. Springer, pp 123–142

Krzisch M, Temprana SG, Mongiat LA et al (2015) Pre-existing astrocytes form functional perisynaptic processes on neurons generated in the adult hippocampus. Brain Struct Funct 220:2027–2042

Kucukdereli H, Allen NJ, Lee AT et al (2011) Control of excitatory CNS synaptogenesis by astrocyte-secreted proteins Hevin and SPARC. Proc Natl Acad Sci U S A 108:E440–E449

Lago-Baldaia I, Fernandes VM, Ackerman SD (2020) More than mortar: glia as architects of nervous system development and disease. Front Cell Dev Biol 8:611269. https://doi.org/10.3389/fcell.2020.611269

Larson VA, Zhang Y, Bergles DE (2016) Electrophysiological properties of NG2(+) cells: matching physiological studies with gene expression profiles. Brain Res 1638:138–160

Lavialle M, Aumann G, Anlauf E et al (2011) Structural plasticity of perisynaptic astrocyte processes involves ezrin and metabotropic glutamate receptors. Proc Natl Acad Sci U S A 108:12915–12919

Lehre KR, Rusakov DA (2002) Asymmetry of glia near central synapses favors presynaptically directed glutamate escape. Biophys J 83:125–134

Liauw J, Hoang S, Choi M, Eroglu C et al (2008) Thrombospondins 1 and 2 are necessary for synaptic plasticity and functional recovery after stroke. J Cereb Blood Flow Metab 28:1722–1732

Lin S-C, Bergles DE (2004) Synaptic signaling between neurons and glia. Glia 47:290–298

Liu H, Leak RK, Hu X (2016) Neurotransmitter receptors on microglia. Stroke Vasc Neurol 1:52–58

Losada-Perez M (2018) Glia: from "just glue" to essential players in complex nervous systems: a comparative view from flies to mammals. J Neurogenet 32:78–91

Lundgaard I, Lu ML, Yang E et al (2017) Glymphatic clearance controls state-dependent changes in brain lactate concentration. J Cereb Blood Flow Metab 37:2112–2124

Luquin S, Naftolin F, Garcia-Segura LM (1993) Natural fluctuation and gonadal hormone regulation of astrocyte immunoreactivity in dentate gyrus. J Neurobiol 24:913–924
Lushnikova I, Skibo G, Muller D et al (2009) Synaptic potentiation induces increased glial coverage of excitatory synapses in CA1 hippocampus. Hippocampus 19:753–762
Lutz SE, Zhao Y, Gulinello M et al (2009) Deletion of astrocyte connexins 43 and 30 leads to a dysmyelinating phenotype and hippocampal CA1 vacuolation. J Neurosci 29:7743–7752
Lyon KA, Allen NJ (2022) From synapses to circuits, astrocytes regulate behavior. Front Neural Circuits 15:786293. https://doi.org/10.3389/fncir.2021.786293
MacDonald A, Lu B, Caron M et al (2021) Single cell transcriptomics of ependymal cells across age, region and species reveals cilia-related and metal ion regulatory roles as major conserved ependymal cell functions. Front Cell Neurosci 15:703951. https://doi.org/10.3389/fncel.2021.703951
Magistretti PJ, Pellerin L, Rothman DL et al (1999) Energy on demand. Science 283:496–497
Martinez FG, Hermel EE, Xavier LL et al (2006) Gonadal hormone regulation of glial fibrillary acidic protein immunoreactivity in the medial amygdala subnuclei across the estrous cycle and in castrated and treated female rats. Brain Res 1108:117–126
Mataga N, Nagai N, Hensch TK (2002) Permissive proteolytic activity for visual cortical plasticity. Proc Natl Acad Sci U S A 99:7717–7721
Mataga N, Mizuguchi Y, Hensch TK (2004) Experience-dependent pruning of dendritic spines in visual cortex by tissue plasminogen activator. Neuron 44:1031–1041
Matsuzaki M, Honkura N, Ellis-Davies GCR et al (2004) Structural basis of long-term potentiation in single dendritic spines. Nature 429:761–766
Matyash V, Kettenmann H (2010) Heterogeneity in astrocyte morphology and physiology. Brain Res Rev 63:2–10
McCarthy MM (2023) Neural control of sexually dimorphic social behavior: Connecting development to adulthood. Annu Rev Neurosci 46:321–339
Mederos S, González-Arias C, Perea G (2018) Astrocyte–neuron networks: A multilane highway of signaling for homeostatic brain function. Front Synaptic Neurosci 10:45. https://doi.org/10.3389/fnsyn.2018.00045
Medvedev N, Popov V, Henneberger C et al (2014) Glia selectively approach synapses on thin dendritic spines. Philos Trans R Soc Lond Ser B Biol Sci 369:20140047. https://doi.org/10.1098/rstb.2014.0047
Mietto BS, Mostacada K, Martinez AM (2015) Neurotrauma and inflammation: CNS and PNS responses. Mediat Inflamm 2015:251204. https://doi.org/10.1155/2015/251204
Miller SJ (2018) Astrocyte heterogeneity in the adult central nervous system. Front Cell Neurosci 12:401. https://doi.org/10.3389/fncel.2018.00401
Miller SJ, Philips T, Kim N et al (2019) Molecularly defined cortical astroglia subpopulation modulates neurons via secretion of Norrin. Nat Neurosci 22:741–752
Miyake A, Hattori Y, Ohta M et al (1996) Rat oligodendrocytes and astrocytes preferentially express fibroblast growth factor receptor-2 and -3 mRNAs. J Neurosci Res 45:534–541
Miyamoto A, Wake H, Ishikawa AW et al (2016) Microglia contact induces synapse formation in developing somatosensory cortex. Nat Commun 7:12540. https://doi.org/10.1038/ncomms12540
Mong JA, Kurzweil RL, Davis AM et al (1996) Evidence for sexual differentiation of glia in rat brain. Horm Behav 30:553–562
Morris GP, Foster CG, Courtney JM et al (2023) Microglia directly associate with pericytes in the central nervous system. Glia 71(8):1847–1869. https://doi.org/10.1002/glia.24371
Murai KK, Nguyen LN, Irie F et al (2003) Control of hippocampal dendritic spine morphology through ephrin-A3/EphA4 signaling. Nat Neurosci 6:153–160
Nägerl UV, Eberhorn N, Cambridge SB et al (2004) Bidirectional activity-dependent morphological plasticity in hippocampal neurons. Neuron 44:759–767
Nakamura S, Todo T, Motoi Y et al (1999) Glial expression of fibroblast growth factor-9 in rat central nervous system. Glia 28:53–65

Nakanishi H, Ni J, Nonaka S, Hayashi Y (2021) Microglial circadian clock regulation of microglial structural complexity, dendritic spine density and inflammatory response. Neurochem Int 142:104905. https://doi.org/10.1016/j.neuint.2020.104905

Navarrete M, Araque A (2010) Endocannabinoids potentiate synaptic transmission through stimulation of astrocytes. Neuron 68:113–126

Navarrete M, Araque A (2014) The Cajal school and the physiological role of astrocytes: a way of thinking. Front Neuroanat 8:33. https://doi.org/10.3389/fnana.2014.00033

Neal-Perry G, Yao D, Shu J et al (2014) Insulin-like growth factor-I regulates LH release by modulation of kisspeptin and NMDA-mediated neurotransmission in young and middle-aged female rats. Endocrinology 155:1827–1837

Neves G, Cooke S, Bliss T (2008) Synaptic plasticity, memory and the hippocampus: a neural network approach to causality. Nat Rev Neurosci 9:65–75

Nguycn AQ, Koeppen J, Woodruff S et al (2020) Astrocytic Ephrin-B1 controls synapse formation in the hippocampus during learning and memory. Front Synaptic Neurosci 12:10. https://doi.org/10.3389/fnsyn.2020.00010

Nishida H, Okabe S (2007) Direct astrocytic contacts regulate local maturation of dendritic spines. J Neurosci 27:331–340

Nonaka S, Nakanishi H (2019) Microglial clearance of focal apoptotic synapses. Neurosci Lett 707:134317. https://doi.org/10.1016/j.neulet.2019.134317

Oberheim NA, Takano T, Han X et al (2009) Uniquely hominid features of adult human astrocytes. J Neurosci 29:3276–3287

Oberlander JG, Erskine MS (2008) Receipt of vaginal-cervical stimulation modifies synapsin content in limbic areas of the female rat. Neuroscience 153:581–593

Ojeda SR, Ma YJ, Lee BJ et al (2000) Glia-to-neuron signaling and the neuroendocrine control of female puberty. Recent Prog Horm Res 55:197–223; discussion 223–224

Ostroff LE, Manzur MK, Cain CK et al (2014) Synapses lacking astrocyte appear in the amygdala during consolidation of Pavlovian threat conditioning. J Comp Neurol 522:2152–2163

Panatier A, Oliet SHR (2006) Neuron-glia interactions in the hypothalamus. Neuron Glia Biol 2:51–58

Pannasch U, Freche D, Dallérac G et al (2014) Connexin 30 sets synaptic strength by controlling astroglial synapse invasion. Nat Neurosci 17:549–558

Paolicelli RC, Bolasco G, Pagani F et al (2011) Synaptic pruning by microglia is necessary for normal brain development. Science 333:1456–1458

Papa M, De Luca C, Petta F et al (2014) Astrocyte-neuron interplay in maladaptive plasticity. Neurosci Biobehav Rev 42:35–54

Papouin T, Dunphy J, Tolman M et al (2017) Astrocytic control of synaptic function. Philos Trans R Soc Lond Ser B Biol Sci 372:20160154. https://doi.org/10.1098/rstb.2016.0154

Parkhurst CN, Yang G, Ninan I, Savas JN et al (2013) Microglia promote learning-dependent synapse formation through brain-derived neurotrophic factor. Cell 155:1596–1609

Parpura V, Zorec R (2010) Gliotransmission: exocytotic release from astrocytes. Brain Res Rev 63:83–92

Patel MR, Weaver AM (2021) Astrocyte-derived small extracellular vesicles promote synapse formation via fibulin-2-mediated TGF-β signaling. Cell Rep 34(10):108829. https://doi.org/10.1016/j.celrep.2021.108829

Perea G, Araque A (2007) Astrocytes potentiate transmitter release at single hippocampal synapses. Science 317:1083–1086

Perea G, Sur M, Araque A (2014) Neuron-glia networks: integral gear of brain function. Front Cell Neurosci 8:378. https://doi.org/10.3389/fncel.2014.00378

Perez-Alvarez A, Navarrete M, Covelo A et al (2014) Structural and functional plasticity of astrocyte processes and dendritic spine interactions. J Neurosci 34:12738–12744. https://doi.org/10.1523/JNEUROSCI.2401-14.2014. Erratum in J Neurosci 34(42):14163

Peters A, Palay SL, Webster H (1991) The fine structure of the nervous system. Oxford University Press, New York

Pickering M, Cumiskey D, O'Connor JJ (2005) Actions of TNF-alpha on glutamatergic synaptic transmission in the central nervous system. Exp Physiol 90:663–670

Pistono C, Bister N, Stanová I et al (2021) Glia-derived extracellular vesicles: role in central nervous system communication in health and disease. Front Cell Dev Biol 8:623771. https://doi.org/10.3389/fcell.2020.623771

Plog BA, Nedergaard M (2018) The glymphatic system in central nervous system health and disease: past, present, and future. Annu Rev Pathol 13:379–394

Pocock JM, Kettenmann H (2007) Neurotransmitter receptors on microglia. Trends Neurosci 30:527–535

Proulx ST, Engelhardt B (2023) Macrophages clear the way for CNS fluid flow. Lancet Neurol 22:194–195

Radomska KJ, Topilko P (2017) Boundary cap cells in development and disease. Curr Opin Neurobiol 47:209–215

Rasia-Filho AA, Fabian C, Rigoti KM et al (2004) Influence of sex, estrous cycle and motherhood on dendritic spine density in the rat medial amygdala revealed by the Golgi method. Neuroscience 126:839-847

Rasia-Filho AA, Xavier LL, dos Santos P et al (2002) Glial fibrillary acidic protein immunodetection and immunoreactivity in the anterior and posterior medial amygdala of male and female rats. Brain Res Bull 58:67–75

Rasia-Filho AA, Dalpian F, Menezes IC et al (2012) Dendritic spines of the medial amygdala: plasticity, density, shape, and subcellular modulation by sex steroids. Histol Histopathol 8:985–1011

Rasia-Filho AA, Guerra KTK, Vásquez CE et al (2021) The subcortical-allocortical-neocortical *continuum* for the emergence and morphological heterogeneity of pyramidal neurons in the human brain. Front Synaptic Neurosci 13:616607. https://doi.org/10.3389/fnsyn.2021.616607

Reddy OC, van der Werf YD (2020) The sleeping brain: harnessing the power of the glymphatic system through lifestyle choices. Brain Sci 10(11):868. https://doi.org/10.3390/brainsci10110868

Reichenbach A, Derouiche A, Kirchhoff F (2010) Morphology and dynamics of perisynaptic glia. Brain Res Rev 63:11–25

Reshef R, Kudryavitskaya E, Shani-Narkiss H et al (2017) The role of microglia and their CX3CR1 signaling in adult neurogenesis in the olfactory bulb. eLife 6:e30809. https://doi.org/10.7554/eLife.30809

Reynolds R, Hardy R (1997) Oligodendroglial progenitors labeled with the O4 antibody persist in the adult rat cerebral cortex in vivo. J Neurosci Res 47:455–470

Richards DA, De Paola V, Caroni P et al (2004) AMPA-receptor activation regulates the diffusion of a membrane marker in parallel with dendritic spine motility in the mouse hippocampus. J Physiol 558:503–512

Río-Hortega P (1919) El "tercer elemento" de los centros nerviosos. III. Naturaleza probable de la microglía. Bol Soc Esp Biol VIII:108–115

Risher WC, Patel S, Kim IH et al (2014) Astrocytes refine cortical connectivity at dendritic spines. eLife 3:e04047. https://doi.org/10.7554/eLife.04047

Robin LM, Oliveira da Cruz JF, Langlais VC et al (2018) Astroglial CB1 receptors determine synaptic D-Serine availability to enable recognition memory. Neuron 98:935–944.e5

Rodríguez-Iglesias N, Sierra A, Valero J (2019) Rewiring of memory circuits: connecting adult newborn neurons with the help of microglia. Front Cell Dev Biol 7:24. https://doi.org/10.3389/fcell.2019.00024

Rusakov DA, Stewart MG (2021) Synaptic environment and extrasynaptic glutamate signals: the quest continues. Neuropharmacology 195:108688. https://doi.org/10.1016/j.neuropharm.2021.108688

Saint-Martin M, Goda Y (2022) Astrocyte-synapse interactions and cell adhesion molecules. FEBS J. https://doi.org/10.1111/febs.16540

Sakry D, Karram K, Trotter J (2011) Synapses between NG2 glia and neurons. J Anat 219:2–7

Sakry D, Neitz A, Singh J et al (2014) Oligodendrocyte precursor cells modulate the neuronal network by activity-dependent ectodomain cleavage of glial NG2. PLoS Biol 12:e1001993. https://doi.org/10.1371/journal.pbio.1001993

Salazar AP, Quagliotto E, Alves J et al (2014) Effect of prior exercise training and myocardial infarction-induced heart failure on the neuronal and glial densities and the GFAP-immunoreactivity in the posterodorsal medial amygdala of rats. Histol Histopathol 29:1423–1435

Schafer DP, Lehrman EK, Kautzman AG et al (2012) Microglia sculpt postnatal neural circuits in an activity and complement-dependent manner. Neuron 74:691–705

Segal M (2005) Dendritic spines and long-term plasticity. Nat Rev Neurosci 6:277–284

Seguella L, Gulbransen BD (2021) Enteric glial biology, intercellular signalling and roles in gastrointestinal disease. Nat Rev Gastroenterol Hepatol 18:571–587

Seil FJ (2001) Interactions between cerebellar Purkinje cells and their associated astrocytes. Histol Histopathol 16:955–968

Semyanov A, Verkhratsky A (2021) Astrocytic processes: from tripartite synapses to the active milieu. Trends Neurosci 44:781–792

Shapson-Coe A, Januszewski M, Berger DR et al (2021) A connectomic study of a petascale fragment of human cerebral cortex. bioRxiv 2021.05.29.446289. https://doi.org/10.1101/2021.05.29.446289

Singh SK, Stogsdill JA, Pulimood NS et al (2016) Astrocytes assemble thalamocortical synapses by bridging NRX1α and NL1 via Hevin. Cell 164:183–196

Sipe GO, Lowery RL, Tremblay M-È et al (2016) Microglial P2Y12 is necessary for synaptic plasticity in mouse visual cortex. Nat Commun 7:10905. https://doi.org/10.1038/ncomms10905

Sipe GO, Petravicz J, Rikhye RV et al (2021) Astrocyte glutamate uptake coordinates experience-dependent, eye-specific refinement in developing visual cortex. Glia 69:1723–1735

Šišková Z, Tremblay M-È (2013) Microglia and synapse: interactions in health and neurodegeneration. Neural Plast 2013:425845. https://doi.org/10.1155/2013/425845

Sosunov A, Wu X, McGovern R et al (2020) Abnormal mitosis in reactive astrocytes. Acta Neuropathol Commun 8:47. https://doi.org/10.1186/s40478-020-00919-4

Spacek J (1985) Three-dimensional analysis of dendritic spines. III. Glial sheath. Anat Embryol 171:245–252

Spacek J, Harris KM (1998) Three-dimensional organization of cell adhesion junctions at synapses and dendritic spines in area CA1 of the rat hippocampus. J Comp Neurol 393:58–68

Squarzoni P, Thion MS, Garel S (2015) Neuronal and microglial regulators of cortical wiring: usual and novel guideposts. Front Neurosci 9:248. https://doi.org/10.3389/fnins.2015.00248

Steward O, Torre ER, Tomasulo R et al (1991) Neuronal activity up-regulates astroglial gene expression. Proc Natl Acad Sci U S A 88:6819–6923

Stewart MG, Popov VI, Kraev IV et al (2014) Structure and complexity of the synapse and dendritic spine. In: Pickel V, Segal M (eds) The synapse. Academic Press, New York, pp 1–20

Sultan S, Li L, Moss J et al (2015) Synaptic integration of adult-born hippocampal neurons is locally controlled by astrocytes. Neuron 88:957–972

Swanson L, Newman E, Araque A et al (2017) The beautiful brain – the drawings of Santiago Ramón y Cajal. Abrams, New York

Szepesi Z, Manouchehrian O, Bachiller S et al (2018) Bidirectional microglia-neuron communication in health and disease. Front Cell Neurosci 12:323. https://doi.org/10.3389/fncel.2018.00323

Takayama F, Zhang X, Hayashi Y et al (2017) Dysfunction in diurnal synaptic responses and social behavior abnormalities in cathepsin S-deficient mice. Biochem Biophys Res Commun 490:447–452

Tanasic S, Mattusch C, Wagner EM et al (2016) Desipramine targets astrocytes to attenuate synaptic plasticity via modulation of the ephrinA3/EphA4 signalling. Neuropharmacology 105:154–163

Tarozzo G, Bortolazzi S, Crochemore C et al (2003) Fractalkine protein localization and gene expression in mouse brain. J Neurosci Res 73:81–88

Theodosis DT, Poulain DA, Oliet SHR (2008) Activity-dependent structural and functional plasticity of astrocyte-neuron interactions. Physiol Rev 88:983–1008
Toni N, Buchs PA, Nikonenko I et al (1999) LTP promotes formation of multiple spine synapses between a single axon terminal and a dendrite. Nature 402:421–425
Tononi G, Cirelli C (2006) Sleep function and synaptic homeostasis. Sleep Med Rev 10:49–62
Trachtenberg JT, Chen BE, Knott GW et al (2002) Long-term in vivo imaging of experience-dependent synaptic plasticity in adult cortex. Nature 420:788–794
Tremblay M-È, Lowery RL, Majewska AK (2010) Microglial interactions with synapses are modulated by visual experience. PLoS Biol 8:e1000527. https://doi.org/10.1371/journal.pbio.1000527
Tremblay M-È, Lecours C, Samson L et al (2015) From the Cajal alumni Achúcarro and Río-Hortega to the rediscovery of never-resting microglia. Front Neuroanat 9:45. https://doi.org/10.3389/fnana.2015.00045
van den Heuvel MP, Scholtens LH, Feldman Barrett L et al (2015) Bridging cytoarchitectonics and connectomics in human cerebral cortex. J Neurosci 35:13943–13948
Vásquez CE, Reberger R, Dall'Oglio A et al (2018) Neuronal types of the human cortical amygdaloid nucleus. J Comp Neurol 526:2776–2801
Ventura R, Harris KM (1999) Three-dimensional relationships between hippocampal synapses and astrocytes. J Neurosci 19:6897–6906
Verkhratsky A, Nedergaard M (2018) Physiology of astroglia. Physiol Rev 98:239–389
Viola GG, Rodrigues L, Américo JC et al (2009) Morphological changes in hippocampal astrocytes induced by environmental enrichment in mice. Brain Res 1274:47–54
Volfovsky N, Parnas H, Segal M et al (1999) Geometry of dendritic spines affects calcium dynamics in hippocampal neurons: theory and experiments. J Neurophysiol 82:450–462
Volterra A, Meldolesi J (2005) Astrocytes, from brain glue to communication elements: the revolution continues. Nat Rev Neurosci 6:626–640
von Bohlen und Halbach O, Unsicker K (2012) Storage and release of nontransmitter signalling molecules from macroglia. In: Kettenmann H, Ransom B (eds) Neuroglia, 3rd edn. Oxford University Press, New York, pp 212–222
VonDran MW, Singh H, Honeywell JZ et al (2011) Levels of BDNF impact oligodendrocyte lineage cells following a cuprizone lesion. J Neurosci 31:14182–14190
Wake H, Moorhouse AJ, Jinno S et al (2009) Resting microglia directly monitor the functional state of synapses in vivo and determine the fate of ischemic terminals. J Neurosci 29:3974–3980
Wang Y, Richter-Landsberg C, Reiser G (2004) Expression of protease-activated receptors (PARs) in OLN-93 oligodendroglial cells and mechanism of PAR-1-induced calcium signaling. Neuroscience 126:69–82
Wang YY, Zhou YN, Jiang L et al (2023) Long-term voluntary exercise inhibited AGE/RAGE and microglial activation and reduced the loss of dendritic spines in the hippocampi of APP/PS1 transgenic mice. Exp Neurol 363:114371. https://doi.org/10.1016/j.expneurol.2023.114371
Weinhard L, di Bartolomei G, Bolasco G et al (2018a) Microglia remodel synapses by presynaptic trogocytosis and spine head filopodia induction. Nat Commun 9:1228. https://doi.org/10.1038/s41467-018-03566-5
Weinhard L, Neniskyte U, Vadisiute A et al (2018b) Sexual dimorphism of microglia and synapses during mouse postnatal development. Dev Neurobiol 78:618–626
Witcher MR, Kirov SA, Harris KM (2007) Plasticity of perisynaptic astroglia during synaptogenesis in the mature rat hippocampus. Glia 55:13–23
Wu Y, Dissing-Olesen L, MacVicar BA et al (2015) Microglia: dynamic mediators of synapse development and plasticity. Trends Immunol 36:605–613
Xu X, Stoyanova EI, Lemiesz AE et al (2018) Species and cell-type properties of classically defined human and rodent neurons and glia. eLife 7:e37551. https://doi.org/10.7554/eLife.37551
Xu-Friedman MA, Regehr WG (2003) Ultrastructural contributions to desensitization at cerebellar mossy fiber to granule cell synapses. J Neurosci 23:2182–2192

Xu-Friedman MA, Harris KM, Regehr WG (2001) Three-dimensional comparison of ultrastructural characteristics at depressing and facilitating synapses onto cerebellar Purkinje cells. J Neurosci 21:6666–6672

Zancan M, Dall'Oglio A, Sarzenski TM et al (2015) Glial and axonal perikaryal coverage and somatic spines in the posterodorsal medial amygdala of male and cycling female rats. J Comp Neurol 523:2127–2137

Zeisel A, Hochgerner H, Lönnerberg P et al (2018) Molecular architecture of the mouse nervous system. Cell 174:999–1014.e22

Zhang Y, Barres BA (2010) Astrocyte heterogeneity: an underappreciated topic in neurobiology. Curr Opin Neurobiol 20:588–594

Zhou J, Sutherland ML (2004) Glutamate transporter cluster formation in astrocytic processes regulates glutamate uptake activity. J Neurosci 24:6301–6306

Zhou Q, Homma KJ, Poo M (2004) Shrinkage of dendritic spines associated with long-term depression of hippocampal synapses. Neuron 44:749–757

Zuo Y, Lin A, Chang P et al (2005) Development of long-term dendritic spine stability in diverse regions of cerebral cortex. Neuron 46:181–189

Chapter 7
Dendritic Spines in Learning and Memory: From First Discoveries to Current Insights

Nicolas Heck and Marc Dos Santos

Abstract The central nervous system is composed of neural ensembles, and their activity patterns are neural correlates of cognitive functions. Those ensembles are networks of neurons connected to each other by synapses. Most neurons integrate synaptic signal through a remarkable subcellular structure called spine. Dendritic spines are protrusions whose diverse shapes make them appear as a specific neuronal compartment, and they have been the focus of studies for more than a century. Soon after their first description by Ramón y Cajal, it has been hypothesized that spine morphological changes could modify neuronal connectivity and sustain cognitive abilities. Later studies demonstrated that changes in spine density and morphology occurred in experience-dependent plasticity during development, and in clinical cases of mental retardation. This gave ground for the assumption that dendritic spines are the particular locus of cerebral plasticity. With the discovery of synaptic long-term potentiation, a research program emerged with the aim to establish whether dendritic spine plasticity could explain learning and memory. The development of live imaging methods revealed on the one hand that dendritic spine remodeling is compatible with learning process and, on the other hand, that their long-term stability is compatible with lifelong memories. Furthermore, the study of the mechanisms of spine growth and maintenance shed new light on the rules of plasticity. In behavioral paradigms of memory, spine formation or elimination and morphological changes were found to correlate with learning. In a last critical step, recent experiments have provided evidence that dendritic spines play a causal role in learning and memory.

Keywords Dendritic spine · Learning and memory · Synaptogenesis · Structural plasticity · History of neuroscience

---

N. Heck (✉)
Laboratory Neurosciences Paris Seine, Sorbonne Université, Paris, France
e-mail: nicolas.heck@sorbonne-universite.fr

M. D. Santos
Department of Neuroscience, Northwestern University Feinberg School of Medicine, Chicago, IL, USA

A. A. Rasia-Filho et al. (eds.), *Dendritic Spines*, Advances in Neurobiology 34,
https://doi.org/10.1007/978-3-031-36159-3_7

7.1 Introduction

Dendritic spines are small protrusions found on dendrites and host the postsynaptic structure receiving excitatory inputs. The density of spines is representative of brain connectivity, and their size correlates with synaptic transmission efficacy. While dendritic spines are highly stable in the adult brain, new spines can appear, and their shapes undergo dynamic changes. Spine remodeling instructs changes in neuronal connectivity that contribute to the adaptive properties of the brain network. Spines are thus considered a privileged locus for cerebral plasticity and therefore relevant for the understanding of cognitive functions.

Providing evidence for the role of dendritic spines in the cerebral process of learning has been a long journey supported by bold hypotheses and technological advances. Seminal ideas brought up at the end of the nineteenth century were remarkably similar to the concepts found in neurobiology at the present day. Following the confirmation, in 1959, that spines host the postsynaptic element, several research programs were deployed, in which the dendritic spine has opened a new window for observing brain plasticity. Live imaging, then, allowed a new picture to emerge, by unraveling the dynamic features of the spines, and the rules that govern synaptic plasticity. Finally, the long thought idea, that dendritic spines contribute to learning and memory, has recently received convincing experimental support. While caution should be taken to not read the terms used in earlier publications with their contemporary meaning, hypotheses and the associated paradigms are better understood in the light of the path that leads to them (Cobb 2020; Morange 2021). Therefore, this chapter offers a historical perspective on the progressive building of our current knowledge that dendritic spines play a causal role in learning and memory.

7.2 The Discovery of Dendritic Spines, and the Early Hypothesis of Morphological Plasticity

7.2.1 First Report, First Question

In 1888, spines were described for the first time by Ramón y Cajal (Ramón y Cajal 1888). Cajal used Golgi staining to observe neuronal morphology in the chicken cerebellum, and wrote that the surface of neurons "appears bristling with points and short spines." Although some researchers disregarded those spines as a staining artifact, other authors published plates showing spines on neurons from different species, including humans (Ramón y Cajal 1899–1904, 1909–1911). Cajal developed several arguments that finally led to the confirmation of his initial description (DeFelipe 2015; García-López et al. 2007; Yuste 2015).

Dendritic spines were not immediately recognized as points of contact by which signal would be transmitted from one neuron to another, which came later with the

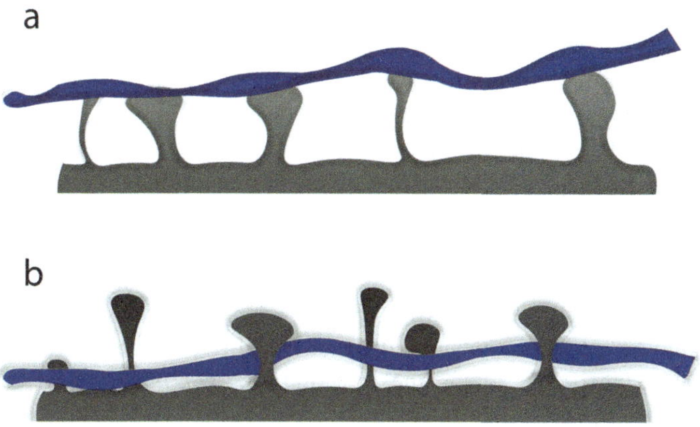

Fig. 7.1 Dendritic spines are protrusions found along the dendrite. After their first description by Cajal in 1888, dendritic spines were thought to increase the global surface of the neuron. It was both proposed the axon would rest on the spine head (**a**) or would find its path in between the spines (**b**)

observation of spiny synapses by electron microscopy (Gray 1959b). The site of transmission, or synapse, a name coined by Stefanowska in 1897a, b, was conceived as a valve, to account for the unidirectionality of nerve impulse and for the synaptic delay (Bennett 1999; Sotelo 2020). It, however, was not simply assumed that synapses would be localized in spines. Cajal viewed the spines as a manner of increasing the total neuronal surface, agreeing with Berkley that they could correspond to charge collectors (Ramón y Cajal 1899–1904, 1909–1911), an idea compatible with axon contacting either the tip or the neck of the spines (Fig. 7.1). Berkley was the first to propose that axon would appose to the tip of the spine, but explicitly claimed that it was more probable that there is no contact so the electrical signal would pass through a gap (Berkley 1896). In most of his writings, Cajal described axons passing in between the spines, explicitly stating axons would be placed in the interspine notches (Ramón y Cajal 1892, 1894a, 1899–1904, 1909–1911). Cajal still described that parallel fibers in the cerebellum rest on spines of Purkinje cells (Ramón y Cajal 1892, 1894a), but also wrote that spines are notches in which parallel fibers are located (Ramón y Cajal 1899–1904, 1909–1911); and he only drew axons apposed on the tip of the spines in his last publication (Ramón y Cajal 1933; García-López et al. 2007).

7.2.2 The Hypothesis of Morphological Plasticity

It is noteworthy that the term "plasticity" was used from the mid-eighteenth century to describe the adaptive properties of living organisms (Jones 2000). In the context of psychology, William James theorized that habit formation could be explained by plasticity (Berlucchi and Buchtel 2009). Plasticity was also considered as revealed by the regenerative capacities of the neural network after lesions. A student of Marinesco, Minea used the term "plasticity" in his thesis to describe the changes in neuronal morphology observed after lesions (Jones 2000; Stahnisch 2003).

A possible role for dendritic spines arose with the question of whether the connections in between neurons would be stable or variable. Pivotal for this idea is the neuronal theory. By the end of the nineteenth century, a burning question was if the brain was composed of a single continuous network (reticular theory, firmly defended by Golgi) or discrete cellular units (neuron doctrine, no less firmly defended by Cajal) (Barbara 2006). Cajal considered spines as free endings, thus being a supplementary argument for the individuality of neurons (Ramón y Cajal 1899–1904, 1909–1911). Against the view of a rigid *continuum*, the acceptance of neuron doctrine opened the gate to imagine that structural changes could modify connectivity and attempts to link such changes with cognition were deployed.

It is Demoor who was the first, in 1896, to use the term "morphological plasticity of neurones" in the title of a publication (Demoor 1896a, b; Jones 2004). Nevertheless, morphological changes related to nerve impulse transmission had been proposed earlier. Actually, the number of published articles with explicit references to other researcher's work attests that the hypothesis of plasticity for neuronal connectivity was matter of debate for the scientific community. Several authors, such as Tanzi, Rabl-Rückhard, Lépine, Duval, Lugaro, Demoor, or Cajal, have debated on the hypothesis that brain connectivity would undergo changes, and proposed possible mechanisms.

7.2.3 Proposed Mechanisms for Changes in Connectivity

In 1893, Tanzi, by making an analogy with the muscle, postulated that the repeated activity at sites of transmission would increase nutrition leading to hypertrophy of the neuronal ramifications that establish connectivity. The hypertrophy would diminish the gap between neurons, thereby facilitating further the transmission of nerve impulses (Tanzi 1893; Berlucchi and Buchtel 2009). Cajal refuted the idea of Tanzi, arguing that the formation of new contacts, rather than facilitation of the transmission of nerve impulses, was required. In his Croonian lecture, Cajal proposed that "acquired intelligence" would be sustained by a "cerebral gymnastic" corresponding to the creation of new contacts between neurons (Ramón y Cajal 1894b), and he specified his arguments later (Ramón y Cajal 1899–1904, 1909–1911). In 1898, Lugaro used the term "plasticity" to combine the nutritive hypertrophy hypothesis of Tanzi with the chemotropism theory of Cajal to account for what he called plasticity of the nervous elements, an idea he proposed earlier (Lugaro 1895, 1898a). Chemotropism was proposed by Cajal as a theory that would explain the guidance of axons during development. Other hypothetical mechanisms were proposed. For example, Cajal proposed that loss of contacts would be the result of the extension of protrusions from glial cells which would intervene in neuronal articulations (a term he often used to mention neuronal contacts) (Ramón y Cajal 1895) but he rejected this idea later (Ramón y Cajal 1899–1904, 1909–1911). Lépine and Duval independently formulated the ameboid theory, according to which ameboid-like morphological changes could happen at the neuronal surface (Lépine 1894; Duval 1895). This idea was based of the observation of Wiedersheim, who first reported that the shape of a cerebral ganglion of a transparent crustacean varied through its lifetime, and thus proposed that neurons can change shape (Wiedersheim 1890). Solvay, in a talk given in 1893, proposed that if nerve cells would possess a kind of plasticity, the electrical activity would modify anastomoses (i.e., neuronal communication) in a uniform or localized, temporarily or persistent, manner (Solvay 1894).

The idea that dendritic spines would participate in morphological plasticity was also proposed. Whereas Cajal did not clearly assume that the tip of the dendritic spines was the privileged locus of the changes in contacts sustaining "cerebral intelligence," he reinforced his hypothesis by arguing that dendritic spines appear gradually through development along intellectual capacities, and that dendritic spines are higher in number through phylogeny, especially in mammals. The general idea of morphological plasticity was subject of experiments by many authors, who described neuronal morphology and dendritic spine density after electrical stimulation, applications of pharmacological agents known to modify brain state (e.g., ether, morphine, chloral hydrate) as well as during sleep or hibernation states. A consensual idea was that spines would retract during periods of inactivity like sleep and hibernation, which was tested at that time (Querton 1898) and much later (Popov et al. 1992). The most convincing results in support of a change in dendritic spine number and shape are probably from the work of Demoor and colleagues at the Solvay Institute in Belgium (Fig. 7.2). Using Golgi staining to observe dendrites and spines,

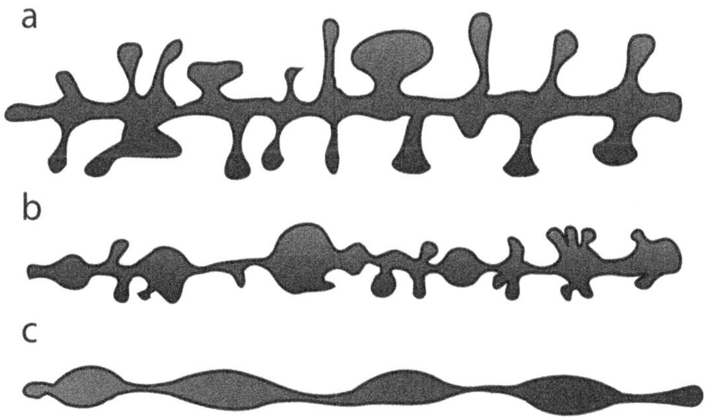

Fig. 7.2 Early experimental evidence of morphological changes in correlation with brain states. Demoor and colleagues provided observations of dendrites from animals subjected to pharmacological agents know to affect brain states. In comparison to naive animals (**a**), they reported reversible spine loss, or spine changes with dendrite blebbing (**b**). However, other authors using similar treatments reported complete loss of spines on dendrite looking like string of pearls, which was interpreted as sign of degeneration (**c**)

Demoor reported a loss of spine in dogs' brain subjected to morphine (Demoor 1896a, b). Importantly, Demoor also observed that when the dog was allowed to recover from morphine administration for 36 h, the number of spines was back to control levels. The work was extended by Stefanoswka and Querton who studied spine density and morphology changes after brain stimulation with electricity or pharmacological agents or hibernation state (Stefanowska 1897a, b; Querton 1898).

Both the ameboid movement theory, which relied on the observation of swelling and varicosities, and the theory of spine changes were criticized based on the knowledge of technical limitations and potential artifacts. While the ameboid theory was abandoned (Black 1981), connectivity changes through spine dynamics were considered, but could not be formally established by unequivocal experimental evidence. For instance, the results from Demoor on dogs subjected to morphine were not found to be reproducible with rabbits (Lugaro 1896). Many authors noted some delay between the treatment and the spine alterations, leading some to conclude that spine changes were the consequence of tissue manipulation rather than a physiological process (discussed in Demoor 1896b, 1898; Querton 1898). The interpretation of Demoor was also confronted with studies using treatments such as starvation (Monti 1895), general paralysis (Azoulay and Klippel 1894), or toxin and alcohol intoxication (Berkley 1896) which induced spine loss and dendrite swelling understood as signs of degeneration (García-López et al. 2007) (Fig. 7.2c). Ether was used for it can have transient, prolonged, or lethal effects, so Stefanowska could argue that she observed spine loss before dendritic swelling appeared (Stefanowska 1900). However, different observations were reported (Lugaro 1898b; Soukhanoff 1898). Cajal, as rightly cited by DeFelipe (2006), considered those works to defend the idea of morphological plasticity of spines, but wrote on the same page that further experiments with better controls were necessary before a definitive conclusion could be drawn (Ramón y Cajal 1899–1904, 1909–1911).

It is again noteworthy that many researchers such as Berkely or Tanzi assumed the connected neurons were not in contact. While the neuron doctrine allowed stemming the idea of connectivity in between neurons, it was also viewed as implying that neurons would be separated by a gap, which would not impair electrical currents to flow through neuronal circuits. That electrical signal could pass all along the neuronal surface precluded to assume that the dendritic spines were discrete foci of signal transmission. Azoulay, translator of Cajal writings at that time, carefully noted that Cajal would employ the terms contact and contiguity in the same way (Azoulay 1895). While Cajal used the term "*contact*" to describe parallel fibers caught "*in the gaps between*" spines (Ramón y Cajal 1892; cited in DeFelipe 2015), it was translated "*by contact, by contiguity*" 2 years later (Ramón y Cajal 1894a). Importantly, all translations by Azoulay were validated by Cajal who took those as opportunities to revise and update his texts. Hence, while dendritic spines were the focus of much attention, and plastic changes were interrogated, it could not be concluded that spine changes would be the explanation of brain states, memory, and cognition.

7.2.4 *Conclusion*

This historical review reveals how the idea of a change in connectivity and the possibility that dendritic spines are its privileged locus was already formulated and subjects of experiments at the end of the nineteenth and very early twentieth century. The hypothesis was of course a matter of debate, given the technical limitations and available background knowledge of that time, but it can be recognized that scientific rigor allowed not to conclude in a peremptory and final way. The aforementioned authors made great achievements in paving the way for a research program according to which network activity would modify synaptic connections to sustain learning and memory.

7.3 Dendritic Spines, Plasticity, and Cognition: Setting Up the Stage

7.3.1 *Synaptic Nature of Dendritic Spines and Paradigms for the Study of Plasticity*

The demonstration that dendritic spines are postsynaptic structures avoided misinterpretations like in earlier works. In the 1950s, electron microscopy entered biology laboratories, which led to the observation of the ultrastructure of synapses (Peters et al. 1976). In 1959, what is today considered the first and definitive proof that dendritic spines are hosting the postsynaptic element faced by presynaptic elements filled by vesicles was provided (Gray 1959a, b). Then, changes in dendritic spines' density would no more fuel hypothetical theories, but would be considered as an indicator of synaptic modification of the neuronal network. Similarly, as the expression of a specific protein is used to identify a cell type, dendritic spine density became a window through which connectivity changes could be reported. The way was then paved to study the potential link between connectivity and learning and other cognitive capabilities.

Obviously, morphological change of synaptic connectivity was not the sole hypothesis, and several cellular and physiological mechanisms were proposed to explain the process of learning. A review covering the trends in this research and the different experimental approaches developed by the year 1968 can be found in Kandel and Spencer (1968).

The hypothesis according to which plastic brain changes involved connectivity changes was tested in four different experimental paradigms: in recovery after lesion, in developmental brain changes induced by interaction with the environment, in correlation with pathologies with learning deficits and mental retardation, and in correlation with synaptic plasticity. Seminal studies are presented below to illustrate those four approaches, with a focus on dendritic spines (Fig. 7.3).

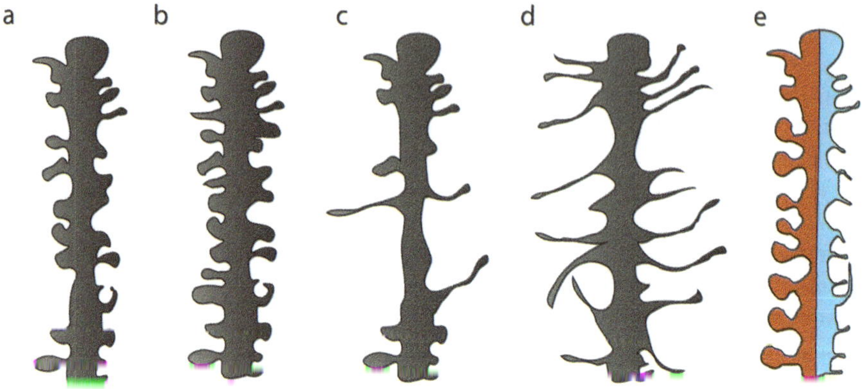

Fig. 7.3 Dendritic spine changes observed in different paradigms of plasticity. In comparison to control animals (**a**), experience-dependent plasticity in enriched environment induced spine density increase (**b**) while sensory deprivation decreased density with smaller spine heads (**c**). Dendrites from brains of infants with mental retardation exhibited abnormal spine densities, and altered spine shapes including very long and thin protrusions (**d**). In hippocampal brain slices, long-term potentiation was found to induce either increase (**e**, red, left side) or decrease (**e**, blue, right side) of spine heads sizes

7.3.2 *Plasticity as Recovery After Lesion*

One approach to tackle potential structural plasticity was the study of recovery from lesions. Those studies, conducted since the beginning of the twentieth century by Nageotte, Marinesco, Cajal, and others, finally led to the seminal publication of Raisman (1969). Following the lesion of afferent pathways, Raisman demonstrated reoccupation of vacant postsynaptic sites by remaining axonal fibers. Either the hippocampal or the hypothalamic fibers projecting onto neurons of the medial nucleus of the septum were subjected to degeneration, and the synaptic profiles on the soma and dendrites of neurons of the septum were observed by electron microscopy. The degeneration of each pathway induced the loss of presynaptic elements in the septum within a few days. However, synapse density seemed restored after some weeks, with presynapses forming multiple synapses, indicating the repopulation of denervated synaptic sites.

In a further experiment using the same circuit from hippocampus and hypothalamus to the septum, animals with long-standing lesions of one pathway were subjected to a lesion of the other pathway and sacrificed after 2 days. The goal was to test whether the remaining pathway undergoes plastic changes to compensate for the loss of synapses induced by the first lesion. Indeed, the first lesion would have been followed by the formation of new synapses, while the short time after the second lesion would allow seeing the loss of presynapses. The presynapses of hippocampal, but not hypothalamic, fibers showed signs of degeneration, hence Raisman acknowledged that both compensations by one pathway as well as nonspecific synapse formation by fibers located in the vicinity of the dendrites could occur. Overall, the anatomical study conducted by Raisman has been considered as a critical step in the study of connectivity and plasticity. It is noteworthy that brain tissue transplantation, obviously studied from a clinical perspective, has been also viewed as an experimental test of the potential for plastic changes of connectivity in the brain (Cotman and Nieto-Sampedro 1984).

7.3.3 *Experience-Dependent Plasticity*

Another approach in the study of neuronal plasticity is to examine brain changes upon experience (i.e., interaction with the environment). Since the visual cortex of each hemisphere receives thalamic afferents conveying information from both eyes, the closure of one eye allows to study whether visual experience shapes the development of the visual pathway. The pioneering work of Hubel and Wiesel unraveled the phenomenon of ocular dominance. They brought evidence, by electrophysiological recordings and observation of thalamic axonal afferents in the visual cortex, that the cortical area that would be innervated by inputs carrying information of the closed eye is decreased at the benefit of the other open eye. Morphological analysis of Golgi-stained neurons of the visual cortex was made following visual deprivation; a decrease in spine density in the mouse visual cortex (Valverde 1967) and increased variability in the size of spine heads in the rabbit visual cortex (Globus and Scheibel 1967) were measured on apical dendrites of pyramidal neurons.

In parallel studies, the brain changes caused by environment enrichment were explored (Bennett et al. 1964). In those conditions, an increase in spine density (Globus et al. 1973) was repeatedly confirmed along extended dendrite length and branching (Volkmar and Greenough 1972). Variety of sensory stimulation applied to rat pups was found to induce an increased spine density too, although it could have been the effect of the stress induced by the stimulation protocol (Schapiro and Vukovich 1970). Dendritic spines could later be directly followed in the brain of living rodent, confirming that spine changes correlate with the experience-dependent plasticity (Holtmaat and Svoboda 2009), including in the context of visual experience and monocular deprivation (Tropea et al. 2010; Zhou et al. 2017). The phenomenon of ocular dominance contributed to the idea that the process at work during development is a competition between active and inactive neural paths. Recent advances in imaging techniques allowed to demonstrate that experience-dependent plasticity is realized at the dendrite level by activity-driven competition between neighboring spines, with the maintenance of active spines and the pruning of the neighboring, less active, spines (Bian et al. 2015).

7.3.4 *Correlates of Intellectual Disability and Learning Deficits in* Postmortem *Brains*

Other results that strengthen the relevance of studying connectivity for the understanding of brain functions came from the observation of altered synapse morphology in clinical cases of mental retardation. Electron microscopy from *postmortem* brains of two patients with psychomotor retardation and learning and memory deficits revealed an alteration of the presynapse (Gonatas 1967; Gonatas and Moss 1975). Swelling of the presynapse with reduced numbers of vesicles and mitochondria, rather than signs of degeneration, was observed in the frontal cortex. Such synaptic alterations, along with preserved neuronal cell body and dendrite, contradicted the general assumption, according to which synapse loss would come as an unspecific consequence of the retrograde degeneration after axonal or cell body lesions. Thus, Gonagas hypothesized a causal relationship between “synaptic lesions” and “psychomotor disorders,” formulated as the “concept of synaptic disease.” Since many studies reported a lack of brain macroanatomical alterations in patients with psychiatric and cognitive disorders, it was then proposed that studying the brain ultrastructure could bring new hypotheses on the etiology of those diseases (Crome 1972).

Other observations confirmed a link between intellectual disability and dendritic spines. Dendrites were observed with optical microscopy on neurons stained with Golgi technique from the brain of children with chromosomal aberrations and mental retardation. In one case, with Patau syndrome, there was a reduced spine density, and the spines were long and tortuous. In another case, with Down syndrome, some dendritic segments were devoid of spines, while others were covered by a high density of very small spines (Marin-Padilla 1972, 1974). Similarly, either dendritic spine loss or abnormally long and thin spines were observed on dendrites from children with mental retardation without apparent chromosomal aberrations (Purpura 1974, 1975). Those observations have since been extended to other diseases such as FragileX syndrome and to murine models of mental disorders (Forrest et al. 2018; Nakai et al. 2018; Penzes et al. 2011), although it remains difficult to establish whether spine alterations are the etiological cause of the disease (Fiala et al. 2002b).

7.3.5 *Morphological Counterpart of Long-Term Potentiation*

The discovery of long-term potentiation (LTP) paved another way for the study of spine plasticity. In 1973, Bliss and Lomo made the demonstration that synaptic transmission efficacy can be increased (Bliss and Lømo 1973). The long-term stability of the potentiation induced by prior synaptic activity was considered as a potential mechanism for memory storage. Bliss and Lomo proposed several mechanisms that would account for that phenomenon, including a change of the morphology of the spine neck, citing theoretical work which calculated that the morphology of the neck would impose a resistance to the synaptic input.

A structural counterpart of long-term plasticity was first reported in 1975 by van Harreveld and Fifkova (1975). The authors acknowledged that changes in the structure of spines could be a mechanism explaining synaptic potentiation and aimed at testing this hypothesis. A tetanic electrical stimulation known to induce an increase in synaptic efficacy was applied in the perforant path, an axonal pathway projecting into the hippocampus. Measurements on electron microscopy plates showed an increase in synaptic area in the proximal dendrites on the molecular layer of the dentate gyrus region. The measurements were done within an hour after stimulation, both on the proximal dendrites, which receive afferents from the perforant, stimulated, pathway, and on the distal dendrites, which receive afferents from the commissural path, in order to confirm the specificity of the effect. The absence of spine changes on the distal dendrites allowed to conclude that structural synaptic changes were restricted to the stimulated path.

The increase in spine size appeared few minutes after stimulation (van Harreveld and Fifkova 1975) and lasted at least 23 h (Fifková and van Harreveld 1977). In their first publication, Fifkova and van Harrefeld postulated that the rapid swelling could be due to "uptake of electrolytes and water," while the long-term stabilization would require "formation of structural proteins." The later observation of the presence of actin in spines provided further support for the dynamic nature of spine morphology (Fifková and Delay 1982; Matus et al. 1982).

The aforementioned studies of Fifkova and van Harreveled were followed by others, in which electrophysiological recordings allowed to confirm synaptic potentiation, and electron microscopy was used to analyze spine density and morphological parameters (for a review summarizing the outcomes of different studies until year 2000, see Yuste and Bonhoeffer (2001)). Although some concluded to an increased spine density while others described constant numbers but change in size, those studies provided strong experimental support in favor of the involvement of structural plasticity in the context of LTP.

An obvious idea was to extend the link between dendritic spines and neuronal plasticity to learning and memory. However, as compared to large connectivity rearrangements that could be observed in paradigms of experience-dependent developmental plasticity or enriched environments, some considered that a single learning task would rather likely induce low or sparse spine formation in restricted dendritic compartments, which would be challenging to observe (Andersen and Trommald 1995). The development and success of that research program are presented in Sects. 7.6 and 7.7.

7.3.6 Conclusion

The confirmation that spines form synapses provided a ground for studies that investigated structural plasticity in different paradigms. The results established that connectivity changes can occur in the brain and correlate with experience, cognitive capacities, and synaptic potentiation. It likely framed later studies guided by the idea that dendritic spines would be involved in learning and memory. It is noteworthy that in both the initial and later studies, electron microscopy was used. The many studies that reported spine change with photonic microscopy could not be interpreted with such confidence without the observations made by electron microscopy. Following the pioneer works aforementioned, a wealth of further studies confirmed and extended the results and considerable knowledge was acquired at the beginning of the 90s (Horner 1993), when new techniques were going to allow the observation of spines in living neurons (see Sect. 7.5).

However, any interpretation of spine density and spine morphology changes requires determining their biological meaning, which is discussed in the next section.

7.4 Beyond Numbers and Shapes

7.4.1 What Can Be Inferred from the Observation of Dendritic Spines

To understand why spine analysis is pivotal to study neuronal plasticity and brain cognition, it is necessary to understand of what morphology of the dendritic spine is indicative. Spine analysis allows for two kinds of evidence. First, the number of spines along the dendrite brings knowledge on an important part of all of synaptic connections. Second, the size of spines is informative on the efficacy of synaptic transmission. Then, any change in one of these two parameters can be correlated with cellular, physiological, and behavioral events. To infer any biological evidence from a defined quantitative parameter, however, requires validation and awareness of its limitations. In the following section, the relationship between spine morphology and its attributed functions is discussed aiming to provide an interplay between spine morphology, learning, and memory.

7.4.2 Dendritic Spines Indicate Excitatory Synapse Density

Electron microscopy played a pivotal role in this field, since it overcomes the limits of optical resolution, thereby unraveling cellular ultrastructure. After the first observation of synapses, Gray published plates in 1959 answering the long-standing question as whether dendritic spines could form synapses. Gray named Type I synapses located on spines, and Type II synapses located on dendritic shaft and cell body (Gray 1959a). Type I synapses departed from Type II by a thickening at the postsynaptic membrane apposed to a presynapse filled with round vesicles (Gray 1959b). Using a different fixation protocol for material preparation, other authors described asymmetrical synapses located on the spine in contrast with symmetrical synapses observed on soma and dendritic shaft (Colonnier 1968; LeVay 1973). Type I and asymmetrical synapses were assumed to be similar and it was proposed they were excitatory, while Type II, symmetrical, synapses would be inhibitory (Uchizono 1965). This assumption based on parallel electrophysiological recordings was finally directly confirmed by immunodetection of glutamate in presynapse (DeFelipe et al. 1988). Finally, only a few percent of spines are not in contact with a presynapse (Arellano et al. 2007), and the vast majority of inhibitory GABAergic inputs do not end on spines (Kwon et al. 2019) (but see Sect. 7.4.3).

Overall, observing and counting spines is equivalent to counting excitatory synapses. Hence, observation at the optical resolution of spine density along the dendrite is to be considered a proper measure of glutamatergic synaptic connectivity. It is noteworthy that another method proposed for assessing the density of connectivity was the observation of the proximity between axons and dendrites. This so-called Peter's rule has been however dismissed when estimations based on Peter's rule were confronted with electron microscopy reconstruction of dendritic segments (Kasthuri et al. 2015; Mishchenko et al. 2010).

7.4.3 Dendritic Spines Bring an Incomplete Picture

A general statement according to which spine density represents glutamatergic synaptic density needs to be considered in the light of further observations, which call for more careful interpretation. Spines dually innervated by glutamatergic and GABAergic boutons are found on pyramidal neurons of the neocortex (Jones and Powell 1969; Kubota et al. 2016), although restricted to the most distal part of the dendrite. More importantly, glutamatergic inputs are not only found on dendritic spines. For example, in striatal neurons, thalamic afferents form synaptic contacts both on spine and on the dendritic shaft (Doig et al. 2010). Therefore, despite an increase in spine density would still correspond to an increase in glutamatergic connectivity, it would not give the whole picture.

Another key point concerns the origin of the afferents' endings on dendritic spines. While a global change in spine density is indicative of connectivity changes, it does not necessarily identify the connected brain regions that undergo such changes. In some cases, a dendritic portion receives an afferent from a single region. For example, pyramidal neurons of the CA1 region of the hippocampus are contacted by mossy fibers of the CA3 region on the proximal part of the apical dendrites, and they receive inputs from the entorhinal cortex on the distal part of the apical dendrites. Inputs can also be distributed in a mixed manner, such as on striatal neurons. In the dorsal striatum, dendritic spines receive afferents from both the neocortex and the thalamus. Those can be distinguished by the presence of different types of vesicular glutamate transporters (VGLUT) when labeled in the presynapses, with VGLUT1 and VGLUT2 being specific for cortical and thalamic presynapses, respectively (Doig et al. 2010; Fremeau et al. 2004). Nevertheless, in the ventral striatum, neurons receive, in addition to thalamic afferents, inputs from neocortex, amygdala, and ventral hippocampus, which are all containing VGLUT1. Last, it is now established that those inputs are not segregated but interleaved along the dendrite (Xia et al. 2020).

In conclusion, spine density can be reasonably considered as a measure of the density of excitatory connectivity, both with optical and electron microscopy, but careful choice of the dendritic segment or other methods allowing to trace the origin of the afferents are mandatory to further investigate connectivity between specific brain regions or neuronal types.

7.1.1 The Meaning of the Morphology

Three-dimensional reconstruction from serial electron microscopy allows to gain precise and reliable picture of the dendritic spine (Harris and Weinberg 2012; Korogod et al. 2015; Tamada et al. 2020). Electron microscopy observations show that dendritic spines exhibit a wide range of size and morphologies along the same dendrite and across brain regions (Parajuli et al. 2020), which contributed to the hypothesis they would be one locus for plastic changes (Harris and Kater 1994). Nevertheless, several studies have demonstrated a strong correlation between the size of the spine and the efficacy of synaptic transmission. A strong correlation between spine head volume and both the area of the postsynaptic density (PSD) and number of vesicles in the presynapse has been measured in neurons from the neocortex (Arellano 2007; Holler et al. 2021), hippocampus (Harris and Stevens 1989; Schikorski and Stevens 1997), striatum (Wilson et al. 1983), and for Purkinje cells of the cerebellum (Harris and Stevens 1988). That correlation was confirmed when measuring in acute slices excitatory postsynaptic potential and the PSD area from neocortical neurons (Holler et al. 2021). In another functional approach, glutamate was uncaged so a single spine would be stimulated while synaptic currents were recorded. In acute slices of the hippocampus (Matsuzaki et al. 2001) and in the neocortex in vivo (Noguchi et al. 2011), a correlation between spine head volume and glutamate-induced synaptic current amplitude was measured at the single-spine level. Another parameter of interest is the morphology of the neck, since it would isolate the spine head from the dendrite. The idea of a biochemical and electrical compartmentalization has been postulated based on theoretical calculation, and recently gained experimental evidence (Adrian et al. 2014; Araya 2014; Cornejo et al. 2022; Yuste 2013). However, electron microscopy and super-resolution techniques are required to properly measure the length and diameter of spine neck.

A classification of dendritic spines based on a morphological criterion has been proposed. According to Peters and Kaiserman-Abramof (1970), spines fall in three categories, namely, stubby, thin, and mushroom. Stubby spines are distinct from others because they do not have a neck, which may have important functional implications since the neck would represent a constraint that isolates the spine head from the dendrite, while the molecular composition of the stubby spine is similar to other spines (Helm et al. 2021). Mushroom spines have a head that is much larger than the diameter of their neck, while thin spines have a head that is not much larger than the neck. However, this is an arbitrary and unjustified categorization, as dendritic spines with necks show a *continuum* of morphological variation (Wilson et al. 1983; Arellano 2007; Tønnesen et al. 2014; Ofer et al. 2021, 2022).

7.4.5 Conclusion

The identification as a glutamatergic synapse combined with the availability of labeling techniques definitively set the spine as an observable unit particularly relevant in neuroscience. Since spines can easily be observed, and so counted, neuroscientists can measure the establishment of connectivity during development, its loss in some pathologies, and test whether or not the number of synapses correlates with learning and other cognitive functions. That measurement, however, does not bring a complete picture. Complementary methods are required for a full description of neuronal connectivity. Furthermore, spine morphology is particularly relevant since the size and shape of spines are indicative of its functional properties. Therefore, the study of morphological changes in the spine allows to explore brain plasticity at the synaptic level, which will be further addressed below.

7.5 Dendritic Spines, Motility, and Stability Compatible with Learning Process

7.5.1 Live Imaging Unravels Spine Dynamics

The many studies that brought invaluable information on the possible role for dendritic spines in plasticity, learning and memory, and cognition, still suffered from a limitation. Those were performed by comparing fixed brains from different experimental conditions, demonstrating a morphological plasticity but dynamic processes remained hidden. The combination of neuronal cultures and brain slices with the development of time-lapse microscopy allowed unprecedented observations, unraveling that dendritic spines are both motile and stable.

7.5.2 Transition from Ex Vivo to In Vivo Spine Imaging

The use of live confocal microscopy for studying spiny neurons began with early postnatal brain slices (Hosokawa et al. 1992; Dailey and Smith 1996; Dunaevsky et al. 1999) and culture models (Fischer et al. 1998; Papa and Segal 1996; Ziv and

Smith 1996). By using live imaging of fluorescent neurons, researchers were able to discover novel facets of spine plasticity that were previously unattainable through staining of *postmortem* brains. Pioneering ex vivo studies demonstrated that spines are highly dynamic under basal conditions, with formation and retraction occurring within several minutes (Bonhoeffer and Yuste 2002). These studies have also demonstrated that filopodia play a crucial role in the development of synapses, with these long and thin protrusions being the initial step for initiating synaptogenesis in immature neurons.

The observation of spine dynamics in immature neurons has raised an important question about whether or not the adult brain also undergoes a high rate of spine remodeling. This is especially relevant because adult animals can form new memories, and if this is correlated with synaptic remodeling, it suggests that new connections can still be formed even when the brain is fully mature. However, for cellular memory traces of previous experiences to be maintained throughout an individual's life span, synapses must also be stable enough.

Longitudinal adult neuron imaging is a challenging task due to the high sensitivity of mature neurons to acute slicing or enzymatic dissociation procedures. In the 1990s, a new type of photonic microscopy was adapted for deep intact brain imaging: two-photon laser scanning microscopy (2PLSM) (Denk et al. 1990). This type of photonic imaging takes advantage of femtosecond pulsed infrared photons, which are less sensitive to light scattering than visible light. Unlike conventional one-photon microscopy, 2PLSM reduces out-of-focus light and lowers phototoxicity, as light emission only occurs when two photons coincidently hit the fluorophores. Based on all these properties, 2PLSM offered the unprecedented capability to acquire fluorescent images up to 1 mm away from the surface of the biological sample, becoming the gold-standard microscopy technique for spine imaging, even 30 years later. In the middle of the 1990s, the first live spines' results using 2PLSM were successfully obtained on brain slices (Yuste and Denk 1995), and in the early 2000s, 2PSLM was then used to visualize neurons in the intact rodent neocortex (Lendvai et al. 2000). The primary obstacle to light penetration is the opacity of the adult cranium, which can be overcome by either replacing the skull with a small glass coverslip or performing skull thinning, thereby allowing 2PLSM to capture images from dendrite in the brain of rodents.

7.5.3 Spontaneous In Vivo Spine Dynamics: From High Motility During Development to Stability in Adult Brain

The 2000s marked a significant milestone since dendritic spines could then be observed in the brain of living rodents over their entire life span (Bhatt et al. 2009). Despite discrepancy between studies using neuronal cultures and brain slices, spine spontaneous dynamics and plastic changes had gained consensus. In vivo time-lapse imaging was offering a mean to unravel those phenomena in the brain (Yuste and Bonhoeffer 2004).

Whereas spines were thought to be transient based on previous ex vivo studies, new results in adult brains started to change this vision by showing that a large proportion of spines can be stable for several weeks in basal conditions. Although initial studies reported contradictory results (Grutzendler et al. 2002; Trachtenberg et al. 2002), further works confirmed the high stability of dendritic spines in the adult brain (Holtmaat et al. 2005; Majewska et al. 2006; Zuo et al. 2005). It appeared that the discrepancy between some studies came from the two different techniques used to image the brain in living rodents. Compared to thinning the cranium, to open the skull and place a glass coverslip induced glia reactivity and increased spine turnover (Xu et al. 2007). More specifically, Grutzendler and colleagues used two-photon imaging through a thinned skull preparation to measure spine dynamics in the mouse visual cortex from 4 weeks (adolescence) to 10 months (adulthood). They showed that filopodia represented around 12% of the spines at 1 month, and this proportion sharply decreased to 1% of the spines at 4 months. They showed for the first time that spines can be stable for a month, a result previously impossible to observe ex vivo. This dramatic change in spines dynamics compared to young neurons observations was replicated and complemented by the work of Zuo et al. (2005) who studied spine dynamics alterations between birth, adolescence, and adulthood in several cortical regions. In the early postnatal days (P14), filopodia represented around 50% of the spines in the neocortex, reminiscent of previous ex vivo observations. The reduction in filopodia proportion after adolescence was perfectly reproduced as well. Nevertheless, one drawback in those studies attempting to capture spontaneous spine dynamics is that they were conducted in anesthetized rodents. Although low doses of anesthetic affect filopodia but not spines within hours (Yang et al. 2011), it seems to increase the rate of spine formation over longer periods of time (Granak et al. 2021). Still, in vivo time-lapse imaging of dendritic spines certainly contributed to acquire knowledge unattainable otherwise, and later imaging studies conducted with awaken rodents yielded similar results. Finally, improvements in imaging technologies will likely bring more detailed information (Fuhrmann et al. 2022). Stimulated emission depletion microscopy that permits super-resolution was recently implemented for in vivo imaging; it confirmed the stability of spines over time, while unraveling spontaneous variations in morphology. It also confirmed that filopodia are rarely present and do not account for a stage of spine formation in the adult brain (Steffens et al. 2021).

Longitudinal observations of spines have been conducted on pyramidal neurons in the neocortex, with different spine turnover rate measured in different cortical areas (Holtmaat et al. 2005; Majewska et al. 2006). The extension to other neuronal types will help to gather a broader view on brain wide network dynamics (Zuo et al. 2005). For instance, stability over months was observed for spines in the olfactory bulb (Mizrahi and Katz 2003), and high spine turnover was found for inhibitory interneurons in the neocortex (Georgiou et al. 2022). Spine stability was also explored in the hippocampus, a brain structure essential for memory formation. In a first study, the cortex above the hippocampus was replaced by a tube to image pyramidal neurons, thereby revealing a negligible turnover rate (Mizrahi et al. 2004). Apical dendrites were imaged only for a few hours, so this study was complemented with long-term imaging showing that most spines are stable during at least 2 weeks (Gu et al. 2014). Interestingly, in the same neurons, the stability of spines on apical dendrites in the *stratum radiatum* CA1 hippocampal region contrasted with a high turnover of spines measured on basal dendrites in the *stratum oriens* of the same neurons (Attardo et al. 2015; Pfeiffer et al. 2018). It remains to be established if this discrepancy reveals that different dendritic compartments of the same neurons have distinct properties, or if results are due to technical improvements the latest study benefited from (Pfeiffer et al. 2018).

7.5.4 Conclusion

Until early adulthood, the mammalian brain is in progressive and constant rewiring to efficiently create short- and long-range connections and circuits adapted to the environment. This is reflected in the prominent presence of long and labile protrusions, the filopodia, in young neurons. 2PLSM revealed that once networks are formed in the adult brain, dendritic spines can potentially become stable during the entire life span, becoming a hypothetical cellular compartment for storage of memory (Fig. 7.4).

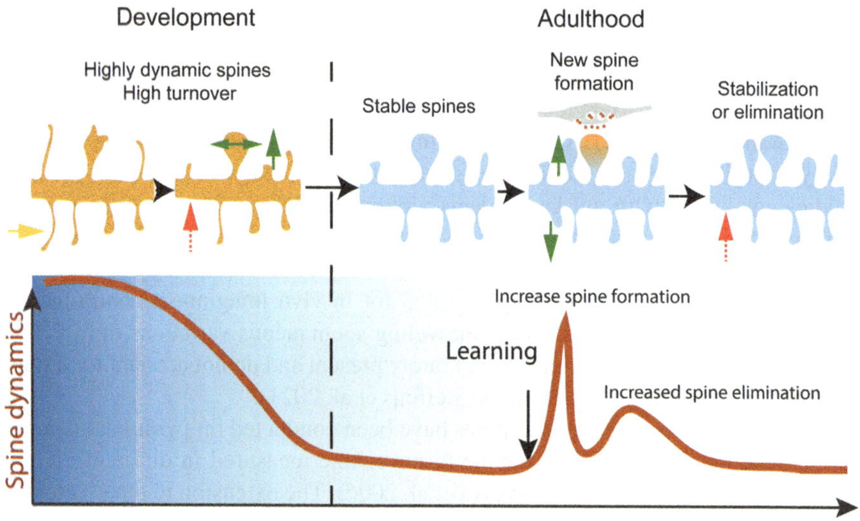

Fig. 7.4 Dendritic spine dynamics from development to adulthood and correlation with learning. During development, dendritic spines are highly dynamic. Motile filopodia (yellow horizontal arrow) are the prime transitory step during spine formation. Spine head enlargement indicates maturation and stability (green horizontal double arrow) Following spine formation (green arrow), spine elimination or pruning (red dotted arrow) participates in brain development. In adult rodents, spines can be stable during the entire life span. Synaptic plasticity induces changes in spine head size and spine formation. Spine formation occurs in cluster (green arrows). Learning is correlated with spine formation. It is followed by elimination since not all new spines are maintained (red dotted arrow)

7.6 Spine Plasticity, Clustered Spinogenesis, and Formation of Synaptic Contacts in Correlation with Learning and Memory

7.6.1 From Synaptic Plasticity to Learning and Memory

The growth and stability of dendritic spines is crucial in shaping the synaptic connections between neurons during brain development, but as the brain circuits mature, the rules governing spinogenesis evolve toward more stable synaptic protrusions in adulthood (see Sect. 7.5). However, despite this process, adult animals can still learn and adapt to new environmental rules, indicating that adult neurons may be capable of rapidly altering their synapses to form new memories. Building on earlier experiments that attempted to correlate LTP with morphological changes of spines (see Sect. 7.3.5), further studies benefited from live imaging to get better insights in structural plasticity and spine changes in correlation with learning and memory. These studies also brought new information on the rules that govern spinogenesis and formation of synaptic contacts.

7.6.2 Structural Synaptic Plasticity

As previously discussed, plastic changes of morphology were actually reported in slices subjected to LTP-inducing protocols (see Sect. 7.3.5). The studies of changes in spine density and morphology could, however, not determine whether the dendrites on which spines were modified belong to neurons that undergo LTP and thus, network effects could not be ruled out. In 1999, two studies established a direct correlation between LTP and spine formation in neurons. LTP was induced specifically on some dendrite branches, either by local superfusion (Engert and Bonhoeffer 1999) or local electrode stimulation (Maletic-Savatic et al. 1999) of one recorded neuron in acute slice of the hippocampus. It allowed to observe spine formation in correlation with LTP, but no changes in spines on the unstimulated dendrite from the same neuron. Furthermore, LTP promoted the stabilization of nascent spines (Hill and Zito 2013), their connection with presynapses (Toni et al. 1999), and spine plasticity was found to be bidirectional, with activity-dependent formation or elimination (Nägerl et al. 2004). The stimulation of single spines by caged glutamate induced both an increase in spine head size and synaptic current amplitude in hippocampal acute slices (Matsuzaki et al. 2004) and in the neocortex in vivo (Noguchi et al. 2019). Synaptic potentiation could thus be elicited in single spines and involves an increase in spine head size. Many studies were then performed to further characterize the spine dynamics in brain slice subjected to LTP (Muller et al. 2000; Segal 2017), building the view that dendritic spines could be a memory unit (Bourne and Harris 2007; Kasai et al. 2010), although such a claim would remain hypothetical until in vivo studies could establish relevant link between spines and memory (Segal 2005).

7.6.3 *Increased Spine Formation/Elimination Rate Following Learning Tasks*

Following initial studies in which spine density increases were observed in animals trained to learn a task (Moser et al. 1994; Patel et al. 1988), the comparison of spine densities between fixed brains from animals that acquired various memory tasks and naive animals became standard experimental design. Nevertheless, in vivo 2PLSM time-lapse imaging has brought more depth and precision to the correlation between dendritic spines changes and learning and memory (Chen et al. 2014; Runge et al. 2020). Motor learning tasks especially highlight the most prominent and consistent results of the impact of learning on spinogenesis. Gross motor skill training, such as accelerating rotarod training (Yang et al. 2009), or finer skills, such as food reaching (Clark et al. 2018) and lever manipulation (Hedrick et al. 2022), correlate with an increase in spinogenesis in the motor cortex as soon as 1 h following the onset of the task. Live imaging of neuronal population activity in the neocortex during motor learning revealed that a reproducible spatiotemporal sequence of activity in a subset of neurons emerged as a result of motor learning, and increase in spine turnover was measured in that neuronal ensemble (Peters et al. 2014). It was confirmed that the dendritic spine formation occurs on neurons activated during learning, and those neurons were reactivated during memory retrieval (Hwang et al. 2022). Moreover, sleep, which is known to play an important role in memory consolidation, was found to promote learning-dependent spine formation (Yang et al. 2014). The generated spines were maintained during subsequent training without additive spinogenesis, when performance reached a plateau (Xu et al. 2009). Furthermore, learning of a new motor task correlated with the formation of a new set of spines. Remarkably, the spine formation rate was correlated with the success of the task (Clark et al. 2018). In mice deficient for the gene coding the receptor Paired Immunoglobulin-Like Receptor B (PirB), increased spine formation rate and spine density correlated with enhanced learning performance. The blockade of the receptor in wild-type animals also increased spine survival and motor learning (Albarran et al. 2021).

Altogether, these results demonstrate the remarkable correlation between spines' structural changes and learning. This raised the question of the role and identity of newly formed synapses induced during learning. The new spines have a lower NMDA receptor-mediated Ca2+ accumulation and they tend to be less stable (Xu et al. 2009; Zito et al. 2009). A phase of increased spine elimination usually occurs after spinogenesis (Fig. 7.4) (Clark et al. 2018; Xu et al. 2009) but despite their reduced stability, newly formed spines are preferentially involved in learning tasks (Qiao et al. 2022). Overall, those results demonstrate a strong correlation between spine formation and motor learning, and reveal that spine formation might be finely tuned for selecting specific neuronal circuits with the learned task.

However, memory formation does not necessarily imply spine formation. It had been proposed that elimination as well as formation of spines could contribute to memory storage (Rosenzweig et al. 1972). That was confirmed in the context of fear associative memory, with the report of higher elimination rate of spines in mouse frontal and motor cortex (Lai et al. 2012; Xu et al. 2019), but higher formation rate in the auditory cortex (Lai et al. 2018). Therefore, spine elimination, as much as formation, should be considered pivotal for explaining learning and memory (Stein and Zito 2019).

Finally, one question remains when dendritic spines are observed: what is the origin of the afferents that contact newly formed spines? Some studies have developed strategies to tackle this question. In the so-called GRASP method, complementary fluorescent protein fragments are expressed in separated presynaptic and postsynaptic identified neurons, so the fluorescent signal is emitted from spines that correspond to synapses established between specified neurons. It allowed to demonstrate in hippocampal CA1 neurons, the formation of spines contacted by axons originating from neurons of the CA3 region following contextual fear conditioning (Choi et al. 2018). Another approach consists in combining dendritic and axonal boutons labeling. The labeling of neurons of the amygdala and the auditory cortex permitted the observation that fear memory association between a sound and a foot-shock induced spines specifically at the amygdala-cortical connections (Yang et al. 2016). Cortical and thalamic axonal boutons contain different vesicular glutamate transporters, so immunostaining post live imaging or promoter-specific expression of fluorescent proteins permits the separation of spines in contact with VGLUT1 cortical boutons from VGLUT2 thalamic boutons. It has been shown that newly formed spines in the motor cortex following a motor learning task create transient connections with cortical presynaptic elements, but stable contacts with thalamic bouton (Sohn et al. 2022). After motor learning, an increased number of synapses from activated neurons of the motor cortex onto striatal neurons were measured (Hwang et al. 2022). In the context of reward-driven learning, increased dopamine levels induced by cocaine were known to increase spine density in the striatum. The analysis of striatal dendrites in mice with labeled boutons showed that both cortical and thalamic afferents contributed to the increased spine density (Dos Santos et al. 2017) and unraveled an increase in the percentage of spines in contact with a dopaminergic bouton (Dos Santos et al. 2018).

7.6.4 From Clustered Dendritic Activity to Clustered Spinogenesis in Adult Neurons

We highlighted the strong correlation between spinogenesis and memory formation in the literature. However, this relationship only provides a partial understanding of the complex nature of synaptic function, as the activity of presynaptic excitatory elements plays a critical role in shaping the overall synaptic activity. To gain a more comprehensive understanding of the characteristics of synapses, researchers have turned to combine spine imaging techniques with measurements of their activity. This approach has led to significant advancements in our understanding of the intricate dynamics underlying synaptic activity. For instance, spines can behave as single units of information processing by responding to unique patterns of sensory stimuli. Indeed, even spines located in the same portion of dendrites can be specifically activated by sensory stimuli. For example, Chen and colleagues showed that different pure tones can reliably activate a specific combination of spines in the same auditory cortex neuron (Chen et al. 2011). These spatially localized dendritic activity areas are also found in the visual cortex neurons when mice or ferrets are presented with different image orientations (Jia et al. 2010; Wilson et al. 2016). Altogether, these findings revealed that synaptic activity across dendritic shafts is not homogeneous but is represented by complex combinations of "hot spots" of high synaptic activity in response to specific stimuli.

These "hot spots" are hypothesized to be linked to localized glutamate release and spine structural plasticity, and glutamatergic signaling has been extensively demonstrated in the literature. In general, local synaptic activity is responsible for an increase in spine size when neurons are depolarized, also known as "spine LTP" (Matsuzaki et al. 2004). Local glutamate application can also induce spine formation (Richards et al. 2005) with localized spinogenesis occurring less than a minute after uncaging glutamate near the dendrite in brain slices (Kwon and Sabatini 2011). Glutamate can also synergize with neuromodulators, such as dopamine, to open a spine plasticity time window in striatal projection neurons (Yagishita et al. 2014). In brain slices, LTP induced spine growth preferentially near activated synapses (De Roo et al. 2008) and repeated activation of single spines induced the shrinkage of neighboring inactivated spines (Oh et al. 2015). Hence, activated spines can form clusters and eliminate neighboring inactive spines. The competition at work between active and inactive synapses was confirmed in vivo in a paradigm of visual plasticity (El-Boustani et al. 2018). Spine clustering (Lu and Zuo 2017), cooperation, and competition (Jenks et al. 2021; Ramiro-Cortés et al. 2014) are thus important rules to understand structural plasticity.

As spinogenesis is tightly correlated to glutamate and neuronal activity, and that activity is spatially clustered in the dendrite, the idea of clustered spine formation following learning tasks was logically tested. These studies yield the results that spinogenesis indeed occurs closer to a neighboring spine than expected by chance (Fu et al. 2012). The precise role of clustered synapse formation remains to be elucidated, however, one of the hypotheses is that these spine clusters increase the density of multispine boutons and therefore, increase the impact of the glutamate release from these boutons (Geinisman et al. 2001). This increased chance of spine coactivation might play an important role in downstream signaling pathways. For example, the nuclear translocation of phosphorylated extracellular signal-regulated kinase (ERK) in pyramidal neurons occurs consistently when a few spines (three to seven) are activated simultaneously (Zhai et al. 2013). The link between local glutamate signaling and clustered spinogenesis in vivo in physiological conditions remained hypothetical until the recent development of genetically encoded glutamate sensors (such as intensity-based glutamate-sensing fluorescent reporter, iGluSnfr) allowing the real-time visualization of single-spine activity in vivo using photonic microscopy (Hedrick et al. 2022). This new tool used in vivo in awake mice confirmed that synapses located in the same dendrites can have largely different strengths, with a few spines receiving significantly more glutamate than their neighbors do. Remarkably, the authors observed that new spines tend to appear near highly active preexisting synapses, proving that the dendritic "hotspots" play the role of spinogenesis attractors.

These results combining elegant novel imaging methods converged toward a model in which learning-dependent spine formation reinforced preexisting neuronal networks in vivo, leading to clustered synapse formation. Hence, learning-dependent synaptogenesis is a finely tuned process that strongly suggests its causal role in memory formation.

7.6.5 Spine Formation, One Step After the Other Toward an Axonal Bouton

How do spines form in the adult brain? Several models have been put forward (Vaughn 1989; Yuste and Bonhoeffer 2004). The filopodia model, that consists of the growth of a long protrusion devoid of head which has strong mobility until contacting an axon and then maturing into a spine with a stabilized head, corresponds to the main mode of spine formation during development. In the adult brain, most in vivo live imaging studies have shown that spines appear without passing through a filopodia stage. Nevertheless, the thinness of filopodia make them difficult to observe, so super-resolution microscopy was necessary to reveal that while they indeed exist, they represent a minority of events and unlikely contribute to main spine formation (Steffens et al. 2021).

Three main models of spine formation have been proposed. In the Sotelo model (Sotelo 1978), spines arise from the dendrite and create contact with an axon, while in the Miller and Peters model, the axon contacts the dendrite to induce spine growth (Miller and Peters 1981). Another proposal was that spine would form by splitting in two (Carlin and Siekevitz 1983), but it was invalidated (Fiala et al. 2002a). The Miller and Peters model was inconsistent with observations of the straight axonal trajectories, and live imaging in brain slices and in vivo definitively confirmed that the spine grows to establish contact with the axon.

Spine formation can be divided into separate steps of growth and stabilization. Those steps seem not to be arbitrary since each is sustained by specific molecular mechanisms. First, membrane bending by Inverse Bin-Amphiphysin-Rvs domain superfamily (I-BAR) proteins makes the spine rise from the dendrite (Saarikangas et al. 2015). Then actin polymerization drives spine growth, and finally the spine is stabilized. Spine growth and stabilization are distinct steps. Indeed, an important observation is many newly formed spines are transient, indicating that spines need to be stabilized for long-term maintenance. One striking difference in between transient and long-lived spines is the presence of the PSD95 protein (Cane et al. 2014). The presence of PSD95 is indicative of persistence and is not a mechanistic explanation, but rather a marker for establishment of neurotransmitter receptors anchored by scaffolding proteins. Spine growth and stabilization are supported by distinct molecular signaling mechanisms. It has been shown in spiny striatal neurons that ERK paves a three-step way for spinogenesis (Dos Santos et al. 2017). Time-lapse analysis demonstrated that first spine growth depends on NMDA glutamatergic receptor-induced ERK activation, but neither translation nor transcription was necessary, indicating that cytoplasmic targets of ERK drive spine growth. Growth was followed by early stabilization for which both ERK-dependent MAP kinase-interacting serine/threonine protein kinase-1 (MNK-1) activity, which known target is the eukaryotic translation initiation factor 4E (eiF4E), and protein synthesis, but not transcription, were essential. Following growth and stabilization, long-term maintenance depends on ERK-instructed gene expression.

For structural plasticity, the process of spine head enlargement was found to involve a Rho-dependent growth phase followed by a cdc42-dependent stabilization (Murakoshi et al. 2011). Further experiments showed an initial phase during which the protein cofilin invaded the spine, to instruct the stabilization phase by forming stable actin-network (Bosch et al. 2014). Those two phases last less than 10 min, and are followed by a protein-synthesis-dependent phase during which postsynaptic density enlarges. The increase of the postsynaptic density necessary for long-term stabilization is paralled by an increase of the presynapse size that matches the spine width (Meyer et al. 2014).

Synaptogenesis requires that spinogenesis makes stable contact with axonal boutons. While the idea that growing spines contact neighbor axons was validated by observations, it ignored whether the new spine instructs the creation of a presynaptic boutons in the contacted axon, or if the spine comes into contact with a preexisting bouton. The emerging picture is that new spines make synapses with axonal boutons already in place. First hints came from observations of multispine boutons and higher stability of axonal boutons as compared to spines (Holtmaat and Svoboda 2009). Electron microscopy after live imaging of spine growth revealed that the associated bouton was in contact with two spines (Hedrick et al. 2022; Knott et al. 2006; Toni et al. 1999). Live dual imaging of spine and boutons in adult striatal slices and neocortex in vivo confirmed that new spines establish contacts with preexisting boutons (Dos Santos et al. 2017; Yang et al. 2016). In the striatum in vivo, increased spine density occurred while the boutons density remained unchanged, and the percentage of spines in contact with a common bouton was increased (Dos Santos et al. 2017). In a murine model of Huntington's disease, a therapeutical approach restored spine density levels and synapse density without changing boutons density, and strikingly the percentage of boutons with more than one active zone was increased (Kacher et al. 2019). Overall, those observations converge to the conclusion that the main mode of formation of synaptic contacts is achieved in the adult brain by the growth of new spines that need to be stabilized and contact preexisting axonal boutons. Nevertheless, other modes of formation could possibly take place depending on neuronal type and specific kind of synapses. Finally, an intriguing morphological specialization of the spine will require more attention in the future. Spinules, thin protrusions that come out of the spine head, have been overlooked because of they lay below the optical resolution. Recent advances in microscopy brought new observations, and the putative role of spinules in synapse formation should be studied (Zaccard et al. 2023).

7.6.6 Conclusion

Twenty years after Donald Hebb proposed his theory famously coined *fire together, wire together*, the discovery of LTP spurred research on connectivity modifications at the level of the dendritic spine (see Sect. 7.3.5). The advent of time-lapse imaging on living neurons and brain slices nailed that dynamic changes occur in potentiated neurons. Since LTP has always been regarded as a potential mechanism for learning, in vivo experiments have been engaged, which confirmed the correlation between spine remodeling and learning and memory. Live imaging in brain slices and in vivo also revealed clustered spinogenesis, with rules of cooperation and competition between neighboring spines, and allowed to decipher how synapses are formed. Following spine growth, transient spines need to be stabilized through distinct molecular mechanisms, and new spines contact preexisting axonal boutons. The remaining question, which has in fact always been the focus, was whether changes in connectivity orchestrated by dendritic spines could be causally linked with the learning process.

7.7 Causal Role of Structural Spine Plasticity in Memory and Disease

7.7.1 An Old Question That Waited for New Technologies

An essential and causal neuronal basis of memory formation and cognition is the formation of neuronal assemblies composed of complex combinations of neurons consistently coactivated during specific tasks or sensory stimuli (Buzsáki 2010). These assemblies are the results of the connections between neurons and, therefore, synapse formation/elimination represents a fundamental mechanism to create and alter neuronal ensembles in response to changes in the environment (Caroni et al. 2012; Holtmaat and Caroni 2016; Josselyn and Tonegawa 2020). The structural changes of dendritic spines have long been considered as implicated in behavioral adaptations, particularly in the context of learning and memory. However, the hypothesis linking spine morphology to cognitive function has been challenging to test, largely due to limitations in methods for specifically modulating the spines that are activated during a learning task (Lu and Zuo 2021; Tsutsumi and Hayashi-Takagi 2021). As a result, the causal relationship between dendritic spine formation and cognitive processes has remained largely correlative in nature.

Some attempts have been made to gain insights in the possible causal link between spine formation and learning process. First, the inhibition of the neuronal activity of the afferents induces both loss of spine formation in postsynaptic neurons and impairment of learning (Hedrick et al. 2022; MacAskill et al. 2014; Yang et al. 2016). Second, heterozygote or knockout mice for C-C chemokine receptor type 5 (ccr5) or PirB exhibit both higher spine turnover rates and improved learning (Albarran et al. 2021; Frank et al. 2018). However, to establish a direct causal link between dendritic spines and learning requires to specifically target the spines involved in learning process.

7.7.2 Evidence for a Direct Causal Link Between Spines and Learning

Remarkable tools were designed to specifically remove spines activated during a learning task, allowing for the first time to decipher the causal roles of the spine as subcellular engram for memory. The laboratory of Haruo Kasai created a genetically expressed chimeric protein to induce a specific shrinkage of recently activated dendritic spines based on a photoactivable Rac-1 (PaRac1) (Fig. 7.5) (Hayashi-Takagi et al. 2015). PaRac1 is expressed using a viral vector and under the control of Arc promoter, an immediate early gene promoter inducing a rapid and local protein expression after neuronal activation. When the neurons express this construct and their spines are stimulated, PaRac1 is locally expressed in potentiated and new synapses; then, illuminating the dendrites induces a specific shrinkage of those spines. These authors proved that removing spines that were newly activated during a motor learning task would completely erase the memory formation of the task. Interestingly, this light-induced erasure of memory did not affect another motor task learned before, proving that each motor task was acquired with the involvement of distinct sets of spines. Hence, spine remodeling is a specific morphological engram

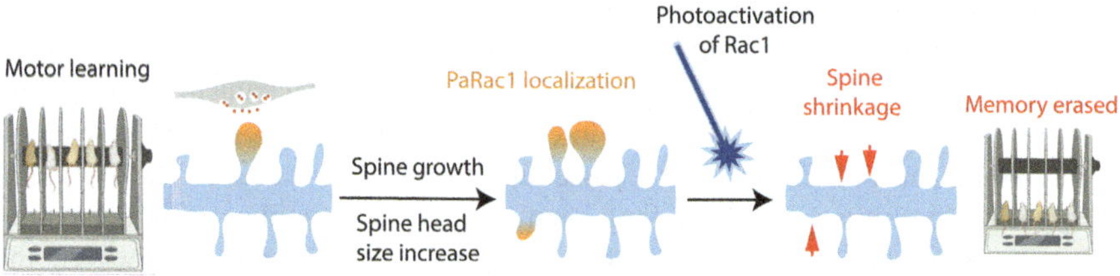

Fig. 7.5 Experimental evidence for a causal role of potentiated and newly formed spines in motor learning. Motor coordination can be evaluated by measuring the time a rodent stays on a rotating rod. Motor learning is correlated with spine formation, and Rac1 is a monomeric G-protein localized in active and new spines. The group of Haruo Kasai developed the synaptic optoprobe PaRac1, a photoactivable form of Rac1. It is expressed in activated neurons of the motor cortex, and localizes in potentiated and new spines. Upon light stimulation, PaRac1 provokes spine shrinkage. The experimentally induced shrinkage of spines erased memory

for memory. More recently, the group of Yasunori Hayashi used a similar approach of light erasure of memory to study the link between episodic memory and spine structural plasticity across the brain. The authors showed that structural plasticity is represented by a sequence of "spine LTP" consecutively occurring first in the hippocampus to then be transferred to the prefrontal cortex (Goto et al. 2021).

The discovery of the causal roles of dendritic spines in fundamental behaviors represents a significant milestone for the field, shedding new light on the underlying mechanisms of normal synaptic function. Notably, the advent of novel research tools that enable precise modulation of spine morphology has opened up new avenues of investigation, particularly with regard to the cellular mechanisms of drug therapies for mental disorders. In the final part of this chapter, we will delve into the latest findings in this area and their implications for our understanding of these complex conditions.

7.7.3 Mental Disorders and Spine Remodeling in the Adult Brain

Brain *postmortem* neuroanatomical studies provided pivotal information about neuronal alterations in mental disorders. Measurement of spine morphology in *postmortem* brains from adult individuals with schizophrenia, Alzheimer's disease, and depression showed a reduction of spine density (Penzes et al. 2011). These observations were replicated in models of induced stress in adult rodents. Most of these studies confirmed that stress paradigms are responsible for a reduction in spine density and synaptic strength in several brain regions, in particular the prefrontal cortex (PFC) (Qiao et al. 2016). Cortical spine loss is hypothesized to mediate long-term effects of stress and anxiety. Indeed, a reduction in synaptic numbers in the PFC might be a major neuronal cause for a lack of cortical feedback essential to mitigate the impact of negative emotions encoded by prelimbic areas (Heller et al. 2013).

Interestingly, recent fast-acting antidepressants, such as ketamine or psilocybin, showed efficient restoration of dendritic spines density in mouse PFC pyramidal neurons (Fig. 7.6) (Li et al. 2010). A study published in 2019 showed the longitudinal impacts on the spines of a cortisol treatment in mice, a pharmacological method used to mimic the physiological effects of stress (Moda-Sava et al. 2019). Mice were implanted with a glass prism in the contralateral PFC to follow spines or neuronal activity in vivo using a genetically encoded calcium indicator. As expected, stressed animals presented a significant reduction in spine density. Remarkably, this structural plasticity is associated with a reduction in neuronal network synchrony in PFC neurons, which suggested that synaptic loss is affecting neuronal ensembles as well. In this model, spine deficit was completely rescued by a single ketamine injection. Additionally, ketamine treatment also restored neuronal network synchrony to control levels. The authors then used the Photoactivable Rac1 described above to

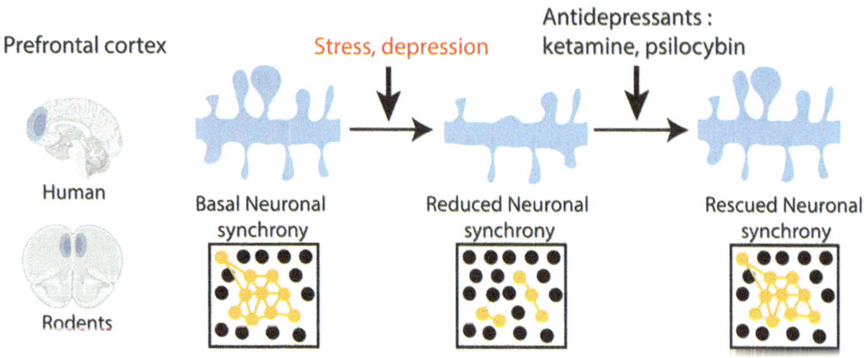

Fig. 7.6 Mental disorders and dendritic spine remodeling in the adult brain. In adult rodents, stress induces a decrease in spine density in the frontal cortex. The decreased spine density corresponds to connectivity loss that leads to altered neuronal network synchrony. Ketamine and psilocybin, two molecules recently tested as treatment against depression, induce spine formation. Using PaRac1 tool, it was shown that antidepressant effects of ketamine rely on its capacity to induce spine formation

specifically eliminate ketamine-induced spines. It allowed to prove that spine formation was not necessary for the rescue of neuronal activity the first days after ketamine treatment, but spine plasticity was required for the maintenance of the network activity rescue. This reveals a novel causal role of spine morphological alterations for the long-term effects of antidepressants.

Lately, high attention was devoted to investigating the impact of psilocybin on spine structural plasticity. Psilocybin is a hallucinogenic targeting the serotonergic receptors of type 2A. Like the ketamine treatment, psilocybin showed significant effects in the treatment of depression in humans. A recently published article showed in a longitudinal in vivo experiment that psilocybin is responsible for a rapid and long-lasting increase in spine density in rodents PFC (Shao et al. 2021). Interestingly, psilocybin-induced spines are stable for at least a month. The spinogenesis is associated with an increase in miniature excitatory postsynaptic currents, revealing an increase in functional excitatory synapses. These recent studies opened promising avenues for new drugs able to promote spine formation in the PFC for treatment-resistant forms of depression.

7.8 General Conclusion

Learning involves dynamic and long-term changes in neural ensembles at different levels such as synaptic plasticity, gene expression, or excitability properties. This chapter has presented, from the birth of the idea, the path that led to empirical evidence that connectivity changes at the level of the dendritic spines are causally involved in learning and memory.

The fact that spines appear distinctly at the surface of neurons gives an opportunity to observe connectivity changes that occur in neuronal networks. Dendritic spines, as motile protrusions, may confer ease to the neural system for modulating its connectivity, and surely provided neuroscientists with a visible mark to follow in their quest to understand the biological substrate of memory. As the present book attests, it is not an exaggeration to consider that dendritic spines are at the center of a research field in itself.

A few years after Cajal turned the attention of the scientific community on dendritic spines, it was proposed that they could participate to dynamic changes in connectivity. Experiments aiming to establish that spine changes would explain brain states and cerebral plasticity were conducted. While both technical limitations and lack of knowledge about the nature of spine precluded definitive conclusions, the dawn of the twentieth century has been the start of a journey that has not ended yet.

The identification, in 1959, of the spine as a component of a synapse obviously gave the information needed to reignite interest. Combining optical and electron microscopy, parallel studies explored the possible link between spines, plasticity, and cognition. Results obtained in the paradigm of experience-dependent plasticity established that spines can undergo changes in numbers and shapes. Clinical cases of intellectual disability contributed to strengthen the idea that synapses and dendritic spines represent a relevant subject of study for understanding cognitive functions.

In 1973, the experimental induction of LTP immediately gave rise to the question whether dendritic spines could be part of the explanation of the phenomenon of synaptic plasticity. Many studies contributed with various but converging results, indicating that structural changes indeed occurred. Twenty years later, the method of live imaging allowed to observe spine dynamics. Structural plasticity at the level of the dendritic spine was then definitively confirmed, and its rules, kinetics, and mechanisms could begin to be unraveled. In 2000, the observation of dendritic spines over time in vivo opened the gate to directly assess the question of whether dendritic spine plasticity could be involved in learning processes. Differences in spine density had been observed since the 90s when comparing naive animals with animals that learned a behavioral task, but live imaging was clearly needed to definitively confirm the correlation between spine dynamics and learning. Both ex vivo and in vivo studies also shed light on the rules of cooperation and competition in spine clusters, and on the mechanisms of spine growth, maintenance, and formation of synaptic contact. In 2015, for the first time, potentiated and new spines could be specifically manipulated and, thus, the more than a century old hypothesis could be tackled: dendritic spines play a causal role in learning and memory.

Acknowledgments We thank Dr. Jean-Gaël Barbara for comments on an earlier version of Sect. 7.1. Figures 7.5 and 7.6 were created with BioRender.com.

References

Adrian M, Kusters R, Wierenga CJ, Storm C, Hoogenraad CC, Kapitein LC (2014) Barriers in the brain: resolving dendritic spine morphology and compartmentalization. Front Neuroanat 8:142. Disponible sur: http://journal.frontiersin.org/article/10.3389/fnana.2014.00142/abstract

Albarran E, Raissi A, Jáidar O, Shatz CJ, Ding JB (2021) Enhancing motor learning by increasing the stability of newly formed dendritic spines in the motor cortex. Neuron 109(20):3298–3311.e4

Andersen P, Trommald M (1995) Possible strategies for finding the substrate for learning-induced changes in the hippocampal cortex. J Neurobiol 26(3):396–402

Araya R (2014) Input transformation by dendritic spines of pyramidal neurons. Front Neuroanat 8:141. Disponible sur: http://journal.frontiersin.org/article/10.3389/fnana.2014.00141/abstract

Arellano JI (2007) Ultrastructure of dendritic spines: correlation between synaptic and spine morphologies. Front Neurosci 1(1):131–143

Arellano JI, Espinosa A, Fairén A, Yuste R, DeFelipe J (2007) Non-synaptic dendritic spines in neocortex. Neuroscience 145(2):464–469

Attardo A, Fitzgerald JE, Schnitzer MJ (2015) Impermanence of dendritic spines in live adult CA1 hippocampus. Nature 523(7562):592–596

Azoulay L (1895) Psychologie histologique et texture du système nerveux. Année Psychol 2:255–294. https://doi.org/10.3406/psy.1895.1538

Azoulay L, Klippel M (1894) Les altérations des cellules de l'écorce cérébrale dans la paralysie générale, étudiées par la méthode de Golgi. C R Seances Soc Biol Fil 46:405–407

Barbara J-G (2006) The physiological construction of the neurone concept (1891–1952). C R Biol 329(5–6):437–449

Bennett MR (1999) The early history of the synapse: from Plato to Sherrington. Brain Res Bull 50(2):95–118. https://doi.org/10.1016/s0361-9230(99)00094-5

Bennett EL, Diamond MC, Krech D, Rosenzweig MR (1964) Chemical and anatomical plasticity of brain: changes in brain through experience, demanded by learning theories, are found in experiments with rats. Science 146(3644):610–619

Berkley HJ (1896) Studies on the lesions produced by the action of certain poisons on the cortical nerve cell. Johns Hopkins Hospital Reports Vol 6. Rep Neurol 3:1–92

Berlucchi G, Buchtel HA (2009) Neuronal plasticity: historical roots and evolution of meaning. Exp Brain Res 192(3):307–319

Bhatt DH, Zhang S, Gan W-B (2009) Dendritic spine dynamics. Annu Rev Physiol 71(1):261–282

Bian W-J, Miao W-Y, He S-J, Qiu Z, Yu X (2015) Coordinated spine pruning and maturation mediated by inter-spine competition for cadherin/catenin complexes. Cell 162(4):808–822

Black SE (1981) Pseudopods and synapses: the amoeboid theories of neuronal mobility and the early formulation of the synapse concept, 1894–1900. Bull Hist Med 55:34–58

Bliss TVP, Lømo T (1973) Long-lasting potentiation of synaptic transmission in the dentate area of the anaesthetized rabbit following stimulation of the perforant path. J Physiol 232(2):331–356

Bonhoeffer T, Yuste R (2002) Spine motility. Phenomenology, mechanisms, and function. Neuron 35(6):1019–1027. https://doi.org/10.1016/s0896-6273(02)00906-6

Bosch M, Castro J, Saneyoshi T, Matsuno H, Sur M, Hayashi Y (2014) Structural and molecular remodeling of dendritic spine substructures during long-term potentiation. Neuron 82(2):444–459

Bourne J, Harris KM (2007) Do thin spines learn to be mushroom spines that remember? Curr Opin Neurobiol 17(3):381–386

Buzsáki G (2010) Neural syntax: cell assemblies, synapsembles, and readers. Neuron 68(3):362–385

Cane M, Maco B, Knott G, Holtmaat A (2014) The relationship between PSD-95 clustering and spine stability *in vivo*. J Neurosci 34(6):2075–2086

Carlin RK, Siekevitz P (1983) Plasticity in the central nervous system: do synapses divide? Proc Natl Acad Sci U S A 80(11):3517–3521

Caroni P, Donato F, Muller D (2012) Structural plasticity upon learning: regulation and functions. Nat Rev Neurosci 13(7):478–490

Chen X, Leischner U, Rochefort NL, Nelken I, Konnerth A (2011) Functional mapping of single spines in cortical neurons in vivo. Nature 475(7357):501–505
Chen C-C, Lu J, Zuo Y (2014) Spatiotemporal dynamics of dendritic spines in the living brain. Front Neuroanat 8:28. Disponible sur: http://journal.frontiersin.org/article/10.3389/fnana.2014.00028/abstract
Choi J-H, Sim S-E, Kim J, Choi DI, Oh J, Ye S et al (2018) Interregional synaptic maps among engram cells underlie memory formation. Science 360(6387):430–435
Clark TA, Fu M, Dunn AK, Zuo Y, Jones TA (2018) Preferential stabilization of newly formed dendritic spines in motor cortex during manual skill learning predicts performance gains, but not memory endurance. Neurobiol Learn Mem 152:50–60
Cobb M (2020) The idea of the brain. Basic Books
Colonnier M (1968) Synaptic patterns on different cell types in the different laminae of the cat visual cortex. An electron microscope study. Brain Res 9(2):268–287
Cornejo VH, Ofer N, Yuste R (2022) Voltage compartmentalization in dendritic spines in vivo. Science 375(6576):82–86
Cotman CW, Nieto-Sampedro M (1984) Cell biology of synaptic plasticity. Science 225(4668):1287–1294
Crome L (1972) Morphological aspects of mental retardation. Proc R Soc Med 65(7):587–588
Dailey ME, Smith SJ (1996) The dynamics of dendritic structure in developing hippocampal slices. J Neurosci 16(9):2983–2994
De Roo M, Klauser P, Muller D (2008) LTP promotes a selective long-term stabilization and clustering of dendritic spines. Sheng M, éditeur. PLoS Biol 6(9):e219
DeFelipe J (2006) Brain plasticity and mental processes: Cajal again. Nat Rev Neurosci 7(10):811–817
DeFelipe J (2015) The dendritic spine story: an intriguing process of discovery. Front Neuroanat 9:14. Disponible sur: http://journal.frontiersin.org/Article/10.3389/fnana.2015.00014/abstract
DeFelipe J, Conti F, Van Eyck SL, Manzoni T (1988) Demonstration of glutamate-positive axon terminals forming asymmetric synapses in cat neocortex. Brain Res 455(1):162–165
Demoor J (1896a) La plasticité morphologique des neurones cérébraux. Arch Biol Brux 14:723–752
Demoor J (1896b) La plasticité morphologique des neurones cérébraux. In: Heger P (ed) Travaux de laboratoire de l'Institut Solvay. Fascicule 1. Institut Solvay, Bruxelles
Demoor J (1898) Le mécanisme et la signification de l'état moniliforme des neurones. In: Heger P (ed) Travaux de laboratoire de l'Institut Solvay, vol 2, Fascicule 1. Institut Solvay, Bruxelles, pp 1–46
Denk W, Strickler JH, Webb WW (1990) Two-photon laser scanning fluorescence microscopy. Science 248(4951):73–76
Doig NM, Moss J, Bolam JP (2010) Cortical and thalamic innervation of direct and indirect pathway medium-sized spiny neurons in mouse striatum. J Neurosci 30(44):14610–14618
Dos Santos M, Salery M, Forget B, Garcia Perez MA, Betuing S, Boudier T et al (2017) Rapid synaptogenesis in the nucleus accumbens is induced by a single cocaine administration and stabilized by mitogen-activated protein kinase interacting kinase-1 activity. Biol Psychiatry 82(11):806–818
Dos Santos M, Cahill EN, Bo GD, Vanhoutte P, Caboche J, Giros B et al (2018) Cocaine increases dopaminergic connectivity in the nucleus accumbens. Brain Struct Funct 223(2):913–923
Dunaevsky A, Tashiro A, Majewska A, Mason C, Yuste R (1999) Developmental regulation of spine motility in the mammalian central nervous system. Proc Natl Acad Sci U S A 96(23):13438–13443
Duval M (1895) Hypothèses sur la physiologie des centres nerveux; théorie histologique du sommeil. C R Soc De Biol 47:74–77
El-Boustani S, Ip JPK, Breton-Provencher V, Knott GW, Okuno H, Bito H et al (2018) Locally coordinated synaptic plasticity of visual cortex neurons in vivo. Science 360(6395):1349–1354
Engert F, Bonhoeffer T (1999) Dendritic spine changes associated with hippocampal long-term synaptic plasticity. Nature 399(6731):66–70

Fiala JC, Allwardt B, Harris KM (2002a) Dendritic spines do not split during hippocampal LTP or maturation. Nat Neurosci 5(4):297–298
Fiala JC, Spacek J, Harris KM (2002b) Dendritic spine pathology: cause or consequence of neurological disorders? Brain Res Rev 39(1):29–54
Fifková E, Delay RJ (1982) Cytoplasmic actin in neuronal processes as a possible mediator of synaptic plasticity. J Cell Biol 95(1):345–350
Fifková E, van Harreveld A (1977) Long-lasting morphological changes in dendritic spines of dentate granular cells following stimulation of the entorhinal area. J Neurocytol 6(2):211–230
Fischer M, Kaech S, Knutti D, Matus A (1998) Rapid actin-based plasticity in dendritic spines. Neuron 20(5):847–854
Forrest MP, Parnell E, Penzes P (2018) Dendritic structural plasticity and neuropsychiatric disease. Nat Rev Neurosci 19(4):215–234
Frank AC, Huang S, Zhou M, Gdalyahu A, Kastellakis G, Silva TK et al (2018) Hotspots of dendritic spine turnover facilitate clustered spine addition and learning and memory. Nat Commun [illegible]
Fremeau RT, Voglmaier S, Seal RP, Edwards RH (2004) VGLUTs define subsets of excitatory neurons and suggest novel roles for glutamate. Trends Neurosci 27(2):98–103
Fu M, Yu X, Lu J, Zuo Y (2012) Repetitive motor learning induces coordinated formation of clustered dendritic spines in vivo. Nature 483(7387):92–95
Fuhrmann M, Gockel N, Arizono M, Dembitskaya Y, Nägerl UV, Pennacchietti F et al (2022) Super-resolution microscopy opens new doors to life at the nanoscale. J Neurosci 42(45):8488–8497
García-López P, García-Marín V, Freire M (2007) The discovery of dendritic spines by Cajal in 1888 and its relevance in the present neuroscience. Prog Neurobiol 83(2):110–130
Geinisman Y, Berry RW, Disterhoft JF, Power JM, Van der Zee EA (2001) Associative learning elicits the formation of multiple-synapse boutons. J Neurosci 21(15):5568–5573
Georgiou C, Kehayas V, Lee KS, Brandalise F, Sahlender DA, Blanc J et al (2022) A subpopulation of cortical VIP-expressing interneurons with highly dynamic spines. Commun Biol 5(1):352
Globus A, Scheibel AB (1967) The effect of visual deprivation on cortical neurons: a Golgi study. Exp Neurol 19(3):331–345. https://doi.org/10.1016/0014-4886(67)90029-5
Globus A, Rosenzweig MR, Bennett EL, Diamond MC (1973) Effects of differential experience on dendritic spine counts in rat cerebral cortex. J Comp Physiol Psychol 82(2):175–181
Gonatas NK (1967) Axonic and synaptic lesions in neuropsychiatric disorders. Nature 214(5086):352–355
Gonatas NK, Moss A (1975) Pathologic axons and synapses in human neuropsychiatric disorders. Hum Pathol 6(5):571–582
Goto A, Bota A, Miya K, Wang J, Tsukamoto S, Jiang X et al (2021) Stepwise synaptic plasticity events drive the early phase of memory consolidation. Science 374(6569):857–863
Granak S, Hoschl C, Ovsepian SV (2021) Dendritic spine remodeling and plasticity under general anesthesia. Brain Struct Funct 226(7):2001–2017
Gray EG (1959a) Electron microscopy of synaptic contacts on dendrite spines of the cerebral cortex. Nature 183(4675):1592–1593. https://doi.org/10.1038/1831592a0
Gray EG (1959b) Axo-somatic and axo-dendritic synapses of the cerebral cortex: an electron microscope study. J Anat 93(Pt 4):420–433
Grutzendler J, Kasthuri N, Gan W-B (2002) Long-term dendritic spine stability in the adult cortex. Nature 420(6917):812–816
Gu L, Kleiber S, Schmid L, Nebeling F, Chamoun M, Steffen J et al (2014) Long-term in vivo imaging of dendritic spines in the hippocampus reveals structural plasticity. J Neurosci Off J Soc Neurosci 34(42):13948–13953
Harris KM, Kater SB (1994) Dendritic spines: cellular specializations imparting both stability and flexibility to synaptic function. Annu Rev Neurosci 17:341–371. https://doi.org/10.1146/annurev.ne.17.030194.002013
Harris K, Stevens J (1988) Dendritic spines of rat cerebellar Purkinje cells: serial electron microscopy with reference to their biophysical characteristics. J Neurosci 8(12):4455–4469

Harris K, Stevens J (1989) Dendritic spines of CA 1 pyramidal cells in the rat hippocampus: serial electron microscopy with reference to their biophysical characteristics. J Neurosci 9(8):2982–2997
Harris KM, Weinberg RJ (2012) Ultrastructure of synapses in the mammalian brain. Cold Spring Harb Perspect Biol 4(5):a005587
Hayashi-Takagi A, Yagishita S, Nakamura M, Shirai F, Wu YI, Loshbaugh AL et al (2015) Labelling and optical erasure of synaptic memory traces in the motor cortex. Nature 525(7569):333–338
Hedrick NG, Lu Z, Bushong E, Singhi S, Nguyen P, Magaña Y et al (2022) Learning binds new inputs into functional synaptic clusters via spinogenesis. Nat Neurosci 25(6):726–737
Heller AS, Johnstone T, Peterson MJ, Kolden GG, Kalin NH, Davidson RJ (2013) Increased prefrontal cortex activity during negative emotion regulation as a predictor of depression symptom severity trajectory over 6 months. JAMA Psychiatry 70(11):1181–1189
Helm MS, Dankovich TM, Mandad S, Rammner B, Jähne S, Salimi V et al (2021) A large-scale nanoscopy and biochemistry analysis of postsynaptic dendritic spines. Nat Neurosci 24(8):1151–1162
Hill TC, Zito K (2013) LTP-induced long-term stabilization of individual nascent dendritic spines. J Neurosci 33(2):678–686
Holler S, Köstinger G, Martin KAC, Schuhknecht GFP, Stratford KJ (2021) Structure and function of a neocortical synapse. Nature 591(7848):111–116
Holtmaat A, Caroni P (2016) Functional and structural underpinnings of neuronal assembly formation in learning. Nat Neurosci 19(12):1553–1562
Holtmaat A, Svoboda K (2009) Experience-dependent structural synaptic plasticity in the mammalian brain. Nat Rev Neurosci 10(9):647–658
Holtmaat AJGD, Trachtenberg JT, Wilbrecht L, Shepherd GM, Zhang X, Knott GW et al (2005) Transient and persistent dendritic spines in the neocortex in vivo. Neuron 45(2):279–291
Horner CH (1993) Plasticity of the dendritic spine. Prog Neurobiol 41(3):281–321
Hosokawa T, Bliss TV, Fine A (1992) Persistence of individual dendritic spines in living brain slices. Neuroreport 3(6):477–480. https://doi.org/10.1097/00001756-199206000-00005
Hwang F-J, Roth RH, Wu Y-W, Sun Y, Kwon DK, Liu Y et al (2022) Motor learning selectively strengthens cortical and striatal synapses of motor engram neurons. Neuron 110(17):2790–2801.e5
Jenks KR, Tsimring K, Ip JPK, Zepeda JC, Sur M (2021) Heterosynaptic plasticity and the experience-dependent refinement of developing neuronal circuits. Front Neural Circuits 15:803401
Jia H, Rochefort NL, Chen X, Konnerth A (2010) Dendritic organization of sensory input to cortical neurons in vivo. Nature 464(7293):1307–1312
Jones EG (2000) NEUROwords. 8. Plasticity and neuroplasticity. J Hist Neurosci 9(1):37–39
Jones EG (2004) Plasticity and neuroplasticity. J Hist Neurosci 13(3):293
Jones EG, Powell TP (1969) Morphological variations in the dendritic spines of the neocortex. J Cell Sci 5(2):509–529. https://doi.org/10.1242/jcs.5.2.509
Josselyn SA, Tonegawa S (2020) Memory engrams: recalling the past and imagining the future. Science 367(6473):eaaw4325
Kacher R, Lamazière A, Heck N, Kappes V, Mounier C, Despres G et al (2019) CYP46A1 gene therapy deciphers the role of brain cholesterol metabolism in Huntington's disease. Brain 142(8):2432–2450
Kandel ER, Spencer WA (1968) Cellular neurophysiological approaches in the study of learning. Physiol Rev 48(1):65–134
Kasai H, Hayama T, Ishikawa M, Watanabe S, Yagishita S, Noguchi J (2010) Learning rules and persistence of dendritic spines: learning rules of dendritic spines. Eur J Neurosci 32(2):241–249
Kasthuri N, Hayworth KJ, Berger DR, Schalek RL, Conchello JA, Knowles-Barley S et al (2015) Saturated reconstruction of a volume of neocortex. Cell 162(3):648–661
Knott GW, Holtmaat A, Wilbrecht L, Welker E, Svoboda K (2006) Spine growth precedes synapse formation in the adult neocortex in vivo. Nat Neurosci 9(9):1117–1124

Korogod N, Petersen CC, Knott GW (2015) Ultrastructural analysis of adult mouse neocortex comparing aldehyde perfusion with cryo fixation. eLife 4:e05793

Kubota Y, Karube F, Nomura M, Kawaguchi Y (2016) The diversity of cortical inhibitory synapses. Front Neural Circuits 10:27. Disponible sur: http://journal.frontiersin.org/Article/10.3389/fncir.2016.00027/abstract

Kwon HB, Sabatini BL (2011) Glutamate induces de novo growth of functional spines in developing cortex. Nature 474(7349):100–104

Kwon T, Merchán-Pérez A, Rial Verde EM, Rodríguez J-R, DeFelipe J, Yuste R (2019) Ultrastructural, molecular and functional mapping of GABAergic synapses on dendritic spines and shafts of neocortical pyramidal neurons. Cereb Cortex 29(7):2771–2781

Lai CSW, Franke TF, Gan W-B (2012) Opposite effects of fear conditioning and extinction on dendritic spine remodelling. Nature 483(7387):87–91

Lai CSW, Adler A, Gan W-B (2018) Fear extinction reverses dendritic spine formation induced by fear conditioning in the mouse auditory cortex. Proc Natl Acad Sci U S A 115(37):9306–9311

Lendvai B, Stern EA, Chen B, Svoboda K (2000) Experience-dependent plasticity of dendritic spines in the developing rat barrel cortex in vivo. Nature 404(6780):876–881

Lépine R (1894) Sur un cas d'hystérie á form particulière. Rev Méd 14:713–728

LeVay S (1973) Synaptic patterns in the visual cortex of the cat and monkey. Electron microscopy of Golgi preparations. J Comp Neurol 150(1):53–85

Li N, Lee B, Liu R-J, Banasr M, Dwyer JM, Iwata M et al (2010) mTOR-dependent synapse formation underlies the rapid antidepressant effects of NMDA antagonists. Science 329(5994):959–964

Lu J, Zuo Y (2017) Clustered structural and functional plasticity of dendritic spines. Brain Res Bull 129:18–22

Lu J, Zuo Y (2021) Shedding light on learning and memory: optical interrogation of the synaptic circuitry. Curr Opin Neurobiol 67:138–144

Lugaro E (1895) Sulle modificazionedelle cellule nervose etc. nei diversi stati funzionali. Lo Sperimentale, t. XLIX

Lugaro E (1896) Nuovi dati e nuovi problemi nella patologia della cellula nervosa. Riv Patol Nerv Ment 1(8):2

Lugaro E (1898a) Le resistenze nell'evoluzione della vita. Riv Mod Cult 1:29–60

Lugaro E (1898b) Sulle modificazioni morfologiche funzionali dei dendriti delle cellule nervose. Riv Patol Nerv Ment 3(8):350–351

MacAskill AF, Cassel JM, Carter AG (2014) Cocaine exposure reorganizes cell type- and input-specific connectivity in the nucleus accumbens. Nat Neurosci 17(9):1198–1207

Majewska AK, Newton JR, Sur M (2006) Remodeling of synaptic structure in sensory cortical areas *in vivo*. J Neurosci 26(11):3021–3029

Maletic-Savatic M, Malinow R, Svoboda K (1999) Rapid dendritic morphogenesis in CA1 hippocampal dendrites induced by synaptic activity. Science 283(5409):1923–1927

Marin-Padilla M (1972) Structural abnormalities of the cerebral cortex in human chromosomal aberrations: a Golgi study. Brain Res 44(2):625–629

Marin-Padilla M (1974) Structural organization of the cerebral cortex (motor area) in human chromosomal aberrations. A Golgi study. I. D1 (13–15) trisomy, Patau syndrome. Brain Res 66(3):375–391

Matsuzaki M, Ellis-Davies GCR, Nemoto T, Miyashita Y, Iino M, Kasai H (2001) Dendritic spine geometry is critical for AMPA receptor expression in hippocampal CA1 pyramidal neurons. Nat Neurosci 4(11):1086–1092

Matsuzaki M, Honkura N, Ellis-Davies GCR, Kasai H (2004) Structural basis of long-term potentiation in single dendritic spines. Nature 429(6993):761–766

Matus A, Ackermann M, Pehling G, Byers HR, Fujiwara K (1982) High actin concentrations in brain dendritic spines and postsynaptic densities. Proc Natl Acad Sci U S A 79(23):7590–7594

Meyer D, Bonhoeffer T, Scheuss V (2014) Balance and stability of synaptic structures during synaptic plasticity. Neuron 82(2):430–443

Miller M, Peters A (1981) Maturation of rat visual cortex. II. A combined Golgi-electron microscope study of pyramidal neurons. J Comp Neurol 203(4):555–573
Mishchenko Y, Hu T, Spacek J, Mendenhall J, Harris KM, Chklovskii DB (2010) Ultrastructural analysis of hippocampal neuropil from the connectomics perspective. Neuron 67(6):1009–1020
Mizrahi A, Katz LC (2003) Dendritic stability in the adult olfactory bulb. Nat Neurosci 6(11):1201–1207
Mizrahi A, Crowley JC, Shtoyerman E, Katz LC (2004) High-resolution in vivo imaging of hippocampal dendrites and spines. J Neurosci Off J Soc Neurosci 24(13):3147–3151
Moda-Sava RN, Murdock MH, Parekh PK, Fetcho RN, Huang BS, Huynh TN et al (2019) Sustained rescue of prefrontal circuit dysfunction by antidepressant-induced spine formation. Science 364(6436):eaat8078
Monti A (1895) Sur les altérations du système nerveux dans l'inanition. Arch Ital Biol 24:347–360
Morange M (2021) A history of biology. Princeton University Press
Moser MB, Trommald M, Andersen P (1994) An increase in dendritic spine density on hippocampal CA1 pyramidal cells following spatial learning in adult rats suggests the formation of new synapses. Proc Natl Acad Sci U S A 91(26):12673–12675
Muller D, Toni N, Buchs P-A (2000) Spine changes associated with long-term potentiation. Hippocampus 10(5):596–604
Murakoshi H, Wang H, Yasuda R (2011) Local, persistent activation of Rho GTPases during plasticity of single dendritic spines. Nature 472(7341):100–104
Nägerl UV, Eberhorn N, Cambridge SB, Bonhoeffer T (2004) Bidirectional activity-dependent morphological plasticity in hippocampal neurons. Neuron 44(5):759–767
Nakai N, Takumi T, Nakai J, Sato M (2018) Common defects of spine dynamics and circuit function in neurodevelopmental disorders: a systematic review of findings from in vivo optical imaging of mouse models. Front Neurosci 12:412
Noguchi J, Nagaoka A, Watanabe S, Ellis-Davies GCR, Kitamura K, Kano M et al (2011) *In vivo* two-photon uncaging of glutamate revealing the structure-function relationships of dendritic spines in the neocortex of adult mice: *in vivo* two-photon glutamate uncaging. J Physiol 589(10):2447–2457
Noguchi J, Nagaoka A, Hayama T, Ucar H, Yagishita S, Takahashi N et al (2019) Bidirectional in vivo structural dendritic spine plasticity revealed by two-photon glutamate uncaging in the mouse neocortex. Sci Rep 9(1):13922
Ofer N, Berger DR, Kasthuri N, Lichtman JW, Yuste R (2021) Ultrastructural analysis of dendritic spine necks reveals a continuum of spine morphologies. Dev Neurobiol 81(5):746–757
Ofer N, Benavides-Piccione R, DeFelipe J, Yuste R (2022) Structural analysis of human and mouse dendritic spines reveals a morphological continuum and differences across ages and species. eNeuro 9(3):ENEURO.0039-22.2022
Oh WC, Parajuli LK, Zito K (2015) Heterosynaptic structural plasticity on local dendritic segments of hippocampal CA1 neurons. Cell Rep 10(2):162–169
Papa M, Segal M (1996) Morphological plasticity in dendritic spines of cultured hippocampal neurons. Neuroscience 71(4):1005–1011
Parajuli LK, Urakubo H, Takahashi-Nakazato A, Ogelman R, Iwasaki H, Koike M et al (2020) Geometry and the organizational principle of spine synapses along a dendrite. eNeuro 7(6):ENEURO.0248-20.2020
Patel SN, Rose SPR, Stewart MG (1988) Training induced dendritic spine density changes are specifically related to memory formation processes in the chick, Gallus domesticus. Brain Res 463(1):168–173
Penzes P, Cahill ME, Jones KA, VanLeeuwen J-E, Woolfrey KM (2011) Dendritic spine pathology in neuropsychiatric disorders. Nat Neurosci 14(3):285–293
Peters A, Kaiserman-Abramof IR (1970) The small pyramidal neuron of the rat cerebral cortex. The perikaryon, dendrites and spines. Am J Anat 127(4):321–355. https://doi.org/10.1002/aja.1001270402

Peters A, Palay SL, Webster H (1976) The fine structure of the nervous system. Saunders, Philadelphia

Peters AJ, Chen SX, Komiyama T (2014) Emergence of reproducible spatiotemporal activity during motor learning. Nature 510(7504):263–267

Pfeiffer T, Poll S, Bancelin S, Angibaud J, Inavalli VK, Keppler K et al (2018) Chronic 2P-STED imaging reveals high turnover of dendritic spines in the hippocampus in vivo. eLife 7:e34700

Popov VI, Bocharova LS, Bragin AG (1992) Repeated changes of dendritic morphology in the hippocampus of ground squirrels in the course of hibernation. Neuroscience 48(1):45–51

Purpura DP (1974) Dendritic spine "dysgenesis" and mental retardation. Science 186(4169):1126–1128. https://doi.org/10.1126/science

Purpura DP (1975) Normal and aberrant neuronal development in the cerebral cortex of human fetus and young infant. In: Buchwald NA, Brazier MAB (eds) Brain mechanisms in mental retardation. Academic Press, pp 141–169

Qiao H, Li MY, Xu C, Chen HB, An SC, Ma XM (2016) Dendritic spines in depression: what we learned from animal models. Neural Plast 2016:8056370

Qiao Q, Wu C, Ma L, Zhang H, Li M, Wu X et al (2022) Motor learning-induced new dendritic spines are preferentially involved in the learned task than existing spines. Cell Rep 40(7):111229

Querton L (1898) Le sommeil hibernal et les modifications des neurones cérébraux. In: Heger P (ed) Travaux de laboratoire de l'Institut Solvay, vol 2, Fascicule 1. Institut Solvay, Bruxelles, pp 1–58

Raisman G (1969) Neuronal plasticity in the septal nuclei of the adult rat. Brain Res 14(1):25–48

Ramiro-Cortés Y, Hobbiss AF, Israely I (2014) Synaptic competition in structural plasticity and cognitive function. Philos Trans R Soc Lond B Biol Sci 369(1633):20130157

Ramón y Cajal S (1888) Estructura de los centros nerviosos de las aves. Rev Trim Histol Normal Patol 1:1–10

Ramón y Cajal S (1892) Nuevo concepto de la histología de los centros nerviosos. Rev Cienc Méd Barc 18:363–376; 457–476; 505–520; 529–540

Ramón y Cajal S (1894a) Les nouvelles idées sur la structure du systeème nerveux chez l'homme et chez les vertébrés (trans: Azoulay L). C. Reinwald & Cie, Paris

Ramón y Cajal S (1894b) The Croonian lecture: la fine structure des centres nerveux. Proc R Soc Lond 55:444–468

Ramón y Cajal S (1895) Algunas conjeturas sobre el mecanismo anatómico de la ideación, asociación y atención. Rev Med Cirug Prác 36:497–508

Ramón y Cajal S (1899–1904) Textura del sistema nervioso del hombre y de los vertebrados. Imprenta y Librería de Nicolás Moya, Madrid

Ramón y Cajal S (1909–1911) Histologie du système nerveux de l'homme et des vertébrés (trans: Azoulay L). Maloine, Paris

Ramón y Cajal S (1933) Neuronismo o reticularismo? Las pruebas objetivas de la unidad anatomica de las celulas nerviosas. Consejo Superior de Investigaciones Cientıficas, Instituto Ramón y Cajal. IX, Madrid

Richards DA, Mateos JM, Hugel S, de Paola V, Caroni P, Gähwiler BH et al (2005) Glutamate induces the rapid formation of spine head protrusions in hippocampal slice cultures. Proc Natl Acad Sci U S A 102(17):6166–6171

Rosenzweig MR, Mollgaard K, Diamond MC, Bennett EL (1972) Negative as well as positive synaptic changes may store memory. Psychol Rev 79(1):93–96

Runge K, Cardoso C, de Chevigny A (2020) Dendritic spine plasticity: function and mechanisms. Front Synaptic Neurosci 12:36

Saarikangas J, Kourdougli N, Senju Y, Chazal G, Segerstråle M, Minkeviciene R et al (2015) MIM-induced membrane bending promotes dendritic spine initiation. Dev Cell 33(6):644–659

Schapiro S, Vukovich KR (1970) Early experience effects upon cortical dendrites: a proposed model for development. Science 167(3916):292–294

Schikorski T, Stevens CF (1997) Quantitative ultrastructural analysis of hippocampal excitatory synapses. J Neurosci 17(15):5858–5867

Segal M (2005) Dendritic spines and long-term plasticity. Nat Rev Neurosci 6(4):277–284
Segal M (2017) Dendritic spines: morphological building blocks of memory. Neurobiol Learn Mem 138:3–9
Shao LX, Liao C, Gregg I, Davoudian PA, Savalia NK, Delagarza K et al (2021) Psilocybin induces rapid and persistent growth of dendritic spines in frontal cortex in vivo. Neuron 109(16):2535–2544.e4
Sohn J, Suzuki M, Youssef M, Hatada S, Larkum ME, Kawaguchi Y et al (2022) Presynaptic supervision of cortical spine dynamics in motor learning. Sci Adv 8(30):eabm0531
Solvay E (1894) Du rôle de l'électricité dans les phénomènes de la vie animale. Hayez, Bruxelles
Sotelo C (1978) Purkinje cell ontogeny: formation and maintenance of spines. In: Progress in brain research. Elsevier, pp 149–170. Disponible sur: https://linkinghub.elsevier.com/retrieve/pii/S0079612308610213
Sotelo C (2020) The history of the synapse. Anat Rec 303(5):1252–1279
Soukhanoff S (1898) Contribution à l'étude des modifications que subissent les prolongements dendritiques des cellules nerveuses sous l'influence des narcotiques. In: Van Gehuchten A (ed) Travaux du laboratoire de neurologie. Fascicule 2. Université de Louvain, Institut Vésale, Louvain
Stahnisch FW (2003) Making the brain plastic: early neuroanatomical staining techniques and the pursuit of structural plasticity, 1910–1970. J Hist Neurosci 12(4):413–435
Stefanowska M (1897a) Les appendices terminaux des dendrites cérébraux et leur différents états physiologiques. Ann Soc R Sci Méd Nat Brux 6:351–407
Stefanowska M (1897b) Les appendices terminaux des dendrites cérébraux. In: Heger P (ed) Travaux de laboratoire de l'Institut Solvay. Fascicule 3. Institut Solvay, Bruxelles
Stefanowska M (1900) Localisation des altérations cérébrales produites par l'éther. In: Heger P (ed) Travaux de laboratoire de l'Institut Solvay, vol 3, Fascicule 3. Institut Solvay, Bruxelles, pp 25–81
Steffens H, Mott AC, Li S, Wegner W, Švehla P, Kan VWY et al (2021) Stable but not rigid: chronic in vivo STED nanoscopy reveals extensive remodeling of spines, indicating multiple drivers of plasticity. Sci Adv 7(24):eabf2806
Stein IS, Zito K (2019) Dendritic spine elimination: molecular mechanisms and implications. Neuroscientist 25(1):27–47
Tamada H, Blanc J, Korogod N, Petersen CC, Knott GW (2020) Ultrastructural comparison of dendritic spine morphology preserved with cryo and chemical fixation. eLife 9:e56384
Tanzi E (1893) I fatti e le induzione nell'odierna istologia del sistema nervoso. Riv Sper Freniatr Med Leg 19:419–472
Toni N, Buchs P-A, Nikonenko I, Bron CR, Muller D (1999) LTP promotes formation of multiple spine synapses between a single axon terminal and a dendrite. Nature 402(6760):421–425
Tønnesen J, Katona G, Rózsa B, Nägerl UV (2014) Spine neck plasticity regulates compartmentalization of synapses. Nat Neurosci 17(5):678–685
Trachtenberg JT, Chen BE, Knott GW, Feng G, Sanes JR, Welker E et al (2002) Long-term in vivo imaging of experience-dependent synaptic plasticity in adult cortex. Nature 420(6917):788–794
Tropea D, Majewska AK, Garcia R, Sur M (2010) Structural dynamics of synapses *in vivo* correlate with functional changes during experience-dependent plasticity in visual cortex. J Neurosci 30(33):11086–11095
Tsutsumi S, Hayashi-Takagi A (2021) Optical interrogation of multi-scale neuronal plasticity underlying behavioral learning. Curr Opin Neurobiol 67:8–15
Uchizono K (1965) Characteristics of excitatory and inhibitory synapses in the central nervous system of the cat. Nature 207(4997):642–643
Valverde P (1967) Apical dendritic spines of the visual cortex and light deprivation in the mouse. Exp Brain Res 3(4):337–352. Disponible sur: http://link.springer.com/10.1007/BF00237559
van Harreveld A, Fifkova E (1975) Swelling of dendritic spines in the fascia dentata after stimulation of the perforant fibers as a mechanism of post-tetanic potentiation. Exp Neurol 49(3):736–749

Vaughn JE (1989) Review: Fine structure of synaptogenesis in the vertebrate central nervous system. Synapse 3(3):255–285

Volkmar FR, Greenough WT (1972) Rearing complexity affects branching of dendrites in the visual cortex of the rat. Science 176(4042):1445–1447

Wiedersheim R (1890) Bewegungserscheinungen im gehirn von leptodora hyalina. Anat Anz 5:673–679

Wilson CJ, Groves PM, Kitai ST, Linder JC (1983) Three-dimensional structure of dendritic spines in the rat neostriatum. J Neurosci 3(2):383–388. https://doi.org/10.1523/JNEUROSCI.03-02-00383.1983

Wilson DE, Whitney DE, Scholl B, Fitzpatrick D (2016) Orientation selectivity and the functional clustering of synaptic inputs in primary visual cortex. Nat Neurosci 19(8):1003–1009

Xia S, Yu J, Huang X, Sesack SR, Huang YH, Schlüter OM et al (2020) Cortical and thalamic interaction with amygdala-to-accumbens synapses. J Neurosci 40(37):7119–7132

Xu H T, Pan F, Yang G, Gan W B (2007) Choice of cranial window type for in vivo imaging affects dendritic spine turnover in the cortex. Nat Neurosci 10(5):549–551

Xu T, Yu X, Perlik AJ, Tobin WF, Zweig JA, Tennant K et al (2009) Rapid formation and selective stabilization of synapses for enduring motor memories. Nature 462(7275):915–919

Xu Z, Adler A, Li H, Pérez-Cuesta LM, Lai B, Li W et al (2019) Fear conditioning and extinction induce opposing changes in dendritic spine remodeling and somatic activity of layer 5 pyramidal neurons in the mouse motor cortex. Sci Rep 9(1):4619

Yagishita S, Hayashi-Takagi A, Ellis-Davies GC, Urakubo H, Ishii S, Kasai H (2014) A critical time window for dopamine actions on the structural plasticity of dendritic spines. Science 345(6204):1616–1620

Yang G, Pan F, Gan W-B (2009) Stably maintained dendritic spines are associated with lifelong memories. Nature 462(7275):920–924

Yang G, Chang PC, Bekker A, Blanck TJJ, Gan W-B (2011) Transient effects of anesthetics on dendritic spines and filopodia in the living mouse cortex. Anesthesiology 115(4):718–726

Yang G, Lai CSW, Cichon J, Ma L, Li W, Gan W-B (2014) Sleep promotes branch-specific formation of dendritic spines after learning. Science 344(6188):1173–1178

Yang Y, Liu D, Huang W, Deng J, Sun Y, Zuo Y et al (2016) Selective synaptic remodeling of amygdalocortical connections associated with fear memory. Nat Neurosci 19(10):1348–1355

Yuste R (2013) Electrical compartmentalization in dendritic spines. Annu Rev Neurosci 36(1):429–449

Yuste R (2015) The discovery of dendritic spines by Cajal. Front Neuroanat 9:18. Disponible sur: http://journal.frontiersin.org/article/10.3389/fnana.2015.00018/abstract

Yuste R, Bonhoeffer T (2001) Morphological changes in dendritic spines associated with long-term synaptic plasticity. Annu Rev Neurosci 24(1):1071–1089

Yuste R, Bonhoeffer T (2004) Genesis of dendritic spines: insights from ultrastructural and imaging studies. Nat Rev Neurosci 5(1):24–34

Yuste R, Denk W (1995) Dendritic spines as basic functional units of neuronal integration. Nature 375(6533):682–684

Zaccard CR, Gippo I, Song A, Geula C, Penzes P (2023) Dendritic spinule-mediated structural synaptic plasticity: implications for development, aging, and psychiatric disease. Front Mol Neurosci 16:1059730

Zhai S, Ark ED, Parra-Bueno P, Yasuda R (2013) Long-distance integration of nuclear ERK signaling triggered by activation of a few dendritic spines. Science 342(6162):1107–1111

Zhou Y, Lai B, Gan W-B (2017) Monocular deprivation induces dendritic spine elimination in the developing mouse visual cortex. Sci Rep 7(1):4977

Zito K, Scheuss V, Knott G, Hill T, Svoboda K (2009) Rapid functional maturation of nascent dendritic spines. Neuron 61(2):247–258

Ziv NE, Smith SJ (1996) Evidence for a role of dendritic filopodia in synaptogenesis and spine formation. Neuron 17(1):91–102

Zuo Y, Lin A, Chang P, Gan W-B (2005) Development of long-term dendritic spine stability in diverse regions of cerebral cortex. Neuron 46(2):181–189

Chapter 8
Steroid Hormone Interaction with Dendritic Spines: Implications for Neuropsychiatric Disease

Maya Frankfurt, Zeinab Nassrallah, and Victoria Luine

Abstract Dendritic spines, key sites for neural plasticity, are influenced by gonadal steroids. In this chapter, we review the effects of gonadal steroids on dendritic spine density in areas important to cognitive function, the hippocampus, and prefrontal cortex. Most of these animal model studies investigated the effects of estrogen in females, but we also include more recent data on androgen effects in both males and females. The underlying genomic and non-genomic mechanisms related to gonadal steroid-induced spinogenesis are also reviewed. Subsequently, we discuss possible reasons for the observed sex differences in many neuropsychiatric diseases, which appear to be caused, in part, by aberrant synaptic connections that may involve dendritic spine pathology. Overall, knowledge concerning the regulation of dendritic spines by gonadal hormones has grown since the initial discoveries in the 1990s, and current research points to a potential role for aberrant spine functioning in many neuropsychiatric disorders.

Keywords Estrogen · Androgen · Dendritic spine · Neural plasticity · Stress · Hippocampus · Prefrontal cortex

M. Frankfurt (✉)
Hofstra Northwell School of Nursing and Physician Assistant Studies, Hempstead, NY, USA
e-mail: maya.frankfurt@hofstra.edu

Z. Nassrallah
Department of Science Education Zucker School of Medicine, 500 Hofstra University, Hempstead, NY, USA

V. Luine
Department of Psychology, Hunter College, New York, NY, USA

A. A. Rasia-Filho et al. (eds.), *Dendritic Spines*, Advances in Neurobiology 34,
https://doi.org/10.1007/978-3-031-36159-3_8

8.1 Introduction

Dendritic spines are an important site of neural plasticity. As such, many factors, including gonadal steroid hormones, which are the primary focus of this chapter, influence spine density. Most of the work reviewed here focuses on the effects of estrogens on dendritic spine plasticity in the context of mediating cognition. Although less well studied, androgens have also been shown to alter spine density and impact cognition, and therefore, they will be reviewed as well. In addition, it is becoming increasingly evident that dendritic spines play a role in neuropsychiatric disorders, and given that gonadal steroids influence spine plasticity, we will speculate on the potential role that gonadal steroids may play in mediating neural dysfunction.

8.2 Dendritic Spines

In general, dendrites are covered extensively by dendritic spines, which, as they are sites for synaptic contact, have a prominent post-synaptic density that contains actin and scaffolding proteins that are activated or deactivated depending on physiological state (Calabrese et al. 2006; Chidambaram et al. 2019). The number of dendritic spines increases with development to a critical point (Urbanska et al. 2012), and following the establishment of connectivity between neurons, dendritic spine turnover actively continues until adulthood when spines achieve relative stability and less turnover (Koleske 2013). In the adult, several distinct dendritic spine subtypes have been described, with thin filopodial types presumed to be immature spines capable of plasticity, and larger, mushroom-shaped spines that are more stable and are the sites of functioning synapses (Bourne and Harris 2007; Von Bohlen Und Halbach 2009).

Although relatively stable in adulthood as compared to development, dendritic spines do exhibit plasticity, including alterations in number and spine subtype, in adult mammals in response to varied stimuli, including denervation/reinnervation (Deller et al. 2006; Parnavelas et al. 1974), hormonal changes (Luine and Frankfurt 2020b; Frankfurt and Luine 2015), drug exposure (Frankfurt et al. 2011; Robinson et al. 2001; Kolb and Gibb 2015), environmental stimuli (Kolb et al. 2003), learning and memory (Luine and Frankfurt 2020c; Kasai et al. 2010a), and stress (Watanabe et al. 1992). Notably, spine plasticity varies during the lifespan. During adolescence, pruning of dendritic spines occurs in the neocortex (Kolb et al. 2012; Holtmaat et al. 2005; Khanal and Hotulainen 2021). Pruning at this stage suggests a refinement of synapses such that weaker connections are eliminated, and stronger ones are maintained. In the aging brain, dendritic spines and synapse density decrease. There are decreases in dendritic spine density in the cortex (Dickstein et al. 2013; Dumitriu et al. 2010) and dendritic spines and axospinous synapses in the hippocampus with aging (Geinisman et al. 1992; Von Bohlen Und Halbach et al.

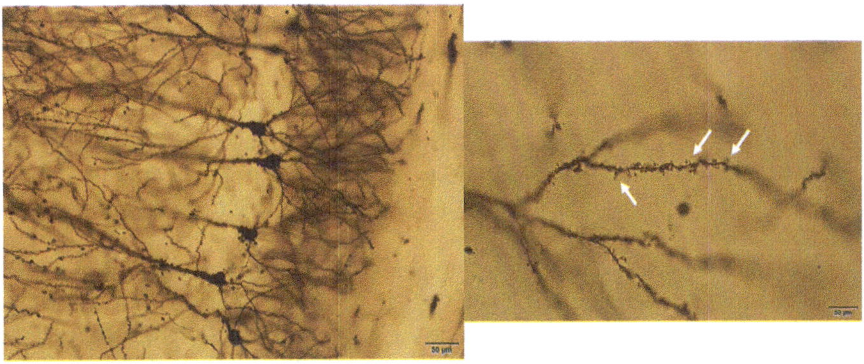

Fig. 8.1 Representative photomicrographs of Golgi-impregnated cells in CA1. Left: Low-power illustrating a single layer of pyramidal cells in CA1. Right: Secondary basal dendrite. Arrows denote spines

2006). Given that during adolescence and aging there is significant change in steroid hormone levels and function, dendritic spine plasticity during these times may be more susceptible to hormonal influences.

Dendritic spine plasticity is essential for learning and memory (Koleske 2013; Chidambaram et al. 2019; Khanal and Hotulainen 2021), which has also been demonstrated to be influenced by gonadal steroids (Luine and Frankfurt 2020a, c; Luine et al. 2018, 2022). The hippocampus and the medial prefrontal cortex (mPFC) are integral to learning and memory (Churchwell and Kesner 2011; Churchwell et al. 2010) and changes in dendritic spine density in these areas play a critical role in these cognitive processes (Jedlicka et al. 2008; Leuner et al. 2003). For this reason, alterations in spine density in the hippocampus and mPFC have been studied more than in other brain regions (Fig. 8.1). Many studies have demonstrated estrogen-dependent enhancements in learning and memory, and these enhancements are associated with increases in spine density on apical and basal dendrites in pyramidal cells in the CA1 region of the hippocampus (CA1) and mPFC in rodents (Luine and Frankfurt 2012, 2013, 2020a, b; Luine 2015, 2016). Therefore, estrogen-induced dendritic spine plasticity has been more extensively studied in the mPFC and CA1 than in other brain regions.

8.2.1 Steroids and Dendritic Spine Plasticity: Estrogens

Early studies demonstrated that dendritic spine density on pyramidal cells in CA1 in gonadally intact female rats fluctuated over the estrous cycle (Woolley et al. 1990; Woolley and McEwen 1992) with the highest levels in proestrus when estrogen levels are also highest. Initial results in the hippocampus were supported by later studies (Kinsley et al. 2006; Gonzalez-Burgos et al. 2005). Alterations in spine density during the estrous cycle have also been demonstrated in other brain regions including the ventromedial nucleus of hypothalamus (Frankfurt et al. 1990; Gonzalez-Burgos et al. 2015), the amygdala (Rasia-Filho et al. 2012), and pyramidal cells in layers III and V of the sensorimotor cortex (Chen et al. 2009). In general, spine density was greatest when estrogen levels were highest apart from the medial nucleus of the amygdala where spine density was lowest on neurons when estrogen levels were high (Rasia-Filho et al. 2012). Alterations in spine density during the estrous cycle in these regions may underlie lordosis and other reproductive behaviors.

The data on intact cycling rats are supported by studies that show a decrease in spine density in ovariectomized (OVX) rats in CA1 (Gould et al. 1990b) that was subsequently restored by administration of estrogen for different time periods (Fig. 8.2), acute (<2 h) to subchronic (2–7 days). In initial studies, subchronic estrogen was shown to reverse the OVX induced decrease in spine density on pyramidal cells in CA1 (Gould et al. 1990b; Luine and Frankfurt 2013). Spine synapses in the hippocampi of OVX monkeys (Leranth et al. 2002) and rats (Woolley and McEwen 1992) are also restored after subchronic estrogen administration. More recently, acute estradiol or estrogen agonists, given for less than 2 h, have been found to induce rapid increases in spine density in gonadectomized female (Inagaki et al. 2012; Luine and Frankfurt 2020a; Phan et al. 2012; Phan et al. 2011; Phan et al. 2015) and male rats (Jacome et al. 2016). A decrease in spine density after OVX has also been shown in CA1 and the mPFC (Wallace et al. 2006). Dendritic spines are decreased in both CA1 and the mPFC in aged females that have lower levels of estrogen (Wallace et al. 2007; Luine et al. 2011). Moreover, when OVX rats are fed a diet low in phytoestrogens, spine density in both CA1 and the mPFC is lower than

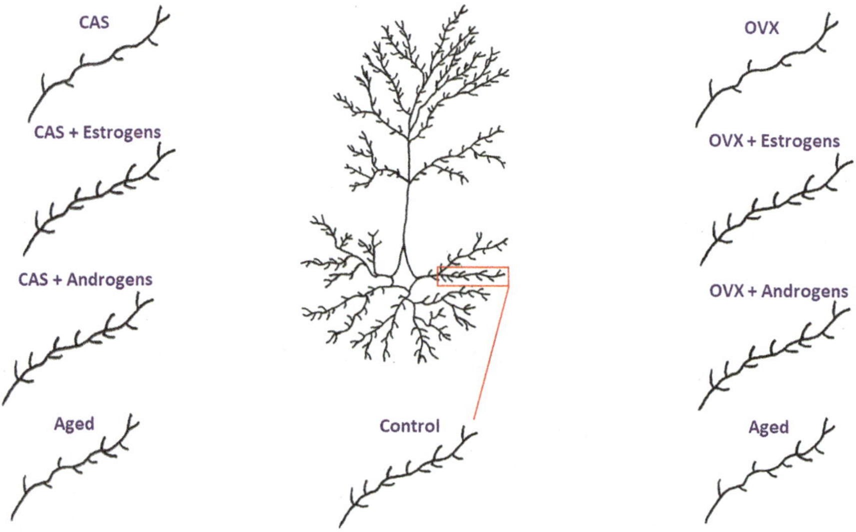

Fig. 8.2 Schematic of the effects of ovariectomy (OVX), castration (CAS), estrogen and androgen replacement, and aging on dendritic spine density on a typical pyramidal cell in CA1

those fed a high phytoestrogen diet (Luine et al. 2006). Finally, chronic exposure to high levels of estrogen during and after pregnancy increases spine density in CA1 pyramidal cells in rats (Kinsley et al. 2006).

As with the estrous cycle, most studies have been done in the hippocampus, but dendritic spine density in other brain regions is also altered when estrogen fluctuates. In the rat, OVX-induced decreases in spine density in the ventromedial nucleus of the hypothalamus (Frankfurt et al. 1990), amygdala (Rasia-Filho et al. 2012), and layers III and V of the somatosensory cortex (Chen et al. 2009) are reversed by estradiol administration (Fig. 8.2). Ye et al. (2019) found that pyramidal cells in layer V of the frontal, motor, and somatosensory cortex in the OVX mouse have decreased spine density that is also reversed when estradiol is given (Ye et al. 2019). The fact that spine density is altered in many brain areas by estrogen illustrates that many neurons are probably sensitive to hormonal alterations, and this understanding may shed light on the observation of sex differences in many neuropsychiatric diseases.

8.2.2 Gonadal Steroids and Dendritic Spine Plasticity: Androgens

Many neurons in the central nervous system are also sensitive to circulating androgens. Although far fewer studies have addressed the interaction between androgens and dendritic spines, it has been clearly demonstrated that various androgens and several androgenic metabolites function similarly to estrogens in terms of their ability to increase spine synapse and dendritic spine density. In gonadectomized male and female rats, both testosterone propionate (TP) and dehydroepiandrosterone (DHEA) increased dendritic spine density on pyramidal cells in CA1 and the mPFC (Luine et al. 2022; Jacome et al. 2016). Similarly, spine synapse density decreases in CA1 after gonadectomy are reversed in rats of both sexes after TP, dihydrotestosterone (DHT), and DHEA administration (Hajszan et al. 2004; Maclusky et al. 2004; Leranth et al. 2004; Atwi et al. 2016). In adult male mouse hippocampus, testosterone (T) increased spine density (Li et al. 2012), and in rats, castration (CAS) reduced, while administration of DHT or estradiol increased spine synapse density in the mPFC (Hajszan et al. 2007). Neurons in the brains of females are also sensitive to androgens as subchronic TP, and DHEA increased spine density on pyramidal cells in the mPFC and CA1 of adult OVX female rats (Luine et al. 2022). Again, as with estrogens, most studies use subchronic treatments, but rapid effects of androgens have also been shown. Acute administration of both T and DHT increases spine density in CA1 in gonadectomized male and female rats (Jacome et al. 2016; Luine et al. 2022) and in hippocampal slices taken from male rats (Murakami et al. 2018). Thus, both spine synapses and synapse density fluctuate with changing androgen levels (Fig. 8.2).

Consistent with the estrogen studies, there are reports of androgens increasing spine density in brain regions other than the hippocampus and mPFC. Syrian hamsters had decreased dendritic spine density in the medial preoptic area 9 weeks after gonadectomy compared to intact male hamsters and gonadectomized hamsters treated with T for 9 weeks (Garelick and Swann 2014). Gonadectomy decreased spine density in the medial preoptic nucleus and medial amygdala, and this effect was reversed by DHT 24 h after injection (Huijgens et al. 2021). Androgen-induced alterations in spine density in these regions may underlie regulation of male sexual behavior.

8.3 Mechanism of Gonadal Steroid Action on Dendritic Spines

Gonadal steroids exert their effects via receptors that mediate both nuclear, genomic, and membrane, non-genomic, mechanisms (Fig. 8.3). Estrogen receptors (ER) α and β are found within cell nuclei and on cell membranes in neurons in many brains regions, while the more recently described G-protein-linked estrogen receptor (GPER) is found on membranes of both neurons and glia (Korol and Pisani 2015; Waters et al. 2015; Torres-Revereron et al. 2020). Chronic effects of steroids are

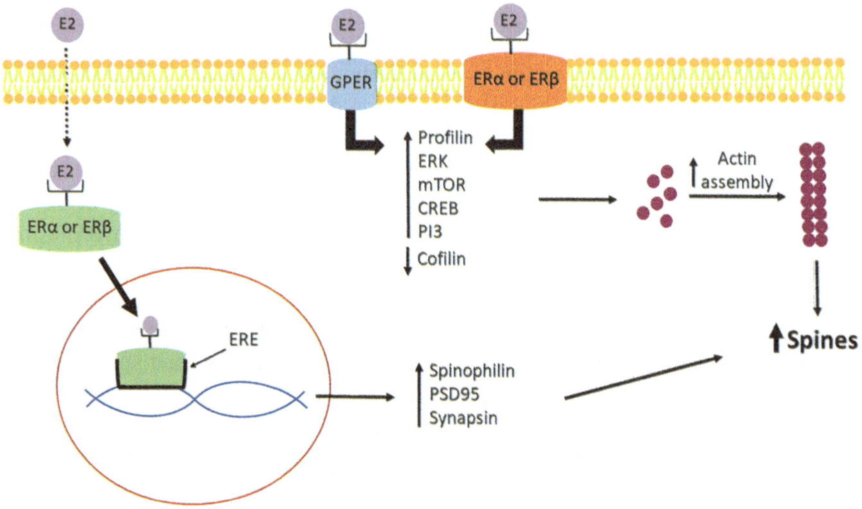

Fig. 8.3 Schematic illustrating the mechanisms underlying estradiol (E_2)-induced spinogenesis via both genomic and non-genomic means. E_2 diffuses across the cell membrane to bind to cytosolic ERs, which then enter the nucleus and bind to the estrogen response element (ERE) inducing the synthesis of synaptic and other proteins. E_2 also binds several membrane receptors, which then alters second messenger systems that result in the polymerization of actin, which increases the number of dendritic spines. Genomic and non-genomic mechanisms may have some degree of interaction in mediating these effects

mediated mainly through nuclear receptors and genomic mechanisms, while binding to membrane receptors mediates the rapid effects through the activation of numerous signaling pathways. Recent studies indicate that interaction between nuclear and membrane receptors may also mediate some steroid effects (Arevalo et al. 2015; Kramár et al. 2013; Luine and Frankfurt 2012). As a membrane receptor, GPER also activates signaling pathways. Which receptors are involved in mediating spine dynamics in neurons and whether the different receptors have an additive effect on steroid-mediated spinogenesis remain to be determined.

Although some differences in effects have been reported, agonists for both ERα and ERβ alter spine density (Murakami et al. 2006; Phan et al. 2011). Studies in OVX mice showed that propyl pyrazole triol (PPT), an ERα agonist, increased dendritic spine density in the stratum radiatum and lacunosum-moleculare of CA1 within 1 h, whereas diarylpropionitrile (DPN), an ERβ agonist, decreased spine density in the lacunosum-moleculare of CA1 (Phan et al. 2011). In the PFC, agonists of the GPER, but not ERα/β receptors, rapidly increased spine density, and the opposite selectivity was found in CA1 (Ye et al. 2019).

Dendritic spine plasticity implies cycling of immature spines to mature spines that make synaptic contact (Ziv and Smith 1996) and changes that existing spines may undergo after exposure to different stimuli (Kasai et al. 2010b; Koleske 2013; Sehgal et al. 2013). This process requires mobilization of many proteins, particularly actin and associated proteins (Penzes and Rafalovich 2012; Hokenson et al. 2021; Koleske 2013). The cycling between filamentous and globular actin is an essential part of spine plasticity and requires interaction with other proteins, including several actin-binding proteins such as cofilin and profilin, which regulate actin polymerization (Basu and Lamprecht 2018; Borovac et al. 2018).

Since spine plasticity is dependent on mobilization of actin and synaptic proteins, it is notable that these proteins have also been shown to be altered by gonadal steroids. Estradiol inactivates cofilin, which is responsible for the disassembly of actin (Kramár et al. 2009). In addition, OVX decreased spine density in the CA1 region of mice in which the expression of cofilin was increased and profilin, which promotes actin polymerization, decreased (Lan et al. 2021). These results may explain how estradiol promotes filamentous actin and spine assembly (Kramár et al. 2009). Estrogen also increases other proteins that are found in the synapses, such as PSD95 and spinophilin (Tang et al. 2004; Maclusky et al. 2005; Lee et al. 2004). Estrogen-induced increases in dendritic spine density have been demonstrated to involve the activation of multiple cell signaling pathways, such as ERK, mTOR (Tuscher et al. 2016), CREB, and PI3, that promote the assembly of actin and protein synthesis and other proteins involved in spine dynamics (Sheppard et al. 2019; Frankfurt and Luine 2015; Fortress et al. 2013; Luine and Frankfurt 2013; Bethea and Reddy 2010; Hansberg-Pastor et al. 2015). Overall, it appears that estrogen acting via multiple pathways influences the assembly of actin and synaptic proteins, which, in turn, increases the number or the maturity of existing spines (Fig. 8.3).

The mechanisms by which androgens influence dendritic spines have been less well studied than estrogens, but the presence of both nuclear and membrane receptors on neurons has also been described for androgens (Atwi et al. 2016;

Chen et al. 2022). Studies to date show that androgens exert similar effects to estrogens in terms of altering cytoskeletal and other proteins in dendritic spines. For example, orchiectomy decreases spine density, actin polymerization, and post-synaptic density thickness in adult male mice (Zhao et al. 2018). In addition, Chen et al. (2022) demonstrated that, in cultured hippocampal neurons from male mice, T promoted the maturation of immature spines and increased synaptic markers PSD 95 and synapsin.

Taken together, the data suggest that steroids can alter dendritic spine density by binding with steroid receptors on neurons and then initiating a series of intracellular events that promote the proteins, which increase the number of dendritic spines. Although most of the studies described here are related to rapid membrane-mediated effects of gonadal steroids, it is interesting to note that both types of receptors appear to mediate similar effects on synaptic proteins. Using antagonists to both nuclear and membrane estrogen receptors, Xing et al. (2018) found that receptor antagonists to ERα, Erβ, and GPER administered to mice decreased PSD-95, spinophilin, spine density, and synaptic density. These results suggest that both genomic and non-genomic receptors play a role in estrogen-induced reorganization of the actin cytoskeleton. There is some evidence for cross talk between the genomic and membrane estrogen receptors, especially given that binding of estrogens to nuclear ERα and ERβ in some circumstances results in alterations in rapid signaling pathways (Kramár et al. 2013; Arevalo et al. 2015; Luine and Frankfurt 2012).

8.4 Dendritic Spine Plasticity and Gonadal Steroids: Potential Clinical Importance

There are clear sex differences in the incidence of some neuropsychiatric diseases (Bangasser and Cuarenta 2021; Seney et al. 2022; Bangasser and Valentino 2014; Schulte Holthausen and Habel 2018; Vegeto et al. 2020). Alzheimer's disease is more prevalent in women, and Parkinson's disease occurs more often in men (Vegeto et al. 2020). Psychiatric disorders such as major depressive disorder and anxiety are more prevalent in women (Bangasser and Cuarenta 2021; Seney et al. 2022; Bangasser and Valentino 2014). Personality disorders, such as paranoid, schizotypal, and narcissistic disorders, are diagnosed more often in men, and borderline histrionic disorders are more common in women (Schulte Holthausen and Habel 2018).

Thus, it is interesting to speculate on the possible clinical importance of gonadal steroid interactions with spines in neural and psychiatric diseases because this information may provide insights into the etiologies and possible treatments for these conditions.

8.4.1 Sex Differences in the Brain

Results of preclinical, clinical, and anatomical studies provide some basis for the sex differences in neuropsychiatric disease. There are reports of sex differences in neural structure in rats (McEwen and Milner 2017; Brandt et al. 2020; Scharfman and Maclusky 2017; Yagi and Galea 2019; Marrocco and McEwen 2016). Nevertheless, sex differences in spine density reports are inconclusive. Female rats in proestrus were found to have greater spine density in CA1 than male rats (Woolley et al. 1990; Shors et al. 2001), and male rats have more thorny excrescences in hippocampal CA3 neurons than female rats (Gould et al. 1990a). However, in other studies no sex differences in spine density were seen in CA1 or the mPFC (Bowman et al. 2015; Gould et al. 1990a). The latter two studies did not consider the estrous stage of the females, and this may account for the different findings among studies. Thus, while sex differences in brain structure exist, data regarding spine density differences are limited and further research is required to determine possible relationship(s) to clinically observed sex differences in diseases.

What about sex differences in brain structure in humans? Imaging studies have shown that men have larger brains, more cortical surface area, and more white matter (except for the corpus callosum) than women, and women have denser gray matter than men (Salminen et al. 2022). Male brains have been shown to have greater ipsilateral connectivity, while female brains have greater commissural connectivity (Ingalhalikar et al. 2014). In the hippocampus, there are sex differences in the size of different hippocampal subregions (Van Eijk et al. 2020). These studies are not conclusive, and it should be noted that there is controversy regarding how real these differences are after being corrected for men's larger brain sizes, sample sizes, and general differences in analysis (for reviews see Hines (2020), Salminen et al. (2022), and Hoggetts and Hausman (2023)).

Therefore, it appears that sex differences in the brain are more subtle than straightforward sexual dimorphisms and may be the result of ongoing developmental exposure during critical periods in the lifespan. One must consider that hormone-induced effects on neurons in adults are the result of multiple effects of hormones at different life stages. These include organizational exposure to gonadal steroids during development and activational exposure starting with adolescence, which may be further influenced by environmental factors (Fig. 8.4), rendering it challenging to correlate levels of steroid hormones with disease (McEwen and Milner 2017). Finally, there is a great deal of variability in preclinical studies with respect to time from gonadectomy to steroid replacement, dose used, length of steroid administration, and age of the animals during the experiment and animal strain, all of which could impact the results. Sex differences in response to stress (see below) are an example of the interaction of factors that may occur when sex differences in disease are manifested. Thus, neural networks seem more important for function than individual differences. Neural networks are connected by spines and synapses, which

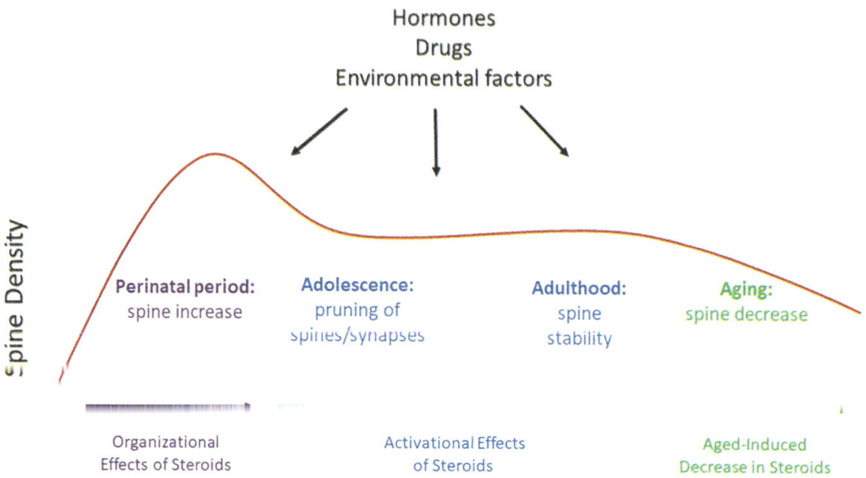

Fig. 8.4 Schematic illustrating the potential interaction between dendritic spine density, hormonal influences, and alterations in environment throughout the lifespan

make them important to study. In the next section, we will review the intersection between spine pathology and gonadal steroids in a few examples to address potential mechanisms that may underlie the sex differences observed clinically.

8.5 Dendritic Spine Plasticity, Gonadal Steroids, and Neuropsychiatric Disorders

8.5.1 Depression

Depression-related alterations in neural plasticity have been studied extensively in animal models subjected to stress because stressed animals exhibit depression-like disturbances, such as anhedonia and alterations in dendritic spines and synapses in the hippocampus and PFC (Leuner and Shors 2013; Yang et al. 2020; Licznerski and Duman 2013). Therefore, stress-induced plasticity in rodents is thought to model what occurs in the human brain with depression.

The brain regions involved in mediating stress-induced responses include the PFC, hippocampus, and amygdala, which have extensive interconnections. Chronic restraint stress has been shown to decrease dendritic spine density in the hippocampus and PFC and increase it in the amygdala (Qiao et al. 2016). Most studies have only been done in male animals, and unfortunately, there are little data from female animals. However, 21 days of chronic restraint stress causes retraction of apical dendrites in the CA3 region of the hippocampus in male, but not female, rats (Galea et al. 1997). In a mouse model in which animals were stressed for 1 h for 6 days using different stressors, only OVX female mice were susceptible to the stress

(Iqbal and Ma 2020). These authors found that OVX female mice had significantly higher corticosterone levels, increased spine density on PFC neurons, increased immobility time of several behavioral tests, and decreased sucrose consumption, which is consistent with anhedonia, in comparison with intact males and sham-operated females. Interestingly, sex differences in behavioral responses to stress have been clearly demonstrated in rats. Chronic restraint stress, 6 h for 21 consecutive days, impairs male performance on several behavioral cognitive tasks and either enhances or has no effect on female cognitive function (Luine et al. 2017; Bowman et al. 2022). In terms of spine density, Shors et al. (2001) found that 24 h after 30 minutes of intermittent stress, spine density in CA1 pyramidal cells was increased in male but decreased in female rats. In the lateral hypothalamic area, there is a sex difference in spine density on putative orexin neurons, and males have less spines than females (Grafe et al. 2019). Following 5 days of 30-minute restraint stress, this sex difference was no longer present, meaning that stress decreased spine density in females only. Following a paradigm of 30 minutes of restraint stress for 5 days, male rats were able to habituate to the stress but female rats did not, and females had significantly higher levels of corticosterone compared to males (Grafe and Bhatnagar 2020). These different stress studies may yield inconsistent results because of the different stress paradigms and behavioral assessments used, but the results do suggest that neural networks related to depression are differentially affected during stress and these changes may help explain sex differences in the incidence of depression.

8.5.2 Schizophrenia

Spine density alterations have also been shown in other diseases (Khanal and Hotulainen 2021). Postmortem Golgi studies have found a decrease in spine density in the dorsolateral prefrontal cortex (DLPFC) and the superior temporal gyrus (Glausier and Lewis 2013; Penzes et al. 2011) of schizophrenic patients, which implies decreased connectivity in regions known to be critical to cognitive function. These authors speculate that dendritic spine plasticity/pruning may be altered during early development and adolescence in schizophrenic patients, time periods in which gonadal steroids influence dendritic spine turnover. Although direct comparisons to the PFC in rodents are difficult, a preclinical study in rats subjected to repeated variable perinatal stress demonstrated a sex difference in the pattern of dendritic development in the PFC (Markham et al. 2013). Dendritic connectivity in both sexes in layer III pyramidal cells of the PFC during adolescence was increased, but maximal growth occurred earlier in female rats and lasted later, into adulthood. Increased spine density was seen in both sexes before puberty, but only females showed pruning of spines in late adolescence. These preliminary results support potential network alterations during a period of gonadal hormone secretions that may explain the observed sex difference in schizophrenia.

8.5.3 *Alzheimer's Disease*

The incidence of neurodegenerative diseases increases with aging. With aging, there are also decreases in spine density (Young et al. 2014; Dumitriu et al. 2010; Walker and Herskowitz 2021; Wallace et al. 2007; Luine et al. 2011). Whether the decreases in spine density in these regions are due to an overall decrease in neuronal number is unclear. However, these changes may be related to the increased incidence of neurodegenerative diseases that is seen with aging. Spine abnormalities, including decreases in number and alterations in spine subtype, have been reported for numerous neurodegenerative diseases that have a cognitive component, such as Alzheimer's disease, Parkinson's disease, and Huntington's disease (for review, see Herms and Dorostkar (2016), and Walker and Herskowitz (2021)). The accumulation of extracellular proteins in Alzheimer's disease appears to interfere with dendritic spines, leading to synaptic loss in both the hippocampus and cortex of patients with Alzheimer's disease (Chidambaram et al. 2019). Given the importance of spine plasticity to the process of learning and memory, it is not surprising that in Alzheimer's disease there are alterations in dendritic spine density (Walker and Herskowitz 2021). Interestingly, Walker and Hershowitz (2021) review the literature that demonstrates that patients with preclinical Alzheimer's disease, who have some signs of the disease but have normal cognitive functions, have higher levels of dendritic spines and synaptic proteins in the hippocampus and PFC than patients who were known to have impaired cognition. This finding implies that dendritic spines may confer resilience to cognitive decline and that decreases in spine density are related to impaired cognition, which is consistent with previous animal studies (Luine and Frankfurt 2020c; Frankfurt and Luine 2015).

8.6 Conclusion

Gonadal steroids exert acute and chronic effects on dendritic spines in pyramidal neurons across the lifespan in both males and females. These effects are mediated by both genomic and non-genomic mechanisms, which influence the assembly of actin and synaptic proteins to promote spinogenesis. Although gonadal steroids have been shown to influence spine density in many brain areas, sex differences have not been adequately investigated, and therefore, it is a challenge to relate differences observed in neuropsychiatric disorders to the basic and clinical data on sex differences to date. This apparent discrepancy may be due to the multifactorial processes and timing during hormonal exposure. Given that there are also spine density changes reported for these disorders, it may become important to consider potential differences in treatment based on sex. However, local alterations in spine density under different conditions imply that alterations in the neural networks may be a critical underlying issue and should be further investigated in relation to potential sex differences.

References

Arevalo MA, Azcoitia I, Gonzalez-Burgos I, Garcia-Segura LM (2015) Signaling mechanisms mediating the regulation of synaptic plasticity and memory by estradiol. Horm Behav 74:19–27

Atwi S, Mcmahon D, Scharfman H, Maclusky NJ (2016) Androgen modulation of hippocampal structure and function. Neuroscientist 22:46–60

Bangasser DA, Cuarenta A (2021) Sex differences in anxiety and depression: circuits and mechanisms. Nat Rev Neurosci 22:674–684

Bangasser DA, Valentino RJ (2014) Sex differences in stress-related psychiatric disorders: neurobiological perspectives. Front Neuroendocrinol 35:303–319

Basu S, Lamprecht R (2018) The role of actin cytoskeleton in dendritic spines in the maintenance of long-term memory. Front Mol Neurosci 11:143

Bethea CL, Reddy AP (2010) Effect of ovarian hormones on genes promoting dendritic spines in laser-captured serotonin neurons from macaques. Mol Psychiatry 15:1034–1044

Borovac J, Bosch M, Okamoto K (2018) Regulation of actin dynamics during structural plasticity of dendritic spines: signaling messengers and actin-binding proteins. Mol Cell Neurosci 91:122–130

Bourne J, Harris KM (2007) Do thin spines learn to be mushroom spines that remember? Curr Opin Neurobiol 17:381–386

Bowman RE, Luine V, Diaz Weinstein S, Khandaker H, Dewolf S, Frankfurt M (2015) Bisphenol-A exposure during adolescence leads to enduring alterations in cognition and dendritic spine density in adult male and female rats. Horm Behav 69:89–97

Bowman R, Frankfurt M, Luine V (2022) Sex differences in cognition following variations in endocrine status. Learn Mem 29:234–245

Brandt N, Löffler T, Fester L, Rune GM (2020) Sex-specific features of spine densities in the hippocampus. Sci Rep 10:11405

Calabrese B, Wilson MS, Halpain S (2006) Development and regulation of dendritic spine synapses. Physiology (Bethesda) 21:38–47

Chen JR, Yan YT, Wang TJ, Chen LJ, Wang YJ, Tseng GF (2009) Gonadal hormones modulate the dendritic spine densities of primary cortical pyramidal neurons in adult female rat. Cereb Cortex 19:2719–2727

Chen H, Qiao D, Si Y, He Z, Zhang B, Wang C, Zhang Y, Wang X, Shi Y, Cui C, Cui H, Li S (2022) Effects of membrane androgen receptor binding on synaptic plasticity in primary hippocampal neurons. Mol Cell Endocrinol 554:111711

Chidambaram SB, Rathipriya AG, Bolla SR, Bhat A, Ray B, Mahalakshmi AM, Manivasagam T, Thenmozhi AJ, Essa MM, Guillemin GJ, Chandra R, Sakharkar MK (2019) Dendritic spines: revisiting the physiological role. Prog Neuro-Psychopharmacol Biol Psychiatry 92:161–193

Churchwell JC, Kesner RP (2011) Hippocampal-prefrontal dynamics in spatial working memory: interactions and independent parallel processing. Behav Brain Res 225:389–395

Churchwell JC, Morris AM, Musso ND, Kesner RP (2010) Prefrontal and hippocampal contributions to encoding and retrieval of spatial memory. Neurobiol Learn Mem 93:415–421

Deller T, Bas Orth C, Vlachos A, Merten T, Del Turco D, Dehn D, Mundel P, Frotscher M (2006) Plasticity of synaptopodin and the spine apparatus organelle in the rat fascia dentata following entorhinal cortex lesion. J Comp Neurol 499:471–484

Dickstein DL, Weaver CM, Luebke JI, Hof PR (2013) Dendritic spine changes associated with normal aging. Neuroscience 251:21–32

Dumitriu D, Hao J, Hara Y, Kaufmann J, Janssen WG, Lou W, Rapp PR, Morrison JH (2010) Selective changes in thin spine density and morphology in monkey prefrontal cortex correlate with aging-related cognitive impairment. J Neurosci 30:7507–7515

Fortress AM, Fan L, Orr PT, Zhao Z, Frick KM (2013) Estradiol-induced object recognition memory consolidation is dependent on activation of mTOR signaling in the dorsal hippocampus. Learn Mem 20:147–155

Frankfurt M, Luine V (2015) The evolving role of dendritic spines and memory: interaction(s) with estradiol. Horm Behav 74:28–36

Frankfurt M, Gould E, Woolley CS, McEwen BS (1990) Gonadal steroids modify dendritic spine density in ventromedial hypothalamic neurons: a Golgi study in the adult rat. Neuroendocrinology 51:530–535

Frankfurt M, Salas-Ramirez K, Friedman E, Luine V (2011) Cocaine alters dendritic spine density in cortical and subcortical brain regions of the postpartum and virgin female rat. Synapse 65:955–961

Garelick T, Swann J (2014) Testosterone regulates the density of dendritic spines in the male preoptic area. Horm Behav 65:249–253

Geinisman Y, Detoledo-Morrell L, Morrell F, Persina IS, Rossi M (1992) Age-related loss of axospinous synapses formed by two afferent systems in the rat dentate gyrus as revealed by the unbiased stereological dissector technique. Hippocampus 2:437–444

Glausier JR, Lewis DA (2013) Dendritic spine pathology in schizophrenia. Neuroscience 251:90–107

Gonzalez-Burgos I, Alejandre Gomez M, Cervantes M (2005) Spine type densities of hippocampal Ca1 neurons vary in proestrus and estrus rats. Neurosci Lett 379:52–54

Gonzalez-Burgos I, Velazquez-Zamora DA, Gonzalez-Tapia D, Cervantes M (2015) A Golgi study of the plasticity of dendritic spines in the hypothalamic ventromedial nucleus during the estrous cycle of female rats. Neuroscience 298:74–80

Gould E, Westlind-Danielsson A, Frankfurt M, McEwen BS (1990a) Sex differences and thyroid hormone sensitivity of hippocampal pyramidal cells. J Neurosci 10:996–1003

Gould E, Woolley CS, Frankfurt M, McEwen BS (1990b) Gonadal steroids regulate dendritic spine density in hippocampal pyramidal cells in adulthood. J Neurosci 10:1286–1291

Grafe LA, Bhatnagar S (2020) The contribution of orexins to sex differences in the stress response. Brain Res 1731:145893

Grafe LA, Geng E, Corbett B, Urban K, Bhatnagar S (2019) Sex- and stress-dependent effects on dendritic morphology and spine densities in putative orexin neurons. Neuroscience 418:266–278

Galea LAM, McEwen BS, Tanapat P, Deak T, Spencer RL, Dhabhar FS (1997) Sex differences in dendritic atrophy of CA3 pyramidal neurons in response to chronic restraint stress. Neuroscience 81:689-697

Hajszan T, Maclusky NJ, Leranth C (2004) Dehydroepiandrosterone increases hippocampal spine synapse density in ovariectomized female rats. Endocrinology 145:1042–1045

Hajszan T, Maclusky NJ, Johansen JA, Jordan CL, Leranth C (2007) Effects of androgens and estradiol on spine synapse formation in the prefrontal cortex of normal and testicular feminization mutant male rats. Endocrinology 148:1963–1967

Hansberg-Pastor V, González-Arenas A, Piña-Medina AG, Camacho-Arroyo I (2015) Sex hormones regulate cytoskeletal proteins involved in brain plasticity. Front Psych 6:165

Herms J, Dosostkar MM (2016) Dendritic spine pathology in neurodegenerative diseases. Ann Rev Pathol 11:221–250

Hines M (2020) Neuroscience and Sex/Gender: Looking Back and Forward. J Neurosci 40:37–43

Hodgetts S, Hausmann M (2023) Sex/gender differences in brain lateralisation and connectivity. Curr Top Behav Neurosci 62:71–99

Hokenson RE, Short AK, Chen Y, Pham AL, Adams ET, Bolton JL, Swarup V, Gall CM, Baram TZ (2021) Unexpected role of physiological estrogen in acute stress-induced memory deficits. J Neurosci 41:648–662

Holtmaat AJ, Trachtenberg JT, Wilbrecht L, Shepherd GM, Zhang X, Knott GW, Svoboda K (2005) Transient and persistent dendritic spines in the neocortex in vivo. Neuron 45:279–291

Huijgens PT, Snoeren EMS, Meisel RL, Mermelstein PG (2021) Effects of gonadectomy and dihydrotestosterone on neuronal plasticity in motivation and reward related brain regions in the male rat. J Neuroendocrinol 33:E12918

Inagaki T, Frankfurt M, Luine V (2012) Estrogen-induced memory enhancements are blocked by acute bisphenol a in adult female rats: role of dendritic spines. Endocrinology 153:3357–3367

Ingalhalikar M, Smith A, Parker D, Satterthwaite TD, Elliott MA, Ruparel K, Hakonarson H, Gur RE, Gur RC, Verma R (2014) Sex differences in the structural connectome of the human brain. Proc Natl Acad Sci U S A 111:823–828
Iqbal J, Ma XM (2020) Impact of subchronic variable stress on ovariectomy and dendritic spine density in prefrontal cortex in mice. Neuroreport 31:213–219
Jacome LF, Barateli K, Buitrago D, Lema F, Frankfurt M, Luine VN (2016) Gonadal hormones rapidly enhance spatial memory and increase hippocampal spine density in male rats. Endocrinology 157:1357–1362
Jedlicka P, Vlachos A, Schwarzacher SW, Deller T (2008) A role for the spine apparatus in LTP and spatial learning. Behav Brain Res 192:12–19
Kasai H, Fukuda M, Watanabe S, Hayashi-Takagi A, Noguchi J (2010a) Structural dynamics of dendritic spines in memory and cognition. Trends Neurosci 33:121–129
Kasai H, Hayama T, Ishikawa M, Watanabe S, Yagishita S, Noguchi J (2010b) Learning rules and persistence of dendritic spines. Eur J Neurosci 32:241–249
Khanal P, Hotulainen P (2021) Dendritic spine initiation in brain development, learning and diseases and impact of bar-domain proteins. Cells 10:2392
Kinsley CH, Trainer R, Stafisso-Sandoz G, Quadros P, Marcus LK, Hearon C, Meyer EA, Hester N, Morgan M, Kozub FJ, Lambert KG (2006) Motherhood and the hormones of pregnancy modify concentrations of hippocampal neuronal dendritic spines. Horm Behav 49:131–142
Kolb B, Gibb R (2015) Plasticity in the prefrontal cortex of adult rats. Front Cell Neurosci 9:15
Kolb B, Gorny G, Soderpalm AH, Robinson TE (2003) Environmental complexity has different effects on the structure of neurons in the prefrontal cortex versus the parietal cortex or nucleus accumbens. Synapse 48:149–153
Kolb B, Mychasiuk R, Muhammad A, Li Y, Frost DO, Gibb R (2012) Experience and the developing prefrontal cortex. Proc Natl Acad Sci U S A 109(Suppl 2):17186–17193
Koleske AJ (2013) Molecular mechanisms of dendrite stability. Nat Rev Neurosci 14:536–550
Korol DL, Pisani SL (2015) Estrogens and cognition: friends or foes?: an evaluation of the opposing effects of estrogens on learning and memory. Horm Behav 74:105–115
Kramár EA, Chen LY, Brandon NJ, Rex CS, Liu F, Gall CM, Lynch G (2009) Cytoskeletal changes underlie estrogen's acute effects on synaptic transmission and plasticity. J Neurosci 29:12982–12993
Kramár EA, Babayan AH, Gall CM, Lynch G (2013) Estrogen promotes learning-related plasticity by modifying the synaptic cytoskeleton. Neuroscience 239:3–16
Lan Z, Meng Z, Lian B, Liu M, Sun T, Sun H, Liu Z, Hu Z, Guo Q, Zhang J (2021) Hippocampal aromatase knockdown aggravates ovariectomy-induced spatial memory impairment, abeta accumulation and neural plasticity deficiency in adult female mice. Neurochem Res 46:1188–1202
Lee SJ, Romeo RD, Svenningsson P, Campomanes CR, Allen PB, Greengard P, McEwen BS (2004) Estradiol affects spinophilin protein differently in gonadectomized males and females. Neuroscience 127:983–988
Leranth C, Shanabrough M, Redmond Jr, DE (2002) Gonadal hormones are responsible for maintaining theintegrity of spine synapses in the CA1 hippocampal subfield of female nonhuman primates. J Comp Neurol 447:34–42
Leranth C, Hajszan T, Maclusky NJ (2004) Androgens increase spine synapse density in the Ca1 hippocampal subfield of ovariectomized female rats. J Neurosci 24:495–499
Leuner B, Shors TJ (2013) Stress, anxiety, and dendritic spines: what are the connections? Neuroscience 251:108–119
Leuner B, Falduto J, Shors TJ (2003) Associative memory formation increases the observation of dendritic spines in the hippocampus. J Neurosci 23:659–665
Li M, Masugi-Tokita M, Takanami K, Yamada S, Kawata M (2012) Testosterone has sublayer-specific effects on dendritic spine maturation mediated by Bdnf and Psd-95 in pyramidal neurons in the hippocampus Ca1 area. Brain Res 1484:76–84
Licznerski P, Duman RS (2013) Remodeling of axo-spinous synapses in the pathophysiology and treatment of depression. Neuroscience 251:33–50

Luine V (2015) Recognition memory tasks in neuroendocrine research. Behav Brain Res 285:158–164
Luine V (2016) Estradiol: mediator of memories, spine density and cognitive resilience to stress in female rodents. J Steroid Biochem Mol Biol 160:189–195
Luine VN, Frankfurt M (2012) Estrogens facilitate memory processing through membrane mediated mechanisms and alterations in spine density. Front Neuroendocrinol 33:388–402
Luine V, Frankfurt M (2013) Interactions between estradiol, Bdnf and dendritic spines in promoting memory. Neuroscience 239:34–45
Luine V, Frankfurt M (2020a) Estrogenic regulation of memory: the first 50 years. Horm Behav 121:104711
Luine V, Frankfurt M (2020b) Estrogenic regulation of object memory, spatial memory, and spinogenesis. In: Frick KM (ed) Estrogens and memory: basic research and clinical implications. Oxford University Press, New York
Luine VN, Frankfurt M (2020c) Estrogenic regulation of recognition memory and spinogenesis. In: Frick KM (ed) Estrogens and memory. Oxford University Press, New York
Luine V, Attalla S, Mohan G, Costa A, Frankfurt M (2006) Dietary phytoestrogens enhance spatial memory and spine density in the hippocampus and prefrontal cortex of ovariectomized rats. Brain Res 1126:183–187
Luine VN, Wallace ME, Frankfurt M (2011) Age-related deficits in spatial memory and hippocampal spines in virgin, female Fischer 344 rats. Curr Gerontol Geriatr Res 2011:316386
Luine V, Gomez J, Beck K, Bowman R (2017) Sex differences in chronic stress effects on cognition in rodents. Pharmacol Biochem Behav 152:13–19
Luine V, Serrano P, Frankfurt M (2018) Rapid effects on memory consolidation and spine morphology by estradiol in female and male rodents. Horm Behav 104:111–118
Luine V, Mohan G, Attalla S, Jacome L, Frankfurt M (2022) Androgens enhance recognition memory and dendritic spine density in the hippocampus and prefrontal cortex of ovariectomized female rats. Neuroscience. https://doi.org/10.1016/j.neuroscience.2022.06.002
Maclusky NJ, Hajszan T, Leranth C (2004) Effects of dehydroepiandrosterone and flutamide on hippocampal Ca1 spine synapse density in male and female rats: implications for the role of androgens in maintenance of hippocampal structure. Endocrinology 145:4154–4161
Maclusky NJ, Luine VN, Hajszan T, Leranth C (2005) The 17alpha and 17beta isomers of estradiol both induce rapid spine synapse formation in the Ca1 hippocampal subfield of ovariectomized female rats. Endocrinology 146:287–293
Markham JA, Mullins SE, Koenig JI (2013) Periadolescent maturation of the prefrontal cortex is sex-specific and is disrupted by prenatal stress. J Comp Neurol 521:1828–1843
Marrocco J, McEwen BS (2016) Sex in the brain: hormones and sex differences. Dialogues Clin Neurosci 18:373–383
McEwen BS, Milner TA (2017) Understanding the broad influence of sex hormones and sex differences in the brain. J Neurosci Res 95:24–39
Murakami G, Tsurugizawa T, Hatanaka Y, Komatsuzaki Y, Tanabe N, Mukai H, Hojo Y, Kominami S, Yamazaki T, Kimoto T, Kawato S (2006) Comparison between basal and apical dendritic spines in estrogen-induced rapid spinogenesis of Ca1 principal neurons in the adult hippocampus. Biochem Biophys Res Commun 351:553–558
Murakami G, Hojo Y, Kato A, Komatsuzaki Y, Horie S, Soma M, Kim J, Kawato S (2018) Rapid nongenomic modulation by neurosteroids of dendritic spines in the hippocampus: androgen, oestrogen and corticosteroid. J Neuroendocrinol 30. https://doi.org/10.1111/jne.12561
Parnavelas JG, Lynch G, Brecha N, Cotman CW, Globus A (1974) Spine loss and regrowth in hippocampus following deafferentation. Nature 248:71–73
Penzes P, Rafalovich I (2012) Regulation of the actin cytoskeleton in dendritic spines. Adv Exp Med Biol 970:81–95
Penzes P, Cahill ME, Jones KA, Vanleeuwen JE, Woolfrey KM (2011) Dendritic spine pathology in neuropsychiatric disorders. Nat Neurosci 14:285–293
Phan A, Lancaster KE, Armstrong JN, Maclusky NJ, Choleris E (2011) Rapid effects of estrogen receptor alpha and Beta selective agonists on learning and dendritic spines in female mice. Endocrinology 152:1492–1502

Phan A, Gabor CS, Favaro KJ, Kaschack S, Armstrong JN, Maclusky NJ, Choleris E (2012) Low doses of 17β-estradiol rapidly improve learning and increase hippocampal dendritic spines. Neuropsychopharmacology 37:2299–2309
Phan A, Suschkov S, Molinaro L, Reynolds K, Lymer JM, Bailey CD, Kow LM, Maclusky NJ, Pfaff DW, Choleris E (2015) Rapid increases in immature synapses parallel estrogen-induced hippocampal learning enhancements. Proc Natl Acad Sci U S A 112:16018–16023
Qiao H, Li MX, Xu C, Chen HB, An SC, Ma XM (2016) Dendritic spines in depression: what we learned from animal models. Neural Plast 2016:8056370
Rasia-Filho AA, Dalpian F, Menezes IC, Brusco J, Moreira JE, Cohen RS (2012) Dendritic spines of the medial amygdala: plasticity, density, shape, and subcellular modulation by sex steroids. Histol Histopathol 27:985–1011
Robinson TE, Gorny G, Mitton E, Kolb B (2001) Cocaine self-administration alters the morphology of dendrites and dendritic spines in the nucleus accumbens and neocortex. Synapse 39:257–266
Salminen LE, Tubi MA, Bright J, Thomopoulos SI, Wieand A, Thompson PM (2022) Sex is a defining feature of neuroimaging phenotypes in major brain disorders. Hum Brain Mapp 43:500–542
Scharfman HE, Maclusky NJ (2017) Sex differences in hippocampal area Ca3 pyramidal cells. J Neurosci Res 95:563–575
Schulte Holthausen B, Habel U (2018) Sex differences in personality disorders. Curr Psychiatry Rep 20:107
Sehgal M, Song C, Ehlers VL, Moyer JR Jr (2013) Learning to learn – intrinsic plasticity as a metaplasticity mechanism for memory formation. Neurobiol Learn Mem 105:186–199
Seney ML, Glausier J, Sibille E (2022) Large-scale transcriptomics studies provide insight into sex differences in depression. Biol Psychiatry 91:14–24
Sheppard PAS, Choleris E, Galea LAM (2019) Structural plasticity of the hippocampus in response to estrogens in female rodents. Mol Brain 12:22
Shors TJ, Chua C, Falduto J (2001) Sex differences and opposite effects of stress on dendritic spine density in the male versus female hippocampus. J Neurosci 21:6292–6297
Tang Y, Janssen WG, Hao J, Roberts JA, Mckay H, Lasley B, Allen PB, Greengard P, Rapp PR, Kordower JH, Hof PR, Morrison JH (2004) Estrogen replacement increases spinophilin-immunoreactive spine number in the prefrontal cortex of female rhesus monkeys. Cereb Cortex 14:215–223
Torres-Revereron A, Brake W, Milner TA (2020) Estrogen receptor distribution in the hippocampus and prefrontal cortex. In: Frick KM (ed) Estrogens and memory: basic research clinical implications. Oxford University Press, New York
Tuscher JJ, Luine V, Frankfurt M, Frick KM (2016) Estradiol-mediated spine changes in the dorsal hippocampus and medial prefrontal cortex of ovariectomized female mice depend on ERK and mTOR activation in the dorsal hippocampus. J Neurosci 36:1483–1489
Urbanska M, Swiech L, Jaworski J (2012) Developmental plasticity of the dendritic compartment: focus on the cytoskeleton. Adv Exp Med Biol 970:265–284
Van Eijk L, Hansell NK, Strike LT, Couvy-Duchesne B, De Zubicaray GI, Thompson PM, Mcmahon KL, Zietsch BP, Wright MJ (2020) Region-specific sex differences in the hippocampus. NeuroImage 215:116781
Vegeto E, Villa A, Della Torre S, Crippa V, Rusmini P, Cristofani R, Galbiati M, Maggi A, Poletti A (2020) The role of sex and sex hormones in neurodegenerative diseases. Endocr Rev 41:273–319
Von Bohlen Und Halbach O (2009) Structure and function of dendritic spines within the hippocampus. Ann Anat 191:518–531
Von Bohlen Und Halbach O, Zacher C, Gass P, Unsicker K (2006) Age-related alterations in hippocampal spines and deficiencies in spatial memory in mice. J Neurosci Res 83:525–531
Walker CK, Herskowitz JH (2021) Dendritic spines: mediators of cognitive resilience in aging and Alzheimer's disease. Neuroscientist 27:487–505

Wallace M, Luine V, Arellanos A, Frankfurt M (2006) Ovariectomized rats show decreased recognition memory and spine density in the hippocampus and prefrontal cortex. Brain Res 1126:176–182

Wallace M, Frankfurt M, Arellanos A, Inagaki T, Luine V (2007) Impaired recognition memory and decreased prefrontal cortex spine density in aged female rats. Ann N Y Acad Sci 1097:54–57

Watanabe Y, Gould E, McEwen BS (1992) Stress induces atrophy of apical dendrites of hippocampal Ca3 pyramidal neurons. Brain Res 588:341–345

Waters EM, Thompson LI, Patel P, Gonzales AD, Ye HZ, Filardo EJ, Clegg DJ, Gorecka J, Akama KT, McEwen BS, Milner TA (2015) G-protein-coupled estrogen receptor 1 is anatomically positioned to modulate synaptic plasticity in the mouse hippocampus. J Neurosci 35:2384–2397

Woolley CS, McEwen BS (1992) Estradiol mediates fluctuation in hippocampal synapse density during the estrous cycle in the adult rat. J Neurosci 12:2549–2554

Woolley CS, Gould E, Frankfurt M, McEwen BS (1990) Naturally occurring fluctuation in dendritic spine density on adult hippocampal pyramidal neurons. J Neurosci 10:4035–4039

[illegible] brane estrogen receptor antagonists induce similar mTORC2 activation-reversible changes in synaptic protein expression and actin polymerization in the mouse hippocampus. CNS Neurosci Ther 24:495–507

Yagi S, Galea LAM (2019) Sex differences in hippocampal cognition and neurogenesis. Neuropsychopharmacology 44:200–213

Yang M, Luo CH, Zhu YQ, Liu YC, An YJ, Iqbal J, Wang ZZ, Ma XM (2020) 7, 8-Dihydroxy-4-methylcoumarin reverses depression model-induced depression-like behaviors and alteration of dendritic spines in the mood circuits. Psychoneuroendocrinology 119:104767

Ye Z, Cudmore RH, Linden DJ (2019) Estrogen-dependent functional spine dynamics in neocortical pyramidal neurons of the mouse. J Neurosci 39:4874–4888

Young ME, Ohm DT, Dumitriu D, Rapp PR, Morrison JH (2014) Differential effects of aging on dendritic spines in visual cortex and prefrontal cortex of the rhesus monkey. Neuroscience 274:33–43

Zhao J, Bian C, Liu M, Zhao Y, Sun T, Xing F, Zhang J (2018) Orchiectomy and letrozole differentially regulate synaptic plasticity and spatial memory in a manner that is mediated by Src-1 in the hippocampus of male mice. J Steroid Biochem Mol Biol 178:354–368

Ziv NE, Smith SJ (1996) Evidence for a role of dendritic filopodia in synaptogenesis and spine formation. Neuron 17:91–102

Chapter 9
Morphological Features of Human Dendritic Spines

Josué Renner and Alberto A. Rasia-Filho

Abstract Dendritic spine features in human neurons follow the up-to-date knowledge presented in the previous chapters of this book. Human dendrites are notable for their heterogeneity in branching patterns and spatial distribution. These data relate to circuits and specialized functions. Spines enhance neuronal connectivity, modulate and integrate synaptic inputs, and provide additional plastic functions to microcircuits and large-scale networks. Spines present a *continuum* of shapes and sizes, whose number and distribution along the dendritic length are diverse in neurons and different areas. Indeed, human neurons vary from aspiny or "relatively aspiny" cells to neurons covered with a high density of intermingled pleomorphic spines on very long dendrites. In this chapter, we discuss the phylogenetic and ontogenetic development of human spines and describe the heterogeneous features of human spiny neurons along the spinal cord, brainstem, cerebellum, thalamus, basal ganglia, amygdala, hippocampal regions, and neocortical areas. Three-dimensional reconstructions of Golgi-impregnated dendritic spines and data from fluorescence microscopy are reviewed with ultrastructural findings to address the complex possibilities for synaptic processing and integration in humans. Pathological changes are also presented, for example, in Alzheimer's disease and schizophrenia. Basic morphological data can be linked to current techniques, and perspectives in this research field include the characterization of spines in human neurons with specific transcriptome features, molecular classification of cellular diversity, and electrophysiological identification of coexisting subpopulations of cells. These data would enlighten how cellular attributes determine neuron type-specific connectivity and brain wiring for our diverse aptitudes and behavior.

J. Renner
Department of Basic Sciences/Physiology and Graduate Program in Biosciences, Universidade Federal de Ciências da Saúde de Porto Alegre, Porto Alegre, RS, Brazil

A. A. Rasia-Filho (✉)
Department of Basic Sciences/Physiology and Graduate Program in Biosciences, Universidade Federal de Ciências da Saúde de Porto Alegre, Porto Alegre, RS, Brazil

Graduate Program in Neuroscience, Universidade Federal do Rio Grande do Sul, Porto Alegre, RS, Brazil

A. A. Rasia-Filho et al. (eds.), *Dendritic Spines*, Advances in Neurobiology 34,
https://doi.org/10.1007/978-3-031-36159-3_9

Keywords Synapse · Synaptic plasticity · Postsynaptic processing · Neural networks · Morphology and function · Microscopy · Spinal cord · Brainstem · Thalamus · Amygdala · Cerebral cortex · Neuropathology · Alzheimer´s disease

> *"When we say that the nervous system contains independent functional and morphological entities called neurons, we are saying much more than what appears at first sight. For, unlike the cells in most other tissues, the components of the nervous system are not equivalent and are not interchangeable parts. Each neuron is unique, and its singularity resides in its specific position in the nervous system. That position is given by its peculiar synaptic connections with other neurons and, either directly or indirectly, with the periphery. As the pattern of these connections is reflected in a rigorous fashion by the form of the neuron, its shape is its most properly neural feature. Thus, the form of a neuron provides the key to its role in the nervous system"* (Peters et al. 1991)

The study of dendritic spines in human neurons is a relatively recent occurrence adding to our knowledge about the "architectonic units of living things" (Peters et al. 1991). Spines were not initially recognized as actual cellular components, and several limitations still exist for visualizing these tiny specialized units with nanodomains in our nervous system. Nevertheless, there is a marvelous history of evolution for the emergence, development, functioning, and plasticity of dendritic spines and neural networks adapted for a myriad of different living species. The evolution of the brain structure, with specialized cells and circuits, provides cues on how complex processing and emergent functional properties let us be *Homo sapiens.* Humans have remarkable phylogenetic differences and ontogenetic development of dendrites, spines, and synapses. There are multiple implications for dendritic morphology for neural circuitry functioning. Spines are specialized elements that usually receive one axonal terminal forming an asymmetric synapse reflecting the excitatory input integrated within a time-space window. Multisynaptic spines also exist and add more complexity to this scenario. Dendritic spines are found from the spinal cord dorsal and ventral horns' neurons to higher-order neocortical areas, although showing heterogeneous numbers, shapes, and sizes. These morphological findings reflect the synaptic ultrastructure, pre- and postsynaptic features, and relevant functions for each neuron in its microenvironment (together with glial cells, extracellular matrix, and vasculature) and every neural circuit. Spines can be normally stable or plastic elements and altered in number, morphology, and ultrastructural composition along healthy aging processes or neuropathological conditions. Here, we discuss these topics and the findings that make dendrites, spines, and synapses key elements for the complexity of our brain and behavior, some showing unique properties in our species compared to other ones. Various examples of human neurons ranging from a very sparse presence to a high density of spines with varied shapes and sizes are presented with their likely functional relevance. Addenda were included to expand some key points with their original descriptions. In the end, perspectives are commented on methodological approaches aiming to expand our comprehension of the integrative role of dendritic spines in different functional networks and brain areas.

9.1 The Evolved Brain Structure: Cells and Circuits

Evolution and genetics are fundamental to unraveling the extraordinary development of what life on the Earth is and how organisms are organized for survival and reproduction while coping with a diversity of challenging conditions coming from a complex and contingent world (Schmidt-Nielsen 1997; Mayr 2001; Catania 2017; Leopold et al. 2019; Keough et al. 2023). Life lies in each detail of the orchestration of morphology, biochemistry, biophysics, and physiology. It relies on matter and energy, and the emergence of activity within a range of biological probabilistic possibilities after the integration of unitary functions that generate one sole body (Schrödinger 1992; Purves et al. 2001; see also Junker 2007). Cellular membrane, few ions relatively abundant in Earth´s crust, and an impressive variety of molecules and proteins compose different subcellular and cellular functional levels. Distinct neurogenic, immune, and developmental genes formed the cellular organization of ctenophores (Moroz et al. 2014) and the likely origin(s) of neurons (Burkhardt 2022). Evolution also provided emergent properties for integrated specialized nerve cells and the generation and modulation of complex functions and behaviors in different species. Emergent properties of nerve cells are more than just the sum of the properties of their parts. While function relies on the properties of each composing element, when assembled, the organized conjunct has more properties and can elaborate higher levels of activity (such as the isolated parts of an airplane that cannot fly but the conjunct does fly).

In addition, the nervous system is organized to elaborate behaviors and maintain a range of homeostatic adjustments, balancing change and stability, despite continuous multi-faceted external and internal variations, dealing with the acquisition, maintenance, development, and flexibility of adaptedness (Cannon 1939; Mayr 2001; Rasia-Filho 2006; Michael et al. 2009; Wefelmeyer et al. 2016; Rasia-Filho et al. 2018; Leopold et al. 2019). The morphological diversity and functional organization of neurons and glial cells reflect this major cellular specialization and connectivity achievements. Species-specific neural features involve ontogenetic changes, the animal´s notion of inanimate objects and who are the "others" (e.g., conspecifics, neutral cues, preys, or predators), elaboration of behavioral strategies for survival (Purves et al. 2001), and experience-dependent memory and learning (Kandel and LeDoux 2021; Shohamy et al. 2021). Epigenetic modulations of cellular functioning provide additional variability for cells and networks in individuals among the population (Kandel and LeDoux 2021; Burton and Greer 2022).

There are similarities in areas and connectivity in the central nervous system (CNS) of primates (e.g., Amiez et al. 2021). On the other hand, some patterns of synaptic organization are characteristic of each cortical area and show differences between species (DeFelipe 2011; Hunt et al. 2022). The nervous system´s connectional and functional organization increased along with the primate evolution (Holstege and Subramanian 2016; Sierpowska et al. 2022), and humans are considered outliers in terms of encephalization quotient (Herculano-Heuzel 2012). Interestingly, the most cognitively able brain is not the largest one

(Herculano-Heuzel 2012), but particular genetic features may have enhanced neurodevelopment, neuronal and glial structure, synaptic processing (e.g., for the glutamatergic one) in our species (Oberheim et al. 2009; Xu et al. 2018; Hodge et al. 2019; Beaulieu-Laroche et al. 2021; Berg et al. 2021; Viscardi et al. 2021; Pinson et al. 2022; An et al. 2023; Keough et al. 2023). It is likely that an increase in brain size with a folded cerebral cortex, the relative number and specialization of nerve cells, the connectional organization and type of synaptic processing, and the plasticity and metaplasticity of different networks formed the route that led us to *Homo sapiens* (based on Cajal 1894; Azevedo et al. 2009; DeFelipe 2011; Geschwind and Rakic 2013; Marín-Padilla 2014; Bruner et al. 2017b; Van Essen et al. 2018; Rasia-Filho et al. 2021; Schmidt and Polleux 2022). These improvements might have developed our ability to interpret and manipulate the external milieu, elaborate on more complex social relationships, and develop intellectual creativity leading to language, successful and progressively more complex inventions, our culture and arts, and the ability to transmit and judge information across generations[1] (Creutzfeldt 1995; Mayr 2001; DeFelipe 2011; Freiwald 2020; Rasia-Filho et al. 2021).

The integrated processing of information involves morphologically and functionally heterogeneous neurons and glial cells organized according to the cytoarchitectonic features of each region in the CNS. Dendrites and spines will be detailed in the next sections, but cell body and axonal features deserve attention as well. The neuronal cell body includes the nucleus, the perikaryon with organelles, and the plasma membrane with biophysical properties to integrate signals coming from dendrites and from direct axosomatic synaptic contacts. The neuronal input/output function not only depends on the somatodendritic morphology but also on the axon initial segment (AIS) origin, length, and position (Höfflin et al. 2017). "Axon-carrying dendrites" were discovered in CA1 pyramidal neurons of mice (Thome et al. 2014). Their occurrence and proportions are variable and type-specific in humans (Wahle et al. 2022). For example, in the CA1 hippocampal pyramidal neurons of adult humans, the axon emerges from the soma or from the initial portion of a basal dendrite (Benavides-Piccione et al. 2020). Changes in the length and position of the AIS location for action potential (AP) generation can impact the neuronal overall level of excitability and output code within neural circuits (Wefelmeyer et al. 2016). In layer V thick-tufted pyramidal neurons, the axon hillock location relative to the soma or a dendrite is finely tuned and related to the somatodendritic capacitive load (Hamada et al. 2016).

It is not completely known how axonal structural plasticity and changes in excitatory and inhibitory contacts upon dendritic shafts and spines cooperate to homeostatically adapt activity at the single-cell level. However, there is a continuous

[1] "Clearly, only humans sit around the fire (or dinner table) to tell each other jokes and stories about past glories or future plans, and only a human would eagerly read what Owen wrote about primate brains 140 years ago. Moreover, only humans use general engineering skills to overcome environmental challenges that other animals can solve solely through evolution by natural selection. These are, of course, merely some of the major differences between human" and other animals (Striedter 2004).

coordination of the formation, maintenance, turnover, strength, and plasticity of the presynaptic and postsynaptic partnered components (Stuart et al. 1999; Bourne and Harris 2007; Wefelmeyer et al. 2016; Kasai et al. 2021). Axonal arbor structural remodeling may involve presynaptic boutons dynamics and the possibility for changes related to postsynaptic plasticity (Wefelmeyer et al. 2016). Synaptic nanomodules underlie the organization and plasticity of discrete and aligned modules of pre- and postsynaptic proteins, whose number scales linearly with spine size (Hruska et al. 2018). Neural cells would then combine varied routes for synaptic plasticity and modulate their input–output relation and excitability in plastic circuits.

9.1.1 Dendritic Morphology

Dendrites have an arborized aspect that greatly increases the membrane available for synaptic contacts, information integration, and network functioning (Stuart et al. 1999). Dendrites represent the highest neuron's receptive surface area (approximately 93%; Torikai et al. 1996), receive most synaptic contacts, compartmentalize information, and integrate inputs along an extensive arbor (Kubota et al. 2007; Spruston et al. 2013). Accordingly, dendrites process most excitatory, and to a lesser extent inhibitory, synaptic inputs usually terminating on dendritic spines and shafts, respectively (Peters et al. 1991; Pannese 2015; see further results in Brusco et al. 2014).

Neurons differ in how their dendrites branch within the neuropil volume and how they receive incoming synaptic information from different afferent sources (Ramón-Moliner 1962; Rollenhagen and Lübke 2013; Benavides-Piccione et al. 2020). The geometry, extension, and three-dimensional (3D) spatial projection of dendrites critically determine the mode of neuronal connectivity within neural circuits (Peters et al. 1991; Rollenhagen and Lübke 2013; Guerra et al. 2023), that is, dendrites can receive a great number of afferents from a wide variety of sources and be highly integrative or, on the other hand, receive inputs from only one or at most a few sources and process a limited range of information (Peters et al. 1991). Afterward, dendrites integrate these inputs and provide synaptic responses of slower and faster time courses, filtering or amplificating signals (Bullock 1979; Stuart et al. 1999; Segev et al. 2003). These response features vary with (1) the pattern of activity in the afferent fibers; (2) the structure of the dendritic tree and location of the synaptic input; (3) the types and actions of synaptic transmitters receptors; (4) the intrinsic membrane properties and slopes of passive and active input–output curves; and (5) the biophysical and biochemical compartmentalizations or cooperativity effects of spines across dendritic segments (Bullock 1979; Peters et al. 1991; Harvey et al. 2008; Sjöström et al. 2008; Spruston et al. 2013; Sala and Segal 2014).

Furthermore, the dendritic number, branching pattern, and size of the field covered by the arborization identify a spectrum of neuronal types and subdivisions (Ramón-Moliner 1962). Morphological diversity of dendrites occurs even within the same neuronal class (Cembrowski and Spruston 2019; Benavides-Piccione et al.

2021; Rasia-Filho et al. 2021). Dendrites adapt their shape to the particularities of the surrounding neuropil, including the tissue volume available, the local cellular density and package, and the spatial distribution of afferent axons, some with specific domains for connectivity (Morishima and Kawaguchi 2006; Larriva-Sahd 2014; Wang et al. 2018; Rasia-Filho et al. 2021). For example, cortical layers II/III (supragranular) pyramidal neurons do not have enough space and cannot have a long apical dendrite like layer V thick-tufted ones, whereas these latter cells differ in their apical and basal branching aspects according to their location within layer V (Morishima and Kawaguchi 2006). Indeed, in the rat frontal cortex, two populations of layer V pyramidal neurons projecting to the striatum differ in their dendritic morphology. Superficial layer V neurons show tufted or slender apical dendrites in layer I, but the same type of neuron in the deeper layer V has a reduced or absent apical tuft (Morishima and Kawaguchi 2006). Morphological differences between layers II/III and V pyramidal neurons mean connectional and electrophysiological particularities with functional implications, as described below.

The first attempts to classify neurons based on "dendroarchitectonic" organization also looked for a correlation between morphology and function, as follows: "The radiate dendritic pattern may well be related to an input of heterogeneous origin and/or to the presence of relatively widely spaced afferent terminal fibers, whereas afferent connections of more homogeneous origin, composed of closely spaced axons that frequently terminate in dense clusters or other specialized endings (relate to) the tufted dendritic pattern. This type of relationship may appear in two different ways: (1) axonal fibers distributed in one or more planes or 'floors' parallel to one another and traversed by dendrites with 'linear' orientation. (2) Dendrites with 'planar' orientation lying in one or more planes parallel to one another and traversed by a stream of parallel axons" (Ramón-Moliner 1962; Fig. 9.1).

Currently, neuron types have been classified into different subsets based on the cell body shape, dendritic number, branching pattern, presence, distribution, density and shape of spines, and axonal ramification and projection. It is possible to recognize the existence of general morphological types and intermediate forms of neurons, although virtually no two cells display matching dendritic trees (Ramón-Moliner 1962; see also a comment in Ascoli 2015). For example, dendritic heterogeneity is evident in pyramidal neurons (Cajal 1909–1911). The morphology of these cells differ along subcortical to neocortical areas or within the same cortical area and show distinct electrophysiological properties within the same or across cortical layers in humans (Benavides-Piccione et al. 2020, 2021; Moradi Chameh et al. 2021; Planert et al. 2021; Rasia-Filho et al. 2021).

Dendrites (with different patterns of branching, number of collaterals in each order, length, and preferential spatial extension within the neuropil volume) show phylogenetic and ontogenetic characteristics to receive a varied number of input pathways and harness them as sources of learning possibilities (Cajal 1909–1911; Sjöström et al. 2008; Benavides-Piccione et al. 2020). There is an activity-dependent reciprocal loop between synaptic plasticity and dendritic excitability (Sjöström et al. 2008). To uncover the structural organization of dendrites, spines, axons, glial

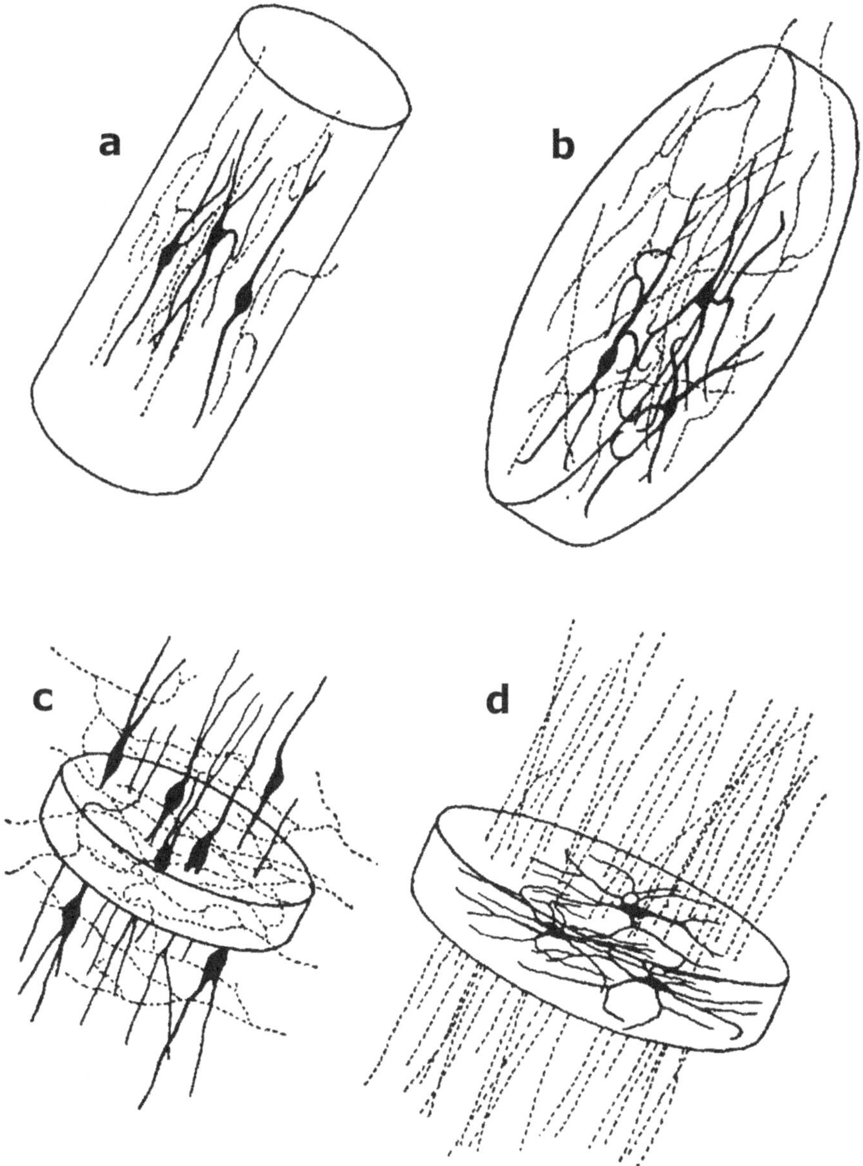

Fig. 9.1 Spatial orientation and contact of dendrites (emerging from the cell body and drawn in black) with axons (dashed lines) in the neuropil volume can show a (**a**) parallel linear (unidimensional) orientation, (**b**) parallel planar (bi- to three-dimensional) orientation, (**c**) linear orientation of dendrites perpendicular to the planar orientation of axons, and (**d**) linear orientation of axons perpendicular to the planar orientation of dendrites. (Legend adapted and figure reproduced from Ramón-Moliner (1962) under CCC RightsLink® license #5383310575621, originally published by John Wiley & Sons, Inc)

cells, and synaptic plasticity in the human nervous system is fundamental to understanding our wide neural wiring diagram and varied functional displays.

9.1.1.1 Wiring Properties Involving Dendritic Spines

Adapted to every CNS area, morphological and functional relationships make axons reach dendritic targets which, in turn, can also actively look for inputs and alter shafts and spine structure for the appropriate wiring of each microcircuit and large network. Dendrites and spines can be both under structural "renovation" and stability (Leopold et al. 2019, see the preceding chapters in this book). For example, both basal and apical dendritic architecture of CA3 and CA1 hippocampal pyramidal neurons show a spatial orientation and functional characteristic related to the local laminar connectivity and pathway of information processing (Andersen et al. 2007). The functional roles of dendrites for synaptic processing and integration were expanded by the existence of protruding spines forming postsynaptic multifunctional units (Shepherd 1996; Yuste 2010; Fig. 9.2). The multitude of spines added many more possibilities for processing synaptic inputs, adjoining a set of functional properties from micron to nanoscale dimensions and different time windows in dendritic arbors. Spines isolate each input and alter the impact of input potentials, expanding the integration and information processing of each neuron within local microcircuits and larger networks (Wefelmeyer et al. 2016). Inputs occur on dendritic spine postsynaptic density (PSD), whose structure and composition connect to a rich intraspine signaling machinery (Calabrese et al. 2006; Yuste 2010; Cohen 2013; Sala and Segal 2014; Nakahata and Yasuda 2018; Kasai et al. 2021). In these spiny cells, the synaptic signals reach different locations along the dendritic tree and find spines with diverse density, sizes, shapes, and stable or dynamic (plastic) features.

Fig. 9.2 Cortical synaptic organization. (**a**) Golgi-impregnated spiny stellate cell (shown in *a'* with the corresponding camera lucida drawing in *b'*) from lamina IV of the cat visual cortex. Both ascending and descending axons (ax) are observed in the neuropil. The vertically ascending axons are in specific synaptic relation with apical dendrites of pyramidal cells forming "synaptic cartridges." (*b'*-*e'*) There are multiple (longitudinal, oblique, and parallel) directions for the axonal fibers and ramifications (arrows) close to spiny dendrites (arrowheads). (**b**) Schematic diagram showing cortical inhibitory interneurons (in black) and pyramidal neurons. Note the spatial distribution of branched axons and the points of contact onto dendritic shafts and spines. (*a'*-*c'*) Drawings made from electron microscopic images showing synaptic types including an *en passant* synapse and an axon terminal contacting the same spine (*a'*), *en passant* axospinous and axodendritic synapses (onto a dendritic shaft, *b'*), and axosomatic synapses (*c'*) from different cortical layers (I-VI). Note the neuropil volume in this schematic representation. ap.den., apical dendrite; a.r., asymmetric membrane contact and round synaptic vesicles—putative excitatory synapses; a.t.c., axonal tuft cells; c.b.c., columnar basket cell; ch.c., chandelier cells; den, dendrite; l.b.c., large basket cell; pyr, cell body of a pyramidal neuron; s.b.c., small basket cells; s.f., symmetric membrane contact and flattened synaptic vesicles—putative inhibitory synapses. (Legends adapted and figures reprinted from Szentágothai (1978) under CCC RightsLink® license #1268521-1, originally published by Proceedings of the Royal Society of London, Series A, Mathematical and Physical Sciences)

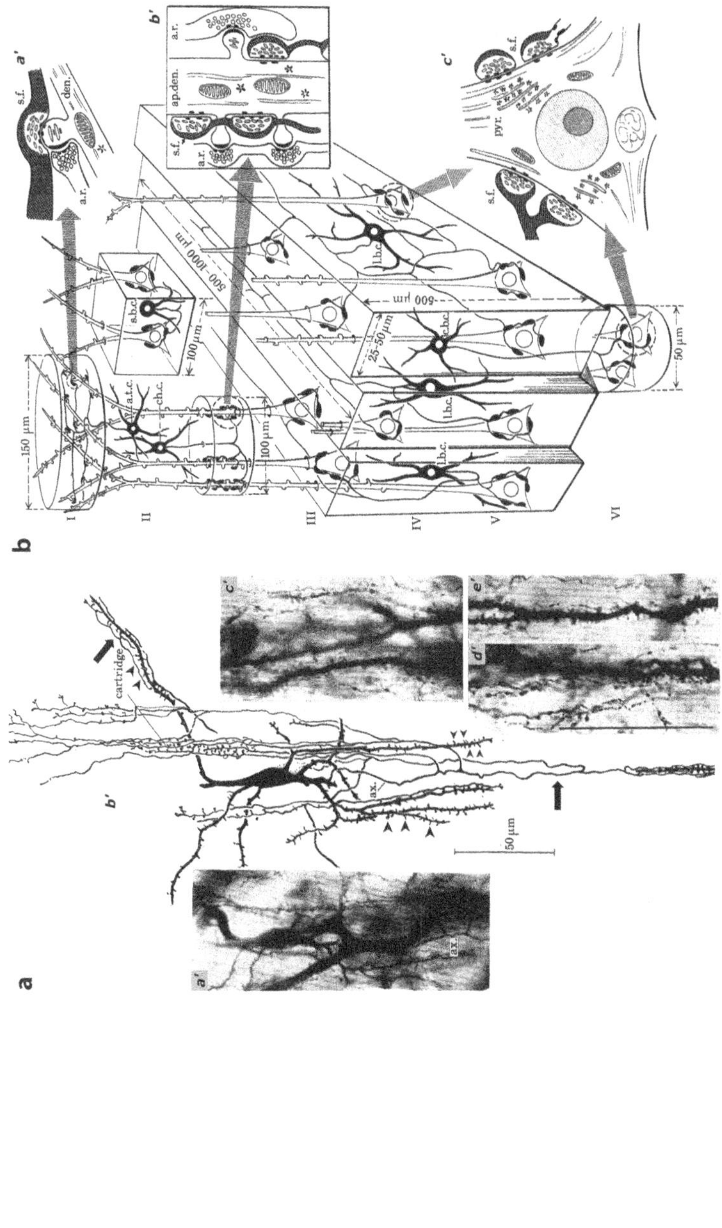

Fig. 9.2 (caption on p. 374)

Synaptic connectivity differs across neuronal types for optimal degrees of wiring and type of plasticity (Litwin-Kumar et al. 2017). Dendritic spines can modulate synaptic processing in multiple dendritic domains, reflecting different patterns of cellular connectivity (see Fig. 1 in Kubota et al. 2016), such as for memory encoding by hippocampal cells (Andersen et al. 2007; González-Ramírez et al. 2014; Larriva-Sahd 2014; Fig. 9.3). Furthermore, cortical pyramidal neurons with distinct basal and apical dendritic domains and synaptic receptive fields can (1) compartmentalize signals; (2) integrate input signals for synchronized transmission of information; (3) involve passive and active membrane properties and modulate synaptic strength; (4) generate anterograde depolarizations, be modulated by strategic inhibitory inputs, or be invaded by retrograde APs; (5) depending on intrinsic membrane properties, impose transient voltage changes that affect the dynamics of ionic diffusion caused by synaptic inputs; and (6) promote different impacts on the somatic excitability and neuronal firing pattern according to the frequency, time, and relative distance of synaptic inputs from the soma (Andersen et al. 2007; Sjöström et al. 2008; Yuste 2010; Spruston et al. 2013; Almog and Korngreen 2014). All these possibilities provide much more computational capabilities for the dynamic processing of information in spiny dendrites within circuits. In conjunction, the shape and function of human nerve cells reflect a complex development over millions of years of multicellular evolution.

Dendritic spines can be one of the key advances in the evolution of the nervous system and part of the way neural networks generate more elaborated emergent functional properties for neural networks (Yuste 2010; see also Brandon and Coss 1982; DeFelipe 2011; Rasia-Filho et al. 2021; Redgrave and Costa 2021). Dendritic spines maximize connectivity by accommodating a great number of distributed synapses (most excitatory) along the dendritic surface although using a minimal volume and containing a huge variety of molecules inside (Yuste 2010; Saga and Segal 2014; Ammassari-Teule et al. 2021; Helm et al. 2021). At the same time that dendritic spines can sample a wider choice of axons, they shorten the wire and condense input processing (Yuste 2010). By providing circuits with increased flexibility and computational power, different spiny neurons likely keep their functional influence distinct from one another by combining maximum connectivity with the functional individuality and modulation of each connection (Yuste 2010; Araya 2014).

Spines can establish contact with an *en passant* fibers, receive one single axon terminal, or be contacted by one or more axonal boutons from the same or different axons (Arellano et al. 2007b; Dall´Oglio et al. 2015; Shapson-Coe et al. 2021; Fig. 9.2). In the mouse cerebral cortex, most afferent axons contact just one spine in a dendritic segment, but a fewer and variable proportion of axons can contact two, three, or more spines of the same dendrite (see Fig. 6 in Kasthuri et al. 2015). Afterward, spines can integrate the action of several molecules in intraspine nanodomains, show varied tunable biophysical properties, promote biochemical compartmentalization, or allow the diffusion of small signaling messengers into the parent

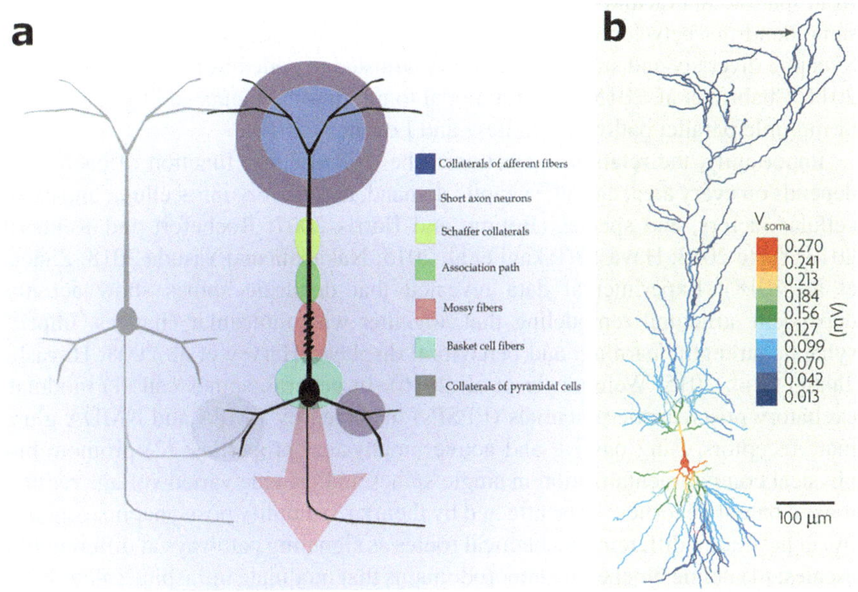

Fig. 9.3 (**a**) Schematic diagram of dendritic domains of a hippocampal pyramidal cell related to the distribution of different input sources (indicated by colors and to the right of this image) terminating along apical and basal segments. (Legend adapted and figure reprinted from Larriva-Sahd (2014) under CC BY license and Copyright © 2014 Larriva-Sahd). (**b**) Simulated attenuation of excitatory postsynaptic potentials (EPSP) on a CA1 pyramidal neuron depending on the dendritic location of synaptic inputs. The passive properties of dendrites make an excitatory synapse of fixed synaptic conductance (0.3 nS) proximal to the cell body promotes somatic amplitudes of 0.2–0.3 mV (colored yellow–red), whereas, in distal dendrites, the amplitude can be less than 0.02 mV (dark blue). The arrow points to a location where the local synaptic potential is approximately 13 mV and the resultant somatic EPSP is around 0.014 mV because of the small diameter and high impedance of distal dendrites. The presence of active properties in dendrites makes that synaptically induced EPSPs modulate the dynamics of voltage-gated channels, exhibit a nonlinear increased response, and lead to the occurrence of dendritic spikes. (Legend adapted and figure reprinted from Spruston (2008) under CCC RightsLink® license #5438800696086, originally published by Springer Nature)

dendrite (see the first chapter of this book).[2] In other words, on a spine-by-spine basis (Oray et al. 2006), there can be a high degree of synaptic modulation arising

[2] "In particular, the activity of one spine can modulate the plasticity of neighboring spines through the mutual sharing of plasticity-related proteins or through the activation of synchronized synaptic inputs. These changes occur across different time scales and typically result in the spatial organization of spines into groups or clusters. This phenomenon suggests that spines do not act as single functional units but are part of a complex network that organizes spines in groups to optimize the connectivity patterns between dendrites and surrounding axons" (Mijalkov et al. (2021) and references therein).

from spatiotemporal and functional heterogeneity among individual synapses on the same dendrite, between different neurons, and across and between brain regions. Synaptic diversity and strength are jointly adjusted to code information (Chen et al. 2011; Chabrol et al. 2015) from unimodal to multimodal inputs and from restricted or multiple parallel pathways (Soltesz and Losonczy 2018).

Importantly, the relationship between the structure and function of each spine depends on every area, circuit, synaptic demand, modulatory intracellular and extracellular factors, and species (Bourne and Harris 2007; Rochefort and Konnerth 2012; Yuste 2013; Hayashi-Takagi et al. 2015; Nakahata and Yasuda 2018; Zancan et al. 2018). Experimental data revealed that dendritic spines show activity-dependent structural remodeling that, together with molecular changes, impacts synaptic strength, learning, and behavioral displays ([illegible] et al. [illegible]; Hayashi-Takagi et al. 2015; Wefelmeyer et al. 2016). In general, spines can (1) modulate excitatory postsynaptic potentials (EPSPs) mediated by AMPA and NMDA glutamate receptors using passive and active biophysical properties; (2) promote biophysical compartmentalization in single spines and provide varied voltage regimes toward parent dendrites (or be affected by them); (3) amplify postsynaptic responses by orchestrating different biochemical routes as signaling pathways at different timescales; (4) define biochemical microdomains that modulate intraspine calcium levels, actin phosphorylation, and second messengers availability, among other possibilities; (5) participate in the synaptic input modulation along diverse dendritic segments, including voltage cooperativity and molecular crosstalk between neighbor spines (with different properties along proximal to distal branches, main shafts, and collateral ones); (6) integrate synaptic signals within a spatiotemporal window for linear and nonlinear impacts on the somatic voltage and neuronal firing output; and (7) maintain a certain degree of isolation for each input or form groups making that each input have an activity-dependent synaptic weight and plasticity in integrated networks (Chen et al. 2011; Chen and Sabatini 2012; Spruston et al. 2013; Tønnesen and Nägerl 2016; Lu and Zuo 2017; Cornejo et al. 2022; see the first chapter of this book).

Spines may not be uniformly distributed along different dendrites; mixed spine types occur along the same dendritic segment, and spines can undergo plastic changes in their turnover, number, shape, size, spatial location, and clustering pattern (Wefelmeyer et al. 2016; Kastellakis and Poirazi 2019; Mijalkov et al. 2021; Rasia-Filho et al. 2021). Although general features identify neuronal types, there may exist intra-individual and interindividual variability in morphology, connectivity, electrophysiology, and function for the same cell class within networks.[3] In this

[3] Neuroanatomical and functional heterogeneity may exist within cortical areas and their subdivisions (e.g., see Vogt 2015) which refers to the surface location, macroscopic aspect, and dimension (e.g., Scheperjans et al. 2008; Bruner et al. 2014, 2017a, b); the corresponding function depending on experience, emotion, and behavior (e.g., in musicians, Gaser and Schlaug 2003; Omigie 2016; Bouhali et al. 2020); and variable gray matter width, heterogeneous cytoarchitectonic width, cellular composition, and neuropil package (Pandya et al. 2015; Mai et al. 2016; Triarhou, 2009 and the von Economo and Koskinas' atlas of cytoarchitectonics of the adult human cerebral cortex;

regard, some spines can also show dynamic features depending on the type of stimuli received and their functional role (González-Burgos et al. 2017). Spines can grow, retract, and shrink with a turnover rate (i.e., a balance between formation and elimination) in a time window that can vary between neurons and brain areas (Toni et al. 1999; Wefelmeyer et al. 2016). Indeed, results depend on the sampling and experimental procedures, and synaptic structure and function can be dynamically modulated at the single-spine level.

There are still many gaps in the elucidation of dendritic spines morphology and function in an inherently highly complex brain like ours. In the next sections, we describe and illustrate relevant findings regarding dendritic spines across different areas of the human CNS. These data address the long pathways that still need to be paved for understanding human spines at the same time that can direct future efforts with clinical implications.

9.2 Phylogenetic Specialization of Dendrites and Spines in Humans

Transcriptome analyses of invertebrate neurons revealed a complexity of synaptic and intracellular signaling pathways comparable to that of vertebrates, suggesting the existence of diverse ancestral presynaptic and postsynaptic pathways for neuronal functions and plasticity (Moroz 2011). On the other hand, the human brain is larger and contains more neurons than expected for a nonprimate mammal of its body size (Azevedo et al. 2009). Although human dendrites and spines have many evolutionary conserved features, they also show unique properties to transform synaptic inputs into complex functions within evolved circuits and brain areas.

Our brain composes only 2% of the adult body's weight but has a high metabolic and energy demand requiring over 15% of the cardiac output continuously (Leopold 2009; see further comments in Herculano-Houzel 2012). Our large cerebral cortex (82% of total brain mass) holds only 19% of all brain neurons (Azevedo et al. 2009). Therefore, more than building on an area by adding more and more neurons in a restricted brain volume, it is likely that cellular morphological and functional specializations had to be developed for wiring new connections and to generate higher mental abilities (DeFelipe 2011; Bianchi et al. 2013; Paredes et al. 2016; Rasia-Filho et al. 2021; Hunt et al. 2022). In other words, neural circuits evolved with additional functional features and increased complexity for information processing from more specialized cells. Species-specific variations in these features are found at the finest level of cell type and circuitry distinction (Striedter 2004; Shepherd and Rowe 2017; Eyal et al. 2016; Van Essen et al. 2018; Hodge et al. 2019; Hunt et al.

note a distinctive layer IV in motor cortex discussed by Yamawaki et al. (2014); and for data indicating that "cortical functions should rather focus on circuits specified by functional cell type composition than mere laminar location," see Guy and Staiger (2017)).

2022; Schmidt and Polleux 2022). Discrete and continuous morphological and functional variations may coexist underlying cell-type diversity within networks (BRAIN Initiative Cell Census Network (BICCN) 2021). In this way, a cell type with heterogeneous elements would accomplish an additional set of functions. "In the case of cell types that repeat across space, such within-cell-type heterogeneity could facilitate the simultaneous execution of distinct computations through the same apparent circuitry" (Cembrowski and Spruston 2019). This is the type of subtle but meaningful neuronal specialization that would support progressively more complex and dynamic networks (Lodato and Arlotta 2015).

For example, pyramidal cell diversity enables parallel information processing in the hippocampus (Soltesz and Losonczy 2018), and human cells in this area show four different branching patterns of their spiny dendrites (Benavides-Piccione et al. 2020). The "within-cell type heterogeneity may provide the hippocampus the intrinsic flexibility that is needed to meet the diverse and variable demands of the external world" (Cembrowski and Spruston 2019). They are also related to our personal memories, cognitive processes, and social behavior, which are all parts of our self-consciousness processes and daily life. These properties would also be applied to glial cells and their participation in synaptic processing. Morphological heterogeneity of astrocytes occurs in the human temporal cortex (Hodge et al. 2019), and human-specific subtype and different astrocytes were demonstrated in the human cerebral cortex (Oberheim et al. 2009; Matyash and Kettenmann 2010; see also the molecular and adapted morphological diversity of astrocytes in Endo et al. (2022), and the description that approximately half of the synapses have an adjacent glial process in the mouse somatosensory cortex in Kasthuri et al. 2015).

We display characteristic genetic architecture phenotype (Grasby et al. 2020) and complex features in cortical layers constituted by different cyto-, myelo-, receptor, and synaptic properties across allocortical and isocortical areas (Palomero-Gallagher and Zilles 2019). In the human temporal cortex layers II/III, which greatly expanded during evolution for increased cortico-cortical connectivity, pyramidal neurons are electrophysiologically heterogeneous and form five subtypes of cells with notable within-individual variability (Planert et al. 2021). Moreover, in the human middle temporal gyrus, layer V pyramidal extratelencephalic- and intratelencephalic-projecting neurons are morphologically distinct. The former type has an apical tuft terminating at the pial surface, a higher apical and basal dendrites total length and dendritic branches, a larger average diameter of the apical dendrite, and a greater total apical and basal dendrite surface area than the latter one (Kalmbach et al. 2021).

9.2.1 Some Differences Between Humans and Other Commonly Studied Species

Understanding the emergence of human higher cognition involves the elucidation of the evolutionary reason for the divergence in gene expression patterns in the human brain and the features that determine neuronal diversity and specialization in our species (Geschwind and Rakic 2013; Hodge et al. 2019, 2020; Kalmbach et al. 2021; Schmidt and Polleux 2022). For example, human pyramidal neurons may have unique morphological and functional properties compared to other species (Mohan et al. 2015; Eyal et al. 2016). However, this is not an easy task. There may exist approximately 16 billion cortical neurons, 61 billion non-neuronal cells, and 180 areas per hemisphere bounded by specific cellular, functional, connectional, and topographic features in the human cerebral cortex (Azevedo et al. 2009; Glasser et al. 2016). For a brief comparison between species, the sea hare *Aplysia californica*, from which fundamental data on the cellular biology of learning and memory were obtained, has a nervous system composed of nine ganglia with approximately 10,000 neurons (Moroz 2011; Liang et al. 2019).

Rats and mice are some of the most studied animal models in the literature. Comparatively, there are morphological differences between the developing human and mouse neocortex (Geschwind and Rakic 2013), as well as in variations for the cortical microanatomical structure (thickness, layers, number of neurons, and synaptic profiles) between humans, rats, and mice (DeFelipe 2011). Large pyramidal extratelencephalic-projecting neurons in temporal cortex layer V are relatively abundant in mice, followed by macaques, and then humans, as evaluated using RNA FISH probes against conserved marker genes (Kalmbach et al. 2021). Neuron density decreased with brain enlargement making the number of synapses per neuron significantly higher as a function of brain expansion in neocortical areas of primates, including humans (Sherwood et al. 2020). There are also species-specific differences in the serotonin (5-HT) receptors and likely differences in the modulation of cortical activity in humans under both normal functioning and psychiatric disorders[4] (Hodge et al. 2019; Kalmabach et al. 2021; see a relevant discussion in Rust and LeDoux 2023).

[4] In the human and mouse temporal cortex, "the most-divergent gene families include neurotransmitter receptors, ion channels, extracellular matrix elements, and cell-adhesion molecules" (Hodge et al. 2019). There are "species differences in the expression of genes encoding 5-HT1 receptor family subunits. In mouse layer V extratelencephalic-projecting neurons, HTR1A and HTR1F were the dominantly expressed subunits, whereas, in human layer V extratelencephalic-projecting neurons, HTR1E (which is absent in the mouse genome) and HTR1F were highly expressed, with little HTR1A expression. These data suggest that human and rodent layer V extratelencephalic-projecting neurons likely share some similar distinctive intrinsic membrane properties and responses to neuromodulation in comparison to neighboring layer V intratelencephalic-projecting neuron types. In contrast, cross-species gene expression differences among layer V extratelencephalic-projecting neurons in mouse versus human highlight areas of potential phenotypic divergence" (Kalmbach et al. 2021). Therefore, extrapolations from animal models data on neuronal morphological, connectional, and functional features have to be done carefully when assuming similar implications to the human cerebral cortex (Hodge et al. 2019). Moreover, it is apparent "that genome-wide quantitative differences in expression profiles between species must also be considered when assessing the fine-tuned functional properties of a given cell type in different species... even classically defined, cerebellar cell types differ between mouse and human by expression of hundreds of orthologous genes... granule cell or astrocyte gene expression profiles can vary between species, or even in individual cells of a type, without losing their cell type identity. Studies of sixteen human *postmortem* brains revealed gender-specific transcriptional differences, cell-specific molecular responses to aging, and the induction of a shared, robust response to an unknown external event evident in three donor samples" (Xu et al. 2018).

Recent transcriptomic data also identified a highly diverse set of excitatory and inhibitory neuron types in the human middle temporal gyrus (Hodge et al. 2019). These data showed not only well-conserved cellular architecture across species that enables the matching of homologous types and predictions of properties of human cell types (Hodge et al. 2019, 2020) but also marked differences between homologous human and mouse cell types in proportions, laminar distributions, gene expression, and morphology (Hodge et al. 2019). The cumulative effects of such differences in the cellular patterning of genes relevant to neuronal signaling and connectivity might have shaped many differences in human cortical circuit function (Hodge et al. 2019) further tuned by a wide repertoire of plastic changes involving learning, emotions, creativity, culture, etc.

As noted above, humans display dendrites, spines, and synapses with crucial differences compared to other animals (Gioia et al. 1998; Elston and DeFelipe 2002; Schmidt and Polleux 2022). The structure of the human cerebral cortex characteristically show (1) larger neurons with more elaborated spiny dendritic trees, notably in (but not restricted to) the prefrontal cortex;[5] (2) thicker layers II/III, an overall larger neuropil, and more spacing between pyramidal neurons, which are the most abundant cells that compose the cortical gray matter; (3) expansion of pathways connecting cortical regions, including those from subcortical areas and between multimodal association areas; (4) more synaptic connections per cell (15,000–30,000 for layer II/III pyramidal neurons); (5) larger white matter volume, myelination, and increased connectivity between primary and unimodal association areas as well as between higher-order multimodal association areas; and (6) a high number and variety of local circuits neurons, most of which are inhibitory interneurons that critically control the pyramidal cells excitability (Elston et al. 2001; DeFelipe et al. 2002; DeFelipe 2011; Bianchi et al. 2013; Geschwind and Rakic 2013; Ardesch et al. 2019; Schmidt and Polleux 2022).

Many features characterize human cells as integrative devices with particular functional properties. Human neocortical pyramidal neurons show larger dendritic length and increased branch complexity with longer segments than mice, marmosets, and macaques (Elston et al. 2001; Mohan et al. 2015). Compared to chimpanzees, human cortical layer III pyramidal neurons are significantly longer and display more branched dendritic arbors in the primary somatosensory (Broadmann area, BA 3b), primary motor (BA 4), prestriate visual (BA 18), and prefrontal (BA 10) cortex, where exists the greatest dendritic complexity (Bianchi et al. 2013). Human pyramidal neurons from layers II/III of the temporal cortex have threefold larger dendritic length and increased branch complexity than similar cells in macaques and mice (Mohan et al. 2015). Likewise, an unusual subpopulation of calcium-binding

[5]The frontal, parietal, and temporal association cortices are larger in humans relative to those of other primates (Kolb and Whishaw (2021); see also Striedter (2004) for a critical discussion and findings on thalamic nuclei and cerebellar hemispheres, and Rasia-Filho et al. (2021) for a discussion on amygdaloid nuclei development with allocortex and neocortex evolved functions). The precuneus in the posteromedial parietal cortex is another cortical area with marked expansion in our species (Bruner et al. 2017a, b; see also Messina et al. 2023).

protein calretinin-positive pyramidal neurons is more abundant in the superficial part of layer V in the anterior cingulate cortex (ACC) of humans than in other primates (Hof et al. 2001).

Human layer V pyramidal neurons also display distinct compartmentalized responses and disrupted coupling between soma and distal dendritic domains compared to rodents (Beaulieu-Laroche et al. 2018). To compensate for the increased size of human dendritic arbors and attenuated signal integration over large distances, (1) neurons in deeper cortical layers show nonlinear properties for the summation of activity in multiple dendritic segments, whereas (2) those in superficial cortical layers reduced membrane capacitance (i.e., less depolarizing charge and fewer coactivated synapses are required for somatic firing) together with an increased propagation speed of APs for an enhanced signal transfer (Schmidt and Polleux 2022 and references therein). In addition, (3) human cortical neurons receive a higher density of synapses per dendritic segment when compared to various other species of primates or mice (Elston et al. 2001; Benavides-Piccione et al. 2002; Sherwood et al. 2020); (4) human synapses are structurally different compared to other species (Molnár et al. 2016; Yakoubi et al. 2019a, b, Rollenhagen et al. 2020) and provide higher information transfer rate by quickly recovering from depression after presynaptic AP train (Testa-Silva et al. 2014); (5) dendritic spines show many shapes and sizes expanding the possibilities for synaptic processing and plasticity (Yuste 2013; Araya et al. 2014; Dall´Oglio et al 2015); (6) human pyramidal neurons have a class of calcium-mediated graded dendritic APs that would classify linearly non-separable inputs, extending the signal processing/integration of connections and the repertoire of computations available to each cell (Gidon et al. 2020); and (7) human membrane properties amplify synaptically induced NMDA-dependent depolarizations (Hunt et al. 2022)[6] and significantly enable synaptic charge-transfer from dendrites to soma and axon (Eyal et al. 2016).

[6] Furthermore, "dendrites of layer II/III human pyramidal cortical neurons are more excitable, generating multiple dendritic Ca^{2+} spikes upon current injection. Dendritic APs in apical tuft dendrites were also found to be sharply tuned to specific input strengths... As a result, when inputs exceed optimal input strength, the amplitude of the dendritic Ca^{2+} AP is reduced. A striking consequence of this change in electrical properties of distal dendrites is that it enables human cortical pyramidal neurons to execute XOR logical operations in apical tuft dendrites, thereby extending the computational repertoire beyond simple AND/OR operations... As such, superficial and deep layer cortical pyramidal neurons may have evolved distinct mechanisms in response to the growing size of the cortex that depends on the computational role they play in the circuit. Each of these mechanisms may have enhanced the computational power in distinct ways. By increasing and tuning dendritic excitability of layer II/III pyramidal neurons, distinct logical operations can be performed on a dendritic level that otherwise would require the implementation of a complex neuronal circuit. In contrast, the isolated nature of distal dendrites in deep layer cortical neurons may provide a distinct compartment for parallel processing of information" (Schmidt and Polleux 2022).

9.2.2 Evidence for Specialized Synaptic Processing in Humans

Hunt et al. (2022) demonstrated that human synaptic connections are threefold stronger[7] and generate more reliable transmission than in mice when studying layer II/III spiny pyramidal neurons in the middle temporal gyrus (Brodmann area 21), part of the distributed cortical language circuitry. Associating compartmental models, *ex vivo* electrophysiological recordings, and morphology of postsynaptic cells, including synapses on the dendritic spine head, pyramidal-to-pyramidal connections in humans are associated with large AMPA- and NMDA-based conductances. The NMDA receptor activation in human neurons showed increased amplitude and prolonged decay of unitary excitatory EPSPs. This finding is important for neuronal excitability that, from a theoretical approach to quantifying properties of spinous synapses, indicates that a human-specific dendritic spine would generate synaptic conductance and voltage change threefold to fourfold larger than in mice (Hunt et al. 2022). Furthermore, considering a variety of spine morphologies and spine–neck values - which relates to spine–neck length and resistance, higher spine–head impedance, and decoupling of the spine head (and site of synaptic contact) from the spine base -, the unitary EPSP amplitude and the recruited NMDA conductance are larger in human versus mouse spines. Assuming that NMDA-dependent recurrent excitation might support persistent activity and working memory, these particular cortical microcircuits and spine properties would contribute to semantic and language processing stronger or unique to our species (Hunt et al. 2022; see further comment on the organization of neuronal circuits in the human brain in Schmidt and Polleux 2022).

Biophysical features of cortical layer V pyramidal neurons also differentiate our cells from other mammalian species. Human dendrites are "outliers" to other species' functional properties by exhibiting a high input resistance, a distinct suprathreshold behavior characterized by small and narrow spikes, low voltage-gated

[7] Accordingly, "the stronger connections in human are not due to a larger number of synaptic contacts. Rather, it is explained by the larger presynaptic active zones and PSDs in human that may allow higher release probability as well as more neurotransmitter release and binding (Benavides-Piccione et al. 2002; Yakoubi et al. 2019b), ultimately leading to larger synaptic conductance at human synapses" (Hunt et al. 2022). Specifically, "beside similarities, human synaptic boutons, although comparably small (approximately 5 μm), differed substantially in several structural parameters, such as the shape and size of active zones, which were on average 2 to 3-fold larger than in experimental animals. The total pool of synaptic vesicles exceeded that in experimental animals by approximately 2 to 3-fold, in particular the readily releasable and recycling pool by approximately 2 to 5-fold, although these pools seemed to be layer-specifically organized" (Rollenhagen et al. 2020). See additional relevant data comparing synaptic structure in humans and rats in Molnár et al. (2016). Importantly, layer-specific synaptic transmission structural parameter differences in temporal lobe layers IV and V, with larger values in layer IV, are suggestive of different neurotransmission efficacy, strength, and plasticity between human cortical layers (Yakoubi et al. 2019a, b). Human layer IV excitatory synaptic boutons may act as "amplifiers" of signals from the sensory periphery to integrate, synchronize, and modulate intra- and extracortical synaptic activity (Yakoubi et al. 2019b).

potassium and hyperpolarization-activated and cyclic nucleotide-gated channel (HCN)-mediated conductances (Beaulieu-Laroche et al. 2021). These data indicate that layer V pyramidal neurons in humans have a unique biophysical makeup for dendritic computations (Beaulieu-Laroche et al. 2021). The human larger dendritic trees of pyramidal neurons would track the activity of synaptic inputs with higher temporal precision to enable efficient information transfer from inputs to output within cortical circuits (Goriounova et al. 2019). These specializations in evolved neocortical cells and networks form a fundamental piece in the neural processing that brings about the spectrum of human behaviors. For neuronal populations, the confluence of molecular microscale architectural attributes (cell-type composition, morphology, and configuration in local circuits) and macroscale connectome architecture may be closely related to spatial patterns of evolutionary expansion, gene expression, intracortical myelin, cortical thickness, and laminar profiles - including spine number - and relate to different temporal dynamics and computations across cortical regions (Shafiei et al. 2020).[8]

In the human allocortex, hippocampal pyramidal neurons show a longer apical dendritic shaft, greater mean values of diameter, surface area, and volume, increased length of dendritic segments after branching, and more complex branching patterns than in mice (Benavides-Piccione et al. 2020; Fig. 9.4). Among other functions, the hippocampal neurons' structure and their synaptic plasticity are relevant for memory formation, consolidation, and retrieval, which are associated with interconnected brain areas to support our high levels of cognitive functions (Andersen et al. 2007; Basu and Siegelbaum 2015). The basal dendritic structure is more complex as a function of the distance from the soma in humans (Benavides-Piccione et al. 2020), which suggests that different dendritic domains in these neurons contribute to our functional features.

On the other hand, we do not have yet comparative data of this kind for human pyramidal neurons in subcortical regions, for example, those composing the amygdaloid complex. These neurons relate to the beginning of the great limbic lobe for higher cortical sensory perception and emotional elaboration, including interpretation of facial expressions and fear, and to evoke complex social behaviors in our species (Rasia-Filho et al. 2021; Guerra et al. 2023; see relevant data and concepts in Heimer et al. 2008; Rolls 2015; Freiwald 2020; Šimić et al. 2021; Rust and LeDoux 2023).

[8] For more information on human multimodal, higher-order processing and connected large-scale networks for the default mode and resting state, sensorimotor, executive control, subcortical and cortical emotional processing, detection of salience and attention, task-related and (abstract) cognitive activity, basal ganglia circuitry, and social behavioral display, see Goulden et al. (2014), van den Heuvel et al. (2015), Margulies et al. (2016), Ng et al. (2016), Diano et al. (2017), Bagarinao et al. (2019), Leopold et al. (2019), Bagarinao et al. (2020), Deming and Koenigs (2020), Shafiei et al. (2020), Bruton (2021), Hidalgo-Lopez et al. (2021), and Kolb and Whishaw (2021).

9.2.3 *Human Dendritic Spines*

As described above, spines are relevant to connectional, electrophysiological, and a large functional repertoire for the synaptic computations in dendrites. Cortical pyramidal neurons possess spines distributed from proximal to distal branches, along main shafts and collaterals (Feldman 1984; Ramaswamy and Markram 2015; Rasia-Filho et al. 2021). Only a small percentage of the axonal inputs to a neuron establish more than one synapse with the postsynaptic cell in the human cerebral cortex (Shapson-Coe et al. 2021). Therefore, spines can isolate inputs and also integrate the information from a larger number of inputs along the dendritic arbor (Yuste 2010). Compared among cortical areas, spine densities and branching patterns of layer III pyramidal cells increase from primary to higher-order cortical areas. This increase represents 2.0-fold more spines in the middle temporal area, 11 fold more spines in the inferior temporal cortex, and 16-fold more spines in the prefrontal cortex compared to the primary visual area V1 of the macaque monkey (Elston and DeFelipe 2002 and references therein). Cross-species comparisons demonstrated that neurons in the human prefrontal cortex have at least 23 times more dendritic spines than those in V1 of the macaque monkey, which likely reflects differences in the number of excitatory inputs to different cortical areas and species-specific differences in cortical function (Elston and DeFelipe 2002).

Human dendritic spines are also larger and longer and show higher densities in hippocampal CA1 pyramidal neurons and layers II/III pyramidal neurons in the temporal cortex than in mice (Benavides-Piccione et al. 2002, 2020; see also Eyal et al. (2018), and different parietal cortex spine data in mouse in Kasthuri et al. 2015; Fig. 9.4). In addition, subcortical and cortical human spines display a diversity of shapes and sizes coexisting along various dendritic segments, some notably convoluted and with complex forms that are not usually found so abundant in corresponding areas of mice and rats (Rasia-Filho et al. 1999; Brusco et al. 2010, 2014;

→

Fig. 9.4 Representative examples of CA1 hippocampal pyramidal cells from human (**a**) and mouse (**b**). The morphology of cells was studied using Lucifer Yellow microinjection, confocal scanning, and fluorescence microscopy. Neuronal reconstructions are presented at the same magnification to illustrate differences in cell size. The main apical dendrite (in black), apical collateral dendrites (in blue), basal arbor (in red), and axon (if traced, in green) are shown with the prevalence percentage of each morphological dendritic pattern found in the studied samples. (**c**) Schematic representation showing the different main apical branching patterns of these human neurons (exemplified in (**a**) and measured within the first 200 μm): *a'*, 0 - no bifurcation; *b'*, 1 bifurcation; *c'*, 2 bifurcations; and *d'*, 3 bifurcations. (Legend adapted and Figure reproduced from Benavides-Piccione et al. (2020) under CCC RightsLink® license #5383911347030, originally published by Oxford University Press). (**d**, **e**) Photomicrograph of horizontally projecting dendrites of neurons injected with Lucifer Yellow from the human (**d**) and mouse (**e**) temporal cortex. At higher magnification, spines along basal dendritic segments in human (**f**) and mouse (**g**) pyramidal cells show morphological differences. Note the smaller size of mice spines compared to larger and longer ones of humans. Scale bar = 45 μm in (**d**) and (**e**); 10 μm in (**f**) and (**g**). (Legend adapted and figures reprinted from Benavides-Piccione et al. (2002) under CCC RightsLink® license #5383920674577, originally published by Springer Nature)

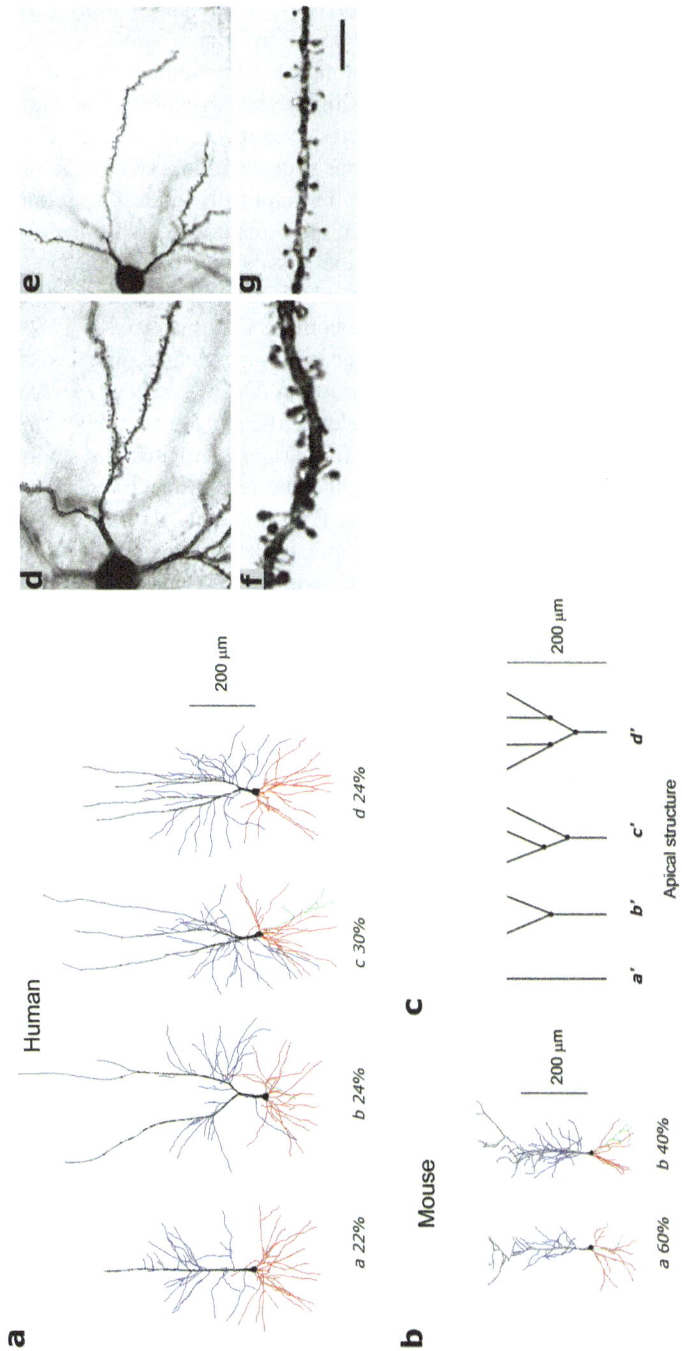

Fig. 9.4 (*caption on p. 386*)

Dall'Oglio et al. 2015; Becker et al. 2017). Taking into account the knowledge available from other species, these numerical and morphological features imply functional differences for the synaptic processing elaborated by human spines.

At the same time, cooperativity and actions among neighbor spines can lead to a wide range of modulatory and computational possibilities key to additional modulation of synaptic strength, integration, and plasticity (Nakahata and Yasuda 2018). Some spines alter the functioning of other neighbor spines within a spatiotemporal window for activity-dependent voltage changes and synaptically induced spreading of signaling molecules along parent dendrites. All these features increase the possibilities of neuronal computations and expand the possibilities of graded modulated functioning of synapses, spines, and dendrites for each neuron, more than a simple deterministic condition and with more possibilities in humans.

In addition, it is assumed that the head diameter in humans, larger spines would also relate to the extension of the PSD, the proportion of AMPA and NMDA glutamate receptors in the PSD, the subcellular cytoskeleton, the molecular composition, and the organelles present in the spine (e.g., actin for the spine structure and smooth endoplasmic reticulum (SER) to modulate intraspine and parent dendrite Ca^{2+} levels; Yuste 2010, 2013). The spine head diameter, PSD area, and its macular and perforated aspects relate to connection stability and structural plasticity in mushroom spines (Arellano et al. 2007a), occurring with selectivity after NMDA receptor-mediated long-term potentiation and depending on SER presence (Borczyk et al. 2019). Stubby and mushroom spines show similar average protein copy number and topology for PSD composition in cultured hippocampal neurons of rats, but proteins related to synaptic strength, spine dynamics, ion channels, endocytosis cofactors, cytoskeletal structure, signaling and trafficking, secretory proteins, and ribosomes are more evident in mushroom spines (Helm et al. 2021).

Human thin spines can show diverse sizes (Benavides-Piccione et al. 2013; Dall´Oglio et al. 2015; Vásquez et al. 2018), some displaying a long thin neck that ends in a bulbous head (Yuste 2013). Supposing a similar functioning in humans as in other species, the spine neck length and resistance can impose compartmentalization for both synaptically mediated electrical signaling and coupling with the adjacent dendrite (Tønnesen and Nägerl 2016). It can also affect the diffusion rate of calcium and small molecules from spines to the parent dendrite. Although long-necked spines can be found in rats (Rasia-Filho et al. 1999), they would provide further dynamic possibilities in the context of the human processing of information. Some human dendritic spines have neck lengths of about 30% longer and 100% more volume than in the somatosensory cortex of mice (DeFelipe 2011 and references therein). Dendritic spines that have longer necks may not generate significant depolarization at the soma, suggesting that EPSPs may be filtered in the spine neck. That may imply that long spines may be "electrically silent" and may be held in "reserve." By changing the spine neck to a shorter one, there might be a "plugging in" and, thereby, a fast circuit switching (Yuste 2013; see also Araya et al. 2014). Human neurons have higher spine densities and many spines with long necks, possibly indicating increased synaptic connectivity and plasticity (Yuste 2013). Following Cajal's descriptions of spiny human neurons and the quest to understand the physical basis of human intelligence, spines displaying abnormally long necks have been seen in young patients with intellectual developmental disorders (Yuste 2013; see also von Bohlen und Halbach 2010 and further data below).

Ramified spines and multiform spines likely indicate a design for the existence of multisynaptic sites with microdomains/nanodomains that would interact when modulating information processing (Chen and Sabatini 2012; Stewart et al. 2014; Reberger et al. 2018). Multisynaptic spines would also relate to homeostatic plasticity and modulation of synaptic activity and demand levels (Wefelmeyer et al. 2016). Human spines of these types display a multitude of complex aspects with various stalk diameters, convoluted shapes, filamentous or thick parts, and various bulbous parts or endings (Dall´Oglio et al. 2015; Correa-Júnior et al. 2020; Rasia-Filho et al. 2021; Fuentealba-Villarroel et al. 2022; Guerra et al. 2023; note the aspect of spines in the "unipolar" neuron in Vásquez et al. 2018).

Because neighbor spines of varying shapes and sizes exist in the same dendritic segment, the morphological heterogeneity of spines even in a small portion of the dendritic shaft supports the possibility that synaptic strength is regulated at the level of every single spine (Frick and Johnston 2005; Arellano et al. 2007a, b, Chen et al. 2011; Lee et al. 2012; Rasia-Filho et al. 2021). These spine features provide a high computational capacity and activity-dependent regulation of synaptic strength for each neuron. Moreover, the presence of different spines in human pyramidal neurons is consistent with recent theories of synaptic learning. These synapses, which show gradation in states and are connected by plastic transitions, would confer an increase in storage capacity in neural networks (Lee et al. 2012; Dall'Oglio et al. 2015 and references therein). In humans, the synaptic processing of sensory, motor, emotional, thinking, and cognitive information reached a higher level adapted for our species-specific social behaviors. The presence, distribution, density, and shape of these spines are clear indications of neuronal connectivity (Cooke and Woolley 2005; Chen et al. 2011) with varied plasticity in each brain area (Toni et al. 1999; Hayashi-Takagi et al. 2015; Mohan et al. 2015; Bucher et al. 2020) and improved structural and encoding capabilities properties. These features would also relate to single-cell RNA-sequencing datasets that revealed particularities in gene expression, morphology, proportions, and laminar distributions of cell types in our cerebral cortex (Hodge et al. 2019) and species-specific differences in key molecules that regulate synaptic plasticity (Beed et al. 2020).

Axodendritic and axospinous synapses coexisting on the same dendritic segment and at different distances from the soma modulate the resultant neuronal excitability (Megías et al. 2001; Kubota et al. 2007; Spruston et al. 2013; Bucher et al. 2020). The phylogenetic development of spiny neurons has an evolutionary value in terms of increased connectivity and integrated functions by providing more computational possibilities and increasing complexity for synaptic processing in assembled cells. Spines add more plasticity to synaptic transmission, serving as time-space encoding and decoding devices for a moment-to-moment modulation of information processing. This kind of activity can be different along the human lifespan.

9.3 Ontogenetic Development and Changes in Dendritic Spines in Humans

Ontogenetic effects were reported for human cortical structure and, specifically, for dendrites and spines. Age relates to the development and dynamic reorganization of neuronal circuitries and synaptic connectivity (Jacobs et al. 1997; Dickstein et al. 2007; Petanjek et al. 2011, 2019; see Chap. 4 in this book). In this regard, the human cerebral cortex shows neoteny (delayed development) and heterochrony (different times for maturation) in circuits related to higher function elaboration (Geschwind and Rakic 2013; see additional data in Leopold et al. 2019). Large-scale networks involving brain connectivity and function have been recognized in the infant's brain (or their earliest forms, Smyser et al. 2010) to be matured with age and further changed in the elderly (Bagarinao et al. 2019 and references therein). Pyramidal neurons in cortical multimodal areas receive and process afferences from both local cortical non-pyramidal cells and from a broad range of synaptic inputs at higher association levels of integrative processing. These cells have longer and more branched dendrites as well as more spines than in areas that process a specific modality of activity (Jacobs et al. 2001; González-Burgos et al. 2019; Kolb and Whishaw 2021).

However, the study of the human brain's ontogenetic morphological and functional features is one of the most challenging tasks for our species. This is because many variables would modulate the fine-tuned shape and activity of neurons from the beginning of our development and continuing along the lifespan. Neuroglial plasticity relates to each personal history of life and culture, and this is valid for approximately 8 billion people living today. We all belong to the same species, and there is abundant genetic variation within humans making individuals display particular phenotypes and abilities (Mayr 2001). Neuroanatomical phenotypes can be partly heritable as well (Panizzon et al. 2009; Kremen et al. 2010). For example, from a large study in 51- to 59-year-old male twins, approximately 70% of the variance in the size of subcortical regions is determined by genetic factors (Kremen et al. 2010). The cortical thickness of prefrontal areas was among the most highly heritable, although individual-specific and not shared environmental factors can account for over 50% of the variance in the thickness of cortical regions (Kremen et al. 2010).

Cortical gray and white matter thinning and thickening change in different ways and regions over the course of children's development[9] and into old age (Kolb and

[9] "In the course of studying changes in gray matter, it has also been possible to distinguish group differences between healthy children and those displaying neurodevelopmental disorders (e.g., Giedd and Rapoport 2010). Fortunately, cortical abnormalities in children with neurodevelopmental disorders may not be permanent. Shaw et al. (2007) followed about 500 children, some typically developing and others with attention-deficit/hyperactivity disorder (ADHD). They found that reduced volume of gray matter in the prefrontal cortex was not permanent but rather reflected a delay in cortical development by about 2 ½ years, suggesting that ADHD is characterized by a delay rather than a deviance in cortical development... The association between delayed cortical

Whishaw 2021). A reduction in gray matter volume begins at 6–7 years of age and continues through adolescence, while white matter tracts and volume enlarge in the same period, likely related to neuron and synaptic pruning for improved functions in more efficiently connected circuits (Kolb and Whishaw 2021). As part of a brain developmental route, progressive changes with a reduction in gray matter density begin in primary areas (dorsal parietal and sensorimotor regions) and spread to secondary (spatial and language skills maturing at 11–13 years) and tertiary regions, such as the prefrontal cortex, maturing in late adolescence and continuing into adulthood (Kolb and Whishaw 2021). Later, in the human Heschl gyrus, *planum temporale*, primary visual cortex, gyrus parahippocampus, anterior insula, amygdala, and hippocampus, gray matter thickness decreased significantly with aging without differences between the left and right hemispheres (Profant et al. 2020). It is assumed that parallel changes in neuronal morphology accompany the reduction of the neuropil in each area.

Moreover, developmental changes in cortical volume vary by sex (Raznahan et al. 2011). The cortical surface area, which reflects complex interactions between brain size-related changes in cortical surface (or convex hull area), the degree of cortical sulcation (calculated as a gyrification index), and the area of cortex hidden in sulci may be adjusted in our species (Raznahan et al. 2011). Brain volumes can be larger in women in the prefrontal and medial paralimbic cortices (precentral gyrus, frontoorbital cortex, superior frontal, and lingual gyri), whereas large volumes in the medial and frontomedial cortex, angular gyrus, amygdala, and hypothalamus can be found in men (Goldstein et al. 2001; Kolb and Whishaw 2021). Interestingly, when assessed with near-infrared spectroscopy, the human prefrontal cortex responds to social/emotional facial expressions and shows sex differences for stimulus-induced selective activation of the cardiac activity (Fogazzi et al. 2020). Males and females respond differently to happy, disgusted, and fearful facial expressions, and higher sympathetic and lower parasympathetic activity occurred in young women when consciously and unconsciously processing negative emotions (Fogazzi et al. 2020).

9.3.1 Human Dendritic Spines Change from Prenatal to Elderly

The development, extension, pruning, and remodeling of human dendrites and spines occur prenatally and postnatally (see details in normal and pathological conditions below and illustrations in Fig. 9.25 in this chapter). Dendritic spines are absent or exist at very low density when cortical pyramidal neurons are developing

development and ADHD is a novel hypothesis that will guide both research and treatment for the foreseeable future" (Kolb and Whishaw 2021). It is also open to determine the role of dendritic spines in this condition.

in the human fetus, while the spine number increases and spine shapes change gradually along the final of the gestational period and after birth (Feldman 1984 and references therein). When differentiating into this specific neuron type, pyramidal cells begin the process of growing dendrites and the axon to establish synapses with other cells; dendrites emerge as simple processes and then ramify, and dendritic branches begin to form spines (Cajal 1909–1911; Kolb and Whishaw 2021).

Morphological data unraveled the gradual development of cortical spines in the human brain. The period of pyramidal dendritic differentiation and development in the visual cortex of the human fetus occurs between the 6th and 8th month of gestational age, which is relatively late when compared to the motor cortex (Feldman 1984). Distinct sequences of cortical ontogenesis follow the arrival of different afferent fibers to the developing cerebral cortex. The integration of incoming fibers and local cellular connectivity induces the formation and organization of the cortical layers and the development and maturation of cortical efferent neurons (Marín-Padilla 1970; see the pattern and distribution of axons around the cell body and proximal dendrites of developing pyramidal cells in Marín-Padilla 1974). Pyramidal cells show initially only a rudimentary smooth apical dendritic shaft, although the visual cortex can present visually evoked potentials at this stage (Purpura 1975b; Feldman 1984). At the 8th month, developing Meynert cells show lengthened apical dendrites, various basal dendritic sprouts, and abundant dendritic spines (Purpura 1975b; Feldman 1984). At this point, human spines are a "long, thin process, often with conspicuous varicosities (whose) trajectory may display prominent kinks or bends" (Feldman 1984). In the motor cortex, a small number of such thin processes can be observed as early as the 5th month, mushroom and stubby spines will appear later, but long thin spines still predominate (see additional data in Chap. 4 in this book).

Spines are evident in supragranular pyramidal and bitufted neurons of the precentral gyrus in a 1-month-old human infant (Cajal 1909–1911). However, the development of mature types of short spines will advance into postnatal life (Purpura 1975a, b; Feldman 1984; see also the postnatal development of spines in layer III pyramidal neurons in the monkey prefrontal cortex in Anderson et al. 1995). The dendritic fields of pyramidal neurons display progressively more branches and gradually occupy more space in the neuropil of the human cerebral cortex. This finding occurs around Broca's area at about 2 years of age, a finding that parallels the development of language in our species (Lenneberg 1967; Kolb and Whishaw 2021; see also the cytoarchitectural features of the orbital and frontal inferior gyri in a 6-year-old child in Fig. 4 from DeFelipe 2011). During childhood and puberty, there is an overproduction of dendritic spines in the human prefrontal cortex, remaining high until the third decade and reducing afterward (Petanjek et al. 2008; Sedmak et al. 2018). In adulthood, there are more spines at intermediate distances between proximal and most distal branches in both apical and basal dendrites of layer III pyramidal neurons of the adult cingulate cortex (Benavides-Piccione et al. 2013). No systematic variation in spine morphologies is evident along the dendritic distance from the soma (Benavides-Piccione et al. 2013). Furthermore, adult human pyramidal neurons may depart from the general description that proximal dendritic segments are spine-free areas, showing spines distributed even at proximal primary shafts (e.g., along the initial 50 µm from the soma; Luengo-Sanchez et al. 2018; Rasia-Filho et al. 2021; see also the presence of these spines in pyramidal-like neurons below).

Throughout the human lifespan, neocortical areas show age effects on dendrites and spines in a nonuniform manner. Importantly, impairments in the complexity of spiny dendrite arborization, dendritic length, and spine numbers during normal aging and in neurodegenerative diseases can occur due to distinct mechanisms (Dickstein et al. 2007; see also a recent review by Aguilar-Hernández et al. 2023). Some spiny neurons may be well preserved in aged (>80 years) persons with adequate cognitive performance (Buell and Coleman 1981; Gefen et al. 2018). For example, Golgi-impregnated layer II pyramidal neurons from the parahippocampal gyrus (between the occipitotemporal sulcus and the apex of the gyrus) were studied in adults (mean age 51 years) and normal-aged individuals (mean age 79), and both were compared to patients with senile dementia (SD, mean age 76; Buell and Coleman 1981). Differences among the studied groups were found in apical and basal dendrites. Normal-aged individuals showed longer apical and basal dendrites and more branched apical dendrites. The greatest differences between groups were seen in the apical trees' terminal segments at distances intermediate from the soma, rather than at the proximal or distal extremes of the dendritic tree. Shrunken and atrophic dendritic trees were found in all cases (more in SD ones). These data indicate that the normal aging cortex contains regressing and dying neurons although surviving and growing neurons predominate normally (Buell and Coleman 1981).

Of individuals aged 69–102 years, six out of nine of them with clinical signs of senile deterioration, hippocampal pyramidal neurons and dentate gyrus granule cells showed changes correlated more clearly with the degree of antecedent psychomotor pathology than with chronologic age (Scheibel et al. 1976). Accordingly, in both cell types, spine loss followed a progressive if patchy course, peripheral dendritic segments usually denuded first, and basal dendritic shafts were affected before the degeneration was found in the apical shaft. Another scenario was found in the adjacent entorhinal cortex. In patients with age of 91, 92, and 102 years with mild-to-moderate emotional and intellectual changes, there was a relative persistence of a relatively high number of dendritic spines in pyramidal neurons until late in a degenerative fragmentation of the apical shaft (Scheibel et al. 1976).

In the human neocortex, basal dendrites of supragranular pyramidal cells were studied in the superior temporal gyrus (Wernicke's area) of both hemispheres and from men and women (18–79 years; Jacobs and Scheibel 1993). Data showed an age-related decrease in total dendritic length making that the interhemispheric dendritic asymmetries (greater values in the left hemisphere) were no longer significant in individuals over 50 years of age (Jacobs and Scheibel 1993). Data were also obtained from Golgi-impregnated neurons in the left prefrontal (supramodal frontopolar region, BA 10) and occipital (visual association region, BA 18) cortex. Regional differences and age-related changes in the spine density of basal dendrites of supragranular (most in upper layer III) small- to medium-sized pyramidal neurons were studied in neurologically normal individuals ranging from 14 to 106 years old (Jacobs et al. 1997). Total dendritic length and dendritic spine number (summing all spines on dendritic segments) were higher in BA 10 than in BA 18. Dendritic spine density was greater in distal over proximal segments. A decrease of approximately 10% in total dendritic length values with age was found in both areas. There was a marked age-related decrease in dendritic spine density. Values showed a higher decrease until the 4th decade, providing an average of 46% fewer spines in both cortical areas from the younger group to the older group (Jacobs et al. 1997; Fig. 9.5). Aging also relates to dendritic atrophy, which starts approximately at 50 years of age, and loss of synapses in the posterior superior temporal gyrus (Anderson and Rutledge 1996). In this area, age was also negatively correlated with the number of primary basal dendrites, the total number of dendritic endings/branches, the total dendritic length, and the number of spines in supragranular pyramidal neurons in men from 21 to 71 years (Anderson and Rutledge 1996).

Likewise, data obtained from over 8,900 3D reconstructed spines along 6.35 mm of main apical and basal dendrites of layer III pyramidal neurons in the cingulate cortex (BA 24) showed differences when compared two men aged 40 and 85 (Benavides-Piccione et al. 2013). Spine density was higher in apical dendrites than basal dendrites in both cases. However, in the older individual, basal dendritic spine density (at approximately 70–130 μm from the soma) was significantly lower than in the middle-aged individual, with the density increasing to 1.2–1.3 spines/μm near 90 μm and remaining similar along intermediate to distal dendritic length. In the middle-aged individual, dendritic spine density increased to a maximum of nearly 1.8 spines/μm at 110 μm and slightly decreased along the remaining length of the basal dendrite. In apical dendrites of the middle-aged person, spine density slightly increased to a maximum of 4.3 spines/μm, whereas values were 2.2 spines/μm (both at 160 μm) in the older person (Benavides-Piccione et al. 2013). Apical dendrites from an individual at 40 years had longer spines than those in basal dendrites, a finding not found at 85 years. Moreover, a significant reduction in spine densities was found for the aged individual, with a loss of small and short spines in basal dendrites and long spines in apical dendrites (Benavides-Piccione et al. 2013).

In the human motor cortex of 14- to 96-year-old individuals, basal dendrites of both layers III and V pyramidal neurons decreased in number with advanced age, with marked differences between old and young groups in the deeper layer (Nakamura et al. 1985). Apical dendrites of layer III pyramidal cells in the human motor cortex show an age-related loss of spines (Watanabe 1981). In addition, in patients aged 74–102 years, layer V giant pyramidal neurons (Betz cells) of the motor cortex show aging and senescence changes represented by progressive loss of dendrite spines and shortening, disruption, and progressive disappearance of the normal long and circumferentially branched basal dendrites (Scheibel et al. 1977). From seven patients studied (three of them reported having mild or moderate senile changes), 75% or more of Betz neurons showed alterations by the eighth decade of life compared to less than 30% of the more numerous surrounding non-Betz pyramidal cells (Scheibel et al. 1977). These data would be related to motor impairments related to aging in humans. Comparatively, in the primary motor cortex of aged mice, apical

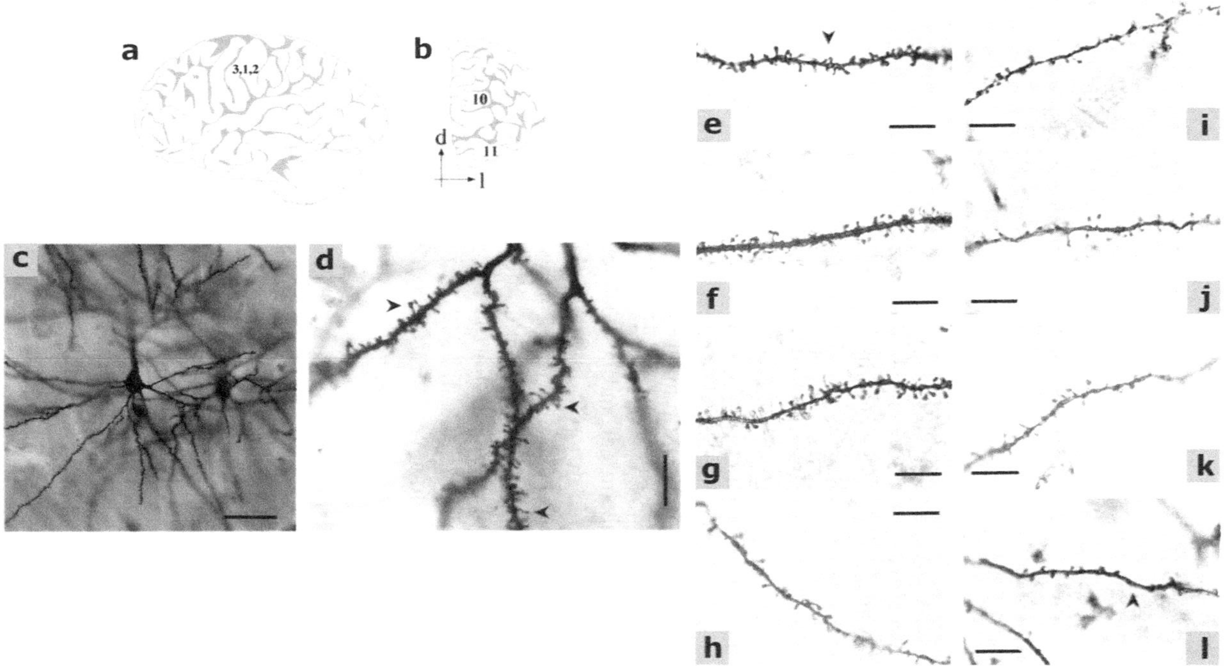

Fig. 9.5 (**a**) Lateral and (**b**) frontal views of the human left hemisphere illustrating the relative position of Brodmann's areas (BA) 3, 1, and 2 considered "low-integration regions" (in **a**), and BA10 and 11 classified as "high-integration region" (in **b**). d dorsal, l lateral. (**c**) Photomicrograph of Golgi-impregnated supragranular pyramidal neurons from the BA11 and (**d**) from BA 3-1-2. Note the varied number of spines with different shapes and sizes along the dendritic length, some with long and thin forms (arrowheads in (**d**)). Images (**c**, **d**) are from an 11-year-old child. Scale bar = 50 μm in (**c**) and 10 μm in (**d**). (Legends adapted and figures reprinted from Jacobs et al. (2001) under CCC RightsLink® license #5384350157140, originally published by Oxford University Press). (**e**–**l**) Photomicrographs of dendritic spines from the BA10 of four subjects younger than 50 years of age [14-year-old male (**e**), 23-year-old male (**f**), 32-year-old female (**g**), and 48-year-old female (**h**)] and four subjects older than 50 years of age [56-year-old male (**i**), 73-year-old male (**j**), 81-year-old female (**k**), and 106-year-old female (**l**)]. Note the general decrease in dendritic spines with increasing age (arrowheads in (**e**) and (**l**)). Scale bar = 10 μm. (Legends adapted and figures reprinted from Jacobs et al. (1997) under CCC RightsLink® license #5384350780357, originally published by John Wiley & Sons, Inc)

dendrites of layer V pyramidal neurons may be in a continuous state of instability and attempts at compensation, exhibiting an increased spine number with elevated turnover, short-term stabilization, and decreased survival (Davidson et al. 2020).

It remains to be determined if similar processes found in mice also occur in the motor cortex and associated areas in our species. If so, it will be important to identify at which point the net result of spine formation and removal tends to be a progressive age-related reduction in spine density affecting motor abilities, executive functions, cognitive performance, and social behavior display. To probe causal relationships is still challenging in humans because multi-faceted spine functions depend on the neuronal subpopulations, intraspine subcellular molecular and structural composition, and the multiple synaptic demands in different networks with short- or long-term demands, stability, or ongoing activity dependent plasticity. Notably, pyramidal neurons show a reduction in dendritic length and complexity with additional changes in spine number, shape, and size along the human lifespan. The ontogenetic, morphological, and functional impact of these dendritic spines' changes on their corresponding synaptic sites can vary and be site-specific. For example, brain structural and functional impairments in aged individuals would relate to various behavioral deficits even in those persons with no neurodegenerative diseases; at the same time, some cognitive and emotional regulation would remain intact (Bagarinao et al. 2019 and references therein).

For emotional processing, subcortical and cortical regions crucial for the flow of information transfer among different functional brain networks ("connector hubs") would be affected by aging (Di Lorenzo Alho et al. 2016; Bagarinao et al. 2020). Age-related changes would alter the brain's intrinsic connectivity and overall connection strength for the whole-brain, short-range, and long-range connections (Bagarinao et al. 2020). Hub regions negatively associated with age include the following: (1) the medial precentral/midcingulate gyri and insula for both short-range and long-range connections' strength, and (2) the angular, middle frontal, and posterior cingulate gyri for long-range connections´ strength. Increases in whole-brain connectivity with age were found in the caudate, thalamus, cerebellum, hippocampus, and temporal pole (Bagarinao et al. 2020). These areas, which are also part of multiple large-scale brain networks (e.g., related to default mode, primary processing, visuospatial, salience, and executive control) are relevant for healthy aging and may be impaired in neurodegenerative diseases that cause progressive deficits in integrated brain functions.

Aging can alter the connectivity of the brain (Bassignana et al. 2022) and, possibly, the cortical neuropil structure and the spines' participation in synaptic transmission within circuits. Nevertheless, the pattern and impact of age-related reorganization of large-scale functional networks would account for individual variability in general cognitive performance (Bagarinao et al. 2019). It is currently open to be determined whether the remodeling of dendritic geometry, spine morphology, and connectivity would serve as a kind of adaptive response to progressive aging effects in humans (Walker and Herskowitz 2021). These structural changes might compensate, within certain possible degrees and considerable interindividual variation, the "wear and tear" burden on the rate of brain aging upon network and neuronal functioning (McEwen 2007; Gefen et al. 2018; Rasia-Filho et al. 2018; WHO 2022).

In this regard, recent functional magnetic resonance imaging (fMRI) and multilevel data analyses from healthy adult men and women (from 21 to 86 years) helped to examine different spatial and functional organization of large-scale resting state networks in the human brain across the lifespan (Bagarinao et al. 2019). Results showed that (1) all canonical resting state networks, except the left and right executive control and dorsal default mode networks, negatively correlated with age; (2) aging relates to weaker within-network and higher between-network connectivity, indicating the reorganization of functional patterns toward more integrated network topology and increased global efficiency;[10] (3) the integrity of primary processing networks including visual and (lateral) sensorimotor networks, as well as the visuospatial network, would contribute to normal cognitive performance; and (4) connectivity among higher-level cognitive networks (including executive control, salience, basal ganglia, and dorsal default mode networks) was minimally affected by age-related connectivity alterations, which would be also relevant for individuals preserving general cognitive performance during healthy aging (Bagarinao et al. 2019).

9.4 Dendritic Spines in Human Cortical and Subcortical Areas Show Heterogeneous Morphological Features

Human cortical neurons are neither simply scaled-up versions of mammalian neurons nor generalizable building blocks of cortical networks between mammals (Mohan et al. 2015; Luebke 2017; Beaulieu-Laroche et al. 2021; Schmidt and Polleux 2022). Multiple evolutionary changes would have increased the number of neurons in the mammalian cerebral cortex and affected the average neuronal cell size and further elaborations of the dendritic and axonal arborization (Herculano-Houzel et al. 2014; Herculano-Houzel 2019). The specialization of our cortical circuits appears to be critical for our higher or more abstract mental abilities (DeFelipe 2011). This is a process involving a genetic background, a subcellular molecular and cellular structural and functional organization associated with learning-related plasticity, and network improvements (Nakahata and Yasuda 2018; Schmidt and Polleux 2022). Hence, neuron morphology relates to connectivity and synaptic patterns with nonrandom wiring in the cerebral cortex (Udvary et al. 2022).

[10] "Topologically direct interconnections between spatially remote brain regions will increase the efficiency of information processing, which is expected to yield benefits in terms of adaptive behavior. Brain networks can therefore be said to negotiate an economical trade-off between minimizing physical connection cost and maximizing topological value" (Bullmore and Sporns 2012).

It has been difficult to synthesize a large number of details into principles for understanding the brain (Leopold et al. 2019). Indeed, recent approaches evidenced the complexity of a connectomic study obtained from a rapidly preserved human surgical fragment of the temporal cortex (1 mm^3) imaged using a high-speed multi-beam scanning electron microscopy (EM) and 3D image reconstruction (Shapson-Coe et al. 2021). This petascale connectomics dataset contained 57,216 cells and nearly 133 million synapses in a 1.4-petabyte volume. Most frequently, there were excitatory spiny neurons (69% of the neuronal population), excitatory synapses (76%), and the synaptic drive onto spiny neurons was excitatory (70%; Shapson-Coe et al. 2021). In another study using a realist model and rodent data, dense digital reconstruction of a 0.3 mm^3 cortical circuit exhibited approximately 31,000 neurons of 55 morphological cell types and 37 million excitatory and inhibitory synapses (Gal et al. 2021). Note the number of spiny neurons and the amount of synapses existing in a cortical minuscule portion (see also the discussion on how to analyze these complex data in Kasthuri et al. (2015)), which needs to be considered when developing computational models of human circuits function.

Not all human neurons show dendritic spines, as will be described below (Braak and Braak 1984a, b, 1986). On the other hand, spiny neurons are described throughout the human CNS,[11] from the spinal cord to high-order cortical areas. Morphological heterogeneity within each neuron type may exist for the dendritic branching pattern and spine density. Nevertheless, dendritic spines are usually in a *continuum* of shapes and sizes in these neurons. Pleomorphic (i.e., with different forms) spines may vary in number and distribution along dendritic segments in the same neuron. Considering the particular connectivity of each area, both neighbor and relatively distant spines have a crucial function in the synaptic establishment and information processing relevant to each local microcircuit and to large networks (Chen et al. 2011; Rochefort and Konnerth 2012; Kasthuri et al. 2015).

Jacobs et al. (2001) provided examples of the likely relationship between neuronal morphology and information processing. These authors studied Golgi-impregnated basal dendrites and spines of supragranular pyramidal cells in the left primary cortex (somatosensory, BA3-1-2; motor, BA4), unimodal cortex (Wernicke's area, BA22; Broca's area, BA44), heteromodal cortex (supplementary motor area, BA6β; angular gyrus, BA39), and supramodal cortex (superior frontopolar zone, BA10; inferior frontopolar zone, BA11). Primary and unimodal areas were

[11] There are open questions about the density and types of dendritic spines - if they indeed exist - in the morphologically heterogeneous populations of enteric neurons in our species. While submucosal neurons can be nondendritic, myenteric cells display around 20 very short dendrites arising from the cell body. These neurons were named Dogiel "spiny" (or with a "thorny" aspect) type I neurons. Considering the axonal and dendritic shapes, other Dogiel types I neurons were classified as "stubby," with short and partly stubby or partly lamellar dendrites, or "hairy" cells, with short and very thin dendrites. Dogiel type II neurons can show up to 16 dendrites, but these processes "look like" and behave as axons conducting APs. Types III and IV neurons are long dendritic, uniaxonal neurons. Type V neurons appear as unipolar cells with a single stem process projecting from the cell body. From this primary shaft, the single axon and various long, branched, tapering dendrites emerge (Brehmer 2021).

considered "low-integrative regions," whereas heteromodal and supramodal areas were "high-integrative regions." Morphological parameters evaluated were total dendritic length, mean segment length, dendritic segment count, and dendritic spine number and density. Even with interindividual variation, data from regions related to early stages of cortical sensory processing (primary cortex) and unimodal ones were consistently less complex than in those regions involved in the later stages of information processing (heteromodal and supramodal ones). In other words, total dendritic length and dendritic spine number in BA10 were greater than that in BA3-1-2 (31% and 69%, respectively), which indicates a broader sampling of afferent information upon dendrites and spines in the higher-order prefrontal area (Jacobs et al. 2001). A large dendritic arbor would enable further processing and provide more associative and learning properties. For example, in the monkey frontal eye field, the need to integrate large numbers of inputs for purposeful eye movements relate to basal dendrites of layer III pyramidal neurons. These neurons show the most complex branching pattern in the visual cortex and nearly 30% more dendritic spines than cellular counterparts in the parietal eye field areas (Elston and Rosa 1998).

Moreover, the basal dendrites of supragranular pyramidal neurons from the left human anterior insular gyri (the secondary gyrus brevis and the precentral gyrus) and posterior insular gyrus (the postcentral gyrus) revealed a specialized structural organization for this type of heteromodal cortex (Anderson et al. 2009). The insular cortex contains transitional zones from agranular–granular cytoarchitectonics, that is, from primitive allocortex to full-fledged isocortex along its anteroposterior extension. These insular portions have extensive interconnections with multisensory cortices, limbic structures (including amygdaloid nuclei), hypothalamic and brainstem nuclei to integrate multimodal interoceptive, emotional processing, and visceral reactions and to elaborate a "sense of self by relating the internal milieu with the external world" (Anderson et al. 2009 and references therein; see also Mesulam and Mufson 1982; Wattendorf et al. 2016). Interestingly, the human insular secondary gyrus brevis was more complex in terms of total dendritic length, dendritic segment count, and high dendritic spine number than either the precentral or postcentral insular gyri (Anderson et al. 2009). Compared to other neocortical areas, insular total dendritic length value was less than those found in heteromodal and supramodal "high-integration" areas (BA 6β, 10, 11, and 39), but greater than those in primary and unimodal "low-integration" areas (BA 3-1-2, 4, 22, and 44). The insular dendritic spine number was higher than both low- and high-integration regions. Greater values were found in the secondary gyrus brevis compared to both low and high-integration regions, whereas the precentral and postcentral insular gyri had more spines than the low-integration regions and similar values compared to the high-integration regions. Therefore, along with its anteroposterior and connectional organization, insular pyramidal cells may be a type of supragranular neuron with shorter dendrites provided with more spines to process convergent heteromodal stimuli (Anderson et al. 2009). It is important to consider that insular neurons can have particular synaptic integration strategies for very complex stimuli in our species (Anderson et al. 2009; Wattendorf et al. 2016; see also Khan et al.

2023). The emergent properties and synaptic processing executed by these circuits, synapses, spines, and dendrites may be one of our most complex. For example, the human insula is strongly activated when "you see the person you are in love with, try to listen to your own heartbeat, suffer from a headache, or crave for a chocolate cookie" (Gogolla 2017).

9.4.1 *Multiple Possibilities for Modulation of Human Dendritic Spines*

There are complex issues and implications for the study of brain areas and functions in humans. Due to clear ethical and technical reasons, we do not have the same amount of information about human dendritic spines as currently available from other species. Some reports on humans were also limited to the number of cases and kind of samples studied. However, descriptive morphological data constitute the foundation for further understanding of the nature of the human brain, its cellular and circuitry´s functional organization for the display of complex behaviors (DeFelipe 2022). This kind of knowledge forms the basis for further progress on neuroanatomy with a potential clinical implication. They can also be linked with precise imaging techniques and current genetics approaches (Grasby et al. 2020; Yuste et al. 2020; Kiwitz et al. 2022).[12]

From technical reasons to marked neuroanatomical differences in development and ultimate identification of borders, some areas impose more difficulties than others do for their study in humans. For example, evident phylogenetic discrepancies

[12] Important current data are available from the following sources: (1) cellular-resolution atlases of the human brain (e.g., Ding et al. 2016; Mai et al. 2016); (2) human brain anatomy using modified Brodmann or gyral annotation and gene expression data in 3D images (Allen Brain Atlas, https://human.brain-map.org/static/brainexplorer; https://atlas.brainmap.org/atlas?atlas=265297126#atlas=265297126&plate=112360888&structure=10390&x=40320&y=46976&zoom=-7&resolution=124.49&z=3; and, https://atlas.brainmap.org/atlas?atlas=138322605#atlas=138322605&plate=112360888&structure=10390&x=40320&y=46976&zoom=-7&resolution=124.49&z=3); (3) atlas of transcriptional features of the mid-gestational human brain (Miller et al. 2014); (4) segmentation of volumetric brain MRIs of infants (first 2 years of life; de Macedo Rodrigues et al. 2015); (5) connectomic study of adult human cerebral cortex on a petascale fragment (e.g., Shapson-Coe et al. 2021; available to peruse online and accessible with the Neuroglancer browser interface); (6) multimodal cell census and atlas of the mammalian primary motor cortex, including human data, and integrating neuronal multi-layered molecular genetic and spatial information with multi-faceted phenotypic properties (BRAIN Initiative Cell Census Network (BICCN) 2021); (7) the "BigBrain" model based on human cell body-stained and 3D-reconstructed sections as an anatomical brain model at a spatial resolution of 20 μm to be associated with ultra-highfield fMRI (with "possibility to measure laminar brain activity as well as identifying functional subdivisions of subcortical and cortical structures"; Kiwitz et al. (2022), and references therein; see also https://interactive-viewer.apps.hbp.eu/, https://ebrains.eu/, and https://ebrains.eu/service/voluba/); and (8) the methodological approach for studying synaptic and non-synaptic profiles of nearby axons and dendritic spines in Kasthuri et al. (2015), for example.

exist for the human hypothalamus and its subdivisions (Saper 2012).[13] Let us also consider the following:

1. It may be more difficult to discern a clear relationship between structure and function in cortical associative areas (Scheibel et al. 1990). That is because there can be a large degree of interindividual variation in dendritic structure complexity in these areas. It is important to have extensive information about the life history, the nature of the work and other social activities, and personal abilities of individuals to describe who serve as subjects for this type of study (Scheibel et al. 1990; Jacobs et al. 1993; see how the functional architecture of the brain is modulated by literacy in Petersson et al. 2001). In addition, there are functionally phenotypic subtypes of pyramidal neurons within cortical layers (Berg et al. 2021; Moradi Chameh et al. 2021; Planert et al. 2021) not detected by histological techniques. Other locally non-spiny neuronal types coexist intermingled with pyramidal cells and do modulate large network inputs and local microcircuitries, influencing selectively dendrites and spines across cortical layers (Kubota et al. 2007, 2016).[14]
2. Cortical areas are currently recognized as a "mosaic" with higher interconnectional complexity, functional networks, and multiple cortical finer-grained parcellations than previously thought (Glasser et al. 2016; Kolb and Whishaw 2021). The neuronal structure is influenced by both local and extrinsic connectivity, and then, the connectional strength is very likely affected and can be reciprocally affected by the dendritic geometry (length, diameter, and branching pattern) associated with all dendritic spine morphological and functional features. Rather than a simple junction of one cortical module with another, there can be a dynamic interplay with feedback and feedforward loops between and among the operation of different subcortical and cortical regions (Kolb and Whishaw 2021). This kind of network organization suggests that the topography of a cortical area and specific functions can be variable between individuals at the same time that subpopulations of neurons with different features can be intermingled within the same area (Anderson and Rutledge 1996). For instance, the face area of the motor cortex also contains neurons that innervate the adrenal

[13] Immunohistochemical characterization of hypothalamic neurons and description of species differences in neuroanatomical subdivisions can be found in Saper (2012). Examples of Golgi-impregnated spiny neurons in hypothalamic nuclei obtained from other species were depicted by Cajal (1909–1911), and the history of neuroendocrinology since the descriptions in *De humanis corporis fabrica* by Vesalius can be found in Kreier and Swaab (2021). See data on pituitary alterations in humans in Baltazar-Gaytan et al. (2019).

[14] For example, horizontal cells in layers I and II, double bouquet cells in layers II and III, chandelier and neurogliaform cells in layers II to IV, basket cells in layers III and IV, spiny stellate cells in layer II and aspiny stellate cells in layer IV, Martinotti cells in layer VI, fusiform stellate cells in layer VI, and a varied group of pleomorphic neurons identified as "modified pyramidal cells" in layer VI, among other varieties of non-spiny or sparsely spinous non-pyramidal heterogeneous cells found in the cerebral cortex of various species (see further data in Lodato and Arlotta (2015) and Kubota et al. (2016) for the complexity of these local cells; and, in addition, Braak 1980; Feldman 1984; Wahle 1993; Pearson 1995; Rudy et al. 2011; Leopold et al. 2019).

medulla, which may modulate sympathetic responses elicited in concert with facial emotions (Leopold et al. 2019).

Subtle but distinct neural processing may coexist in the same area. Cortical responses with laminar synaptic elaboration and cellular functioning differ if a hand movement is executed or just imagined (Persichetti et al. 2020)[15]. In this regard, the precentral cortical gyrus contains multiple and partially overlapping sites whose activation occurs for the motor programming of the wrist and single finger movements (Matelli et al. 2004). Cortical cellular modules coexist side-by-side in the human primary motor cortex (M1; Persichetti et al. 2020), and approximately one million axons in each pyramid compose the corticospinal tract in humans (Schoenen and Grant 2004). By testing the responses of the identified human hand-selective region in the M1, executed movements of finger tapping enhanced activity in both the superficial layers (II-III), which receive cortico-cortical input, and the deep layers (Vb-VI) that send output to the spinal cord. When the task was imagining the movement, a laminar-specific evoked response remained restricted to the superficial circuitry of adult individuals (Persichetti et al. 2020). These neuroanatomical and functional data relate not only to extrinsic and intrinsic cortical circuitries but also to plastic and learning effects for neural repetition improvements (i.e., repetition priming) in behavior (Persichetti et al. 2020). Second, in sexually active women (aged 18–45 years), the somatotopically ordered representation of the female clitoris lies in BAs 1, 2, and 3a/3b, dorsolateral in the postcentral gyrus (S1) adjacent to the representation of the hips and upper legs, and not in the mesial wall of the precentral lobe as hitherto considered (Knop et al. 2022). Individual variability of the precise location of the clitoral region, and sensory-tactile stimulation of the clitoral region evoked bilateral neural activations in S1 in most women or, in some cases, activation in either the right or the left hemisphere only. Structural plasticity findings involving local connectivity showed a positive correlation between the cortical thickness of the individually mapped left-hemispheric genital field and the frequency of sexual intercourse within the past 12 months (Knop et al. 2022).

3. The dendroarchitecture and the functional dendritic domains are associated with spine computations for synaptic processing and integration. A higher number of spines in dendrites imply many forms of synaptic establishment and modulation, long-term neuronal plasticity, learning, and memory, varied biochemical cascades, and molecular interactions modulated by the synaptic demands (Yuste 2010; Sala and Segal 2014; Nakahata and Yasuda 2018). Interestingly, differences between the left and right hemispheres exist for morphological features of human cortical neurons. For example, in seven out of nine subjects, supragranular pyramidal neurons of the posterior superior temporal gyrus posterior to Heschl's gyrus showed longer dendrites, more dendritic branch endings, and

[15] See also data on human cortical stimulus-related hemodynamic changes and the modulation made by task difficulty, arousal, and behavioral performance in Oelschlägel et al. (2022) and Burlingham et al. (2022), respectively. Commentaries on results from alert-behaving monkeys can be found in Sirotin and Das (2009) further discussed in Leopold (2009).

more dendritic spines in the right hemisphere than in the left hemisphere, where it corresponds to Wernicke's area (Anderson and Rutledge 1996).

4. There may also exist differences between males and females in dendritic spine number in subcortical areas revealed by experimental and morphological data using rats (e.g., in the medial amygdaloid nucleus, Rasia-Filho et al. (2012) and references therein) or in human cortical neurons. For example, there was a significant interaction between sex and insular gyri for total dendritic length and dendritic spine number. Men showed supragranular pyramidal neurons longer and with more spines in basal dendrites than the corresponding area in women (Anderson et al. 2009). On the other hand, women exhibited slightly greater basal dendritic values and variability in supragranular pyramidal cells in Wernicke's area, related to language comprehension, than men (Jacobs et al. 1993).
5. Finally, it has to be considered that "snapshots" of dendritic spines number and shape likely represent a moment within a dynamic time window for spine formation, stabilization, persistence, or shrinkage (von Bohlen und Halbach 2009; Davidson et al. 2020; Kasai et al. 2021; see also the first chapter of this book). For different neuronal subpopulations or the morphological and functional heterogeneity within the same class of neurons, it is conceivable that subtle differences can exist *in vivo* and involve a variable proportion of intermingled, neighbor dendritic spines (Chen et al. 2011). These spines can be stable or plastic and their number, shape, and composition can adapt to physiological demands. Plastic dendritic spines can show functional selectivity and be sensitive to specific stimuli along the same segment, but spine structure–function couplings of such kind cannot be assessed with current techniques in humans.

We have not reached the ultimate interpretation of the morphological features of all nerve cells in the human CNS. Nevertheless, even considering various technical limitations, there are relevant reports reflecting decades of study of human spiny neurons and their likely integrated functions across CNS areas and functional circuits. Some examples of human neurons will be presented in the following sections with representative figures, maintaining the original description and classification of cells and spines. To be more concise, data on the axonal aspect of each neuron will be omitted most frequently. For additional data, the interested reader is directed to relevant morphological descriptions that can be found in the references indicated in the text.

9.4.2 Spinal Cord

Neurons showing many spines and others with few spines form part of different afferent and efferent pathways in the human spinal cord (Schoenen 1982a, b; Fig. 9.6). It is plausible to consider that neurons that possess more dendritic spines would modulate differently the axonal inputs they receive than those neurons whose

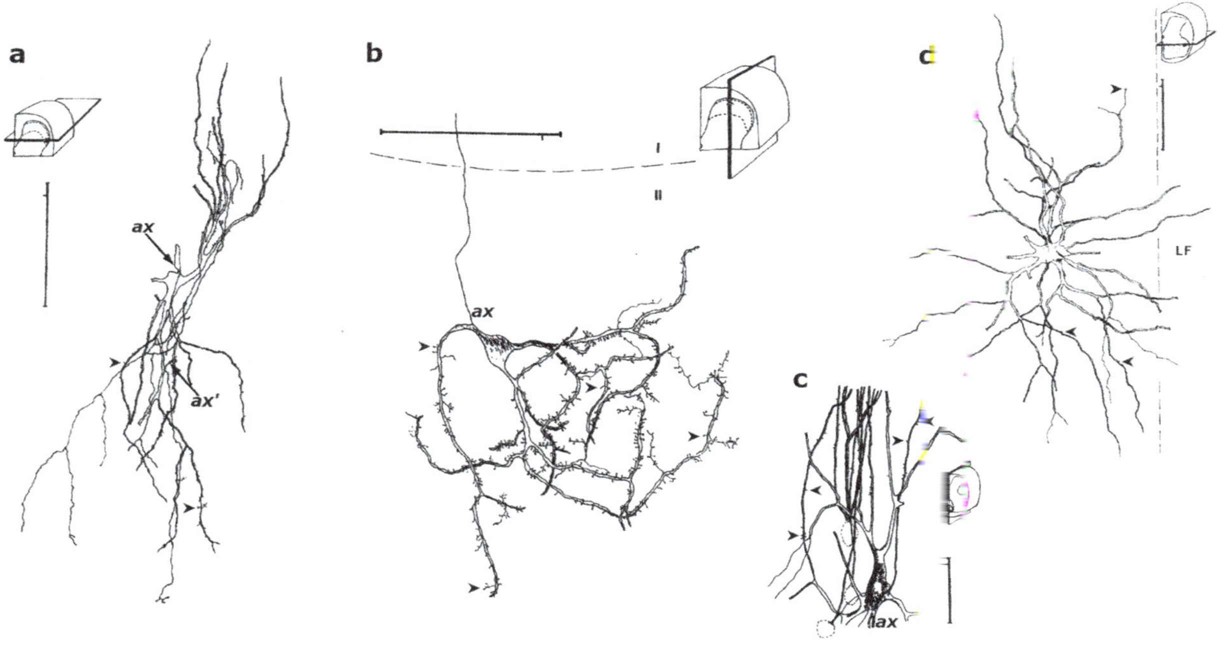

Fig. 9.6 Camera lucida drawings of Golgi-impregnated neurons from the human dorsal (**a**, **b**) and ventral (**c**, **d**) horns of the spinal cord. (**a**) Islet cell in lamina II in the cervical enlargement of a 2-month-old infant. This cell has two axons, one (*ax*) emerging from the perikaryon and another (*ax'*) from a secondary dendrite. Note the presence of a few spines (arrowheads) in this neuron and compare it to those more abundant in a (**b**) curly cell in the lamina II in the lumbar enlargement of a 1-day-old neonate. (**c**) Ventromedial motoneuron in the lumbar enlargement of a 2-year-old child. (**d**) Dorsolateral motoneuron in the lumbar enlargement of a 28-week-old fetus. The impregnation of dendritic spines (arrowheads) was variable in the studied samples. LF lateral funiculus. The plane of cutting where each neuron was observed is indicated by a scheme of the corresponding part of the spinal cord (drawn in perspective at the top of each image). Scale bar = 100 μm (one subdivision at 10 μm is shown). (Legends adapted and figures reprinted (**a**, **b**) from Schoenen (1982b) under CCC RightsLink® license #5384201195677, originally published by Elsevier; and (**c**, **d**) from Schoenen (1982a) under CCC RightsLink® license #5384330581077, originally published by John Wiley & Sons, Inc)

afferents make synapses directly upon spine-free dendritic shafts. Smooth dendritic segments are equally important for synaptic integration (Kubota et al. 2016), and, theoretically, the axodendritic contacts would transform fast and less modulated synaptic inputs into voltage changes in the target dendritic segment. On the other hand, axospiny synapses let the synaptic input more modulated and would balance inputs' strength and plasticity with more possibilities of information processing and integration. It is a counterintuitive possibility to be proved that some circuits need a more direct, rapid, and less modulated effect on dendrites and neuronal excitability, whereas others evoke responses integrating spines' processing for weighing the afferent pattern activity, intensity, and plasticity.

Considering the spinal cord physiology, local afferent pathways coding noxious stimulation can induce motor reflex and hemodynamic responses (Cravo et al. 1995). At the same time, plastic excitatory responses in postsynaptic targeted neurons in the dorsal horn can be induced by the intensity and chronicity of stimulation involving, the neurotransmitter, neuromodulator, and neurotrophic factor released (Basbaum 2021; Gardner 2021). In the ventral horn of the spinal cord, proper and fine-tuned muscle contraction and relaxation are orchestrated by different motoneurons, which accommodate, balance the impact, and compute several ongoing inputs from various central motor control areas (Scott and Kalaska 2021). The morphology of each neuron in the spinal cord reflects the circuitries' properties and synaptic modulation they compute and the cellular role in specific intrinsic or projecting pathways. This rationale applies to other brain areas and essentially reinforces that morphology and function are two sides of the same token, although it has to be demonstrated in detail in each studied area. Even though these concepts may appear obvious, the myriad of neuronal shapes and different circuitries in humans are not indeed.

For example, the morphology of Golgi-impregnated neurons in the human spinal cord was obtained during 26 weeks of gestational age, adult life, and into old age by Shoenen (1982a, b). The dorsal horn has neurons with distinct dendroarchitectonic and neurochemical features, dendrites with preferential orientation in one or more spatial planes, and varied spine densities. Dorsal horn neurons receive superficial, deep, and visceral somatosensory inputs from approximately 2–2.5 million fibers in human adult dorsal roots on each side (Schoenen and Grant 2004). Lamina I contains a uniform population of large, poorly ramified neurons with disk-shaped dendritic arbor and spiny branches projecting ventrally into the adjacent lamina II, some reaching layers III and IV or the white matter. These neurons are moderately covered with "sessile" or "pedicled" dendritic spines, more numerous on the ventral dendrites (Schoenen 1982b).

Four neuronal types compose the human lamina II (the *substantia gelatinosa*): (1) "Islet cells," which display a cylindrical "parasol pine" dendritic tree shape or primary dendrites on two opposite poles of the soma with sparse and irregularly distributed dendritic spines in small bunches (Fig. 9.6a); (2) "filamentous cells," small-sized neurons characterized by multiple filiform spiny dendrites (spines are more numerous on these filiform dendritic branches than on the thick ones); (3) "curly cells" with the most complex, twisted, spine-rich dendrites, and spines with a thin stubby-necked aspect covering the entire dendritic tree, except the initial segments of primary dendrites (Fig. 9.6b); and (4) "stellate cells," which display a simple dendritic tree architecture and various primary straight dendrites (some curved and recurrent) with spine-poor branches extending into laminae I and III, or with branches covered by numerous spines on thin collateral branches, but also irregularly distributed over main shafts of the dendritic tree (Shoenen 1982b).

Lamina III shows a mixed population of "antenna-like cells" with an asymmetric dendritic tree, cone-shaped dendritic aspect, and numerous short-necked dendritic spines. Local "radiate cells" have a radial dendritic tree and spherical territory. Lamina IV neurons are medium- or large-sized "antenna-like cells" with an asymmetric dendritic tree, some reaching the neuropil of lamina II and I, and richly covered with dendritic spines on dorsal dendrites (Shoenen 1982b; see also Sengul and Watson 2012).

These dorsal spinal cord neurons relate to afferent, intrinsic, and efferent pathways. For example, lamina II neurons integrate segmental and suprasegmental influences in the modulation of nociceptive transmission (Sengul and Watson 2012), curly cells have an axon projection directly into lamina I, islet cells might be inhibitory interneurons in nociceptive processing, whereas stellate cells might integrate inputs to lamina II and transfer them to the deeper lamina III to V (Shoenen 1982b). Lamina I neurons receive inputs coming from peripheral nociceptors, which form multiple contacts with local dendrites, as well as other inputs from non-noxious mechanic stimuli. These neurons' axons and those from laminae IV and V compose the ascending sensory spinothalamic tract (Shoenen 1982b; Murray 1995; Schoenen and Grant 2004; Basbaum 2021).

Ventral motoneurons have a particular dendroarchitectonics and organization in columns developed for multiple connections and integrated functions (Shoenen 1982a). One somatic motoneuron, "the final common pathway" of Sherrington, receives the convergence of inputs of approximately 10,000 axon terminals from descending pathways of different sources (Murray 1995). Furthermore, although the motoneuron cell body can be localized into a specific ventral lamina, dendrites can extend into the intermediate zone and the dorsal horn as well as reach the white matter (Murray 1995). This structural organization relates to an enormous receptive field (see drawings of representative spiny neurons from various species in Cajal (1909–1911)). Other motoneurons are propriospinal interneurons, such as most lamina VIII neurons, and coordinate motor activity including long pathways that connect upper and lower spinal cord segments for forelimb and hindlimb movements (e.g., the reciprocal arm and leg swing in walking; Sengul and Watson 2012).

Four motor neuron columns were described in the human lumbar spinal cord (Shoenen 1982a). The ventromedial column motoneurons have one or two thick primary dendrites arising from the dorsal surface of the cell body forming tight bundles, and ventral primary dendrites that are more numerous (three to five), which ramify in a fanlike arrangement and display more spines (Fig. 9.6c). Vertical dendritic bundles are long and narrow-meshed, extending over 2–3 mm in the ventrodorsal axis. Longitudinal dendritic bundles are composed of branches of variable thicknesses and orders from neighboring neurons and from cells several hundred micrometers distant from each other in the rostrocaudal axis. Motoneurons of the central column have a radial, multipolar dendritic tree predominantly oriented longitudinally and with sparse spines, whose primary to tertiary branches form large transverse bundles or thin longitudinal plexus. Ventrolateral column motoneurons present a radial dendritic tree. Most dendrites have a longitudinal orientation and form bundles, which also occur in the transverse plane, forming the largest dendritic domain rostrocaudally in the spinal cord. Motoneurons of the dorsolateral column display a rich multipolar dendritic tree, mostly branched out in the transverse plane and some forming small bundles of short dendrites that can extend into the white matter (Fig. 9.6d). Dendritic spines are scarce, although relatively more numerous on proximal dendrites than on distal branches (Shoenen 1982a).

These data indicate that motoneurons likely form most synapses directly on dendritic shafts to compute information from multiple descending motor pathways and propriospinal fibers. Integrated with their intrinsic membrane properties, the synaptic impact of different inputs along the dendritic length within specific time windows will determine the resultant firing pattern of these cells to control and execute phasic and tonic muscle contraction. This is a complex task. In macaques, M1 corticomotoneuronal (CM) cells project monosynaptically to spinal motor neurons and control a single hand muscle, although showing an extensive overlap of axonal branches with the distribution of other M1 neurons that control different hand muscles. Moreover, a single CM axon terminal arborizes extensively in the ventral horn of one or various segments of the spinal cord. The terminal ramifications form synapses with spinal cord motorneuron pools of different intrinsic hand muscles and with different combinations of spinal cord interneuronal networks, which can also activate different combinations of agonist and antagonist (flexor and extensor) muscles (Scott and Kalaska 2021 and references therein). In this example, spinal motoneuron dendrites with heterogeneous spines integrate programmed, learned, feedback, and feedforward synaptically mediated codes to display the fine control necessary for movements of the distal parts of the limbs (Sengul and Watson 2012; Scott and Kalaska 2021; see also the effects of electrical stimulation of dorsal roots C3 to T1 on the arm and hand motor control in two patients with chronic post-stroke hemiparesis in Powell et al. 2023).

9.4.3 Brainstem and Cerebellum

There are few morphological descriptions of human brainstem spiny neurons. Brainstem morphology, itself, is diverse reflecting its varied functions, neuroactive elements, and intrinsic neural pathways. The anatomical organization of the human brainstem is a complex combination of neuronal groups and cell areas with differing cytoarchitecture (Koutcherov et al. 2004; Paxinos et al. 2012; see also Ding et al. 2016). Descriptions and delineations of brainstem nuclei have been based on topography, cytoarchitecture using Nissl staining, myeloarchitecture, connectivity, and chemoarchitectonic analysis of reactivity patterns to acetylcholinesterase, tyrosine hydroxylase, NADPH diaphorase, bombesin, substance P, and neuropeptide Y, among other histochemical markers (Koutcherov et al. 2004; Paxinos et al. 2012). Findings from other mammals (e.g., mice, rats, cats, dogs, and rabbits) are mentioned for comparative reasons and to provide some insights into expected human data.

Different types of neurons were found along the various brainstem areas, some with sparse spines (Cajal 1909–1911). For example, (1) the hypoglossal nucleus (XII cranial nerve) of the newborn mouse and cat presents large cell bodies with long, thick, winding, and typically spiny dendrites radiating in all directions, some ending within the nuclear borders and others extending outside to lie parallel or perpendicular to the fibers associated with the central pathway of the vagus and glossopharyngeal nerves; (2) the dorsal motor nucleus of the vagus nerve (DMV, X cranial nerve) of newborn cat contains relatively large motor neurons with short primary dendrites that give rise to branched segments, all covered with a moderate to a large density of spines (some spines also protruding from the cell body). These cells are intermingled with relatively small neurons with smooth or spine-poor dendritic segments and short axons, whereas other cells, similarly shaped, show a moderate to a large number of spines and axons grouped to form the motor root of the vagus nerve; in humans, the DMV is formed by two zones, one is superficial and showing more widely spaced cells, and another is deeper and show more densely packed cells; (3) neurons in the nucleus *ambiguus* of the cat medulla have thick dendrites classified with moderate density to "spine-laden" segments; (4) in the newborn mouse, neurons in the lateral vestibular nucleus of Deiters show stellate, multipolar cell body and long, spiny dendrites that divide repeatedly. In contrast, neurons in the vestibular ganglion show a fusiform cell body and two straight, predominantly smooth or spine-poor dendrites arising from opposite somatic poles; (5) in the ventral cochlear nucleus of a neonatal cat, fusiform cells display one or two short primary dendrites (some with more very short primary shafts), which are covered with numerous and pleomorphic spines in intermediate to distal shafts and branches (other shapes for spiny neurons are found in the dorsal cochlear nucleus); (6) in the cat and rabbit dorsal nucleus of the raphe, fusiform, triangular, or stellate neurons display very spiny dendrites (also found covering the cell body of some cells) extending along the dorsal to ventral transversal plane; and (7) the rabbit interpeduncular nucleus have neurons with an ovoid, fusiform, or triangular cell body and two to four thick primary shafts which branch sparingly and, after a rather long course, dendrites have a number of thick and short spines, occasionally ramifying as a bouquet of spiny tangled branches or two or three terminal branches (Cajal 1909–1911; Cajal 1995 translated from the original work).

The spatial orientation of dendrites and the distribution, number, and type of spines would be related to the local cellular packing density, neuroanatomical disposition of afferent pathways, and the role of each area in hierarchical or parallel circuits receiving axon terminals or "en passage" collateral fibers. Then, the dendroarchitecture and connectivity of these spiny neurons would suggest the existence of neuronal populations with different coding strategies for processing specific stimuli. Likewise, they would indicate whether the same neurons elaborate more than one type of information by dealing with different spatiotemporal inputs computed along the dendritic tree. Again, the scenario can be complex. Ultrastructural data demonstrated that a single dendritic spine in the cochlear nuclei might receive nine axon terminals and form symmetric synapses containing vesicles of different sizes (and assumed different transmitter content) in the adult rat (Peters et al. 1991).

Moreover, the functional implication of basic neuroanatomical and molecular research of human brainstem nuclei can be highly significant by revealing the patterns of innervation and action upon target sites for the projections from the serotoninergic raphe nuclei, noradrenergic cells in the *locus coeruleus*, and dopaminergic neurons in the ventral tegmental area in the mesocorticolimbic circuitry, to cite a few examples (see also Paxinos et al. 2012; Benarroch 2018; Horn et al. 2020; Saper and Elmquist 2021).

Let us consider another example of an important integrative brainstem area, the nucleus of the solitary tract in the dorsal medulla (NTS; Paxinos et al. 2012; Zoccal et al. 2014; Holstein 2020; Saper and Elmquist 2021). Local neurons are functionally organized to receive inputs and participate in the baroreflex and chemoreflex control of blood pressure, respiratory function, wake-sleep cycle, integrated vestibular responses, esophageal-mediated vago-vagal gastric reflexes, taste perception, and activation of hypothalamic neuroendocrine secretion. In adult rats, NTS neurons receiving viscerosensory inputs and those that project to the DMV display morphologic characteristics correlated with the type of afferent synaptic information they process (Glatzer et al. 2003). Three types of local neurons were identified with one to six primary dendrites "relatively aspiny," that is, with scarce spines of different shapes, some protruding from the cell body and proximal dendrites (see Figs. 3D and 6 in Glatzer et al. 2003). "Group I" neurons have a small soma, three or fewer dendrites, and a relatively small proximal dendritic arbor; "Group II" neurons have a larger cell body area, four or more dendrites branching near the soma, and a larger dendritic arbor than group I neurons; and "Group III" neurons share some morphological similarities with Group II cells, except that they show longer primary dendrites and shafts branching further away from the soma (Glatzer et al. 2003).

Interestingly, there is a transient increase in the number of dendritic spines in NTS neurons of rats (Vincent and Tell 1999). At birth, there are no such "appendages," but the number of pleomorphic spines (some thin and long) covering secondary dendrites reaches a maximum at postnatal day 12 and decreases until adulthood to remain at a very low density (see Figs. 1 and 5 in Vincent and Tell 1999). These findings suggest a functional adjustments with dendritic segments adapting the number of spines and promoting synaptic stabilization in a gradual way. Moreover, contacts are rather made on smooth parts of the dendrites of NTS neurons. In adult hamsters, dendrites are sparsely to moderately covered with pedunculated, filiform, claw-shaped, small, thin pedunculated, and short, blunt spines along the extension and spatial orientation of branches in different neuronal subpopulations and NTS subdivisions (see Figs. 13 and 15 in Whitehead 1988). These morphological data have to be linked to the way NTS neurons are receiving information from multiple sources and integrating different functional responses, which likely present some homology in humans (see further data in Ran et al. 2022).

Further morphological data may also shed light on the synaptic organization of other brainstem structures that elaborate sympathetic and parasympathetic homeostatic (and allostatic) responses, micturition, nociception, alertness, defensive and anxiety-like behaviors, vestibular, auditory, visual, and other sensory and motor cranial nerves functions. Various areas along the brainstem are also connected with subcortical/basal ganglia/cerebellar and cortical circuits to modulate positive and negative emotional expressions in our and other species (Calanchini et al. 2016; Leopold et al. 2019). Groups of cells in regions or defined nuclei and their pathways participate in motor control, including adjustments of spinal reflexes, muscle tone, posture, and voluntary movements (for Nissl-stained soma and primary dendrites of magnocellular red nucleus neurons in humans; see Onodera and Hicks 2009). These emotional and motor circuits provide more complex behavioral displays than those generated by neuronal ensembles near cranial nerves in the brainstem reticular formation called "pattern generators" for stereotyped innate responses (Saper and Elmquist 2021). Indeed, considering an evolutionary perspective, facial expressions are a relevant part of our social communication and social intelligence (Schmidt and Cohn 2001; see additional data on the amygdaloid function in this chapter). There is a complex descending cortical command to the brainstem motoneurons composing the facial nerve (VII cranial nerve) to control 17-paired mimetic facial muscles. The cortical motor control of facial expressions has a distributed neural representation in each brain hemisphere, involving the primary motor cortex, the ventral lateral premotor cortex, and the supplementary motor area on the medial wall for the voluntary control of facial expressions, and the rostral and caudal cingulate cortex to elaborate our emotional expressions (Müri 2016).[16]

The human mesencephalic periaqueductal gray (PAG) is part of the neural circuits activated by nociception and for analgesia (Basbaum 2021) as well as for basic emotions (grief, joy, rage, and fear—see Fig. 3 in Panksepp 2016) or maternal love (Zeki 2007). The PAG neurons are relevant components of the human capacity for speech, forming a species-specific circuit for our emotional expressions and vocalization (Holstege and Subramanian 2016).[17] It is connected with the superior col-

[16] "In humans, the motor nucleus of the facial nerve is the largest of all motor nuclei of the brainstem... The comparative anatomy of the facial musculature and of the central nervous apparatus that controls facial movements suggests that, in some primates, group size, facial motor control, and primary visual cortex evolved with the same pattern... Species living in relatively large social groups tend to have relatively large facial motor nuclei, and species with enlarged facial nuclei and facial mobility have rather large primary visual cortices... Great apes and humans have facial motor cortices that are thicker and richer in local circuitry, their facial movements have the highest degree of dependence on the primary motor cortex... (and) more pronounced direct cortico-facial projections" (Müri 2016 and references therein).

[17] As described by Holstege and Subramanian (2016), human speech needs the activation of two motor systems: one generates vocalization by activating the prefrontal - PAG - nucleus retroambiguus (NRA)–motoneuronal pathway, and the other modulates vocalization into words and sentences by activating the motor cortex and the corticobulbar fibers to specific muscles. The PAG has a central role in the "emotional motor system" when vocalizations also express emotions (e.g.,

liculus and modulates aversion and defensive behaviors (Laemle 1979; Brandão et al. 1994; see additional functional implications in Carrive and Morgan 2012).

Golgi-impregnated PAG cells were studied in two men (28 and 45 years of age) and were classified as "fusiform" or "multipolar" neurons (Gioia et al. 1998). Fusiform neurons have two primary dendrites emerging at opposite somatic poles, which are straight and smooth but ramify to secondary spiny dendrites. Multipolar neurons have more primary dendrites branching in all directions and at variable lengths. These branches provide a radial aspect for the dendritic field. The primary dendrites are usually smooth, and the density of spines increases in dendrites of second or higher order. Multipolar neurons are the spiniest local ones, have an extensive receptive dendritic surface, and are classified as projection cells (Gioia et al. 1998). Additional morphological descriptions on dendrites and spines of Golgi-impregnated PAG neurons were obtained from fetuses to adult humans (Laemle 1979). In these samples, most neurons were small to medium-sized and were named (1) "vertical" cells with one or two (occasionally three) primary dendrites; (2) "stellate" cells with four or more randomly oriented primary dendrites and radial dendritic field; (3) "horizontal" cells show one or two primary dendrites, less spiny dendrites than the other local cells but with spines tending to be in clusters, and having a preferential mediolateral dendritic field; and (4) "other types," which display a triangular cell body resembling pyramidal and "inverted pyramidal" cells and three primary dendrites that ramify sparingly to generate spiny branches. Both densely and sparsely spinous cells were observed, and, notably, pleomorphic spines were found even in fetal tissue. Some dendrites crossed the midline and extended into the contralateral PAG or were directed to the superior colliculus. These data indicate that local cells have multiple receptive fields and process inputs even outside the PAG boundaries (Laemle 1979).

The human superior colliculus receives visual inputs and presumably orients the head and eyes toward salient stimuli in the environment (Tardif et al. 2005; Vanderah and Gould 2021). Five types of neurons were identified in its deep layers: (1) horizontal cells; (2) vertical cells; (3) pyramidal cells; (4) inverted pyramidal cells; and (5) stellate cells (Laemle 1983). In the superficial layers of the human superior colliculus, neurons were classified as (1) small marginal cells with a locally oriented dendritic field; (2) horizontal cells presumed to be interneurons; (3) piriform cells; (4) stellate cells with multiple radiating dendrites; (5) narrow field vertical cells, with dendrites vertically oriented and cylindrical fields; and (6) wide-field vertical cells, with dendrites extending obliquely toward the surface and forming a broad field (Laemle 1981; May 2006).

crying and laughter in humans). "The PAG receives strong projections from higher limbic regions and from the anterior cingulate, insula, and orbitofrontal cortical areas. In turn, the PAG has strong access to the caudal medullary NRA. The NRA is the only cell group that has direct access to the motoneurons involved in vocalization, i.e., the motoneuronal cell groups innervating soft palate, pharynx, and larynx as well as the diaphragm, intercostal, abdominal, and pelvic floor muscles. Together they determine the intraabdominal, intrathoracic, and subglottic pressure, the control of which is necessary for generating vocalization. Only humans can speak, because, via the lateral component of the volitional or somatic motor system, they are able to modulate vocalization into words and sentences" (Holstege and Subramanian 2016).

Another brainstem area with relevant morphological data to be linked with functional and clinical implications is the substantia nigra (SN, van Domburg and ten Donkelaar 1991).[18] The human SN presents an interindividual variability in shape, asymmetry, and volume in normal individuals (50–91 years of age; Di Lorenzo Alho et al. 2016). Comparing the SN *pars compacta, pars reticulata, and pars diffusa*, the former contains large neuromelanin-rich neurons (Braak and Braak 1986; Di Lorenzo Alho et al. 2016). Cells usually superimpose each other in the heterogeneous SN parts, although showing distinct morphological, pigmentation, and histochemical features (Braak and Braak 1986; Halliday et al. 2012).

Three types of neurons were identified in the SN of individuals 20–93 years old: "Type I," mainly found in *pars compacta* (but also in *pars diffusa* and less in *pars reticulata*), is a medium sized to a large cell with neuromelanin granules in the cell body. They are tyrosine hydroxylase immunopositive projecting neurons with a triangular or polygonal cell body shape or, in a few cases, spindle, bitufted soma (Braak and Braak 1986; Halliday et al. 2012; Fig. 9.7a, b). Type I dopaminergic cells have two to six thick and sparsely branching dendrites, frequently as one large and other smaller primary dendrites with a straight extent and smooth aspect, or with some segments with small-sized spines (Halliday et al. 2012; Fig. 9.7a). Extended thick dendrites form small bundles into the *pars reticulata*, but not beyond nuclear boundaries. Neurons in *pars diffusa* may send dendrites into the perirubral formation. The heterogeneous distribution of dendritic spines is evident in these type I neurons. Spines can be sparse along dendritic shafts, forming isolated patches of grouped spines, or show a high density of pleomorphic spines, some as "thorny appendages" (Braak and Braak 1986, Fig. 9.7a, b).

"Type II" neurons, mainly found in the *pars reticulata* and *pars diffusa* (less in *pars compacta*), are usually non-pigmented and have a variable size (usually slightly smaller than type I cells). The cell body shape is determined by the number of primary dendrites. Slender and elongated cells with two dendrites at opposite cell body poles are most commonly observed followed by other multipolar forms with more

[18] The morphology, neurochemical profile, and number of affected SN *pars compacta* neurons relate the pathophysiology of Parkinson's disease (PD; Gibb and Lees 1991). Recently, single-nucleus RNA-sequencing profiling of human SN *pars compacta* dopaminergic neurons identified a population that is selectively vulnerable and degenerates in PD (Kamath et al. 2022). These neurons have a unique, very large, dense, and widely spread axonal arbor architecture targeting the neostriatum (Matsuda et al. 2009), which might impose a chronic high energetic demand and likely damaging oxidative stress (Bolam and Pissadaki 2012; see additional comments in Giguère et al. (2018) and recent findings in an animal model in Ferreira et al. (2020)). There are various intracellular molecular changes in human SN dopaminergic neurons critical to the neurodegeneration and symptomatic manifestations of PD (Halliday et al. 2005), but the loss of neurons in this disease is not confined to the SN. Morphological and functional impairments include motor and nonmotor circuits, and PD progression alters the olfactory bulb, other brainstem areas, temporal mesocortex, and neocortical areas (Braak et al. 2003; Bolam and Pissadaki 2012; Giguère et al. 2018). For example, the pedunculopontine nucleus in the ponto-mesencephalic tegmentum is part of the mesencephalic locomotor region and may be involved in sleep and cognitive disturbances (French and Muthusamy 2018) at different stages of PD progression (Braak et al. 2003; see also Rietdijk et al. 2017). Many patients also suffer from hallucinations and delusions.

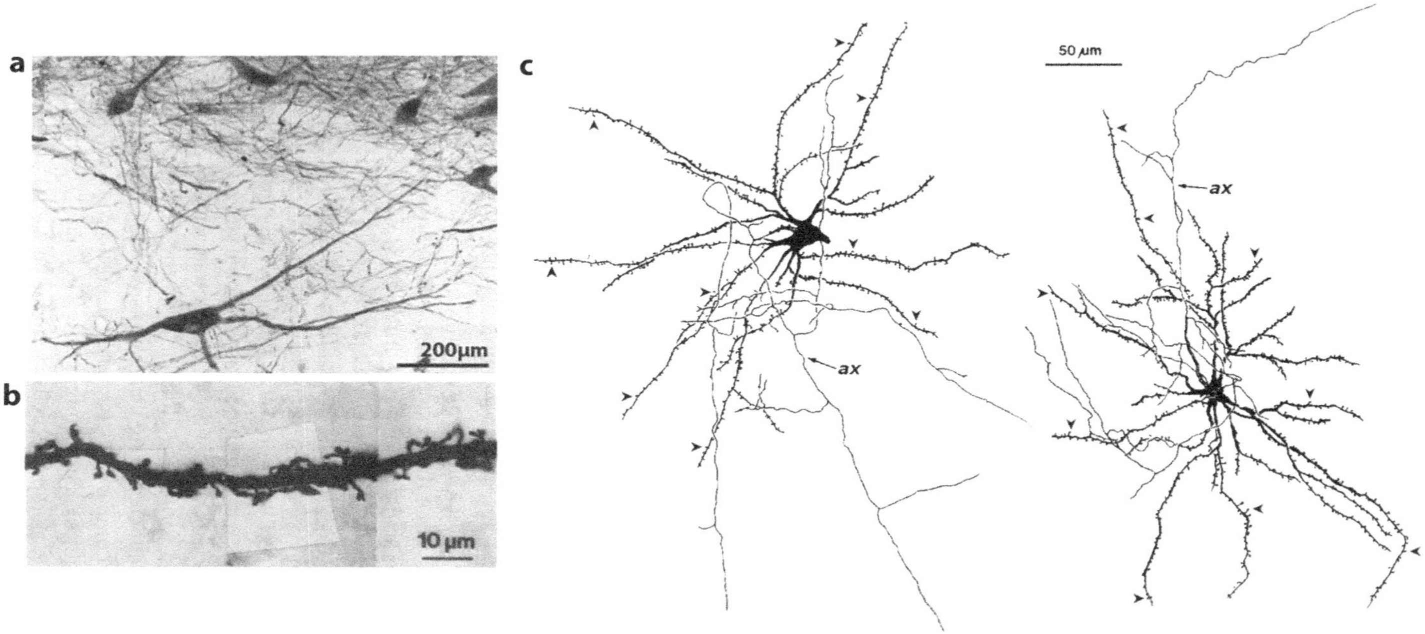

Fig. 9.7 (**a**) Typical medium-sized to large tyrosine hydroxylase-immunoreactive dopaminergic neuron found in the human substantia nigra *pars compacta*. These neurons characteristically contain neuromelanin pigment and are projection cells. [Image originally published in "The Human Nervous System," 3rd edition, by Halliday et al. "Substantia Nigra, Ventral Tegmental Area, and Retrorubral Fields," p. 439–455, reproduced under license #221112-000190, Copyright Elsevier (2012)]. (**b**) Golgi-impregnated dendritic segment from the same type of dopaminergic neuron as in (**a**), also classified as a "type I" cell. Although such neurons can show many smoothly contoured thick dendrites, some dendrites also display spines and "spine-like protrusions" as shown in this image. (Figure reproduced from Braak and Braak (1986), originally published and copyrighted by Springer Verlag). (**c**) Camera lucida drawing of Golgi-impregnated medium-sized spiny neurons of the human striatum. Note the highly ramified arbors with branches radiating in all directions. Spines (arrowheads) are found after the initial segments of the main dendritic shafts and along various collateral branches. Spines also occur in the axon hillock (*ax*, pointed by an arrow in the neuron at the center of the figure). (Legend and figure reprinted from Braak and Braak (1982b) under CCC RightsLink® license #5386640793281, originally published by Springer Verlag)

than two primary dendrites and some collateral branches near the cell body. Smooth primary dendrites give rise to intermediate and distal dendritic segments with a low density of isolated spines or clusters of spines. "Type III" neurons, in all SN portions, have usually a small cell body with varied shapes. A few thin primary dendrites radiate in various directions, which, as a rule, show very few branches and are spineless shafts. These cells show lipofuscin granules that did not tend to agglomerate (Braak and Braak 1986).

Two types of non-pigmented, non-dopaminergic SN neurons are immunoreactive to the calcium-binding proteins calretinin and parvalbumin. Calretinin-containing neurons have two primary dendrites emerging from opposite somatic poles and are observed surrounded by pigmented cell clusters only in *pars compacta*. Parvalbumin-containing neurons concentrate in *pars reticulata* and are small multipolar cells with round or triangular cell bodies and dendrites with few branching points (Halliday et al. 2012; see Damier et al. 1999 for calbindin immunohistochemistry distribution within the human SN).

The human dentate nucleus, one of the cerebellar central nuclei, consists of 6–10 irregular rows of closely packed cells that, compared to other mammals, have smaller neurons with thinner, shorter, and less branched cells (Cajal 1909–1911). Peters et al. (1991) depicted a Golgi-impregnated small neuron in the dentate nucleus of the cerebellum. This cell has an ovoid cell body and multiple primary dendrites. The thickest primary shafts radiate and immediately branch around the cell body. Thin primary branches are straight or tortuous and show few branching points. Dendrites of different branching orders are found close to the cell body with a radial orientation. There are few spines with a heterogeneous distribution. Some dendritic segments are smooth. Other dendritic segments have isolated spines, spines at a low density along the dendritic length, or pleomorphic spines in distal parts (Peters et al. 1991).

Purkinje cells in the human cerebellar cortex display thick and short proximal primary, secondary, and tertiary dendrites devoid of spines or showing isolated spines. Phylogenetic differences are found in climbing fibers and dendrites of Purkinje cells in humans. In these latter, after proximal ramifications, further branched dendrites have spines along their extensions, which are so close that they almost touch. These spines are abundant and of varied sizes, which include short and thick or long and thin spines, as depicted by Cajal (1909–1911).

In the human cerebellar cortex granular layer, three types of large nerve cells were identified in samples from 37- to 55-year-old individuals (Braak and Braak 1983a): (1) Most cells are "Type I" neurons, also found in the Purkinje cell layer, and correspond to the Golgi cells with a rounded or polygonal cell body. Only a few dendrites usually arise from various sites of the cell body, radiating in all directions with few branching points, and forming a globular dendritic domain. These dendrites show little reduction in diameter along their extension and collateral branches often do not taper significantly compared to the parent dendrite. Very few spines are found along the dendritic length of these Golgi cells. They are "small kinks and twists and even single spines." (2) "Type II" cells (found even within the white matter) show a fusiform, triangular, or polygonal cell body, generally with a few primary dendrites. These dendrites are remarkably long and straight, extending up to

700 μm away from the soma, and rarely branching or, distally, ramifying up to two times. Horizontal dendritic shafts are often densely covered with "stalked" (thin) spines. Finally, (3) "Type III" cells are heterogeneous large multipolar neurons with varied shapes, including spindle and triangular aspects, and broad distribution (also within the Purkinje cell layer and the white matter) but a small number. They display two to eight primary dendrites, which show different features. For example, some dendrites are long straight shafts that do not ramify or, when dividing, generate secondary branches distally. Other cells display a dense dendritic arborization close to the cell body whereas others have a straight dendrite that ramifies profusely distally. Intermediate morphological aspects for branching patterns are observed as well. Notably, some dendrites project toward and extend into the deep portions of the molecular layer and the Purkinje cell layer. Dendrites are smoothly contoured for the most part, although some "small appendages" and spines are occasionally observed (Braak and Braak 1983a). After this description, another type of monodendritic neuron was reported in the human cerebellar granule cell layer. It is a calretinin-immunoreactive bipolar cell with a characteristic spherical, ovoid, or more elongated cell body, a single very thick and short primary dendrite with no preferential orientation terminating in a closeby tuft of short and fine arborizations with small swellings or leaflets, and a thin axon emerging from the cell body (Braak and Braak 1993).

These data address the functional organization of the afferent, intrinsic, and efferent cerebellar circuitries in humans (see fundamental data in Voogd and Ruigrok 2012). It is noteworthy that the dendrites of Purkinje cells are spiny, but the presence and density of spines are variable in different granular layer neurons. Purkinje cells compute and rather modulate a huge amount of afferent synaptic inputs along the dendritic tree before determining their firing pattern as the cerebellar cortex efferent pathway. With the number of contacts ranging between 200,000 and 1 million per cell, the location and kind of synaptic inputs along the Purkinje cell dendritic tree are controlled, integrated with intrinsic membrane properties, and determine the cellular excitability and pattern of firing output (Llinás and Sugimori 1980; Bastian and Lisberger 2021). On the other hand, granular cells can integrate information along the extension of long dendrites, some without ramification, which suggests direct synapses on dendritic shafts and distinct cable properties based on such structural dendritic organization. In our species, the cerebellum reaches maturity after birth (as the neocortex), enabling further motor skills during childhood as well as continuing to show plastic properties for memory and learning of planned motor behavior and other nonmotor functions along our lifespan (Diamond 2000; Bastian and Lisberger 2021; Kolb and Whishaw 2021).

9.4.4 Thalamus and Basal Ganglia

Multiple spiny multipolar neurons are found in the thalamic "somatosensory nucleus" of cats, rabbits (Cajal 1909–1911), rats, and mice (Scheibel and Scheibel 1967). In humans, thalamic neurons in the superior region, including the anterior nuclear group and the dorsal superficial nucleus, show round to polygonal cell bodies and are typically isodendritic, some with marked ramification close to the soma (Mai and Forutan 2012). There are principal/projection cells and interneurons in different thalamic nuclei (Braak and Weinel 1985). In the medial region, represented by the mediodorsal nucleus and subdivisions, there are characteristic very large, polygonal or flattened cells in the paralaminar (lateral) component; the paratenial nucleus of the periventricular formation contains medium-sized triangular neurons with thin and long processes; and neurons in the inferior pulvinar nucleus in the posterior thalamus show a predominance of smaller cells with a polygonal cell body and various thin proximal processes (Mai and Forutan 2012).

The nuclei composing the lateral region of the thalamus are functionally involved with processing and relaying relevant information for sensory and motor modulation in humans (Mai and Forutan 2012). As an initial hypothesis, it would be expected that neurons in this region would display a variety of spines for the synaptic computation of multiple inputs and for sending a modulated output firing pattern toward the next circuitry steps. Surprisingly, it is not the case for the dendrites in the adult human neocortex-dependent nuclei of the lateral thalamus. These neurons are devoid of spines or show very few long and thin protrusions ("appendages") or stalks without spine heads (Braak and Braak 1984a). In other words, (1) "Type I" neurons have a polygonal cell body and various radiating primary dendrites (three to seven), which branch profusely and generate a dense and mostly globular dendritic field. According to the territory where these cells are located, they can also display a bush-like dendritic arbor with branches mainly oriented in one preferred direction. Typical spines are absent on dendrites. Rather, thread-like appendages are present, which display elongated stalks lacking a terminal knob. (2) "Type 2" neurons show cell bodies smaller than type I cells, elongated, spindle-shaped, or polygonal with some primary dendrites (e.g., four to five), which ramify sparsely and proximally providing an ellipsoidal dendritic domain with few branches. These local circuit cells have smooth shafts or, along the extension of the dendritic tree, isolated or very few spines are observed in restricted dendritic segments. (3) "Type 3" neurons have polygonal medium- to large-sized cell bodies and thick, short primary dendrites that ramify close to the cell body. Dendrites have a straight or slightly curved course extending in all directions. These cells do not show spines or other appendages (Braak and Braak 1984a). In this same thalamic area, medium-sized multipolar neurons with a fair number of typical dendritic spines were observed forming small cellular islands near the boundaries of the intralaminar nuclei. These cells were named "ectopic neurons" by Braak and Braak (1984a).

Visual processing pathways activate and are modulated by the thalamus and metathalamus. Novelty seeking and interaction with the world by exploring visual objects involves the zona *incerta* in primates (Ogasawara et al. 2022), an extension of the reticular nucleus of the thalamus related to visceral functions, arousal, and attention (Zrinzo and Hyam 2018). In the visual metathalamic pathway relay station, the lateral geniculate nucleus process visual information modulated by selective attention. It integrates retinal and brainstem nuclei inputs with top-down signals from corticothalamic feedback projections (Reynolds et al. 2013). Neuronal types in the lateral geniculate nucleus show smooth dendrites or few spines in adult humans. "Type 1" projecting cells (the most abundant) are heterogeneous medium- to large-sized polygonal and spindle-shaped cells with two to ten radiating primary dendrites of varied features. The variety of shapes for these multipolar neurons is notable. Some cells have two thick primary dendrites emerging from one pole of the cell body and the axon hillock at the other pole. Dendrites can have isolated or clustered "excrescences" (i.e., protrusions with a blunt aspect or with a long and thin gnarled outline, some of them resembling thin, wide, and large complex spines) placed close to proximal branching points. Few pleomorphic dendritic spines were found isolated along the extension of the dendritic tree and rarely exist along thin distal segments. Another type I neuron possesses thin, radiating, and smoothly contoured dendrites with few dendritic excrescences or thorny appendages. "Type 2" and "Type 3" cells (interneurons) usually have a small round, ovoid, triangular, or spindle-shaped cell body. They have two to four primary dendrites, some with a tufted aspect, others with sparse branching and straight, smooth dendrites forming an ellipsoidal domain. Dendritic spines are not found along the dendritic shafts and branches, except for the occurrence of rare irregularly distributed thorny appendages. In their distal portions, Types 2 and 3 neurons have dendrites whose appearance resembles thin axon-like structures (Braak and Braak 1984b; Braak and Bachmann 1985). These and other thalamic components, such as its centromedian nucleus, surely deserve more attention and in-depth histological and functional studies (Ilyas et al. 2019).

The basal ganglia is part of a circuit involving the cerebral cortex, hippocampus, amygdaloid complex, thalamus, and brainstem nuclei to elaborate, execute, and learn new planned, motivated behaviors requiring executive function, motor adjustments, cognitive, and emotional processing (Smith and Bolam 1990; Haber et al. 2012; Horak and Earhart 2021). The cytoarchitecture and multiple functional circuitries of the striatum have been investigated in humans and other species (Cajal 1909–1911; Smith and Bolam 1990; Haber et al. 2012; Lanciego et al. 2012). Input projections (sensorimotor, associative, and limbic), intrinsic connections, and outputs of the basal ganglia form a series of reentrant parallel projecting functional loops under within and outside modulatory actions (Redgrave and Costa 2021).

The neostriatum receives inputs from cortical and subcortical areas that innervate local medium spiny neurons, including dopaminergic neurons from the SN (Smith and Bolam 1990). There are important features involving the synaptic organization and functioning of this circuit. First, inputs from the cerebral cortex and the central lateral thalamic nucleus appear to make few individual contacts but extend their action upon many striatal neurons, whereas those from other areas (e.g., the parafascicular thalamic nucleus) might do the opposite for each cell make various contacts to influence few local neurons (Redgrave and Costa 2021). Second, the complex axonal arbor of each SN projecting cell innervates a significant volume of striosome and matrix compartments of the neostriatum (Matsuda et al. 2009). In rats, it is estimated that approximately 75,000 striatal neurons might be innervated by a single dopaminergic neuron, and one striatal neuron might be under the influence of 95–194 of these dopaminergic cells (Matsuda et al. 2009). In humans, it is estimated that one single dopaminergic neuron from the SN *pars compacta* gives rise to between 1 and 2.4 million synapses at the level of the striatum (Bolam and Pissadaki 2012). Third, there are different patterns of synaptic innervation and organization on the spines and dendrites of striatal neurons (Smith et al. 1994, 2004). Striatal medium spiny projection neurons are targeted by corticostriatal, thalamostriatal, and dopaminergic input axons. Corticostriatal inputs converge on the head of dendritic spines, most nigrostriatal inputs converge on the neck of dendritic spines, and most thalamic terminals from caudal intralaminar nuclei form asymmetric synapses on dendritic shafts (in rats and monkeys; see complete data in Smith and Bolam (1990), Smith et al. (2004)). Furthermore, thalamostriatal afferent fibers generate strong synaptic inputs in proximal dendrites and perikarya of striatal cholinergic interneurons, while corticostriatal inputs are scarce on distal dendrites (Smith et al. (2004) and references therein). Thalamic and dopaminergic inputs usually are not nearby or have synaptic relationships when innervating the same postsynaptic target in the sensorimotor territory of the striatum in monkeys (Smith et al. 1994).

These data indicate morphologically tuned synaptic processing adjusted for each input source as well as the integration of varied information by medium spiny neurons before projecting an output code toward parallel striatal functional loops. The importance of these integrated actions is reflected in the fact that, besides the components of the classical "direct" and "indirect" circuits, other relevant output pathways connect the striatum with the superior colliculus, periaqueductal gray, pedunculopontine, parabrachial nuclei, and nucleus *accumbens* (Smith and Bolam 1990; Redgrave and Costa 2021). The involvement of various brain areas and circuits with the striatal medium-sized spiny neurons provides research avenues to investigate the diversity and magnitude of clinical manifestations, including emotional disturbances, in the course of neurodegenerative Parkinson´s and Huntington´s disease, for example.

Cajal (1909–1911) and Braak and Braak (1982b) detailed the morphology of neuronal types in the striatum of humans. Cajal described the morphology of three main neuronal types, namely, (1) particularly in the caudate nucleus, a more or less spherical to polygonal small to medium-sized neuron with a short axon or with an axon projecting throughout a broad region of adjacent striatal nucleus or nuclei. These cells have a large number of spiny, moderately branched, radiating, and rather short dendrites; (2) large cells (with fusiform, triangular, or polygonal soma and rare Nissl granules) and rare giant stellate cells (with Nissl bodies in their cytoplasm) with a characteristic long axon and long, stout, and moderately branched dendrites covered with many spines; and (3) "dwarf or neurogliaform neurons" with a small spherical soma, several short varicose dendrites, and short axon (Cajal (1995)—see Fig. 325 for his drawings of neuronal types from a human infant brain).

Braak and Braak (1982b) reported the following five types of Golgi-impregnated neurons in the striatum of adult individuals (44–52 years old): (1) "Type I," the most abundant cell, is the typical medium-sized neuron with spiny dendrites. Four to 10 primary spine-free dendrites emerge from a spindle-shaped or polygonal cell body. Close to the soma, dendrites ramify and various second-order branches follow a relatively straight course. Secondary and further orders branches show many thin spines with different sizes. Mushroom and stubby spines were also observed (Fig. 9.7c; see also Fig. 2 in Braak and Braak 1982b). Taking into account the local connectional organization, "dendrites radiate in all directions, but in places, especially close to thicker bundles of myelinated fibers, they are more or less strictly oriented in parallel to the fascicles." Also noteworthy, the axons of these cells can show a few stubby "appendages" proximally, can ramify within the dendritic domain of the parent neuron, and then project forward (Fig. 9.7c). The identification of such Golgi-impregnated axons implies that projection fibers and local ramified collaterals lack a myelin sheath. (2) "Type II" is a medium- to large-sized cell with long intertwining dendrites covered with spines of uncommon shape as well as somatic spines. The cell body is triangular, spindle- or horn-shaped from where emerge a few very thick primary dendrites. These dendrites ramify repeatedly and show a gnarled varicose aspect. Thin terminal branches have a curved course and form a small dendritic domain. There is a moderate number of pleomorphic spines, including "stubby protuberances" and "a fair number of spicular appendages with elongate pedicles." Perikarya and proximal dendrites are covered with "knobbed spines" (see Figs. 3 and 4 in Braak and Braak 1982b). (3) "Type III" is a large spindle-shaped or multipolar neuron with thick extended dendrites, few branching points, and isolated spines along intermediate to distal dendrites. (4) "Type IV" is a large aspiny cell with a rounded cell body and richly branching tortuous thick and thin primary dendrites. These neurons have a dense radial dendritic domain made by many relatively short dendrites close to the cell body. Although considered "aspiny," slender terminal dendritic branches have a gnarled appearance and varicosities and a "few isolated peduncular spines." (5) "Type V" are small- to medium-sized round aspiny cells with only a few relatively thick dendrites that ramify close to the soma. Beaded branches and fine terminal collaterals form a small ellipsoidal dendritic domain. Although dendrites are rather smooth, a "few isolated and minute thorns" can also be observed along the dendritic arbor (Braak and Braak 1982b).

Braak and Braak (1982a) also described five types of neurons in the human claustrum. Among large and small aspiny or "almost aspiny" nerve cells, it is significant the predominant occurrence of local Golgi-impregnated "pyramidal-like" spiny neurons - with no spines at proximal dendritic shafts and an evident number of pleomorphic spines along intermediate dendritic segments - or "modified pyramidal cells" (see Figs. 1–3 in Braak and Braak 1982a). These spiny neurons (likely projecting ones) would integrate reciprocal circuits with various cerebral cortex areas as well as would modulate information from amygdaloid nuclei in claustrocortical pathways (Fernández-Miranda et al. 2008).

9.4.5 Amygdala

The amygdala (or amygdaloid complex) is not a structural or functional unit; rather, it is composed of heterogeneous nuclei and subnuclei with particular phylogenetic, cytoarchitecture, connectional, neurochemical, and functional features (Brodal 1981; de Olmos et al. 1985; Gloor 1997; Swanson and Petrovich 1998; Rasia-Filho et al. 2000; DiMarino et al. 2016; Petrulis 2020), including "extended" areas outside previously assumed anatomical borders (de Olmos 2004; Heimer et al. 2008; Yilmazer-Hanke 2012). The amygdaloid complex nuclei have both cortical and subcortical origins (Medina and Abellán 2012; Akhmadeev and Kalimullina 2015; Olucha-Bordonau et al. 2015) with pallial (most nuclei and their subdivisions) and subpallial components forming parallel circuits. These latter relate to the medial amygdaloid nucleus (MeA) and the central nucleus in the "extended amygdala" (CeA; Martínez-García et al. 2007)

The specialization of neuronal subpopulations in the amygdaloid complex likely provided more possibilities for sensory stimuli processing, for triggering somatic responses from primary emotions, and for associating them with complex social cues together with the gradual expansion of cortical circuitries (Janak and Tye 2015; Diano et al. 2017; Koroster et al. 2017; Rasia-Filho et al. 2021; Šimić et al. 2021; Underwood et al. 2021). In this regard, the development of different amygdaloid nuclei would have contributed to important species-specific social behaviors, for example: (1) to further perceive and elaborate visual and auditory cues; (2) to attribute meanings and emotional valence to these stimuli; (3) to modulate new memories and cognitive abilities; and (4) to expand the behavioral repertoire for complex social interactions between individuals, including judgments of facial expressions and emotional vocalization contributing to behavioral displays of parenting, empathy, happiness, fear, or disgust (Adolphs 2003; Heimer et al. 2008; Rutishauser et al. 2011, 2015; Diano et al. 2017; Grisendi et al. 2019). That might have provided human brain networks with a higher level of information processing from cellular to network levels and from subcortical to cortical areas, with new emergent functional properties and adaptive responses.

The MeA, one of the phylogenetically oldest parts of the amygdaloid complex (Johnston 1923), has a relatively small area in humans (de Olmos 2004; Schumann and Amaral 2005). It contains five types of multipolar spiny neurons (Dall'Oglio et al. 2013). Golgi-impregnated neuronal types were morphologically classified in humans as (1) fusiform neurons with usually two (at most three) primary dendrites arising from opposite cell body poles, very long and straight dendrites with few branching points (Fig. 9.8a) or more branched shafts with a tufted aspect (see Fig. 2E in Dall'Oglio et al. (2015)); (2) neurons with a pear-shaped cell body, two primary dendrites emerging approximately at a right angle, branching sparingly, and showing very long shafts (Fig. 9.8b); (3) angular neurons with three primary dendrites that, despite the triangular cell body shape, do not display the characteristics of cortical-like pyramidal neurons (e.g., there is no evidence of one main thick apical shaft and two thinner and more branched basal shafts; instead, there are two main thick and long shafts with collaterals or three primary dendrites with almost similar thickness and branching pattern); (4) stellate neurons with a round or ovoid cell body and various primary dendrites radiating in multiple directions; and (5) rectangular neurons with four or more primary dendrites. The density of dendritic spines is variable in these human MeA neurons (Dall'Oglio et al. 2013). Close to the MeA, the human CeA possesses various neuronal types as well. All of them are spiny cells (unpublished observation).

The axonal morphology in the human MeA neuropil indicates various possibilities for synaptic contacts with local dendrites and spines. For example, there are "en passage" varicose fibers (probably for an ample "alarming" response evoked by inputs) as well as fibers with simple collateral prolongments with a terminal aspect. Other axonal collaterals display complex shapes, multiple branching patterns, and a bulbous ending that was further ultrastructurally related to a glomerular synaptic structure and function (Fig. 9.8c; see data in Dall'Oglio et al. 2013, 2015). Axons were observed crossing perpendicularly a restricted segment of spiny dendrites (Fig. 9.8d) and eventually having a terminal bouton close to a spine (Fig. 9.8e). These synaptic features are relevant to the human MeA role in neural networks that elaborate emotions (Blood and Zatorre 2001; Liberzon et al. 2003; Gamer et al. 2010), responding to aversive olfactory stimulation (Zald and Pardo 1997), to gonadal hormone-mediated premenstrual increase in stress-induced negative affect (Ossewaarde et al. 2013), and for the perception and processing of both aversive stimulus evoked by fearful faces and the positive valence of happy faces (Gamer et al. 2010).

In addition, the amygdaloid complex is related to the emergence and morphological heterogeneity of pyramidal neurons found in an anatomical and functional integrated subcortical–allocortical–neocortical *continuum* in the human brain (Rasia-Filho et al. 2021). The search for the emergence of pyramidal neurons led to the interface between subcortical and cortical nuclei of the amygdaloid complex and cerebral cortex transitional areas (Gloor 1997; Heimer et al. 2008; Insausti et al. 2017; McDonald and Duque 2022). Transition areas and nuclei with cellular components suggestive of a primitive cortex are found in the amygdaloid complex of primates (Amaral et al. 1992). In rats, the pallial amygdala is considered an initial allocortical structure characterized by superficial layered cortical areas and deep non-layered parts (Olucha-Bordonau et al. 2015). Likewise, a cytoarchitectonic organization in distinct layers was proposed for the human cortical amygdaloid nucleus (CoA; reviewed in Vásquez et al. (2018)). The CoA composes the "superficial cortex-like amygdaloid region" and is associated with the olfactory tract pathways together with other adjacent areas (see a critical discussion in Yilmazer-Hanke 2012). In conjunction, it would form a cortical transition area having the piriform and entorhinal cortices in the rostral part and the caudal piriform cortex and lateral entorhinal cortex in the caudal part (Olucha-Bordonau et al. 2015).

By composing the superficial amygdaloid nuclei and at the beginning of the limbic lobe, the CoA participates in somatosensory processing, associative learning, and distinct aspects of emotional processes (Heimer et al. 2008; Goossens et al. 2009; Bach et al. 2011; Yilmazer-Hanke 2012; Bzdok et al. 2013). For example, the CoA is connected to the prosubiculum, the dysgranular temporo-polar and perirhinal cortices, the agranular insula, and the lateral orbitofrontal cortex (Gloor 1997). The CoA neurons would also integrate multiple stimuli, as evidenced by connectivity-based parcellation of the superficial amygdaloid group, to modulate olfactory and emotional information in the context of social interaction and for approach-avoidance behaviors (Bzdok et al. 2013).

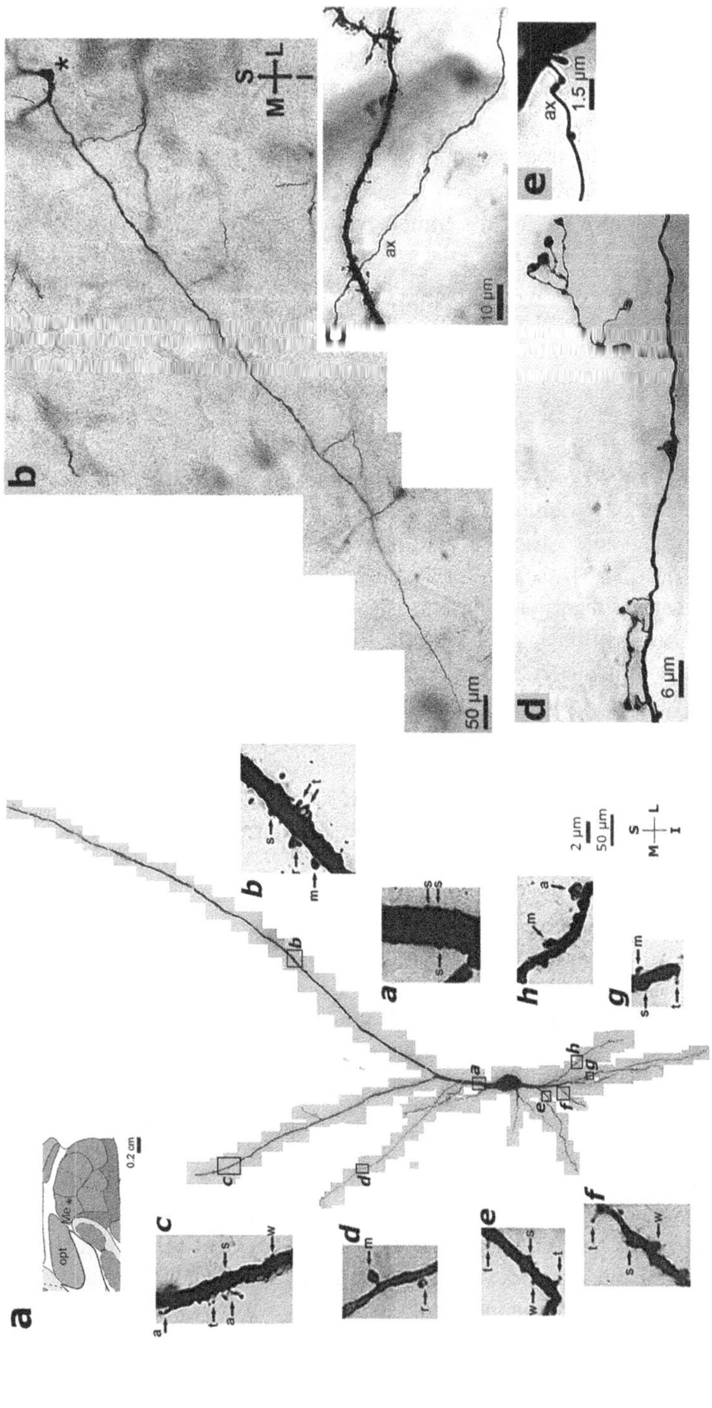

Fig. 9.8 (*caption on p. 423*)

Based on morphological criteria, ten Golgi-impregnated neuronal types were described in the human CoA, although three of them show some similarities to a pyramidal-like shape (Vásquez et al. 2018, 3D reconstructions available as Supplementary Images in Guerra et al. 2023). All are spiny neurons, but show differences in spine distribution and density along dendritic shafts (e.g., there is an impressive spine density in the unipolar neuron type shown in Fig. 15 from Vásquez et al. (2018)). Every type of spine is found in human CoA neurons. The 3D reconstruction of dendritic spines evidenced various shapes and sizes of stubby/wide, thin, mushroom, ramified, and transitional or atypical/polymorphic spines (Fig. 9.9a–e). The heterogeneous distribution, number, shape, and grouping aspect of spines along dendrites in the same neuron are exemplified by the pear-shaped cell in this brain nucleus (Fig. 9.9a–e; Vásquez et al. 2018). Note the "coralline excrescence" aspect of the highly packed dendritic spines with "numerous thin protrusions, velamentous expansions, and tendrils" (Fig. 9.9e, e'; classified according to Fiala and Harris 1999) and spinule protruding from different spines (e.g., Fig. 9.9c, d'; for structure, regulatory mechanisms, functions, and role of spinules in synaptic plasticity see Spacek and Harris 2004; Petralia et al. 2018; Zaccard et al. 2020).

Main dendritic shafts and collaterals in CoA neurons have longitudinal, transverse, and oblique orientations with a rather "planar" aspect within the neuropil volume. The number of neuronal types and their morphology suggest that cellular subpopulations developed to process differently the information from extrinsic and intrinsic circuits. In addition, the packing density and the existence of intermingled different neurons would allow afferent axons to contact more than one CoA neuron type to process the salience of the incoming stimulus. The diversity of spines also suggests that local cells might have different plastic possibilities for synaptic

Fig. 9.8 (**a**) Top: schematic diagram of a coronal section of the adult human brain showing the medial amygdaloid nucleus (Me) close to the optic tract (opt) 12 mm posterior to the midpoint of the anterior commissure. (Adapted from Mai et al. 2008). The asterisk indicates the location of a Golgi-impregnated large multipolar fusiform neuron shown as a composition of multiple digitized microscopic images obtained under light microscopy. Note the long dendrites and branching pattern, as well as the presence, distribution, density, and shape of dendritic spines. Spines are shown from samples (boxes *a* to *h*) along the dendritic length and corresponding inserts at higher magnification. Spines were classified as stubby (s), wide (w), thin (t), mushroom (m), ramified (r), or transitional/atypical (a, such as a double-spine in *c* and a spine with a racemose appearance in *h*). Photoshop CS3 adjustments (Adobe Systems, USA). I inferior, L lateral, M medial, S superior. Bars = 50 μm for the general view of the neuron and 2 μm for the inserts. (Legend adapted and figure reprinted from Dall'Oglio et al. (2015) under CCC RightsLink® license #5433161449807, originally published by John Wiley & Sons, Inc). (**b**) Reconstructed digitized microscopic image of another human medial amygdaloid Golgi-impregnated multipolar spiny neuron with a rounded, pear-shaped cell body, two thick primary dendrites, and few but long dendritic branches. (**c**) Straight axons cross perpendicularly dendrites in the Me neuropil, (**d**) show varicosities and collateral fibers with convoluted branching patterns and large bulbous endings, or (**e**) a small terminal bouton close to a spine head. (Legend adapted and figure reprinted from Dall'Oglio et al. (2013) under CCC RightsLink® license #5433161269561, originally published by John Wiley & Sons, Inc)

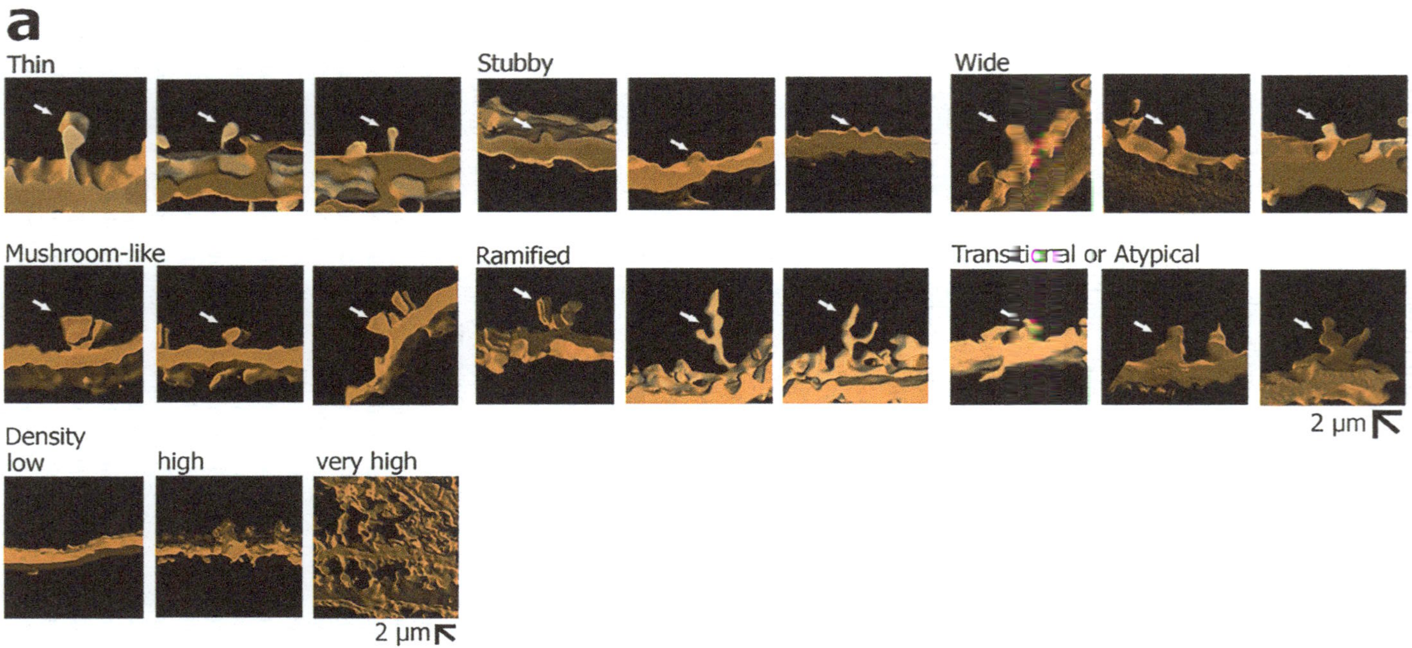

Fig. 9.9 (**a**) Three-dimensional (3D) reconstruction of Golgi-impregnated dendritic spines from the human cortical amygdaloid nucleus. Brightfield microscopy images were submitted to a segmentation model for the 3D reconstruction of spines. (Algorithm described in Reberger et al. (2018)). Spines were classified into different types, and the number of pleomorphic spines varies between dendritic segments, ranging from a low number of isolated spines to a very high density of clustered spines. (**b**) Golgi-impregnated pear-shaped neuron from the human posterior part of the cortical amygdaloid nucleus (PCO), 9.3 mm posterior to the midpoint of the anterior commissure. (Top, schematic diagram adapted from Mai et al. 2008). opt: optic tract. (Center) Reconstructed two-dimensional microscopic image showing the cell body and branched spiny dendrites. Examples of 3D reconstructed dendritic spines are shown in the inserts at higher magnification and different rotation angles (boxes *a-e* and corresponding inserts). Rotated segments are indicated by an apostrophe (*c'*, *c"*, *d'*, and *e'*).

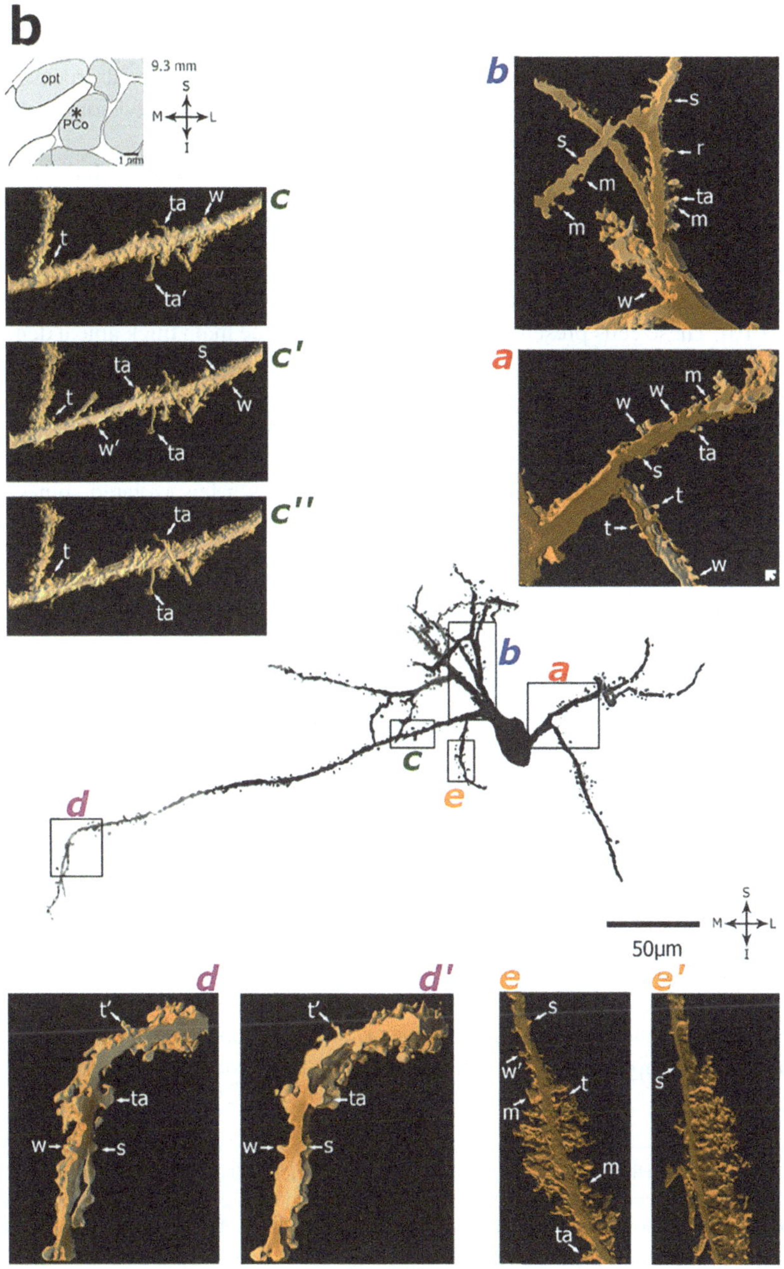

Fig. 9.9 (continued) The dendritic spine density varied from low (*a*) and moderate (*b* and *c*) to very high (*e*). Spines (arrows) were classified as stubby (s), wide (w), thin (t), mushroom-like (m), ramified (r), or transitional/atypical ones (ta). The presence of a spinule is indicated by an apostrophe in the corresponding spine (e.g., ta' in *c*, t' in *d* and *d'*, and w' in *e*). All images are from adult individuals. I inferior; L lateral; M medial; S superior. Scale bar = 50 μm for the neuron shape and 2 μm for the higher-magnification inserts (shown in (**a**)). (Legend adapted and figure reprinted from Vásquez et al. (2018) under CCC RightsLink® license #5384490072219, originally published by John Wiley & Sons, Inc)

modulation and computational possibilities, translating differently these input demands (Guerra et al. 2023). It remains to be determined how the CoA and adjacent areas participate in these circuits, from olfactory and emotional processing to cortical default mode network and cortical "dorsal attention subdivision" functions (Caparelli et al. 2017; Sylvester et al. 2020).

As mentioned above, the occurrence of spiny pyramidal neurons has an evolutionary and ontogenetic value *per se* in terms of increased connectivity and integrated functions in neural circuits (Spruston 2008; DeFelipe 2011; Marín-Padilla 2014; Sedmak et al. 2018; Petanjek et al. 2019; Rasia-Filho et al. 2021). The human CoA contains "pyramidal-like" neurons (Vásquez et al. 2018; Guerra et al. 2023; Fig. 9.10). These cells present a triangular cell body, one main thick apical dendrite, and two basal dendrites of a similar thickness at their emerging points. The apical dendrite ramifies close to the cell body and the main apical shaft is directed to the CoA surface. Both apical and basal dendrites have variable branching points and lengths. Some dendrites extend considerable distances away from the soma. Proximal dendritic shafts usually are smooth or show a few isolated spines. The density of spines increases along the dendritic segments (Fig. 9.10a, b). Pleomorphic spines include complex, atypical/polymorphic shapes, and spines with spinule (Fig. 9.10c, d, spines are shown at different rotation angles after using a supervised machine learning-based algorithm for 3D reconstructions of Golgi-impregnated human neurons).

It is noteworthy that the CoA and the components of the amygdaloid basolateral nuclei form the largest nuclear group within the amygdaloid complex. They progressed most in size along with the evolution that led to primates, displaying a coordinated development with the neocortex, and possessing more than 50% of all neurons in the human amygdaloid complex (Stephan et al. 1987; Gloor 1997; Schumann and Amaral 2005; Chareyron et al. 2011; Yilmazer-Hanke 2012).[19] Compared to rats, the large volume of the monkey amygdala relates to a greater expansion of the basolateral complex neuropil with a higher number of glial cells and more extensive dendritic and axonal arborization (Chareyron et al. 2011; Morgan and Amaral 2014). The lateral, basal, and basomedial (BM) nuclei are the primary targets of cortical and subcortical afferent projections to the amygdaloid complex in primates (Kelly and Stefanacci 2009). Inputs to the lateral amygdala come from the hippocampal formation, thalamic, and neocortical modality-specific

[19] The functional participation in cortical circuits is not restricted to a few amygdaloid nuclei. Due to multimodal input convergence from multiple parallel circuits, directly or indirectly, amygdaloid nuclei of different origins and in "a degree of mosaic evolution" and function can connect higher-order cortical areas (Hirata et al. 2009; Yilmazer-Hanke 2012; Kedo et al. 2018). For example, "being part of an organized neural network that projects to the bed nucleus of the stria terminalis and to various hypothalamic and brainstem nuclei, the CeA and MeA subnuclei also participate in social and defensive reactions against innate and learned threats with neuroendocrine, behavioral, and sympathetic/parasympathetic responses to fearful/defensive and stressful stimuli... The MeA projects to the periallocortical, paleocortex, and archicortex, as well as to the insular agranular cortex and ventromedial prefrontal cortex... These data indicate that the MeA also participates, although with varied magnitude, in parallel circuits with different parts of the evolved neocortex for social and emotional processing in our species" (Rasia-Filho et al. 2021 and references therein)

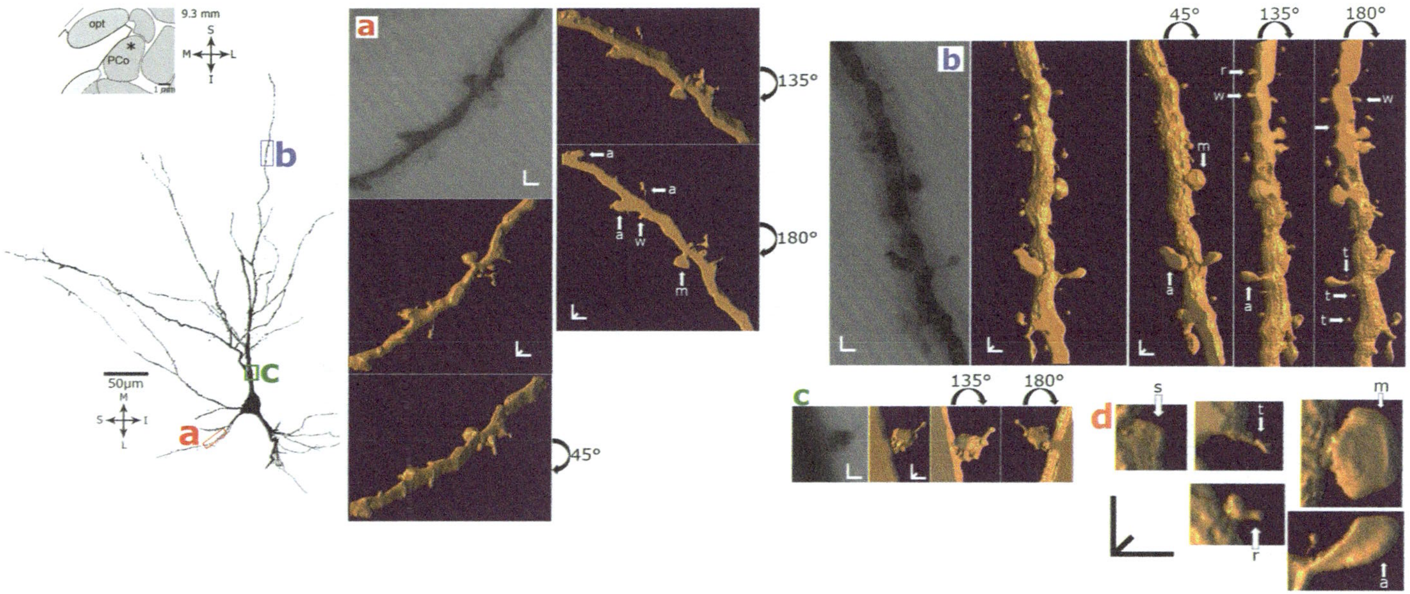

Fig. 9.10 Three-dimensional (3D) reconstruction of Golgi-impregnated dendritic spines from the human cortical amygdaloid nucleus. Brightfield microscopy images were submitted to a supervised machine learning-based approach for the 3D reconstruction of spines. (Described in Guerra et al. (2023)). (Left top) Schematic diagram of a coronal section of the human brain showing the posterior part of the cortical amygdaloid nucleus (PCO), 9.3 mm posterior to the midpoint of the anterior commissure. (Adapted from Mai et al. (2008)). opt: optic tract. (Left center) Two-dimensional reconstructed image illustrating the neuronal morphology under light microscopy. Boxes (**a**–**c**) indicate sampled segments in which 3D reconstruction of spiny proximal basal (**a**) and distal apical (**b**) dendrites showed varied spines number and shapes along the dendritic length. Curved arrows indicate that the 3D reconstructed images were rotated and are shown at different observation angles (**a**–**c**). In the inserts, dendritic spines (arrows) were classified as stubby (s), wide (w), thin (t), mushroom-like (m), ramified (r), or "atypical"/transitional ones (a), including a protruding spinule (**c**). Some examples of dendritic spines shapes and sizes are provided after 3D reconstruction at higher magnification in (**d**). Images are from an adult individual. Scale bar = 50 μm for the neuron shape and 1 μm for the dendritic spines in (**a**–**d**). I inferior, L lateral, M medial, S superior. (Legend adapted and figure reprinted from Guerra et al. (2023) under license #5441540588231 from CCC RightsLink® originally published by John Wiley & Sons, Inc.)

sensory processing areas to display responses to signal danger as quickly as possible and to initiate defensive behaviors without necessarily requiring additional neocortical processing (Janak and Tye 2015). In addition, the attention associated with the response to stimulus novelty and the amygdalo-hippocampal communication during the encoding of emotional stimuli can be translated into memory engrams and cognition (Rutishauser et al. 2015). The complexity of processing involving the human basolateral amygdala is further exemplified by its implication in late-life depressive symptom severity, associated with the dentate gyrus/hippocampal CA3 field and the lateral entorhinal cortex, during emotional episodic memory (Leal et al. 2017) and in the emotional content of dreams (Blake et al. 2019).

The human BM contains pyramidal-like (Fig. 9.11) and pyramidal (Fig. 9.12) neurons (Rasia-Filho et al. 2021, 3D reconstructions of these BM neurons available as Supplementary Figures in this reference). Pyramidal-like neurons show the apical dendrite ramifying close to the cell body while pyramidal neurons have a long apical shaft with a distal main ramification, and both may display different spatial orientations in the neuropil. These cells have branched basal dendrites and varied pleomorphic spines along the dendritic main shafts and collaterals (Figs. 9.10 and 9.11). Pyramidal-like neurons (or "modified pyramidal neurons") were also found in the human basolateral amygdaloid nucleus. They also show no preferential dendritic orientation within the neuropil and have apical and basal dendrites displaying several pleomorphic spines in secondary and further branched dendrites (see Fig. 5 in Braak and Braak 1983b). As observed from Golgi-impregnated samples, the spiny neurons of the accessory basal, basal, and lateral nucleus show different sizes, larger in the dorsolateral portions and gradually smaller in the ventromedial direction (Braak and Braak 1983b). Pyramidal neurons were also found in the lateral nucleus of the amygdaloid complex in humans (Sorvari et al. 1996).

The occurrence of morphologically and functionally specialized spiny pyramidal neurons and their organization in cortical areas might have represented a crucial form to develop circuits with proper wiring length and space (based on Guy and Staiger 2017; see also Cajal 1894; Chklovskii et al. 2002; Buckmaster 2017; Mercer and Thomson 2017; Narayanan et al. 2017). The morphological complexity of human pyramidal neurons varies along their emergence in areas initially considered as subcortical ones (e.g., some specific amygdaloid nuclei and the claustrum)[20] or transitional areas toward the allocortical, hippocampal regions, and the neocortical layers II-VI, with small pyramidal neurons in the upper layers II/III and large pyramidal neurons in the deep layer V (Ramaswamy and Markram 2015; Soltesz and Losonczy 2018; Cembrowski and Spruston 2019; Rasia-Filho et al. 2021), as shown in the following section.

[20] For the description of putative pyramidal neurons in the human basal nucleus of Meynert; see Mesulam et al. (1983), Saper and Chelimsky (1984), Morrison et al. (1987), and Liu et al. (2015). In rats, see the description of pyramidal-like neurons and Fig. 1F in Nakajima et al. (1985).

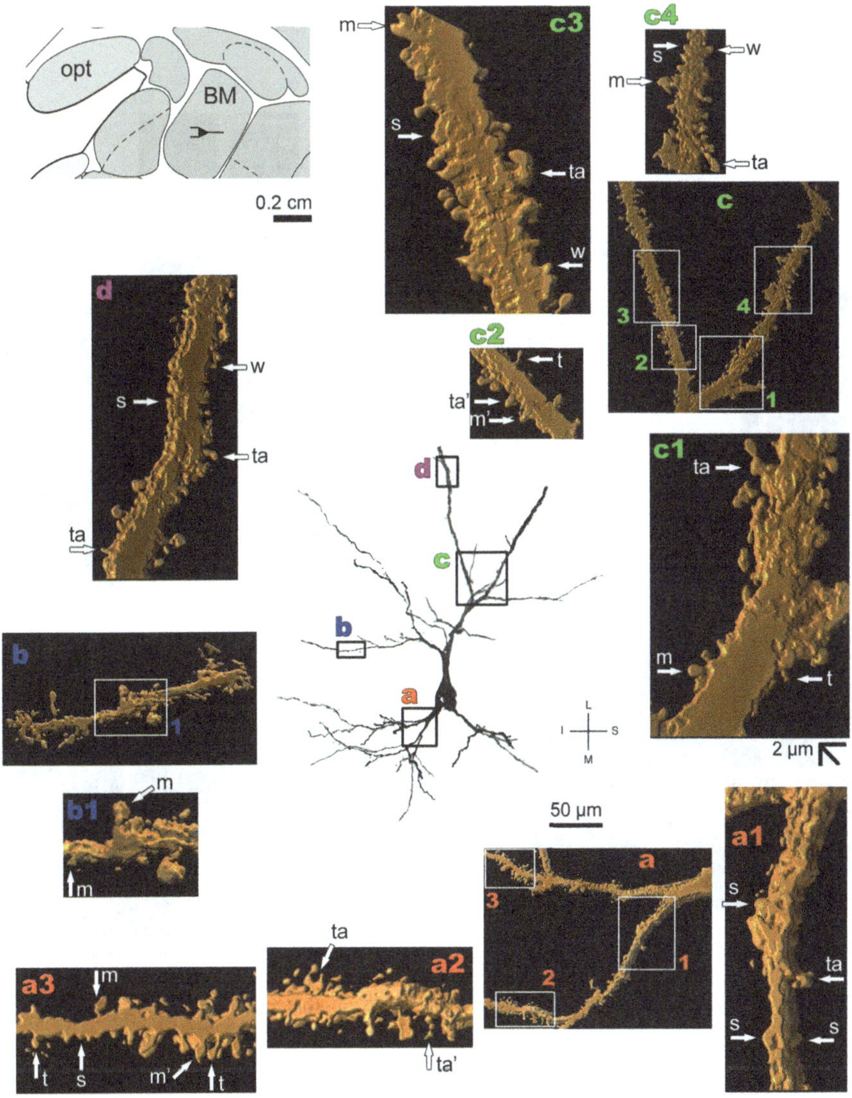

Fig. 9.11 Golgi-impregnated pyramidal-like neuron from the human basomedial amygdaloid nucleus (BM). (Top) Schematic diagram of a coronal section showing the human BM, 9.3 mm posterior to the midpoint of the anterior commissure. (Adapted from Mai et al. (2008)). opt, optic tract. (Center) Two-dimensional reconstructed image illustrating the neuronal morphology under light microscopy. The distribution, density, and shape of three-dimensional-reconstructed dendritic spines are shown in the inserts at higher magnification (**a**–**d**). Letters and numbers represent sampled parts of the same dendritic segment (**a1**–**a3**, **b1**, **c1**–**c4**). Note the high density of spines in the proximal basal dendrite (**a2**, **a3**) and along the intermediate to distal apical dendrites (**c1**, **c3**, **c4**, **d**). Spines were classified as stubby (s), wide (w), thin (t), mushroom-like (m), or transitional/atypical ones (ta). The presence of a spinule is indicated by the apostrophe attached to the corresponding spine (ta' in **a2** and **c2**, and m' in **a3** and **c2**). I inferior, L lateral, M medial, S superior. Images are from an adult individual. Scale bars = 50 µm for the neuron shape and 2 µm for the higher-magnification inserts (shown in **c1**). (Legend adapted and figure reprinted from Rasia-Filho et al. (2021) under CC BY license and Copyright © 2021 Rasia-Filho, Guerra, Vásquez, Dall'Oglio, Reberger, Jung and Calcagnotto)

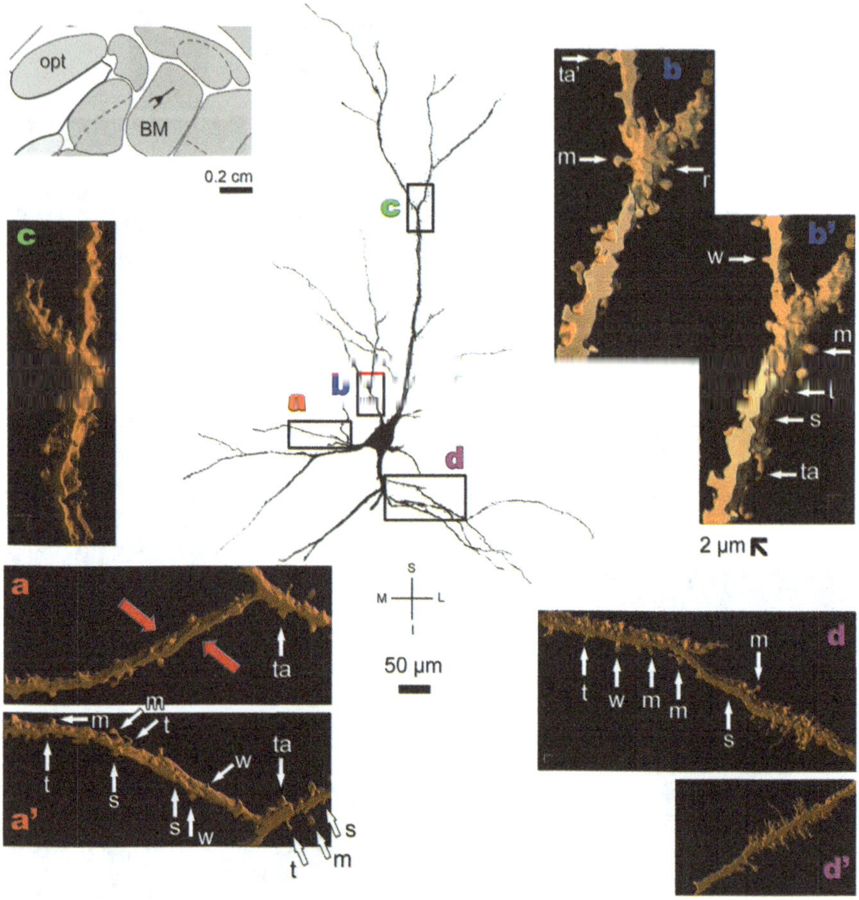

Fig. 9.12 Golgi-impregnated pyramidal neuron from the human basomedial amygdaloid nucleus (BM). (top) Schematic diagram of a coronal section showing the human BM, 9.3 mm posterior to the midpoint of the anterior commissure. (Adapted from Mai et al. (2008)). opt, optic tract. (Center) Two-dimensional reconstructed image illustrating the neuronal morphology under light microscopy. The distribution, density, and shape of three-dimensional-reconstructed dendritic spines are demonstrated in the inserts at higher magnification (**a**–**d**). The apostrophe over these letters represents an image that was rotated in space after reconstruction to show various pleomorphic dendritic spines (**a'**, **b'**, **d'**). Note that the same dendritic segment can have one side covered by spines and another side smooth (red arrows in **a**) There is a moderate (**a**, **a'**) to high (**d**, **d'**) density of spines in basal dendrites, proximal collaterals (**b**, **b'**), and distal apical dendrites (**c**). Spines were classified as stubby (s), wide (w), thin (t), mushroom-like (m), ramified (r), or transitional/atypical ones (ta). The presence of a spinule is indicated by the apostrophe attached to the corresponding spine (ta' in **b**). Images are from an adult individual. I inferior, L lateral, M medial, S superior. Scale bars = 50 μm for the neuron shape and 2 μm for the higher-magnification inserts (shown in **b'**). (Legend adapted and figure reprinted from Rasia-Filho et al. (2021) under CC BY license and Copyright © 2021 Rasia-Filho, Guerra, Vásquez, Dall'Oglio, Reberger, Jung, and Calcagnotto)

9.4.6 Hippocampus and Neocortex

An evident finding in the cerebral cortex is the abundance and functional specialization of pyramidal neurons (70–85% of all cells in the gray matter; Nieuwenhuys 1994; DeFelipe 2011; Mohan et al. 2015; Eyal et al. 2016; Kolb and Whishaw 2021). Pyramidal-like or other variations of "classic" pyramidal cells might have increased the synaptic processing and computational properties of evolved neural networks across various species (Spruston 2008; Rasia-Filho et al. 2021). For human evolution, pyramidal neurons' structure, functional properties, and increased connections may represent a fundamental advance for the elaboration of our higher nervous capacities (referred to as "the psychic cells" by Cajal 1894). Much about hippocampal neurons and neocortical pyramidal neurons was discussed in previous sections of this chapter. Further data on hippocampal formation's cellular components, connections, electrophysiology, and relevant functions can be found in other relevant works (e.g., Andersen et al. 2007; Insausti and Amaral 2012; Kandel and LeDoux 2021). Here, some representative 3D reconstructed images illustrate such neurons and the diversity of dendritic spines in human cells.

Gloor (1997) included the prepiriform–periamygdaloid area as part of the evolving mammalian allocortex, homologous with the ventral portion of the lateral pallium of amphibians and reptiles. Following the allocortical and neocortical *continuum* for the occurrence and development of pyramidal cells, the human periallocortex (i.e., the presubiculum, parasubiculum, and entorhinal cortex) contain pyramidal neurons (Insausti et al. 2017). Likewise, pyramidal cells are key elements for the hippocampal formation functional organization and display different morphological features locally (Andersen et al. 2007; Soltesz and Losonczy 2018; Rasia-Filho et al. 2021). The intrinsic circuit for information flow involves a serial pathway with several collateral and feedback projections connecting the entorhinal cortex, dentate gyrus, CA3 and CA1 fields, and subiculum (Wyss and van Groen 1995). The human CA3 hippocampal region contains pyramidal-like neurons or "short cortical pyramidal neurons" with a shape and dendritic field adapted to the position of these cells in a relatively small tissue volume (Fig. 9.13). These cells show a short apical dendrite ramifying proximally and oriented to the medial surface of the brain. Two main thick basal dendrites generate various branches close to the cell body. Several small spines are found in proximal basal shafts, whereas other pleomorphic spines are found along with the extension of both basal and apical branches. These CA3 neurons display thorny excrescences (Fig. 9.13a–c, but also found in the human MeA as reported by Dall'Oglio et al. 2015), a complex spine type with multiple PSDs and synaptic microdomains (3D reconstructed in Stewart et al. 2014; Reberger et al. 2018).[21]

[21] The CA3 pyramidal neurons are critical for encoding, storage, and recall of associative memory along the mouse transverse (or proximodistal) CA3 axis (Sun et al. 2017), showing distinct projections to CA1 cells for processing nonspatial information in rats (Nakamura et al. 2013). Mice CA3 pyramidal neurons show heterogeneous morphology, afferent connectivity, and electrophysiology. Regular-spiking pyramidal neurons have complex spines (thorny excrescences) that receive monosynaptic mossy-fiber input and were found at any position along the radial axis of the CA3 *stratum pyramidale*. On the other hand, intrinsically bursting pyramidal neurons lack thorny excrescences ("athorny" cells), receive no or very few mossy-fibers input, and preferentially occupy distal CA3 parts closer to *stratum oriens* (Hunt et al. 2018).

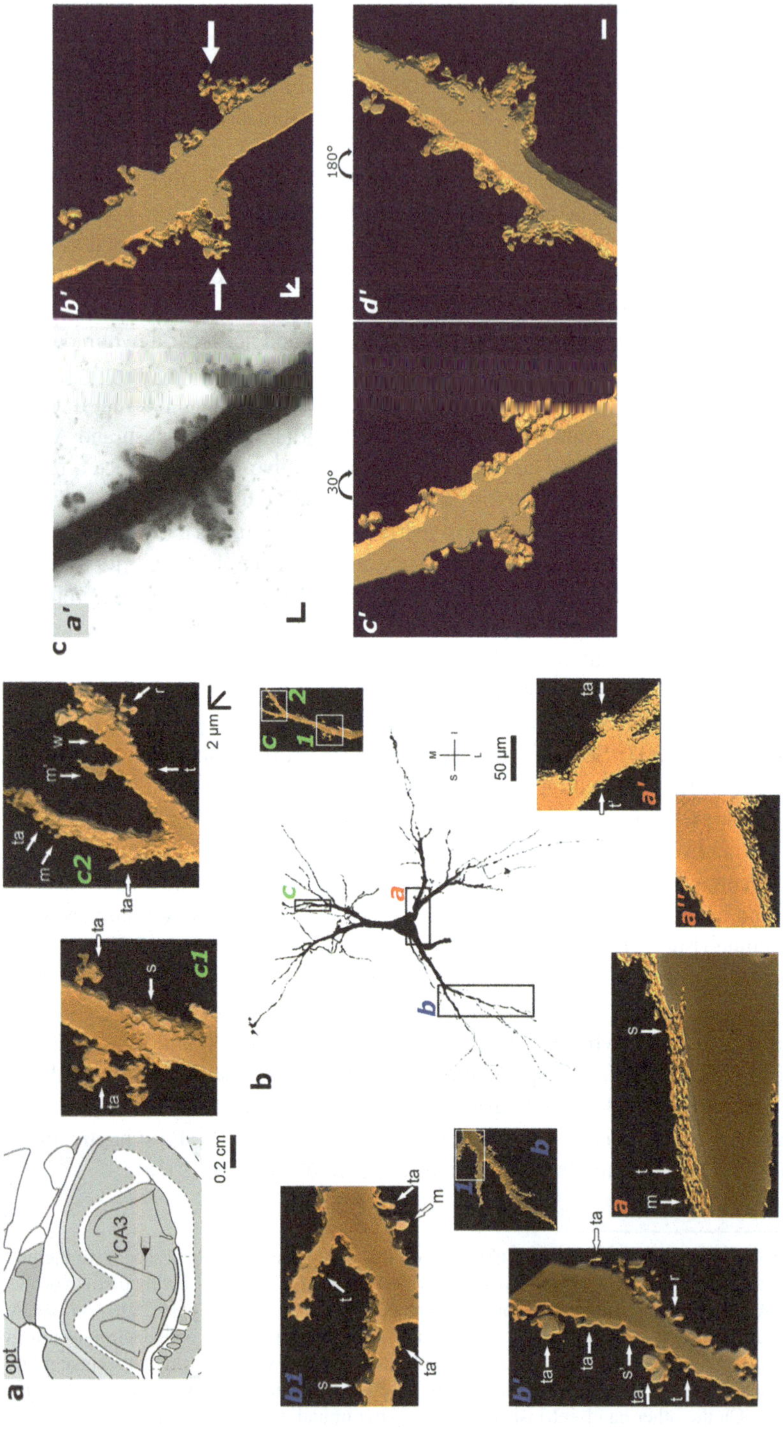

Fig. 9.13 *(caption on p. 433)*

The human CA1 hippocampal field shows a variety of pyramidal neuron shapes with different dendroarchitectonics and a variety of spine shapes and sizes, as revealed by the 3D reconstruction of Golgi-impregnated neurons under brightfield microscopy (Figs. 9.14 and 9.15; see relevant data based on fluorescent images for the dendroarchitecture of hippocampal cells in Benavides-Piccione et al. (2020)). Some CA1 neurons display basal and apical dendrites with a relatively short aspect and various branching points (Fig. 9.14), while other cells in deeper positions have straight and very long (at the order of millimeters) apical dendrites with scarce ramification (Fig. 9.15). Basal dendrites emerge at different points along the triangular or ovoid cell body. In both short and long CA1 pyramidal neurons, the apical dendrite is oriented to the surface of this brain area. These pyramidal cells coexist within the same neuropil volume and receive inputs at different dendritic fields (Benavides-Piccione et al. 2020). Basal and apical dendrites display an impressive number of pleomorphic spines, some with spinule (Figs. 9.14a–f and 9.15a–f). The basal dendrites of a long pyramidal neuron can show a huge density of pleomorphic spines along proximal to distal segments (Fig. 9.15a). Likewise, very long apical branches are covered by a huge density of all types of spines densely intermingled in the same dendritic segment (Fig. 9.15b–f, 3D reconstructions of human CA3 and CA1 hippocampal neurons available as Supplementary Figures in Rasia-Filho et al. 2021).

Morphological data suggest that hippocampal pyramidal neurons have a multitude of modulatory possibilities for the integration of information. Indeed, after noting the density of spines per cell, one has also to consider that there are approximately 16 million and 2.7 million neurons in human CA2-3 and CA1 hippocampal areas, respectively (West and Gundersen 1990). Synaptic inputs occur more on pleomorphic spines than on spine-free dendritic shafts in very long dendrites, which opens many possibilities to elaborate hypotheses on the morphological and functional relationships. For example, (1) how different spiny dendritic domains and

Fig. 9.13 Golgi-impregnated short pyramidal neuron from the human CA3 hippocampal region. (**a**) Schematic diagram of a coronal section showing the human CA3 region, 16 mm posterior to the midpoint of the anterior commissure. (Adapted from Mai et al. 2008). opt, optic tract. (**b**) Two-dimensional recon structed image illustrating the neuronal morphology under light microscopy. The distribution, density, and shape of three-dimensional-reconstructed dendritic spines are demonstrated in the inserts at higher magnification (*a-c*). Letters and numbers represent sampled parts of the same dendritic segment (*b1*, *c1*, *c2*). The apostrophe over the letters represents an image that was rotated in space after reconstruction to show various pleomorphic dendritic spines (*a'*, *a"*, *b'*). There is a high density of small spines in the proximal basal dendrite (*a*) and a moderate to high density of pleomorphic spines in intermediate dendritic segments of basal (*b*, *b1*, *b'*) and apical dendrites (*c1*, *c2*). Spines were classified as stubby (s), wide (w), thin (t), mushroom-like (m), ramified (r), or transitional/atypical ones (ta). Note the presence and shape of thorny excrescences in both basal (ta in *a'*) and apical (ta in *c1*) dendrites. The presence of a spinule is indicated by the apostrophe attached to the corresponding spine (t' in *a'*, s' in *b'*, and m' in *c2*). I inferior, L lateral, M medial, S superior. Scale bars = 50 μm for the neuron shape and 2 μm for the higher-magnification inserts (shown in *c2*). (Legend adapted and figure reprinted from Rasia-Filho et al. (2021) under CC BY license and Copyright © 2021 Rasia-Filho, Guerra, Vásquez, Dall'Oglio, Reberger, Jung, and Calcagnotto). (**c**) Three-dimensional reconstruction (using a supervised machine learning-based approach) of a thorny excrescence from the same neuron shown in (**a**). Note the convoluted form of this dendritic spine (arrows) and the multiple heads and stalks, from the original Golgi data (*a'*) to the final reconstruction and observation at different rotation angles (*b'*-*d'*). Images are from an adult individual. Scale bar = 2 μm

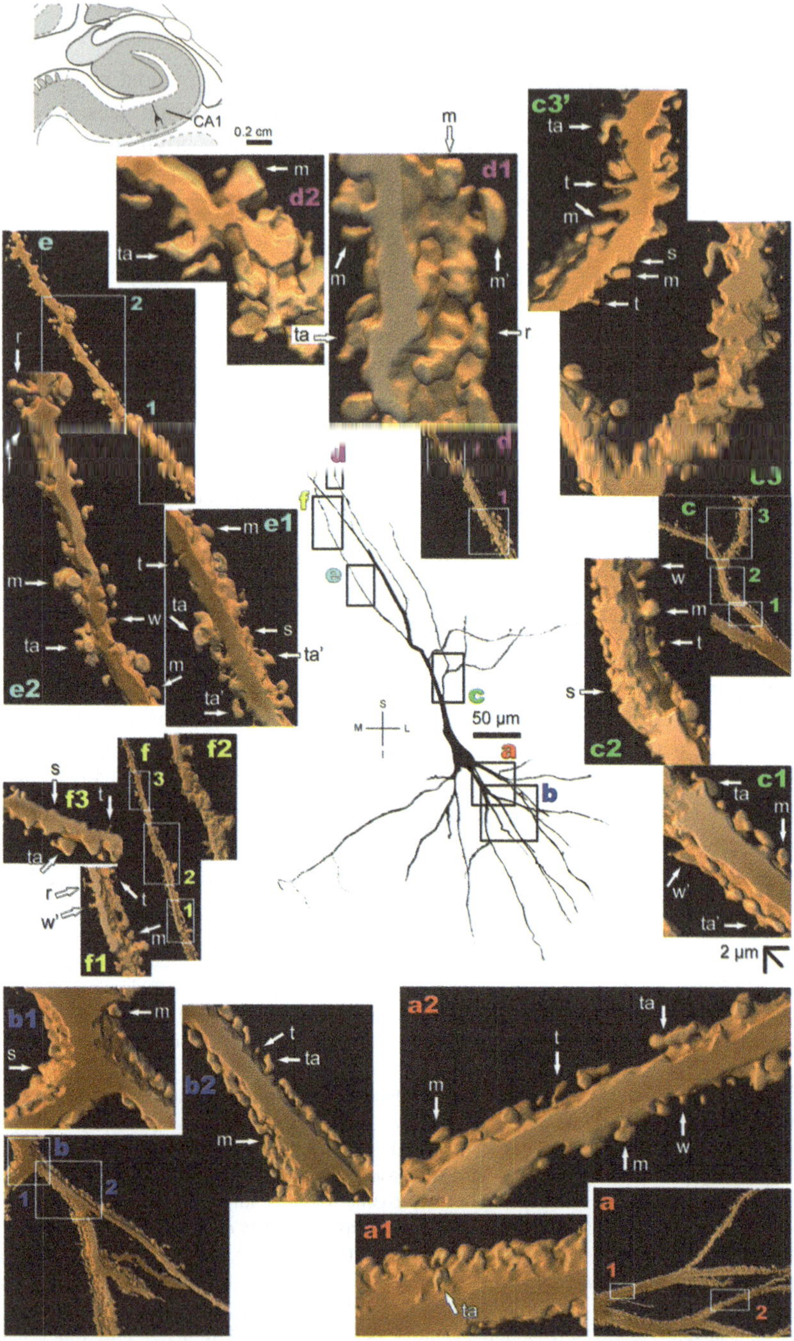

Fig. 9.14 Golgi-impregnated small pyramidal neuron from the human CA1 hippocampal region. (Top) Schematic diagram of a coronal section showing the human CA1 region, 22.6 mm posterior to the midpoint of the anterior commissure. (Adapted from Mai et al. (2008)). opt, optic tract. (Center) Two-dimensional reconstructed image illustrating the neuronal morphology under light microscopy. The distribution, density, and shape of three-dimensional-reconstructed dendritic spines are demonstrated in the inserts at higher magnification (**a–f**). Letters and numbers represent

synapses might be compartmentalizing the processing of information and what are the possibilities of coactivation and cooperativity between neighbor spines; (2) how distant synaptic inputs impact the function of other segments when signals are transmitted anterogradely along the dendritic tree and what is the resultant "code" elaborated from multiple activated sites; and (3) what is the balance between stable and plastic spines related to the need for processing new tasks and associating them with older ones within circuits and different brain areas. Considering that hippocampal spines are involved with memory formation, stable spines would serve as a common substrate for output activities that code this action. Otherwise, a certain percentage of spines should always be in a constant turnover to allow enough space in the membrane for processing new stimuli and determining the "weight" of these new inputs that occur every day along the lifespan (see a parallel discussion in Leopold et al. (2019)). It is also possible that small dendritic spines in these neurons would participate in long-term potentiation induction while large, thin, and mushroom spines would modulate different steps for long-term memory and circuitry plasticity (Matsuzaki et al. 2004; Bourne and Harris 2007; Kasai et al. 2010; Yuste 2013).

The turnover and functional implications of spines can be site- and task-specific. The dynamics of appearance and shrinkage of spines can be fast and in the order of minutes after activation in the hippocampus (Toni et al. 1999), but the loss of motor cortex spines many hours after inducing memory for a task impairs the execution of a learned behavior (Hayashi-Takagi et al. 2015). Then, the image of these spines likely represents the conjunct of spines needed to balance between the maintenance of long-lasting functional connections and the available synaptic sites with functional individuality to receive new, plastic network demands, keeping accurate integration of inputs and individual synaptic tuning. Such neuron is one among several others whose morphology and integrated function elaborate the memories and self-identity of each unique person. Interestingly, the principles controlling spine formation, retreat, or shape reconfiguration reflect how the adult brain remains adaptable and, with spines, has a capacity for experience-dependent alterations that seems "endless" (Leopold et al. 2019).

Cortical neurons are not randomly interconnected within the neuropil volume; rather, wiring specificity can result from the genetically defined identity of its constituents, morphological properties, electrical activity (Udvary et al. 2022), and activity- and circuitry-dependent features. It is then comprehensible why pyramidal

Fig. 9.14 (continued) sampled parts of the same dendritic segment (**a1**, **a2**, **b1**, **b2**, **c1**–**c3**, **d1**, **d2**, **e1**, **e2**, **f1**–**f3**). The apostrophe over the letters represents an image that was rotated in space after 3D reconstruction to show various pleomorphic dendritic spines (**c3'**). Note an intermediate to a high density of pleomorphic spines in the proximal segments of the basal (**a**, **b**) and apical (**c**) dendrites, which continue toward distal segments (**d**–**f**). Spines were classified as stubby (s), wide (w), thin (t), mushroom-like (m), ramified (r), or transitional/atypical ones (ta). Note the occurrence of spines with different shapes along the same dendritic segment, some relatively apart from the others (e.g., in **a2**, **e2**) or in groups (e.g., **b2**, **c3**, **d1**, **d2**). The presence of a spinule is indicated by the apostrophe attached to the corresponding spine (w' and ta' in **c1**, m' in **d1**, ta' in **e1**, and w' in **f1**). Images are from an adult individual. I inferior, L lateral, M medial, S superior. Scale bars = 50 μm for the neuron shape and 2 μm for the higher-magnification inserts (shown in **c1**). (Legend adapted and figure reprinted from Rasia-Filho et al. (2021) under CC BY license and Copyright © 2021 Rasia-Filho, Guerra, Vásquez, Dall'Oglio, Reberger, Jung, and Calcagnotto)

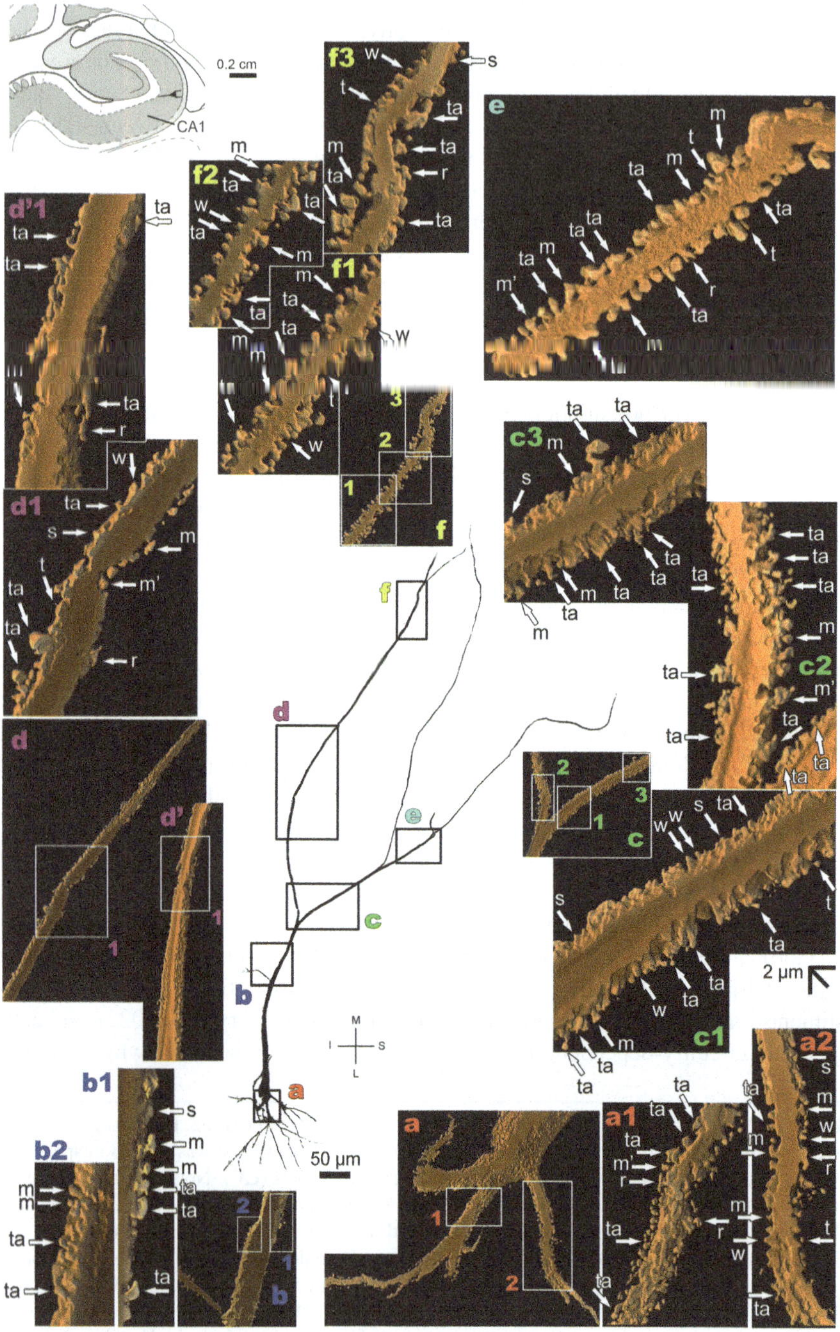

Fig. 9.15 Golgi-impregnated long pyramidal neuron from the human CA1 hippocampal region. (Top) Schematic diagram of a coronal section showing the human CA1 region, 22.6 mm posterior to the midpoint of the anterior commissure. (Adapted from Mai et al. (2008)). opt, optic

neurons vary in shape and properties within and between cortical areas, layers, and networks of different processing complexities (Moradi Chameh et al. 2021; Planert et al. 2021; Rasia-Filho et al. 2021). For example, pyramidal cell morphology shows interindividual variability and is not the same across the human anterior temporal lobe, as evidenced by different patterns of dendritic complexity and spine number (Benavides-Piccione et al. 2021). The greatest spine density was seen in the smallest and least complex layer III cells in the posterior temporal region T2c. On the other hand, neurons displaying the greatest size and high number of dendritic spines were found in the temporal pole. The estimated number of basal dendritic spines per distance from the soma showed a maximum number located around 90 μm, reaching a peak of approximately 1,000 spines, and an estimated total number of ~13,000 spines. Pyramidal cells in layer V were larger in size and number of spines (~ 1,200 spines at 110 μm and an estimated total number of 22,000 spines) than those in layer III in two individuals (Benavides-Piccione et al. 2021). An example of a Golgi-impregnated layer V long pyramidal neuron from the human temporal lobe is shown in Fig. 9.16. This cell has basal dendrites ramifying horizontally or directed to the adjacent layer VI, a long and straight apical dendrite with a main vertical shaft oriented to the superficial layers and some collateral branches. A moderate density of pleomorphic spines can be observed even in proximal segments of basal dendrites (Fig. 9.16a), whereas, in the apical dendrite, spines show a moderate-to-high density from proximal to intermediate segments (Fig. 9.16b, c) and a moderate density distally (Fig. 9.16d). Spines may show heterogeneous shapes and sizes along with the dendritic length, including various atypical/polymorphic forms (the 3D reconstruction of this layer V neocortical neuron is available as Supplementary Figures in Rasia-Filho et al. 2021).

Other neuron types compose the cortical cytoarchitecture and are found nearby large pyramidal neurons in layer V of higher-order human neocortical areas. An example is the "spindle-shaped" neuron in the posteromedial precuneus (PC) region (Fuentealba-Villaroel et al. 2022; Fig. 9.17). These cells are not so abundant and do

Fig. 9.15 (continued) tract. (Center) Two-dimensional reconstructed image illustrating the neuronal morphology under light microscopy. The distribution, density, and shape of three-dimensional-reconstructed dendritic spines are demonstrated in the inserts at higher magnification (**a–f**). Letters and numbers represent sampled parts of the same dendritic segment (**a1**, **a2**, **b1**, **b2**, **c1–c3**, **d1**, **f1–f3**). The apostrophe over the letters represents an image that was rotated in space after 3D reconstruction to show various pleomorphic dendritic spines (**d'**, **d'1**). Note the multitude of types and very high density of pleomorphic spines in the proximal segments of the basal dendrites (**a1**), moderate-to-high spine density in the proximal to intermediate segments of the apical dendrite (**b1**, **b2**), and very high density of spines in intermediate apical segments (**c1–c3**). Toward the distal parts of the apical dendritic branches, spines show an intermediate to high (**d1**, **d'1**, **e**) and a very high density of spines close to final shafts (**f1–f3**). Dendritic spines were classified as stubby (s), wide (w), thin (t), mushroom-like (m), ramified (r), or transitional/atypical ones (ta). Note that spines of different shapes and sizes coexist along the same dendritic segments (e.g., **a1**, **c1–c3**, **e**, **f1–f3**). The presence of a spinule is indicated by the apostrophe attached to the corresponding spine (m' in **a1**, **d1**, **e**). Images are from an adult individual. I inferior, L lateral, M medial, S superior. Scale bars = 50 μm for the neuron shape and 2 μm for the higher-magnification inserts (shown in **c1**). (Legend adapted and figure reprinted from Rasia-Filho et al. (2021) under CC BY license and Copyright © 2021 Rasia-Filho, Guerra, Vásquez, Dall'Oglio, Reberger, Jung, and Calcagnotto)

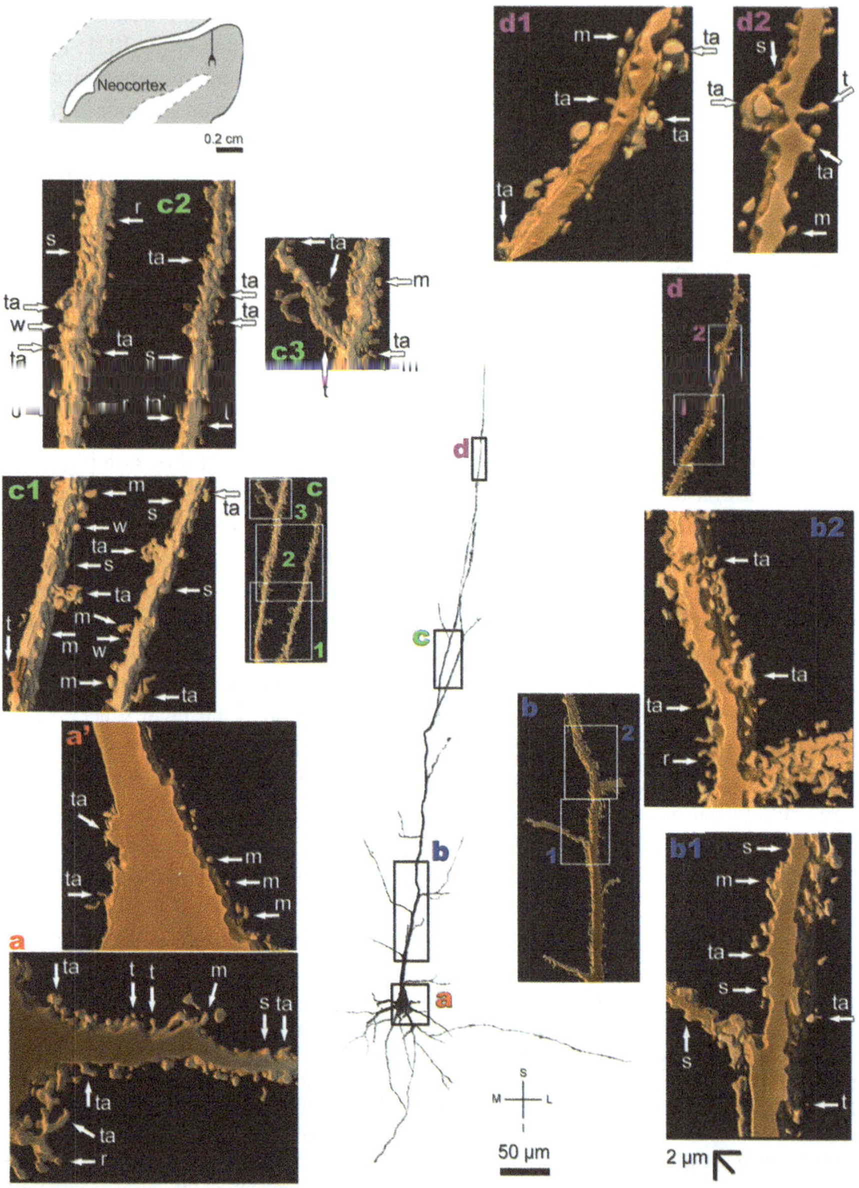

Fig. 9.16 Golgi-impregnated large pyramidal neuron from layer V of the human anterolateral temporal lobe. (Top) Schematic diagram of a coronal section showing the corresponding neocortical region, 9.3 mm posterior to the midpoint of the anterior commissure. (Adapted from Mai et al. (2008)). opt, optic tract. (Center) Two-dimensional reconstructed image illustrating the neuronal morphology under light microscopy. The distribution, density, and shape of 3D-reconstructed dendritic spines are demonstrated in the inserts at higher magnification (**a–d**). Letters and numbers represent sampled parts of the same dendritic segments (**b1**, **b2**, **c1–c3**, **d1**, **d2**). The apostrophe (in **a'**) represents an image that was rotated in space after 3D reconstruction to show various somatic

not have the same morphology compared to pyramidal ones. Spiny spindle-shaped cells were found in layer V merging with layer VI in the PC of adult humans and have a characteristic vertical, elongated cell body, two main longitudinally oriented ascending and descending dendrites with varied branching aspect and radial orientation. Pleomorphic spines with a sparse to moderate density cover proximal to distal segments. Some spines show atypical/polymorphic forms, including convoluted shapes with different neck thicknesses and/or multiple bulbous structures (Fig. 9.17f2 right). Other cells in the same location display a similar cell body shape, straight dendritic shafts devoid of main ramifications in both ascending and descending branches, a moderate density of small dendritic spines, or few spines along the dendritic shafts (Fuentealba-Villaroel et al. 2022). There are no electrophysiological recordings or transcriptomic profile studies of these cells currently, but it is likely that these cells also participate in higher-order neural processing circuitries involving the PC. For example, the human PC modulates sensorimotor, visual, cognitive/associative information, episodic memory retrieval, consciousness, self-centered mental imagery, and empathy (Cavanna 2007; Margulies et al. 2009, 2016; Zebarjadi et al. 2021; Lyu et al. 2023; see also Messina et al. 2023). However, based on their distinct dendritic shape, spindle-shaped neurons probably process synaptic inputs in distinct ways than the adjacent pyramidal cells.

The cell body shape and the display of longitudinal primary dendrites at opposite somatic poles of "spindle-shaped" neurons in the human PC are intriguing. That is because they resemble the general somatic and primary dendrites' shape of layer V von Economo neurons (VENs) from the human ACC (Fig. 9.18; see a critical discussion in Banovac et al. 2019, 2021 and Fuentealba-Villaroel et al. 2022), and both the PC and the ACC can be part of a general intelligence circuit (Bruton 2021). In this regard, the human VENs have been described in restricted cortical areas with some features that are not similar to layer V pyramidal neurons. VENs have phylogenetic, ontogenetic, immunohistochemical, and electrophysiological particularities as well as a vulnerability in diseases with social and emotional deficits (Allman et al. 2005, 2010, 2011a, b; Stimpson et al. 2011; Raghanti et al. 2015; Cobos and Seeley 2015; Dijkstra et al. 2018; Jacot-Descombes et al. 2020; see also Yang et al. 2019 and Hodge et al. 2020 for transcriptomic data). For example, VENs are more numerous in the ACC and frontoinsular cortex of humans than in apes (Allman et al. 2010). VENs are rarely seen during gestation (i.e., VENs are not identified morphologically until gestational week 36), but their number increases during the first

Fig. 9.16 (continued) and dendritic spines. Note the moderate density of pleomorphic spines in the proximal segments of the basal dendrites (**a**) and along the apical dendrite (**b1**). Moderate-to-high density of dendritic spines is observed toward intermediate (**b2**) to distal (**c1**–**c3**) segments of the apical dendrite. Moderate spine density is observed at distal apical dendritic segments (**d1**, **d2**). Some spines were observed in the cell body (**a'**). Dendritic spines were classified as stubby (s), wide (w), thin (t), mushroom-like (m), ramified (r), or transitional/atypical ones (ta). Dendritic spines of very different shapes and sizes coexist along the same dendritic segments (e.g., **a**, **b2**, **c1**, **d1**). Images are from an adult individual. I inferior, L lateral, M medial, S superior. Scale bars = 50 μm for the neuron shape and 2 μm for the higher-magnification inserts (shown in **b1**). (Legend adapted and figure reprinted from Rasia-Filho et al. (2021) under CC BY license and Copyright © 2021 Rasia-Filho, Guerra, Vásquez, Dall'Oglio, Reberger, Jung, and Calcagnotto)

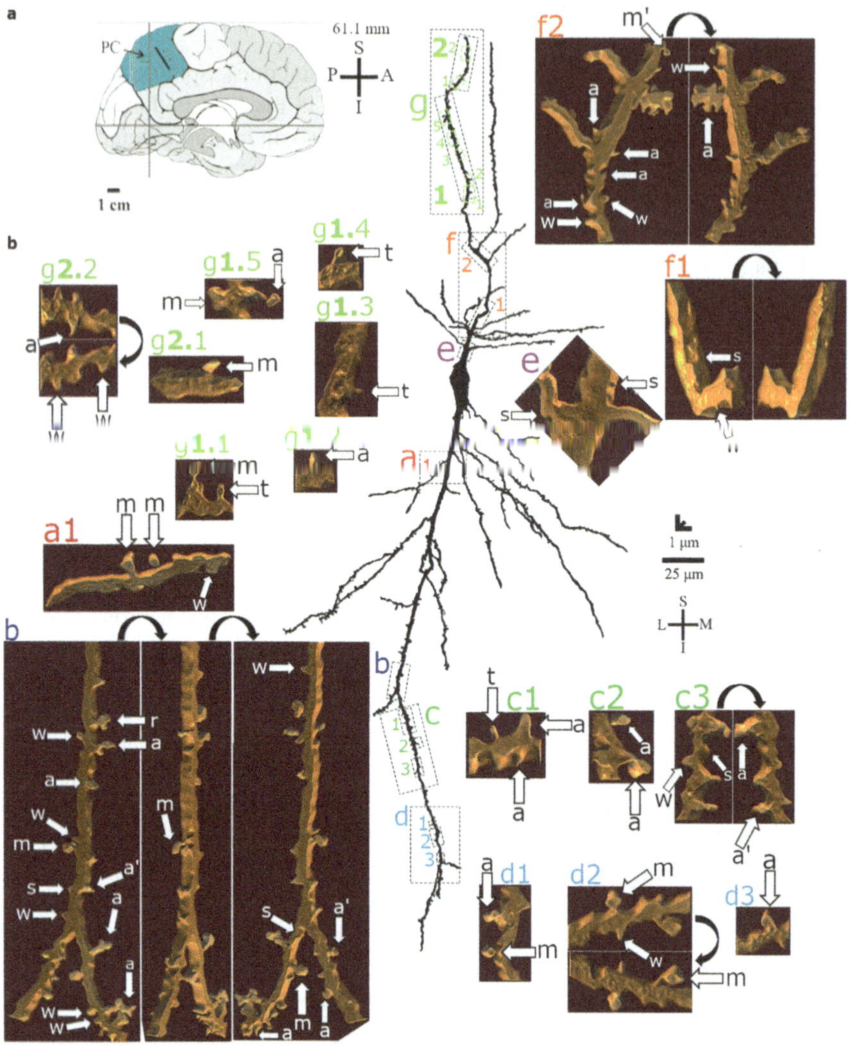

Fig. 9.17 Golgi-impregnated spindle-shaped neuron from layer V in transition to layer VI in the human precuneus (central region), part of the posteromedial cerebral cortex. (Top) Schematic diagram of a coronal section showing the corresponding neocortical region, 61.1 mm posterior to the midpoint of the anterior commissure. (Adapted from Mai et al. (2008, 2016)). opt, optic tract. (Center) Two-dimensional reconstructed image illustrating the neuronal morphology under light microscopy. The distribution, density, and shape of 3D-reconstructed dendritic spines are demonstrated in the inserts at higher magnification (**a–g**). Letters, numbers, and subdivisions represent sampled parts of the same dendritic segment (**c1-3**, **d1-3**, **f1-2**, **g1**.1-2.2). Curved arrows indicate that the 3D reconstructed images were rotated and are shown at different observation angles (**b**, **c3**, **d2**, **f1**, **f2**, **g2**.2). Note the distribution of a low-to-moderate density of pleomorphic spines along the dendritic segments. Spines were classified as stubby (s), wide (w), thin (t), mushroom (m), ramified (r), with a transitional (t), or atypical aspect (a). The presence of a spinule is indicated by the apostrophe attached to the corresponding spine (a' in **b** and **c3**). Images are from an adult individual. I inferior, L lateral, M medial, S superior. Scale bars = 25 μm for the neuron shape and 1 μm for the inserts. (Legend adapted and figure reprinted from Fuentealba-Villarroel et al. (2022) under

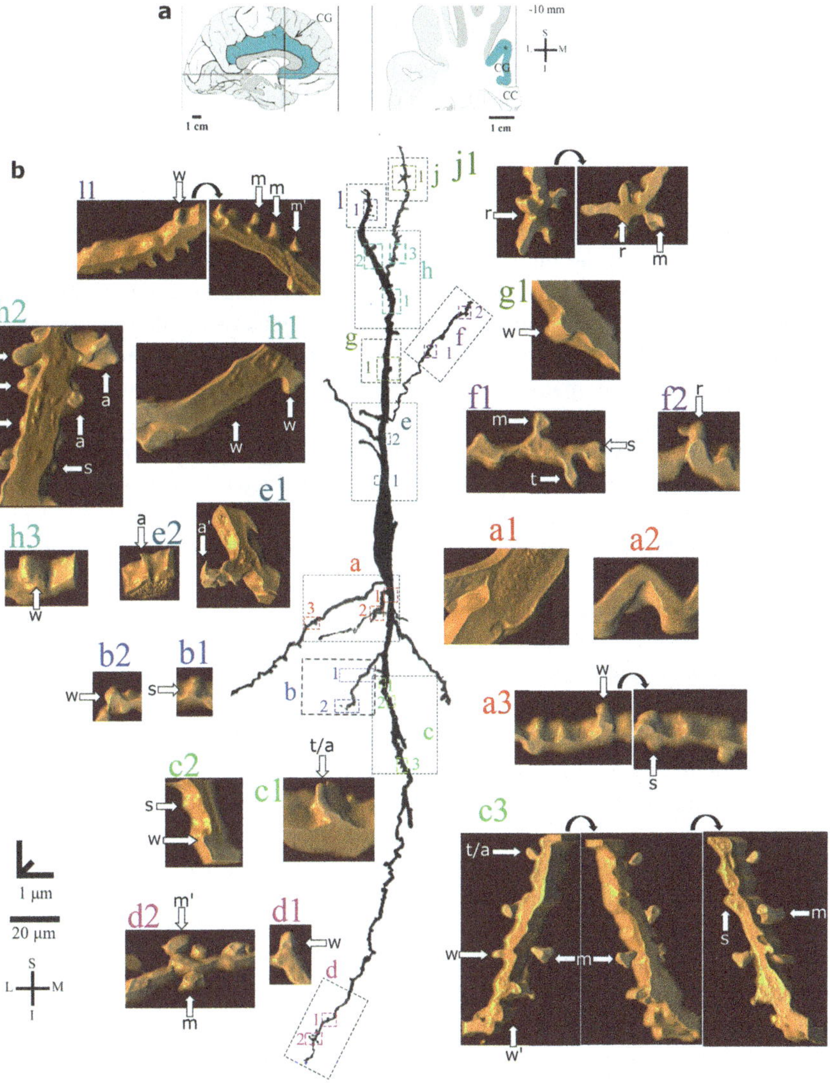

Fig. 9.18 Golgi-impregnated von Economo neuron from layer V in the human cingulate cortex. (Top) Schematic diagram of a coronal section showing the corresponding neocortical region, 10 mm anterior to the midpoint of the anterior commissure. (Adapted from Mai et al. 2008). opt, optic tract. (Center) Two-dimensional reconstructed image illustrating the neuronal morphology under light microscopy. The distribution, density, and shape of 3D-reconstructed dendritic spines are demonstrated in the inserts at higher magnification (**a–l**). Letters, numbers, and subdivisions represent sample parts of the same dendritic segment (**a1-3**, **b1**, **b2**, **c1-3**, **d1**, **d2**, **e1**, **e2**, **f1**, **f2**, **h1-3**). Curved arrows indicate that the 3D reconstructed images were rotated and are shown at different observation angles (**a3**, **c3**, **j1**, **l1**). Note the intermediate density of spines along the dendritic length, but the variety of spine shapes in main descending and ascending shafts and collaterals. Spines were classified as stubby (s), wide (w), thin (t), mushroom (m), ramified (r), with a transitional (t/), or atypical aspect (a). The presence of a spinule is indicated by the apostrophe attached to the corresponding spine (w' in **c3**, m' in **d2**, a' in **e1**, m' in **l1**). Images are from an adult individual. I inferior, L lateral, M medial, S superior. Scale bars = 20 μm for the neuron shape and 1 μm for the inserts. (Legend adapted and figure reprinted from Correa-Júnior et al (2020) under CC BY license and Copyright © 2020 Correa-Júnior, Renner, Fuentealba-Villarroel, Hilbig, and Rasia-Filho)

8 months after birth, decrease afterward, and reach stable values at 4–8 years (Allman et al. 2010, 2011a). The number of VENs remains constant along the lifespan in individuals with average cognition, and a higher density of VENs exists in the ACC of individuals (≥age 80) who show outstanding memory abilities (Gefen et al. 2015). Based on transcriptomic data, human VENs may be projecting cells to extratelencephalic, subcortical targets (Hodge et al. 2020), whereas, in gorillas, VENs also project to the inferior frontal gyrus, inferotemporal cortex, hippocampus, septum, and amygdala (Allman et al. 2010). VENs may also innervate the midbrain periaqueductal gray and the parabrachial nucleus of monkeys (discussed in Evrard et al. 2012), likely associating cortical processing with emotions and the control of sympathetic/parasympathetic sites in the brainstem and spinal cord (Cobos and Seeley 2015, Jacot-Descombes et al. 2020).

The human ACC VENs were reported in a morphological *continuum* from sparsely branched (Watson et al. 2006) to more extensively ramified cells with varied collateral branches with distinct distribution, density, and shape of dendritic spines (Correa-Júnior et al. 2020). VENs can display a brush-like aspect for the descending dendrites from where emerges an axon (Banovac et al. 2019, depicted in layer V of a 1-month-old human infant insular cortex by Cajal (1909–1911) and shown in Banovac et al. (2021)). If VENs show more or less branched ascending and descending dendrites, these morphological features imply a higher surface for synaptic processing and plasticity modulated by pleomorphic spines within an extended neuropil volume. Furthermore, the firing output of VENs would be affected by a selective location of the axon along the dendritic architecture (Banovac et al. 2019, 2021). Jointly with some particular intrinsic properties of putative VENs' membrane (as described in the human frontoinsular cortex, Hodge et al. 2020), the heterogeneity of dendritic arbor ramification and spine number suggest a heterogeneous capacity between VENs for input sampling, computational power, and elaboration of output codes as projection neurons (Correa-Júnior et al. 2020; 3D reconstruction of human PC and ACC neurons are available as Supplementary Figures in Fuentealba-Villarroel et al. 2022 and Correa-Júnior et al. 2020, respectively).

In addition, the human cingulate cortex (from 62- to 91-year-old individuals) shows large pyramidal cells of Betz located in layer Vb of the primitive gigantopyramidal field at the bottom of the cingulate sulcus (Braak and Braak 1976). These Golgi-impregnated cells have a pyramidal or pear-shaped cell body, short and slender (sometimes thick) branched basal dendrites, a varied number of dendrites emerging from the lateral surface of the soma, and a moderate number of spines with "thread-like stems and rounded knobs." Spines are evenly distributed along the whole length of the dendrite. Proximal parts of dendrites are covered with spiny appendages, the soma is often covered with a small number of spiny projections with large rounded knobs, and the thick main shaft has numerous spines at proximal segments, which continues with "numerous long-stalked, bent, and twisted spines closely spaced together." At most distal parts, the apical dendrite bifurcates repeatedly. These dendrites have "sessile and stalked spines ending in a knob" and occasionally "small clusters of spines may be followed by short zones almost devoid of spiny projections" (Braak and Braak 1976; see also the description and images of dendrites with a high density of pleomorphic spines of Betz cells in the human precentral gigantopyramidal field in this same work).

Together, the cingulate cortex microcircuits exemplify how coexisting different types of neurons with discrepant proportions can integrate multimodal inputs and elaborate coherent output displays. The cingulate cortex is an interesting part of the "proisocortex" in the paralimbic cortex, a phylogenetically older part compared to the neocortex (Braak 1979; Pandya et al. 2015; Kolb and Whishaw 2021). However, it is much more than a primitive stage of cortical evolution and comprises neural networks for high-order attentional processes/executive functions with sensory processing, flexible control of the motor, and affective information for goal-directed and exploratory behavior within complex social interactions (Allman et al. 2001; Vogt 2015; Medalla et al. 2022). The human cingulate cortex elaborates cognition and premotor planning with motivational features, social awareness, and emotions such as love, trust, empathy, deception, guilt, and fear, as well as evokes heart rate and arterial pressure, respiratory, and gastrointestinal responses (Allman et al. 2001, 2010, 2011a; Watson et al. 2006; Xuan et al. 2016 and references therein).

It is also noteworthy that morphologically and biophysically distinct pyramidal neurons in the ACC project to the dorsal premotor cortex and the amygdaloid nuclei in rhesus monkeys. Layer V pyramidal cells projecting to amygdaloid nuclei show greater excitability, apical dendritic complexity, spine densities, and diversity of inhibitory inputs than those targeting the dorsal premotor cortex (Medalla et al. 2022). The projecting neurons in the ACC also stimulate two pathways simultaneously to generate human speech: one via PAG for vocalization and another via Broca´s area to convert vocalization into words and sentences (Holstege and Subramanian 2016).

9.5 Human Spine Features Revealed by Further Microscopic Techniques

Besides the "classic" Golgi-impregnation procedure mentioned above, other methodological approaches corroborated and expanded the study of dendritic spines in humans (see also Chap. 2 in this book). One of them is the use of DiI, a fluorescent dye of the carbocyanine family, under confocal microscopy. Fine-powdered dye DiI extracellularly diffuses over intact cellular membranes and reveals fine 3D features of dendritic spine morphology. Spine connectivity can be assessed when associated with immunohistochemical markers for presynaptic proteins, such as synaptophysin (Kim et al. 2007; Rasia-Filho et al. 2010). Confocal images using DiI dye allow the visualization of human MeA dendritic spines with varied stubby/wide, thin, mushroom, ramified, or intermediate/atypical forms within a *continuum* of sizes, either isolated or grouped, and from proximal to distal dendritic branches (Fig. 9.19). Large and thin spines with a gemmule appearance and filopodium found in this brain area suggest additional possibilities for spinogenesis in adult individuals (Dall'Oglio et al. 2015; Fig. 9.19f).

The variety of heterogeneous spines in the human MeA includes thin spines (Fig. 9.19a) and stubby/wide spines (Fig. 9.19b) ranging from small to relatively large forms. The head sizes of mushroom spines can also vary and display a large and flat appearance (Fig. 9.19c, first to sixth images) or irregular surface (Fig. 9.19c, last image). Ramified spines show a single large stalk branching into two heads

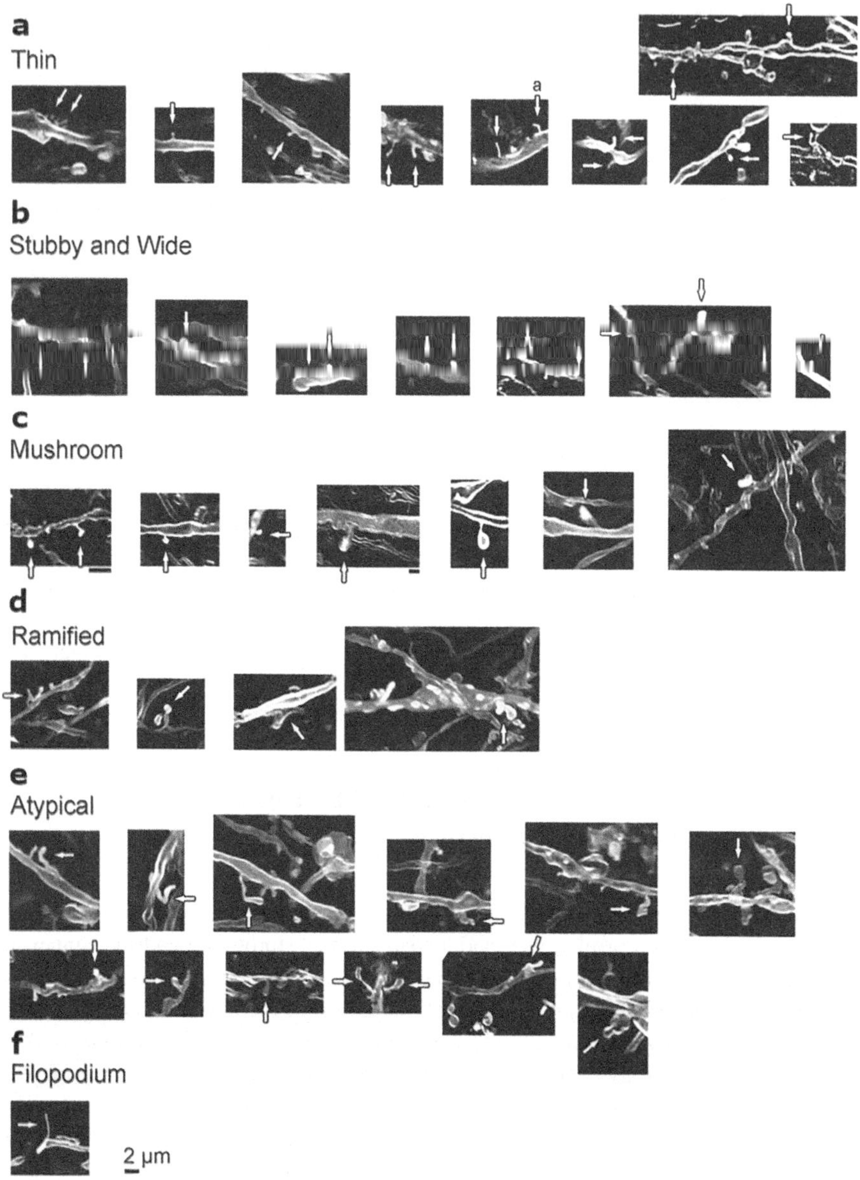

Fig. 9.19 Three-dimensional reconstructed fluorescent confocal microscopy image (using extracellularly applied DiI dye) of dendritic spines from the human medial amygdaloid nucleus. Differently shaped spines (arrows) were classified as (**a**) thin, (**b**) stubby/wide, (**c**) mushroom, (**d**) ramified, and (**e**) transitional or atypical forms. Note the varied shapes and sizes within each spine class. (**f**) A filopodium is also shown with an elongated neck and no evident head. Images are from an adult individual. Scale bar = 2 μm. (Legend adapted and figure reprinted from Dall'Oglio et al. (2015) under CCC RightsLink® license #5385561144316, originally published by John Wiley & Sons, Inc)

(Fig. 9.19d, first and second images), a single stalk giving rise to a branch with no identified spine head (Fig. 9.19d, third image), or a large stalk and separated heads (Fig. 9.19d, last image). Some transitional/atypical spines (Fig. 9.19e) might be intermediate forms of previously classified spines, considering the combination of the length and thickness of the spine neck and the size and shape of the spine head (Fig. 9.19e, first to third images in the second row). Other atypical/polymorphic spines include the following: (1) double spine (Fig. 9.19a, fifth image, marked "a"); (2) spines with a single, large, elongated, and thick protrusion (Fig. 9.19e, first and second images in the first row); (3) spines with an enlarged finger-like neck and a fusiform head (Fig. 9.19e, third image in the first row); (4) spines with curved or "twisted" endings (Fig. 9.19e, fourth and fifth images in the first row); (5) branched spine with various bulbous protrusions (Fig. 9.19e, last image in the second row); (6) large spine resembling a simple racemose aspect (Fig. 9.19e, last image in the first row); (7) spines with a thin, elongated neck resembling a gemmule aspect with an atypical elongated head (Fig. 9.19e, fourth image in the second row, left); (8) spines with a single bulbous protrusion giving rise to small protruding heads attached to a thick "neck" (Fig. 9.19e, fifth image in the second row); and (9) thorny excrescence with lobed protrusions (Fig. 9.19e, last image in the second row; descriptions based on Fiala and Harris 1999). The synaptic relevance for some of these spines was revealed by transmission electron microscopy (TEM, described below).

Likewise, intracellular injection of Lucifer Yellow allowed 3D reconstructions (Figs. 9.20a and 9.21) and quantifiable morphological data for supragranular pyramidal dendrites and spines of the adult human cingulate cortex, CA1 hippocampal field, and temporal cortex (Benavides-Piccione et al. 2013; Eyal et al. 2018; Benavides-Piccione et al. 2020, 2021). Using *postmortem* human brains from two men (aged 40 and 85), 3D fluorescent images under confocal microscopy revealed the dendritic spine density of layer III pyramidal neurons of the human anterior cingulate gyrus (Benavides-Piccione et al. 2013). Accordingly, pleomorphic spines were found along the dendritic length of both basal (Fig. 9.20b, note the lack of spines in the proximal segment) and apical (Fig. 9.20c, 100 μm away from the soma) shafts, with no variation in spine morphologies related to distance from the soma, but with smaller spines in basal dendrites than in the apical one. On the other hand, there is a high mean density of spines in apical dendrites than in basal dendrites. In the basal dendrite of the younger person, dendritic spine density reached a maximum of approximately 1.9 spines/μm at a 110-μm distance from the soma and then slightly decreased distally. In the apical dendrite, spine density increased to a maximum of 4.3 spines/μm at a 160-μm distance from the soma (Benavides-Piccione et al. 2013).

This same procedure unraveled the heterogeneity in the dendritic arbor of pyramidal neurons in the CA1 field of the adult human hippocampal formation, which is more complex than in rats (Benavides-Piccione et al. 2020; Fig. 9.4a). Examples of reconstructed fluorescent dendritic spines of layer III pyramidal cells from the human temporal cortex (Brodmann's area 20) are shown in Fig. 9.21. These are the spiny neurons identified with morphological, biophysical, and computational

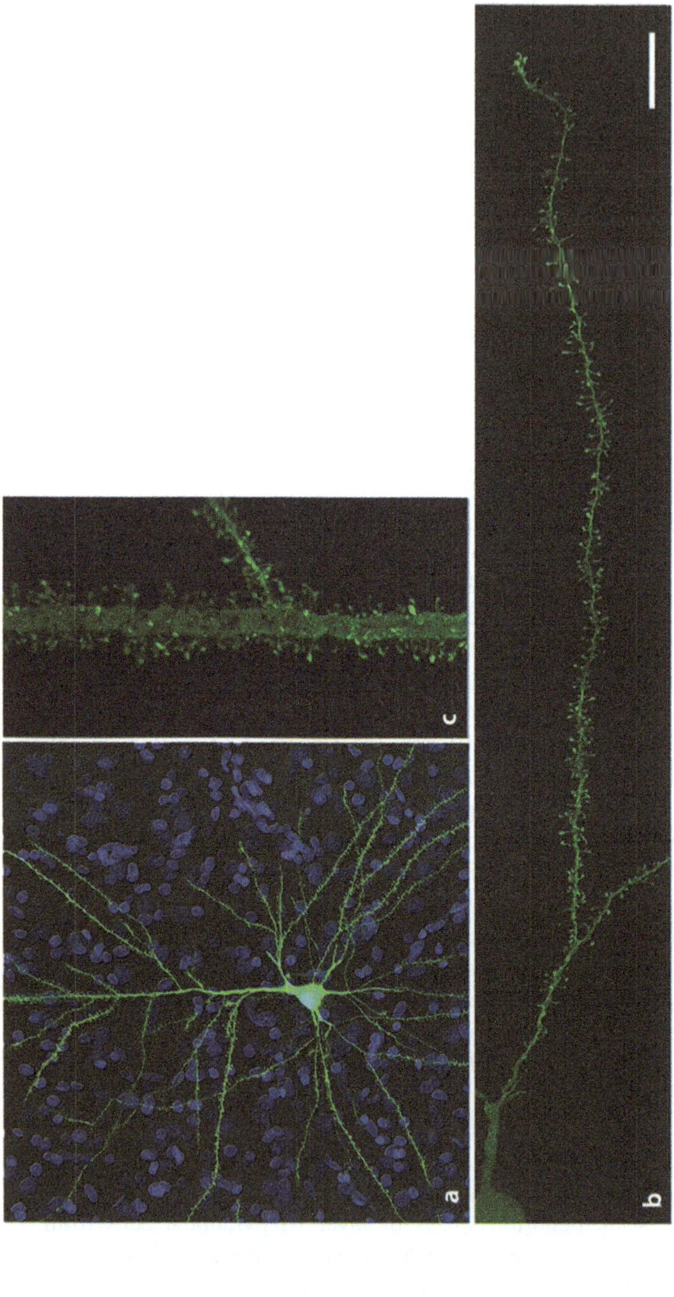

Fig. 9.20 (**a**) Confocal microscopy image "z" projection of a pyramidal neuron (injected with Lucifer Yellow) located in layer III of the adult human cingulate cortex. DAPI staining (in blue) identifies cell bodies. High-magnification images are provided for (**b**) a ramified basal dendrite and (**c**) an apical dendritic segment. In (**b**), note virtually no spines at the first proximal micrometers (left), the progressive increase of pleomorphic spines along intermediate parts (center), and spines remaining visible at distal segments (right). In (**c**), note the density of spines covering the main apical dendritic shaft (100 μm away from the soma) as well as along the collateral branch. Scale bar = 40 μm in (**a**); 13 μm in (**b**); and 7 μm in (**c**). (Legend adapted and figure reprinted from Benavides-Piccione et al. (2013) under CCC RightsLink® license #5385651024769, originally published by Oxford University Press)

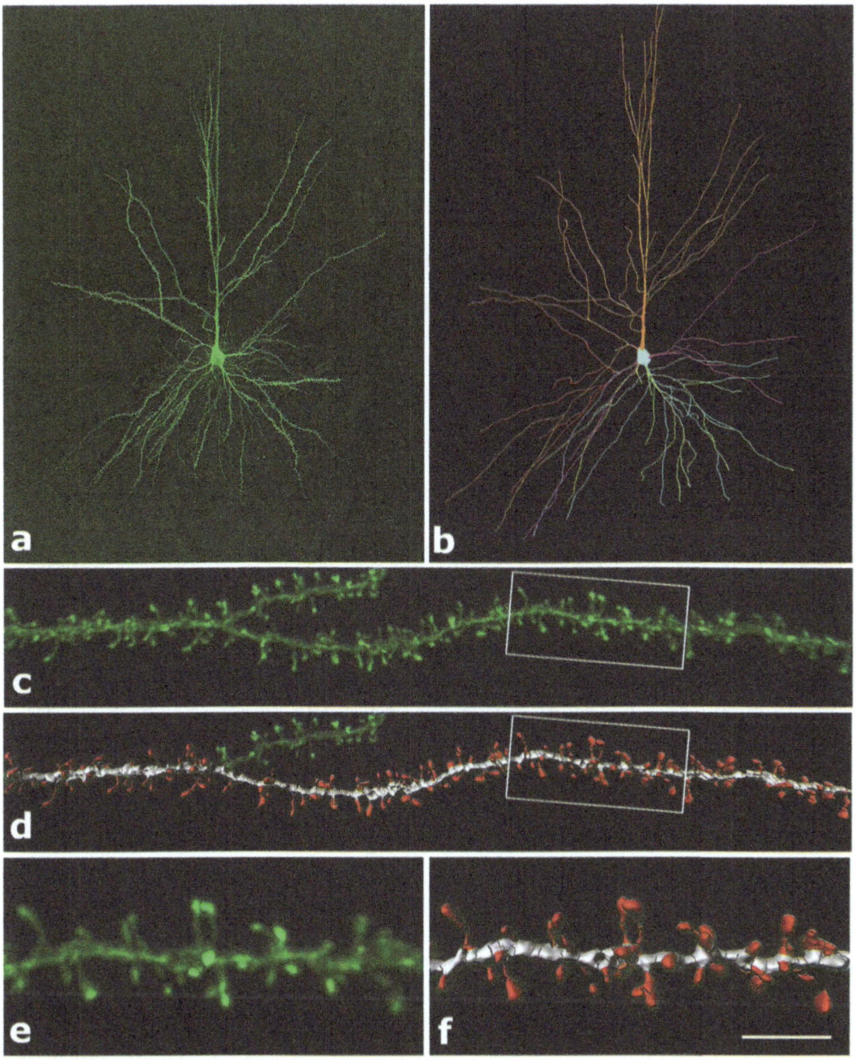

Fig. 9.21 (**a**) Confocal microscopy image "z" projection of a pyramidal neuron (injected with Lucifer Yellow) located in layer III of the human temporal cortex. (**b**) Three-dimensional reconstruction of the neuron shown in (**a**) showing the apical dendritic arbor in orange and the varied basal dendritic arborization with other colors. (**c**) Confocal microscopy image showing a horizontally projecting labeled spiny basal dendrite. (**d**) Reconstructed dendritic spines (red) and dendritic shaft (white). (**e**, **f**) Higher-magnification images of the dendritic segment are indicated by boxed areas in (**c**) and (**d**). Note the presence and distribution of intermingled pleomorphic spines, some with long and thin shapes. Scale bar = 110 μm (in **a** and **b**), 10 μm (in **c** and **d**), and 4.5 μm (in **e** and **f**). (Legend adapted and figure reprinted from Eyal et al. (2018) under CC BY license and Copyright © 2018 Eyal, Verhoog, Testa-Silva, Deitcher, Benavides-Piccione, DeFelipe, de Kock, Mansvelder, and Segev)

distinct properties in our species (Eyal et al. 2018). The morphology of Lucifer Yellow injected pyramidal neurons was studied in layers IIIa and Vb of the antero-lateral middle and inferior temporal gyri (Brodmann's areas 20, 21, and 38) from patients neurosurgically treated for pharmacoresistant temporal lobe epilepsy (Benavides-Piccione et al. 2021). Findings suggest a relationship between dendritic branching and the number of spines likely aiming for balanced neuronal excitability (Benavides-Piccione et al. 2021). Moreover, data reinforce that differences in dendritic structure and density of spines are regulated at every single cell, brain area, and person, providing a high degree of possible arrays and probabilistic possibilities of functioning for each component of neural circuits, for information processing demands, and flexible, adaptable responses to new conditions.

9.5.1 Topology and Ultrastructure of Human Dendritic Spines

Another relevant issue is whether dendritic spines emerge isolated or grouped for synaptic purposes. Spines were already reported to form functional clusters within micrometers along the dendritic segment (Lu and Zuo 2017; Kastellakis and Poirazi 2019), and clustered synaptic modifications facilitate dendritic computation in learning and memory (Ma and Zuo 2022). For example, the apical terminal dendrites of layer III pyramidal neurons of the monkey prefrontal cortex display spatially (nonrandom) clusters of mushroom and stubby spines (Yadav et al. 2012). "The clustering of synapses may emerge from synapses receiving similar input, or via many processes that allow for crosstalk between nearby synapses within a dendritic branch, leading to cooperative plasticity. Clustered synapses can act in concert to exploit maximally the nonlinear integration potential of the dendritic branches in which they reside. Their main contribution is to facilitate the induction of dendritic spikes and dendritic plateau potentials, which provide advanced computational and memory-related capabilities to dendrites and single neurons" (Kastellakis and Poirazi 2019). As will be mentioned in the following section, some human spines can be lost in clusters in Alzheimer's disease (Mijalkov et al. 2021).

On the other hand, when synaptic wiring is random, the dimension of a representation formed by many sparsely connected neurons can be higher than that of a smaller number of densely connected elements (Litwin-Kumar et al. 2017). If this-possibility is adapted to dendritic spines, multiple connections in different spines along dendrites would provide a high-dimensional representation and output pattern to different ensembles of inputs. It is also possible that both clustered and scattered random contacts can occur on dendrites, allowing different but complementary strategies for the dendritic integration of information. Providing more alternatives for the dendritic mechanisms recruited for postsynaptic summation might give an exceptional repertoire for spiny dendritic segments associate signals linearly and nonlinearly and modulate cellular excitability (Tran-Van-Minh et al. 2015).

Dendritic spines with a helical topology means that spines are distributed in regular linear arrays on the shaft perimeter, even though helixes would show no particular rotational preference (Yuste 2010). Helical spine ordering was observed in distal dendrites of cerebellar Purkinje cells of fish and mice (O´Brien and Unwin 2006). Such short-pitch helical paths would maximize the chance for different spines to interact with different passing axons and facilitate synaptic interactions (O´Brien and Unwin 2006). The existence of a similar pattern was tested in 3D reconstructed apical and basal spiny dendrites of layer III pyramidal neurons from the frontal, temporal, and cingulate cortex in humans (40 and 85 years, Morales et al. 2014). Using Fourier and spatial statistical analyses of spine insertion points, some evidence for helical patterns was present in a minority of samples. Nevertheless, most spines followed a random positioning principle and did not show a helical pattern in apical or basal dendrites or among spines of different volumes and lengths in these human samples (Morales et al. 2014; Fig. 9.22).

Finally, TEM and additional modern techniques revealed the fine-tuned synaptic processing that reflects the complexity of dendritic and spine structural and functional organization in humans. As mentioned above, human synapses can show unique properties and synaptic heterogeneities can occur in cortical and subcortical areas related to cellular specializations and circuits. Some basic principles can be found, but particular arrangements imply the "role" of each synapse and its integration in each cell. For example, the synaptic organization of the human temporal lobe neocortex was recently reported using high-resolution transmission, focused ion beam scanning, and electron microscopic tomography (Rollenhagen et al. 2020). This approach provided quantitative 3D models of synapses (overall geometry, number, size, shape of active zones, and synaptic vesicles) and insights into the morphological correlates of layer-specifically organized synaptic transmission and plasticity in the human neocortex (Rollenhagen et al. 2020; see additional data in this reference).

Ultrastructural data from human MeA neurons also demonstrated how common and particular features coexist in the synaptic organization of a brain area. Axonal synaptic contacts can form asymmetric and symmetric contacts directly on dendritic shafts and on differently shaped spines (Dall'Oglio et al. 2015; Figs. 9.23 and 9.24). Axon terminals formed isolated (Fig. 9.23a) or multiple asymmetric contacts (Fig. 9.23b), and the presynaptic terminals of asymmetric synapses contained round, small, electron-lucent vesicles (e.g., Fig. 9.23a, c, d) whereas symmetric presynaptic terminals exhibited small or pleomorphic vesicles, some intermingled with large, dense-core vesicles (e.g., Fig. 9.23e, g, top).

Dendritic spines of different shapes and sizes showed axospinous monosynaptic or, in some cases, multisynaptic sites with contacts located at the spine head and neck (Fig. 9.23f–m). These synapses were usually of the asymmetric type (e.g., Fig. 9.23f, h). Spines could also form axospinous symmetric contacts (e.g., Fig. 9.23e). Synaptic terminals contacted different parts of stubby/wide spines (e.g., at the base, Fig. 9.23c; at the upper part, Fig. 9.23d; or at the lateral part of the spine,

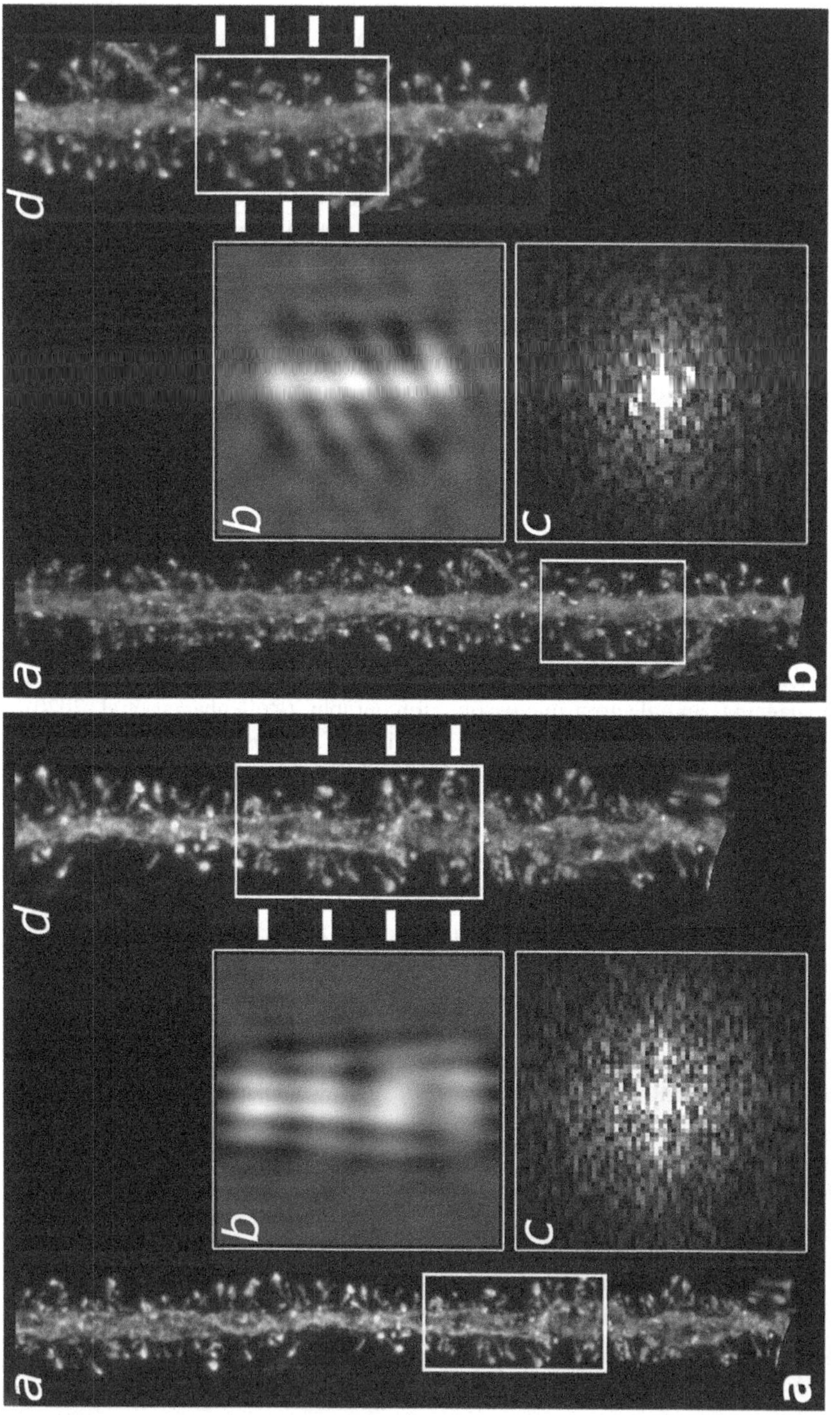

Fig. 9.22 Helical or random positions of dendritic spines were tested using confocal microscopy images of Lucifer Yellow-injected pyramidal neurons from layer III of the adult human frontal, temporal, and cingulate cortex. Fourier analysis and spatial statistics were used to analyze spine position along the apical and basal

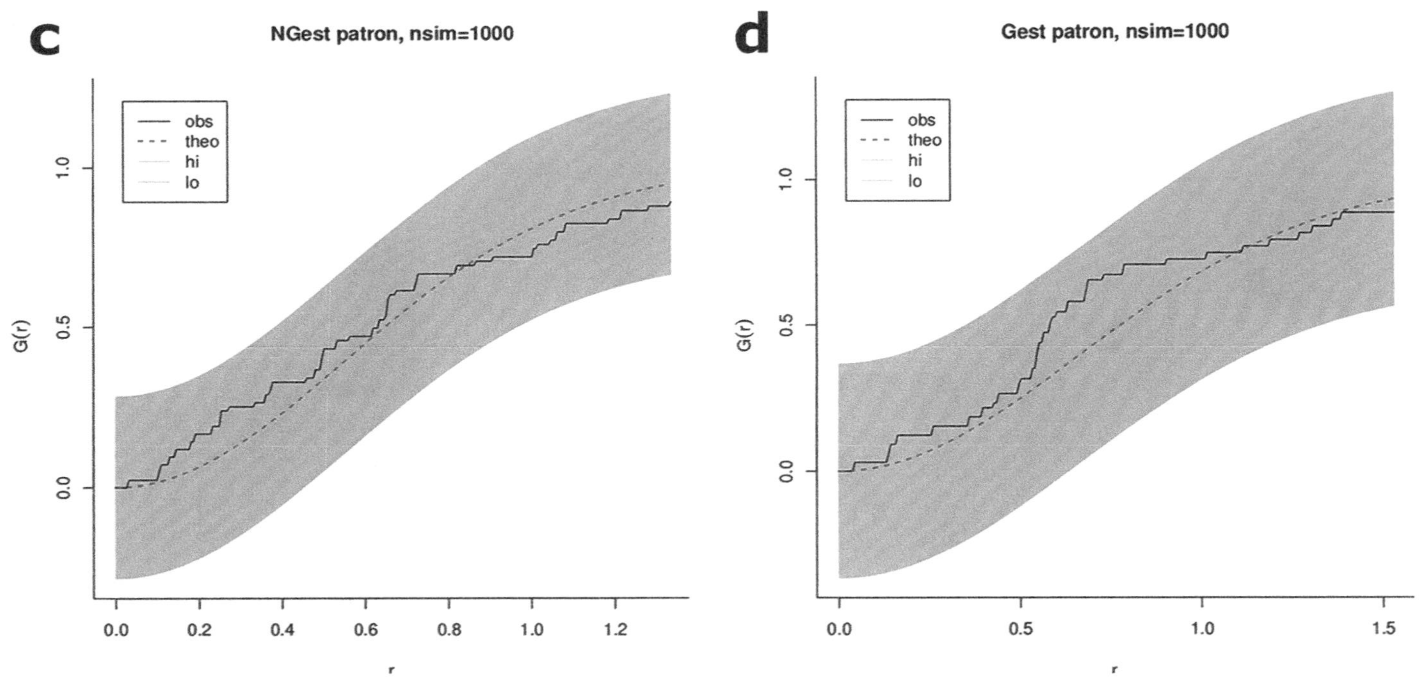

Fig. 9.122 (continued) dendrites. (**a**, **b**) "Two examples of dendritic segments from apical dendrites (**a**), showing regions in which helical distributions were apparent (white boxed areas) after the corresponding Fourier and inverse Fourier transforms (**b**, **c**). (**d**) Higher-magnification images of the same dendritic segments indicate a possible frequency of regular spacing along the lateral insertion of dendritic spines (dashes). (**c**, **d**) Monte Carlo envelope tests of these examples. Graphs represent the allowed random band (shaded), the theoretic (theo) distribution function [$G_{CSR}(r)$, red dotted line], and the estimated [$\hat{G}(r)$] obtained from the patterns analyzed (obs, black line) using the proposed edge correction. Because examples did not surpass the shaded envelope (high and low, hi and lo), the null hypothesis of complete spatial randomness was not rejected." (Legend and figure reprinted from Morales et al. (2014) under CC BY license and Copyright Morales, Benavides-Piccione, Dar, Fernaud, Rodríguez, Anton-Sanchez, Bielza, Larrañaga, DeFelipe, Yuste, doi: 10.1523/JNEUROSCI.1085-14.2014)

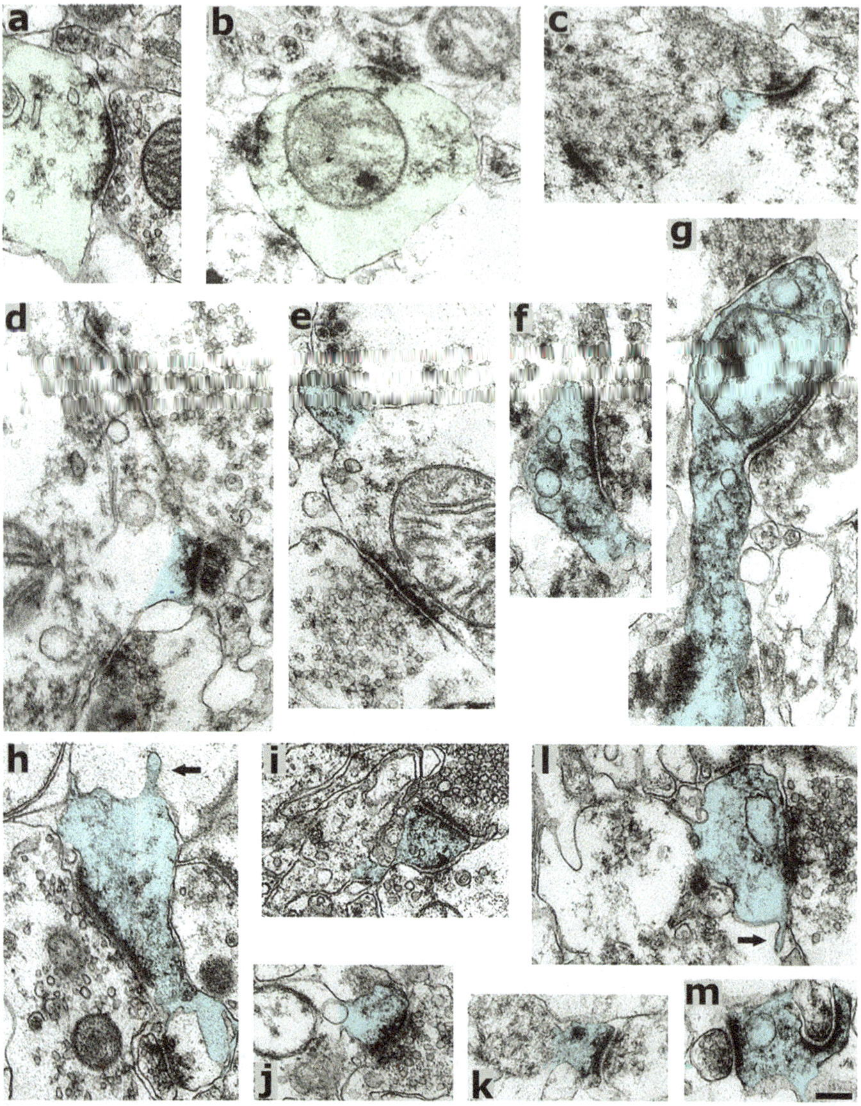

Fig. 9.23 Ultrastructure of dendrites (slightly highlighted in green), spines (blue), and synapses in the human medial amygdaloid nucleus. (**a**) Isolated or (**b**) multiple, adjacent axon terminals making asymmetric synapses upon dendritic shafts were found in the neuropil. Dendritic spines were classified as (**c**, **d**) stubby, (**e**) wide, (**f**) thin with small or (**g**) long necks, (**h**) with a transitional aspect, (**i**–**k**) mushroom shape with small to (**l**) big head and macular postsynaptic densities. A spine with a transitional aspect between mushroom and ramified shapes is shown in (**m**). Spinules are indicated by an arrow in (**h**) and (**l**). Synapses showed asymmetric (e.g., in (**c**, **d**, **f**, **j**, **k**)) or symmetric contacts (in **e** and **i**). Note that the multisynaptic spine receives two asymmetric contacts ((**g**), one at the spine head and another at the neck) and a symmetric contact (at the top of the spine head). Synapses were found at the spine base (**c**), neck (**g**), and head (**i**). Images are from adult individuals. Scale bar = 0.2 μm. (Legend adapted and figures reprinted from Dall'Oglio et al. (2013) under CCC RightsLink® license #5385630478535, and Dall'Oglio et al. (2015) under CCC RightsLink® license #5385561144316, both originally published by John Wiley & Sons, Inc)

Fig. 9.23e). Thin spines ranged in size from small (Fig. 9.23f) to long-necked ones (Fig. 9.23g). This latter can be multisynaptic and show both asymmetric and symmetric contacts at opposite sites on the spine head, and an asymmetric contact at the spine neck (Fig. 9.23g). An example of transitional shaped spine, with morphological features between thin and mushroom types, displayed a protruding spinule that, in this case, showed no evident PSD (Fig. 9.23h). Mushroom spines with variable sizes had small-to-large spine heads (Figs. 9.23i–m and 9.24a), and both symmetric and asymmetric macular synapses (Fig. 9.23i) with asymmetric contacts at the spine head (Fig. 9.23j,k). Mushroom spines also displayed a large head with a perforated PSD (Fig. 9.24a). Spinules were found in multisynaptic mushroom spines (Fig. 9.23l). Spines with other transitional shapes could be observed (with mushroom to ramified features) in some cases (Fig. 9.23m; other atypical/polymorphic examples are shown in Fig. 9.24b, c).

Other interesting ultrastructural findings draw attention to the human MeA: (1) an axonal envelopment around the perimeter of the spine head close to a symmetric synapse (Fig. 9.24d); (2) reciprocal synapses (Fig. 9.24e, top); (3) a multisynaptic spine protruding from an axon showing an unusual spinule-like double-membrane evagination (Fig. 9.24e, bottom and right); (4) a serial axo-axo-dendritic synapse (Fig. 9.24f, bottom); (5) an "*en passant*" axospinous synapse (Fig. 9.24f, top); and (6) the close apposition of cellular membranes suggestive of a "gap-like" junction (Fig. 9.24g and insert; Dall'Oglio et al. 2015). These axonal features in the human MeA neuropil (Fig. 9.8c–e) indicate that connections can be made directly on a specific end-target element or evoke more diffuse synaptic relationships with various cells at the same time. For this latter purpose, the human MeA neuropil shows coarse and finely beaded axons, reinforcing the occurrence of *en passage* synapses, as well as axonal collateral branches forming terminal boutons ranging from simple short appendages to multiple bulbous forms (Dall'Oglio et al. 2013). Bulbous endings can form a glomerular-like synaptic organization, that is, a bulbous axonal vesicle-filled ending making various asymmetric contacts with multiple dendrites at the same time (see Fig. 9J in Dall'Oglio et al. 2013). Taking into account the occurrence of reciprocal (axo–axonal connection) and serial (axo–axo–dendritic connection) synapses, it is very likely that input pathways can generate divergent processing of information in the human MeA. It is hypothesized that such an arrangement would rapidly activate various MeA neurons in parallel and promote a widespread response to the emotionally valenced stimulus and a prompt behavioral display (Dall'Oglio et al. 2015). How this integrated synaptic organization can vary between individuals and according to personal experience is currently difficult to determine and prove, but encourages interesting hypotheses on circuitry functioning.

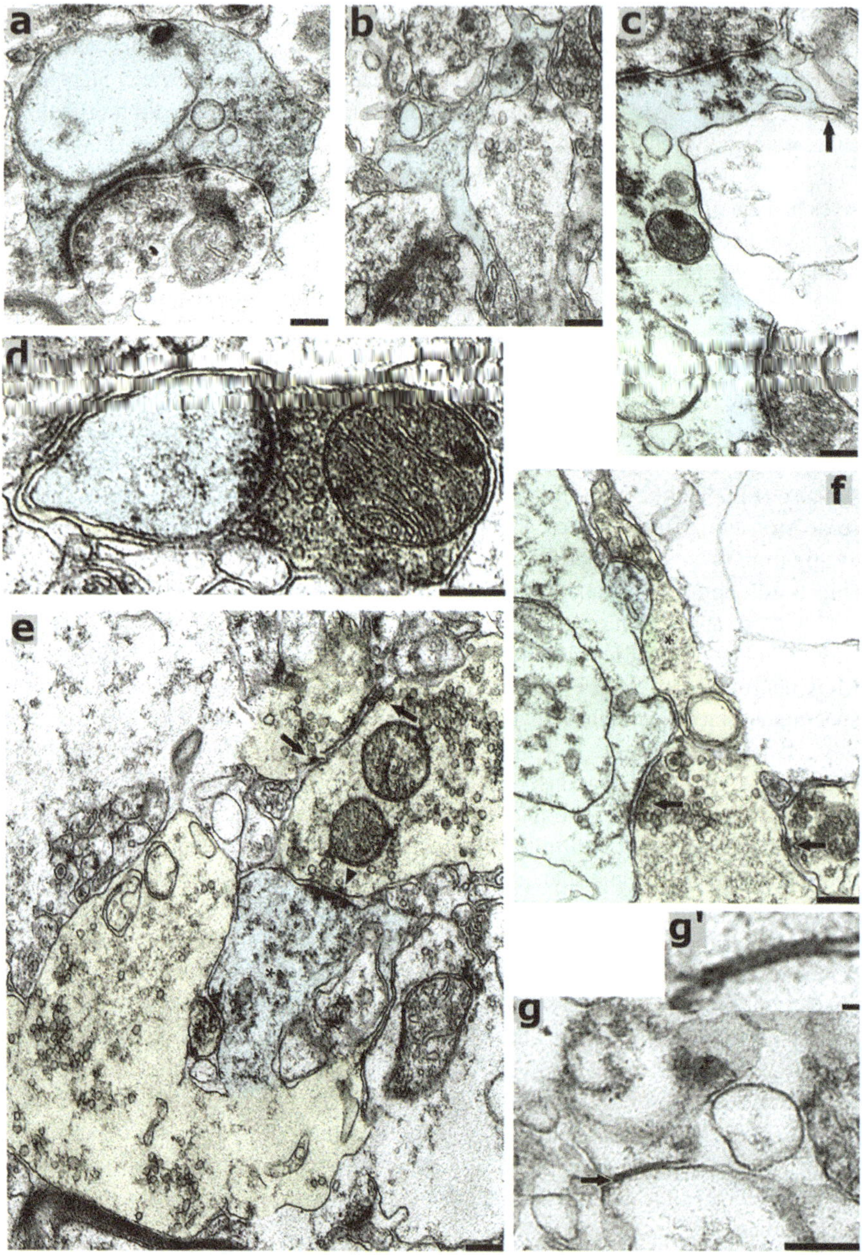

Fig. 9.24 Ultrastructural organization of dendrites (slightly highlighted in green), spines (blue), axons (yellow), and synapses in the human medial amygdaloid nucleus. Some examples of spines synaptic and shape heterogeneity are presented in (**a–c**). (**a**) Mushroom spine with a typical large head and a perforated postsynaptic density contacting an axonal bouton with vesicles. (**b**) Atypical spines with ramification and protrusions or (**c**) with an elongated and wide shape and a spinule (arrow). Spines can form complex synaptic arrangements, such as (**d**) an axodendritic asymmetric

9.6 Some Examples of Altered Dendritic Spines in Human Neuropathological Conditions

Various morphological findings on dendritic spines have been obtained from human brain tissue with no evident neuropathological damage or noticeable neurological and psychiatric disorders. The study of spines in normal conditions might serve as the "baseline" for further comparisons with pathological ones. Again, this can be a complex task. First, it is highly expected that some degree of variability in spine number and types can normally occur within the same neuronal subpopulation and brain area. The study of human spines usually involves uncontrollable variables that may influence results (see a parallel discussion in Falougy et al. 2019), which imposes difficulties to determine a suitable sample size considering intra-individual and interindividual variability (see also Forrest et al. 2018). Second, altered dendritic spines can occur in different pathological conditions and since early neurodevelopmental periods to puberty and old age. At the same time, mental health disorders can show different and graded signs and symptoms. The study of dendritic spines and brain connectivity has also to consider differences in men and women as a biological variable in neuropsychiatric research and related to clinical evidence (Joel and McCarthy 2017; Rubinow and Schmidt 2019; Sheppard et al. 2019; Arnold 2020; Hidalgo-Lopez et al. 2021).

Despite these inherent difficulties, neurological and psychiatric disorders have been associated with cross-scale disruptions in the synaptic organization, spine structure, functioning of individual cell types, and wiring of microcircuits in the human brain (Hunt et al. 2022 and references therein). Dendritic spine abnormalities have been classified as "pathologies of spine distribution" and "pathologies of ultrastructure" (Fiala et al. 2002). The former affects the number, morphology, and loci of spine origin on the neuron. In other words, there can be increases and decreases in spine density, alterations in spine size and distortion of spine shape, dendritic swellings with concomitant loss of spines, and ectopic spines (Fiala et al. 2002). The latter involves distortion of the ultrastructure and the subcellular organelles within dendritic spines, including electron-dense spines, altered endoplasmic reticulum, hypertrophied spine apparatus, multivesicular bodies and cytoskeleton, with higher spine volume in giant spines, and formation of aberrant synapse-like connections with axonless spines and axon-free PSDs (Fiala et al. 2002). The

---

Fig. 9.24 (continued) contact with an axon that envelopes the spine head, (**e**) axo-axonal reciprocal synapse (arrows close to synaptic vesicles), and an axospinous asymmetric contact (arrowhead) in an axonal spine (asterisk) with an atypical "spinule-like" double-membrane evagination (right), and (**f**) serial axo-axo-dendritic synapse (arrows) and an *en passant* axospinous synapse (asterisk). In (**g**), the close appositions of membranes resemble a gap-like junction (arrow; at high magnification in the insert **g′**). Images are from adult individuals. Scale bar = 0.2 μm, except for the inserts where scale bar = 20 nm. (Legend adapted and figure reprinted from Dall'Oglio et al. (2015) under CCC RightsLink® license #5385561144316, originally published by John Wiley & Sons, Inc)

study of the synaptic signaling cascades and the integrated molecular pathways that lead to irregular dendrites and altered spine structure–function coupling is key to unraveling the different groups of neuropsychiatric disorder-associated risk factors and patients' clinical characteristics (Penzes et al. 2011; Chassefeyre et al. 2015; Hrvoj-Mihic et al. 2017; Forrest et al. 2018; Kommaddi et al. 2018; Runge et al. 2020; Heinze et al. 2022).

Genetic and environmental risk factors can alter putative trajectories of dendritic spines morphogenesis and pruning during typical development or in neuropsychiatric disorders (Forrest et al. 2018). Converging on a subset of networks and pathways, both hypoconnected and hyperconnected cells disrupt short- and long-range networks in the brain, affecting different psychological domains and how symptoms are displayed (Forrest et al. 2018). Dysregulation of dendritic and spine structure and dynamics occur in various neurodevelopmental, neuropsychiatric, and neurodegenerative disorders. The list of spinopathies includes the affected brain following malnutrition, alcohol or toxin exposure, infection, maternal phenylketonuria, autism spectrum disorder (ASD), Fragile X syndrome, Rett syndrome, Williams syndrome, Creutzfeldt–Jakob disease, frontal dementia, Alzheimer's disease (AD), Pick's disease, Parkinson's disease, Huntington's disease, schizophrenia, bipolar disorder, epilepsy, hypoxic or ischemic insults, traumatic lesions, and brain tumors.[22]

It is not completely clear whether altered spines are the cause, consequence, compensatory response, or a combination of these possibilities in each studied disorder and spectrum of clinical presentation (Fiala et al. 2002; Penzes et al. 2011). Although it is currently unsettled whether morphological changes in spines are associated with the time course of specific symptoms, "spine alterations may be relevant no matter where in the pathological cascade of a disease they occur" (Penzes et al. 2011). Indeed, pioneer works with Golgi-impregnated neurons reported remarkable morphological abnormalities of dendritic spines in cases of children clinically diagnosed with intellectual disability compared to children with normal cortical development and behavior. Human neocortical pyramidal neurons display spines whose number normally increases from the fetal to infancy periods, as mentioned previously (Fig. 9.25a). Genetic disorders that cause intellectual disability (e.g., Patau syndrome and Down syndrome) are associated with abnormal distribution, number, and shapes of dendritic spines in neocortical neurons along the postnatal development (Marín-Padilla 1972, Fig. 9.25a). Neurons have long fine spines with multiple varicosities on pedicles, entangled spines, and expanded terminal heads (Purpura 1974, Fig. 9.25b).

The abnormal morphology of dendritic spines is associated with neuropsychiatric disorders such as ASD and schizophrenia. Besides other affected brain areas (e.g., the frontal, parietal, and temporal lobes), there is a high dendritic spine density with more immature shapes in the human lateral amygdaloid nucleus in ASD

[22] For relevant data, see Baloyannis et al. (2001), Fiala et al. (2002), Penzes et al. (2011), Jiang et al. (2013), Overk and Masliah (2014), Maiti et al. (2015), Herms and Dorostkar (2016), Hrvoj-Mihic et al. 2017, Forrest et al. (2018), Chidambaram et al. (2019), Runge et al. (2020), Bączyńska et al. (2021), and Parajuli and Koike (2021), for example.

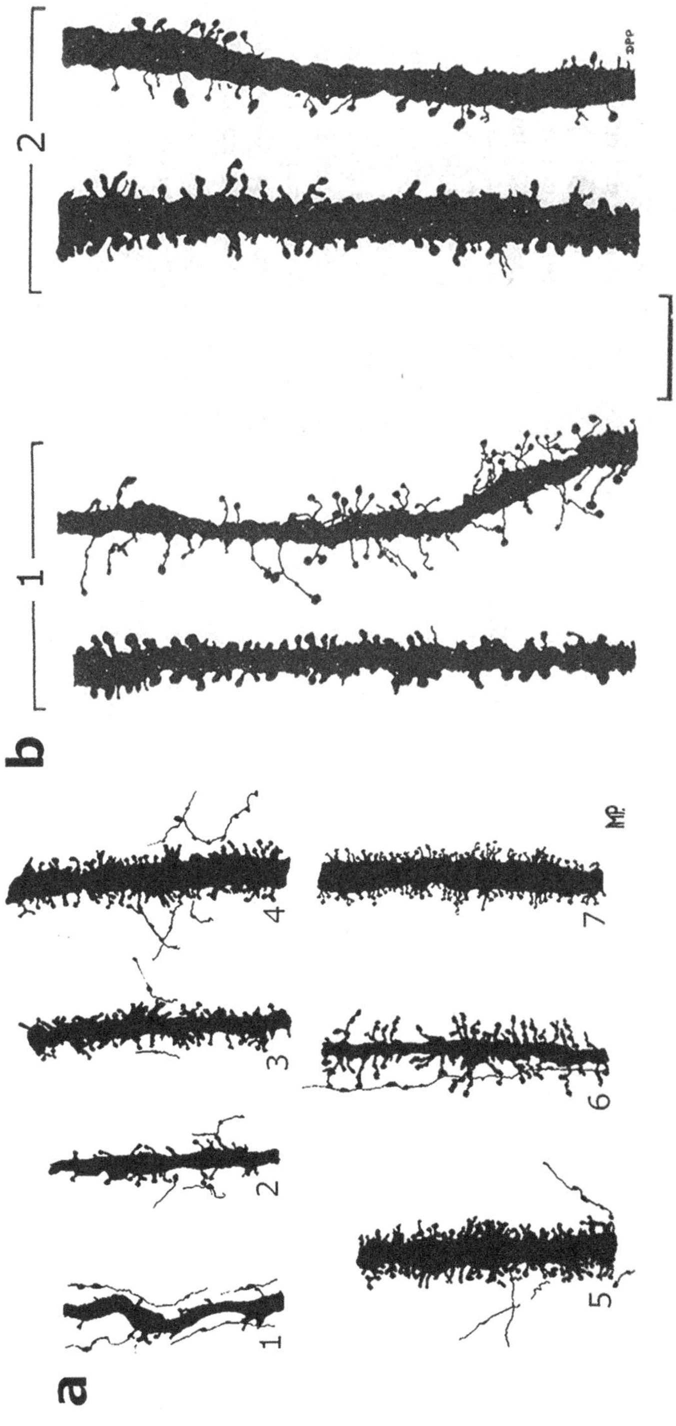

Fig. 9.25 (*caption on p. 459*)

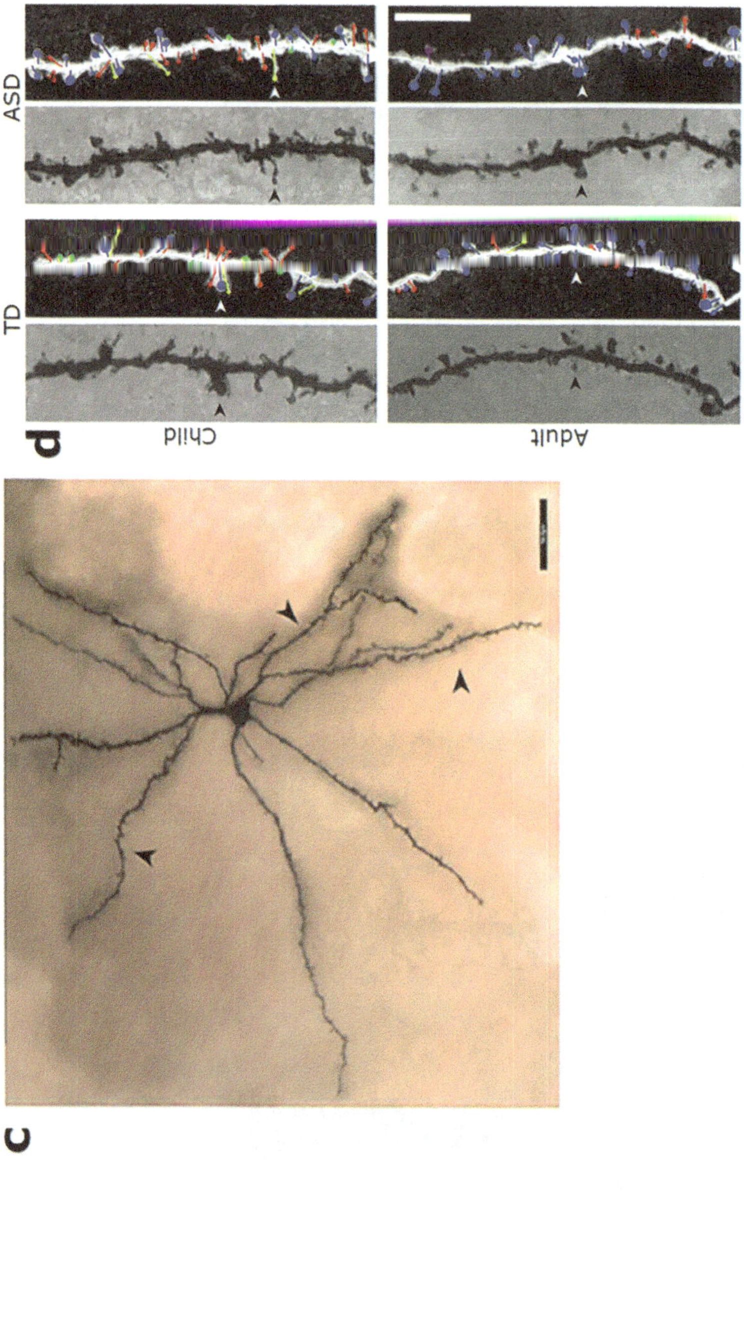

Fig. 9.25 (*caption on p. 459*)

individuals than in control individuals (Weir et al. 2018, Fig. 9.25c). This finding might be associated with some ASD symptoms, such as disruption in information processing and deficits in social interactions because the integrity of the amygdaloid nuclei and their connections are relevant for emotional processing, memory and cognition, anxiety and stress regulation, and proper social behavior display (Rasia-Filho et al. 2000; Šimić et al. 2021). Moreover, Golgi-impregnated pyramidal neurons from the dorsolateral prefrontal cortex display a reduced density of dendritic spines in subjects diagnosed with schizophrenia (under treatment with antipsychotic agents) compared to control data (Glantz and Lewis 2000). Reduced spine density was also found in the parietal cortex in these cases (Garey et al. 1998). Spines in higher-order cortical areas process synaptic information from multiple circuitries for executive functions and proper social behavior display, some of which are specific to humans. Then, deficits in synaptic transmission in these areas might account for some of the clinical findings of these patients (see more data in Lewis and González-Burgos 2008; Jaaro-Peled et al. 2010; Penzes et al. 2011; Orellana and Slachevsky 2013).

Fig. 9.25 (**a**) Camera lucida drawings of Golgi-impregnated spiny apical dendrites of layer V pyramidal neurons (crossing the territory of layer III) in the human motor cerebral cortex. Data were obtained along normal development and from genetic disorders. The apical dendritic segments are from (1) a 5-month fetus; (2) a 7-month fetus; (3) a newborn; (4) a 2-month-old infant; (5) an 8-month-old infant; (6) a newborn girl with 13–15 trisomy (Patau syndrome); and (7) an 18-month-old girl with 21 trisomy (Down syndrome). Images illustrate the morphological characteristics of the human dendritic spines during prenatal and early postnatal cortical development (1–5). The number of spines normally increases with age. The spines from two infants with proven chromosomal trisomies and intellectual disability have abnormal distribution, number, and shapes (6, 7). (Legend adapted and figure reprinted from Marín-Padilla (1972) under CCC RightsLink® license #5387100803473, originally published by Elsevier). (**b**) Camera lucida drawing of Golgi-impregnated dendritic segments of medium-sized pyramidal neurons of the human motor cortex in normal conditions (infant and child) or cases with intellectual disability. (1, left) Normal aspect for the dendritic spines of a 6-month-old infant. Three basic types of spines were identified: thin, stubby, and mushroom-like spines. (1, right) Dendritic spines from a 10-month-old infant with intellectual disability displaying abnormally long spines and some entangled spines with prominent terminal heads. Multiple varicosities on the pedicles of these long, thin spines were occasionally found, but mushroom-like and stubby spines were rarely observed in this case. (2, left) Apical dendrite of a normal 7-year-old child. (2, right) Apical dendritic segment from a 12-year-old child with severe intellectual disability. The paucity of mushroom-like and stubby spines and the presence of thin spines with abnormally expanded terminal heads are characteristics of this case. Scale bar = 10 μm. (Legend adapted and figure reprinted from Purpura (1974) under CCC RightsLink® license #5387120853098, originally published by Science, The American Association for the Advancement of Science). (**c**) Two-dimensional photomicrograph of a Golgi-impregnated principal neuron from the human lateral amygdaloid (La) nucleus. Arrows indicate second-order dendrites where dendritic spines were studied in control (typical development, TD) and autism spectrum disorder (ASD) cases from youth to adulthood. (**d**) Representative high-magnification images of second-order dendrites and spines (arrowheads) of La neurons. Each Golgi-impregnated figure is accompanied by a representative trace image. Spine density in child ASD cases was higher compared to age-matched TD cases and adult ASD cases. Young ASD cases also display spines that tend to be more immature than TD cases. Blue: mushroom spine; red: thin spine; yellow: filopodial spine; green: stubby spine; purple: branched spine. Scale bar = 50 μm (in **c**) and 10 μm (in **d**). (Legend adapted and figures reprinted from Weir et al. (2018) under CCC RightsLink® license #5387140851871, originally published by John Wiley & Sons, Inc)

9.6.1 Additional Data on Alzheimer's Disease

Different experimental and imaging approaches for the study of the net loss of dendritic spines in AD opened a new frontier for research. For example, human hippocampal CA1 pyramidal neurons with and without tau pathology were imaged using intracellularly Lucifer Yellow fluorescent dye and confocal microscopy (Mijalkov et al. 2021; Fig. 9.26a). The number of dendritic spines is reduced in hippocampal pyramidal neurons and dentate granule cells in AD cases (Fiala et al. 2002; Penzes et al. 2011). In this case, the presence of tau pathology determines the loss of dendritic spines in clusters (not at random) in neocortical pyramidal neurons, letting the remaining spines in smaller, more tightly packed, and isolated groups along dendritic segments (Mijalkov et al. 2021, Fig. 9.26a). Dendrites passing through or near fibrillar β-amyloid deposition show a progressive reduction in dendritic number, shaft diameter, and length also related to local axonal abnormalities and permanent disruption of neuronal connections (Tsai et al. 2004). At least part of the cognitive dysfunction in AD may be related to this degree of disruption of spatial location of dendritic spines, their function and connectivity, affecting the modulated synaptic organization and plasticity from subcortical to allocortical- and neocortical-associated regions (Heimer et al. 2008; Overk and Masliah 2014; Rasia-Filho et al. 2021).

Fig. 9.26 (**a**) Human hippocampal CA1 pyramidal neurons with and without tau pathology in Alzheimer's disease (AD). Confocal microscopy images of merged images (*a'*, *d'*) following intracellularly injected Lucifer Yellow fluorescent dye (green, *b'*, *e'*) and immunostaining with phospho-tau AT8 (red, *c'*, *f'*). Neurons (*a'-f'*) and their basal dendrites and spines (*g'-m'*) are shown in different tau pathology conditions. Dendrites from cell body neurons immunostained by either phospho-tauAT8 or phospho-tauPHF-1 were classified as "SomaTau+" dendrites, whereas those not immunostained by either of these antibodies were labeled as "SomaTau-" ones. (*a'-c'*, *g'*, *i'*) Neurons and dendrites from a cell body free of phospho-tau AT8 and (*d'-f'*, *h'-j'*) with phospho-tau AT8 in an intermediate stage of neurofibrillary pathology. Note the different number of dendritic spines between dendrites SomaTau- and SomaTau+ (lower in this latter) in the zoomed-in views of dendrites and spines (from *g'* to *j'*). The high-resolution dendritic segment in *k'* demonstrates the spines along the white line shown in *j'*. Red points indicate spines on the dendritic shaft (*l'* and *m'*), their three-dimensional spatial distribution, and insertion points (*m'*). Dendritic spines were lost in clusters. Scale bar (shown in *m'*) = 12 μm (*a'-f'*), 5 μm (*g'-j'*), and 3 μm (*k'-m'*). (Legend adapted and figure reprinted from Mijalkov et al. (2021), https://doi.org/10.1038/s41598-021-91726-x, under CC BY 4.0 copyright license). (**b**) Focused ion beam/scanning electron microscopy images and postsynaptic target identification in the *stratum pyramidale* and *stratum radiatum* in the hippocampal CA1 field of control and AD cases. (*a'*) Dendritic shaft (blue) with a protruding dendritic spine (purple). Axospinous asymmetric synapse (Asym, white arrowhead) established on the head of the spine. (*b'-e'*) Axospinous asymmetric synapse followed through the stack of images. (**c**, **d**) The percentage of axospinous and axodendritic asymmetric synapses (**c**) and symmetric synapses (**d**) in the *stratum pyramidale* (str pyr) and *stratum radiatum* (str rad) of CA1 in both control and AD cases. Lower values were found in AD cases. Scale bar in E = 500 nm (*a'*) and 450 nm (*b'-e'*). (Legend adapted and figure reprinted from Montero-Crespo et al. (2021) under CCC RightsLink® license #5387231262727, originally published by Oxford University Press)

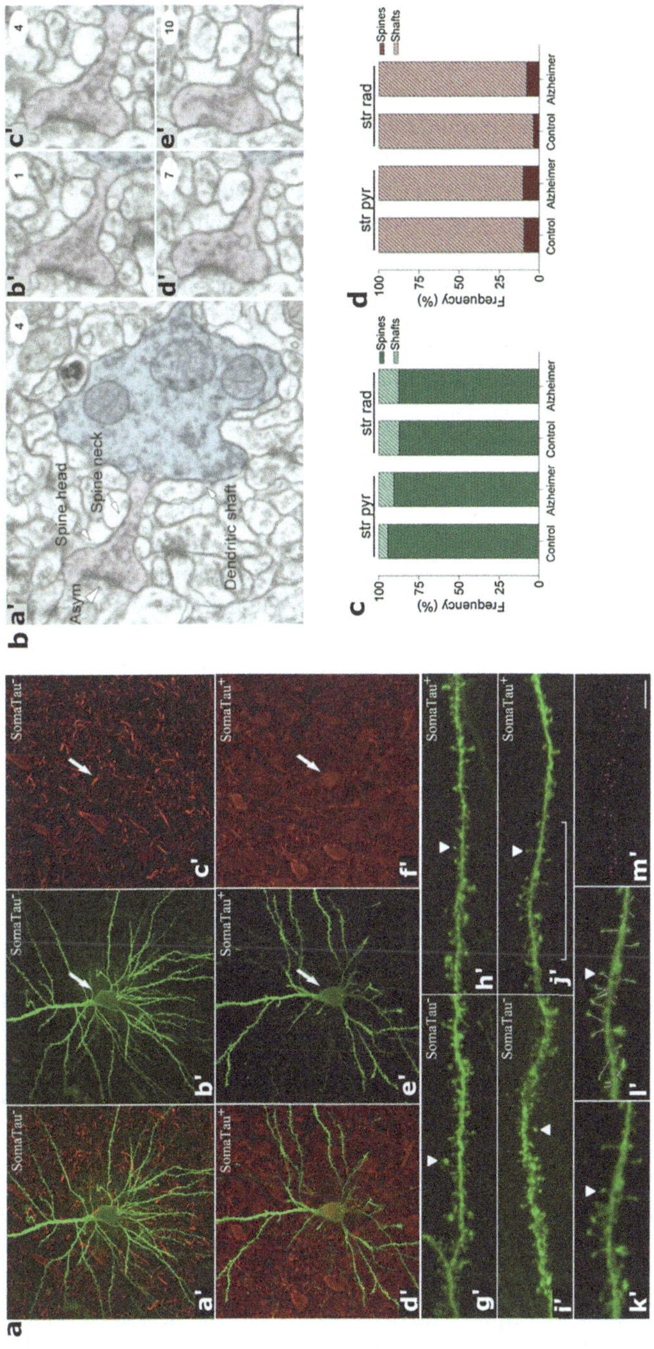

Fig. 9.26 (*caption on p. 460*)

In the same regard, recent 3D analysis of focused ion beam/scanning electron microscopy (FIB/SEM) images identified changes in the synaptic organization in the *stratum pyramidale* and *radiatum* in the hippocampal CA1 field of subjects diagnosed with AD (Fig. 9.26b). Axonal synapses on spine heads showed an asymmetric aspect that was the most abundant type, multisynaptic spines were observed in small numbers, and axospinous symmetric synapses were scarce in both strata (Montero-Crespo et al. 2021). Compared to controls, AD cases have a lower proportion of axospinous asymmetric synapses (with a diminished proportion of perforated PSD) and a higher frequency of axodendritic asymmetric synapses in the *stratum pyramidale* (Fig. 9.26c). In the *stratum radiatum*, a lower proportion of axodendritic symmetric synapses is found in AD cases compared to controls (Fig. 9.26d). In both strata, a lower proportion of large asymmetric synapses (mainly axospinous and macular synapses) exist in AD cases (Montero-Crespo et al. 2021).

The human medial entorhinal cortex (EC) is another critical area related to cognitive deficits in the course of AD (Domínguez-Álvaro et al. 2021a, b). The 3D analysis of FIB/SEM images served for the description of the synapses in the neuropil of medial EC in layers II and III from human brain autopsies (Domínguez-Álvaro et al. 2021a). The proportion of asymmetric and symmetric synapses on the dendritic shaft, on the spine head or neck, and the number and type of single or multiple axospinous contacts were determined for both layers in normal subjects (Fig. 9.27). In both layers, most synapses were of the asymmetric type presenting a macular morphology. Approximately half of the asymmetric contacts were made on dendritic shafts and half were on spine heads (usually one single axonal contact per spine). The majority of symmetric contacts were found on dendritic shafts and the remaining on spine heads (Domínguez-Álvaro et al. 2021a). In AD patients, there was a significantly lower synaptic density in both layers, asymmetric synapses were larger and more complex ("horseshoe-shaped") in layer II, and asymmetric synapses were smaller and simple (fragmented and macular were more frequent, whereas perforated was less frequent) in layer III compared to control individuals (Domínguez-Álvaro et al. 2021b).

Fig. 9.27 (*Left and Center*, **a–d**) Synaptic morphology analysis and identification of postsynaptic target distribution in the neuropil of layers II (**a**, **b**) and III (**c**, **d**) of the human medial entorhinal cortex (EC) under focused ion beam/scanning electron microscopy. (**a**) Distribution of asymmetric synapses (AS, green, considered excitatory) and symmetric synapses (SS, red, considered inhibitory) contacting postsynaptic targets (spine head and neck or dendritic shaft) in EC layer II. (**b**) Schematic representation of the proportions of AS and SS on different postsynaptic targets in EC layer II. (**c**) Distribution of AS and SS contacting different postsynaptic targets in EC layer III. (**d**) Schematic representation of the proportions of AS and SS on different postsynaptic targets in EC layer III. In (**b**) and (**d**), note - from left to right - the most frequent type (AS on dendritic shafts) to the least frequent type (SS on spine necks) of contact. (*Right*) Schematic representation of single excitatory synapses and multiple excitatory ones with or without inhibitory synapses upon the head of dendritic spines in EC layers II and III. Percentages of each synaptic type are indicated. Note that one single excitatory contact on the spine head is the most frequent type. Synapses on the necks and other combinations were rarely observed (less than 1%) and were not included. (Legend and figure slightly adapted and reprinted from Domínguez-Álvaro et al. (2021a) under CCC RightsLink® license #5386740499269, originally published by Oxford University Press)

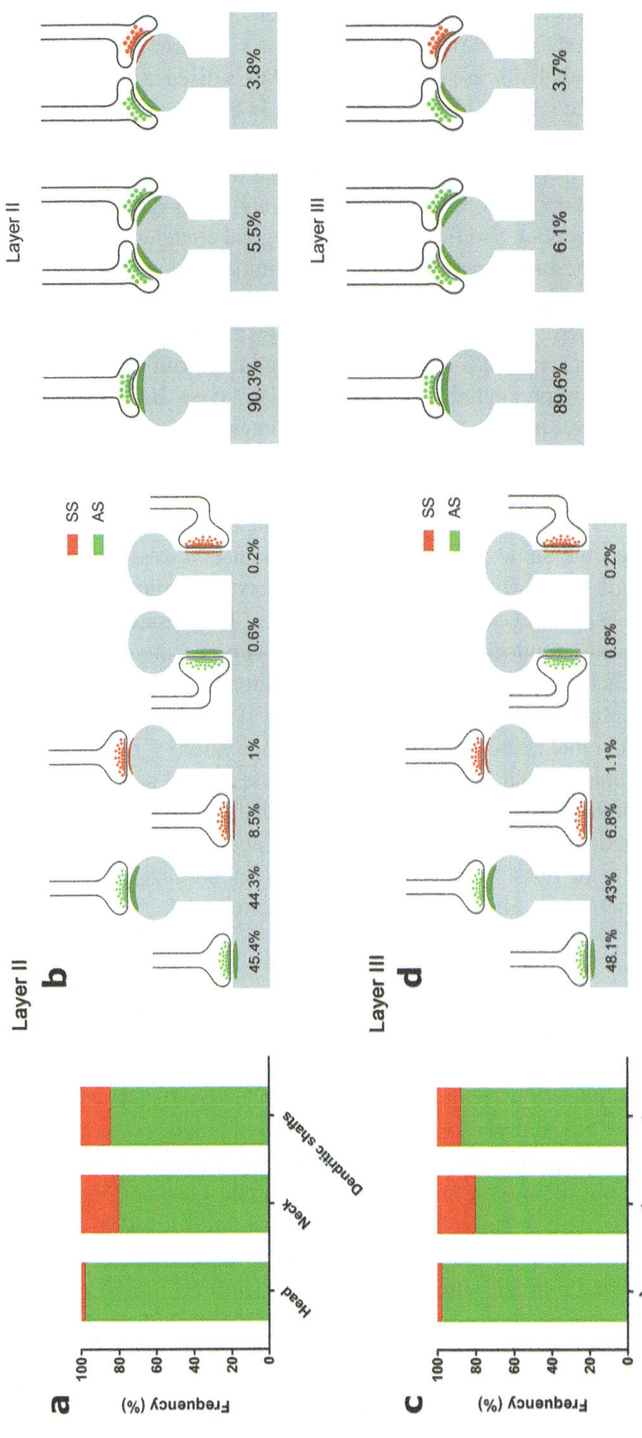

Fig. 9.27 (*caption on p. 462*)

A reduction in the number of apical dendritic spines on cortical pyramidal neurons was also reported in AD (three patients), Parkinson's disease and dementia (one patient), Creutzfeldt–Jakob disease (four patients), Pick's disease (one patient), and dementia paralytica (one patient; Catalá et al. 1988). In conjunction, these data indicate that altered excitatory synapses on dendritic spines are likely associated with impaired synaptic and information processing in areas and neurons vulnerable to tau pathology, β-amyloid deposition, AD progression, and other dementia cases. Structural, metabolic, perfusional, and functional pathological changes can be associated with cognitive decline in AD patients, which can initiate many years before the unequivocal clinical diagnosis of a symptomatic disorder.

On the other hand, cognitive resilience to AD pathology can occur in older individuals with abnormal brain levels of β-amyloid plaques and neurofibrillary tangles (Boros et al. 2017). Golgi-impregnated dendritic spines in layers II/III pyramidal neurons from the dorsolateral prefrontal cortex showed interesting results in these cases compared to age-matched pathology-free controls and AD cases. Spine density and, specifically, thin and mushroom spines were reduced in AD cases. Mean spine head diameter also showed lower values in AD. On the other hand, cognitively resilient persons showed the same spine density as normal controls, few stubby spines (but a higher length for this type of spine), and reduced thin spine head diameter compared to controls and AD cases (Boros et al. 2017). These findings suggest that maintenance of certain dendritic spine types, remodeling of their shape, and modulation of synaptic activity would protect older individuals with AD pathology from developing dementia (Boros et al. 2017; see also Walker and Herskowitz 2021). The study of vulnerable neuronal subtypes, the participation of glia and other cells in the local inflammatory and oxidative stress, and the altered neural and circuitry excitability are fundamental steps to elucidate the pathogenesis of AD. They will improve our knowledge about neuroprotection (Lourenço et al. 2019) or timely diagnosis and, hopefully, the possibility to halt progressive neurodegenerative processes and impaired functions in affected brain areas and circuitries.

9.7 Final Remarks and Perspectives

As evidenced by the existence of heterogeneous "aspiny" neurons or cells with very sparse spines along the CNS, morphological data indicate that spines may not be crucial for every information processing in all human neural circuits. Examples are found from the spinal cord to various brainstem, diencephalic, and cortical areas. Axons terminating directly upon smooth dendritic shafts might promote "direct" and less modulated synaptic responses than those made upon spines. In both aspiny and spiny cells, the neuronal input/output ratio depends on the presynaptic elements' activity, the synaptic transmission, and all the postsynaptic responses, including the dendritic tree's biophysical and biochemical attributes. Non-spiny neurons might have more variability in the EPSPs occurring at different dendritic locations and, within certain ranges, with no "low-pass filters" for synaptic transmission represented by spines (adapted from Tsay and Yuste 2004; Gulledge et al. 2012). Circuits may have cells devoid of spines or with sparse spines aiming for a cellular activity pattern that would show varied amplitudes for the dendritic EPSPs (related to the resting potential, input resistance, and membrane time constant) and, accordingly, input/output ratios for fast firing patterns depending on timely synaptic demands.

On the other hand, dendritic spines are evident in neurons coexisting with other aspiny cells along the CNS and, notably, in "higher-order" brain areas. For example, most neurons of evolved cortical areas that integrate unimodal to multimodal information are spiny ones. Although spines are not always needed to connect pre- and postsynaptic sides, when doing so they provide many more modulatory and plastic possibilities for signal integration upon dendrites. Spines enhance wiring possibilities, maximize connectivity with several different axons, and form networks with increased flexibility and high computational power. Spines can modulate the strength of synaptic inputs along different dendritic distances and branching orders. Spines also expand the complexity of synaptic processing by adding varied molecular pathways for activity-dependent integration and learning along dendritic segments. Various projecting neurons in the CNS are spiny, which implies that they compute numerous different and dynamic inputs, modulate, and integrate them before sending an output code onward within each respective neural circuit. The heterogeneity of spine number, distribution, shapes and sizes indicate a myriad of malleable computational possibilities for these neurons and networks. Some complementary possibilities are commented on in this final analysis.

Spiny pyramidal neurons have significant phylogenetic, ontogenetic, and functional development (Rasia-Filho et al. 2021 and references therein; see also Chaps. 4, 7, and 8 in this book). The excitability of cortical pyramidal cells is determined by thousands of spines along hundreds of micrometers of main dendritic shafts and collaterals with spatiotemporal integrative synaptic properties. From approximately 100 trillion spines in the human cortex (Kasai et al. 2021), nearly 99.5% of all spines lie in pyramidal neurons (Kubota et al. 2016) organizing the ongoing synaptic transmission from multiple neurochemical circuits (Palomero-Gallagher and Zilles 2019). Aspiny and spiny interneurons compose local microcircuits (Kubota et al. 2016; Foggetti et al. 2019). Ten or more cortical GABAergic non-pyramidal cell subtypes with a unique form of axonal arborization and innervation pattern control the excitation of these pyramidal neurons (Kubota et al. 2016). Pyramidal neurons and circuits have notable human-specific features within the expanded cortical layers II/III. Spines reflect past and ongoing synaptic demands for each neuronal type, area, and network. Based on this scenario, it is plausible to suppose that we will still discover more cellular and connectional particularities in our species when looking for areas with a higher development in primates and for our distinguished behaviors.

Here, one insight deserves additional consideration, that is, when mentioning that "dendritic spines are postsynaptic units" we are also indicating the existence of three "worlds" of functional specialization and evolution working together (i.e., *dendrites* plus *spines* plus *synapses*). Each component has its complexity and active role in modulating information that, jointly, provide much more functional possibilities to information processing (Sjöström et al. 2008; Sala and Segal 2014; Nakahata and Yasuda 2018; Rollenhagen et al. 2020; Schmidt and Polleux 2022). The timely and spatially integrated actions by these elements generate multiple properties for the beginning of emergent features in assembled cells within neural networks. They are also associated with neural organization occurring on a full range of scales, from synaptic structure and individual proteins to changes in long-distance connectivity and gray and white matter volumes (Mancuso et al. 2014).

Indeed, spines have the capacity to modulate the magnitude of the EPSPs and associate them with the geometry and functional properties of the dendritic arbor. It is then important to determine the actual impact of spines on neuronal activity. Spines can integrate the inputs received and transform synaptic inputs into a variety of linear and nonlinear voltage signals in the parent dendrite (Spruston 2008). Depending on voltage dynamics in spines (Cornejo et al. 2022) and the specific form of dendritic integration (i.e., from summed EPSPs to brief or prolonged dendritic spikes), dendrites can show a single AP, trains and bursts of multiple APs (Losonczy and Magee 2006). Even a partial fusion of active spines or changes in spine branch positions would increase synaptic signal transfer (Rusakov et al. 1996). For example, glutamate evokes local dendritic spikes in rat CA1 pyramidal neurons. Somatic voltage responses show a nonlinear increase when such stimulations are performed in rapid succession at multiple spots either clustered or distributed along the dendritic segment, although the dendritic spike threshold is lower for the distributed input than for the clustered one (Losonczy and Magee 2006; Spruston 2008). These synchronous and sized input patterns would be represented by a tightly clustered stimulation of ~20 inputs within ~6 ms (i.e., 20 synapses would represent ~30% of all synapses within 20 μm in this specific dendritic segment), whereas distributed stimuli would be represented by the activation of spines randomly covering approximately 67% of a single oblique dendrite (along ~60 μm; Losonczy and Magee 2006). By modeling human temporal layer II–III pyramidal neurons, a somatic AP would be generated (with 50% probability) by 134 ± 28 simultaneously activated synapses with large AMPA and NMDA conductances per synaptic contact, with EPSPs at the spine head of 12.7 ± 4.6 mV, and at spine base of 9.7 ± 5.0 mV (Eyal et al. 2018). Based on the extension of the pyramidal neuron dendritic arbor (in μm and with different domains) and the number and location of pleomorphic spines, the activation of a relatively a few number of dendritic spines provides neurons with many different possibilities to elaborate a flexible input–output code within neural circuits.

Let us consider again the likely integrative action made by a human long and densely spined CA1 hippocampal neuron (as shown in Fig. 9.15) in the context of our current knowledge about this area and spines' features. The CA1 area has a relevant role in encoding memories along with integrated and widespread circuits that endow us with a clear notion of who we are, where we stay, and the time we live in. Dendrites are capable of compartmentalization combined with amplification, inhibition, and neuromodulation of inputs for routing and multiplexing streams of information (Payeur et al. 2019). Spines are multifunctional integrative units (Shepherd 1996) with a turnover rate (formation and shrinkage), varied number, and a *continuum* of shapes and sizes that compute and modulate the network state with multiple spatiotemporal axonal inputs reaching every single cell. Pre- and postsynaptic elements may show associated structure and function in terms of connective strength (Leopold et al. 2019), with the former adjusting the speed of learning and postsynaptic firing rates, whereas postsynaptic plasticity can also amplify the response range (Mizusaki et al. 2018). Contextual information processing and different types of local computations and organizational gradients (Shafiei et al. 2020) as well as synaptic-induced voltage transients and molecular cascades beginning in the intraspine microenvironment can spread to neighbor spines and within the parent dendrite and onward (Harvey et al. 2008; Sala and Segal 2014; Cornejo et al. 2022). Together with glial cells and extracellular matrix, it is formed a system with a very large range of possible activities, from the nanoscale to major networks functioning, using both stable and plastic characteristics. The specialization of our neural cells and the expanded connectivity of the human neuropil can involve unique elements (Cajal 1909–1911; DeFelipe 2011; Schmidt and Polleux 2022). Approximately 10% of the proteins encoded in the human genome can be found composing synapses, their diversity and individual function (Grant and Fransén 2020) in complex PSD compositions (Cohen 2013). Over 500 different types of molecules modulate the structure and function of dendritic spines (Ammassari-Teule et al. 2021; see also Calabrese et al. 2006 and Helm et al. 2021). The *dendrites* plus *spines* plus *synapses* and their rich network features provide emergent functions expanding our capacity for learning and adaptive responses. This is part of the complexity elaborated by one single neuron in a conjunct of millions of other neural cells composing one unique person.

There are still many unexplored human brain areas, and more data are needed for the characterization of human neurons in an integrated functional context. New experimental frontiers offer exciting perspectives for the detailed characterization of individual neuronal profiles within areas and networks (González-Burgos et al. 2019; Gouwens et al. 2019). They involve "multi-omics" and multiple molecular signatures associated with computational approaches (Luo et al. 2022), and multi-neuron patch-clamp recordings and morphology assayed from the same cells in neurosurgically resected human brain tissue (Hodge et al. 2020; Berg et al. 2021; Planert et al. 2021). The number of neuronal types can be impressive. For example, by revealing conserved cell types with divergent features, 24 excitatory and 45 inhibitory transcriptomically distinct neuron types were identified in the middle temporal gyrus in humans (Hodge et al. 2019). Non-neuronal cell types had the most-divergent expression of genes comparing human and mouse temporal cortex (Hodge et al. 2019). Joint analyses of the methylome, transcriptome, chromatin accessibility, and conformation revealed 63 cell finely defined populations in the human frontal cortex, including 19 excitatory neuronal subtypes, 33 inhibitory neuronal subtypes, and 11 non-neuronal cell subtypes (Luo et al. 2022). In addition, human cortical pyramidal neurons in supragranular layers show electrophysiological characteristics that support their subdivision into five different functional subpopulations (Berg et al. 2021; Planert et al. 2021; see relevant mice data in Gouwens et al. 2019; Fig. 9.28a, b). The characterization of dendritic spines in these specific neuronal subpopulations is highly desirable (as done to describe gonadotropin-releasing hormone cells, Campbell et al. 2005; see also the work of González-Burgos et al. 2019).

Furthermore, neuron morphology is a major source of wiring specificity and nonrandom cortical network topology in addition to genetically defined cellular identity and molecular recognition of cell type and activity (Udvary et al. 2022). However, axodendritic overlap does not predict connectivity. Spines from one dendrite can be completely intermingled with the spines of other dendrites and various axons can be close to a spine, but most juxtapositions of the trajectories of excitatory axons with dendritic spines do not establish synapses in the mouse cerebral cortex (Kasthuri et al. 2015; Udvary et al. 2022). On the other hand, "the greater the cellular diversity, the broader are the distributions of connection probabilities" (Udvary et al. 2022; Fig. 9.28c). This is a finding that can advance with paired recordings from synaptically coupled neurons and by combining electrophysiological properties with 3D structural and molecular profiles to characterize both pre- and postsynaptic neurons at the light and electron microscopic level in local brain microcircuits, including human samples (Qi et al. 2020; see also Kubota et al. 2018 and Leopold et al. 2019).

Complementary approaches are needed toward a unified, consensual neuronal classification based on a high-throughput single-cell transcriptomic-based taxonomy, building a "probabilistic definition" of every cell type along brain regions (Yuste et al. 2020; see also Peng et al. 2021). We will need to critically evaluate intra- and inter-type variability in cellular phenotypes along different ages, brain areas and their synaptic demands (Lodato and Arlotta 2015; Fuentealba-Villarroel et al. 2022) as well as taking into account that transcriptional profiles and cell-specific molecular responses can vary between individuals (Xu et al. 2018). "The existence of cell states, spatial gradients of phenotypes and mixtures of differences and similarities in cross-species comparisons present challenges to a discrete and categorical perspective on defining cell types. Prematurely adopting an inflexible definition of types will obscure the significance of observed phenotypic variability and its biological interpretation…" while might exist a "… core and intermediate cells or the description of a cell type as a continuous trajectory in transcriptomic space" (Yuste et al. 2020). When doing so with advancing techniques, additional perspectives will emerge for relating human-specific genomic features with morphology, connectivity of specific neuronal subpopulations, dendritic spines, and synaptic structure and composition (Kommaddi et al. 2018; Xu et al. 2018; Helm et al. 2021; Fuentealba-Villarroel et al. 2022; Hunt et al. 2022; Schmidt and Polleux 2022; see the modular nanomolecular architecture of synaptic proteins described in mice and rats by Hruska et al. 2018). For example, intraspine nanodomains with biochemical cascades and proteins can remodel the dendritic spine structure. Current approaches can evaluate the polymerization/depolymerization ratio of actin in normal and pathological conditions (Fig. 9.29). Integrating foundational knowledge and clinical findings can open new possibilities to understand the

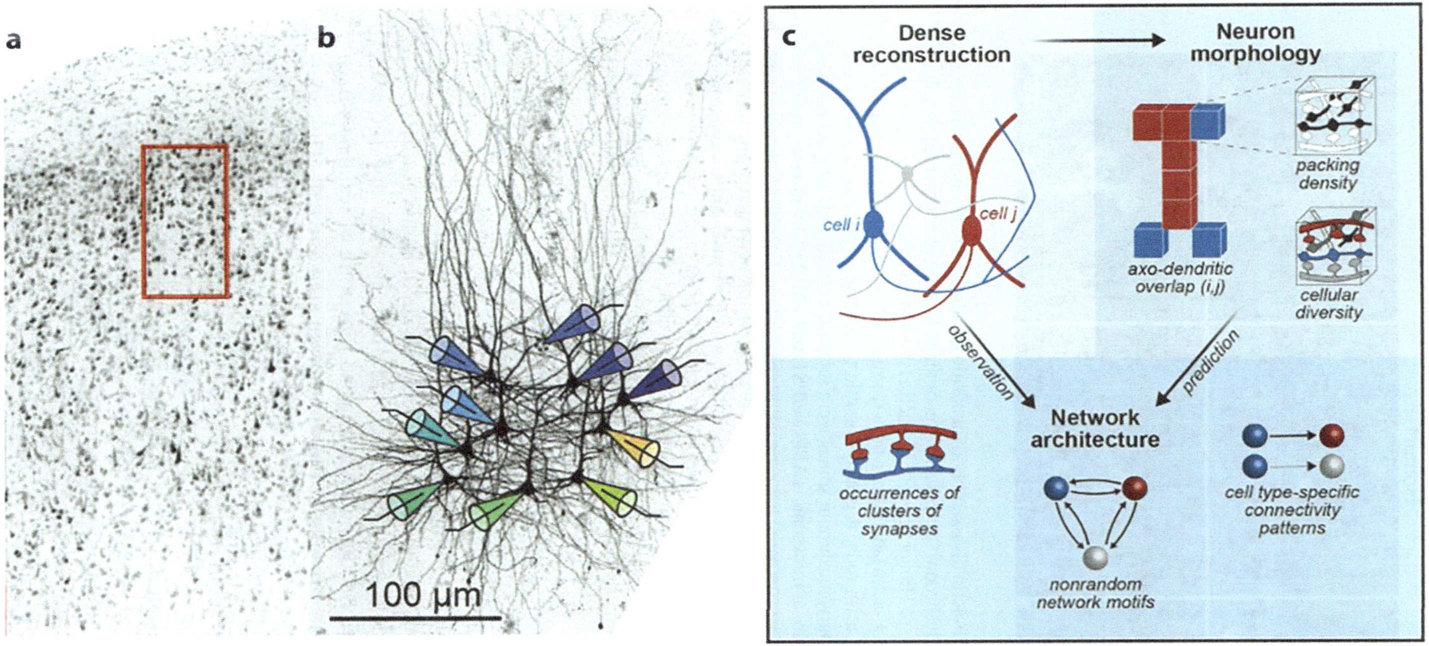

Fig. 9.28 High-throughput whole-cell patch-clamp recordings can be performed in human cortical brain tissue *ex vivo* providing hundreds of pyramidal neurons for electrophysiological and morphological study. (**a**) Nissl staining of human cortical cell layers. Cells from layer II and III cells (indicated by a rectangle) were sampled for further study. (**b**) Colored patch pipettes illustrate the multi-neuron patch-clamp approach. Neighboring pyramidal neurons were electrophysiologically recorded and, afterward, biocytin stained to unravel their morphology. (Legend adapted and Figure reprinted from the original work by Planert et al. (2021) https://doi.org/10.1101/2021.11.08.467668; copyright CC BY 4.0 International license). (**c**) To test the impact of neuron morphology on network connectivity, reconstructed 3D geometry and cytoarchitecture of a brain area of interest were obtained and the morphology and classification of neurons inside a selected volume were determined (in this case, neurons are labeled *i* in blue and *j* in red). Afterward, within brain subvolumes, the axodendritic overlaps were identified. Note the representation of neuron *j* and the axonal projections of neuron *i* contacting it related to the local packing density of presynaptic axons (black), postsynaptic dendrites, and spines (white) as well as the cellular diversity (i.e., the number of neurons from which the presynaptic and postsynaptic elements originate). Predictions and empirical observations are then compared for the existence of clusters of synapses, nonrandom network motifs, and cell-type-specific connectivity patterns. (Figure reprinted from Udvary et al. (2022) under CCC RightsLink® license #5440750664829, originally published by Elsevier)

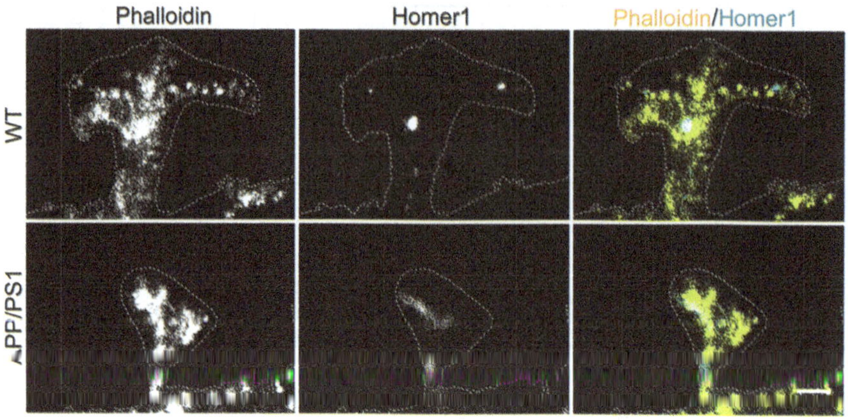

Fig. 9.29 β-amyloid mediates F-actin disassembly in dendritic spines leading to cognitive deficits in Alzheimer's disease (AD). Mushroom spines can be recognized by their characteristic morphology and the intraspine presence of Homer1 puncta and actin (merged images shown, respectively, in cyan and yellow on the right). F-actin labeling can be performed with Alexa Fluor 647-phalloidin. In addition to the detection of actin rods, it is also possible to determine whether the fibers are parallel to each other. Accordingly, mushroom spines were selected to study radiating actin structures within the spine head in a transgenic (APPswe/PS1ΔE9, APP/PS1) mouse model of AD. Mean anisotropy values were compared between wild-type (WT) and APP/PS1 mice to quantify changes in F-actin organization. Direct stochastic optical reconstruction microscopy revealed the occurrence of F-actin depolymerization in dendritic spines, which is causal and not consequent to decreased spine density in AD and leads to a breakdown of the nano-organization of outwardly radiating F-actin rods in primary cortical neuronal culture. (Legend adapted and figure reprinted from Kommaddi et al. (2018) under CC BY license and Copyright Kommaddi, Das, Karunakaran, Nanguneri, Bapat, Ray, Shaw, Bennett, Nair, Ravindranath; doi: 10.1523/JNEUROSCI.2127-17.2017)

dysregulation of spine dynamics in neuropathologies (Runge et al. 2020), which can also support predictive models aiming for novel therapies for neurological and psychiatric disorders (e.g., Paoletti et al. 2013; Beutel et al. 2020; Hunt et al. 2022).

These morphological data would then help to elaborate how specific cells bring about different functional outputs and a broad repertoire of responses from a multitude of axonal inputs arriving at the same brain area and cell. This complex task looks like finding the functional feature of each cell or brain area as a piece of a mosaic. One realizes that these neuronal and glial subpopulations in different organized circuits and timely interconnected areas function as a synaptically coded and responsive "orchestra" able to perform various meaningful "symphonies". Each node (and there are multiple nodes with intermingled subsets of cells, distinct neural temporal dynamics, and connectional strengths) receives "weighted" inputs to be integrated for an output code, both displaying a repertoire of flexible possibilities within the limits and properties of each circuit (Li et al. 2017; McCarthy 2023). At the cellular level, the number, spatial distribution, and strength of each synaptic contact along a widespread receptive field are modulated by the interaction of each axon upon dendritic segments and pleomorphic spines. Perpendicular axons may establish few contacts with the dendrites of one neuron but may contact many neurons; otherwise, parallel projections may selectively innervate a few neurons and strongly influence these targets (Yuste 2010). Together, the "orchestra" plays with the activity-dependent intensity and source of inputs combined with variable amplitudes of functional output codes from the embedded components. Graded "ups" and "downs" of cellular excitability all over time mean that cells and areas will be activated, and others will be less activated or even deactivated (Newman 1999; Fig. 9.30a). These dynamic switches form different codes that represent the interpretation of what is being processed (Busler et al. 2019; e.g., see functional data from young men submitted to sexually arousing films in Redouté et al. (2000)and neural correlates of romantic love and affection in women and men in Zeki (2007), Acevedo et al. (2012), de Boer et al. (2012)). Dendritic spines are part of this scenario modulating excitatory neural processing and elaborating the initial steps for these output codes. It is expected that different types of cellular and network computations manifest with different time-series features of coordinated gradients of activity within and between brain areas (Shafiei et al. 2020). In the end, these functional codes established by interacting neural populations "sound" as different instruments being played by musicians in an orchestra. The sequence and intensity performed by each "group of instruments" are determined by the connectivity of their elements under a formidable functional and temporal fine-tune (Fig. 9.30b; see also Fig. 9.3 in Leopold et al. 2019 and further recent data on brain networks).

Species-specific interpretation of the "world," (un)conscious elaborations, memories and learning possibilities, emotions, and social behaviors depend on the distributed and coordinated activity of these neural populations, their codes at multiple spatial and temporal scales, and the integration of information in local and long-range networks. The interconnection of elements can be extensive within circuits (Fig. 9.30b; Glasser et al. 2016) progressing toward eventual "actions" in a multimodal integrated organization (Bruton 2021; Kolb and Whishaw 2021). Under physiological conditions, these actions emerge from a dynamic pattern of activation across the network rather than an "on" or "off" state of any one of its nodes (Newman 1999). A sequence of behaviors relates to a particular temporal pattern of activity of associated areas in the whole network (Fig. 9.30a), and subtle changes in this activity would mean different states and emergent properties, such as the few microvolts that evoke different firing rhythm patterns of thalamic relay neurons coding slow-wave sleep or wakefulness (McCormick and Pape 1990). In this same regard, cortical motor neurons are activated when processing information by observing or executing movements in rhesus macaques. However, rather than existing functionally distinct subpopulations of cells to perform these activities, motor neuron populations exhibit both a commonly shared computation and a unique task-dependent and dynamic coordinated temporal activation for one or another action (Jiang et al. 2020). At this point, an outstanding question is how microscopic-level neural

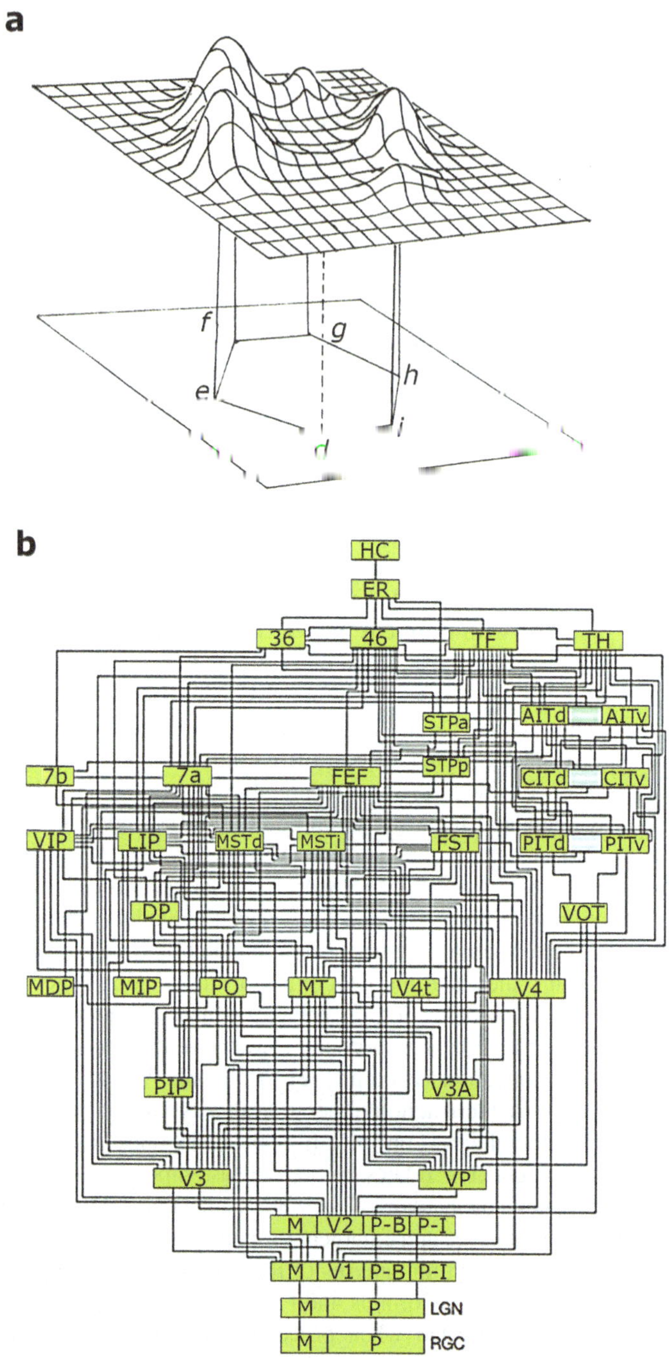

Fig. 9.30 (**a**) A hypothetical three-dimensional graph representing the magnitude of activation or inhibition/deactivation of integrated brain areas. These areas are organized as a functional network for the display and to modulate homeostatic activity or behaviors. Areas are indicated by letters (*d-i*) at the bottom of the image. The present synchronous pattern of activity represents an output code in the network and will determine a specific response. Changing the spatiotemporal ampli

coding mechanisms interact with macroscopic scale aspects for evoking our most complex neural circuitry activities, including consciousness, abstract thoughts, creativity, and feelings (Creutzfeldt 1995; Timo-Iaria and Valle 1995; Panzeri et al. 2015; LeDoux and Brown 2017; see an interesting discussion on perceiving itch or pain using multiple combinatorial codes from the same sensory receptors in Gardner, 2021).

The intra-individual and interindividual differences associated with the structural plasticity of human neurons, dendrites, spines, and synapses imply a large repertoire of computational possibilities for the human brain. Cellular morphology provides clues for functional implications of local connectivity and the levels of information processing, and a foundational knowledge that is continuously refined as new information is acquired (adapted from Rust and LeDoux 2023). Moreover, basic morphological data can be associated with probabilistic cytoarchitectonic maps and 3D-reconstructed images to interpret ultra-high field fMRI results in normal and pathological conditions (Kiwitz et al. 2022). Spines evolved to expand the plasticity of networks' functioning as miniature elements with multi-integrative capacity (Yuste 2010). It is not surprising that the number of cells and the level of complexity of the human cerebral cortex reflects its capacity to execute over 1.2 Zetta logical operations per second (Georgiev et al. 2020). Unraveling the dynamic role of dendritic spines for synaptic processing in humans is a task that needs multiple complementary experimental approaches linking morphology to physiology. This endeavor currently resides on the astonishing scale of around 10^{15} connections in the human brain (Rasia-Filho 2022; contextualized by DeWeerdt 2019). Nevertheless, spines can provide varied modulatory possibilities for synaptic transmission serving homeostatic and flexible requirements, showing a varied number and shape with phylogenetic and ontogenetic implications for network performance, and providing relevant cues to be investigated in human neurological and psychiatric conditions.

Fig. 9.30 (continued) tude of activity of one or various components would change the resultant output code and, consequently, evoke a different response display. (Figure reprinted and slightly modified from Newman (1999) under CCC RightsLink® license #5438910072828, originally published by John Wiley and Sons). (**b**) Multiple, intertwined streams for visual processing in the primate cerebral cortex. From the bottom to the top, visual areas and other associated nonvisual areas are indicated by rectangles (highlighted in yellow) and connecting pathways by lines, most of which are reciprocal ones. This schematic representation includes magnocellular (M) and parvocellular (P) retinal ganglion cell (RGC) and layers of the lateral geniculate nucleus (LGN) as the afferent pathway to visual area 1 [V1, with further split into the parvo-blob (P-B) and parvo-interblob (PI)], progressing sequentially to neighbor visual cortical areas (V2, V3, VP, and on) plus nonvisual areas [BAs 7a and 7b, perirhinal area 36, middle frontal area 46, entorhinal cortex (ER), and hippocampal complex (HC)]. Note the connectional complexity of such circuitries and the possibility of feedback and feedforward loops affecting the dynamics of information processing within it. It is important to include spines and synapses needed in each step for the elaboration of memories, emotions, social judgments of facial expressions, and empathy, for example. (Figure reprinted and slightly modified from Felleman and Van Essen (1991) under CCC RightsLink® license #5438980718898, originally published by Oxford University Press)

Fig. 9.31

P.S.1: "To write is to weave" (Eduardo Galeano) to transform imagination into symbolic representations of words. Imagination also makes us prone to find "dendritic spines" (and a "PSD" indicated by an asterisk) in unsuspected places (Titiqaqa Lake, border of Perú and Bolivia, seen from a plane window; Fig. 9.31).

P.S.2: Last but not least, the authors are thankful (and dedicate this chapter) to all persons who donated samples from their beloved relatives for the study of the human brain by scientists they may have never known before. Our gratitude and respect are due not only because of such an altruistic attitude while grieving but also to their deep trust in science and the best future for other persons and our humanity.

References

Acevedo BP, Aron A, Fisher HE et al (2012) Neural correlates of long-term intense romantic love. Soc Cogn Affect Neurosci 7:145–159

Adolphs R (2003) Is the human amygdala specialized for processing social information? Ann N Y Acad Sci 985:326–340

Aguilar-Hernández L, Alejandre R, Morales-Medina JC et al (2023) Cellular mechanisms in brain aging: Focus on physiological and pathological aging. J Chem Neuroanat 128:102210. https://doi.org/10.1016/j.jchemneu.2022.102210

Akhmadeev AV, Kalimullina LB (2015) Paleoamygdala: the morphogenesis of nuclear-type, paleocortical and intermediate formations in the perial of postnatal development in rats. Rus J Develop Biol 46:27–32

Allman JM, Hakeem A, Erwin JM et al (2001) The anterior cingulate cortex. The evolution of an interface between emotion and cognition. Ann N Y Acad Sci 935:107–117

Allman JM, Watson KK, Tetreault NA et al (2005) Intuition and autism: a possible role for von Economo neurons. Trends Cogn Sci 9:367–373

Allman JM, Tetreault NA, Hakeem AY et al (2010) The von Economo neurons in frontoinsular and anterior cingulate cortex in great apes and humans. Brain Struct Funct 214:495–517

Allman JM, Tetreault NA, Hakeem AY et al (2011a) The von Economo neurons in the frontoinsular and anterior cingulate cortex. Ann N Y Acad Sci 1225:59–71

Allman JM, Tetreault NA, Hakeem AY et al (2011b) The von Economo neurons in apes and humans. Am J Hum Biol 23:5–21
Almog M, Korngreen A (2014) A quantitative description of dendritic conductances and its application to dendritic excitation in layer 5 pyramidal neurons. J Neurosci 34:182–196
Amaral DG, Price JL, Pitkänen A et al (1992) Anatomical organization of the primate amygdaloid complex. In: Aggleton JP (ed) The Amygdala: neurobiological aspects of emotion, memory, and mental dysfunction. Wiley-Liss, New York, pp 1–66
Amiez C, Sallet J, Novek J et al (2021) Chimpanzee histology and functional brain imaging show that the paracingulate sulcus is not human-specific. Commun Biol 4:54. https://doi.org/10.1038/s42003-020-01571-3
Ammassari-Teule M, Sala C, Segal M (2021) Editorial - Dendritic spines: from biophysics to neuropathology. Front Synaptic Neurosci 13:652117. https://doi.org/10.3389/fnsyn.2021.652117
An NA, Zhang J, Mo F et al (2023) De novo genes with an lncRNA origin encode unique human brain developmental functionality. Nat Ecol Evol. https://doi.org/10.1038/s41559-022-01925-6
Andersen P, Morris R, Amaral D et al (2007) The Hippocampus book. Oxford University Press, New York
Anderson B, Rutledge V (1996) Age and hemisphere effects on dendritic structure. Brain 119:1983–1990
Anderson SA, Classey JD, Condé F et al (1995) Synchronous development of pyramidal neuron dendritic spines and parvalbumin-immunoreactive chandelier neuron axon terminals in layer III of monkey prefrontal cortex. Neuroscience 67:7–22
Anderson K, Bones B, Robinson B et al (2009) The morphology of supragranular pyramidal neurons in the human insular cortex: a quantitative Golgi study. Cereb Cortex 19:2131–2144
Araya R (2014) Input transformation by dendritic spines of pyramidal neurons. Front Neuroanat 8:141. https://doi.org/10.3389/fnana.2014.00141
Araya R, Vogels TP, Yuste R (2014) Activity-dependent dendritic spine neck changes are correlated with synaptic strength. Proc Natl Acad Sci USA 111:E2895-E2904
Ardesch DJ, Scholtens LH, Li L et al (2019) Evolutionary expansion of connectivity between multimodal association areas in the human brain compared with chimpanzees. Proc Natl Acad Sci U S A 116:7101–7106
Arellano JI, Benavides-Piccione R, DeFelipe J et al (2007a) Ultrastructure of dendritic spines: correlation between synaptic and spine morphologies. Front Neurosci 1:131–143. https://doi.org/10.3389/neuro.01.1.1.010.2007
Arellano JI, Espinosa A, Fairén A et al (2007b) Non-synaptic dendritic spines in neocortex. Neuroscience 145:464–469
Arnold AP (2020) Sexual differentiation of brain and other tissues: five questions for the next 50 years. Horm Behav 120:104691. https://doi.org/10.1016/j.yhbeh.2020.104691
Ascoli GA (2015) Trees of the brain, roots of the mind. The MIT Press
Azevedo FA, Carvalho LR, Grinberg LT et al (2009) Equal numbers of neuronal and nonneuronal cells make the human brain an isometrically scaled-up primate brain. J Comp Neurol 513:532–541
Bach DR, Behrens TE, Garrido L et al (2011) Deep and superficial amygdala nuclei projections revealed in vivo by probabilistic tractography. J Neurosci 31:618–623
Bączyńska E, Pels KK, Basu S et al (2021) Quantification of dendritic spines remodeling under physiological stimuli and in pathological conditions. Int J Mol Sci 22:4053. https://doi.org/10.3390/ijms22084053
Bagarinao E, Watanabe H, Maesawa S et al (2019) Reorganization of brain networks and its association with general cognitive performance over the adult lifespan. Sci Rep 9:11352. https://doi.org/10.1038/s41598-019-47922-x
Bagarinao E, Watanabe H, Maesawa S et al (2020) Aging impacts the overall connectivity strength of regions critical for information transfer among brain networks. Front Aging Neurosci 12:592469. https://doi.org/10.3389/fnagi.2020.592469
Baloyannis SJ, Manolidis SL, Manolidis LS (2001) The acoustic cortex in frontal dementia. Acta Otolaryngol 121:289–292

Baltazar-Gaytan E, Aguilar-Alonso P, Brambila E et al (2019) Increased cell number with reduced nitric oxide level and augmented superoxide dismutase activity in the anterior-pituitary region of young suicide completers. J Chem Neuroanat 96:7–15

Banovac I, Sedmak D, Džaja D et al (2019) Somato-dendritic morphology and axon origin site specify von Economo neurons as a subclass of modified pyramidal neurons in the human anterior cingulate cortex. J Anat 235:651–669

Banovac I, Sedmak D, Judaš M et al (2021) Von Economo neurons – primate-specific or commonplace in the mammalian brain? Front Neural Circuits 15:714611. https://doi.org/10.3389/fncir.2021.714611

Basbaum AI (2021) Pain. In: Kandel ER, Koester JD, Mack SH, Siegelbaum SA (eds) Principles of neural science. McGraw Hill, New York, pp 470–495

Bassignana G, Lacidogna G, Bartolomeo P et al (2022) The impact of aging on human brain network target controllability. Brain Struct Funct. https://doi.org/10.1007/s00429-022-02584-w

Bastian AJ, Lisberger SG (2021) The cerebellum. In: Kandel ER, Koester JD, Mack SH, Siegelbaum SA (eds) Principles of neural science. McGraw Hill, New York, pp 908–931

Basu J, Siegelbaum SA (2015) The corticohippocampal circuit, synaptic plasticity, and memory. Cold Spring Harb Perspect Biol 7(11):a021733. https://doi.org/10.1101/cshperspect.a021733

Beaulieu-Laroche L, Toloza EH, Van der Goes MS et al (2018) Enhanced dendritic compartmentalization in human cortical neurons. Cell 175:643–651

Beaulieu-Laroche L, Brown NJ, Hansen M et al (2021) Allometric rules for mammalian cortical layer 5 neuron biophysics. Nature 600:274–278

Becker RO, Rasia-Filho AA, Giovenardi M (2017) Selective deletion of the oxytocin gene remodels the number and shape of dendritic spines in the medial amygdala of males with and without sexual experience. Neurosci Lett 660:155–159

Beed P, Ray S, Velasquez LM et al (2020) Species-specific differences in synaptic transmission and plasticity. Sci Rep 10:16557. https://doi.org/10.1038/s41598-020-73547-6

Benarroch EE (2018) Brainstem integration of arousal, sleep, cardiovascular, and respiratory control. Neurology 91:958–966

Benavides-Piccione R, Ballesteros-Yáñez I, DeFelipe J et al (2002) Cortical area and species differences in dendritic spine morphology. J Neurocytol 31:337–346

Benavides-Piccione R, Fernaud-Espinosa I, Robles V et al (2013) Age-based comparison of human dendritic spine structure using complete three-dimensional reconstructions. Cereb Cortex 23:1798–1810

Benavides-Piccione R, Regalado-Reyes M, Fernaud-Espinosa I et al (2020) Differential structure of hippocampal CA1 pyramidal neurons in the human and mouse. Cereb Cortex 30:730–752

Benavides-Piccione R, Rojo C, Kastanauskaite A et al (2021) Variation in pyramidal cell morphology across the human anterior temporal lobe. Cereb Cortex 31:3592–3609

Berg J, Sorensen SA, Ting JT et al (2021) Human neocortical expansion involves glutamatergic neuron diversification. Nature 598:151–158. Erratum in Nature (2022) 601:E12. https://doi.org/10.1038/s41586-021-04322-4

Beutel T, Dzimiera J, Kapell H et al (2020) Cortical projection neurons as a therapeutic target in multiple sclerosis. Expert Opin Ther Targets 24:1211–1224

Bianchi S, Stimpson CD, Bauernfeind AL et al (2013) Dendritic morphology of pyramidal neurons in the chimpanzee neocortex: regional specializations and comparison to humans. Cereb Cortex 23:2429–2436

Blake Y, Terburg D, Balchin R et al (2019) The role of the basolateral amygdala in dreaming. Cortex 113:169–183

Blood AJ, Zatorre RJ (2001) Intensely pleasurable responses to music correlate with activity in brain regions implicated in reward and emotion. Proc Natl Acad Sci U S A 98:11818–11823

Bolam JP, Pissadaki EK (2012) Living on the edge with too many mouths to feed: why dopamine neurons die. Mov Disord 27:1478–1483. https://doi.org/10.1002/mds.25135

Borczyk M, Sliwińska MA, Caly A et al (2019) Neuronal plasticity affects correlation between the size of dendritic spine and its postsynaptic density. Sci Rep 9:1693. https://doi.org/10.1038/s41598-018-38412-7

Boros BD, Greathouse KM, Gentry EG et al (2017) Dendritic spines provide cognitive resilience against Alzheimer's disease. Ann Neurol 82:602–614

Bouhali F, Mongelli V, Thiebaut de Schotten M et al (2020) Reading music and words: the anatomical connectivity of musicians' visual cortex. Neuroimage 212:116666. https://doi.org/10.1016/j.neuroimage.2020.116666

Bourne JN, Harris KM (2007) Do thin spines learn to be mushroom spines that remember? Curr Opin Neurobiol 17:381–386

Braak H (1979) Pigment architecture of the human telencephalic cortex. V. Regio anterogenualis. Cell Tissue Res 204:441–451

Braak H (1980) Architectonics of the human telencephalic cortex. Springer-Verlag, Berlin

Braak H, Bachmann A (1985) The percentage of projection neurons and interneurons in the human lateral geniculate nucleus. Hum Neurobiol 4:91–95

Braak H, Braak E (1976) The pyramidal cells of Betz within the cingulate and precentral gigantopyramidal field in the human brain. A Golgi and pigment architectonic study. Cell Tissue Res 172:103–119

Braak H, Braak E (1982a) Neuronal types in the claustrum of man. Anat Embryol 163:447–460

Braak H, Braak E (1982b) Neuronal types in the striatum of man. Cell Tissue Res 227:319–342

Braak E, Braak H (1983a) On three types of large nerve cells in the granular layer of the human cerebellar cortex. Anat Embryol 166:67–86

Braak H, Braak E (1983b) Neuronal types in the basolateral amygdaloid nuclei of man. Brain Res Bull 11:349–365

Braak H, Braak E (1984a) Neuronal types in the neocortex-dependent lateral territory of the human thalamus. Anat Embryol 169:61–72

Braak H, Braak E (1984b) Neuronal types in the lateral geniculate nucleus of man. A Golgi-pigment study. Cell Tissue Res 237:509–520

Braak H, Braak E (1986) Nuclear configuration and neuronal types of the nucleus niger in the brain of the human adult. Hum Neurobiol 5:71–82

Braak E, Braak H (1993) The new monodendritic neuronal type within the adult human cerebellar granule cell layer shows calretinin-immunoreactivity. Neurosci Lett 154:199–202

Braak H, Weinel U (1985) The percentage of projection neurons and local circuit neurons in different nuclei of the human thalamus. J Hirnforsch 26:525–530

Braak H, Del Tredici K, Rüb U et al (2003) Staging of brain pathology related to sporadic Parkinson's disease. Neurobiol Aging 24:197–211

BRAIN Initiative Cell Census Network (BICCN) (2021) A multimodal cell census and atlas of the mammalian primary motor cortex. Nature 598:86–102

Brandão ML, Cardoso SH, Melo LL et al (1994) Neural substrate of defensive behavior in the midbrain tectum. Neurosci Biobehav Rev 18:339–346

Brandon JG, Coss RG (1982) Rapid dendritic spine stem shortening during one-trial learning: the honeybee's first orientation flight. Brain Res 252:51–61

Brehmer A (2021) Classification of human enteric neurons. Histochem Cell Biol 156:95–108

Brodal A (1981) Neurological anatomy. Oxford University Press, New York

Bruner E, Rangel de Lázaro G, de la Cuétara JM et al (2014) Midsagittal brain variation and MRI shape analysis of the precuneus in adult individuals. J Anat 224:367–376

Bruner E, Pereira-Pedro AS, Chen X et al (2017a) Precuneus proportions and cortical folding: a morphometric evaluation on diverse human sample. Ann Anat 211:120–128

Bruner E, Preuss TM, Chen X et al (2017b) Evidence for expansion of the precuneus in human evolution. Brain Struct Funct 222:1053–1060

Brusco J, Dall'Oglio A, Rocha LB et al (2010) Descriptive findings on the morphology of dendritic spines in the rat medial amygdala. Neurosci Lett 483:152–156

Brusco J, Merlo S, Ikeda ÉT et al (2014) Inhibitory and multisynaptic spines, and hemispherical synaptic specialization in the posterodorsal medial amygdala of male and female rats. J Comp Neurol 522:2075–2088

Bruton OJ (2021) Is there a "g-neuron"? Establishing a systematic link between general intelligence (g) and the von Economo neuron. Intelligence 86:101540. https://doi.org/10.1016/j.intell.2021.101540

Bucher M, Fanutza T, Mikhaylova M (2020) Cytoskeletal makeup of the synapse: shaft versus spine. Cytoskeleton 77:55–64

Buckmaster PS (2017) Comparative biology and species effects on expression of epilepsy. In: Pitkänen A, Buckmaster PS, Galanopoulou AS et al (eds) Models of seizures and epilepsy. Academic Press, pp 7–19

Buell SJ, Coleman PD (1981) Quantitative evidence for selective dendritic growth in normal human aging but not in senile dementia. Brain Res 214:23–41

Bullmore E, Sporns O (2012) The economy of brain network organization. Nat Rev Neurosci 13:336–349

Bullock TH (1979) Evolving concepts of local integrative operations in neurons. In: Schmitt FO, Worden FG (eds) The Neurosciences, Fourth Study Program, The MIT Press, Cambridge, pp 10–19

Burkhardt P (2022) Ctenophores and the evolutionary origin(s) of neurons. Trends Neurosci 45:878–880

Burlingham CS, Ryoo M, Roth ZN et al (2022) Task-related hemodynamic responses in human early visual cortex are modulated by task difficulty and behavioral performance. eLife 11:e73018. https://doi.org/10.7554/eLife.73018

Burton NO, Greer EL (2022) Multigenerational epigenetic inheritance: transmitting information across generations. Semin Cell Dev Biol 127:121–132

Busler JN, Yanes JA, Bird RT et al (2019) Differential functional patterns of the human posterior cingulate cortex during activation and deactivation: a meta-analytic connectivity model. Exp Brain Res 237:2367–2385

Bzdok D, Laird AR, Zilles K et al (2013) An investigation of the structural, connectional, and functional subspecialization in the human amygdala. Hum Brain Mapp 34:3247–3266

Cajal SR (1894) New ideas on the structure of the nervous system in man and vertebrates. Reinwald & Cie, Paris

Cajal SR (1909–1911) Histologie du système nerveux de l'homme et des vertébrés. Maloine, Paris

Cajal SR (1995) Histology of the nervous system of man and vertebrates (translated by Swanson N, Swanson LW). Oxford University Press, New York

Calabrese B, Wilson MS, Halpain S (2006) Development and regulation of dendritic spine synapses. Physiology 21:38–47

Calanchini C, Bassetti C, Celio MR (2016) Special edition on positive emotions. J Comp Neurol 524:1529–1531

Campbell RE, Han SK, Herbison AE (2005) Biocytin filling of adult gonadotropin-releasing hormone neurons in situ reveals extensive, spiny, dendritic processes. Endocrinology 146:1163–1169

Cannon W (1939) The Wisdom of the body. Norton, New York

Caparelli EC, Ross TJ, Gu H et al (2017) Graph theory reveals amygdala modules consistent with its anatomical subdivisions. Sci Rep 7:14392. https://doi.org/10.1038/s41598-017-14613-4

Carrive P, Morgan MM (2012) Periaqueductal gray. In: Mai JK, Paxinos G (eds) The human nervous system, 3rd edn. Elsevier Academic Press, Amsterdam, pp 367–400

Catalá I, Ferrer I, Galofré E et al (1988) Decreased numbers of dendritic spines on cortical pyramidal neurons in dementia. A quantitative Golgi study on biopsy samples. Hum Neurobiol 6:255–259

Catania KC (2017) Behavioral pieces of neuroethological puzzles. J Comp Physiol A Neuroethol Sens Neural Behav Physiol 203:677–689

Cavanna AE (2007) The precuneus and consciousness. CNS Spectr 12:545–552

Cembrowski MS, Spruston N (2019) Heterogeneity within classical cell types is the rule: lessons from hippocampal pyramidal neurons. Nat Rev Neurosci 20:193–204

Chabrol FP, Arenz A, Wiechert MT et al (2015) Synaptic diversity enables temporal coding of coincident multisensory inputs in single neurons. Nat Neurosci 18:718–727

Chareyron LJ, Banta Lavenex P, Amaral DG et al (2011) Stereological analysis of the rat and monkey amygdala. J Comp Neurol 519:3218–3239

Chassefeyre R, Martínez-Hernández J, Bertaso F et al (2015) Regulation of postsynaptic function by the dementia-related ESCRT-III subunit CHMP2B. J Neurosci 35:3155–3173

Chen Y, Sabatini BL (2012) Signaling in dendritic spines and spine microdomains. Curr Opin Neurobiol 22:389–396

Chen X, Leischner U, Rochefort NL et al (2011) Functional mapping of single spines in cortical neurons in vivo. Nature 475:501–505

Chidambaram SB, Rathipriya AG, Bolla SR et al (2019) Dendritic spines: revisiting the physiological role. Prog Neuropsychopharmacol Biol Psychiatry 92:161–193

Chklovskii DB, Schikorski T, Stevens CF (2002) Wiring optimization in cortical circuits. Neuron 34:341–347

Cobos I, Seeley WW (2015) Human von Economo neurons express transcription factors associated with layer V subcerebral projection neurons. Cereb Cortex 25:213–220

Cohen RS (2013) The postsynaptic density. In: Pfaff DW (ed) Neuroscience in the 21st century. Springer, New York, pp 403–437

Cooke BM, Woolley CS (2005) Sexually dimorphic synaptic organization of the medial amygdala. J Neurosci 25:10759–10767

Cornejo VH, Ofer N, Yuste R (2022) Voltage compartmentalization in dendritic spines in vivo. Science 375:82–86

Correa-Júnior ND, Renner J, Fuentealba-Villarroel F et al (2020) Dendritic and spine heterogeneity of von Economo neurons in the human cingulate cortex. Front Synapt Neurosci 12:25. https://doi.org/10.3389/fnsyn.2020.00025

Cravo SL, Lopes OU, Fraga CA et al (1995) Cardiovascular adjustments in limb retraction provoked by noxious stimulation in decerebrate and spinal cats. Evidence for a somatotopic organization. Braz J Med Biol Res 28:385–396

Creutzfeldt OD (1995) Cortex Cerebri: performance, structural and functional organisation of the cortex. Oxford University Press, Oxford

Dall'Oglio A, Xavier LL, Hilbig A et al (2013) Cellular components of the human medial amygdaloid nucleus. J Comp Neurol 521:589–611

Dall'Oglio A, Dutra AC, Moreira JE et al (2015) The human medial amygdala: structure, diversity, and complexity of dendritic spines. J Anat 227:440–459

Damier P, Hirsch EC, Agid Y et al (1999) The substantia nigra of the human brain. I. Nigrosomes and the nigral matrix, a compartmental organization based on calbindin D(28K) immunohistochemistry. Brain 122:1421–1436

Davidson AM, Mejía-Gómez H, Jacobowitz M et al (2020) Dendritic spine density and dynamics of layer 5 pyramidal neurons of the primary motor cortex are elevated with aging. Cereb Cortex 30:767–777

de Boer A, van Buel EM, Ter Horst GJ (2012) Love is more than just a kiss: a neurobiological perspective love and affection. Neuroscience 201:114–124

de Macedo RK, Ben-Avi E, Sliva DD et al (2015) A FreeSurfer-compliant consistent manual segmentation of infant brains spanning the 0–2 year age range. Front Hum Neurosci 9:21. https://doi.org/10.3389/fnhum.2015.00021

de Olmos JS (2004) Amygdala. In: Paxinos G, Mai J (eds) The human nervous system, 2nd edn. Elsevier, San Diego, pp 739–868

de Olmos J, Alheid GF, Beltramino CA (1985) Amygdala. In: Paxinos G (ed) The rat nervous system, vol 1. Academic Press, Australia, pp 223–334

DeFelipe J (2011) The evolution of the brain, the human nature of cortical circuits, and intellectual creativity. Front Neuroanat 5:29. https://doi.org/10.3389/fnana.2011.00029

DeFelipe J (2022) Manifesto of a neuroanatomist. Front Neuroanat 16:931547. https://doi.org/10.3389/fnana.2022.931547

DeFelipe J, Alonso-Nanclares L, Arellano JI (2002) Microstructure of the neocortex: comparative aspects. J Neurocytol 31:299–316

Deming P, Koenigs M (2020) Functional neural correlates of psychopathy: a meta-analysis of MRI data. Transl Psychiatry 10:133. https://doi.org/10.1038/s41398-020-0816-8

DeWeerdt S (2019) How to map the brain/Deep connections. Nature 571:S6–S8

Di Lorenzo Alho AT, Suemoto CK, Polichiso L et al (2016) Three-dimensional and stereological characterization of the human substantia nigra during aging. Brain Struct Funct 221:3393–3403

Diamond A (2000) Close Interrelation of motor development and cognitive development and of the cerebellum and prefrontal cortex. Child Dev 71:44–56

Diano M, Tamietto M, Celeghin A et al (2017) Dynamic changes in amygdala psychophysiological connectivity reveal distinct neural networks for facial expressions of basic emotions. Sci Rep 7:1–13

Dickstein DL, Kabaso D, Rocher AB et al (2007) Changes in the structural complexity of the aged brain. Aging Cell 6:275–284

Dijkstra AA, Lin LC, Nana AL et al (2018) von Economo neurons and fork cells: a neurochemical signature linked to monoaminergic function. Cereb Cortex 28:131–144

DiMarino V, Etienne Y, Niddam M (2016) The Amygdaloid nuclear complex. Springer, Cham

Ding S-L, Royall JJ, Sunkin SM (2016) Comprehensive cellular-resolution atlas of the adult human brain. J Comp Neurol 524:3127–3481

Domínguez-Álvaro M, Montero-Crespo M, Blazquez-Llorca L et al (2021a) 3D ultrastructural study of synapses in the human entorhinal cortex. Cereb Cortex 31:410–425

Domínguez-Álvaro M, Montero-Crespo M, Blazquez-Llorca L et al (2021b) 3D Analysis of the synaptic organization in the entorhinal cortex in Alzheimer's disease. eNeuro 8(3):ENEURO.0504-20.2021. https://doi.org/10.1523/ENEURO.0504-20.2021

Elston GN, DeFelipe J (2002) Spine distribution in cortical pyramidal cells: a common organizational principle across species. Prog Brain Res 136:109–133

Elston GN, Rosa MGP (1998) Complex dendritic fields of pyramidal cells in the frontal eye field of the macaque monkey: comparison with parietal areas 7a and LIP. Neuroreport 9:127–131

Elston GN, Benavides-Piccione R, DeFelipe J (2001) The pyramidal cell in cognition: a comparative study in human and monkey. J Neurosci 21:RC163. https://doi.org/10.1523/jneurosci.21-17-j0002.2001

Endo F, Kasai A, Soto JS et al (2022) Molecular basis of astrocyte diversity and morphology across the CNS in health and disease. Science 378:6619. https://doi.org/10.1126/science.adc9020

Evrard HC, Forro T, Logothetis NK (2012) von Economo neurons in the anterior insula of the macaque monkey. Neuron 74:482–489

Eyal G, Verhoog MB, Testa-Silva G et al (2016) Unique membrane properties and enhanced signal processing in human neocortical neurons. Elife 5:e16553. https://doi.org/10.7554/eLife.16553

Eyal G, Verhoog MB, Testa-Silva G et al (2018) Human cortical pyramidal neurons: from spines to spikes via models. Front Cell Neurosci 12:181. https://doi.org/10.3389/fncel.2018.00181

Falougy HE, Filova B, Ostatnikova D et al (2019) Neuronal morphology alterations in autism and possible role of oxytocin. Endocr Regul 53:46–54

Feldman ML (1984) Morphology of the neocortical pyramidal neuron. In: Jones EG, Peters A (eds) Cerebral cortex. Plenum Press, New York, pp 107–121

Felleman DJ, Van Essen DC (1991) Distributed hierarchical processing in the primate cerebral cortex. Cereb Cortex 1:1–47

Fernández-Miranda JC, Rhoton AL Jr, Kakizawa Y et al (2008) The claustrum and its projection system in the human brain: a microsurgical and tractographic anatomical study. J Neurosurg 108:764–774

Ferreira AFF, Binda KH, Singulani MP et al (2020) Physical exercise protects against mitochondria alterations in the 6-hidroxydopamine rat model of Parkinson's disease. Behav Brain Res 387:112607. https://doi.org/10.1016/j.bbr.2020.112607

Fiala JC, Harris KM (1999) Dendrite structure. In: Stuart G, Spruston N, Häusser M (eds) Dendrites. Oxford University Press, New York, pp 1–34

Fiala JC, Spacek J, Harris KM (2002) Dendritic spine pathology: cause or consequence of neurological disorders? Brain Res Rev 39:29–54

Fogazzi DV, Neary JP, Sonza A et al (2020) The prefrontal cortex conscious and unconscious response to social/emotional facial expressions involve sex, hemispheric laterality, and selective activation of the central cardiac modulation. Behav Brain Res 393:112773. https://doi.org/10.1016/j.bbr.2020.112773

Foggetti A, Baccini G, Arnold P et al (2019) Spiny and non-spiny parvalbumin-positive hippocampal interneurons show different plastic properties. Cell Rep 27:3725–3732.e5. https://doi.org/10.1016/j.celrep.2019.05.098

Forrest MP, Parnell E, Penzes P (2018) Dendritic structural plasticity and neuropsychiatric disease. Nat Rev Neurosci 19:215–234

Freiwald WA (2020) Social interaction networks in the primate brain. Curr Opin Neurobiol 65:49–58

French IT, Muthusamy KA (2018) A review of the pedunculopontine nucleus in Parkinson's disease. Front Aging Neurosci 10:99. https://doi.org/10.3389/fnagi.2018.00099

Frick A, Johnston D (2005) Plasticity of dendritic excitability. J Neurobiol 64:100–115

Fuentealba-Villarroel FJ, Renner J, Hilbig A et al (2022) Spindle-shaped neurons in the human posteromedial (precuneus) cortex. Front Synaptic Neurosci 13:769228. https://doi.org/10.3389/fnsyn.2021.769228

Gal E, Amsalem O, Schindel A et al (2021) The role of hub neurons in modulating cortical dynamics. Front Neural Circuits 15:718270. https://doi.org/10.3389/fncir.2021.718270

Gamer M, Zurowski B, Buchel C (2010) Different amygdala subregions mediate valence-related and attentional effects of oxytocin in humans. Proc Natl Acad Sci U S A 107:9400–9405

Gardner EP (2021) Receptors of the somatosensory system. In: Kandel ER, Koester JD, Mack SH, Siegelbaum SA (eds) Principles of Neural Science. McGraw Hill, New York, p 408–434

Garey LJ, Ong WY, Patel TS et al (1998) Reduced dendritic spine density on cerebral cortical pyramidal neurons in schizophrenia. J Neurol Neurosurg Psychiatry 65:446–453

Gaser C, Schlaug G (2003) Brain structures differ between musicians and non-musicians. J Neurosci 23:9240–9245

Gefen T, Peterson M, Papastefan ST et al (2015) Morphometric and histologic substrates of cingulate integrity in elders with exceptional memory capacity. J Neurosci 35:1781–1791

Gefen T, Papastefan ST, Rezvanian A et al (2018) von Economo neurons of the anterior cingulate across the lifespan and in Alzheimer's disease. Cortex 99:69–77

Georgiev DD, Kolev SK, Cohen E et al (2020) Computational capacity of pyramidal neurons in the cerebral cortex. Brain Res 1748:147069. https://doi.org/10.1016/j.brainres.2020.147069

Geschwind DH, Rakic P (2013) Cortical evolution: judge the brain by its cover. Neuron 80:633–647

Gibb WR, Lees AJ (1991) Anatomy, pigmentation, ventral and dorsal subpopulations of the substantia nigra, and differential cell death in Parkinson's disease. J Neurol Neurosurg Psychiatry 54:388–396

Gidon A, Zolnik TA, Fidzinski P et al (2020) Dendritic action potentials and computation in human layer 2/3 cortical neurons. Science 367:83–87

Giedd JN, Rapoport JL (2010) Structural MRI of pediatric brain development: what have we learned and where are we going? Neuron 67:728–734

Giguère N, Burke Nanni S, Trudeau LE (2018) On cell loss and selective vulnerability of neuronal populations in Parkinson's disease. Front Neurol 9:455. https://doi.org/10.3389/fneur.2018.00455

Gioia M, Tredici G, Bianchi R (1998) Dendritic arborization and spines of the neurons of the cat and human periaqueductal gray: a light, confocal laser scanning, and electron microscope study. Anat Rec 251:316–325

Glantz LA, Lewis DA (2000) Decreased dendritic spine density on prefrontal cortical pyramidal neurons in schizophrenia. Arch Gen Psychiatry 57:65–73

Glasser MF, Coalson TS, Robinson EC et al (2016) A multi-modal parcellation of human cerebral cortex. Nature 536:171–178

Glatzer NR, Hasney CP, Bhaskaran MD et al (2003) Synaptic and morphologic properties in vitro of premotor rat nucleus tractus solitarius neurons labeled transneuronally from the stomach. J Comp Neurol 464:525–539

Gloor P (1997) The temporal lobe and limbic system. Oxford University Press, New York

Gogolla N (2017) The insular cortex. Curr Biol 27:R580–R586

Goldstein JM, Seidman LJ, Horton NJ et al (2001) Normal sexual dimorphism of the adult human brain assessed by in vivo magnetic resonance imaging. Cereb Cortex 11:490–497

González-Burgos I, Velázquez-Zamora DA, González-Tapia D et al (2017) Plasticity of dendritic spines. Not only for cognitive processes. In: Heinbockel T (ed) Synaptic plasticity. IntechOpen, pp 153–166

González-Burgos G, Miyamae T, Krimer Y et al (2019) Distinct properties of layer 3 pyramidal neurons from prefrontal and parietal areas of the monkey neocortex. J Neurosci 39:7277–7290

González-Ramírez MM, Velázquez-Zamora DA, Olvera-Cortés ME et al (2014) Changes in the plastic properties of hippocampal dendritic spines underlie the attenuation of place learning in healthy aged rats. Neurobiol Learn Mem 109:94–103

Goossens L, Kukolja J, Onur OA et al (2009) Selective processing of social stimuli in the superficial amygdala. Hum Brain Mapp 30:3332–3338

Goriounova NA, Heyer DB, Wilbers R et al (2019) Large and fast human pyramidal neurons associate with intelligence. Elife 7:e41714. https://doi.org/10.7554/eLife.41714

Goulden N, Khusnulina A, Davis NJ et al (2014) The salience network is responsible for switching between the default mode network and the central executive network: replication from DCM. Neuroimage 99:180–190

Gouwens NW, Sorensen SA, Berg J et al (2019) Classification of electrophysiological and morphological neuron types in the mouse visual cortex. Nat Neurosci 22:1182–1195

Grant SG, Fransén E (2020) The synapse diversity dilemma: molecular heterogeneity confounds studies of synapse function. Front Synaptic Neurosci 12:590403. https://doi.org/10.3389/fnsyn.2020.590403

Grasby KL, Jahanshad N, Painter JN et al (2020) The genetic architecture of the human cerebral cortex. Science 367(6484):eaay6690. https://doi.org/10.1126/science.aay6690. Erratum in: Science (2021) 374(6566):eabm7211

Grisendi T, Reynaud O, Clarke S et al (2019) Processing pathways for emotional vocalizations. Brain Struct Funct 224:2487–2504

Guerra KTK, Renner J, Vásquez CE et al (2023) Human cortical amygdala dendrites and spines morphology under open-source 3D reconstruction procedures. J Comp Neurol 531:344–365

Gulledge AT, Carnevale NT, Stuart GJ (2012) Electrical advantages of dendritic spines. PloS One 7(4):e36007. https://doi.org/10.1371/journal.pone.0036007

Guy J, Staiger JF (2017) The functioning of a cortex without layers. Front Neuroanat 11:54. https://doi.org/10.3389/fnana.2017.00054

Haber SN, Adler A, Bergman H (2012) The basal ganglia. In: Mai JK, Paxinos G (eds) The human nervous system, 3rd edn. Elsevier Academic Press, Amsterdam, pp 678–738

Halliday GM, Ophof A, Broe M et al (2005) Alpha-synuclein redistributes to neuromelanin lipid in the substantia nigra early in Parkinson's disease. Brain 128:2654–2664

Halliday G, Reyes S, Double K (2012) Substantia nigra, ventral tegmental area, and retrorubral fields. In: Mai JK, Paxinos G (eds) The human nervous system, 3rd edn. Elsevier Academic Press, Amsterdam, pp 439–455

Hamada MS, Goethals S, de Vries SI et al (2016) Covariation of axon initial segment location and dendritic tree normalizes the somatic action potential. Proc Natl Acad Sci U S A 113:14841–14846

Harvey CD, Yasuda R, Zhong H et al (2008) The spread of Ras activity triggered by activation of a single dendritic spine. Science 321:136–140

Hayashi-Takagi A, Yagishita S, Nakamura M et al (2015) Labeling and optical erasure of synaptic memory traces in the motor cortex. Nature 525:333–338

Heimer L, Van Hoesen GW, Trimble M et al (2008) Anatomy of neuropsychiatry – the new anatomy of the Basal Forebrain and its implications for neuropsychiatric illness. Academic Press, San Diego

Heinze A, Schuldt C, Khudayberdiev S et al (2022) Functional interdependence of the actin regulators CAP1 and cofilin1 in control of dendritic spine morphology. Cell Mol Life Sci 79:558. https://doi.org/10.1007/s00018-022-04593-8

Helm MS, Dankovich TM, Mandad S et al (2021) A large-scale nanoscopy and biochemistry analysis of postsynaptic dendritic spines. Nat Neurosci 24:1151–1162

Herculano-Houzel S (2012) Brain evolution. In: Mai JK, Paxinos G (eds) The human nervous system, 3rd edn. Elsevier Academic Press, Amsterdam, pp 2–13

Herculano-Houzel S (2019) Life history changes accompany increased numbers of cortical neurons: a new framework for understanding human brain evolution. Prog Brain Res 250:179–216

Herculano-Houzel S, Manger PR, Kaas JH (2014) Brain scaling in mammalian evolution as a consequence of concerted and mosaic changes in numbers of neurons and average neuronal cell size. Front Neuroanat 8:77. https://doi.org/10.3389/fnana.2014.00077

Herms J, Dorostkar MM (2016) Dendritic spine pathology in neurodegenerative diseases. Annu Rev Pathol 11:221–250

Hidalgo-Lopez E, Zeidman P, Harris T et al (2021) Spectral dynamic causal modelling in healthy women reveals brain connectivity changes along the menstrual cycle. Commun Biol 4:954. https://doi.org/10.1038/s42003-021-02447-w

Hirata T, Li P, Lanuza GM et al (2009) Identification of distinct telencephalic progenitor pools for neuronal diversity in the amygdala. Nat Neurosci 12:141–149

Hodge RD, Bakken TE, Miller JA et al (2019) Conserved cell types with divergent features in human versus mouse cortex. Nature 573:61–68

Hodge RD, Miller JA, Novotny M et al (2020) Transcriptomic evidence that von Economo neurons are regionally specialized extratelencephalic-projecting excitatory neurons. Nat Commun 11:1172. https://doi.org/10.1038/s41467-020-14952-3

Hof PR, Nimchinsky EA, Perl DP et al (2001) An unusual population of pyramidal neurons in the anterior cingulate cortex of hominids contains the calcium-binding protein calretinin. Neurosci Lett 307:139–142

Höfflin F, Jack A, Riedel C et al (2017) Heterogeneity of the axon initial segment in interneurons and pyramidal cells of rodent visual cortex. Front Cell Neurosci 11:332. https://doi.org/10.3389/fncel.2017.00332

Holstege G, Subramanian HH (2016) Two different motor systems are needed to generate human speech. J Comp Neurol 524:1558–1577

Holstein GR (2020) Morphophysiological organization of vestibulo-autonomic pathways. In: Fritzsch B (ed) The senses: a comprehensive reference, 2nd edn. Elsevier, pp 432–444

Horak FB, Earhart GM (2021) Posture. In: Kandel ER, Koester JD, Mack SH, Siegelbaum SA (eds) Principles of neural science. McGraw Hill, New York, pp 883–907

Horn AKE, Němcová V, ten Donkelaar HJ et al (2020) The reticular formation and the neuromodulatory systems. In: Clinical neuroanatomy. Springer, pp 257–307. https://doi.org/10.1007/978-3-030-41878-6_5

Hruska M, Henderson N, Le Marchand SJ et al (2018) Synaptic nanomodules underlie the organization and plasticity of spine synapses. Nat Neurosci 21:671–682

Hrvoj-Mihic B, Hanson KL, Lew CH et al (2017) Basal dendritic morphology of cortical pyramidal neurons in Williams syndrome: prefrontal cortex and beyond. Front Neurosci 11:419. https://doi.org/10.3389/fnins.2017.00419

Hunt DL, Linaro D, Si B et al (2018) A novel pyramidal cell type promotes sharp-wave synchronization in the hippocampus. Nat Neurosci 21:985–995

Hunt S, Leibner Y, Mertens EJ et al (2022) Strong and reliable synaptic communication between pyramidal neurons in adult human cerebral cortex. Cereb Cortex 8:bhac246. https://doi.org/10.1093/cercor/bhac246

Ilyas A, Pizarro D, Romeo AK et al (2019) The centromedian nucleus: anatomy, physiology, and clinical implications. J Clin Neurosci 63:1–7

Insausti R, Amaral DG (2012) Hippocampal formation. In: Mai JK, Paxinos G (eds) The Human Nervous System. Academic Press, Amsterdam, pp 896–942

Insausti R, Muñoz-López M, Insausti AM et al (2017) The human periallocortex: layer pattern in presubiculum, parasubiculum and entorhinal cortex. A Review. Front Neuroanat 11:84. https://doi.org/10.3389/fnana.2017.00084

Jaaro-Peled H, Ayhan Y, Pletnikov MV et al (2010) Review of pathological hallmarks of schizophrenia: comparison of genetic models with patients and nongenetic models. Schizophr Bull 36:301–313

Jacobs B, Scheibel AB (1993) A quantitative dendritic analysis of Wernicke's area in humans. I. Lifespan changes. J Comp Neurol 327:83–96

Jacobs B, Schall M, Scheibel AB (1993) A quantitative dendritic analysis of Wernicke's area in humans. II. Gender, hemispheric, and environmental factors. J Comp Neurol 327:97–111

Jacobs B, Driscoll L, Schall M (1997) Life-span dendritic and spine changes in areas 10 and 18 of human cortex: a quantitative Golgi study. J Comp Neurol 386:661–680

Jacobs B, Schall M, Prather M et al (2001) Regional dendritic and spine variation in human cerebral cortex: a quantitative Golgi study. Cereb Cortex 11:558–571

Jacot-Descombes S, Keshav N, Brosch CMS et al (2020) von Economo neuron pathology in familial dysautonomia: quantitative assessment and possible implications. J Neuropathol Exp Neurol 79:1072–1083

Janak PH, Tye KM (2015) From circuits to behavior in the amygdala. Nature 517:284–292

Jiang M, Ash RT, Baker SA et al (2013) Dendritic arborization and spine dynamics are abnormal in the mouse model of MECP2 duplication syndrome. J Neurosci 33:19518–19533

Jiang X, Saggar H, Ryu SI et al (2020) Structure in neural activity during observed and executed movements is shared at the neural population level, not in single neurons. Cell Rep 32(6):108006. https://doi.org/10.1016/j.celrep.2020.108006. Erratum in: Cell Rep. 2020 Sep 8;32(10):108148

Joel D, McCarthy MM (2017) Incorporating sex as a biological variable in neuropsychiatric research: where are we now and where should we be? Neuropsychopharmacology 42:379–385

Johnston JB (1923) Further contributions to the study of the evolution of the forebrain. J Comp Neurol 35:337–481

Junker T (2007) Ernst Mayr (1904–2005) and the new philosophy of biology. J Gen Phil Sci 38:1–17

Kalmbach BE, Hodge RD, Jorstad NL et al (2021) Signature morpho-electric, transcriptomic, and dendritic properties of human layer 5 neocortical pyramidal neurons. Neuron 109:2914–2927.e5. https://doi.org/10.1016/j.neuron.2021.08.030

Kamath T, Abdulraouf A, Burris SJ et al (2022) Single-cell genomic profiling of human dopamine neurons identifies a population that selectively degenerates in Parkinson's disease. Nat Neurosci 25:588–595

Kandel ER, LeDoux J (2021) Cellular mechanisms of implicit memory storage and the biological basis of individuality. In: Kandel ER, Koester JD, Mack SH, Siegelbaum SA (eds) Principles of neural science. McGraw Hill, New York, pp 1312–1338

Kasai H, Fukuda M, Watanabe S et al (2010) Structural dynamics of dendritic spines in memory and cognition. Trends Neurosci 33:121–129

Kasai H, Ziv NE, Okazaki H et al (2021) Spine dynamics in the brain, mental disorders and artificial neural networks. Nat Rev Neurosci 22:407–422

Kastellakis G, Poirazi P (2019) Synaptic clustering and memory formation. Front Mol Neurosci 12:300. https://doi.org/10.3389/fnmol.2019.00300

Kasthuri N, Hayworth KJ, Berger DR et al (2015) Saturated reconstruction of a volume of neocortex. Cell 162:648–661

Kedo O, Zilles K, Palomero-Gallagher N et al (2018) Receptor-driven, multimodal mapping of the human amygdala. Brain Struct Funct 223:1637–1666

Kelly R, Stefanacci L (2009) Amygdala: structure and circuitry in primates. In: Squire LR (ed) Encyclopedia of neuroscience. Academic Press, Oxford, pp 341–345

Keough KC, Whalen S, Inoue F et al (2023) Three-dimensional genome rewiring in loci with human accelerated regions. Science 380(6643). https://doi.org/10.1126/science.abm1696

Kerestes R, Chase HW, Phillips ML et al (2017) Multimodal evaluation of the amygdala's functional connectivity. Neuroimage 148:219–229
Khan H, Gamble DT, Rudd A et al (2023) Structural and functional brain changes in acute Takotsubo syndrome. J Am Coll Cardiol HF (Epublished). https://doi.org/10.1016/j.jchf.2022.11.001
Kim BG, Dai HN, McAtee M et al (2007) Labeling of dendritic spines with the carbocyanine dye DiI for confocal microscopic imaging in lightly fixed cortical slices. J Neurosci Methods 162:237–243
Kiwitz K, Brandstetter A, Schiffer C et al (2022) Cytoarchitectonic maps of the human metathalamus in 3D space. Front Neuroanat 16:837485. https://doi.org/10.3389/fnana.2022.837485
Knop AJJ, Spengler S, Bogler C et al (2022) Sensory-tactile functional mapping and use-associated structural variation of the human female genital representation field. J Neurosci 42:1131–1140
Kolb B, Whishaw IQ (2021) Fundamentals of human neuropsychology, 8th edn. Worth Publishers, New York
Kommaddi RP, Das D, Karunakaran S et al (2018) Aβ mediates F-actin disassembly in dendritic spines leading to cognitive deficits in Alzheimer's disease. J Neurosci 38:1085–1099
Koutcherov Y, Huang X-F, Halliday G et al (2004) Organization of human brain stem nuclei. In: Paxinos G, Mai J (eds) The human nervous system, 2nd edn. Academic Press, San Diego, pp 267–320
Kreier F, Swaab DF (2021) History of hypothalamic research: "The spring of primitive existence". In: Swaab DF, Kreier F, Lucassen PJ, Salehi A, Buijs RM (eds) Handbook of clinical neurology. Elsevier, pp 7–43
Kremen WS, Prom-Wormley E, Panizzon MS et al (2010) Genetic and environmental influences on the size of specific brain regions in midlife: the VETSA MRI study. Neuroimage 49:1213–1223. Erratum in: Neuroimage 2010 49:3499–3502
Kubota Y, Hatada S, Kondo S et al (2007) Neocortical inhibitory terminals innervate dendritic spines targeted by thalamocortical afferents. J Neurosci 27:1139–1150
Kubota Y, Karube F, Nomura M et al (2016) The diversity of cortical inhibitory synapses. Front Neural Circuits 10:27. https://doi.org/10.3389/fncir.2016.00027
Kubota Y, Sohn J, Kawaguchi Y (2018) Large volume electron microscopy and neural microcircuit analysis. Front Neural Circuits 12:98. https://doi.org/10.3389/fncir.2018.00098
Laemle LK (1979) Neuronal populations of the human periaqueductal gray, nucleus lateralis. J Comp Neurol 186:93–107
Laemle LK (1981) A Golgy study of cellular morphology in the superficial layers of superior colliculus of man, Saimiri, and Macaca. J Hirnforsch 22:253–263
Laemle LK (1983) A Golgi study of cell morphology in the deep layers of the human superior colliculus. J Hirnforsch 24:297–306
Lanciego JL, Luquin N, Obeso JÁ (2012) Functional neuroanatomy of the basal ganglia. Cold Spring Harb Perspect Med 2(12):a009621. https://doi.org/10.1101/cshperspect.a009621
Larriva-Sahd JA (2014) Some predictions of Rafael Lorente de Nó 80 years later. Front Neuroanat 8:147. https://doi.org/10.3389/fnana.2014.00147
Leal SL, Noche JA, Murray EA et al (2017) Disruption of amygdala-entorhinal-hippocampal network in late-life depression. Hippocampus 27:464–476
LeDoux JE, Brown R (2017) Emotions as higher-order states of consciousness. Proc Natl Acad Sci U S A 114:E2016–E2025
Lee KFH, Soares C, Béïque JC (2012) Examining form and function of dendritic spines. Neural Plast 2012:704103. https://doi.org/10.1155/2012/704103
Lenneberg E (1967) Biological foundations of language. Wiley, New York
Leopold DA (2009) Pre-emptive blood flow. Nature 457:387–388
Leopold DA, Strick PL, Bassett DS et al (2019) Functional architecture of the cerebral cortex. In: Singer W, Sejnowski TJ, Rakic P (eds) The Neocortex. Strüngmann Forum reports (Lupp JR, series ed. 27). MIT Press, Cambridge, pp 141–164
Lewis DA, González-Burgos G (2008) Neuroplasticity of neocortical circuits in schizophrenia. Neuropsychopharmacology 33:141–165

Li Y, Mathis A, Grewe BF et al (2017) Neuronal representation of social information in the medial amygdala of awake behaving mice. Cell 171:1176–1190.e17

Liang JJH, Cole BE, Rankin CH (2019) Habituation. In: Choe JC (ed) Encyclopedia of animal behavior, 2nd edn. Academic Press, pp 411–422

Liberzon I, Phan KL, Decker LR et al (2003) Extended amygdala and emotional salience: a PET activation study of positive and negative affect. Neurospsychopharmacol 28:726–733

Litwin-Kumar A, Harris KD, Axel R et al (2017) Optimal degrees of synaptic connectivity. Neuron 93:1153–1164

Liu AK, Chang RC, Pearce RK et al (2015) Nucleus basalis of Meynert revisited: anatomy, history and differential involvement in Alzheimer's and Parkinson's disease. Acta Neuropathol 129:527–540

Llinás R, Sugimori M (1980) Electrophysiological properties of in vitro Purkinje cell dendrites in mammalian cerebellar slices. J Physiol 305:197–213

Lodato S, Arlotta P (2015) Generating neuronal diversity in the mammalian cerebral cortex. Annu Rev Cell Dev Biol 31:699–720

Losonczy A, Magee JC (2006) Integrative properties of radial oblique dendrites in hippocampal CA1 pyramidal neurons. Neuron 50:291–307

Lourenço MV, Frozza RL, de Freitas GB et al (2019) Exercise-linked FNDC5/irisin rescues synaptic plasticity and memory defects in Alzheimer's models. Nat Med 25:165–175

Lu J, Zuo Y (2017) Clustered structural and functional plasticity of dendritic spines. Brain Res Bull 129:18–22

Luebke JI (2017) Pyramidal neurons are not generalizable building blocks of cortical networks. Front Neuroanat 11:11. https://doi.org/10.3389/fnana.2017.00011

Luengo-Sanchez S, Fernaud-Espinosa I, Bielza C et al (2018) 3D morphology-based clustering and simulation of human pyramidal cell dendritic spines. PLoS Comput Biol 14:e1006221. https://doi.org/10.1371/journal.pcbi.1006221

Luo C, Liu H, Xie F et al (2022) Single nucleus multi-omics identifies human cortical cell regulatory genome diversity. Cell Genom 2:100106. https://doi.org/10.1016/j.xgen.2022.100107

Lyu D, Stieger JR, Xin C et al (2023) Causal evidence for the processing of bodily self in the anterior precuneus. Neuron, https://doi.org/10.1016/j.neuron.2023.05.013

Ma S, Zuo Y (2022) Synaptic modifications in learning and memory - a dendritic spine story. Semin Cell Dev Biol 125:84–90

Mai JK, Forutan F (2012) Thalamus. In: Mai JK, Paxinos G (eds) The human nervous system, 3rd edn. Elsevier Academic Pres, Amsterdam, pp 618–677

Mai JK, Paxinos G, Voss T (2008) Atlas of the human brain, 3rd edn. Academic Press, New York

Mai JK, Majtanik M, Paxinos G (2016) Atlas of the human brain, 4th edn. Elsevier, New York

Maiti P, Manna J, Ilavazhagan G, Rossignol J, Dunbar GL (2015) Molecular regulation of dendritic spine dynamics and their potential impact on synaptic plasticity and neurological diseases. Neurosci Biobehav Rev 59:208–237

Mancuso JJ, Cheng J, Yin Z et al (2014) Integration of multi scale dendritic spine structure and function data into systems biology models. Front Neuroanat 8:130. https://doi.org/10.3389/fnana.2014.00130

Margulies DS, Vincent JL, Kelly C et al (2009) Precuneus shares intrinsic functional architecture in humans and monkeys. Proc Natl Acad Sci U S A 106:20069–20074

Margulies DS, Ghosh SS, Goulas A et al (2016) Situating the default-mode network along a principal gradient of macroscale cortical organization. Proc Natl Acad Sci 113:12574–12579

Marín-Padilla M (1970) Prenatal and early postnatal ontogenesis of the human motor cortex: a golgi study. I. The sequential development of the cortical layers. Brain Res 23:167–183

Marín-Padilla M (1972) Structural abnormalities of the cerebral cortex in human chromosomal aberrations: a Golgi study. Brain Res 44:625–629

Marín-Padilla M (1974) Three-dimensional reconstruction of the pericellular nests (baskets) of the motor (area 4) and visual (area 17) areas of the human cerebral cortex. A Golgi study. Z Anat Entwicklungsgesch 144:123–135

Marín-Padilla M (2014) The mammalian neocortex new pyramidal neuron: a new conception. Front Neuroanat 7:51. https://doi.org/10.3389/fnana.2013.00051
Martínez-García F, Novejarque A, Lanuza E (2007) Evolution of the amygdala in vertebrates. In: Kaas JH (ed) Evolution of nervous systems. a comprehensive reference. Elsevier, Oxford, pp 255–334
Matelli M, Luppino G, Geyer S et al (2004) Motor cortex. In: Paxinos G, Mai J (eds) The human nervous system. Academic Press, San Diego, pp 973–996
Matsuda W, Furuta T, Nakamura KC et al (2009) Single nigrostriatal dopaminergic neurons form widely spread and highly dense axonal arborizations in the neostriatum. J Neurosci 29:444–453
Matsuzaki M, Honkura N, Ellis-Davies GC et al (2004) Structural basis of long-term potentiation in single dendritic spines. Nature 429:761–766
Matyash V, Kettenmann H (2010) Heterogeneity in astrocyte morphology and physiology. Brain Res Rev 63:2–10
May PJ (2006) The mammalian superior colliculus: laminar structure and connections. Prog Brain Res 151:321–378
Mayr E (2001) What evolution is. Basic Books/Perseus Books Group, New York
McCarthy MM (2023) Neural control of sexually dimorphic social behavior: Connecting development to adulthood. Annu Rev Neurosci 46:321–339
McCormick DA, Pape HC (1990) Properties of a hyperpolarization-activated cation current and its role in rhythmic oscillation in thalamic relay neurones. J Physiol 431:291–318
McDonald AJ, Duque A (2022) Specific neuronal subpopulations in the amygdala of macaque monkeys express high levels of nonphosphorylated neurofilaments. Brain Res 1777:147767. https://doi.org/10.1016/j.brainres.2021.147767
McEwen BS (2007) Physiology and neurobiology of stress and adaptation: central role of the brain. Physiol Rev 87:873–904
Medalla M, Chang W, Ibañez S et al (2022) Layer-specific pyramidal neuron properties underlie diverse anterior cingulate cortical motor and limbic networks. Cereb Cortex 32:2170–2196
Medina L, Abellán A (2012) Subpallial structures. In: Watson C, Paxinos G, Puelles L (eds) The mouse nervous system. Academic Press, Cambridge, pp 173–220
Megías M, Emri Z, Freund T et al (2001) Total number and distribution of inhibitory and excitatory synapses on hippocampal CA1 pyramidal cells. Neuroscience 102:527–540
Mercer A, Thomson AM (2017) Cornu Ammonis regions-antecedents of cortical layers? Front Neuroanat 11:83. https://doi.org/10.3389/fnana.2017.00083
Messina A, Cucci G, Crescimanno C et al (2023) Clinical anatomy of the precuneus and pathogenesis of the schizophrenia. Anat Sci Int. https://doi.org/10.1007/s12565-023-00730-w
Mesulam M-M, Mufson EJ (1982) The insula of the old world monkey. III: efferent cortical output and comments on function. J Comp Neurol 212:38–52
Mesulam M-M, Mufson EJ, Levey AI et al (1983) Cholinergic innervation of cortex by the basal forebrain: cytochemistry and cortical connections of the septal area, diagonal band nuclei, nucleus basalis (Substantia innominata), and hypothalamus in the rhesus monkey. J Comp Neurol 214:170–197
Michael J, Modell H, McFarland J et al (eds) (2009) The "core principles" of physiology: what should students understand? Adv Physiol Educ 33:10–16
Mijalkov M, Volpe G, Fernaud-Espinosa I et al (2021) Dendritic spines are lost in clusters in Alzheimer's disease. Sci Rep 11:12350. https://doi.org/10.1038/s41598-021-91726-x
Miller JA, Ding SL, Sunkin SM (2014) Transcriptional landscape of the prenatal human brain. Nature 508:199–206
Mizusaki BEP, Li SSY, Costa RP et al (2018) Pre- and postsynaptically expressed spiking-timing-dependent plasticity contribute differentially to neuronal learning. bioRxiv Preprint at https://doi.org/10.1101/450825
Mohan H, Verhoog MB, Doreswamy KK et al (2015) Dendritic and axonal architecture of individual pyramidal neurons across layers of adult human neocortex. Cereb Cortex 25:4839–4853

Molnár G, Rózsa M, Baka J et al (2016) Human pyramidal to interneuron synapses are mediated by multi-vesicular release and multiple docked vesicles. Elife 5:e18167. https://doi.org/10.7554/eLife.18167

Montero-Crespo M, Domínguez-Álvaro M, Alonso-Nanclares L et al (2021) Three-dimensional analysis of synaptic organization in the hippocampal CA1 field in Alzheimer's disease. Brain 144:553–573

Moradi Chameh H, Rich S, Wang L et al (2021) Diversity amongst human cortical pyramidal neurons revealed via their sag currents and frequency preferences. Nat Commun 12:2497. https://doi.org/10.1038/s41467-021-22741-9

Morales J, Benavides-Piccione R, Dar M et al (2014) Random positions of dendritic spines in human cerebral cortex. J Neurosci 34:10078–10084. https://doi.org/10.1523/JNEUROSCI.1085-14.2014

Morgan JT, Amaral DG (2014) Comparative analysis of the dendritic organization of principal neurons in the lateral and central nuclei of the rhesus macaque and rat amygdala. J Comp Neurol 522:689–716

Morishima M, Kawaguchi Y (2006) Recurrent connection patterns of corticostriatal pyramidal cells in frontal cortex. J Neurosci 26:4394–4405

Moroz LL (2011) Aplysia. Curr Biol 21:R60–R61. https://doi.org/10.1016/j.cub.2010.11.028

Moroz LL, Kocot KM, Citarella MR et al (2014) The ctenophore genome and the evolutionary origins of neural systems. Nature 510:109–114

Morrison JH, Lewis DA, Campbell MJ et al (1987) A monoclonal antibody to non-phosphorylated neurofilament protein marks the vulnerable cortical neurons in Alzheimer's disease. Brain Res 416:331–336

Müri RM (2016) Cortical control of facial expression. J Comp Neurol 524:1578–1585

Murray M (1995) Spinal cord. In: Conn PM (ed) Neuroscience in Medicina. J.B. Lippincott Co., Philadelphia, pp 197–209

Nakahata Y, Yasuda R (2018) Plasticity of spine structure: local signaling, translation and cytoskeletal reorganization. Front Synaptic Neurosci 10:29. https://doi.org/10.3389/fnsyn.2018.00029

Nakajima Y, Nakajima S, Obata K et al (1985) Dissociated cell culture of cholinergic neurons from nucleus basalis of Meynert and other basal forebrain nuclei. Proc Natl Acad Sci U S A 82:6325–6329

Nakamura S, Akiguchi I, Kameyama M et al (1985) Age-related changes of pyramidal cell basal dendrites in layers III and V of human motor cortex: a quantitative Golgi study. Acta Neuropathol 65:281–284

Nakamura NH, Flasbeck V, Maingret N et al (2013) Proximodistal segregation of nonspatial information in CA3: preferential recruitment of a proximal CA3-distal CA1 network in nonspatial recognition memory. J Neurosci 33:11506–11514

Narayanan RT, Udvary D, Oberlaender M (2017) Cell type-specific structural organization of the six layers in rat barrel cortex. Front Neuroanat 11:91. https://doi.org/10.3389/fnana.2017.00091

Newman SW (1999) The medial extended amygdala in male reproductive behavior: a node in the mammalian social behavior network. Ann N Y Acad Sci 877:242–257

Ng KK, Lo JC, Lim JKW et al (2016) Reduced functional segregation between the default mode network and the executive control network in healthy older adults: a longitudinal study. Neuroimage 133:321–330

Nieuwenhuys R (1994) The neocortex. An overview of its evolutionary development, structural organization and synaptology. Anat Embryol 190:307–337

Oberheim NA. Takano T, Han X et al (2009) Uniquely hominid features of adult human astrocytes. J Neurosci 29:3276–3287

O'Brien J, Unwin N (2006) Organization of spines on the dendrites of Purkinje cells. Proc Natl Acad Sci U S A 103:1575–1580

Oelschlägel M, Polanski WH, Morgenstern U et al (2022) Characterization of cortical hemodynamic changes following sensory, visual, and speech activation by intraoperative optical imaging utilizing phase-based evaluation methods. Hum Brain Mapp 43:598–615

Ogasawara T, Sogukpinar F, Zhang K et al (2022) A primate temporal cortex-zona incerta pathway for novelty seeking. Nat Neurosci 25:50–60
Olucha-Bordonau FE, Fortes-Marco L, Otero-García M et al (2015) Amygdala: structure and function. In: Paxinos G (ed) The rat nervous system. Academic Press, San Diego, pp 441–490
Omigie D (2016) Basic, specific, mechanistic? Conceptualizing musical emotions in the brain. J Comp Neurol 524:1676–1686
Onodera S, Hicks TP (2009) A comparative neuroanatomical study of the red nucleus of the cat, macaque and human. PloS One 4(8):e6623. https://doi.org/10.1371/journal.pone.0006623
Oray S, Majewska A, Sur M (2006) Effects of synaptic activity on dendritic spine motility of developing cortical layer V pyramidal neurons. Cereb Cortex 16:730–741
Orellana G, Slachevsky A (2013) Executive functioning in schizophrenia. Front Psych 4:35. https://doi.org/10.3389/fpsyt.2013.00035
Ossewaarde L, van Wingen GA, Rijpkema M et al (2013) Menstrual cycle-related changes in amygdala morphology are associated with changes in stress sensitivity. Hum Brain Mapp 34:1187–1193
Overk CR, Masliah E (2014) Pathogenesis of synaptic degeneration in Alzheimer's disease and Lewy body disease. Biochem Pharmacol 88:508–516
Palomero-Gallagher N, Zilles K (2019) Cortical layers: cyto-, myelo-, receptor- and synaptic architecture in human cortical areas. Neuroimage S1053-8119:30682–30691. https://doi.org/10.1016/j.neuroimage.2017.08.035
Pandya DN, Seltzer B, Petrides M et al (2015) Cerebral cortex – architecture, connections, and the dual origin concept. Oxford University Press, New York
Panizzon MS, Fennema-Notestine C, Eyler LT et al (2009) Distinct genetic influences on cortical surface area and cortical thickness. Cereb Cortex 19:2728–2735
Panksepp J (2016) The cross-mammalian neurophenomenology of primal emotional affects: from animal feelings to human therapeutics. J Comp Neurol 524:1624–1635
Pannese E (2015) Neurocytology. Springer International, Basel
Panzeri S, Macke JH, Gross J et al (2015) Neural population coding: combining insights from microscopic and mass signals. Trends Cogn Sci 19:162–172
Paoletti P, Bellone C, Zhou Q (2013) NMDA receptor subunit diversity: impact on receptor properties, synaptic plasticity and disease. Nat Rev Neurosci 14:383–400
Parajuli LK, Koike M (2021) Three-dimensional structure of dendritic spines revealed by volume electron microscopy techniques. Front Neuroanat 15:627368. https://doi.org/10.3389/fnana.2021.627368
Paredes MF, Sorrells SF, Garcia-Verdugo JM et al (2016) Brain size and limits to adult neurogenesis. J Comp Neurol 524:646–664
Paxinos G, Huang X, Sengul G et al (2012) Organization of brainstem nuclei. In: Mai JK, Paxinos G (eds) The human nervous system, 3rd edn. Elsevier Academic Pres, Amsterdam, pp 260–327
Payeur A, Béïque JC, Naud R (2019) Classes of dendritic information processing. Curr Op Neurobiol 58:78–85
Pearson C (1995) Cerebral cortex. In: Williams PL, Bannister LH, Berry MM, Collins P, Dyson M, Dussek JE, Ferguson MWJ (eds) Gray´s anatomy. Churchill Livingstone, New York, pp 1141–1186
Peng H, Xie P, Liu L et al (2021) Morphological diversity of single neurons in molecularly defined cell types. Nature 598:174–181
Penzes P, Cahill ME, Jones KA et al (2011) Dendritic spine pathology in neuropsychiatric disorders. Nat Neurosci 14:285–293
Persichetti AS, Avery JA, Huber L et al (2020) Layer-specific contributions to imagined and executed hand movements in human primary motor cortex. Curr Biol 30:1721–1725.e3
Petanjek Z, Judas M, Kostovic I et al (2008) Lifespan alterations of basal dendritic trees of pyramidal neurons in the human prefrontal cortex: a layer-specific pattern. Cereb Cortex 18:915–929
Petanjek Z, Judaš M, Šimic G et al (2011) Extraordinary neoteny of synaptic spines in the human prefrontal cortex. Proc Natl Acad Sci U S A 108:13281–13286

Petanjek Z, Sedmak D, Džaja D et al (2019) The protracted maturation of associative layer IIIC pyramidal neurons in the human prefrontal cortex during childhood: a major role in cognitive development and selective alteration in autism. Front Psych 14:122. https://doi.org/10.3389/fpsyt.2019.00122

Peters A, Palay SL, Webster H (1991) The fine structure of the nervous system. Oxford University Press, New York

Petersson KM, Reis A, Ingvar M (2001) Cognitive processing in literate and illiterate subjects: a review of some recent behavioral and functional neuroimaging data. Scand J Psychol 42:251–267

Petralia RS, Wang YX, Mattson MP et al (2018) Invaginating structures in mammalian synapses. Front Synaptic Neurosci 10:4. https://doi.org/10.3389/fnsyn.2018.00004

Petrulis A (2020) Structure and function of the medial amygdala. In: Urban JH, Rosenkranz JA (eds) Handbook of behavioral neuroscience. Elsevier, Amsterdam, pp 39–61

Pinson A, Xing L, Namba T et al (2022) Human TKTL1 implies greater neurogenesis in frontal neocortex of modern humans than Neanderthals. Science 377(6611):eabl6422. https://doi.org/10.1126/science.abl6422

Planert H, Mittermaier FX, Grosser S et al (2021) Intra-individual physiomic landscape of pyramidal neurons in the human neocortex. bioRxiv preprint at https://doi.org/10.1101/2021.11.08.467668

Powell MP, Verma N, Sorensen E et al (2023) Epidural stimulation of the cervical spinal cord for post-stroke upper-limb paresis. Nat Med. https://doi.org/10.1038/s41591-022-02202-6

Profant O, Škoch A, Tintěra J et al (2020) The influence of aging, hearing, and tinnitus on the morphology of cortical gray matter, amygdala, and hippocampus. Front Aging Neurosci 12:553461. https://doi.org/10.3389/fnagi.2020.553461

Purpura DP (1974) Dendritic spine "dysgenesis" and mental retardation. Science 186:1126–1128

Purpura DP (1975a) Dendritic differentiation in human cerebral cortex: normal and aberrant developmental patterns. Adv Neurol 12:91–134

Purpura DP (1975b) Morphogenesis of visual cortex in the preterm infant. In: Brazier MAB (ed) Growth and development of the brain. Raven Press, New York, pp 33–49

Purves WK, Sadava D, Orians GH et al (2001) Life: the science of biology. W.H. Freeman and Co., New York

Qi G, Yang D, Ding C et al (2020) Unveiling the synaptic function and structure using paired recordings from synaptically coupled neurons. Front Synaptic Neurosci 12:5. https://doi.org/10.3389/fnsyn.2020.00005

Raghanti MA, Spurlock LB, Uppal N et al (2015) von Economo neurons. In: Zilles K, Amunts K (eds) Brain mapping: an Encyclopedic reference. Elsevier, Amsterdam, pp 81–91

Ramaswamy S, Markram H (2015) Anatomy and physiology of the thick-tufted layer 5 pyramidal neuron. Front Cell Neurosci 9:233. https://doi.org/10.3389/fncel.2015.00233

Ramón-Moliner E (1962) An attempt at classifying nerve cells on the basis of their dendritic patterns. J Comp Neurol 119:211–227

Ran C, Boettcher JC, Kaye JA et al (2022) A brainstem map for visceral sensations. Nature 609:320–326

Rasia-Filho AA (2006) Is there anything "autonomous" in the nervous system? Adv Physiol Educ 30:9–12

Rasia-Filho AA (2022) Unraveling brain microcircuits, dendritic spines, and synaptic processing using multiple complementary approaches. Front Physiol 13:831568. https://doi.org/10.3389/fphys.2022.831568

Rasia-Filho AA, Londero RG, Achaval M (1999) Effects of gonadal hormones on the morphology of neurons from the medial amygdaloid nucleus of rats. Brain Res Bull 48:173–183

Rasia-Filho AA, Londero RG, Achaval M (2000) Functional activities of the amygdala: an overview. J Psychiatry Neurosci 25:14–23

Rasia-Filho AA, Brusco J, Rocha LB et al (2010) Dendritic spines observed by extracellular DiI dye and immunolabeling under confocal microscopy. Nat Protoc/Protoc Exch http://www.nature.com/protocolexchange/protocols/1890 Protocol Exchange (2010) https://doi.org/10.1038/nprot.2010.153.

Rasia-Filho AA, Dalpian F, Menezes IC et al (2012) Dendritic spines of the medial amygdala: plasticity, density, shape, and subcellular modulation by sex steroids. Histol Histopathol 8:985–1011
Rasia-Filho AA, Andrejew R, Belló-Klein A (2018) Integrating concepts of resilience from cellular functioning to human behavior. In: Manu A (ed) Amygdala: mechanisms, structure and role in disease. Nova Science Publishers, Hauppauge, pp 1–30
Rasia-Filho AA, Guerra KTK, Vásquez CE et al (2021) The subcortical-allocortical-neocortical *continuum* for the emergence and morphological heterogeneity of pyramidal neurons in the human brain. Front Synaptic Neurosci 13:616607. https://doi.org/10.3389/fnsyn.2021.616607
Raznahan A, Shaw P, Lalonde F et al (2011) How does your cortex grow? J Neurosci 31:7174–7177
Reberger R, Dall'Oglio A, Jung CR et al (2018) Structure and diversity of human dendritic spines evidenced by a new three-dimensional reconstruction procedure for Golgi staining and light microscopy. J Neurosci Methods 293:27–36
Redgrave P, Costa RM (2021) The basal ganglia. In: Kandel ER, Koester JD, Mack SH, Siegelbaum SA (eds) Principles of neural science. McGraw Hill, New York, pp 932–952
Redouté J, Stoléru S, Grégoire M-C et al (2000) Brain processing of visual sexual stimuli in human males. Hum Brain Mapp 11:162–177
Reynolds JH, Gottlieb JP, Kastner S (2013) Attention. In: Squire LR, Berg D, Bloom FE et al (eds) Fundamental neuroscience, 4th edn. Academic Press, pp 989–1007
Rietdijk CD, Perez-Pardo P, Garssen J et al (2017) Exploring Braak's hypothesis of Parkinson's disease. Front Neurol 8:37. https://doi.org/10.3389/fneur.2017.00037
Rochefort NL, Konnerth A (2012) Dendritic spines: from structure to in vivo function. EMBO Rep 13:699–708
Rollenhagen A, Lübke JHR (2013) Dendrites: a key structural element of neurons. In: Pfaff DW (ed) Neuroscience in the 21st century. Springer, New York, pp 179–217
Rollenhagen A, Walkenfort B, Yakoubi R et al (2020) Synaptic organization of the human temporal lobe neocortex as revealed by high-resolution transmission, focused ion beam scanning, and electron microscopic tomography. Int J Mol Sci 21:5558. https://doi.org/10.3390/ijms21155558
Rolls ET (2015) Limbic systems for emotion and for memory, but no single limbic system. Cortex 62:119–157
Rubinow DR, Schmidt PJ (2019) Sex differences and the neurobiology of affective disorders. Neuropsychopharmacology 44:111–128
Rudy B, Fishell G, Lee S et al (2011) Three groups of interneurons account for nearly 100% of neocortical GABAergic neurons. Dev Neurobiol 71:45–61
Runge K, Cardoso C, de Chevigny A (2020) Dendritic spine plasticity: function and mechanisms. Front Synaptic Neurosci 12:36. https://doi.org/10.3389/fnsyn.2020.00036
Rusakov DA, Stewart MG, Korogod SM (1996) Branching of active dendritic spines as a mechanism for controlling synaptic efficacy. Neuroscience 75:315–323
Rust NC, LeDoux JE (2023) The tricky business of defining brain functions. Trends Neurosci 46:3–4
Rutishauser U, Tudusciuc O, Neumann D et al (2011) Single-unit responses selective for whole faces in the human amygdala. Curr Biol 21:1654–1660
Rutishauser U, Mamelak AN, Adolphs R (2015) The primate amygdala in social perception – insights from electrophysiological recordings and stimulation. Trends Neurosci 38:295–306
Sala C, Segal M (2014) Dendritic spines: the locus of structural and functional plasticity. Physiol Rev 94:141–188
Saper CB (2012) Hypothalamus. In: Mai JK, Paxinos G (eds) The human nervous system, 3rd edn. Elsevier Academic Press, Amsterdam, pp 548–583
Saper CB, Chelimsky TC (1984) A cytoarchitectonic and histochemical study of nucleus basalis and associated cell groups in the normal human brain. Neuroscience 13:1023–1037
Saper CB, Elmquist JK (2021) The brain stem. In: Kandel ER, Koester JD, Mack SH, Siegelbaum SA (eds) Principles of neural science. McGraw Hill, New York, pp 981–1009

Scheibel ME, Scheibel AB (1967) Structural organization of nonspecific thalamic nuclei and their projection toward cortex. Brain Res 6:60–94

Scheibel ME, Lindsay RD, Tomiyasu U et al (1976) Progressive dendritic changes in the aging human limbic system. Exp Neurol 53:420–430

Scheibel ME, Tomiyasu U, Scheibel AB (1977) The aging human Betz cell. Exp Neurol 56:598–609

Scheibel A, Conrad T, Perdue S et al (1990) A quantitative study of dendrite complexity in selected areas of the human cerebral cortex. Brain Cogn 12:85–101

Scheperjans F, Eickhoff SB, Hömke L et al (2008) Probabilistic maps, morphometry, and variability of cytoarchitectonic areas in the human superior parietal cortex. Cereb Cortex 18:2141–2157

Schmidt KL, Cohn JF (2001) Human facial expressions as adaptations: evolutionary questions in facial expression research. Am J Phys Anthropol Suppl 33:3–24

Schmidt ERE, Polleux F (2022) Genetic mechanisms underlying the evolution of connectivity in the human cortex. Front Neural Circuits 15:787164. https://doi.org/10.3389/fncir.2021.787164

Schmidt-Nielsen K (1997) Animal physiology: adaptation and environment. Cambridge University Press

Schoenen J (1982a) Dendritic organization of the human spinal cord: the motoneurons. J Comp Neurol 211:226–247

Schoenen J (1982b) The dendritic organization of the human spinal cord: the dorsal horn. Neuroscience 7:2057–2087

Schoenen J, Grant G (2004) Spinal cord: connections. In: Paxinos G, Mai J (eds) The human nervous system, 2nd edn. Academic Press, San Diego, pp 233–249

Schrödinger E (1992) What is life? with mind and matter with autobiographical sketches. Cambridge University Press

Schumann CM, Amaral DG (2005) Stereologial estimation of the number of neurons in the human amygdaloid complex. J Comp Neurol 491:320–329

Scott SH, Kalaska JF (2021) Voluntary movement: motor cortices. In: Kandel ER, Koester JD, Mack SH, Siegelbaum SA (eds) Principles of neural science. McGraw Hill, New York, pp 815–859

Sedmak D, Hrvoj-Mihiæ B, Džaja D et al (2018) Biphasic dendritic growth of dorsolateral prefrontal cortex associative neurons and early cognitive development. Croat Med J 59:189–202

Segev I, Rinzel J, Shepherd GM (2003) The theoretical foundation of dendritic function. The MIT Press

Sengul G, Watson C (2012) Spinal cord: regional anatomy, cytoarchitecture and chemoarchitecture. In: Mai JK, Paxinos G (eds) The human nervous system. Elsevier Academic Press, Amsterdam, pp 186–232

Shafiei G, Markello RD, Vos de Wael R et al (2020) Topographic gradients of intrinsic dynamics across neocortex. Elife 9:e62116. https://doi.org/10.7554/eLife.62116

Shapson-Coe A, Januszewski M, Berger DR et al (2021) A connectomic study of a petascale fragment of human cerebral cortex. bioRxiv 2021.05.29.446289 Preprint at https://doi.org/10.1101/2021.05.29.446289

Shaw P, Eckstrand K, Sharp W et al (2007) Attention-deficit/hyperactivity disorder is characterized by a delay in cortical maturation. Proc Natl Acad Sci U S A 104:19649–19654

Shepherd GM (1996) The dendritic spine: a multifunctional integrative unit. J Neurophysiol 75:2197–2210

Shepherd GM, Rowe TB (2017) Neocortical lamination: insights from neuron types and evolutionary precursors. Front Neuroanat 11:100. https://doi.org/10.3389/fnana.2017.00100

Sheppard PAS, Choleris E, Galea LAM (2019) Structural plasticity of the hippocampus in response to estrogens in female rodents. Mol Brain 12:22. https://doi.org/10.1186/s13041-019-0442-7

Sherwood CC, Miller SB, Karl M et al (2020) Invariant synapse density and neuronal connectivity scaling in primate neocortical evolution. Cereb Cortex 30:5604–5615

Shohamy D, Schacter DL, Wagner AD (2021) Learning and memory. In: Kandel ER, Koester JD, Mack SH, Siegelbaum SA (eds) Principles of neural science. McGraw Hill, New York, pp 1291–1311

Sierpowska J, Bryant KL, Janssen N (2022) Comparing human and chimpanzee temporal lobe neuroanatomy reveals modifications to human language hubs beyond the frontotemporal arcuate fasciculus. Proc Natl Acad Sci U S A 119(28):e2118295119. https://doi.org/10.1073/pnas.2118295119
Šimić G, Tkalčić M, Vukić V et al (2021) Understanding emotions: origins and roles of the amygdala. Biomolecules 11:823. https://doi.org/10.3390/biom11060823
Sirotin YB, Das A (2009) Anticipatory haemodynamic signals in sensory cortex not predicted by local neuronal activity. Nature 457:475–479
Sjöström PJ, Rancz EA, Roth A et al (2008) Dendritic excitability and synaptic plasticity. Physiol Rev 88(2):769–840
Smith AD, Bolam JP (1990) The neural network of the basal ganglia as revealed by the study of synaptic connections of identified neurones. Trends Neurosci 13:259–265
Smith Y, Bennett BD, Bolam JP et al (1994) Synaptic relationships between dopaminergic afferents and cortical or thalamic input at the single cell level in the sensorimotor territory of the striatum in monkey. J Comp Neurol 344:1–19
Smith Y, Raju DV, Pare JF et al (2004) The thalamostriatal system: a highly specific network of the basal ganglia circuitry. Trends Neurosci 27:520–527
Smyser CD, Inder TE, Shimony JS et al (2010) Longitudinal analysis of neural network development in preterm infants. Cereb Cortex 20:2852–2862
Soltesz I, Losonczy A (2018) CA1 pyramidal cell diversity enabling parallel information processing in the hippocampus. Nat Neurosci 21:484–493
Sorvari H, Miettinen R, Soininen H et al (1996) Parvalbumin-immunoreactive neurons make inhibitory synapses on pyramidal cells in the human amygdala: a light and electron microscopic study. Neurosci Lett 217:93–96
Spacek J, Harris KM (2004) Trans-endocytosis via spinules in adult rat hippocampus. J Neurosci 24:4233–4241
Spruston N (2008) Pyramidal neurons: dendritic structure and synaptic integration. Nat Rev Neurosci 9:206–221
Spruston N, Häusser M, Stuart G (2013) Information processing in dendrites and spines. In: Squire LR, Berg D, Bloom FE et al (eds) Fundamental neuroscience. Academic Press, Waltham, pp 231–260
Stephan H, Frahm HD, Baron G (1987) Comparison of brain structure volumes in insectivores and primates. VII. Amygdaloid components. J Hirnforsch 28:571–584
Stewart MG, Popov VI, Kraev IV et al (2014) Structure and complexity of the synapse and dendritic spine. In: Pickel V, Segal M (eds). The Synapse Academic Press, Kidlington, pp 1–20
Stimpson CD, Tetreault NA, Allman JM et al (2011) Biochemical specificity of von Economo neurons in hominoids. Am J Hum Biol 23:22–28
Striedter GF (2004) Brain evolution. In: Paxinos G, Mai J (eds) The human nervous system. Academic Press, San Diego, pp 3–21
Stuart G, Spruston N, Häusser M (1999) Dendrites. Oxford University Press, New York
Sun Q, Sotayo A, Cazzulino AS et al (2017) Proximodistal heterogeneity of hippocampal CA3 pyramidal neuron intrinsic properties, connectivity, and reactivation during memory recall. Neuron 95:656–672.e3
Swanson L, Petrovich G (1998) What is the amygdala? Trends Neurosci 21:323–331
Sylvester CM, Yu Q, Srivastava AB et al (2020) Individual-specific functional connectivity of the amygdala: a substrate for precision psychiatry. Proc Natl Acad Sci U S A 117:3808–3818
Szentágothai J (1978) The neuron network of the cerebral cortex: a functional interpretation. Proc R Soc Lond B 201:219–248
Tardif E, Delacuisine B, Probst A et al (2005) Intrinsic connectivity of human superior colliculus. Exp Brain Res 166:316–324
Testa-Silva G, Verhoog MB, Linaro D et al (2014) High bandwidth synaptic communication and frequency tracking in human neocortex. PLoS Biol 12:e1002007. https://doi.org/10.1371/journal.pbio.1002007

Thome C, Kelly T, Yanez A et al (2014) Axon-carrying dendrites convey privileged synaptic input in hippocampal neurons. Neuron 83:1418–1430

Timo-Iaria C, Valle AC (1995) The functional role of the conscious process. Ciência e Cultura/J Braz Assoc Adv Sci 47:221–234

Toni N, Buchs PA, Nikonenko I et al (1999) LTP promotes formation of multiple spine synapses between a single axon terminal and a dendrite. Nature 402:421–425

Tønnesen J, Nägerl V (2016) Dendritic spines as tunable regulators of synaptic signals. Front Psych 7:101. https://doi.org/10.3389/fpsyt.2016.00101

Torikai H, Hayashi F, Tanaka K et al (1996) Recruitment order and dendritic morphology of rat phrenic motoneurons. J Comp Neurol 366:231–243

Tran-Van-Minh A, Cazé RD, Abrahamsson T et al (2015) Contribution of sublinear and supra-linear dendritic integration to neuronal computations. Front Cell Neurosci 9:67. https://doi.org/10.3389/fncel.2015.00067

Triarhou LC (2009) von Economo and Koskinas' atlas of cytoarchitectonics of the adult human cerebral cortex. Karger, Basel

Tsai J, Grutzendler J, Duff K et al (2004) Fibrillar amyloid deposition leads to local synaptic abnormalities and breakage of neuronal branches. Nat Neurosci 7:1181–1183

Tsay D, Yuste R (2004) On the electrical function of dendritic spines. Trends Neurosci 27:77–83

Udvary D, Harth P, Macke JH et al (2022) The impact of neuron morphology on cortical network architecture. Cell Rep 39:110677. https://doi.org/10.1016/j.celrep.2022.110677

Underwood R, Tolmeijer E, Wibroe J et al (2021) Networks underpinning emotion: a systematic review and synthesis of functional and effective connectivity. Neuroimage 243:118486. https://doi.org/10.1016/j.neuroimage.2021.118486

van den Heuvel MP, Scholtens LH, Feldman Barrett L et al (2015) Bridging cytoarchitectonics and connectomics in human cerebral cortex. J Neurosci 35:13943–13948

van Domburg PHMF, ten Donkelaar HJ (1991) The Human Substantia Nigra and Ventral Tegmental area. Advances in anatomy embryology and cell biology. Springer, Berlin, Heidelberg

Van Essen DC, Donahue CJ, Glasser MF (2018) Development and evolution of cerebral and Cerebellar Cortex. Brain Behav Evol 91:158–169

Vanderah TW, Gould DJ (2021) Nolte's the human brain, 8th edn. Elsevier, pp 396–434

Vásquez CE, Reberger R, Dall'Oglio A et al (2018) Neuronal types of the human cortical amygdaloid nucleus. J Comp Neurol 526:2776–2801

Vincent A, Tell F (1999) Postnatal development of rat nucleus tractus solitarius neurons: morphological and electrophysiological evidence. Neuroscience 93:293–305

Viscardi LH, Imparato DO, Bortolini MC et al (2021) Ionotropic receptors as a driving force behind human synapse establishment. Mol Biol Evol 38:735–744

Vogt BA (2015) Mapping cingulate subregions. In: Toga AW (ed) Brain mapping: an Encyclopedic reference. Academic Press, Oxford, pp 325–339

von Bohlen und Halbach O (2009) Structure and function of dendritic spines within the hippocampus. Ann Anat 191:518–531

von Bohlen und Halbach O (2010) Dendritic spine abnormalities in mental retardation. Cell Tissue Res 342:317–323

Voogd J, Ruigrok TJH (2012) Cerebellum and precerebellar nuclei. In: Mai JK, Paxinos G (eds) The human nervous system, 3rd edn. Elsevier Academic Press, Amsterdam, pp 471–545

Wahle P (1993) Differential regulation of substance P and somatostatin in martinotti cells of the developing cat visual cortex. J Comp Neurol 329:519–538

Wahle P, Sobierajski E, Gasterstädt I et al (2022) Neocortical pyramidal neurons with axons emerging from dendrites are frequent in non-primates, but rare in monkey and human. Elife 11:e76101. https://doi.org/10.7554/eLife.76101

Walker CK, Herskowitz JH (2021) Dendritic spines: mediators of cognitive resilience in aging and Alzheimer's disease. Neuroscientist 27:487–505

Wang Y, Ye M, Kuang X et al (2018) A simplified morphological classification scheme for pyramidal cells in six layers of primary somatosensory cortex of juvenile rats. IBRO Rep 5:74–90

Watanabe H (1981) Aging changes of apical dendritic spines in the area 4 human pyramidal cells with the rapid Golgi method. Clin Neurol 121:895–902
Watson KK, Jones TK, Allman JM (2006) Dendritic architecture of the von Economo neurons. Neuroscience 141:1107–1112
Wattendorf E, Westermann B, Lotze M et al (2016) Insular cortex activity and the evocation of laughter. J Comp Neurol 524:1608–1615
Wefelmeyer W, Puhl CJ, Burrone J (2016) Homeostatic plasticity of subcellular neuronal structures: from inputs to outputs. Trends Neurosci 39:656–667
Weir RK, Bauman MD, Jacobs B et al (2018) Protracted dendritic growth in the typically developing human amygdala and increased spine density in young ASD brains. J Comp Neurol 526:262–274
West MJ, Gundersen HJ (1990) Unbiased stereological estimation of the number of neurons in the human hippocampus. J Comp Neurol 296:1–22
Whitehead MC (1988) Neuronal architecture of the nucleus of the solitary tract in the hamster. J Comp Neurol 276:547–572
WHO (2022) World Health Organization – Ageing and health. Available at https://www.who.int/news-room/fact-sheets/detail/ageing-and-health. Accessed 27 Dec 2022
Wyss JM, van Groen T (1995) The limbic system. In: Conn PM (ed) Neuroscience in medicine. J.B. Lippincott, Philadelphia, pp 321–337
Xu X, Stoyanova EI, Lemiesz AE et al (2018) Species and cell-type properties of classically defined human and rodent neurons and glia. Elife 7:e37551. https://doi.org/10.7554/eLife.37551
Xuan B, Mackie MA, Spagna A et al (2016) The activation of interactive attentional networks. Neuroimage 129:308–319
Yadav A, Gao YZ, Rodriguez A et al (2012) Morphologic evidence for spatially clustered spines in apical dendrites of monkey neocortical pyramidal cells. J Comp Neurol 520:2888–2902
Yakoubi R, Rollenhagen A, von Lehe M et al (2019a) Quantitative three-dimensional reconstructions of excitatory synaptic boutons in layer 5 of the adult human temporal lobe neocortex: a fine-scale electron microscopic analysis. Cereb Cortex 29:2797–2814
Yakoubi R, Rollenhagen A, von Lehe M et al (2019b) Ultrastructural heterogeneity of layer 4 excitatory synaptic boutons in the adult human temporal lobe neocortex. Elife 8:e48373. https://doi.org/10.7554/eLife.48373
Yamawaki N, Borges K, Suter BA et al (2014) A genuine layer 4 in motor cortex with prototypical synaptic circuit connectivity. Elife 3:e05422. https://doi.org/10.7554/eLife.05422
Yang L, Yang Y, Yuan J et al (2019) Transcriptomic landscape of von Economo neurons in human anterior cingulate cortex revealed by microdissected-cell RNA sequencing. Cereb Cortex 29:838–851
Yilmazer-Hanke DM (2012) Amygdala. In: Mai JK, Paxinos G (eds) The human nervous system, 3rd edn. Elsevier Academic Press, Amsterdam, pp 759–834
Yuste R (2010) Dendritic spines. MIT Press, Cambridge
Yuste R (2013) Electrical compartmentalization in dendritic spines. Ann Rev Neurosci 36:429–449
Yuste R, Hawrylycz M, Aalling N et al (2020) A community-based transcriptomics classification and nomenclature of neocortical cell types. Nat Neurosci 23:1456–1468
Zaccard CR, Shapiro L, Martin-de-Saavedra MD et al (2020) Rapid 3D enhanced resolution microscopy reveals diversity in dendritic spinule dynamics, regulation, and function. Neuron 107:522–537.e6
Zald DH, Pardo JV (1997) Emotion, olfaction, and the human amygdala: amygdala activation during aversive olfactory stimulation. Proc Natl Acad Sci U S A 94:4119–4124
Zancan M, da Cunha RSR, Schroeder F et al (2018) Remodeling of the number and structure of dendritic spines in the medial amygdala: from prepubertal sexual dimorphism to puberty and effect of sexual experience in male rats. Eur J Neurosci 48:1851–1865
Zebarjadi N, Adler E, Kluge A et al (2021) Rhythmic neural patterns during empathy to vicarious pain: beyond the affective-cognitive empathy dichotomy. Front Hum Neurosci 15:708107. https://doi.org/10.3389/fnhum.2021.708107

Zeki S (2007) The neurobiology of love. FEBS Lett 581:2575–2579

Zoccal DB, Furuya WI, Bassi M et al (2014) The nucleus of the solitary tract and the coordination of respiratory and sympathetic activities. Front Physiol 5:238. https://doi.org/10.3389/fphys.2014.00238

Zrinzo L, Hyam JA (2018) Deep brain stimulation for movement disorders. In: Ellenbogen RG, Sekhar LN, Kitchen ND, Brito da Silva H (eds) Principles of neurological surgery. Elsevier, pp 781–798

Index

A. A. Rasia-Filho et al. (eds.), *Dendritic Spines*, Advances in Neurobiology 34,
https://doi.org/10.1007/978-3-031-36159-3

SPRINGER NATURE

GPSR Compliance

The European Union's (EU) General Product Safety Regulation (GPSR) is a set of rules that requires consumer products to be safe and our obligations to ensure this.

If you have any concerns about our products, you can contact us on ProductSafety@springernature.com

In case Publisher is established outside the EU, the EU authorized representative is:

Springer Nature Customer Service Center GmbH
Europaplatz 3
69115 Heidelberg, Germany

The manufacturer's authorised representative in the EU is Springer Nature Customer Service Centre GmbH, Europaplatz 3, 69115 Heidelberg, Germany. If you have any concerns regarding our products, please contact ProductSafety@springernature.com

Printed and bound by CPI Group (UK) Ltd, Croydon, CR0 4YY
30/09/2026
02225509-0015